国外优秀数学著作
原版系列

A Course in Analysis

—Vol. III, Measure and Integration Theory, Complex-Valued Functions of a Complex Variable

分析学教程

——第3卷，测度与积分理论，复变量的复值函数

（英文）

[英] 尼尔斯·雅各布（Niels Jacob）
[英] 克里斯蒂安·P. 埃文斯（Kristian P. Evans）
著

哈尔滨工业大学出版社
HARBIN INSTITUTE OF TECHNOLOGY PRESS

黑版贸审字 08 - 2020 - 189 号

图书在版编目(CIP)数据

分析学教程. 第 3 卷, 测度与积分理论, 复变量的复值函数 = A Course in Analysis: Vol. Ⅲ, Measure and Integration Theory, Complex - Valued Functions of a Complex Variable: 英文 / (英) 尼尔斯·雅各布 (Niels Jacob), (英) 克里斯蒂安·P. 埃文斯 (Kristian P. Evans) 著. —哈尔滨: 哈尔滨工业大学出版社, 2023.3

ISBN 978 - 7 - 5767 - 0613 - 0

Ⅰ. ①分… Ⅱ. ①尼… ②克… Ⅲ. ①数学分析 - 英文 Ⅳ. ①O17

中国国家版本馆 CIP 数据核字(2023)第 030373 号

FENXIXUE JIAOCHENG: DI-SAN JUAN, CEDU YU JIFEN LILUN, FUBIANLIANG DE FUZHI HANSHU

World Scientific

策划编辑　刘培杰　杜莹雪
责任编辑　刘家琳
封面设计　孙茵艾
出版发行　哈尔滨工业大学出版社
社　　址　哈尔滨市南岗区复华四道街 10 号　邮编 150006
传　　真　0451 - 86414749
网　　址　http://hitpress. hit. edu. cn
印　　刷　哈尔滨博奇印刷有限公司
开　　本　720 mm × 1 000 mm　1/16　印张 51.5　字数 890 千字
版　　次　2023 年 3 月第 1 版　2023 年 3 月第 1 次印刷
书　　号　ISBN 978 - 7 - 5767 - 0613 - 0
定　　价　118.00 元

(如因印装质量问题影响阅读,我社负责调换)

Preface

A detailed description of the content of Volume III of our Course in Analysis will be provided in the introduction. Here we would like to take the opportunity to thank those who have supported us in writing this volume. We owe a debt of gratitude to James Harris who has typewritten the majority of the manuscript. Thanks for typewriting further parts are expressed to Saroj Limbu and James Morgan. Huw Fry, James Harris and Elian Rhind undertook a lot of proofreading for which we are grateful. We also want to thank the Department of Mathematics, the College of Science and Swansea University for providing us with funding for typewriting.

It turned out that R. Schilling was working on the second edition of his book "Measure, Integration and Matingales" [75] while we were working on Part 6. This led to many interesting discussions between Dresden and Swansea from which we could benefit a lot and for which we are grateful.

Finally we want to thank our publisher, in particular Tan Rok Ting and Ng Qi Wen, for a pleasant collaboration.

Niels Jacob
Kristian P. Evans
Swansea, January 2017

Introduction

The third volume of our "Course in Analysis" covers two topics indispensable for every mathematician whether specializing in analysis or not. Part 6 discusses Lebesgue's theory of integration and first consequences in the theory of a real-valued function of a real variable. Part 7 introduces the theory of complex-valued functions of one complex variable - traditionally just called the "the theory of functions".

The advanced theory of real-valued functions, Fourier analysis, functional analysis, the theory of dynamical systems, partial differential equations or the calculus of variations are all topics in analysis depending on Lebesgue's integration theory. But in addition, probability theory, parts of geometry (fractal sets) and more applied subjects such as information theory or optimization needs a proper understanding of Lebesgue integration. Moreover, many mathematicians will agree that the structure of the real line is one of the most complicated structures, if not the most complicated structure of all, we have to deal with. Many problems leading to a better understanding of $\mathbb{R}$ and in fact leading to deep developments in the foundation of mathematics (mathematical logic and set theory) we encounter when relating the topological structure of $\mathbb{R}$ as induced by the Euclidean metric to the problem of defining and determining the size of subsets of $\mathbb{R}$. Often the underlying model of set theory determines the relation between topology and measure theory. In our treatise we always assume ZF as underlying model of set theory but we do not spend much time on investigating the problems mentioned above, they are the topic of different and more advanced courses.

We will describe the content of Part 6 in more detail below, however we would like to mention the influence of [11] in our presentation. The forerunner of H. Bauer's monograph [11], i.e. [10], was the standard text book on measure theory in Germany for many decades. When [10] was split into two books the first named author was heavily involved in proof reading and discussing the material.

The theory of functions is key to every other advanced theory in analysis and at the same time it is needed as a tool in many other applied mathematical disciplines such as mechanics or in fields such as electrical engineering or physics. But much more holds: holomorphic functions, i.e. complex differentiable complex-valued functions of a complex variable enter into many fields in pure mathematics such as number theory, algebraic geometry, representation theory, differential geometry or combinatorics and many others. It is fair to say that without a proper knowledge of function theory no undergraduate education in pure or applied mathematics can be viewed as being complete. A more detailed discussion of Part 7 follows below.

Before going into the details we have to add two remarks. Firstly, originally we also planned to include in our third volume Part 8: Fourier analysis. While writing Part 6 and 7 it emerged that a better strategy is to add more (advanced or specialized) material from the two theories covered in Part 6 and 7 already here and not in later parts where they will be needed, e.g. differentiability properties of real-valued functions, Sard's theorem, dense subsets in L^p-spaces and the Friedrichs mollifier, or hypergeometric function, elliptic integrals and elliptic functions, just to mention a few. This will of course lead to some alteration of the previous plan of arranging the entire material. Secondly, the reader will notice a different mode of referring to the literature. We now sometimes deal with topics which admit quite different approaches and representations. Clearly we are influenced by authors who dealt with this material before and of course we want to and we have to give fair credit where appropriate. The fact that we have meanwhile reached more advanced material is also taken into account when in some but not many proofs we leave straightforward calculations to the reader, something we have strictly avoided in the first two volumes. Some of our problems are now more involved and some of the solutions are more brief, again a reflection of the fact that we now address more (mathematically) matured readers.

In Chapter 1 we introduce σ-fields, their generators and measures, and Chapter 2 is devoted to the Carathéodory extension theorem. A discussion of the Lebesgue-Borel measure and of the Hausdorff measure as well as the Hausdorff dimension follows in Chapter 3. In particular the Cantor set is treated in great detail. Measurable mappings are the topic of Chapter 4. The standard approach to define the Lebesgue integral with respect to a measure is developed in Chapter 5, and Chapter 6 starts to handle measures with densities. We prove the Radon-Nikodym theorem as we prove the transformation theorem for Lebesgue integrals. The role of sets of measure zero and almost everywhere statements are treated in Chapter 7 along with the main convergence results, especially the dominated convergence theorem. We also look at spaces of p-fold integrable functions. These results are then applied in Chapter 8 to prove typical theorems about the interchanging of limits such as the continuity or the differentiability of parameter dependent integrals. We prove Jensen's inequality and discuss the relation between the Lebesgue integral and the Riemann integral. By this we fill in some of the gaps left by dealing with the Riemann integral in higher dimensions in Volume II. This discussion includes improper integrals in particular as well as the introduction of the L^p-spaces. Product integrations and most of all the theorems of Tonelli and Fubini are the main content of Chapter 9. As an important application we look at integration with respect to the distribution function and we give some examples of theoretical interest. We also provide a complete proof of Minkowski's integral inequality. From our point of view the content of Chapters 1-9 forms the core of any module on measure and integration theory suitable for the purposes of analysis as well as probability theory. Chapters 10-12 deal with topics which are interesting by themselves but they are also major tools in areas such as Fourier analysis, functional analysis or (partial) differential equations. Chapter 10 is devoted to the convolution of functions and measures and we prove the density of continuous functions with compact support in the spaces $L^p(\mathbb{R}^n)$, $1 \leq p < \infty$. Further we handle the Friedrichs mollifier which will turn out to become a first class tool in many later considerations. Finally, we have a first look at convolution operators. Lebesgue's theory of differentiation is our topic in Chapter 11. After having proved the Vitali covering theorem we introduce absolutely continuous functions and functions of bounded variation and study their relations. The key results are a new version of the fundamental theorem of calculus and Lebesgue's differentiation theorem, the proof of which is given by making use of the Hardy-Littlewood maximal function. Eventually in Chapter 12 we discuss three special results which will become important later: a version of Sard's theorem on the measure of the critical points of a differentiable mapping, Lusin's theorem which is followed by a first discussion of weak convergence and the Kolmogorov-Riesz theorem which characterises relative compact sets in L^p-spaces.

Our treatment of complex-valued functions of a complex variable starts with a brief recollection of the complex numbers including convergence of sequences and series. In Chapter 14 we embark onto a small digression and summarize obvious properties of complex-valued functions defined on an arbitrary set and Chapter 15 is devoted to the geometry of the plane and complex numbers. We also include the Riemann sphere and the stereographic projection. A chapter on complex-valued functions of a complex variable follows this where we handle continuity, convergence and uniform convergence and we prove Abels' theorem as well as the Cauchy-Hadamard theorem for power series. Our examples of power series include the binomial series and the Gaussian hypergeometric series. We close this chapter with a first look at Möbius transformations. Complex differentiation is handled in Chapter 17 where all the standard results are shown including those related to the Cauchy-Riemann differential equations. The key notion is that of a holomorphic function and we also have a first encounter with biholomorphic functions as well as harmonic functions. Next is a discussion of some important holomorphic functions, e.g. the exponential function, trigonometrical functions and their inverses and hyperbolic functions and their inverses. Most important in Chapter 18 are the investigations into the logarithmic function. Topological notions such as connectivity, simple connectivity or homology play a crucial part in function theory and Chapter 19 treats some of the related questions before in Chapter 20 we introduce line integrals. For defining line integrals we can rely on our considerations in Chapter II.15 as we can when deriving their basic properties. Often we just need to change the notation, i.e. we use complex numbers. Once this is done the theory of complex-valued functions of a complex variable offers many new results when put in context of line integrals over (simply) closed curves. We introduce the notion of a primitive and relate properties of line integrals to the existence of primitives. Chapter 21 is a central chapter in the theory with a detailed discussion of Cauchy's integral theorem, Cauchy's integral formulae and many of their applications including standard estimates for derivatives of holomorphic functions and the relation to Taylor expansions. In Chapter 22 we continue our investigations by looking at power series, holomorphy and holomorphy domains, and applications to differential equations. An important result is Riemann's theorem on removable singularities and the uniqueness theorem for holomorphic functions. We develop the theory further in Chapter 23 by discussing the boundary maximum principle, the Lemma of Schwarz, Liouville's theorem and the approximation theorem of Runge. Meromorphic functions and the Laurent expansions are investigated in Chapter 24 including the classification of their singularities and a first version of the residue theorem as well as the argument principle and Rouché's theorem. In Chapter 25 we give a general version of the residue theorem. This more theoretical consideration is followed by many very concrete applications, e.g. the evaluation of certain (improper) integrals. We take up our discussion of the Γ-function from Volume I and now look at $\Gamma(z)$ as a meromorphic function in Chapter 26. We also introduce Dirichlet series, in particular the Riemann ζ-function and prove some of their properties. This enables us to eventually formulate the Riemann hypothesis - we believe that the reader will understand that we do not provide its proof here. Mathematicians may encounter elliptic integrals only late in their education, physicists meet them quite early. Our discussion of elliptic integrals and elliptic functions in Chapter 27 starts with the integrals and then we move on to the Jacobi elliptic functions. We show that they are double periodic functions and we begin to investigate double periodic functions by proving the three Liouville theorems. Then we develop, up to a certain point, the theory of the Weierstrass $\wp$-function. This chapter serves as a first introduction but it covers more material (as the previous chapter) than is usually included in a first course on function theory. Chapter 28 provides a proof of the Riemann mapping theorem, a result which must be included in every course, however we do not discuss many applications to conformal mappings such as Schwarz-Christoffel mappings. The final chapter of Part 7 is devoted to power series in several complex variables. The aim is to point out that in several (real and complex) variables the domain of convergence of a power series is a much more delicate question as it is in one (real or complex) dimension.

Two of the appendices handle some topological questions, partly in relation to measure theory. In Appendix III we discuss Möbius transformations further and the geometry of the extended plane. The final appendix is devoted to Bernoulli numbers which we have encountered in several places before.

As in the previous volumes we have provided solutions to all the ca. 275 problems. Moreover, in particular in Part 7, we needed quite a substantial number of figures (ca. 90). All these figures were done by the second named author using LaTeX. Finally a remark about referring to Volume I or II: Theorem II.4.20 means Theorem 4.20 in Volume II and (I.3.12) stands for formula (3.12) in Volume I. Problems marked with a * are more challenging.

Contents

List of Symbols

$\mathbb{N}$	natural numbers
$\mathbb{N}_0 = \mathbb{N} \cup \{0\}$,	
$\mathbb{N}_0^n$	multi-indices (see more in Volume II)
$\mathbb{Z}$	integers
$\mathbb{Q}$	rational numbers
$\mathbb{R}$	real numbers
$\mathbb{R}_+ = \{x \in \mathbb{R} \mid x \geq 0\}$	
$\overline{\mathbb{R}} = \mathbb{R} \cup \{-\infty, \infty\}$	
$\mathbb{D}$	unit disc in $\mathbb{C}$
$\mathbb{H}$	upper half plane in $\mathbb{C}$
$B_r(x) = \{y \in \mathbb{R} \mid \|x - y\| < r\}$	
$D_\rho(z_0) = \{z \in \mathbb{C} \mid |z - z_0| < \rho\}$	
$S^{n-1} = \{x \in \mathbb{R}^n \mid \|x\| = 1\}$	
$A_{r,R}(z_0) = B_r(z_0) \backslash \overline{B_r(z_0)}$	annulus in $\mathbb{C}$
$A_{0,R} = B_R(z_0) \backslash \{z_0\}$	punctured disc in $\mathbb{C}$
$a \vee b = \max(a, b)$	
$a \wedge b = \min(a, b)$	
$\delta_{kl} = \begin{cases} 1, & k = l \\ 0, & k \neq l \end{cases}$	Kronecker delta
$e = \exp(1)$	
$i, \quad i^2 = -1$	
γ	Euler constant
B_k	Bernoulli number
$\beta_{2k} = (-1)^{k-1} B_{2k}$	
$n!$	n factorial
$\binom{n}{k}$	binomial coefficient for $n, k \in \mathbb{N}$
$\binom{a}{k} = \frac{a(a-1)(a-2) \cdot \ldots \cdot (a-k+1)}{k!}$	general binomial coefficents

$(a)_n = a(a+1)(a+2) \cdot \ldots \cdot (a+n-1)$ Pochhammer symbol

$M(n; \mathbb{R})$ $n \times n$-matrices with real elements

$GL(n; \mathbb{R})$ general linear group over $\mathbb{R}$ in dimension n

$SL(n; \mathbb{R})$ special linear group over $\mathbb{R}$ in dimension n

$O(n)$ orthogonal group in $\mathbb{R}^n$

$SO(n)$ special orthogonal group in $\mathbb{R}^n$

$GL(n; \mathbb{C})$ general linear group over $\mathbb{C}$ in dimension n

$SL(n; \mathbb{C})$ special linear group over $\mathbb{C}$ in dimension n

$U(n)$ unitary group in $\mathbb{C}^n$

$SU(n)$ special unitary group in $\mathbb{C}^n$

$\mathrm{Aut}(G)$ automorphism group of a domain $G \subset \mathbb{C}$

$z = x + iy$ complex number

$\bar{z} = x - iy$ conjugate complex number

$\mathrm{Re}\, z = x$ real part of z

$\mathrm{Im}\, z = y$ imaginary part of z

$|z|$ absolute value or modulus of z

$z^\alpha = z_1^{\alpha_1} \cdot \ldots \cdot z_n^{\alpha_n}, \ z \in \mathbb{C}^n, \ \alpha \in \mathbb{N}_0^n$

f, g, h generic symbol for mappings or functions

$f(A)$ image of A under f

$f^{-1}(A')$ pre-image of A' under f

$f^+ = f \vee 0$ positive part of f

$f^- = (-f) \vee 0$ negative part of f

$\Gamma(f)$	graph of f
$\operatorname{supp} f$	support of f
$\operatorname{crit}(f)$	critical points of f
$\Lambda(f)$	Lebesgue set of f
s_f	singularity function of f
$\operatorname{Sing}(f)$	singularities of a meromorphic function
$\operatorname{res}(f, z_0)$	residue of f at z_0
$\operatorname{Pol}(f)$	poles of f
$N(f)$	zeroes of f
J_f	Jacobi matrix of f
$\det J_f$	Jacobi determinant of f
$f = u + iv$	decomposition of f into real and imaginary parts
$f * g$	convolution of f and g
$T^c_{(a_n)}(z) = \displaystyle\sum_{k=0}^{\infty} a_k (z - c)^k$	
$\operatorname{pr}_j : A_1 \times \cdots \times A_N \to A_j$	projection
$\emptyset$	empty set
$\mathcal{P}(\Omega)$	power set of Ω
$A^{\complement}$	complement of the set A
$A_k \uparrow A$	$A_k \subset A_{k+1}$ and $A = \bigcup A_k$
$V/\sim$	equivalence classes in V with respect to the equivalence relation "$\sim$"
$(X, \|\cdot\|)$	normed space
(X, d)	metric space
$(X, \mathcal{O}_X)$	topological space

$\mathcal{O}_X$	topology, open sets in a topological space
$C(x)$	connectivity component of x in a topological space
$\mathcal{C}_X$	closed sets in a topological space
$\mathcal{O}_n$	open sets in $\mathbb{R}$
$\mathcal{C}_n$	closed sets in $\mathbb{R}^n$
$\mathcal{K}_n$	compact sets in $\mathbb{R}^n$
$\mathcal{I}_{r,n}, \mathcal{I}_{l,n}$	right open, left open intervals in $\mathbb{R}^n$
$\mathcal{F}^{(n)}$	figures in $\mathbb{R}^n$
$\mathcal{B}^{(n)}, \mathcal{B}(\mathbb{R}^n)$	Borel σ-field in $\mathbb{R}^n$
$\overline{\mathcal{B}}^{(1)}, \mathcal{B}(\overline{\mathbb{R}})$	Borel σ-field in $\overline{\mathbb{R}}$
$\sigma(\mathcal{E})$	σ-field generated by $\mathcal{E}$
$\delta(\mathcal{E})$	Dynkin system generated by $\mathcal{E}$
$\mathring{G}$	interior of a set G
$\overline{G}$	closure of a set G
∂G	boundary of a set G
$\mathrm{diam}(A) = \sup\{d(x,y) \mid x, y \in A\}$	diameter of a set A
$\mathrm{dist}(A, B)$	distance of A and B
$\mathrm{dist}(A, x) = \mathrm{dist}(A, \{x\})$	
$\mathrm{dist}(x, A) = \mathrm{dist}(\{x\}, A)$	
$\mathrm{vol}_n(K)$	volume of a set $K \subset \mathbb{R}^n$
$g_T = \det T^*T$	Gram determinant
$\{f = g\} = \{x \in X \mid f(x) = g(x)\}$	
$\{f \leq g\} = \{x \in X \mid f(x) \leq g(x)\}$	
$\{f < g\} = \{x \in X \mid f(x) < g(x)\}$	
$\{f \geq g\} = \{x \in X \mid f(x) \geq g(x)\}$	
$\{f > g\} = \{x \in X \mid f(x) > g(x)\}$	

$\mathcal{A}$	generic σ-field
$\mathcal{A}_{\Omega'} = \Omega' \cap \mathcal{A}$	trace σ-field
$(\Omega, \mathcal{A})$	measurable space
$(\Omega, \mathcal{A}, \mu)$	measure space
$(\Omega, \tilde{\mathcal{A}}, \tilde{\mu})$	completion of $(\Omega, \mathcal{A}, \mu)$ with respect to μ
$\mathcal{N}_\mu$	null sets of $(\Omega, \mathcal{A}, \mu)$
$\displaystyle\bigotimes_{j=1}^{N} \mathcal{A}_j = \mathcal{A}_1 \otimes \cdots \otimes \mathcal{A}_N$	product σ-field
$\mathcal{A}/\mathcal{A}' -$ measurable	$f : \Omega \to \Omega'$ measurable with respect to $\mathcal{A}, \mathcal{A}'$
$f^{-1}(\mathcal{A}') = \{f^{-1}(A') \mid A' \in \mathcal{A}'\}$	for $f : \Omega \to \Omega'$
S_x, S_y	$x-, y-$section of $S \subset X \times Y$
μ, ν	generic measures
ϵ_{w_0}	unit measure, Dirac measure at w_0
$\epsilon = \epsilon_0$	Dirac measure at $0 \in \mathbb{R}^n$
$\lambda^{(n)}$	Lebesgue-Borel and Lebesgue measure in $\mathbb{R}^n$
$\lambda_G^{(n)}$	restriction of the Lebesgue-Borel measure to G
β_p^N	binomial distribution
π_α	Poisson distribution
c_α	Cauchy distribution (density)
$g_{\sigma,\alpha}$	Gauss distribution (density)
$\mu_{\Omega'}(A) = \mu(\Omega' \cap A)$	trace of a measure
$T(\mu)$	image measure
$\mu - $ a.e.	μ-almost everywhere
$\sim_\mu$	μ-a.e. equivalence
μ^*	generic outer measure

$\mu \otimes \nu$	product measure
$\mu * \nu$	convolution of measures
$\mu \ll \nu$	absolute continuity of measures
$\mathcal{H}_\alpha$	Hausdorff measure
$\dim_H A$	Hausdorff dimension of A
$\mathcal{H}_A^\delta$	outer Hausdorff measure
P_X	distribution of a random variable
$E(X)$	expectation of a random measure

D_+, D_-, D^+, D_-	Dini numbers or Dini derivatives
$\dfrac{df}{dx}(x) = f'(x)$	derivative of a function of a real variable
$\dfrac{d^k f}{dx^k} = f^{(k)}(x)$	higher order derivatives of a function of a real variable
$\dfrac{df}{dz}(z) = f'(z)$	derivative of a function of a complex variable
$\dfrac{d^k f(z)}{dz^k} = f^{(k)}(z)$	higher order derivative of a function of a complex variable
$\partial^\alpha f, \partial_z^\alpha f$	partial derivates ($\alpha \in \mathbb{N}_0^n$)
$u_x = \dfrac{\partial u}{\partial x}$	
Δ or Δ_2	Laplacian in $\mathbb{R}^2$
Δ_n	Laplacian in $\mathbb{R}^n$
$P(\zeta, z)$	Poisson kernel for the disc in $\mathbb{R}^2$ or $\mathbb{C}$

γ	generic curve (in $\mathbb{C}$ or $\mathbb{R}^2$ or $\mathbb{R}^n$)
γ^{-1}	reversed curve
$\mathrm{tr}(\gamma)$	trace of a curve

$\gamma_1 \oplus \gamma_2$	sum of two curves
ind_γ	index of a curve in $\mathbb{C}$
$\mathrm{int}\,\gamma$	interior of a curve in $\mathbb{C}$
$\mathrm{ext}\,\gamma$	exterior of a curve in $\mathbb{C}$
$Z(t_0, \ldots, t_m)$	partitions of $[t_0, t_m]$ or $[a, b]$, $a = t_0$ and $b = t_m$
$V_Z(g)$, $V_Z(\gamma)$	variation of a function or a curve with respect to the partition $Z(t_0, \ldots, t_m)$
$V(g)$, $V(\gamma)$	(total) variation of a function or a curve
l_γ	length of a curve
$\displaystyle\int_\gamma f(z)\,\mathrm{d}z$	line integral in $\mathbb{C}$
$\gamma_1 \sim \gamma_2$	equivalence of curves
$[\gamma]$	equivalence class of γ
$[\gamma_1] \oplus [\gamma_2] := [\gamma_1 \oplus \gamma_2]$	
$[\gamma_1]^{-1} := [\gamma_1^{-1}]$	
$\gamma = a_1\gamma_1 \tilde{+} a_2\gamma_2 \tilde{+} \cdots \tilde{+} a_M\gamma_M$	cycle
$\simeq \gamma = (-a_1)\gamma_1 \tilde{+} (-a_2)\gamma_2 \tilde{+} \cdots \tilde{+} (-a_M)\gamma_M$	
S_{x_0,x_1}, $[x_0, x_1]$, $\overline{x_0 x_1}$	line segment connecting x_0 and x_1
$\Delta(z_1, z_2, z_3)$	closed triangle with vertices $z_1, z_2,$ $z_3 \in \mathbb{C}$
$\partial\Delta(z_1, z_2, z_3)$	boundary of $\Delta(z_1, z_2, z_3)$
$\|\cdot\|_\infty$	supremum norm
$\|f\|_{\infty,G} = \sup\limits_{x \in G}|f(z)|$	
$\|f\|_{\infty,\mathrm{tr}(\gamma)} = \sup\limits_{z \in \mathrm{tr}(\gamma)}|f(z)|$	
$\|\cdot\|_{L^p} = \|\cdot\|_p$	norm in $L^p(\Omega)$
$N_p(u) = \left(\displaystyle\int_\Omega |u(w)|^p\,\mu(\mathrm{d}w)\right)^{\frac{1}{p}}$	
$N_\infty(u)$	essential supremum of u

$A_r u(x)$	average or mean value of u over $B_r(x)$
$M_r(u)(x)$	average operator
$M(f)(x)$	Hardy-Littlewood maximal function
$J_\epsilon(f)$	Friedrichs mollifier applied to f
$L = L(w_1, w_2)$	periodicity lattice of a double periodic function
$P = P(w_1, w_2)$	fundamental parallelogram of a double periodic function
$\mathrm{ord}(f)$	order of an elliptic function
$P(a, \vec{r})$	polydisc in $\mathbb{C}^n$
$\partial_d P(a, \vec{r})$	distinguished boundary of a polydisc in $\mathbb{C}^n$
$S(\Omega)$	simple functions
$\mathcal{H}(G)$	holomorphic functions
$AC([a, b])$	absolutely continuous functions
$BV([a, b])$	functions of bounded variation
$C(X),\ C(X; \mathbb{C})$	continuous (real-/complex-valued) functions on X
$C_b(X),\ C_b(X; \mathbb{C})$	bounded elements in $C(X),\ C(X; \mathbb{C})$
$C_0(X),\ C_0(X; \mathbb{C})$	elements with compact support in $C(X),\ C(X; \mathbb{C})$
$C_\infty(\mathbb{R}^n)$	continuous functions on $\mathbb{R}^n$ vanishing at infinity
$C^k(X),\ C^k(X; \mathbb{C})$	k-times continuously differentiable elements in $C(X),\ C(X; \mathbb{C})$
$C^\infty(U) = \bigcap_{k \in \mathbb{N}} C^k(U)$	
$C_b^\infty(U) = \bigcap_{k \in \mathbb{N}} C_b^k(U)$	

$$C_0^\infty(U) = \bigcap_{k\in\mathbb{N}} C_0^k(U)$$

$\mathcal{L}^p(\Omega)$	p-fold integrable functions
$\mathcal{L}^\infty(\Omega) = \mathcal{L}^p/\sim \mu, \quad 1 \le p \le \infty$	
$L^1_{loc}(\Omega)$	locally integrable functions (equivalence classes)
χ_A	characteristic function of the set A
exp	exponential function
sin	sine function
cos	cosine function
tan	tangent function
cot	cotangent function
arcsin	inverse sine function
arccos	inverse cosine function
arctan	inverse tangent function
arccot	inverse cotangent function
sinh	hyperbolic sine function
cosh	hyperbolic cosine function
tanh	hyperbolic tangent function
coth	hyperbolic cotangent function
arsinh	inverse hyperbolic sine function
arcosh	inverse hyperbolic cosine function
artanh	inverse hyperbolic cotangent function
arcoth	inverse hyperbolic cotangent function
ln	real logarithmic function $\ln : (0, \infty) \to \mathbb{R}$
$\log z = \ln|z| + i\varphi$	principal branch of the complex logarithmic function

$$\tilde{\ln}(1 + z) = \sum_{k=1}^{\infty} \frac{(-1)^{k-1}}{k} z^k$$

$$\tilde{L}(z) = \tilde{\ln}(1 - z)$$

$$L_{a,b}(z) = b + \tilde{L}\left(\frac{z}{a}\right)$$

$$L(z) = \frac{1}{2}\ln\left(x^2 + y^2\right) + i\arctan\frac{y}{z}$$

$$B_a(z) = \sum_{k=0}^{\infty} \binom{a}{k} z^k \qquad\qquad \text{binomial series}$$

$B(z, w)$	beta function
$D_f(z)$	Dirichlet series associated with f
E_α	Mittag-Leffler function
$E_{\alpha\beta}$	generalised Mittag-Leffler function
$_2F_1$	Gauss hypergeometric series or functions
$_pF_q$	generalised hypergeometric series or function
$G_n = G_n(L) = G_n(w_1, w_2)$	Eisenstein series
J_l	Bessel function of order l
am	amplitude of an elliptic integral

$$\left.\begin{array}{l} \text{sn} \\ \text{cn} \\ \text{dn} \end{array}\right\} \qquad\qquad \text{Jacobi elliptic functions}$$

$\mathcal{P}(z)$	Weierstrass $\wp$-function
$\Gamma(z)$	Γ-function
$\zeta(z)$	ζ-function
$\mu(n)$	Möbius function

$$\psi(z) = \frac{\Gamma'(z)}{\Gamma(z)}$$

Part 6: Measure and Integration Theory

1 A First Look at σ-Fields and Measures

At this stage of our Course much has already been learnt about integration. In Part 1 we introduced the Riemann integral for real-valued functions of one real variable, in Part 2 we provided the underlying theory including all proofs. Then, in Part 4, we investigated integrals of real-valued functions of several real variables, followed by line and surface integrals of vector fields in Part 5. A motivation for introducing integrals was the problem to determine the area (volume) of a figure bounded by the graphs of certain functions. A further central topic was the fundamental theorem of calculus which links the derivative of a function to the integral. Powerful as these classical results are, eventually they lead to many open questions; here are some of these:

- We can integrate certain discontinuous functions, but we cannot (yet) prove the fundamental theorem for them. It would be desirable to have necessary and sufficient conditions the fundamental theorem to hold.

- We can define the length of a rectifiable curve and integrals over rectifiable curves, but we lack an analogous theory for surfaces or more generally hypersurfaces.

- How do we define a volume integral for functions $f : G \to \mathbb{R}$, $G \subset \mathbb{R}^n$, without depending on topological notions, i.e. ∂G?

- Can we derive more satisfactory results for interchanging limits (continuity, differentiability, sequences) and integrals?

As it turns out, within the Riemann (Darboux, Jordan) approach to define an integral we cannot resolve our problems in a convincing manner. Only after set theory was developed, and relying on the first ideas of E. Borel, the mathematician H. Lebesgue in [54] succeeded to construct a new theory of integration which allows us to resolve many of the problems we have with the Riemann integral. In addition this integral coincides with the Riemann integral for continuous functions defined on compact non-degenerate cells $K \subset \mathbb{R}^n$, $n \geq 1$, but it is defined on a much larger class of domains than the Riemann integral. The fundamental idea of Lebesgue was first to handle the problem of defining the volume (length, area) of a set. Starting with a non-empty set Ω he considered families $\mathcal{A} \subset \mathcal{P}(\Omega)$ of its subsets, $\mathcal{P}(\Omega)$ denoting the power set of Ω, which satisfy certain "natural" conditions. Then he assigned a volume, or more generally a "mass" or a "measure" to each set $A \in \mathcal{A}$.

3

The elements of $\mathcal{A}$ are called **measurable** sets and the process of assigning a "measure to A" is formalized by introducing a mapping $\mu : \mathcal{A} \to [0, \infty]$ called the **measure** and $\mu(A)$ is then by definition the volume/mass/measure of A. Of course μ has to fulfil certain conditions. The first observation is the generality of this approach. In principle on one set Ω we can have quite a lot of different families of measurable sets, and in addition on one family $\mathcal{A}$ of measurable sets we can have different measures. It was ca. 25 years after Lebesgue had published his work that A. N. Kolmogorov [49] realised that Lebesgue's thoery of measures and integrals provides the frame needed for probability theory. Thus the generality of the proposed approach was and is one of its advantages. At the same time it allows in the concrete setting of $\mathbb{R}$ or $\mathbb{R}^n$ to resolve most of the problems we encounter when using the Riemann integral. The definition of the Riemann integral (in $\mathbb{R}^n$) depends much on the topological structure of $\mathbb{R}^n$ induced by the Euclidean metric. In Chapter II.19 we have seen that Jordan measurable sets are characterised by properties of their boundary, and the boundary of a set is a topological notion. By separating the concept of measurable sets from topology, Lebesgue gained his advantage. However as a consequence we now have to investigate the relations of measurable sets belonging to $\mathcal{A}$ with the topological structure we have equipped Ω with. The surprise was that this is seriously non-trivial and deeply depends on the underlying model of set theory. This is a rather interesting topic but we cannot discuss it within our Course and refer to [9] or [46].

Before going into the details, a further general remark might be helpful. Due to the importance of measure theory in probability theory, nowadays many textbooks, e.g. [11], [15], [21], [75] or [85], on measure and integration concentrate on the general (abstract) theory and give some emphasis on those parts of relevance for probability theory. On the other hand, when studying real-valued functions other aspects of measure and integration theory are of more interest and there are also some good text books in this direction, e.g. [14], [47], [50], [59], [70], [71], [84], [87] or [94]. In our presentation of the topic, while still aiming to serve the needs of probabilists, we will emphasise the real-variable theory approach.

In case of $\mathbb{R}^n$ we want to integrate functions, say continuous functions to begin with, defined on a non-degenerate compact cell $K := \bigtimes_{j=1}^{n} [a_j, b_j]$, $a_j < b_j$. To achieve this within Lebesgue's approach we need to define a

4

family of measurable sets which contains all non-degenerate compact cells K. Furthermore we want to have that the measure of a compact cell $K = \bigtimes_{j=1}^{n}[a_j, b_j]$ is its natural "Euclidean" volume $\mathrm{vol}_n(K) = \prod_{j=1}^{n}(b_j - a_j)$. If for a moment $\mathcal{C}_n$ denotes all non-degenerate compact cells in $\mathbb{R}^n$, we are searching for a family $\mathcal{A}$ of subsets of $\mathbb{R}^n$ containing all bounded Jordan measurable sets, hence in particular $\mathcal{C}_n$, and for a mapping $\mu : \mathcal{A} \to [0, \infty]$ such that $\mu|_{\mathcal{C}_n} = \mathrm{vol}_n$, i.e. $\mu(K) = \mathrm{vol}_n(K) = \prod_{j=1}^{n}(b_j - a_j)$ for $K = \bigtimes_{j=1}^{n}[a_j, b_j]$. From μ we would further expect

$$\mu(\emptyset) = 0;$$
$$\mu(G) \geq 0, \quad G \subset \mathbb{R}^n;$$
$$\mu\left(\bigcup_{j=1}^{\infty} G_j\right) = \sum_{j=1}^{\infty} \mu(G_j) \text{ if } G_l \cap G_k = \emptyset \text{ for } k \neq l;$$
$$\mu(G + x) = \mu(G), \quad x \in \mathbb{R}, \quad G \subset \mathbb{R}^n, \quad \text{(translation invariance)};$$

and

$$\mu(R(G)) = \mu(G), \quad G \subset \mathbb{R}^n, \quad R \in SO(n), \quad \text{(rotation invariance)}.$$

In particular for $G_1 \cap G_2 = \emptyset$ we expect

$$\mu(G_1 \cup G_2) = \mu(G_1) + \mu(G_2)$$

and this implies for $G_1 \subset G$ that

$$\mu(G \backslash G_1) = \mu(G) - \mu(G_1)$$

since $G = (G \backslash G_1) \cup G_1$ and $(G \backslash G_1) \cap G_1 = \emptyset$. For a single point $c \in \mathbb{R}^n$ we expect

$$\mu(\{c\}) = 0$$

which implies, say for $n = 1$, that

$$\mu([a, b]) = \mu([a, b)) = \mu((a, b]) = \mu((a, b)) = b - a.$$

From this we deduce, since $\mathbb{Q}$ is countable, that

$$\mu(\mathbb{Q} \cap [a, b]) = 0 \text{ and } \mu([a, b] \backslash \mathbb{Q}) = \mu([a, b]),$$

i.e. we may take away from $[a, b]$ an infinite, but countable set, and the "length" remains unchanged. It turns out that on $\mathcal{P}(\mathbb{R}^n)$ a mapping μ with all these properties does not exist. We refer to Appendix II where we will discuss some of the underlying problems.

Let us collect some ideas for the conditions a family $\mathcal{A}$ of measurable subsets $A \subset \Omega$, $\Omega \neq \emptyset$ but otherwise arbitrary, should satisfy. It is natural to assume that Ω itself is a measurable set, i.e. $\Omega \in \mathcal{A}$. Moreover, if $A_1, A_2 \in \mathcal{A}$ are two measurable sets we expect that $A_1 \cup A_2$ is measurable too, which then clearly extends to finitely many sets, i.e. $A_j \in \mathcal{A}$, $j = 1, \ldots, N$, implies $A_1 \cup \ldots \cup A_n \in \mathcal{A}$. Since $\Omega = A \cup A^{\complement}$, it is now natural to require also $A^{\complement} \in \mathcal{A}$ for $A \in \mathcal{A}$. However we know that finite unions of cells or hyper-rectangles can never lead to balls, ellipsoids, etc, i.e. more general Jordan measurable sets. Lebesgue's insight was that denumerable unions will fit the bill, i.e. he added the condition that $\bigcup_{j \in \mathbb{N}} A_j \in \mathcal{A}$ holds for $(A_j)_{j \in \mathbb{N}}$, $A_j \in \mathcal{A}$ (which we have already stated above). Before giving a formal definition, we recall that a set is called **countable** if it is the bijective image of a mapping with domain $\mathbb{N}$, while a set is called **denumerable** if it is either finite or countable. By definition the empty set $\emptyset$ is finite. With these preparations we now start to develop Lebesgue's theory.

Definition 1.1. *Let $\Omega \neq \emptyset$ be a set. A family $\mathcal{A}$ of subsets of Ω is called a σ-field or σ-algebra in Ω if*

$$\Omega \in \mathcal{A}; \tag{1.1}$$

$$A \in \mathcal{A} \; \text{implies} \; A^{\complement} \in \mathcal{A}; \tag{1.2}$$

$$A_j \in \mathcal{A}, \; j \in \mathbb{N}, \; \text{implies} \; \bigcup_{j \in \mathbb{N}} A_j \in \mathcal{A}. \tag{1.3}$$

*The elements of $\mathcal{A}$ are called **measurable sets** and the pair $(\Omega, \mathcal{A})$ is called a **measurable space**.*

Remark 1.2. A. If we always choose $A_j = A_N$ in (1.3) for $j \geq N$ we find that $A_1, \ldots, A_N \in \mathcal{A}$ implies $\bigcup_{j=1}^{N} A_j \in \mathcal{A}$. Since $\Omega^{\complement} = \emptyset$ it follows also that $\emptyset \in \mathcal{A}$ for every σ-field.
B. In the case that we consider on Ω several σ-fields $\mathcal{A}_1, \ldots, \mathcal{A}_M$, we may, and may have to speak of $\mathcal{A}_j$-**measurable sets**.

Example 1.3. A. The power set $\mathcal{P}(\Omega)$ is obviously a σ-field as is the set $\mathcal{A} := \{\emptyset, \Omega\}$.

B. Let $\Omega \neq \emptyset$ be any set and define

$$\mathcal{A} := \left\{ A \in \mathcal{P}(\Omega) \,\big|\, A \text{ or } A^{\complement} \text{ is denumerable} \right\}.$$

Then $\mathcal{A}$ is a σ-field in Ω, see Problem 1.

C. Let $(\Omega, \mathcal{A})$ be a measurable space and $\Omega' \subset \Omega$ a non-empty subset. We define the **trace** or **trace σ-field** of $\mathcal{A}$ in Ω' by

$$\mathcal{A}_{\Omega'} := \Omega' \cap \mathcal{A} := \left\{ \Omega' \cap A \,\big|\, A \in \mathcal{A} \right\}. \tag{1.4}$$

Note that this construction is completely analogous to that of the trace or relative topology, see Proposition II.1.24, and the proof, which we leave as an exercise, goes analogously. The notation $\Omega' \cap \mathcal{A}$ is quite common, hence we shall adopt this, but of course $\Omega' \cap \mathcal{A}$ cannot be interpreted as an intersection of two sets.

D. Let Ω and Ω' be two non-empty sets and $\mathcal{A}'$ a σ-field in Ω'. Further let $f : \Omega \to \Omega'$ be a mapping. For $A' \subset \Omega'$ we denote as usual by $f^{-1}(A') \subset \Omega$ the pre-image of A' under f. The family

$$f^{-1}(\mathcal{A}') := \left\{ f^{-1}(A') \,\big|\, A' \in \mathcal{A}' \right\} \tag{1.5}$$

is a σ-field in Ω. Indeed, we have $f^{-1}(\Omega') = \Omega$, thus $\Omega \in f^{-1}(\mathcal{A}')$ and for $A' \in \mathcal{A}'$ it follows that $(f^{-1}(A'))^{\complement} = \Omega \backslash f^{-1}(A') = f^{-1}(\Omega') \backslash f^{-1}(A') = f^{-1}(\Omega' \backslash A') = f^{-1}(A'^{\complement})$ and therefore $A \in f^{-1}(\mathcal{A}')$ implies $A^{\complement} \in f^{-1}(\mathcal{A}')$. Finally let $(A_j)_{j \in \mathbb{N}}$ be a countable collection of elements of $f^{-1}(\mathcal{A}')$. We can find $A'_j \in \mathcal{A}'$ such that $A_j = f^{-1}(A'_j)$ and therefore we find

$$\bigcup_{j \in \mathbb{N}} A_j = \bigcup_{j \in \mathbb{N}} f^{-1}(A'_j) = f^{-1}\left(\bigcup_{j \in \mathbb{N}} A'_j \right)$$

and since $\bigcup_{j \in \mathbb{N}} A'_j \in \mathcal{A}'$ we conclude that $\bigcup_{j \in \mathbb{N}} A_j \in f^{-1}(\mathcal{A}')$ implying that $f^{-1}(\mathcal{A}')$ is a σ-field.

Given a measurable space $(\Omega, \mathcal{A})$, since $\emptyset = \Omega^{\complement}$ it follows that

$$\emptyset \in \mathcal{A} \tag{1.6}$$

7

and for $A_j \in \mathcal{A}$, $j \in \mathbb{N}$, we find $A_j^{\complement} \in \mathcal{A}$ and therefore

$$\bigcap_{j \in \mathbb{N}} A_j = \bigcup_{j \in \mathbb{N}} A_j^{\complement} \in \mathcal{A} \tag{1.7}$$

which also includes the statement that

$$\bigcap_{j=1}^{N} A_j \in \mathcal{A} \text{ for } A_j \in \mathcal{A}, \ j = 1, \ldots, N. \tag{1.8}$$

Moreover, since $A \backslash B = A \cap B^{\complement}$ we have

$$A \backslash B \in \mathcal{A} \text{ for } A, B \in \mathcal{A}. \tag{1.9}$$

Our first theorem gives a powerful tool to construct σ-fields.

Theorem 1.4. *The intersection $\bigcap_{j \in I} \mathcal{A}_j$ of an arbitrary family $(\mathcal{A}_j)_{j \in I}$ of σ-fields in Ω is a σ-field in Ω.*

Proof. First we note that

$$\bigcap_{j \in I} \mathcal{A}_j = \left\{ A \in \mathcal{P}(\Omega) \,\middle|\, A \in \mathcal{A}_j \text{ for all } j \in I \right\}.$$

Since $\Omega \in \mathcal{A}_j$ for all $j \in I$ it follows that $\Omega \in \bigcap_{j \in I} \mathcal{A}_j$. Furthermore, if $A \in \bigcap_{j \in I} \mathcal{A}_j$ then $A \in \mathcal{A}_j$ for all $j \in I$, hence $A^{\complement} \in \mathcal{A}_j$ for all $j \in I$ implying that $A^{\complement} \in \bigcap_{j \in I} \mathcal{A}_j$. Eventually, suppose that $A_k \in \mathcal{A}_j$ for all $j \in I$ and $k \in \mathbb{N}$, which yields by (1.7) for all $j \in I$ that $\bigcap_{k \in \mathbb{N}} A_k \in \mathcal{A}_j$ and therefore $\bigcap_{k \in \mathbb{N}} A_k \in \bigcap_{j \in I} \mathcal{A}_j$. $\qquad \square$

This theorem justifies

Definition 1.5. *Let $\mathcal{E} \subset \mathcal{P}(\Omega)$, $\Omega \neq \emptyset$. The σ-field*

$$\sigma(\mathcal{E}) := \bigcap \left\{ \mathcal{A} \,\middle|\, \mathcal{E} \subset \mathcal{A} \text{ and } \mathcal{A} \text{ is a } \sigma\text{-field in } \Omega \right\} \tag{1.10}$$

*is called the σ-**field generated by** $\mathcal{E}$ and we call $\mathcal{E}$ a **generator** of $\sigma(\mathcal{E})$.*

Note that since $\mathcal{E} \subset \mathcal{P}(\Omega)$ and $\mathcal{P}(\Omega)$ is a σ-field we take an intersection over a non-empty system and therefore $\sigma(\mathcal{E})$ is indeed a σ-field in Ω.

Example 1.6. A. If $\mathcal{E}$ is a σ-field in Ω then $\mathcal{E} = \sigma(\mathcal{E})$, in particular we have $\sigma(\sigma(\mathcal{E})) = \sigma(\mathcal{E})$.
B. For $\mathcal{E} = \{A\}$, $A \subset \Omega$, we find $\sigma(\mathcal{E}) = \{\emptyset, A, A^\complement, \Omega\}$.
C. If $\mathcal{E}_1 \subset \mathcal{E}_2$ then $\sigma(\mathcal{E}_1) \subset \sigma(\mathcal{E}_2)$. However it may happen that $\mathcal{E}_1 \subset \mathcal{E}_2$, $\mathcal{E}_1 \neq \mathcal{E}_2$ but $\sigma(\mathcal{E}_1) = \sigma(\mathcal{E}_2)$ as we will see soon below in Corollary 1.8.

Our next goal is to find a "good" σ-field on $\mathbb{R}$ which contains all intervals. An easy way to introduce such a σ-field is to take the σ-field generated by all intervals. However we want to give more emphasis on topological notions and start with

Definition 1.7. *The σ-field generated by all open subsets $\mathcal{O}_1$ of the real line is called the **Borel σ-field** in $\mathbb{R}$ and is denoted by $\mathcal{B}^{(1)}$ or $\mathcal{B}(\mathbb{R})$, hence*

$$\sigma(\mathcal{O}_1) = \mathcal{B}^{(1)}. \tag{1.11}$$

*The elements of $\mathcal{B}^{(1)}$ are called the **Borel sets** of $\mathbb{R}$.*

Corollary 1.8. *The Borel σ-field $\mathcal{B}^{(1)}$ is generated by all open intervals.*

Proof. From Example 1.6.C we deduce that the σ-field $\mathcal{A}$ generated by all open intervals is contained in $\mathcal{B}^{(1)}$. By Theorem I.19.27 we know that for every open set $A \subset \mathbb{R}$ there exists a denumerable family $I_j \subset \mathbb{R}$, $j \in J \subset \mathbb{N}$, of open intervals such that $A = \bigcup_{j \in J} I_j$. This implies however for every open set $A \subset \mathbb{R}$ that $A \in \mathcal{A}$ which yields

$$\mathcal{B}^{(1)} = \sigma(\mathcal{O}_1) \subset \sigma(\mathcal{A}) = \mathcal{A} \subset \mathcal{B}^{(1)}.$$

$\square$

The next step is to determine further generators of $\mathcal{B}^{(1)}$. For this let us collect some simple observations about generators of σ-fields, compare with Problem 2.

Lemma 1.9. *Let $\Omega \neq \emptyset$ be a set and $\mathcal{E}, \mathcal{E}_1, \mathcal{E}_2 \subset \mathcal{P}(\Omega)$.*
A. *If we denote by $\mathcal{E}^\complement$ the system*

$$\mathcal{E}^\complement := \left\{ E^\complement \mid E \in \mathcal{E} \right\}$$

then we have

$$\sigma(\mathcal{E}^\complement) = \sigma(\mathcal{E}). \tag{1.12}$$

B. *For* $\mathcal{I} := \left\{ F = \bigcup_{j \in J} E_j \,\middle|\, E_j \in \mathcal{E},\, J \subset \mathbb{N} \right\}$ *it follows that*

$$\sigma(\mathcal{I}) = \sigma(\mathcal{E}). \tag{1.13}$$

C. *If* $\sigma(\mathcal{E}_1) = \sigma(\mathcal{E}_2)$ *then*

$$\sigma(\mathcal{E}_1 \cup \mathcal{E}_2) = \sigma(\mathcal{E}_1). \tag{1.14}$$

We use the following notations:

$$
\begin{aligned}
\mathcal{O}_n & \quad \text{all open sets in } \mathbb{R}^n, & (1.15) \\
\mathcal{C}_n & \quad \text{all closed sets in } \mathbb{R}^n, & (1.16) \\
\mathcal{K}_n & \quad \text{all compact sets in } \mathbb{R}^n, & (1.17) \\
\mathcal{I}_r & := \left\{ [a,b) \,\middle|\, a < b,\, a,b \in \mathbb{R} \right\}, & (1.18) \\
\mathcal{I}_l & := \left\{ (a,b] \,\middle|\, a < b,\, a,b \in \mathbb{R} \right\}, & (1.19) \\
\mathcal{I}_{r,n} & := \mathcal{I}_r^n, \quad \mathcal{I}_{l,n} := \mathcal{I}_l^n. & (1.20)
\end{aligned}
$$

Corollary 1.10. *We have*

$$\mathcal{B}^{(1)} = \sigma(\mathcal{C}_1) = \sigma(\mathcal{K}_1). \tag{1.21}$$

Proof. Since $\mathcal{C}_1 = \mathcal{O}_1^{\complement}$ the first equality follows from Lemma 1.9.A. Moreover, since $\mathcal{K}_1 \subset \mathcal{C}_1$ we have $\sigma(\mathcal{K}_1) \subset \sigma(\mathcal{C}_1)$. For $C \in \mathcal{C}_1$ we have $C = \bigcup_{k \in \mathbb{N}} (C \cap [-k, k])$, but $C \cap [-k, k]$ is compact implying by Lemma 1.9.B that $\sigma(\mathcal{C}_1) \subset \sigma(\mathcal{K}_1)$. $\square$

Corollary 1.11. *The following holds*

$$\mathcal{B}^{(1)} = \sigma(\mathcal{I}_r) = \sigma(\mathcal{I}_l). \tag{1.22}$$

Proof. For $a < b$ we have

$$[a, b) = \bigcap_{j \in \mathbb{N}} \left(a - \frac{1}{j}, b \right)$$

and

$$(a, b] = \bigcap_{j \in \mathbb{N}} \left(a, b + \frac{1}{j} \right).$$

This implies
$$\sigma(\mathcal{I}_r) \subset \mathcal{B}^{(1)} \text{ and } \sigma(\mathcal{I}_l) \subset \mathcal{B}^{(1)}.$$
Now for an open interval $I = (a, b)$ we know that
$$(a, b) = \bigcup \{[c, d) \,|\, [c, d) \subset (a, b) \text{ and } c, d \in \mathbb{Q}\}$$
and
$$(a, b) = \bigcup \{(c, d] \,|\, (c, d] \subset (a, b) \text{ and } c, d \in \mathbb{Q}\}.$$
Since $\mathbb{Q}$ is countable, we deduce from Corollary 1.8
$$\mathcal{B}^{(1)} \subset \sigma(\mathcal{I}_r) \text{ and } \mathcal{B}^{(1)} \subset \sigma(\mathcal{I}_l).$$

$\square$

Remark 1.12. The proof of Corollary 1.11 yields even more, namely that $\mathcal{B}^{(1)}$ is already generated by all half-open intervals (either from the right or from the left) with rational end points. For $a, b \in \mathbb{Q}$, $a < b$, we have $[a, b) = (-\infty, b) \backslash (-\infty, a)$ which implies that
$$\mathcal{B}^{(1)} = \sigma\left(\{[a, b) \subset \mathbb{R} \,|\, a < b, \, a, b \in \mathbb{Q}\}\right) \subset \sigma\left(\{(-\infty, c) \subset \mathbb{R} \,|\, c \in \mathbb{Q}\}\right).$$
However $(-\infty, c) = \bigcup_{\substack{k \in \mathbb{N} \\ -k < c}} [-k, c)$ and therefore we deduce that
$$\sigma\left(\{(-\infty, c) \subset \mathbb{R} \,|\, c \in \mathbb{Q}\}\right) \subset \sigma(\mathcal{I}_r) = \mathcal{B}^{(1)},$$
i.e. we have proved $\mathcal{B}^{(1)} = \sigma\left(\{(-\infty, c) \subset \mathbb{R} \,|\, c \in \mathbb{Q}\}\right)$. Since for $d \in \mathbb{R}$ it follows that $(-\infty, d) = \bigcup_{\substack{c \in \mathbb{Q} \\ d < c}} (-\infty, c)$ we find further that $\mathcal{B}^{(1)} = \sigma(\{(-\infty, d) \subset \mathbb{R} \,|\, d \in \mathbb{R}\})$. With analogous arguments it follows
$$\begin{aligned} \mathcal{B}^{(1)} &= \sigma\left(\{(c, \infty) \subset \mathbb{R} \,|\, c \in \mathbb{Q}\}\right) = \sigma\left(\{(d, \infty) \subset \mathbb{R} \,|\, d \in \mathbb{R}\}\right) \\ &= \sigma\left(\{[c, \infty) \subset \mathbb{R} \,|\, c \in \mathbb{Q}\}\right) = \sigma\left(\{[d, \infty) \subset \mathbb{R} \,|\, d \in \mathbb{R}\}\right) \\ &= \sigma\left(\{(-\infty, c] \subset \mathbb{R} \,|\, c \in \mathbb{Q}\}\right) = \sigma\left(\{(-\infty, d] \subset \mathbb{R} \,|\, d \in \mathbb{R}\}\right). \end{aligned}$$

In later considerations we will benefit from the fact that we are free to choose any of these generators, for different purposes different choices will be convenient. Note that we need to consider intervals with rational end points since we have to rely on countable operations. Of course we can replace $\mathbb{Q}$ by any countable and dense set.

The next step is to study the Borel σ-field in $\mathbb{R}^n$:

Definition 1.13. *The σ-field $\mathcal{B}^{(n)} := \mathcal{B}(\mathbb{R}^n) := \sigma(\mathcal{O}_n)$ is called the **Borel** σ-field in $\mathbb{R}^n$. Its elements we call the **Borel sets** in $\mathbb{R}^n$.*

We are of course interested in further generators of $\mathcal{B}^{(n)}$, however most important is to understand the relation of $\mathcal{B}^{(n)}$ to $\mathcal{B}^{(1)}$. Clearly, we do **not** expect $\mathcal{B}^{(n)}$ to be the n-times Cartesian product of $\mathcal{B}^{(1)}$. In fact, as we will see in Problem 4, the product of two σ-fields is in general not a σ-field. When studying analysis in $\mathbb{R}^n$ we often could reduce theorems to analogous results in $\mathbb{R}$. For example a sequence $(x^{(k)})_{k \in \mathbb{N}}$, $x^{(k)} = \left(x_1^{(k)}, \ldots, x_n^{(k)} \right) \in \mathbb{R}^n$, converges to $x = (x_1, \ldots, x_n) \in \mathbb{R}^n$ if and only if the sequences $\left(x_j^{(k)} \right)_{k \in \mathbb{N}}$, $1 \le j \le n$, converges to x_j in $\mathbb{R}$. A way to understand such a result is by looking at the projections $\mathrm{pr}_j : \mathbb{R}^n \to \mathbb{R}$, $x \mapsto x_j$. It turns out that projections are also central to understanding certain σ-fields on products of measurable spaces and for this we need some preparations. As a first step we extend Example 1.3.D.

Example 1.14. Let $(\Omega_j, \mathcal{A}_j)_{j \in J}$ be an arbitrary family of measurable spaces. Further let $\Omega \ne \emptyset$ be a set and $f_j : \Omega \to \Omega_j$, $j \in J$. Then

$$\sigma \left(\{ f_j \mid j \in J \} \right) := \sigma \left(\bigcup_{j \in J} f_j^{-1}(\mathcal{A}_j) \right) \tag{1.23}$$

is a σ-field in Ω as the reader may easily check by inspection. Thus we may use families of mappings to construct σ-fields. Note that $\sigma \left(\{ f_j \mid j \in J \} \right)$ is constructed in such a way that for every $j \in J$ the pre-image of a measurable set $A_j \in \mathcal{A}_j$ under f_j is a measurable set in $\sigma \left(\{ f_j \mid j \in J \} \right)$, i.e. $f_j^{-1}(A_j) \in \sigma \left(\{ f_j \mid j \in J \} \right)$ for all $A_j \in \mathcal{A}_j$ and all $j \in J$. In addition $\sigma \left(\{ f_j \mid j \in J \} \right)$ is the smallest σ-field in Ω with this property. This observation will soon become very important.

To clarify the situation further we give

Definition 1.15. *Let $(\Omega, \mathcal{A})$ and $(\Omega', \mathcal{A}')$ be two measurable spaces and $f : \Omega \to \Omega'$ be a mapping. We call f **measurable** or $\mathcal{A}/\mathcal{A}'$ **measurable** if the pre-image $f^{-1}(A')$ of every measurable set $A' \in \mathcal{A}'$ is measurable in $\mathcal{A}$, i.e. $A' \in \mathcal{A}'$ implies $f^{-1}(A') \in \mathcal{A}$.*

Remark 1.16. A. We can now characterise $\sigma \left(\{ f_j \mid j \in J \} \right)$ in Example 1.14 as the smallest σ-field in Ω for which all mappings $f_j : \Omega \to \Omega_j$ are measurable.

B. Note the analogy of measurability and continuity. If $(X, \mathcal{O}_X)$ and $(Y, \mathcal{O}_Y)$ are two topological spaces then $f : X \to Y$ is continuous if and only if the pre-image of an open set in $\mathcal{O}_Y$ is open in $\mathcal{O}_X$, i.e. $f^{-1}(U_Y) \in \mathcal{O}_X$ for every $U_Y \in \mathcal{O}_Y$, see Definition II.3.24.

C. Since for pre-images we have $(g \circ f)^{-1}(A) = f^{-1}(g^{-1}(A))$ it follows that the composition of measurable mappings is measurable.

It is helpful to note that we can check measurability of a mapping on a generator.

Lemma 1.17. *Let $(\Omega, \mathcal{A})$ and $(\Omega', \mathcal{A}')$ be two measurable spaces and $f : \Omega \to \Omega'$ a mapping. Further let $\mathcal{E}'$ be a generator of $\mathcal{A}'$, i.e. $\mathcal{A}' = \sigma(\mathcal{E}')$. The $\mathcal{A}/\mathcal{A}'$-measurability of f is equivalent to the condition that the pre-image of every $E' \in \mathcal{E}'$ belongs to $\mathcal{A}$, i.e. $f^{-1}(E') \in \mathcal{A}$ for $E' \in \mathcal{E}'$.*

Proof. If f is measurable then it follows by definition that $f^{-1}(E') \in \mathcal{A}$ for all $E' \in \mathcal{E}'$. Now assume that $f^{-1}(E') \in \mathcal{A}$ for all $E' \in \mathcal{A}'$ and consider $\tilde{\mathcal{A}}' := \{A' \subset \Omega' \,|\, f^{-1}(A') \in \mathcal{A}\}$. Clearly we have $\mathcal{E}' \subset \mathcal{A}'$. Moreover, $\tilde{\mathcal{A}}'$ is a σ-field in Ω' since

i) $\Omega' \in \tilde{\mathcal{A}}'$ because $f^{-1}(\Omega') = \Omega \in \mathcal{A}$,

ii) $f^{-1}\left(A'^{\complement}\right) = f^{-1}(A')^{\complement} \in \mathcal{A}$ for $A' \in \mathcal{A}'$,

iii) $f^{-1}\left(\bigcup_{j=1}^{\infty} A'_j\right) = \bigcup_{j=1}^{\infty} f^{-1}(A'_j) \in \mathcal{A}$ for $A'_j \in \mathcal{A}'$.

Thus it follows that
$$\mathcal{A}' = \sigma(\mathcal{E}') \subset \sigma(\tilde{\mathcal{A}}') = \tilde{\mathcal{A}}',$$
i.e. $f^{-1}(A') \in \mathcal{A}$ for all $A' \in \mathcal{A}'$ which means that f is $\mathcal{A}/\mathcal{A}'$-measurable. $\square$

Corollary 1.18. *Every continuous mapping $f : \mathbb{R}^m \to \mathbb{R}^n$ is $\mathcal{B}^{(m)}/\mathcal{B}^{(n)}$-measurable.*

Proof. By the definition of $\mathcal{B}^{(n)}$ the open sets $\mathcal{O}_n$ are a generator of $\mathcal{B}^{(n)}$ and for $U \in \mathcal{O}_n$ it follows from the continuity of f that $f^{-1}(U) \in \mathcal{O}_m \subset \mathcal{B}^{(m)}$. Now Lemma 1.17 yields the result. $\square$

We return to $\mathcal{B}^{(n)}$ and try to identify $\mathcal{B}^{(n)}$ as a type of product since the underlying set $\mathbb{R}^n$ is a product. Again we will see a certain analogy to topological spaces. It is worth handling products of general measurable spaces first.

Let $(\Omega_j, \mathcal{A}_j)_{j=1,\ldots,N}$ be a finite family of measurable spaces and let $\Omega := \Omega_1 \times \cdots \times \Omega_N$ be the product of the sets $\Omega_1, \ldots, \Omega_N$. We can define the jth projection by

$$\mathrm{pr}_j : \Omega \to \Omega_j, \ \mathrm{pr}_j(\omega) = \mathrm{pr}_j(\omega_1, \ldots, \omega_N) = \omega_j. \tag{1.24}$$

Definition 1.19. *For the finite family of measurable spaces $(\Omega_j, \mathcal{A}_j)_{j=1,\ldots,N}$ the σ-field*

$$\mathcal{A} := \bigotimes_{j=1}^{N} \mathcal{A}_j := \mathcal{A}_1 \otimes \cdots \otimes \mathcal{A}_N := \sigma\left(\{\mathrm{pr}_j \,|\, j = 1, \ldots, N\}\right) \tag{1.25}$$

*is called the **product σ-field** in $\Omega = \Omega_1 \times \cdots \times \Omega_N$.*

Thus at the moment we have on $\mathbb{R}^n$ two σ-fields, the Borel σ-field $\mathcal{B}^{(n)}$ and the product σ-field induced by n copies of $(\mathbb{R}, \mathcal{B}^{(1)})$ with the help of the projections. Our aim is to identify these two σ-fields. The following theorem tells us how to construct a generator for $\bigotimes_{j=1}^{N} \mathcal{A}_j$ given generators $\mathcal{E}_j$ of $\mathcal{A}_j$, i.e. $\sigma(\mathcal{E}_j) = \mathcal{A}_j$, and as it turns out this will also resolve our problem.

Theorem 1.20. *Let $(\Omega_j, \mathcal{A}_j)_{j=1,\ldots,N}$ be a finite family of measurable spaces with generators $\mathcal{E}_j$ of $\mathcal{A}_j$, $\sigma(\mathcal{E}_j) = \mathcal{A}_j$. Suppose that for every j, $1 \leq j \leq N$, the generator $\mathcal{E}_j$ contains a sequence $(E_k^{(j)})_{k\in\mathbb{N}}$ such that $E_k^{(j)} \subset E_{k+1}^{(j)}$ and $\bigcup_{k\in\mathbb{N}} E_k^{(j)} = \Omega_j$. Then the product σ-field $\bigotimes_{j=1}^{N} \mathcal{A}_j$ is generated by*

$$\left\{ E_1 \times \cdots \times E_N \,\middle|\, E_j \in \mathcal{E}_j, \ 1 \leq j \leq N \right\}.$$

Proof. The statement we have to prove is that for a σ-field $\mathcal{A}$ in Ω the following holds: every projection pr_j is $\mathcal{A}/\mathcal{A}_j$-measurable if and only if $E_1 \times \cdots \times E_N \in \mathcal{A}$ for arbitrary $E_j \in \mathcal{A}_j$. By Lemma 1.17 each pr_j is $\mathcal{A}/\mathcal{A}_j$-measurable if $E_j \in \mathcal{A}_j$ implies $\mathrm{pr}_j^{-1}(E_j) \in \mathcal{A}$. In this case it follows that $E_1 \times \cdots \times E_N = \mathrm{pr}_1^{-1}(E_1) \cap \cdots \cap \mathrm{pr}_N^{-1}(E_N) \in \mathcal{A}$. Suppose now that $E_1 \times \cdots \times E_N \in \mathcal{A}$ for all $E_j \in \mathcal{A}_j$. We deduce that

$$F_k := E_k^{(1)} \times \cdots \times E_k^{(j-1)} \times E_j \times E_k^{(j+1)} \times \cdots \times E_k^{(N)} \in \mathcal{A}.$$

14

By construction we have $F_k \subset F_{k+1}$ and

$$\bigcup_{k\in\mathbb{N}} F_k = \Omega_1 \times \cdots \times \Omega_{j-1} \times E_j \times \Omega_{j+1} \times \cdots \times \Omega_N = \mathrm{pr}_j^{-1}(E_j),$$

and hence $\mathrm{pr}_j^{-1}(E_j) \in \mathcal{A}$ for all $1 \leq j \leq N$ and $E_j \in \mathcal{A}_j$. $\qquad\square$

Corollary 1.21. *The following holds*

$$\mathcal{B}^{(n)} = \sigma\left(\left\{\mathrm{pr}_j \,\middle|\, j = 1, \ldots, n\right\}\right).$$

Proof. We take for $j = 1, \ldots, n$ as Ω_j the set $\mathbb{R}$, i.e. $\Omega_j = \mathbb{R}$, as $\mathcal{A}_j$ we choose $\mathcal{B}^{(1)}$ and as $\mathcal{E}_j$ we take $\mathcal{O}_1$. Since $(-k, k) \in \mathcal{O}_1$, $(-k, k) \subset (-k - 1, k + 1)$ and $\bigcup_{k=1}^{\infty}(-k, k) = \mathbb{R}$, Theorem 1.20 yields that the system $\left\{U_1 \times \cdots \times U_n \,\middle|\, U_j \in \mathcal{O}_1\right\}$ generates $\bigotimes_{j=1}^{n} \mathcal{B}^{(1)} = \sigma\left(\left\{\mathrm{pr}_j \,\middle|\, j = 1, \ldots, n\right\}\right)$. We now want to show that this system also generates $\mathcal{B}^{(n)}$. We are done if we can prove that $J := \left\{(a_1, b_1) \times \cdots \times (a_n, b_n) \,\middle|\, a_j < b_j,\, a_j,\, b_j,\, j = 1, \ldots, n\right\}$ generates $\mathcal{B}^{(n)}$. Clearly $\sigma(J) \subset \mathcal{B}^{(n)}$. On the other hand, the inequality $\|x\|_\infty \leq \|x\|_2 \leq \sqrt{n}\|x\|_\infty$ for $x \in \mathbb{R}^n$ implies (for more details see Problem 5) that every open set $U \in \mathcal{O}_n$ has the representation $U = \bigcup_{I \in J,\, I \subset U} I$ which yields $\sigma(J) \subset \mathcal{B}^{(n)}$, i.e. $\mathcal{B}^{(n)} = \bigotimes_{j=1}^{n} \mathcal{B}^{(1)}$. $\qquad\square$

We provide more generators of $\mathcal{B}^{(n)}$ in the following

Exercise 1.22. *(Problem 6) Prove that $\mathcal{B}^{(n)}$ is generated by $\mathcal{K}_n$, $\mathcal{C}_n$ as well as by $\mathcal{I}_{r,n}$, $\mathcal{I}_{l,n}$ and each of the following systems:*

$$\left\{(a, b) \subset \mathbb{R} \,\middle|\, a < b,\, a, b \in \mathbb{Q}\right\}^n \ \text{ and } \ \left\{(a, b) \subset \mathbb{R} \,\middle|\, a < b,\, a, b \in \mathbb{R}\right\}^n;$$
$$\left\{[a, b] \subset \mathbb{R} \,\middle|\, a < b,\, a, b \in \mathbb{Q}\right\}^n \ \text{ and } \ \left\{[a, b] \subset \mathbb{R} \,\middle|\, a < b,\, a, b \in \mathbb{R}\right\}^n.$$

As already mentioned the Borel σ-field will be of utmost importance to us, however

$$\mathcal{B}^{(n)} \neq \mathcal{P}(\mathbb{R}^n), \tag{1.26}$$

a result which is non-trivial, its proof needs the Axiom of Choice (or one of its equivalent statements). Indeed there are several surprises in measure theory and their understanding needs more involved set theory. Some of these results we will discuss in Appendix II.

We now return to the problem of defining a measure and it is advantageous to first handle the general case before turning to $\mathbb{R}^n$ and our original question.

Definition 1.23. *Let $(\Omega, \mathcal{A})$ be a measurable space. A mapping $\mu : \mathcal{A} \to [0, \infty]$ is called a **measure** on $\mathcal{A}$ (or $(\Omega, \mathcal{A})$) if*

$$\mu(\emptyset) = 0 \tag{1.27}$$

and

$$\mu\left(\bigcup_{j \in \mathbb{N}} A_j\right) = \sum_{j=1}^{\infty} \mu(A_j) \tag{1.28}$$

*for every sequence $(A_j)_{j \in \mathbb{N}}$, $A_j \in \mathcal{A}$, of mutually disjoint sets, i.e. $A_k \cap A_l = \emptyset$ for $k \neq l$. The triple $(\Omega, \mathcal{A}, \mu)$ is called a **measure space**.*

Before we give examples and explore the consequences of this definition we need to fix the arithmetic with $+\infty$ and $-\infty$. By $\overline{\mathbb{R}}$ we denote the set $[-\infty, +\infty]$ and $\overline{\mathbb{R}}_+ = [0, \infty]$. For $a \in \mathbb{R}$ we set

$$a + \infty = +\infty, \quad a - \infty = -\infty; \tag{1.29}$$

as well as

$$+\infty + \infty = +\infty, \quad -\infty - \infty = -\infty. \tag{1.30}$$

Further, for $a \in (0, \infty]$ we define

$$a \cdot (+\infty) = +\infty \text{ and } a \cdot (-\infty) = -\infty \tag{1.31}$$

and for $a \in [-\infty, 0)$

$$a(+\infty) = -\infty \text{ and } a(-\infty) = +\infty. \tag{1.32}$$

However the terms

$$\infty - \infty, \quad (+\infty) \cdot (-\infty), \quad (-\infty) \cdot (-\infty), \quad (-\infty) \cdot (+\infty)$$

and

$$0 \cdot (+\infty), \quad 0 \cdot (-\infty)$$

are not defined (although we later on need to define in some special cases $0 \cdot (+\infty) = 0$). Often we write ∞ for $+\infty$.

Let $(\Omega, \mathcal{A}, \mu)$ be a measure space. We call μ a **finite** measure and $(\Omega, \mathcal{A}, \mu)$ a **finite measure space** if $\mu(\Omega) < \infty$. For a finite measure it follows for every $A \in \mathcal{A}$

$$\mu(A) \leq \mu(A) + \mu(\Omega \backslash A) = \mu(\Omega) < \infty,$$

i.e. $\mu(A) < \infty$ for all $A \in \mathcal{A}$. In the case where $\mu(\Omega) = 1$ we call μ a **probability measure** and $(\Omega, \mathcal{A}, \mu)$ a **probability space**.

Example 1.24. Let $(\Omega, \mathcal{A})$ be an arbitrary measure space and $\omega_0 \in \Omega$ fixed. Define $\epsilon_{\omega_0} : \mathcal{A} \to [0, \infty]$ by

$$\epsilon_{\omega_0}(A) := \begin{cases} 1, & \omega_0 \in A \\ 0, & \omega_0 \notin A. \end{cases} \tag{1.33}$$

Then ϵ_{ω_0} is a probability measure called the **unit mass at** ω_0 or the **Dirac measure at** ω_0. Clearly $\epsilon_{\omega_0}(\emptyset) = 0$. Further for a countable collection $(A_j)_{j \in \mathbb{N}}$, $A_j \in \mathcal{A}$, of mutually disjoint sets we find

$$\epsilon_{\omega_0}\left(\bigcup_{j \in \mathbb{N}} A_j\right) = \begin{cases} 1, & \text{if } \omega_0 \in A_{j_0} \text{ for some } j_0 \in \mathbb{N} \\ 0, & \text{if } \omega_0 \notin A_{j_0} \text{ for some } j_0 \in \mathbb{N} \end{cases}$$

which entails

$$\epsilon_{\omega_0}\left(\bigcup_{j \in \mathbb{N}} A_j\right) = \sum_{j \in \mathbb{N}} \epsilon_{\omega_0}(A_j).$$

Since $\epsilon_{\omega_0}(\Omega) = 1$ it follows that ϵ_{ω_0} is always a probability measure.

The following construction is quite universal and gives rise to many probability measures, indeed to all discrete (probability) measures.

Example 1.25. Let $(\Omega, \mathcal{A})$ be a measurable space and $(\alpha_k)_{k \in \mathbb{N}_0}$ a sequence of non-negative numbers $\alpha_k \in \mathbb{R}$, $\alpha_k \geq 0$, such that $\sum_{k=0}^{\infty} \alpha_k = \alpha < \infty$, i.e. the series is absolutely convergent. For $k \in \mathbb{N}_0$ pick ω_k and define

$$\mu := \sum_{k=0}^{\infty} \alpha_k \epsilon_{\omega_k}. \tag{1.34}$$

Then μ is a finite measure on $(\Omega, \mathcal{A})$ and

$$\nu := \sum_{k=0}^{\infty} \frac{\alpha_k}{\alpha} \epsilon_{\omega_k} \tag{1.35}$$

is a probability measure. Obviously, since $\mu(\Omega) = \alpha$, if μ is a measure then ν is a probability measure. Since $\epsilon_{\omega_k}(\emptyset) = 0$ for all $k \in \mathbb{N}_0$ we find $\mu(\emptyset) = \sum_{k=0}^{\infty} \alpha_k \epsilon_{\omega_k}(\emptyset) = 0$. Moreover, for a sequence $(A_l)_{l \in \mathbb{N}}$ of mutually disjoint elements of $\mathcal{A}$ we find

$$\mu\left(\bigcup_{l\in\mathbb{N}} A_l\right) = \sum_{k\in\mathbb{N}_0} \alpha_k \epsilon_{\omega_k}\left(\bigcup_{l\in\mathbb{N}} A_l\right)$$

$$= \sum_{k\in\mathbb{N}_0} \alpha_k \sum_{l=1}^{\infty} \epsilon_{\omega_k}(A_l)$$

$$= \sum_{l=1}^{\infty}\left(\sum_{k\in\mathbb{N}_0} \alpha_k \epsilon_{\omega_k}(A_l)\right)$$

$$= \sum_{l=1}^{\infty} \mu(A_l).$$

Example 1.26. The **Poisson distribution** π_α, $\alpha > 0$, on $(\mathbb{R}, \mathcal{B}^{(1)})$ is given by

$$\pi_\alpha := \sum_{k=0}^{\infty} e^{-\alpha} \frac{\alpha^k}{k!} \epsilon_k \tag{1.36}$$

and since $\sum_{k=0}^{\infty} \frac{\alpha^k}{k!} = e^\alpha$ it is a probability measure. (In Chapter 5 we will understand the term "distribution".) In the case where $\alpha_k = 0$ for all but finitely many values $\alpha_{k_0}, \ldots, \alpha_{k_N} > 0$ the measure μ in (1.34) is given as a finite sum

$$\mu = \sum_{j=0}^{N} \alpha_{k_j} \epsilon_{\omega_{k_j}}. \tag{1.37}$$

An example of a corresponding probability measure is the **Bernoulli** or **binomial distributions** with parameter N and $p \in [0,1]$ defined by

$$\beta_p^N := \sum_{k=0}^{N} \binom{N}{k} p^k (1-p)^{N-k} \epsilon_k, \tag{1.38}$$

note that $\sum_{k=0}^{N} \binom{N}{k} p^k (1-p)^{N-k} = (p + (1-p))^N = 1$.

In the following let $(\Omega, \mathcal{A}, \mu)$ be an arbitrary measure space. For $A, B \in \mathcal{A}$ we always have $A \cup B = A \cup (B \setminus A)$ and $A \cap (B \setminus A) = \emptyset$ implying

$$\mu(A \cup B) = \mu(A) + \mu(B \setminus A) \tag{1.39}$$

and since $B = (A \cap B) \cup (B \backslash A)$ it follows that

$$\mu(B) = \mu(A \cap B) + \mu(B \backslash A). \tag{1.40}$$

Adding (1.39) and (1.40) we arrive at

$$\mu(A \cup B) + \mu(A \cap B) + \mu(B \backslash A) = \mu(A) + \mu(B) + \mu(B \backslash A),$$

and if $\mu(B \backslash A) < \infty$, which holds, for example, for every finite measure, we have

$$\mu(A \cup B) + \mu(A \cap B) = \mu(A) + \mu(B). \tag{1.41}$$

However, for $\mu(B \backslash A) = \infty$ both $\mu(A \cup B)$ and $\mu(B)$ must be equal to $+\infty$ and hence (1.41) holds in all cases. In particular we have for $A \subset B$

$$\mu(B) = \mu(A \cap B) + \mu(B \backslash A) = \mu(A) + \mu(B \backslash A)$$

implying the **monotonicity of measures**:

$$A \subset B \text{ implies } \mu(A) \leq \mu(B). \tag{1.42}$$

Let us return to the problem of defining on $\mathbb{R}^n$, better on $\mathcal{B}^{(n)}$, a measure $\lambda^{(n)}$ with the properties mentioned earlier on, in particular with the property that for all compact cells $K := \bigtimes_{j=1}^{n}[a_j, b_j]$, $a_j < b_j$, we have

$$\lambda^{(n)}(K) = \prod_{j=1}^{n}(b_j - a_j). \tag{1.43}$$

We can easily extend (1.43) to $A \in \mathcal{I}_r^n \cup \mathcal{I}_l^n$ or all open cells $\mathring{K}$, but we can not easily extend this definition to $\mathcal{B}^{(n)}$. The general question we have to address is the following: given $\Omega \neq \emptyset$ and a system $\mathcal{G} \subset \mathcal{P}(\Omega)$ of subsets. On $\mathcal{G}$ a mapping $\nu : \mathcal{G} \to [0, \infty]$ is defined. When can we extend ν to a measure on $\sigma(\mathcal{G})$?

The answer is given by Carathéodory's extension theorem which we will discuss in the next chapter.

Problems

1. a) Prove that for a non-empty set Ω a σ-field is given by $\mathcal{A} := \{A \in \mathcal{P}(\Omega)|A$ or A^C is denumerable$\}$.

    b) Prove that the trace $\mathcal{A}_{\Omega'} := \Omega' \cap \mathcal{A}$ as defined in (1.4) is a σ-field in Ω'.

2. Prove Lemma 1.9.

3. Let (X, d) be a metric space and denote by $\mathcal{B}(X) = \sigma(X)$ the Borel σ-field in X, i.e. the σ-field generated by all open sets $\mathcal{O}_X$ in X. Prove that $\mathcal{B}(X)$ is also generated by all closed sets in X. Let $\mathcal{D} = \{B_r(x) \subset X | r > 0$ and $x \in X\}$ be the system of all open balls in X and suppose that X is separable, i.e. X has a countable dense subset. Does D generate $\mathcal{B}(X)$?

4. Let $(\Omega_j, \mathcal{A}_j), j = 1, 2$, be two measurable spaces. Prove that in general $A_1 \times A_2$ is not a σ-field in $\Omega_1 \times \Omega_2$.

5. Let $U \subset \mathbb{R}^n$ be open. Prove that $U = \bigcup_{I \subset J, I \subset U} I$, where J is defined as in the proof of Corollary 1.21.

6. Solve Exercise 1.22.

7. a) Let $f : \mathbb{R} \to \mathbb{R}$ be a bijective continuous mapping and choose on $\mathbb{R}$ the Borel σ-field $\mathcal{B}^{(1)}$. What can we say about $\sigma(f)$?

    b) Consider the three measurable spaces $(\Omega_1, \mathcal{A}_1) := (\mathbb{R}, \mathcal{P}(\mathbb{R}))$, $(\Omega_2, \mathcal{A}_2) := (\mathbb{R}, \mathcal{B}^{(1)})$ and $(\Omega_3, \mathcal{A}_3) := (\mathbb{R}, \{\mathbb{R}, \emptyset, A, A^C\})$ where $A \neq \emptyset$ is any set in $\mathbb{R}$. Further, always consider on $\mathbb{R}$ the Euclidean topology. For $1 \leq k, l \leq 3$ we can consider continuous mappings $g_{kl} : \mathbb{R} \to \mathbb{R}$. When is a generic map g_{kl} $\mathcal{A}_k/\mathcal{A}_l$-measurable.

8. Let $(\Omega_j, \mathcal{A}_j), j = 1, 2, 3$, be measurable spaces and $f : \Omega_1 \to \Omega_2, g : \Omega_2 \to \Omega_3$ two measurable mappings. Prove that $g \circ f : \Omega_1 \to \Omega_3$ is measurable. Now prove that $f : \mathbb{R} \to \mathbb{R}^n$ is $\mathcal{B}^{(1)}/\mathcal{B}^{(n)}$-measurable if and only if $f_j : \mathbb{R} \to \mathbb{R}$ is $\mathcal{B}^{(1)}/\mathcal{B}^{(1)}$-measurable, where $f = (f_1, \ldots, f_n)$.

9. a) Let $(\Omega, \mathcal{A})$ be the measurable space from Example 1.3.B. Prove that $\nu : \mathcal{A} \to [0, 1], \nu(A) = \begin{cases} 0, & A \text{ is denumerable} \\ 1, & A^C \text{ is denumerable} \end{cases}$ is a probability measure.

20

b) On $(\mathbb{Z}, \mathcal{P}(\mathbb{Z}))$ consider $\mu : \mathcal{P}(\mathbb{Z}) \to \overline{\mathbb{R}}$, $\mu(A) = \sum_{k \in \mathbb{Z}} \epsilon_k(A)$. Show that on $\mathcal{P}(\mathbb{Z})$ and $\mu(A + m) = \mu(A)$ for all $m \in \mathbb{Z}$ and $A \in \mathcal{P}(\mathbb{Z})$.

c) Let $(\Omega, \mathcal{A}, \mu_j)$, $j \in \mathbb{N}$, be probability spaces. Prove that $(\Omega, \mathcal{A}, \mu)$ where $\mu = \sum_{j=1}^{\infty} \frac{1}{2^j} \mu_j$ is a further probability space.

10.   a) Let $(\Omega, \mathcal{A}, \mu)$ be a measure space and $\Omega' \in \mathcal{A}$. Prove that $\mu_{\Omega'}$ defined by $\mu_{\Omega'}(A) := \mu(\Omega' \cap A)$ is a further measure on $\mathcal{A}$.

b) Let $(\Omega, \mathcal{A}, \mu)$ be a measure space and $\tilde{\mathcal{A}} \subset \mathcal{A}$ a σ-field. Show that $\mu|_{\tilde{\mathcal{A}}}$ is a measure on $\tilde{\mathcal{A}}$.

2 Extending Pre-Measures. Carathéodory's Theorem

While in the last chapter we introduced the framework for eventually resolving our problem of defining a measure and integral for real-valued functions defined on suitable domains in $\mathbb{R}^n$, we still cannot apply this framework to this concrete situation, i.e. $(\mathbb{R}^n, \mathcal{B}^{(n)})$. The measure $\lambda^{(n)}$ we are seeking on $\mathcal{B}^{(n)}$ should satisfy

$$\lambda^{(n)}(K) = \prod_{j=1}^{n}(b_j - a_j)$$

for every compact cell $K = \bigtimes_{j=1}^{n}[a_j, b_j] \subset \mathbb{R}^n$, but at the moment it is not defined on $\mathcal{B}^{(n)}$ but only on (closed, open and half-open) hyper-rectangles. Thus we need to find a way to extend certain mappings defined on a subset of $\mathcal{B}^{(n)}$ (or more generally of a σ-field $\mathcal{A}$) to measures on $\mathcal{B}^{(n)}$ (or the σ-field $\mathcal{A}$). This problem will be solved by Carathéodory's theorem which we are going to prove in this chapter.

First we want to explore further what we actually know about the set-function which we want to extend to the measure $\lambda^{(n)}$ on $\mathcal{B}^{(n)}$. In our discussion of the Riemann integral in Volume II we learnt that partitions are helpful and even important, however we cannot find partitions of non-degenerate compact cells into non-degenerate compact cells. This does not even work for compact intervals. Note that a partition of a set Ω consists of **mutually disjoint** subsets. For this reason we switch from non-degenerate compact cells to elements of $\mathcal{I}_{r,n}$ (we could also have chosen $\mathcal{I}_{l,n}$). Recall that $A \in \mathcal{I}_{r,n} = (\mathcal{I}_r)^n$ if

$$A = \bigtimes_{j=1}^{n}[a_j, b_j), \quad a_j < b_j \text{ and } a_j, b_j \in \mathbb{R}, 1 \le j \le n.$$

First we note that for $A \in \mathcal{I}_{r,n}$ such that $a_j < b_j$ for all $1 \le j \le n$ it follows that $\overline{A}$ is a non-degenerate compact cell and further that the "natural" volume for A and $\overline{A}$ is equal and given by $\prod_{j=1}^{n}(b_j - a_j)$, i.e. for $\lambda^{(n)}$ we will have $\lambda^{(n)}(\overline{A}) = \lambda^{(n)}(A)$.

By switching from compact cells to elements of $\mathcal{I}_{r,n}$ we gain that we can find partitions of A into elements of $\mathcal{I}_{r,n}$: each family of partitions of $[a_j, b_j)$ induces a partition of A into elements of $\mathcal{I}_{r,n}$, compare with II.18. We now introduce the n-**dimensional figures** $\mathcal{F}^{(n)}$ as finite unions of elements belonging to $\mathcal{I}_{r,n}$:

$$\mathcal{F}^{(n)} := \left\{ F \in \mathbb{R}^n \,\middle|\, F = \bigcup_{k=1}^{M} A_k,\ A_k \in \mathcal{I}_{r,n} \right\}. \tag{2.1}$$

Here are some examples of elements on $\mathcal{F}^{(2)}$:

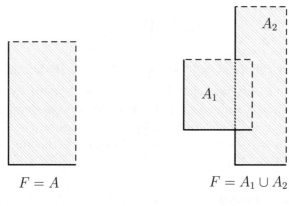

$F = A$

Figure 2.1

$F = A_1 \cup A_2$

Figure 2.2

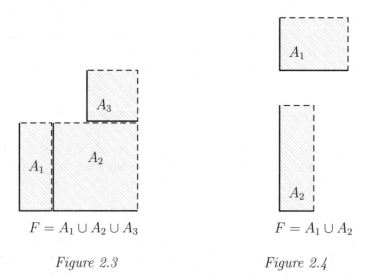

$F = A_1 \cup A_2 \cup A_3$

Figure 2.3

$F = A_1 \cup A_2$

Figure 2.4

Given $F \in \mathcal{F}^{(n)}$ such that $F = \bigcup_{k=1}^{M} A_k$ for mutually disjoint sets $A_k \in \mathcal{I}_{r,n}$ we can easily define

$$\lambda^{(n)}(F) := \sum_{k=1}^{M} \lambda^{(n)}(A_k). \tag{2.2}$$

By standard arguments we often used in II.18 and which we can easily modify to this situation (see Problem 1) it is clear that the definition of $\lambda^{(n)}(F)$ is independent of the choice of the partition $(A_k)_{k=1,\dots,M}$. In order to study the structure of $\mathcal{F}^{(n)}$ let us introduce some helpful notation. For $a, b \in \mathbb{R}^n$ we write

$$a \leq b \text{ if and only if } a_j \leq b_j \text{ for all } 1 \leq j \leq n,$$

and

$$[a, b) := \underset{j=1}{\overset{n}{\times}} [a_j, b_j).$$

Given partitions $a_j = a_j^{(0)} < a_j^{(1)} < \dots < a_j^{(k_j)} = b_j$ of $[a_j, b_j)$,

$$[a_j, b_j) = \bigcup_{l=0}^{k_j-1} \left[a_j^{(l)}, a_j^{(l+1)} \right),$$

we obtain a partition of $[a, b)$. More precisely we find with $\alpha = (\alpha_1, \dots, \alpha_n)$, $\alpha_j \in \{0, \dots, k_j - 1\}$ and $Q := \{\alpha \in \mathbb{N}_0^n \mid \alpha_j \in \{0, \dots, k_j - 1\}\}$ that

$$[a, b) = \bigcup_{\alpha \in Q} [a^{(\alpha)}, a^{(\alpha+1)})$$

where $[a^{(\alpha)}, a^{(\alpha+1)}) := \underset{j=1}{\overset{n}{\times}} \left[a_j^{(\alpha_j)}, a_j^{(\alpha_j+1)} \right)$. By construction we have for $\alpha \neq \alpha'$ that $[a^{(\alpha)}, a^{(\alpha+1)}) \cap [a^{(\alpha')}, a^{(\alpha'+1)}) = \emptyset$. We note further that $[c, d) \subset [a, b)$ if and only if $a_j \leq c_j$ and $d_j \leq b_j$ for $j = 1, \dots, n$. Now let $[c, d) \subset [a, b)$. Then a partition of $[a_j, b_j)$ is given by the points of $\{a_j, c_j, d_j, b_j\}$ and therefore

$$[a, b) = \bigcup [a^{(\alpha)}, a^{(\alpha+1)})$$

with $a_j^{(0)} = a_j$, $a_j^{(1)} = c_j$, $a_j^{(2)} = d_j$ and $a_j^{(3)} = b_j$. Since $[c, d)$ is one of the sets in this union it follows that $[a, b) \backslash [c, d)$ is a (disjoint) union of elements in $\mathcal{I}_{r,n}$, hence $[a, b) \backslash [c, d)$ is a figure for $[c, d) \subset [a, b)$. This is almost the proof of

Lemma 2.1. *For $[a, b), [c, d) \in \mathcal{I}_{r,n}$ it follows that $[a, b) \cap [c, d) \in \mathcal{F}^{(n)}$ and $[a, b) \backslash [c, d) \in \mathcal{F}^{(n)}$.*

Proof. With $a \vee c := (\max(a_1, c_1), \ldots, \max(a_n, c_n))$ and $b \wedge d := (\min(b_1, d_1), \ldots, \min(b_n, d_n))$ we find that $[a, b) \cap [c, d) = [a \vee c, b \wedge d)$ or otherwise $[a, b) \cap [c, d) = \emptyset$, i.e. $[a, b) \cap [c, d) \in \mathcal{F}^{(n)}$ since $\emptyset = [a, a) \in \mathcal{F}^{(n)}$. Since $[a, b) \backslash [c, d) = [a, b) \backslash ([a, b) \cap [c, d))$, and since we know by now that $[a, b) \cap [c, d) \in \mathcal{F}^{(n)}$, in order to prove $[a, b) \backslash [c, d) \in \mathcal{F}^{(n)}$ we may assume that $[c, d) \subset [a, b)$. But then we can apply our preceeding considerations. $\square$

Lemma 2.2. *Every figure $F \in \mathcal{F}^{(n)}$ is a finite union of mutually disjoint elements belonging to $\mathcal{I}_{r,n}$.*

Proof. Let $F = \bigcup_{k=1}^{N} A_k$, $A_k \in \mathcal{I}_{r,n}$. We choose $A \in \mathcal{I}_{r,n}$ and such that $F \subset A$. Now the intervals $[a_{j,k}, b_{j,k})$, $1 \leq j \leq n$ and $1 \leq k \leq N$, induce a partition of the intervals $[a_j, b_j)$ where $A_k = \bigtimes_{j=1}^{n}[a_{j,k}, b_{j,k})$ and $A = \bigtimes_{j=1}^{n}[a_j, b_j)$, see Figure 2.5.

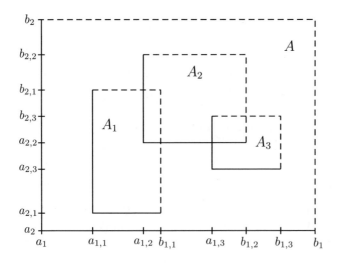

Figure 2.5

Hence we obtain a partition $(K_\alpha)_{\alpha=1,\ldots,M}$, $K_\alpha \in \mathcal{I}_{r,n}$ of A. We observe that

$$F = \bigcup_{k=1}^{N} A_k = \bigcup_{k=1}^{N} \bigcup_{\alpha=1}^{M} (K_\alpha \cap A_k)$$

and $K_\alpha \cap A_k \in \mathcal{I}_{r,n}$. Moreover, for $\alpha \neq \alpha'$ it follows that $(K_\alpha \cap A_k) \cap (K_{\alpha'} \cap A_k) = \emptyset$, and by the construction of the sets K_α, if $(K_\alpha \cap A_k) \cap (K_\alpha \cap A_{k'}) \neq \emptyset$ then they must coincide which implies the lemma. $\square$

Definition 2.3. *Let $\Omega \neq \emptyset$ be a set. We call $\mathcal{R} \subset \mathcal{P}(\Omega)$ a **ring** in Ω if*

$$\emptyset \in \mathcal{R} \tag{2.3}$$

$$A, B \in \mathcal{R} \ \text{implies} \ A \backslash B \in \mathcal{R}; \tag{2.4}$$

$$A, B \in \mathcal{R} \ \text{implies} \ A \cup B \in \mathcal{R}. \tag{2.5}$$

*If in addition $\Omega \in \mathcal{R}$ then $\mathcal{R}$ is called an **algebra** in Ω.*

Remark 2.4. A. Since $A \cap B = A \backslash (A \backslash B)$ it follows for a ring $\mathcal{R}$ that

$$A, B \in \mathcal{R} \ \text{implies} \ A \cap B \in \mathcal{R}. \tag{2.6}$$

B. A family $\mathcal{R} \subset \mathcal{P}(\Omega)$ is an algebra if and only if $\Omega \in \mathcal{R}$, $A \in \mathcal{R}$ implies $A^\complement \in \mathcal{R}$ and $A, B \in \mathcal{R}$ implies $A \cup B \in \mathcal{R}$. Indeed, for an algebra these conditions are obvious since $A^\complement = \Omega \backslash A$. Conversely, since $\Omega^\complement = \emptyset$ and $A \backslash B = A \cap B^\complement = (A^\complement \cup B)^\complement$ we deduce that these conditions imply (2.3) - (2.5).

Now we claim

Theorem 2.5. *The n-dimensional figures $\mathcal{F}^{(n)}$ form a ring.*

Proof. Since $[a, a) = \emptyset$ it follows that $\emptyset \in \mathcal{F}$. Moreover, for $F_1, F_2 \in \mathcal{F}^{(n)}$ with $F_1 = \bigcup_{k=1}^{M} A_k$, $F_2 = \bigcup_{l=1}^{N} B_l$, where $A_k, B_l \in \mathcal{I}_{r,n}$, we obtain

$$F_1 \cup F_2 = \bigcup_{k=1}^{M} \bigcup_{l=1}^{N} (A_k \cup B_l) \in \mathcal{F}^{(n)}.$$

In order to see that $F_1 \backslash F_2 \in \mathcal{F}^{(n)}$ for $F_1, F_2 \in \mathcal{F}^{(n)}$ we now assume that in the representations of F_1 and F_2 used above the families $(A_k)_{1 \leq k \leq M}$ and $(B_l)_{1 \leq l \leq N}$ consist of mutually disjoint sets $A_k, B_l \in \mathcal{I}_{r,n}$, which in light of Lemma 2.2 can be done. This implies

$$F_1 \backslash F_2 = \bigcup_{k=1}^{M} \left(\bigcap_{l=1}^{N} A_k \backslash B_l \right)$$

and it remains to prove that $\bigcap_{l=1}^{N} (A_k \backslash B_l) \in \mathcal{F}^{(n)}$. We already know that $A_k \backslash B_l \in \mathcal{F}^{(n)}$, hence we need to prove that the intersection of two, hence finitely many figures is a figure. But for two figures F_1 and F_2 with the given representation we have

$$F_1 \cap F_2 = \bigcup_{k=1}^{M} \bigcup_{l=1}^{N} (A_k \cap B_l)$$

and $A_k \cap B_l \in \mathcal{F}^{(n)}$ by Lemma 2.1, and the theorem follows. $\qquad\square$

Since $\mathcal{F}^{(n)}$ is not a σ-field, we cannot define a measure on $\mathcal{F}^{(n)}$. However it turns out that $\lambda^{(n)}$ is on $\mathcal{F}^{(n)}$ a pre-measure in the sense of the following:

Definition 2.6. *Let $\mathcal{R}$ be a ring in Ω.*
A. *We call $\mu : \mathcal{F}^{(n)} \to [0, \infty]$ a **content** if*

$$\mu(\emptyset) = 0 \qquad\qquad (2.7)$$

and

$$\mu\left(\bigcup_{l=1}^{N} A_l\right) = \sum_{l=1}^{N} \mu(A_l) \qquad\qquad (2.8)$$

for every finite family $(A_l)_{l=1,\dots,N}$ of mutually disjoint sets $A_l \in \mathcal{R}$.
B. *If (2.7) holds for $\mu : \mathcal{F}^{(n)} \to [0, \infty]$ and*

$$\mu\left(\bigcup_{l=1}^{\infty} A_l\right) = \sum_{l=1}^{\infty} \mu(A_l) \qquad\qquad (2.9)$$

*for every countable family $(A_l)_{l \in \mathbb{N}}$ of mutually disjoint sets $A_l \in \mathcal{R}$ such that $\bigcup_{l=1}^{\infty} A_l \in \mathcal{R}$ then we call μ a **pre-measure** on $\mathcal{R}$ (or on $(\Omega, \mathcal{R})$).*

Exercise 2.7. *If $(\mu_k)_{k \in \mathbb{N}}$ is a sequence of pre-measures on a ring $\mathcal{R}$ and $(a_k)_{k \in \mathbb{N}}$, $a_k \geq 0$, a sequence of non-negative numbers then $\mu := \sum_{k=1}^{\infty} a_k \mu_k$ is a pre-measure on $\mathcal{R}$ too.*

Remark 2.8. A pre-measure on a σ-field is a measure, and if $(\Omega, \mathcal{A}, \mu)$ is a measure space and $\mathcal{R} \subset \mathcal{A}$ a ring, then $\mu|_{\mathcal{R}}$ is a pre-measure.

As in the case of a measure, compare with the end of Chapter 1, we can prove (2.10) - (2.12) in the following:

Lemma 2.9. *Let $\mathcal{R}$ be a ring in Ω and $\mu : \mathcal{R} \to [0, \infty)$ a pre-measure. For $A, B \in \mathcal{R}$ we have*

$$\mu(A \cup B) + \mu(A \cap B) = \mu(A) + \mu(B), \tag{2.10}$$
$$A \subset B \text{ implies } \mu(A) \leq \mu(B), \tag{2.11}$$
$$A \subset B \text{ and } \mu(A) < \infty \text{ implies } \mu(B \setminus A) = \mu(B) - \mu(A), \tag{2.12}$$

and

$$\mu\left(\bigcup_{j=1}^{\infty} A_j\right) \leq \sum_{j=1}^{\infty} \mu(A_j) \tag{2.13}$$

for any, not necessarily mutually disjoint, sequence $(A_j)_{j \in \mathbb{N}}$, $A_j \in \mathcal{R}$ such that $\bigcup_{j=1}^{\infty} A_j \in \mathcal{R}$. Furthermore for $A_0 \in \mathcal{R}$ and a sequence $(A_k)_{k \in \mathbb{N}}$, $A_k \in \mathcal{R}$, we have

$$A_0 \subset \bigcup_{k \in \mathbb{N}} A_k \text{ implies } \mu(A_0) \leq \sum_{k=1}^{\infty} \mu(A_k). \tag{2.14}$$

Proof. It remains to prove (2.14). Since $A_0 = \bigcup_{k \in \mathbb{N}}(A_0 \cap A_k)$ and since by (2.11) $\mu(A_0 \cap A_k) \leq \mu(A_k)$ we may assume that $A_0 = \bigcup_{k \in \mathbb{N}} A_k$. With $B_1 := A_k \setminus \left(\bigcup_{j=1}^{k-1} A_j\right)$, $k > 1$, we find $A_0 = \bigcup_{k \in \mathbb{N}} A_k = \bigcup_{k \in \mathbb{N}} B_k$, but $B_k \cap B_l = \emptyset$ for $k \neq l$, and therefore

$$\mu\left(\bigcup_{k=1}^{N} B_k\right) = \sum_{k=1}^{N} \mu(B_k) \leq \sum_{k=1}^{N} \mu(A_k) \leq \sum_{k=1}^{\infty} \mu(A_k).$$

The sequence $\left(\mu\left(\bigcup_{k=1}^{N} B_k\right)\right)_{k \in \mathbb{N}}$ is increasing with upper bound $\mu\left(\bigcup_{k \in \mathbb{N}} B_k\right)$ $= \mu\left(\bigcup_{k \in \mathbb{N}} A_k\right)$ implying now

$$\mu(A_0) \leq \mu\left(\bigcup_{k \in \mathbb{N}} A_k\right) \leq \sum_{k=1}^{\infty} \mu(A_k).$$

$\square$

Note that (2.14) yields in particular for all $(A_k)_{k \in \mathbb{N}}$, $A_k \in \mathcal{R}$ such that $\bigcup_{k \in \mathbb{N}} A_k \in \mathcal{R}$, that

$$\mu\left(\bigcup_{k \in \mathbb{N}} A_k\right) \leq \sum_{k=1}^{\infty} \mu(A_k). \tag{2.15}$$

The next result deals with sequential continuity properties of a pre-measure and it will be needed to construct the Lebesgue pre-measure $\lambda^{(n)}$ on the ring $\mathcal{F}^{(n)}$.

Theorem 2.10. *Let μ be a pre-measure on the ring $\mathcal{R}$.*
A. *For a sequence $(A_k)_{k \in \mathbb{N}}$, $A_k \in \mathcal{R}$, such that $A_k \uparrow A$, i.e. $A_k \subset A_{k+1}$ and $\bigcup_{k \in \mathbb{N}} A_k = A \in \mathcal{R}$, it follows*

$$\lim_{k \to \infty} \mu(A_k) = \mu(A). \tag{2.16}$$

B. *Let $(A_k)_{k \in \mathbb{N}}$, $A_k \in \mathcal{R}$, be a sequence such that $A_k \downarrow A$, i.e. $A_k \supset A_{k+1}$ and $\bigcap_{k \in \mathbb{N}} A_k = A$. If $\mu(A_k) < \infty$ for all $k \in \mathbb{N}$ then*

$$\lim_{k \to \infty} \mu(A_k) = \mu(A). \tag{2.17}$$

Proof. **A.** With $A_0 := \emptyset$ and $B_k := A_k \backslash A_{k-1}$, $k \in \mathbb{N}$, we now have a sequence $(B_k)_{k \in \mathbb{N}}$ of mutually disjoint sets $B_k \in \mathcal{R}$ such that $A = \bigcup_{k \in \mathbb{N}} B_k$ and $A_k = \bigcup_{j=1}^{k} B_j$. It follows that

$$\mu(A) = \sum_{k=1}^{\infty} \mu(B_k) = \lim_{m \to \infty} \sum_{k=1}^{m} \mu(B_k) = \lim_{m \to \infty} \mu(A_m).$$

B. First we note that

$$\mu(A_1 \backslash A_k) = \mu(A_1) - \mu(A_k)$$

and $A_k \downarrow A$ yields $(A_1 \cap A_k) \uparrow (A_1 \cap A)$ and all these sets belong to $\mathcal{R}$. Now by part A we find

$$\mu(A_1 \backslash A) = \lim_{k \to \infty} (A_1 \backslash A_k) = \mu(A_1) - \lim_{k \to \infty} \mu(A_k),$$

and since $A \subset A_k$ and $\mu(A) < \infty$ we have $\mu(A_1 \backslash A) = \mu(A_1) - \mu(A)$ implying the result. $\qquad \square$

Corollary 2.11. *Let μ be a pre-measure on the ring $\mathcal{R}$. The statement of part B is equivalent to the following: if $(A_k)_{k \in \mathbb{N}}$ is a sequence in $\mathcal{R}$ such that $A_k \downarrow \emptyset$ and $\mu(A_k) < \infty$ for all $k \in \mathbb{N}$ then*

$$\lim_{k \to \infty} \mu(A_k) = 0 \tag{2.18}$$

holds.

Proof. Of course part B of Theorem 2.10 implies the statement of the corollary. Now let $A_k \downarrow A$, $\mu(A_k) < \infty$. Then $A_k \backslash A \downarrow \emptyset$, and since $A_k \backslash A \subset A_k$ it follows that $\mu(A_k \backslash A) < \infty$. Hence, by (2.18) we have

$$\lim_{k \to \infty} \mu(A_k \backslash A) = 0$$

and from $\mu(A_k \backslash A) = \mu(A_k) - \mu(A)$ we now deduce that

$$\lim_{k \to \infty} \mu(A_k) = \mu(A).$$

$\square$

It is helpful to introduce some definitions:

Definition 2.12. A. *For a measure or a pre-measure we call the property*

$$\mu\left(\bigcup_{k=1}^{\infty} A_k\right) = \sum_{k=1}^{\infty} \mu(A_k)$$

*for every countable, mutually disjoint family $(A_k)_{k \in \mathbb{N}}$ of sets for which μ is defined the σ-**additivity** of μ.*
B. *We call a pre-measure μ on a ring $\mathcal{R}$ in Ω (a measure μ on a σ-field $\mathcal{A}$ in Ω) σ-**finite** if there exists a sequence $(A_k)_{k \in \mathbb{N}}$, $A_k \in \mathcal{R}$ $(A_k \in \mathcal{A})$, such that $\mu(A_k) < \infty$ and $\Omega = \bigcup_{k=1}^{\infty} A_k = \Omega$.*
C. *In the situation of Theorem 2.10.A we call μ **continuous from below**, and in the situation of Theorem 2.10.B we call μ **continuous from above**. If Corollary 2.11 holds we call μ **continuous at $\emptyset$** or just $\emptyset$-**continuous**.*

Corollary 2.13. *Let $(\Omega, \mathcal{A})$ be a measurable space and $\mu : \mathcal{A} \to [0, \infty]$ a content. In addition assume that for μ the following holds (continuity from below): for every sequence $(A_k)_{k \in \mathbb{N}}$, $A_k \in \mathcal{A}$, such that $A_k \uparrow A$ it follows that $\lim_{k \to \infty} \mu(A_k) = \mu(A)$. Then μ is a measure.*

Proof. Clearly $\mu(\emptyset) = 0$. Now let $(B_k)_{k \in \mathbb{N}}$ be a sequence of mutually disjoint sets. Define $A_l := B_1 \cup \cdots \cup B_l \in \mathcal{A}$, and note that $A_l \uparrow \bigcup_{k \in \mathbb{N}} B_k = \bigcup_{j \in \mathbb{N}} A_j = A \in \mathcal{A}$. By our assumption and (2.8) we find

$$\mu(A) = \lim_{l \to \infty} \mu(A_l) = \lim_{l \to \infty} \sum_{k=1}^{l} \mu(B_k) = \sum_{k=1}^{\infty} \mu(B_k),$$

i.e.

$$\mu\left(\bigcup_{k\in\mathbb{N}} B_k\right) = \sum_{k=1}^{\infty} \mu(B_k)$$

and we have proved that μ is σ-additive on $\mathcal{A}$, hence a measure. $\square$

In order to solve our major problem we are now going to prove two central results: on the ring $\mathcal{F}^{(n)}$ we can define $\lambda^{(n)}$ as a pre-measure and every σ-finite pre-measure $\tilde{\mu}$ on a ring $\mathcal{R}$ in Ω has a unique extension to a measure μ on $\sigma(\mathcal{R})$. Both results require some lengthy proofs. Let $F \in \mathcal{F}^{(n)}$ and let $F = \bigcup_{k=1}^{M} A_k$ be any representation of F by a finite family $(A_k)_{k=1,\dots,M}$ of mutually disjoint sets $A_k \in \mathcal{I}_{r,n}$. We already know that

$$\lambda^{(n)}(F) := \sum_{k=1}^{M} \lambda^{(n)}(A_k)$$

is independent of the choice of the family $(A_k)_{k=1,\dots,M}$, see the remark following (2.2) or Problem 1.

Theorem 2.14. *On the ring $\mathcal{F}^{(n)}$ a pre-measure is given by $\lambda^{(n)}$.*

Proof. (Following E. Behrends [13].) Clearly we have $\lambda^{(n)}(\emptyset) = 0$. We need to show the σ-additivity of $\lambda^{(n)}$: if $(F_k)_{k\in\mathbb{N}}$ is a family of mutually disjoint figures $F_k \in \mathcal{F}^{(n)}$ with $\bigcup_{k\in\mathbb{N}} F \in \mathcal{F}^{(n)}$ then

$$\lambda^{(n)}\left(\bigcup_{k\in\mathbb{N}} F_k\right) = \sum_{k=1}^{\infty} \lambda^{(n)}(F_k).$$

With $F = \bigcup_{k\in\mathbb{N}} F_k$ and $(F_k)_{k\in\mathbb{N}}$ as above we know that $G_m := F \backslash \bigcup_{l=1}^{m} F_l \in \mathcal{F}^{(n)}$. For the sequence $(G_m)_{m\in\mathbb{N}}$ we have $G_1 \supset G_2 \supset \cdots$ and $\bigcap_{m\in\mathbb{N}} G_m = \emptyset$, i.e. $G_m \downarrow \emptyset$. Moreover we have

$$\lambda^{(n)}(G_m) = \lambda^{(n)}(F) - \sum_{l=1}^{m} \lambda^{(n)}(F_l).$$

Thus if we can prove $\lim_{m\to\infty} \lambda^{(n)}(G_m) = 0$ the σ-additivity of $\lambda^{(n)}$ on $\mathcal{F}^{(n)}$ will follow. Note that we are aiming to prove that $\lambda^{(n)}$ is $\emptyset$-continuous on $\mathcal{F}^{(n)}$. From our construction it follows that $\lambda^{(n)}(G_k) \leq \lambda^{(n)}(G_l)$ for $l \leq k$ and since the decreasing sequence $\left(\lambda^{(n)}(G_k)\right)_{k\in\mathbb{N}}$ is bounded from below by 0 it must have a limit $\delta \geq 0$, $\lim_{k \to \infty} \lambda^{(n)}(G_k) = \delta \geq 0$.

The aim is to prove that $\delta = 0$. We will prove that if $\delta > 0$ then exists a sequence of non-empty compact sets $(K_m)_{m \in \mathbb{N}}$ such that $K_m \supset K_{m+1}$ and $K_m \subset G_m$. From Theorem II.2.15 we deduce that $\bigcap_{m \in \mathbb{N}} K_m \neq \emptyset$, however by construction we have $\bigcap_{m \in \mathbb{N}} K_m \subset \bigcap_{m \in \mathbb{N}} G_m = \emptyset$ which is a contradiction. Assume now that $\delta > 0$. In this case each G_m is a union of elements $F_l \in \mathcal{F}^{(n)}$, $G_m = \bigcup_{l=m+1}^{\infty} F_l \neq \emptyset$, and $F_l = \bigcup_{\nu=1}^{M_l} A_{l\nu}$ where we can assume that $\overline{A_{l\nu}}$ is a non-degenerate compact cell since otherwise $\lambda^{(n)}(A_{l\nu}) = 0$. By shrinking $A_{l\nu}$ to $A'_{l\nu}$ such that $\overline{A'_{l\nu}}$ is still a non-degenerate compact cell we can shrink G_m to G'_m such that $\overline{G'_m} \subset G_m$ and

$$\lambda^{(n)}(G'_m) \geq \lambda^{(n)}(G_m) - \frac{\delta}{2^m}. \tag{2.19}$$

The set $K_m := \overline{G'_1} \cap \cdots \cap \overline{G'_m}$ is compact, $K_m \subset G_m$ and $K_1 \supset K_2 \supset \cdots$. We need to prove that $K_m \neq \emptyset$. For this it is sufficient to prove $G'_1 \cap \cdots \cap G'_m \neq \emptyset$ or equivalently $\lambda^{(n)}(G'_1 \cap \cdots \cap G'_m) > 0$. We show now by induction that

$$\lambda^{(n)}(G'_1 \cap \cdots \cap G'_m) \geq \lambda^{(n)}(G_m) - \delta \left(1 - \frac{1}{2^m}\right) \tag{2.20}$$

which implies

$$\lambda^{(n)}(G'_1 \cap \cdots \cap G'_m) \geq \frac{\delta}{2^m} \tag{2.21}$$

since $\lambda^{(m)}(G_m) \geq \delta$. For $m = 1$ estimate (2.20) is just (2.19). Suppose that (2.20) holds for m. Using (2.10) we obtain

$$\lambda^{(n)}(G'_1 \cap \cdots \cap G'_{m+1}) = \lambda^{(n)}(G'_1 \cap \cdots \cap G'_m) + \lambda^{(n)}(G'_{m+1}) - \lambda^{(n)}((G'_1 \cap \cdots \cap G'_m) \cup G'_{m+1})$$

and we observe

$$\lambda^{(n)}(G'_1 \cap \cdots \cap G'_m) \geq \lambda^{(n)}(G_m) - \delta \left(1 - \frac{1}{2^m}\right)$$

by induction hypothesis,

$$\lambda^{(n)}(G'_{m+1}) \geq \lambda^{(n)}(G_{m+1}) - \frac{\delta}{2^{m+1}}$$

by construction, and since $(G'_1 \cap \cdots \cap G'_m) \cup G'_{m+1} \subset G_m$ we have

$$\lambda^{(n)} \left((G'_1 \cap \cdots \cap G'_m) \cup G'_{m+1} \right) \leq \lambda^{(n)}(G_m),$$

which implies

$$\lambda^{(n)}(G'_1 \cap \cdots \cap G'_{m+1}) \geq \lambda^{(n)}(G_m) - \delta \left(1 - \frac{1}{2^m} \right) + \lambda^{(n)}(G_{m+1}) - \frac{\delta}{2^{m+1}} - \lambda^{(n)}(G_m)$$

$$= \lambda^{(n)}(G_{m+1}) - \delta \left(1 - \frac{1}{2^{m+1}} \right),$$

and hence the theorem is proved. $\qquad\square$

In order to prove our central extension and uniqueness theorem, Carathéodory's theorem, we need some further tools. At certain stages we need to decide of a given system of subsets of Ω extending a given ring $\mathcal{R}$ in Ω that it is already a σ-field, or even more whether it is the σ-field generated by $\mathcal{R}$. As it turns out Dynkin systems are best suited for such a purpose.

Definition 2.15. *We call $\mathcal{D} \subset \mathcal{P}(\Omega)$, $\Omega \neq \emptyset$, a **Dynkin system** in Ω if*

$$\Omega \in \mathcal{D} \tag{2.22}$$

$$A \in \mathcal{D} \text{ implies } A^{\complement} \in \mathcal{D}, \tag{2.23}$$

and for every sequence $(A_k)_{k \in \mathbb{N}}$ of mutually disjoint sets $A_k \in \mathcal{D}$ we have

$$\bigcup_{k \in \mathbb{N}} A_k \in \mathcal{D}. \tag{2.24}$$

Remark 2.16. A. Clearly $\emptyset = \Omega^{\complement} \in \mathcal{D}$ and finite unions of mutually disjoint sets belonging to $\mathcal{D}$ are elements of D.
B. Every σ-field is a Dynkin system.
C. For $A, B \in \mathcal{D}$, $A \subset B$ it follows that $A \cap B^{\complement} = \emptyset \in \mathcal{D}$ and $A \cup B^{\complement} \in \mathcal{D}$ which yields

$$B \backslash A = B \cap A^{\complement} = \left(A \cup B^{\complement} \right)^{\complement} \in \mathcal{D}. \tag{2.25}$$

Theorem 2.17. *A Dynkin system $\mathcal{D}$ in Ω is a σ-field in Ω if and only if $A, B \in \mathcal{D}$ implies that $A \cap B \in \mathcal{D}$.*

Proof. (Compare with H. Bauer [11].) It remains to prove that if $\mathcal{D}$ is a Dynkin system and $A, B \in \mathcal{D}$ implies $A \cap B \in \mathcal{D}$, then $\mathcal{D}$ is a σ-field, i.e. $\Omega \in \mathcal{D}$, $A^\complement \in \mathcal{D}$ for $A \in \mathcal{D}$ and $\bigcup_{k \in \mathbb{N}} A_k \in \mathcal{D}$ for a sequence $(A_k)_{k \in \mathbb{N}}$, $A_k \in \mathcal{D}$. The first two assertions are trivial due to the definition of a Dynkin system. Now let $(A_k)_{k \in \mathbb{N}}$, $A_k \in \mathcal{D}$. We define further

$$A_0' := \emptyset, \quad A_k' := A_1 \cup \cdots \cup A_k.$$

For $k \neq l$ this implies

$$\left(A_{k+1}' \backslash A_k'\right) \cap \left(A_{l+1}' \backslash A_l'\right) = \emptyset.$$

Moreover, since for any two sets $B_1, B_2 \in \mathcal{D}$ we have $B_1 \backslash B_2 \in \mathcal{D}$ and $B_1 \cup B_2 = (B_1 \backslash B_2) \cup B_2$ and $(B_1 \backslash B_2) \cap B_2 = \emptyset$ we deduce that a finite union of elements of $\mathcal{D}$ belongs to $\mathcal{D}$, hence $A_{k+1}' \backslash A_k' \in \mathcal{D}$ which now yields

$$\bigcup_{k \in \mathbb{N}} A_k = \bigcup_{k \in \mathbb{N}} \left(A_{k+1}' \backslash A_k'\right) \in \mathcal{D}$$

proving the theorem. □

The next construction is similar to that of generating a σ-field or a topology.

Proposition 2.18. *For every family $\mathcal{E} \subset \mathcal{P}(\Omega)$, $\Omega \neq \emptyset$, the family*

$$\delta(\mathcal{E}) := \bigcap \left\{ \mathcal{D} \subset \mathcal{P}(\Omega) \, \middle| \, \mathcal{E} \subset \mathcal{D} \text{ and } \mathcal{D} \text{ is a Dynkin system} \right\} \tag{2.26}$$

*is a Dynkin system called the **Dynkin system generated by $\mathcal{E}$**.*

Proof. We refer to Problem 6. □

The usefulness of Dynkin systems is due to

Theorem 2.19. *Let $\mathcal{E} \subset \mathcal{P}(\Omega)$ be a family of subsets of Ω, $\Omega \neq \emptyset$, such that $A, B \in \mathcal{E}$ implies $A \cap B \in \mathcal{E}$. Then we have*

$$\delta(\mathcal{E}) = \sigma(\mathcal{E}), \tag{2.27}$$

i.e. the Dynkin system generated by $\mathcal{E}$ is already the σ-field generated by $\mathcal{E}$.

Proof. (Compare with H. Bauer [11].) It is trivial that $\delta(\mathcal{E}) \subset \sigma(\mathcal{E})$. Thus we have only to prove the converse inclusion. For this we will employ Theorem 2.17 and show that $D, E \in \mathcal{E}$ implies that $D \cap E \in \mathcal{E}$. For $D \in \delta(\mathcal{E})$ we define

$$\mathcal{D}_D := \{Q \in \mathcal{P}(\Omega) \,|\, Q \cap D \in \delta(\mathcal{E})\} \qquad (2.28)$$

and claim that $\mathcal{D}_D$ is a Dynkin system. Since $\Omega \cap D = D$ it follows (2.22) and since $D^{\complement} \in \delta(\mathcal{E})$ we find for $Q \in \mathcal{D}_D$

$$Q^{\complement} \cap D = \left(Q^{\complement} \cup D^{\complement}\right) \cap D = (Q \cap D)^{\complement} \cap D = \left((Q \cap D) \cup D^{\complement}\right)^{\complement} \in \delta(\mathcal{E})$$

implying $Q^{\complement} \in \mathcal{D}_D$, i.e. (2.23). We now prove (2.24) for $\mathcal{D}_D$. For this let $A_k \in \mathcal{D}_D$, $k \in \mathbb{N}$, such that $A_k \cap A_l = \emptyset$ for $k \neq l$. It follows that $A_k \cap D \in \delta(\mathcal{E})$ and $(A_k \cap D) \cap (A_l \cap D) = \emptyset$ for $k \neq l$. Since $\delta(\mathcal{E})$ is a Dynkin system we find

$$\left(\bigcup_{k \in \mathbb{N}} A_k\right) \cap D = \bigcup_{k \in \mathbb{N}} (A_k \cap D) \in \delta(\mathcal{E})$$

implying that $\bigcup_{k \in \mathbb{N}} A_k \in \mathcal{D}_D$, i.e. $\mathcal{D}_D$ is indeed a Dynkin system. For $E \in \mathcal{E}$ we know by our assumption that $E \cap E' \in \mathcal{E}$, for every $E' \in \mathcal{E}$ hence $\mathcal{E} \subset \mathcal{D}_E$ and therefore $\delta(\mathcal{E}) \subset \delta(\mathcal{D}_E) = \mathcal{D}_E$. Further, for $D \in \delta(\mathcal{E})$ and $E \in \mathcal{E}$ we have $E \cap D \in \delta(\mathcal{E})$, i.e. $\mathcal{E} \subset \mathcal{D}_D$ and therefore $\delta(\mathcal{E}) \subset \mathcal{D}_D$ for every $D \in \delta(\mathcal{E})$, which is just a reformulation of the statement we want to prove and the theorem is shown. $\qquad\square$

Following R. Schilling [75], as a first application of Dynkin systems, we prove a uniqueness result for measures.

Theorem 2.20. *Let $(\Omega, \sigma(\mathcal{G}))$ be a measurable space where $\mathcal{G} \subset \mathcal{P}(\Omega)$, $\Omega \neq \emptyset$. Assume that $C_1, C_2 \in \mathcal{G}$ implies $C_1 \cap C_2 \in \mathcal{G}$ and that there exists a sequence of sets $(B_k)_{k \in \mathbb{N}}$ of sets $B_k \in \mathcal{G}$ such that $B_k \uparrow \Omega$. Suppose μ and ν are two measures on $\sigma(\mathcal{G})$ such that $\mu(B) = \nu(B)$ for all $B \in \mathcal{G}$ and that $\mu(B_k) = \nu(B_k) < \infty$ for all k. Then $\mu = \nu$, i.e. $\mu(A) = \nu(A)$ for all $A \in \sigma(\mathcal{G})$.*

Proof. For $k \in \mathbb{N}$ we claim that

$$\mathcal{D}_k := \{A \in \sigma(\mathcal{G}) \,|\, \mu(B_k \cap A) = \nu(B_k \cap A)\}$$

is a Dynkin system. Note that by assumption $\mu(B_k \cap A) < \infty$. Since $\Omega \cap B_k = B_k$ it is clear that $\Omega \in \mathcal{D}_k$. For $A \in \mathcal{D}_k$ we find

$$
\begin{aligned}
\mu\left(B_k \cap A^{\complement}\right) &= \mu(B_k \setminus A) = \mu(B_k) - \mu(B_k \cap A) \\
&= \nu(B_k) - \nu(B_k \cap A) = \nu(B_k \setminus A) \\
&= \nu(B_k \cap A^{\complement}),
\end{aligned}
$$

i.e. $A \in \mathcal{D}_k$ implies $A^{\complement} \in \mathcal{D}_k$. Now let $(A_j)_{j \in \mathbb{N}}$ be a sequence of mutually disjoint sets $A_j \in \mathcal{D}_k$. It follows that

$$
\begin{aligned}
\mu\left(B_k \cap \bigcup_{j \in \mathbb{N}} A_j\right) &= \mu\left(\bigcup_{j \in \mathbb{N}}(B_k \cap A_j)\right) \\
&= \sum_{j=1}^{\infty} \mu(B_k \cap A_j) = \sum_{j=1}^{\infty} \nu(B_k \cap A_j) \\
&= \nu\left(\bigcup_{j \in \mathbb{N}}(B_k \cap A_j)\right) \\
&= \nu\left(B_k \cap \bigcup_{j \in \mathbb{N}} A_j\right),
\end{aligned}
$$

i.e. $\bigcup_{j \in \mathbb{N}} A_j \in \mathcal{D}_k$. By Theorem 2.17 we know that $\delta(\mathcal{G}) = \sigma(\mathcal{G})$ which yields that $\mathcal{G} \subset \mathcal{D}_k$ implies $\delta(\mathcal{G}) = \sigma(\mathcal{G}) \subset \mathcal{D}_k$ for all $k \in \mathbb{N}$. However $\mathcal{D}_k \subset \sigma(\mathcal{G})$ and we deduce for all $k \in \mathbb{N}$ that $\mathcal{D}_k = \sigma(\mathcal{G})$ which implies

$$
\mu(B_k \cap A) = \nu(B_k \cap A) \tag{2.29}
$$

for all $k \in \mathbb{N}$ and all $A \in \sigma(\mathcal{G})$. From Theorem 2.10.A we obtain now as $k \to \infty$ that

$$
\mu(A) = \lim_{k \to \infty} \mu(B_k \cap A) = \lim_{k \to \infty} \nu(B_k \cap A) = \nu(A).
$$

$\square$

Next we want to address the problem of how to extend a pre-measure μ defined on a ring $\mathcal{R}$ over Ω, $\Omega \neq \emptyset$. As a first step we use μ to define a set function μ^* on $\mathcal{P}(\Omega)$ which we later will call the **outer** (Carathéodory)

measure induced by μ. For this let $G \in \mathcal{P}(\Omega)$ be any subset of Ω. By $\mathcal{U}(G)$ we denote the family of all sequences $(A_k)_{k \in \mathbb{N}}$, $A_k \in \mathcal{R}$, such that G is covered by $(A_k)_{k \in \mathbb{N}}$, i.e.

$$G \subset \bigcup_{k \in \mathbb{N}} A_k. \tag{2.30}$$

Now we define $\mu^* : \mathcal{P}(\Omega) \to [0, \infty]$ by

$$\mu^*(G) := \begin{cases} \inf \left\{ \sum_{k=1}^{\infty} \mu(A_k) \,\middle|\, (A_k)_{k \in \mathbb{N}} \in \mathcal{U}(G) \right\}, & \text{if } \mathcal{U}(G) \neq \emptyset \\ +\infty, & \text{if } \mathcal{U}(G) = \emptyset, \end{cases} \tag{2.31}$$

note that $\mu^*(G) \geq 0$ is trivial.

Since the sequence $(A_k)_{k \in \mathbb{N}}$, $A_k = \emptyset$, belongs to $\mathcal{U}(\emptyset)$ and $\mu(\emptyset) = 0$, we deduce

$$\mu^*(\emptyset) = 0. \tag{2.32}$$

Further, if $G_1 \subset G_2$ then $\mathcal{U}(G_2) \subset \mathcal{U}(G_1)$, and therefore it follows that

$$G_1 \subset G_2 \text{ implies } \mu^*(G_1) \leq \mu^*(G_2). \tag{2.33}$$

Finally we claim that for every sequence $(G_l)_{l \in \mathbb{N}}$, $G_l \in \mathcal{P}(\Omega)$, we have

$$\mu^* \left(\bigcup_{l \in \mathbb{N}} G_l \right) \leq \sum_{l=1}^{\infty} \mu^*(G_l). \tag{2.34}$$

In order to prove (2.34) we may assume that $\mu^*(G_l) < \infty$ for all $l \in \mathbb{N}$, in particular we assume $\mathcal{U}(G_l) \neq \emptyset$ for all $l \in \mathbb{N}$. For $\epsilon > 0$ and $l \in \mathbb{N}$ we find now a sequence $(A_{lk})_{k \in \mathbb{N}} \in \mathcal{U}(G_l)$ such that

$$\sum_{k=1}^{\infty} \mu(A_{lk}) \leq \mu^*(Q_l) + \frac{\epsilon}{2^l}.$$

The sequence $(A_{lk})_{l, k \in \mathbb{N}}$ belongs to $\mathcal{U}\left(\bigcup_{l \in \mathbb{N}} G_l \right)$ and therefore

$$\mu^* \left(\bigcup_{l \in \mathbb{N}} G_l \right) \leq \sum_{l,k=1}^{\infty} \mu(A_{lk})$$

$$= \sum_{l=1}^{\infty} \left(\sum_{k=1}^{\infty} \mu(A_{lk}) \right) \leq \sum_{l=1}^{\infty} \mu^*(G_l) + \sum_{l=1}^{\infty} \frac{\epsilon}{2^l}$$

or

$$\mu^* \left(\bigcup_{l \in \mathbb{N}} G_l \right) \le \sum_{l=1}^{\infty} \mu^*(G_l) + \epsilon$$

for all $\epsilon > 0$ implying (2.34).

Definition 2.21. A. *Let $\Omega \ne \emptyset$ and $\mu^* : \mathcal{P}(\Omega) \to [0, \infty]$ be a set function satisfying (2.32) - (2.34). Then μ^* is called an **outer measure** on $\mathcal{P}(\Omega)$.*
B. *For an outer measure μ^* on $\mathcal{P}(\Omega)$ we call $A \subset \Omega$ a μ^*-**measurable set** or **outer measurable set** with respect to μ^* if*

$$\mu^*(G) = \mu^*(G \cap A) + \mu^*(G \cap A^{\complement}) \tag{2.35}$$

holds for all $G \in \mathcal{P}(\Omega)$.

Proposition 2.22. *Let $\Omega \ne \emptyset$ and $\mathcal{R}$ be a ring in Ω as well as μ a pre-measure on $\mathcal{R}$. Denote by μ^* the outer measure associated with μ by (2.31). Then every $A \in \mathcal{R}$ is μ^*-measurable and for $A \in \mathcal{R}$ we have*

$$\mu^*(A) = \mu(A). \tag{2.36}$$

Proof. For $A \in \mathcal{R}$ we have to prove (2.35) and for this we may assume $\mu^*(G) < \infty$, and in particular $\mathcal{U}(G) \ne \emptyset$. Since μ is finite additive we have for $(A_k)_{k \in \mathbb{N}} \in \mathcal{U}(G)$

$$\mu(A_k) = \mu(A_k \cap A) + \mu(A_k \backslash A)$$

implying

$$\sum_{k=1}^{\infty} \mu(A_k) = \sum_{k=1}^{\infty} \mu(A_k \cap A) + \sum_{k=1}^{\infty} (A_k \backslash A).$$

Furthermore $(A_k \cap A)_{k \in \mathbb{N}} \in \mathcal{U}(G \cap A)$ and $(A_k \backslash A)_{k \in \mathbb{N}} \in \mathcal{U}(G \backslash A)$ which yields

$$\sum_{k=1}^{\infty} \mu(A_k) \ge \mu^*(G \cap A) + \mu^*(G \backslash A)$$

for any such a sequence $(A_k)_{k \in \mathbb{N}}$ and this implies

$$\mu^*(G) \ge \mu^*(G \cap A) + \mu^*(G \cap A^{\complement}).$$

Now we consider the sequence $A_1 = G \cap A$, $A_2 = G \backslash A$ and $A_k = \emptyset$ for $k > 2$. Applying (2.34) to this sequence we find

$$\mu^*(G) = \mu^* \left((G \cap A) \cup (G \backslash A)\right) \leq \mu^*(G \cap A) + \mu^*(G \backslash A),$$

and hence (2.35).

In order to prove (2.36) for $A \in \mathcal{R}$ we note that the sequence $(A_k)_{k \in \mathbb{N}}$, $A_1 = A$, $A_k = \emptyset$ for $k > 1$, is an element of $\mathcal{U}(A)$ and therefore we have $\mu(A) \geq \mu^*(A)$. On the other hand, applying (2.14) to A and $(A_k)_{k \in \mathbb{N}}$ we have $\mu(A) \leq \mu^*(A)$, i.e. $\mu(A) = \mu^*(A)$, and the proposition is proven. $\qquad \square$

So far we have seen that a pre-measure μ on a ring $\mathcal{R}$ in $\Omega \neq \emptyset$ can be extended to an outer measure μ^* on $\mathcal{P}(\Omega)$ and all elements of $\mathcal{R}$ are μ^*-measurable. Our final step is to prove that the μ^*-measurable sets form a σ-field in Ω. In this proof we will again make use of Dynkin systems.

Theorem 2.23. *Let μ^* be an outer measure on $\mathcal{P}(\Omega)$, $\Omega \neq \emptyset$. The system $\mathcal{A}^*$ of all μ^*-measurable sets $A \subset \Omega$ is a σ-field and $\mu^*|_{A*}$ is a measure.*

Proof. It is obvious that (2.35) holds for Ω since $G \cap \Omega = G$ and $G \backslash \Omega = \emptyset$, and that $A \in \mathcal{A}^*$ if and only if $A^C \in \mathcal{A}^*$. Next we prove that $\mathcal{A}^*$ is an algebra, i.e. taking into account the previous observation, that $A, B \in \mathcal{A}^*$ implies $A \cup B \in \mathcal{A}^*$. We know that

$$\mu^*(G) = \mu^*(G \cap A) + \mu^*(G \backslash A) \tag{2.37}$$

for $G \in \mathcal{P}(\Omega)$ and $A \in \mathcal{A}^*$, thus we have also

$$\mu^*(G) = \mu^*(G \cap B) + \mu^*(G \backslash A) \tag{2.38}$$

as well as

$$\mu^*(G \cap A) = \mu^* \left((G \cap A) \cap B\right) + \mu^* \left((G \cap A) \backslash B\right)$$

and

$$\mu^*(G \backslash A) = \mu^* \left((G \backslash A) \cap B\right) + \mu^* \left((G \backslash A) \cap B\right)$$

where we replaced in (2.38) G by $G \cap A$ and $G \backslash A$, respectively. Combined with (2.37) this yields

$$\mu^*(G) = \mu^*(G \cap A) + \mu^*(G \backslash A) \tag{2.39}$$
$$= \mu^*(G \cap A \cap B) + \mu^*(G \cap A \cap B^C)$$
$$+ \mu^*(G \cap A^C \cap B) + \mu^*(G \cap A^C \cap B^C)$$

and replacing G by $G \cap (A \cup B)$ we find

$$\mu^*(G \cap (A \cup B)) = \mu^*(G \cap A \cap B) + \mu^*(G \cap A \cap B^{\complement}) + \mu^*(G \cap A^{\complement} \cap B) \quad (2.40)$$

and with (2.39)

$$\mu^*(G) = \mu^*(G \cap (A \cup B)) + \mu^*(G \cap (A \cup B)^{\complement}).$$

Since $G \in \mathcal{P}(\Omega)$ was arbitrary we deduce $A \cup B \in \mathcal{A}^*$. Eventually we want to prove that $\mathcal{A}^*$ is a Dynkin system stable under finite intersections which implies by Theorem 2.19 that $\mathcal{A}^*$ is a σ-field. For this let $(A_k)_{k \in \mathbb{N}}$ be a sequence of mutually disjoint sets $A_k \in \mathcal{A}^*$ and $A := \bigcup_{k \in \mathbb{N}} A_k$. If we choose in (2.40) $A = A_1$ and $B = A_2$ we find using that $A_1 \cap A_2 = \emptyset$ that

$$\mu^*(G \cap (A_1 \cup A_2)) = \mu^*(G \cap A_1) + \mu^*(G \cap A_2),$$

and by induction we obtain

$$\mu^*\left(G \cap \bigcup_{k=1}^{N} A_k\right) = \sum_{k=1}^{N} \mu^*(G \cap A_k)$$

for all $N \in \mathbb{N}$ and $G \in \mathcal{P}(\Omega)$. We know already that $B_N := \bigcup_{k=1}^{N} A_k \in \mathcal{A}^*$ and $G \backslash A \subset G \backslash B_N$, i.e. $\mu^*(G \backslash A) \leq \mu^*(G \backslash B_N)$, therefore we get for all $N \in \mathbb{N}$

$$\mu^*(G) = \mu^*(G \cap B_N) + \mu^*(G \backslash B_N)$$
$$\geq \sum_{k=1}^{N} \mu^*(G \cap A_k) + \mu^*(G \backslash A)$$

and by (2.34) it follows that

$$\mu^*(G) \geq \sum_{k=1}^{\infty} \mu^*(G \cap A_k) + \mu^*(G \backslash A) \geq \mu^*(G \cap A) + \mu^*(G \backslash A).$$

As in the proof of Proposition 2.22 we further conclude that

$$\mu^*(G) = \sum_{k=1}^{\infty} \mu^*(G \cap A_k) + \mu^*(G \backslash A) = \mu^*(G \cap A) + \mu^*(G \backslash A), \quad (2.41)$$

i.e. $A = \bigcup_{k \in \mathbb{N}} A_k \in \mathcal{A}^*$. Thus we have proved that $\mathcal{A}^*$ is a Dynkin system stable under finite intersections, i.e. it is a σ-field. If we choose in (2.41) now $G = A$ then we find

$$\mu^*(A) = \mu^*\left(\bigcup_{k \in \mathbb{N}} A_k\right) = \sum_{k=1}^{\infty} \mu^*(A_k),$$

i.e. μ^* is on $\mathcal{A}^*$ indeed a measure. □

Combining Theorem 2.23 with Theorem 2.20 and Proposition 2.22 we arrive at Carathéodory's theorem.

Theorem 2.24. *Every σ-finite pre-measure μ on a ring $\mathcal{R}$ in Ω, $\Omega \neq \emptyset$, has a unique extension to a measure $\tilde{\mu}$ on $\sigma(\mathcal{R})$.*

Proof. We know by Proposition 2.22 that $\mathcal{R} \subset \mathcal{A}^*$, hence $\sigma(\mathcal{R}) \subset \mathcal{A}^*$ and therefore $\tilde{\mu} := \mu^*|_{\sigma(\mathcal{R})}$ is a measure on $\sigma(\mathcal{R})$. Since by assumption μ is σ-finite on $\mathcal{R}$, $\mathcal{R}$ and μ have all properties required in Theorem 2.20. □

Finally we return to the pre-measure $\lambda^{(n)}$ on $\mathcal{F}^{(n)}$. We know from Theorem 2.14 that $\lambda^{(n)}$ is a σ-finite pre-measure on the ring $\mathcal{F}^{(n)}$. Hence according to Theorem 2.24 it has a unique extension as a measure on $\sigma(\mathcal{F}^{(n)})$. For simplicity we will denote this extension again by $\lambda^{(n)}$.

Lemma 2.25. *For $n \in \mathbb{N}$ we have $\sigma(\mathcal{F}^{(n)}) = \mathcal{B}^{(n)}$.*

Proof. This result can be deduced from Exercise 1.22, but due to its importance we give the proof here (once again). Since $\mathcal{F}^{(n)} \subset \mathcal{B}^{(n)}$ we have $\sigma(\mathcal{F}^{(n)}) \subset \mathcal{B}^{(n)}$. In order to prove the converse inclusion we note first that every non-empty open set $U \subset \mathbb{R}^n$ is a countable union of elements belonging to $\left\{(a,b) \subset \mathbb{R}^n \mid (a,b) = \bigtimes_{j=1}^{n}(a_j, b_j), \, a_j < b_j\right\}$. For example we may consider all sets (a,b) with $a_j, b_j \in \mathbb{Q}$ such that $(a,b) \subset U$. (This is purely a topological problem, see Problem 8, which we have encountered before several times). In addition we have

$$(a,b) = \bigcup_{k \in \mathbb{N}} \left[a + \frac{1}{k}h, b\right), \quad h = (1, 1, \ldots, 1) \in \mathbb{R}^n,$$

which now implies $\mathcal{O}_n \subset \sigma(\mathcal{F}^{(n)})$, hence $\mathcal{B}^{(n)} \subset \sigma(\mathcal{F}^{(n)})$ proving the lemma. □

Theorem 2.26. *On the Borel σ-field $\mathcal{B}^{(n)}$ exist a unique measure $\lambda^{(n)}$ with the property that for all $[a, b) \in \mathcal{I}_{r,n}$ the following holds*

$$\lambda^{(n)}([a, b)) = \bigtimes_{j=1}^{n}(b_j - a_j).$$

Definition 2.27. *The measure $\lambda^{(n)}$ is called the **Lebesgue-Borel measure** on $\mathcal{B}^{(n)}$ or $\mathbb{R}^n$.*

In the next chapter we also introduce a measure called Lebesgue measure. Later on we will (as do many authors) often speak about the Lebesgue measure when we actually mean the Lebesgue-Borel measure.

Problems

1. Let $F \in \mathcal{F}^{(n)}$ be an n-dimensional figure. Prove that $\lambda^{(n)}(F)$ is independent of the partition of F, i.e. if $F = \bigcup_{k=1}^{M} A_k$ and $F = \bigcup_{l=1}^{N} B_l$, $A_k, B_l \in \mathcal{I}_{r,n}$, then $\sum_{k=1}^{M} \lambda^{(n)}(A_k) = \sum_{l=1}^{N} \lambda^{(n)}(B_l)$.

2. Let Ω be a non-denumerable set and define

$$\mathcal{R} := \{A \in \mathcal{P}(\Omega) | A \text{ or } A^{\complement} \text{ is finite}\}.$$

   Prove that $\mathcal{R}$ is a ring in Ω and that $\mu : \mathcal{R} \to [0, \infty]$ defined on $\mathcal{R}$ by

$$\mu(A) := \begin{cases} 0, & A \text{ is finite} \\ 1, & A^{\complement} \text{ is finite} \end{cases}$$

   is a pre-measure on $\mathcal{R}$.

3. Solve Exercise 2.7.

4. Let $\mathcal{R}$ be a ring in Ω and $(\mu_k)_{k \in \mathbb{N}}$ a sequence of pre-measures on $\mathcal{R}$ satisfying $\mu_k(A) \le \mu_{k+1}(A)$ for all $A \in \mathcal{R}$. Prove that by $\mu(A) := \sup_{k \in \mathbb{N}} \mu_k(A)$ a further pre-measure is defined on $\mathcal{R}$.

5. Let $\mathcal{R}$ be a ring in Ω and μ a pre-measure on $\mathcal{R}$. Further let $(A_k)_{k \in \mathbb{N}}$, $A_k \in \mathcal{R}$, be a sequence such that $\mu(A_k) < \infty$ and $\Omega = \bigcup_{k \in \mathbb{N}} A_k$, i.e. $\mathcal{R}$ is σ-finite with respect to μ. Construct a sequence $(\tilde{A}_k)_{k \in \mathbb{N}}$, $\tilde{A}_k \in \mathcal{R}$, such that $\tilde{A}_k \subset \tilde{A}_{k+1}$, $\mu(\tilde{A}_k) < \infty$ and $\bigcup_{k \in \mathbb{N}} \tilde{A}_k = \Omega$.

6. Prove Proposition 2.18.

7. Let Ω be a non-empty set and $\mathcal{G} \subset \mathcal{H} \subset \mathcal{P}(\Omega)$. Show that $\delta(\mathcal{G}) \subset \delta(\mathcal{H})$ and $\delta(\mathcal{G}) \subset \sigma(\mathcal{G})$.

8. Let $U \subset \mathbb{R}^n$ be open. Prove that U is the countable union of open rectangles with rational vertices contained in U.
   Hint: re-interpret Problem 5 of Chapter 1.

3 The Lebesgue-Borel Measure and Hausdorff Measures

The Lebesgue-Borel measure and its extension to the Lebesgue measure is a fundamental object of analysis and in this chapter we investigate more of its properties and extensions. Most of all we are interested in "geometric" properties, for example invariance properties. We need

Lemma 3.1. *Let $B \in \mathcal{B}^{(n)}$ be a Borel set, $x_0 \in \mathbb{R}^n$ and $T \in O(n)$. Then $x + B$ and $T(B)$ are Borel sets too.*

Proof. From Corollary 1.18 we know that continuous mappings $f : \mathbb{R}^n \to \mathbb{R}^n$ are $\mathcal{B}^{(n)}/\mathcal{B}^{(n)}$-measurable. Since $g_{x_0} : \mathbb{R}^n \to \mathbb{R}^n$, $g(y) = y - x_0$ is continuous and $x_0 + B = g_{x_0}^{-1}(B)$, it follows that $x_0 + B \in \mathcal{B}^{(n)}$. Further we know that $T \in O(n)$ is a homeomorphism implying that T^{-1} establishes a bijective correspondence of $\mathcal{O}_n$ with itself, i.e. $T^{-1}U \in \mathcal{O}_n$ for every $U \in \mathcal{O}_n$, and every $V \in \mathcal{O}_n$ is of type $V = T^{-1}U$, $U \in \mathcal{O}_n$. Since $\mathcal{O}_n$ generates $\mathcal{B}^{(n)}$ we now deduce that $B = T^{-1}(T(B))$ is Borel-measurable for every $B \in \mathcal{B}^{(n)}$. $\square$

Theorem 3.2. *The Lebesgue-Borel measure $\lambda^{(n)}$ is translation invariant in the sense that $\lambda^{(n)}(x_0+B) = \lambda^{(n)}(B)$ for all $x_0 \in \mathbb{R}^n$ and $B \in \mathcal{B}^{(n)}$. Moreover, every translation invariant measure μ on $\mathcal{B}^{(n)}$ for which $\mu\left([0,1)^n\right)$ is strictly positive and finite is a positive multiple of $\lambda^{(n)}$, i.e. $0 < \mu\left([0,1)^n\right) < \infty$ implies for some constant $c_\mu > 0$ that $\mu(B) = c_\mu \lambda^{(n)}(B)$ for all $B \in \mathcal{B}^{(n)}$.*

Proof. For $A = \bigtimes_{j=1}^n [a_j, b_j)$ it follows that $x_0 + A = \bigtimes_{j=1}^n [a_j + x_{0j}, b_j + x_{0j})$ and consequently $\lambda^{(n)}(x_0+A) = \prod_{j=1}^n (b_j + x_{0j} - (a_j + x_{0j})) = \prod_{j=1}^n (b_j - a_j) = \lambda^{(n)}(A)$, which extends to all $A \in \mathcal{F}^{(n)}$. On the other hand $\mu_{x_0} : \mathcal{B}^{(n)} \to [0, \infty]$ defined by $\mu_{x_0}(B) := \lambda^{(n)}(x_0 + B)$ is a measure on $\mathcal{B}^{(n)}$ since $x_0 + \emptyset = \emptyset$ and $x_0 + \bigcup_{k \in \mathbb{N}} B_k = \bigcup_{k \in \mathbb{N}}(x_0 + B_k)$. Since on $\mathcal{F}^{(n)}$ the measures μ_{x_0} and $\lambda^{(n)}$ coincide it follows that $\mu_{x_0} = \lambda^{(n)}$ on $\mathcal{B}^{(n)}$. In order to prove the second assertion we first note that it is sufficient to prove it for $\mathcal{F}^{(n)}$, hence for $\mathcal{I}_{r,n}$ and thus for all $A := \bigtimes_{j=1}^n [a_j, b_j)$, $a_j, b_j \in \mathbb{Q}$. For each A we can find $N, m_1, \ldots, m_n \in \mathbb{N}$ such that $N m_j = b_j - a_j$ and therefore we can cover A by $m = \prod_{j=1}^n m_j$ mutually disjoint half-open cubes with side length $\frac{1}{N}$. Each such a cube is a translation of the cube $C_N := \left[0, \frac{1}{N}\right) \times \cdots \times \left[0, \frac{1}{N}\right) \subset \mathbb{R}^n$. Therefore we find by the translation invariance of μ and $\lambda^{(n)}$ that

$$\mu(A) = m\, \mu(C_N) \quad \text{and} \quad \lambda^{(n)}(A) = m\, \lambda^{(n)}(C_N)$$

as well as

$$\mu(C_1) = N^n \mu(C_N) \text{ and } 1 = \lambda^{(n)}(C_1) = N^n \lambda^{(n)}(C_N).$$

Thus $\frac{m}{N^n} = \lambda^{(n)}(A)$ and this applies

$$\mu(A) = m\,\mu(C_N) = \frac{m\,\mu(C_1)}{N^n} = \mu(C_1)\lambda^{(n)}(A)$$

proving the theorem with $c_\mu = \mu(C_1)$. $\qquad\square$

Corollary 3.3. *On $\left(\mathbb{R}^n, \mathcal{B}^{(n)}\right)$ the Lebesgue-Borel measure $\lambda^{(n)}$ is the unique translation invariant measure satisfying $\lambda^{(n)}(C_1) = 1$.*

Remark 3.4. We will use Theorem 3.2 in Appendix II to discuss the existence of non-Borel-measurable sets in $\mathbb{R}^n$, i.e. of sets $A \subset \mathbb{R}^n$, $A \notin \mathcal{B}^{(n)}$.

Invariance of measures is an important property. Before we prove that the Lebesgue-Borel measure is also invariant under the operation of $O(n)$, we want to give an example of a "translation" invariant measure on $\mathbb{Z}^n$.

Example 3.5. (Also see Problem 9.b) in Chapter 1) For a finite set A denote by $\#(A)$ the number of its elements. On $\mathbb{Z}^n$ we consider $\mathcal{P}(\mathbb{Z}^n)$ as σ-field and define the measure

$$\mu_{\mathbb{Z}^n}(A) := \sum_{k \in \mathbb{Z}^n} \epsilon_k(A). \tag{3.1}$$

Clearly this is a measure with $\mu_{\mathbb{Z}^n}(A) = \#(A)$ for a finite set $A \subset \mathbb{Z}^n$ and $\mu_{\mathbb{Z}^n}(A) = +\infty$ otherwise. For $l_0 \in \mathbb{Z}^n$ fixed we can consider the translated set $l_0 + A$ for which we obtain

$$\mu_{\mathbb{Z}^n}(l_0 + A) = \sum_{k \in \mathbb{Z}^n} \epsilon_k(l_0 + A) = \sum_{k - l_0 \in \mathbb{Z}^n} \epsilon_{k-l_0}(A) = \sum_{m \in \mathbb{Z}^n} \epsilon_m(A),$$

i.e. $\mu_{\mathbb{Z}^n}$ is on $(\mathbb{Z}^n, \mathcal{P}(\mathbb{Z}^n))$ a measure invariant under the group translations of $(\mathbb{Z}^n, +)$. Later we will discuss the existence of invariant measures on certain topological groups in more detail.

Our intuition tells us that $\lambda^{(n)}$ is the "correct" object to measure the volume of (certain) subsets of $\mathbb{R}^n$. Therefore for $A \in \mathcal{B}^{(n)}$ the set $T(A)$, $T \in O(n)$, and the set A should have the same volume, i.e. we expect

$$\lambda^{(n)}(T(A)) = \lambda^{(n)}(A) \tag{3.2}$$

for all $A \in \mathcal{B}^{(n)}$ and all $T \in O(n)$. We have already seen in Lemma 3.1 that $T(A) \in \mathcal{B}^{(n)}$ for $A \in \mathcal{B}^{(n)}$ and $T \in O(n)$.

Theorem 3.6. *The Lebesgue-Borel measure $\lambda^{(n)}$ is invariant under the operation of $O(n)$, i.e. (3.2) holds.*

Proof. Let $T \in O(n)$ and $A \in \mathcal{B}^{(n)}$. Since $T^{-1} \in O(n)$ it follows that $T^{-1}(A) \in \mathcal{B}^{(n)}$ and we can define on $\mathcal{B}^{(n)}$

$$\kappa_T^{(n)}(A) := \lambda^{(n)}\left(T^{-1}(A)\right). \tag{3.3}$$

Clearly, $\kappa_T^{(n)}(\emptyset) = 0$ and for a sequence of mutually disjoint sets $A_k \in \mathcal{B}^{(n)}$, $k \in \mathbb{N}$, we have

$$\kappa_T^{(n)}\left(\bigcup_{k \in \mathbb{N}} A_k\right) = \lambda^{(n)}\left(T^{-1}\left(\bigcup_{k \in \mathbb{N}} A_k\right)\right)$$

$$= \lambda^{(n)}\left(\bigcup_{k \in \mathbb{N}} T^{-1}(A_k)\right) = \sum_{k=1}^{\infty} \lambda^{(n)}\left(T^{-1}(A_k)\right)$$

$$= \sum_{k=1}^{\infty} \kappa_T^{(n)}(A_k),$$

where we used $T^{-1}(A_k) \cap T^{-1}(A_l) = \emptyset$ for $A_k \cap A_l = \emptyset$. Hence $\kappa_T^{(n)}$ is a measure on $\mathcal{B}^{(n)}$. Next we prove that $\kappa_T^{(n)}$ is translation invariant. For $x_0 \in \mathbb{R}^n$ and $A \in \mathcal{B}$ we find

$$\kappa_T^{(n)}(x_0 + A) = \lambda^{(n)}(T^{-1}(x_0 + A))$$

$$= \lambda^{(n)}\left((T^{-1}x_0) + T^{-1}(A)\right) = \lambda^{(n)}(T^{-1}(A))$$

$$= \kappa_T^{(n)}(A),$$

where we used that T^{-1} is linear and $\lambda^{(n)}$ is translation invariant. For all $R \in O(n)$ it follows $R\left(B_\rho(0)\right) = B_\rho(0)$, $B_\rho(0) = \{x \in \mathbb{R}^n \mid \|x\| < \rho\}$, and with $C_1 = [0, 1) \times \cdots \times [0, 1) \in \mathbb{R}^n$ we have $R(C_1) \subset B_{\sqrt{n}}(0)$. This implies $\lambda^{(n)}(R(C_1)) \le \lambda^{(n)}\left(B_{\sqrt{n}}(0)\right) < \infty$. Let y_0 be the midpoint of C_1. Then $B_{\frac{1}{4}}(y_0) \subset C_1$ and the translation invariance of $\lambda^{(n)}$ yields

$$0 < \lambda^{(n)}\left(B_{\frac{1}{4}}(0)\right) = \lambda^{(n)}\left(y_0 + B_{\frac{1}{4}}(0)\right) = \lambda^{(n)}\left(B_{\frac{1}{4}}(y_0)\right) \le \lambda^{(n)}(C_1).$$

Thus by Theorem 3.2 it follows that $\kappa_T^{(n)}(A) = C_{\kappa_T^{(n)}}\lambda^{(n)}(A)$, but for $A = B_1(0)$ we must have $\kappa_T^{(n)}(B_1(0)) = \lambda^{(n)}(B_1(0))$ since $B_1(0)$ is invariant under

$T^{-1} \in O(n)$. It follows that $C_{\kappa_T^{(n)}} = 1$, i.e. $\kappa_T^{(n)} = \lambda^{(n)}$ on $\mathcal{B}^{(n)}$ and by replacing A by $T(A)$ in (3.3) we obtain (3.2). $\qquad\square$

A second look at the proof of Theorem 3.6 reveals that (3.3) suggests a universal construction for a measure given a measure and a measurable mapping.

Theorem 3.7. *Let $(\Omega, \mathcal{A})$ and $(\Omega', \mathcal{A}')$ be two measurable spaces and $f : \Omega \to \Omega'$ a measurable mapping. If μ is a measure on $\mathcal{A}$ then $f(\mu) : \mathcal{A}' \to [0, \infty]$ defined by*

$$f(\mu)(A') := \mu(f^{-1}(A')) = \mu\left\{\omega \in \Omega \,\middle|\, f(\omega) \in A'\right\}, \quad A' \in \mathcal{A}', \qquad (3.4)$$

is a measure on $\mathcal{A}'$.

Proof. Clearly $f(\mu)(\emptyset) = 0$ and as in the proof of Theorem 3.6 we conclude for a sequence $(A'_k)_{k\in\mathbb{N}}$ of mutually disjoint sets $A'_k \in \mathcal{A}'$, $k \in \mathbb{N}$, that

$$f(\mu)\left(\bigcup_{k\in\mathbb{N}} A'_k\right) = \mu\left(f^{-1}\left(\bigcup_{k\in\mathbb{N}} A'_k\right)\right) = \sum_{k=1}^{\infty} \mu(f^{-1}(A'_k)) = \sum_{k=1}^{\infty} f(\mu)(A'_k),$$

proving the theorem. $\qquad\square$

Definition 3.8. *The measure $f(\mu)$ on $\mathcal{A}'$ constructed in Theorem 3.7 by (3.4) is called the image of μ under f, or the **image measure** of μ under f. Instead of $f(\mu)$ we often will write μ_f.*

From Remark 1.16.C we deduce immediately

Corollary 3.9. *Let $(\Omega_j, \mathcal{A}_j)$, $j = 1, 2, 3$, be three measurable spaces and $f : \Omega_1 \to \Omega_2$, $g : \Omega_2 \to \Omega_3$ two measurable mappings. For a measure μ on $\mathcal{A}_1$ we have*

$$(g \circ f)(\mu) = g(f(\mu))$$

where $(g \circ f)(\mu)$ is a measure on $\mathcal{A}_3$ and $f(\mu)$ is a measure on $\mathcal{A}_2$.

As usual we call the group generated by the group of translations on $\mathbb{R}^n$ and $O(n)$ the group of **motions** on $\mathbb{R}^n$. This is the group of all transformations on $\mathbb{R}^n$ leaving the Euclidean distance invariant. Combining Corollary 3.9 with Theorem 3.2 and Theorem 3.6 we have

Corollary 3.10. *The Lebesgue-Borel measure $\lambda^{(n)}$ is invariant under all motions on $\mathbb{R}^n$.*

Next we calculate $T(\lambda^{(n)})$ for $T \in GL(n; \mathbb{R})$.

Theorem 3.11. *For all $A \in \mathcal{B}^{(n)}$ and $T \in GL(n; \mathbb{R})$ the following holds*

$$\lambda^{(n)}(T(A)) = |\det T| \, \lambda^{(n)}(A), \tag{3.5}$$

i.e.

$$T(\lambda^{(n)}) = \frac{1}{|\det T|} \lambda^{(n)}. \tag{3.6}$$

Proof. First let $A = \bigtimes_{j=1}^{n} [a_j, b_j)$ and $T = D_k(r)$, $r > 0$, $0 \leq k \leq n$, where

$$D_k(r) = \begin{pmatrix} 1 & & & & & \\ & \ddots & & & & \\ & & 1 & & & \large 0 \\ & & & r & & \\ & & & & 1 & \\ & \large 0 & & & & \ddots \\ & & & & & & 1 \end{pmatrix}$$

with r in position (k, k). It follows that

$$D_k(r)(A) = \bigtimes_{j=1}^{k-1} [a_j, b_j) \times [ra_k, rb_k) \times \bigtimes_{j=k+1}^{n} [a_j, b_j),$$

and therefore

$$\lambda^{(n)}(D_k(r)(A)) = r\lambda^{(n)}(A) = |\det D_k(r)| \, \lambda^{(n)}(A).$$

For $D(r_1, \ldots, r_n) = D_1(r_1) \cdots D_n(r_n) = (r_k \delta_{kl})_{k,l=1,\ldots,n}$, $r_k > 0$, we find

$$\lambda^{(n)}(D(r_1, \ldots, r_n)(A)) = r_1 \cdot \ldots \cdot r_n \lambda^{(n)}(A) = |\det D(r_1, \ldots, r_n)| \, \lambda^{(n)}(A). \tag{3.7}$$

It follows that (3.7) holds for all $A \in \mathcal{F}^{(n)}$ and hence for all $A \in \mathcal{B}^{(n)}$. By the polar decomposition theorem for $T \in GL(n; \mathbb{R})$, Theorem II.A.I.28, we can find $U \in O(n)$ and a symmetric positive definite matrix S such that $T = US$. Further, since S is symmetric, there exists $V \in O(n)$ such that $S = V^{-1}D(r)V$ with $r_1, \ldots, r_n$ being the eigenvalues of S, $r_k > 0$. Using the transitivity of the image measure, i.e. Corollary 3.9, and the invariance of $\lambda^{(n)}$ under the actions of $O(n)$ we now get for all $A \in \mathcal{B}^{(n)}$

$$\begin{aligned}
\lambda^{(n)}(T(A)) &= \lambda^{(n)}\left(UV^{-1}D(r)V(A)\right) \\
&= \lambda^{(n)}(D(r)V(A)) = |\det D(r)|\,\lambda^{(n)}(V(A)) \\
&= |\det D(r)|\,\lambda^{(n)}(A).
\end{aligned}$$

Since

$$\det T = \det U \cdot \det V^{-1} \det D(r) \det V = \det U \det D(r)$$

and since $|\det U| = 1$ we find $|\det T| = \det D(r) = |\det D(r)|$, and (3.5) follows. Taking in (3.5) for A now $T^{-1}(A)$ we arrive at (3.6). $\qquad\square$

Remark 3.12. In Chapter II.16 when starting with our considerations on the Riemann integral in $\mathbb{R}^n$ we discussed parallelotops $P(a_1, \dots, a_n)$ defined for n linearly independent vectors $a_j \in \mathbb{R}^n$ as

$$P(a_1, \dots, a_n) := \left\{ x \in \mathbb{R}^n \,\Big|\, x = \sum_{j=1}^{n} \lambda_j a_j, \; \lambda_j \in [0,1] \right\},$$

and we have seen that with $T \in GL(n; \mathbb{R})$ defined by $Te_j = a_j$ we always have

$$P(a_1, \dots, a_n) = T(P(e_1, \dots, e_n)),$$

but $P(e_1, \dots, e_n) = [0,1] \times \cdots \times [0,1] \subset \mathbb{R}^n$. Thus we arrive again at

$$\left|\lambda^{(n)}(P(a_1, \dots, a_n))\right| = |\det T|\lambda^{(n)}(P(e_1, \dots, e_n)) = |\det T| = |\det (a_1, \dots, a_n)| \tag{3.8}$$

where we used our old notation $T = (a_1, \dots, a_n)$. Clearly (3.8) is nothing but (II.16.8).

Our notion of Jordan measurable sets which was key for developing the theory of Riemann integrals in $\mathbb{R}^n$ depends on the concept of Lebesgue null sets, see Definition II.19.6. Now we can put this concept in a more natural context.

Definition 3.13. *Let* $(\Omega, \mathcal{A}, \mu)$ *be a measure space. By* $\mathcal{N}_\mu$ *we denote the sets of μ-measure zero, or just μ-zero sets, i.e.*

$$\mathcal{N}_\mu := \left\{ A \in \mathcal{A} \,\middle|\, \mu(A) = 0 \right\}. \tag{3.9}$$

A subset B of a set of μ-measure zero is called a μ-null set, compare with Definition I.32 and Definition II.19.6.

Lemma 3.14. A. *Clearly* $\emptyset \in \mathcal{N}_\mu$ *and if* $M \in \mathcal{A}$ *and* $M \subset N$ *for some* $N \in \mathcal{N}_\mu$ *then* $M \in \mathcal{N}_\mu$.
B. *The union of a denumerable family of μ-zero sets is a μ-zero set, i.e. for* $(N_k)_{k \in \mathbb{N}}$, $N_k \in \mathcal{N}_\mu$, *it follows that* $\bigcup_{k \in \mathbb{N}} N_k \in \mathcal{N}_\mu$.
C. *The union of a denumerable family of μ-null sets is a μ-null set.*

Proof. Part A is trivial, but note that the condition $M \in \mathcal{A}$ is essential, otherwise $\mu(M)$ may not be defined. Now, given a sequence $(N_k)_{k \in \mathbb{N}}$, $N_k \in \mathcal{N}_\mu$. For $\epsilon > 0$ we have $\mu(N_k) < \frac{\epsilon}{2^k}$ and therefore

$$\mu\left(\bigcup_{k \in \mathbb{N}} N_k\right) \leq \sum_{k=1}^{\infty} \mu(N_k) < \epsilon \sum_{k=1}^{\infty} \frac{1}{2^k} = \epsilon$$

implying $\mu\left(\bigcup_{k \in \mathbb{N}} N_k\right) = 0$. Part C is obvious from Part B.   □

It is important to realise that two measures on one and the same σ-field $\mathcal{A}$ may have quite different sets of measure zero.

Example 3.15. A. Let $(\Omega, \mathcal{A})$ be a measurable space and $\omega_0 \in \Omega$. We consider the measure space $(\Omega, \mathcal{A}, \epsilon_{\omega_0})$. A set $A \in \mathcal{A}$ belongs to $\mathcal{N}_{\epsilon_{\omega_0}}$ if and only if $\omega_0 \notin A$, i.e. $\mathcal{N}_{\epsilon_{\omega_0}} = \{A \in \mathcal{A} \mid A \subset \{\omega_0\}^{\complement}\}$.
B. For every point $x_0 \in \mathbb{R}^n$ we have $\lambda^{(n)}(\{x_0\}) = 0$, i.e. $\{x_0\} \in \mathcal{N}_{\lambda^{(n)}}$. To see this, for $\epsilon > 0$ choose the cube $C_\epsilon(x_0)$ with centre x_0 and side length ϵ. We have $C_\epsilon(x_0), \{x_0\} \in \mathcal{B}^{(n)}$ and

$$\lambda^{(n)}(\{x_0\}) \leq \lambda^{(n)}(C_\epsilon(x_0)) \leq \epsilon^n$$

implying $\{x_0\} \in \mathcal{N}_{\lambda^{(n)}}$. Combined with Lemma 3.14.B we now find that $\mathbb{N}^n$, $\mathbb{Z}^n$, $\mathbb{Q}^n \in \mathcal{N}_{\lambda^{(n)}}$ for $n \in \mathbb{N}$. This is due to the fact that these are countable sets.
C. When we now take in Part A the space $(\mathbb{R}^n, \mathcal{B}^{(n)})$ as the measurable space $(\Omega, \mathcal{A})$ and $\omega_0 = x_0 \in \mathbb{R}^n$ we see immediately that $\mathcal{N}_{\epsilon_{x_0}} \neq \mathcal{N}_{\lambda^{(n)}}$ since any non-degenerate compact cube K with $x_0 \notin K$ has positive Lebesgue measure, however $\epsilon_{x_0}(K) = 0$.

Recall Definition II.19.6: $A \subset \mathbb{R}^n$ was said to be a Lebesgue null set if for every $\epsilon > 0$ exists a denumerable covering $(K_j)_{j \in \mathbb{N}}$ of A by open cells such that $\sum_{j=1}^{\infty} \lambda^{(n)}(K_j) < \epsilon$. In Problem 7 we will see that for sets $A \in \mathcal{B}^{(n)}$ this definition coincides with Definition 3.13. However in Definition II.19.6 the set A is not necessarily a Borel set. Thus we may have the situation that null sets in the sense of Definition II.19.6 are not elements in $\mathcal{N}_{\lambda^{(n)}}$ since they do not belong to the Borel σ-field. As in general μ-null sets do not necessarily belong to $\mathcal{A}$, in general they do not belong to $\mathcal{N}_\mu$. To cope with this problem we give

Definition 3.16. *A measure space* $(\Omega, \mathcal{A}, \mu)$ *is called* **complete** *if every subset of a set of measure zero belongs to* $\mathcal{A}$ *and hence is a set of measure zero.*

Theorem 3.17. *Let* $(\Omega, \mathcal{A}, \mu)$ *be a measure space. Denote by* $\mathcal{N}_\mu$ *all sets of measure zero and by* $\tilde{\mathcal{N}}_\mu$ *all subsets of* $\mathcal{N}_\mu$, *i.e.* $N \in \tilde{\mathcal{N}}_\mu$ *if there exists* $N_0 \in \mathcal{N}_\mu$ *such that* $N \subset N_0$. *The family*

$$\tilde{\mathcal{A}} := \left\{ A \cup N \mid A \in \mathcal{A} \text{ and } N \in \tilde{\mathcal{N}}_\mu \right\}$$

is a σ-field in Ω *and we have* $\mathcal{A} \subset \tilde{\mathcal{A}}$. *Moreover* $\tilde{\mu} : \tilde{\mathcal{A}} \to [0, \infty]$ *defined by* $\tilde{\mu}(A \cup N) := \mu(A)$ *is a measure defined on* $\tilde{\mathcal{A}}$ *and* $(\Omega, \tilde{\mathcal{A}}, \tilde{\mu})$ *is a complete measure space with* $\tilde{\mu}|_\mathcal{A} = \mu$.

Definition 3.18. *The measure space* $(\Omega, \tilde{\mathcal{A}}, \tilde{\mu})$ *is called the* **completion** *of* $(\Omega, \mathcal{A}, \mu)$.

Proof of Theorem 3.17. First we note that the representation $A \cup N$ for an element is not unique. For $N_0 \in \mathcal{N}_\mu$ we have $(A \backslash N_0) \cup (N \cup N_0) = A \cup N$, $A \backslash N_0 \in \mathcal{A}$ and $N \cup N_0 \subset \tilde{N}$, $\mu(\tilde{N}) = 0$. In the following, when dealing with elements of $\tilde{\mathcal{A}}$ we always start with a representation $A \cup N$ and we take care that results are independent of the special choice. Clearly $A = A \cup \emptyset \in \tilde{\mathcal{A}}$ for all $A \in \mathcal{A}$. Moreover for $N \subset N_0 \in \mathcal{N}_\mu$ we find with $A \in \mathcal{A}$ that

$$(A \cup N)^\complement = A^\complement \cap N^\complement = A^\complement \cap N^\complement \cap \left(N_0^\complement \cup N_0 \right)$$
$$= \left(A^\complement \cap N^\complement \cap N_0^\complement \right) \cup \left(A^\complement \cap N^\complement \cap N_0 \right)$$
$$= \left(A^\complement \cap N_0^\complement \right) \cup \left(A^\complement \cap N^\complement \cap N_0 \right).$$

Since $A^\complement \cap N_0^\complement \in \mathcal{A}$ and $A^\complement \cap N^\complement \cap N_0 \subset N_0 \in \mathcal{N}_\mu$ it follows that $(A \cup N)^\complement \in \tilde{\mathcal{A}}$. Now we consider a sequence $(A_k \cup N_k)_{k \in \mathbb{N}}$ with $A_k \cup N_k \in \tilde{\mathcal{A}}$ with corresponding sequence $(N_{0,k})_{k \in \mathbb{N}}$, $N_k \subset N_{0,k} \in \mathcal{N}_\mu$. It follows that

$$\bigcup_{k \in \mathbb{N}} (A_k \cup N_k) = \left(\bigcup_{k \in \mathbb{N}} A_k \right) \cup \left(\bigcup_{k \in \mathbb{N}} N_k \right)$$

and $\bigcup_{k\in\mathbb{N}} A_k \in \mathcal{A}$ as well as $\bigcup_{k\in\mathbb{N}} N_k \subset \bigcup_{k\in\mathbb{N}} N_{0,k} \in \mathcal{N}_\mu$, hence we have $\bigcup_{k\in\mathbb{N}}(A_k \cup N_k) \in \tilde{\mathcal{A}}$ and $\tilde{\mathcal{A}}$ is a σ-field containing $\mathcal{A}$. Next we want to show that the definition of $\tilde{\mu}$ is independent of the representation. Suppose $A \cup N = B \cup M$ with $A, B \in \mathcal{A}$ and $N \subset N_0$, $M \subset M_0$, $N_0, M_0 \in \mathcal{N}_\mu$. By assumption we have

$$A \subset A \cup N = B \cup M \subset B \cup M_0$$

and

$$B \subset B \cup M = A \cup N \subset A \cup N_0$$

and the monotonicity of μ yields

$$\mu(A) \le \mu(B \cup M_0) \le \mu(B) + \mu(M_0) = \mu(B)$$

as well as

$$\mu(B) \le \mu(A \cup N_0) \le \mu(A) + \mu(N_0) = \mu(A),$$

i.e. $\mu(A) = \mu(B)$ and hence $\tilde{\mu}$ is well defined, i.e. independent of the representation of $\tilde{A} \in \tilde{\mathcal{A}}$. We claim that $\tilde{\mu}$ is a measure on $\tilde{\mathcal{A}}$. The fact that $\tilde{\mu}(\emptyset) = 0$ is trivial and for $(A_k \cup N_k)_{k\in\mathbb{N}}$ of mutually disjoint sets we find

$$\tilde{\mu}\left(\bigcup_{k\in\mathbb{N}}(A_k \cup N_k)\right) = \tilde{\mu}\left(\left(\bigcup_{k\in\mathbb{N}} A_k\right) \cup \left(\bigcup_{k\in\mathbb{N}} N_k\right)\right)$$

$$= \mu\left(\bigcup_{k\in\mathbb{N}} A_k\right) = \sum_{k=1}^{\infty}\mu(A_k) = \sum_{k\in\mathbb{N}}\tilde{\mu}(A_k \cup N_k),$$

where we used that $\bigcup_{k\in\mathbb{N}} N_k \subset N_0 \in \mathcal{N}_\mu$. Thus $\tilde{\mu}$ is a measure on $\tilde{\mathcal{A}}$ extending μ, i.e. $\tilde{\mu}|_{\mathcal{A}} = \mu$. It remains to prove that $(\Omega, \tilde{\mathcal{A}}, \tilde{\mu})$ is complete. For this let $\tilde{A} = A \cup N \in \tilde{\mathcal{A}}$ such that $\tilde{\mu}(\tilde{A}) = 0$ and $\tilde{B} \subset \tilde{A}$. We have to show that $\tilde{B} \in \tilde{\mathcal{A}}$. But $\tilde{\mu}(\tilde{A}) = 0$ implies $\mu(A) = 0$ and we can write $\tilde{A} = \emptyset \cup (A \cup N)$, $A \cup N \subset N_0 \in \mathcal{N}_\mu$. Thus $\tilde{B} \subset \tilde{A} = A \cup N \subset N_0$ and $\tilde{B} = \emptyset \cup \tilde{B}$, $\emptyset \in \mathcal{A}$ and $\tilde{B} \subset N_0 \in \mathcal{N}_\mu$, i.e. $\tilde{B} \in \tilde{\mathcal{A}}$.   □

Definition 3.19. *The completion of $\mathcal{B}^{(n)}$ is the σ-field of all **Lebesgue sets** and denoted by $\mathcal{L}^{(n)}$. The completion of $\lambda^{(n)}$ is called the **Lebesgue measure**, however we will in general still write $\lambda^{(n)}$ instead of $\tilde{\lambda}^{(n)}$.*

Exercise 3.20. *Show that $\tilde{\lambda}^{(n)}$ is again translation invariant.*

We return to sets of measure zero, more precisely to sets of Lebesgue measure zero in $(\mathbb{R}^n, \mathcal{L}^{(n)}, \lambda^{(n)})$. We can apply our results from Part 4 and find that in particular sets of Jordan content zero are Lebesgue sets of measure zero as are the graphs of continuous functions $f : \mathbb{R}^{n-1} \to \mathbb{R}$ or certain boundaries of sets defined by equations, in particular this applies to hyperplanes in $\mathbb{R}^n$ as well as their subsets. (Compare with Remark II.19.7.C, Corollary II.19.13, Proposition II.19.19 and II.19.21, Lemma II.21.3, Proposition II.21.11 and Corollary II.21.12.)

One of our first examples of a σ-field was the trace σ-field, Example 1.3.C. Let $(\Omega, \mathcal{A}, \mu)$ be a measure space and $\Omega' \subset \Omega$ a subset then the trace σ-field $\mathcal{A}_{\Omega'} = \Omega' \cap \mathcal{A}$ was defined as $\{\Omega' \cap A \,|\, A \in \mathcal{A}\}$. We may ask whether we can construct a measure on $\Omega' \cap \mathcal{A}$ starting with μ. In the case where $\Omega' \in \mathcal{A}$, i.e. Ω' is $\mathcal{A}$-measurable, we can define for $A \in \Omega' \cap \mathcal{A}$

$$\mu_{\Omega'}(A) := \mu(\Omega' \cap A) \tag{3.10}$$

and it is trivial that $\mu_{\Omega'}$ is a measure on $\Omega' \cap \mathcal{A}$. However for a non-measurable set $\Omega' \subset \Omega$, i.e. for $\Omega' \notin \mathcal{A}$, in general we can not define a corresponding measure on $\Omega' \cap \mathcal{A}$.

Thus for every Borel set $A \in \mathcal{B}^{(n)}$ the measure $\lambda_A^{(n)}$ is well defined on $A \cap \mathcal{B}^{(n)}$ as $\lambda_A^{(n)}$ is well defined for every $A \in \mathcal{L}^{(n)}$. In particular, for a Lebesgue set N of measure zero (or a Borel set M of measure zero) the measure $\lambda_N^{(n)}$ (and $\lambda_M^{(n)}$) is defined, however by (3.10) it is identically zero. Thus this construction does not allow us to define a "nice" measure on a hyperplane or a C^k-surface in $\mathbb{R}^n$. For the graph of a measurable function we may try to work with the image measure. Let $f : G \to \mathbb{R}$, $G \in \mathcal{B}^{(n-1)}$, be a measurable function and consider on G the measure $\lambda_G^{(n-1)}$. On $\Gamma(f) = \{(x, f(x)) \,|\, x \in G\}$ we may study $F(\lambda_G^{(n-1)})$ where $F : G \to \mathbb{R}^n$, $F(x) = (x, f(x))$. But it remains the general problem to find for Lebesgue measurable subsets of $\mathbb{R}^n$ a natural geometric measure. The Hausdorff measure we are going to study now will give such a measure. At the moment it mainly serves us as a further application of the Carathéodory theorem, however we will make much use of it when returning to the question of how to define the area of a surface (or the area of a submanifold). Most of the results are best studied with context of metric spaces. We start with

Definition 3.21. *Let (X, d) be a metric space and μ^* an outer measure on $\mathcal{P}(X)$. We call μ^* a **metric outer measure** if*

$$\mu^*(A \cup B) = \mu^*(A) + \mu^*(B) \tag{3.11}$$

holds for all $A, B \in \mathcal{P}(X)$ such that $\mathrm{dist}(A, B) > 0$.

Recall that in a metric space $\mathrm{dist}(A, B)$ is defined by

$$\mathrm{dist}(A, B) = \inf \left\{ d(x, y) \,\middle|\, x \in A, \, y \in B \right\},$$

compare with (II.3.2).

The following theorem implies that a metric outer measure restricted to the Borel sets on X, i.e. to $\mathcal{B}(X) := \sigma(\mathcal{O}_X)$ where $\mathcal{O}_X$ denotes the metric topology in X, is a measure.

Theorem 3.22. *The Borel sets of a metric space are measurable with respect to any metric outer measure on X.*

Proof. We will prove that all closed sets are measurable with respect to μ^*, i.e. that

$$\mu^*(G) = \mu^*(G \cap A) + \mu^*(G \cap A^C)$$

holds for all closed sets $A \subset X$, and in fact we may assume that $\mu^*(G) < \infty$. Further it is clear that we only need to prove

$$\mu^*(G) \geq \mu^*(G \cap A) + \mu^*(G \cap A^C)$$

since the converse inequality follows from the sub-additivity of μ^*. For $k \in \mathbb{N}$ we define $G_k := \left\{ x \in A^C \cap G \,\middle|\, \mathrm{dist}(x, A) > \frac{1}{k} \right\} \subset G$. Clearly we have $G_k \subset G_{k+1}$ and the closedness of A implies $A^C \cap G = \bigcup_{k \in \mathbb{N}} G_k$. Moreover $\mathrm{dist}(A \cap G, G_k) \geq \frac{1}{k}$ and since μ^* is an outer metric measure it follows that

$$\mu^*(G) \geq \mu^* \left((A \cap G) \cup G_k \right) = \mu^*(A \cap G) + \mu^*(G_k).$$

If we can show that $\lim_{k \to \infty} \mu^*(G_k) = \mu^*(A^C \cap G)$ then the result will follow. For this let $H_k := G_{k+1} \cap G_k^C$ and consider

$$\mathrm{dist}(H_{k+1}, G_k) = \mathrm{dist}\left(G_{k+1} \cap G_k^C, G_k \right).$$

For $x \in H_{k+1}$, in particular $x \in G_k^\complement$, and $d(x, y) < \frac{1}{k(k+1)}$, it follows that

$$d(y, A) \leq d(x, y) + \operatorname{dist}(x, A) < \frac{1}{k(k+1)} + \frac{1}{k+1} = \frac{1}{k},$$

thus we have

$$\operatorname{dist}(H_{k+1}, G_k) \geq \frac{1}{k(k+1)}. \tag{3.12}$$

Since $H_{2l} \backslash G_{2l-1} \subset G_{2l+1}$ we arrive at

$$\mu^*(G_{2l+1}) \geq \mu^*(H_{2l} \cup G_{2l-1}) = \mu^*(H_{2l}) + \mu^*(G_{2l-1})$$

where we used that μ^* is a metric outer measure and (3.12). By induction we find now

$$\mu^*(G_{2l+1}) \geq \sum_{j=1}^{l} \mu^*(H_{2j})$$

and analogously

$$\mu^*(G_{2l}) \geq \sum_{j=1}^{l} \mu^*(H_{2j-1}).$$

By assumption $\mu^*(G) < \infty$ implying the convergence of the series $\sum_{j=1}^{\infty} \mu^*(H_{2j})$ and $\sum_{j=1}^{\infty} \mu^*(H_{2j-1})$, and hence

$$\mu^*(G_k) \leq \mu^*(A^\complement \cap G) \leq \mu^*(G_k) + \sum_{j=k+1}^{\infty} \mu^*(H_j)$$

which yields for $k \to \infty$ that

$$\lim_{k \to \infty} \mu^*(G_k) = \mu^*(A^\complement \cap G)$$

and the theorem is proved. $\qquad\square$

For a subset $Y \subset X$ of a metric space (X, d) we define its **diameter** as

$$\operatorname{diam}(Y) := \sup \left\{ d(x, y) \,|\, x, y \in Y \right\},$$

see Definition II.2.13. We note that if $\operatorname{diam}(Y) \leq \rho$ then $\operatorname{diam}(\overline{Y}) \leq \rho$, and further if $Y \subset X$ then $Y_\epsilon := \{ x \in X \,|\, d(x, Y) < \epsilon \}$ is an open set and

$\operatorname{diam} Y_\epsilon \leq \operatorname{diam} Y + \epsilon$.

Let $A \subset X$ be any set and $\alpha > 0$, $\delta > 0$. We set

$$\mathcal{H}_\alpha^\delta(A) := \inf \left\{ \sum_{k=1}^\infty (\operatorname{diam} Y_k)^\alpha \,\Big|\, A \subset \bigcup_{k \in \mathbb{N}} Y_k, \operatorname{diam} Y_k \leq \delta \right\}. \qquad (3.13)$$

As δ decreases to 0 the function $\delta \mapsto \mathcal{H}_\alpha^\delta(A)$ is increasing and therefore

$$\mathcal{H}_\alpha^*(A) := \lim_{\delta \to 0} \mathcal{H}_\alpha^\delta(A) \qquad (3.14)$$

exists, maybe as improper limit $+\infty$. In particular we have

$$\mathcal{H}_\alpha^\delta(A) \leq \mathcal{H}_\alpha^*(A). \qquad (3.15)$$

We claim

Theorem 3.23. *By $\mathcal{H}_\alpha^*$ a metric outer measure is defined on $\mathcal{P}(X)$.*

Proof. Clearly, $\mathcal{H}_\alpha^\delta(\emptyset) = 0$, so $\mathcal{H}_\alpha^*(\emptyset) = 0$, and if $A_1 \subset A_2$ a covering of A_2 is also a covering of A_1, so it follows that $\mathcal{H}_\alpha^\delta(A_1) \leq \mathcal{H}_\alpha^\delta(A_2)$ implying that $\mathcal{H}_\alpha^*(A_1) \leq \mathcal{H}_\alpha^*(A_2)$. Next we want to show the sub-additivity of $\mathcal{H}_\alpha^*$. For this let $(A_k)_{k \in \mathbb{N}}$ be a sequence in $\mathcal{P}(X)$ and $(Y_{k,l})_{l \in \mathbb{N}}$ a covering of A_k by sets $Y_{k,l}$ such that $\operatorname{diam}(Y_{k,l}) < \delta$ and

$$\sum_{l=1}^\infty (\operatorname{diam}(Y_{k,l}))^\alpha \leq \mathcal{H}_\alpha^\delta(A_k) = \frac{\epsilon}{2^k}.$$

The family $(Y_{k,l})_{k,l \in \mathbb{N}}$ is a covering of $\bigcup_{k \in \mathbb{N}} A_k$ by sets with diameter less than δ and therefore

$$\mathcal{H}_\alpha^\delta \left(\bigcup_{k \in \mathbb{N}} A_k \right) \leq \sum_{k=1}^\infty \mathcal{H}_\alpha^\delta(A_k) + \sum_{k=1}^\infty \frac{\epsilon}{2^k}$$
$$\leq \sum_{k=1}^\infty \mathcal{H}_\alpha^*(A_k) + \epsilon,$$

which yields

$$\mathcal{H}_\alpha^\delta \left(\bigcup_{k \in \mathbb{N}} A_k \right) \leq \sum_{k=1}^\infty \mathcal{H}_\alpha^*(A_k),$$

and for δ decreasing to 0 we obtain $\mathcal{H}_\alpha^* \left(\bigcup_{k \in \mathbb{N}} A_k \right) \leq \sum_{k=1}^\infty \mathcal{H}_\alpha^*(A_k)$, i.e. the subadditivity of $\mathcal{H}_\alpha^*$. This already implies that $\mathcal{H}_\alpha^*$ is an outer measure. Finally we prove that $\mathcal{H}_\alpha^*$ is a metric outer measure. For this let $A, B \subset X$ such that $\mathrm{dist}(A, B) > 0$. We need to show that $\mathcal{H}_\alpha^*(A \cup B) = \mathcal{H}_\alpha^*(A) + \mathcal{H}_\alpha^*(B)$, and since we already know the sub-additivity of $\mathcal{H}_\alpha^*$ it remains to verify

$$\mathcal{H}_\alpha^*(A \cup B) \geq \mathcal{H}_\alpha^*(A) + \mathcal{H}_\alpha^*(B). \tag{3.16}$$

Let $0 < \epsilon < \mathrm{dist}(A, B)$ and let $A \cup B$ be covered by $(Y_k)_{k \in \mathbb{N}}$ with $\mathrm{diam}(Y_k) \leq \delta < \epsilon$. By $A_k' := A \cap Y_k$ and $B_k' := B \cap Y_k$, $k \in \mathbb{N}$, we now have disjoint covers of A and B, respectively. Therefore we find

$$\sum_{k=1}^\infty (\mathrm{diam}\, A_k')^\alpha + \sum_{k=1}^\infty (\mathrm{diam}\, B_k')^\alpha \leq \sum_{k=1}^\infty (\mathrm{diam}\, Y_k)^\alpha.$$

We now take the infimum over all coverings and then we let δ go to zero to arrive at (3.16). $\qquad \square$

Combining Theorem 2.23 with Theorem 3.22 and Theorem 3.23 we get

Corollary 3.24. *Restricted to $\mathcal{B}^{(n)}$ the metric outer measure $\mathcal{H}_\alpha^*$ is a measure.*

Note that $\mathcal{H}_\alpha^*$ as a measure on $\mathcal{B}^{(n)}$ allows a completion on $\mathcal{L}^{(n)}$, but at this stage we cannot decide whether for certain values of α the family of $\mathcal{H}_\alpha^*$-measurable sets is larger than $\mathcal{B}^{(n)}$ (or $\mathcal{L}^{(n)}$).

Definition 3.25. *The measure*

$$\mathcal{H}_\alpha := \mathcal{H}_\alpha^*\big|_{\mathcal{B}^{(n)}} \tag{3.17}$$

*is called the α-**dimensional Hausdorff measure** on (X, d).*

Lemma 3.26. *The α-dimensional Hausdorff measure on the n-dimensional Euclidean space $\mathbb{R}^n$ is translation invariant as well as invariant under the action of $O(n)$, i.e. for $x \in \mathbb{R}^n$ and $T \in O(n)$*

$$\mathcal{H}_\alpha(x + A) = \mathcal{H}_\alpha(A) \text{ and } \mathcal{H}_\alpha(T(A)) = \mathcal{H}_\alpha(A) \tag{3.18}$$

holds for all $A \in \mathcal{B}^{(n)}$.

Proof. This follows immediately since the metric induced by the Euclidean norm is invariant under translation and orthogonal transformations.   □

Remark 3.27. Lemma 3.26 has a counterpart for a general metric space (X, d). The measure $\mathcal{H}_\alpha$ is on $\mathcal{B}^{(n)}(X)$ invariant under isometries of (X, d), i.e. if $T : X \to X$ satisfies $d(Tx, Ty) = d(x, y)$, then for all $A \in \mathcal{B}(X)$ we have $\mathcal{H}_\alpha(T(A)) = \mathcal{H}_\alpha(A)$.

Corollary 3.28. *There exists a constant $\gamma_n > 0$ such that*

$$\mathcal{H}_n(A) = \gamma_n \lambda^{(n)}(A) \tag{3.19}$$

holds for all $A \in \mathcal{B}^{(n)}$.

Proof. The result will follow from Corollary 3.3 if we can prove for $C_n = [0, 1] \times \cdots \times [0, 1] \subset \mathbb{R}^n$ that $0 < \mathcal{H}_n(C_n) < \infty$. Since for every covering of C_n we have $1 \leq \sum_{k=1}^\infty (\text{diam } Y_k)^n$ the estimate $0 < \mathcal{H}_n(C_n)$ is trivial. If we divide C_n into k^n cubes of side length $\frac{1}{k}$ and choose $\delta > \frac{\sqrt{n}}{k}$ we find $\mathcal{H}_n^\delta(C_n) \leq k^n \left(\frac{\sqrt{n}}{k}\right)^n = n^{\frac{n}{2}}$ implying $\mathcal{H}_n(C) < \infty$.   □

Remark 3.29. It can be shown with tools currently not at our disposal that

$$\gamma_n = \frac{1}{2^n} \lambda^{(n)}(B_1(0)) = \frac{\pi^{\frac{n}{2}}}{2^n \Gamma(\frac{n}{2} + 1)}, \tag{3.20}$$

see [24].

The next result compares $\mathcal{H}_\alpha^*$ and $\mathcal{H}_\beta^*$.

Lemma 3.30. *Let $0 < \alpha < \beta < \gamma$ and assume that $\mathcal{H}_\alpha^*(A) < \infty$, then $\mathcal{H}_\beta^*(A) = 0$. If however $\mathcal{H}_\gamma^*(A) > 0$ then $\mathcal{H}_\beta^*(A) = +\infty$.*

Proof. Clearly, the second statement follows from the first. We note that if $\text{diam}(Y) < \delta$ and $\beta > \alpha$ then

$$(\text{diam}(Y))^\beta = (\text{diam}(Y))^{\beta-\alpha} (\text{diam}(Y))^\alpha \leq \delta^{\beta-\alpha} (\text{diam}(Y))^\alpha$$

which yields

$$\mathcal{H}_\beta^\delta(A) \leq \delta^{\beta-\alpha} \mathcal{H}_\alpha^\delta(A) \leq \delta^{\beta-\alpha} \mathcal{H}_\alpha^*(A).$$

Now, if $\mathcal{H}_\alpha^*(A) > 0$, since $\beta - \alpha > 0$, for $\delta \to 0$ it follows that $\mathcal{H}_\beta^*(A) = 0$.   □

From this lemma we deduce immediately that in the case of $\mathbb{R}^n$ when dealing with $\mathcal{H}_\alpha$ we can always assume that $\alpha \leq n$, otherwise $\mathcal{H}_\alpha$ is identically zero.

Example 3.31. Let $U \subset \mathbb{R}^n$ be an open set and $\alpha < n$, then $\mathcal{H}_\alpha(U) = \infty$ since $\mathcal{H}_n(U) > 0$. The latter statement follows since U must contain a non-degenerate cube. Now we can deduce that every Borel set of $\mathbb{R}^n$ with non-empty interior has for $\alpha < n$ an α-dimensional Hausdorff measure equal to zero.

A further consequence of Lemma 3.30 is that for every Borel set $A \in \mathcal{B}^{(n)}$ there exists a unique $\alpha \geq 0$ such that

$$\mathcal{H}_\beta(A) = \begin{cases} +\infty, & \beta < \alpha, \\ 0, & \alpha < \beta, \end{cases}$$

i.e. we have

$$\alpha = \sup \left\{ \beta > 0 \,\middle|\, \mathcal{H}_\beta(A) = \infty \right\} = \inf \left\{ \gamma > 0 \,\middle|\, \mathcal{H}_\gamma(A) = 0 \right\}. \tag{3.21}$$

Definition 3.32. *For $A \in \mathcal{B}^{(n)}$ the number α uniquely determined by (3.21) is called the **Hausdorff dimension** of A and we write $\alpha = \dim_H A$.*

Remark 3.33. A. If $A \in \mathcal{B}^{(n)}$ has Hausdorff dimension α, it is still possible that $H_\alpha(A) = 0$ or $H_\alpha(A) = +\infty$. In the case where $0 < H_\alpha(A) < \infty$ some authors say that A has strict Hausdorff dimension α, see [84].
B. The Hausdorff dimension of a Borel set $A \in \mathcal{B}^{(n)}$ can be a non-integer number. In such a case A is called a **fractal**.
C. *A priori* there is no relation between the Hausdorff dimension of a Borel set and the dimensions of a subspace of $\mathbb{R}^n$. However the situation will become more clear in the following examples.

Before studying more examples we want to investigate the behaviour of the Hausdorff dimension under certain mappings.

Lemma 3.34. *For $\rho > 0$ define $h_\rho : \mathbb{R}^n \to \mathbb{R}^n$ by $h_\rho(x) = \rho x$. If $A \in \mathcal{B}^{(n)}$ and $\alpha > 0$ then it follows that*

$$\mathcal{H}_\alpha(\rho A) := \mathcal{H}_\alpha(h_\rho(A)) = \rho^\alpha \mathcal{H}_\alpha(A). \tag{3.22}$$

Proof. Let $(Y_k)_{k\in\mathbb{N}}$, $Y_k \subset \mathbb{R}^n$, be a covering of A such that $\mathrm{diam}(Y_k) \leq \delta$. Then $(\rho Y_k)_{k\in\mathbb{N}}$ is a covering of ρA and $\mathrm{diam}(\rho Y_k) \leq \rho\delta$ which yields

$$\mathcal{H}_\alpha^{\rho\delta}(\rho A) \leq \sum_{k=1}^{\infty} (\mathrm{diam}(\rho Y_k))^\alpha = \rho^\alpha \sum_{k=1}^{\infty} (\mathrm{diam}\, Y_k)^\alpha \leq \rho^\alpha \mathcal{H}_\alpha^*(A)$$

implying for Borel sets $\mathcal{H}_\alpha^{\rho\delta}(\rho A) \leq \rho^\alpha \mathcal{H}_\alpha(A)$ and now $\mathcal{H}_\alpha(\rho A) \leq \rho^\alpha \mathcal{H}_\alpha(A)$ follows for $\delta \to 0$. Replacing ρ by $\frac{1}{\rho}$ and A by ρA we get $\mathcal{H}_\alpha(A) = \mathcal{H}_\alpha\left(\frac{1}{\rho}\rho A\right) \leq \frac{1}{\rho^\alpha}\mathcal{H}_\alpha(\rho A)$ or $\rho^\alpha \mathcal{H}_\alpha(A) \leq \mathcal{H}_\alpha(\rho A)$ and the result follows. □

In Lemma 3.34 it was easy to see that $h_\rho(A)$ is a Borel set since $\rho A = h_{\frac{1}{\rho}}^{-1}(A)$ and as a continuous mapping $h_{\frac{1}{\rho}}$ is measurable. For the following results we need to know that the image of a Borel set under a continuous mapping is a Borel set, and this is a non-trivial result. A partial result is easy to get: since the image of a compact set under a continuous mapping is compact we can deduce using $f\left(\bigcup_{k\in\mathbb{N}} A_k\right) = \bigcup_{k\in\mathbb{N}} f(A_k)$ that the image of a countable union of compact sets under a continuous mapping is Borel measurable. For bijective continuous functions $f : \mathbb{R} \to \mathbb{R}$ we prove in Problem 11 that they map Borel sets into Borel sets. The proof of the general result needs knowledge about Suslin sets and goes far beyond what can be reasonably handled in a first course on measure theory. In fact there are many results in measure theory in topological spaces which depends on more advanced results of descriptive set theory. We refer the interested reader to H. Federer [25], K. Jacobs [42] and A. B. Kharazishvili [46], and T. Bartoszynski, H. Judah [9] where special attention is paid to the real line. For reference purpose we quote from [25] result 2.2.13 in the formulation suitable for our needs.

Theorem 3.35. *Denote on $\mathbb{R}^n$ the Euclidean metric by d_E and suppose that (G, d_E), $G \subset \mathbb{R}^n$, is a complete metric subspace of $(\mathbb{R}^n, d_E)$. Then for a continuous function $f : G \to \mathbb{R}^m$ the image of a Borel set $A \subset G$ is a Borel set in $\mathbb{R}^m$. In particular every continuous mapping $f : \mathbb{R}^n \to \mathbb{R}^m$ maps Borel sets of $\mathbb{R}^n$ onto Borel sets in $\mathbb{R}^m$.*

Of course we can take G just to be closed in Theorem 3.35, but we intended to be close to Federer's formulation.

Lemma 3.36. *Let $G \subset \mathbb{R}^n$ be a closed set and $f : G \to \mathbb{R}^m$ be a **Hölder continuous** mapping with **Hölder exponent** $s \in (0, 1]$, i.e.*

$$\|f(x) - f(y)\| \leq c \|x - y\|^s \tag{3.23}$$

for all $x, y \in G$ and some $c > 0$. For every Borel set $A \in \mathcal{B}^{(n)}$, $A \subset G$, we have

$$\mathcal{H}_{\frac{\alpha}{s}}(f(A)) \leq c^{\frac{\alpha}{s}} \mathcal{H}_\alpha(A). \tag{3.24}$$

Proof. Let $(Y_k)_{k \in \mathbb{N}}$, $Y_k \subset \mathbb{R}^n$, $\operatorname{diam}(Y_k) \leq \delta$, be a covering of A. The Hölder condition (3.23) yields $\operatorname{diam}(f(A \cap Y_k)) \leq c\,(\operatorname{diam}(Y_k))^s$, i.e. $(f(A \cap Y_k))_{k \in \mathbb{N}}$ is a covering of $f(A)$ by sets with diameter less or equal to $c\delta^s$. Since

$$\sum_{k=1}^\infty (\operatorname{diam}(f(A \cap Y_k)))^{\frac{\alpha}{s}} \leq c^{\frac{\alpha}{s}} \sum_{k=1}^\infty (\operatorname{diam}(Y_k))^\alpha$$

we find

$$\mathcal{H}_{\frac{\alpha}{s}}^{c\delta^s}(f(A)) \leq c^{\frac{\alpha}{s}} \mathcal{H}_s^\delta(A)$$

and in the limit $\delta \to 0$ we have $\mathcal{H}_{\frac{\alpha}{s}}(f(A)) \leq c^{\frac{\alpha}{s}} \mathcal{H}_\alpha(A)$. $\qquad\square$

Corollary 3.37. *For a Lipschitz continuous mapping $f : G \to \mathbb{R}^m$ and $A \in \mathcal{B}^{(n)}$, $A \subset G$, the following holds*

$$\mathcal{H}_\alpha(f(A)) \leq c^\alpha \mathcal{H}_\alpha(A). \tag{3.25}$$

This estimate applies in particular if G is arcwise connected and f is a C^1-mapping with bounded derivative, since such a mapping is by the mean-value theorem Lipschitz continuous.

Corollary 3.38. *In the situation of Lemma 3.36 we have*

$$\dim_H(f(A)) \leq \frac{1}{s} \dim_H A. \tag{3.26}$$

Proof. By (3.34) we have $\mathcal{H}_{\frac{\alpha}{s}}(f(A)) \leq c^{\frac{\alpha}{s}} \mathcal{H}_\alpha(A)$ which implies for $\alpha > \dim_H A$ that $\mathcal{H}_{\frac{\alpha}{s}}(f(A)) = 0$, i.e. $\dim_H f(A) \leq \frac{\alpha}{s}$ for all $\alpha > \dim_H A$, which in turn yields $\dim_H f(A) \leq \frac{1}{s} \dim_H A$. $\qquad\square$

In particular, for Lipschitz continuous mappings we have in the situation of Lemma 3.36

$$\dim_H f(A) \leq \dim_H A. \tag{3.27}$$

Corollary 3.39. *Let $G \subset \mathbb{R}^n$ be closed and $f : G \to \mathbb{R}^n$, be a **bi-Lipschitz mapping**, i.e. for $0 < c_1 \leq c_2$ we have*

$$c_1 \|x - y\| \leq \|f(x) - f(y)\| \leq c_2 \|x - y\| \tag{3.28}$$

for all $x, y \in G$. If $A \in \mathcal{B}^{(n)}$, $A \subset G$, then we have the equality $\dim_H f(A) = \dim_H A$.

Proof. First we note that f must be injective and so we can denote by f^{-1} the inverse of $f : G \to f(G)$. From (3.28) we deduce further that $f^{-1} : f(G) \to G$ is also Lipschitz continuous. Hence we have $\dim_H f(A) \leq \dim_H A$ as well as $\dim_H A = \dim_H f^{-1}(f(A)) \leq \dim_H f(A)$. □

Example 3.40. The embedding $j : \mathbb{R}^n \to \mathbb{R}^m$, $n < m$, $j(x_1, \ldots, x_n) = (x_1, \ldots, x_n, 0, \ldots, 0)$ is injective and its "inverse" in the sense of the proof of Corollary 3.39 is the projection $\pi_{(n,m)} : \mathbb{R}^m \to \mathbb{R}^n$, $\pi_{(n,m)}(y_1, \ldots, y_m) = (y_1, \ldots, y_n)$. Both mappings are trivially Lipschitz continuous and therefore the Hausdorff dimension of every Borel set $A \in \mathcal{B}^{(n)}$ is equal to the Hausdorff dimension of $j(A) \in \mathcal{B}^{(m)}$. In particular if A has a non-empty interior $\mathring{A} \subset \mathbb{R}^n$ then $\dim_H j(A) = n$. This applies for example to non-degenerate cubes in $\mathbb{R}^n$.

The next proposition will eventually pave the way to define in Volume VI integrals over sub-manifolds. Analogous to Definition II.27.1.B we call for a linear mapping $T : \mathbb{R}^k \to \mathbb{R}^n$

$$g_T := \det(T^*T) \tag{3.29}$$

the **Gram determinant** of T. Since T^*T is a symmetric positive semi-definite $k \times k$-matrix $\sqrt{g_T}$ is well defined.

Proposition 3.41. *Let $k \leq n$ and $A \in \mathcal{B}^{(k)}$. Further let $T : \mathbb{R}^k \to \mathbb{R}^n$ be a linear mapping. Then we have*

$$\mathcal{H}_k(T(A)) = \sqrt{g_T}\mathcal{H}_k(A). \tag{3.30}$$

Proof. (Following G. Folland [27]) In the case where $k = n$ it follows that $\mathcal{H}_n = \gamma_n \lambda^{(n)}$, see Corollary 3.28, and $g_T = (\det T)^2$. Now (3.30) follows from Theorem 3.11. For $k < n$ we can find a rotation $R \in O(n)$ such that $R(T(\mathbb{R}^n)) \subset \{(x_1, \ldots, x_n) \in \mathbb{R}^n \,|\, x_{k+1} = \cdots = x_n = 0\}$ for which we write $R(T(\mathbb{R}^n)) \subset \mathbb{R}^k \times \{0\} \subset \mathbb{R}^n$. With $S := RT$ we have $S^*S = T^*R^*RT = T^*T$ and the rotation invariance of the Hausdorff measure yields $g_S = g_T$ as well as $\mathcal{H}_k(S(A)) = \mathcal{H}_k(T(A))$. Identifying $\mathbb{R}^k \times \{0\}$ with $\mathbb{R}^k$ we observe that S maps $\mathbb{R}^k$ into itself and S^*S does not change under this identification. Hence we may apply the result already shown for the case $k = n$ to obtain (3.30). □

We end this chapter by discussing a Borel set $C \subset \mathbb{R}$ which has fractional Hausdorff dimension. The set C is the **Cantor set** which we have already met in Chapter I.32. We recall its construction. Let $C_0 := [0,1]$, define $C_1 := [0,1] \setminus \left(\frac{1}{3}, \frac{2}{3}\right) = \left[\frac{0}{3}, \frac{1}{3}\right] \cup \left[\frac{2}{3}, \frac{3}{3}\right]$ and now we continue in the following way: C_{N+1} is obtained by removing the 2^N open middle intervals of length 3^{-N} from the intervals forming C_N. The Cantor set is then defined by

$$C := \bigcap_{N=0}^{\infty} C_N, \tag{3.31}$$

which is by Theorem I.32.4 a compact non-denumerable null set with respect to $\lambda^{(1)}$. The set C_N is obtained by taking from $[0,1]$ away $2^N - 1$ open intervals which we denote by J_l, $1 \leq l \leq 2^N - 1$ and with $J_l = (a_l, b_l)$ we assume that $b_l < a_{l+1}$. We set $\overline{J}_0 := \{0\}$ and $\overline{J}_{2^N} := \{1\}$. The set C_N consists of 2^N closed intervals $K_k = [c_k, d_k]$ and we order these intervals such that $c_1 = 0$, $c_k = b_{k-1}$ and $d_k = a_k$, $d_{2^N} = 1$. We define now the function $F_N : [0,1] \to \mathbb{R}$ in the following way

$$F_N(x) = \begin{cases} 0, & x = 0 \\ \frac{l}{2^N}, & x \in \overline{J}_l, \, l = 1, \dots, 2^N - 1 \\ g_l(x), & x \in \overset{\circ}{C}_l, \, l = 1, \dots, 2^N \\ 1, & x = 0 \end{cases} \tag{3.32}$$

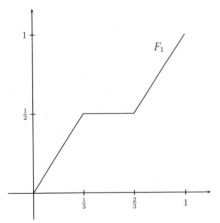

Figure 3.1

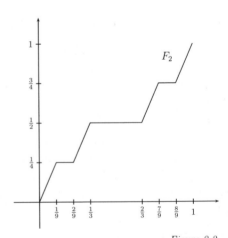

Figure 3.2

where the graph $\Gamma(g_l)$ is the line segment connecting $(c_l, F_N(b_{l-1}))$ with $(d_l, F_N(a_l))$. For $N = 1$ and $N = 2$ the functions F_1 and F_2 are drafted in Figure 3.1 and Figure 3.2 above.

By construction $F_N : [0,1] \to [0,1]$ is continuous and monotone increasing and further we have

$$|F_{N+1}(x) - F_N(x)| \leq \frac{1}{2^{N+1}}. \tag{3.33}$$

The triangle inequality yields for $M \in \mathbb{N}$ that

$$|F_{N+M}(x) - F_N(x)| \leq \sum_{j=1}^{M} |F_{N+j}(x) - F_{N+j-1}(x)| \leq \sum_{j=1}^{M} \frac{1}{2^{N+j}}.$$

Taking the supremum over $x \in [0,1]$ we find

$$\|F_{N+M} - F_N\|_\infty \leq \frac{1}{2^N}, \tag{3.34}$$

i.e. $(F_N)_{N \in \mathbb{N}}$ is a Cauchy sequence in $C([0,1])$, hence it has a limit $F \in C([0,1])$, i.e.

$$F(x) = \lim_{N \to \infty} F_N(x) \quad \text{(uniform convergence)}. \tag{3.35}$$

Moreover $F(0) = 0$ and $F(1) = 1$, F is increasing since for $y < x$ we have $F(y) = \lim_{N \to \infty} F_N(y) \leq \lim_{N \to \infty} F_N(x) = F(x)$, F is constant on each interval of the complement of C, and hence F maps C onto $[0,1]$.

Definition 3.42. *The function F defined by (3.35) is called the **Cantor-Lebesgue function** or by some authors the **Lebesgue singular function**.*

Lemma 3.43. *The Cantor-Lebesgue function is Hölder continuous with Hölder exponent $s = \frac{\ln 2}{\ln 3}$.*

Proof. The function F is the uniform limit of the functions F_N which are piecewise linear and F_N increases in an interval of length $\frac{1}{3^N}$ by at most $\frac{1}{2^N}$. Thus the slope of F_N is bounded by $\left(\frac{3}{2}\right)^N$ which yields

$$|F_N(x) - F_N(y)| \leq \left(\frac{3}{2}\right)^N |x - y|. \tag{3.36}$$

Passing in (3.34) to the limit $M \to \infty$ we find

$$|F(x) - F_N(x)| \leq \frac{1}{2^N}. \tag{3.37}$$

Combining (3.36) and (3.37) gives for $x, y \in [0, 1]$

$$|F(x) - F(y)| \leq |F_N(x) - F_N(y)| + |F(x) - F_N(x)| + |F(y) - F_N(y)|$$

$$\leq \left(\frac{3}{2}\right)^N |x - y| + \frac{2}{2^N} = \frac{1}{2^N}\left(3^N|x - y| + 2\right).$$

For $x, y \in [0, 1]$ fixed, $x \neq y$, we now choose N such that $1 \leq 3^N|x - y| \leq 3$ which is always possible since $|x - y| \leq 1$. Hence we arrive at

$$|F(x) - F(y)| \leq \frac{c}{2^N}$$

with some $c \in [3, 5]$. Since $3^\gamma = 2$ means $e^{\gamma \ln 3} = e^{\ln 2}$, we find with $s = \frac{\ln 2}{\ln 3}$ that $3^s = 2$, and it follows that

$$|F(x) - F(y)| = \frac{c}{3^{sN}}$$

and taking into account that $3^{-N} \leq |x - y|$ we eventually get

$$|F(x) - F(y)| \leq c'|x - y|^s.$$

$\square$

Now we can prove

Theorem 3.44. *The Hausdorff dimension of the Cantor set C is*

$$\dim_H(C) = \frac{\ln 2}{\ln 3}. \tag{3.38}$$

Proof. Recall that $C = \bigcap_{N \in \mathbb{N}} C_N$ and each C_N is the union of 2^N intervals of length 3^{-N}. Given $\delta > 0$ we choose M such that $3^{-M} < \delta$ and the intervals belonging to C_M cover C. For $\alpha > 0$ we have

$$\mathcal{H}_\alpha^\delta(C) \leq 2^M (3^{-M})^\alpha.$$

For $\alpha = \frac{\ln 2}{\ln 3}$ it follows that $2^M (3^{-M})^\alpha = 1$, i.e. $\mathcal{H}_{\frac{\ln 2}{\ln 3}}(C) \leq 1$.

On the other hand, combining Lemma 3.36 with Lemma 3.43 and using that $F(C) = [0, 1]$ we get $\mathcal{H}_1([0, 1]) \leq c\mathcal{H}_{\frac{\ln 2}{\ln 3}}(C)$, and both estimates imply $\dim_H(C) = \frac{\ln 2}{\ln 3}$. $\square$

Remark 3.45. A. Our arguments leading to the proof of (3.38) are much influenced by the presentation of E. M. Stein and R. Shakarchi [84].
B. We will meet the Cantor-Lebesgue function again in Chapter 11 when discussing differentiability and the fundamental theorem of calculus within Lebesgue's theory of integration.
C. Fractals are quite popular (and important) objects, partly due to the beautiful pictures that can be created using so called self-similar fractals. A good first reading is K. Falconer [23].

Problems

1.   a) On the set $\{1, \ldots, n\}$ we consider the power set as its σ-field and we define the measure $\mu := \sum_{k=1}^{n} \epsilon_k$. Prove that μ is invariant under the operation of the symmetric group S_n, i.e. $\mu(\sigma(A)) = \mu(A)$ for every permutation $\sigma : \{1, \ldots, n\} \to \{1, \ldots, n\}$ and every subset A of $\{1, \ldots, m\}$.

   b) Now consider on $\{1, \ldots, n\}$ the measure

$$\nu := \sum_{\substack{k \leq n \\ k \text{ even}}} \frac{1}{2} \epsilon_k + \sum_{\substack{l \leq n \\ l \text{ odd}}} \frac{1}{3} \epsilon_l$$

which has the power set as its σ-field. Prove that in general ν is not invariant under S_n. However, if $\sigma \in S_n$ maps even numbers onto odd numbers, then σ leaves ν invariant.

2. On $(\mathbb{Z}, \mathcal{P}(\mathbb{Z}))$ consider the measure $\mu = \sum_{k \in \mathbb{Z}} \epsilon_k$. For $T : \mathbb{Z} \to \mathbb{Z}$, $k \mapsto T(k) = k^2$ find the image measure $T(\mu)$ on $\mathcal{P}(\mathbb{Z})$.

3. Consider the mapping $T : \mathbb{R}^3 \to \mathbb{R}^3$, $T(x, y, z) = (ax, by, cz)$, where $a > b > c > 0$ are given and $\mathbb{R}^3$ is equipped with the Borel σ-field $\mathcal{B}^{(3)}$. Prove that T is measurable and find the image of $B_1(0) \subset \mathbb{R}^3$ under T as well as $T(\lambda^{(3)})$. Now find the volume of the ellipsoid $\mathcal{E} := \left\{ (x, y, z) \in \mathbb{R}^3 \,\middle|\, \frac{x^2}{9} + \frac{y^2}{16} + \frac{z^2}{25} < 1 \right\}$, i.e. find $\lambda^{(n)}(\mathcal{E})$.

4.   a) For the Bernoulli distribution β_p^N as well as for the Poisson distribution π_α find all sets of measure zero.

b) Give an example of a measurable mapping $T : \mathbb{R}^n \to \mathbb{R}^n$ and a Borel set $A \in \mathcal{B}^{(n)}$ such that $\lambda^{(n)}(A) > 0$ but $\lambda^{(n)}(T(A)) = 0$.

c) Let $T : \mathbb{R} \to \mathbb{R}$ be any mapping. Prove that $\lambda(T(\mathbb{Q})) = 0$.

5. Give an example of two measure spaces $(\Omega_1, \mathcal{A}_1, \mu_1)$ and $(\Omega_2, \mathcal{A}_2, \mu_2)$ and a measurable mapping $h : \Omega_1 \to \Omega_2$ such that h maps a non-trivial set of μ_1-measure zero to a set of strictly positive μ_2-measure, i.e. $A \in \mathcal{A}_1$, $A \neq \emptyset$ and $\mu_1(A) = 0$ but $\mu_2(h(A)) > 0$.

6. The mapping $\varphi : \mathbb{R} \to \mathbb{R}^2$, $t \mapsto \varphi(t) = (\cos t, \sin t)$ maps $\mathbb{R}$ onto the unit circle $S^1 \subset \mathbb{R}^2$. Prove that $\varphi(\lambda^{(1)})|_{S^1}$ is invariant under rotation.

7. Prove that a Borel set $A \subset \mathbb{R}^n$ which is a null set in the sense of Definition II.19.6 is a set of measure zero in the sense of Definition 3.13. (Note that the assumption that A is a Borel set, i.e. measurable, is crucial.)

8. Let $(\Omega, \mathcal{A}, \mu)$ be a complete measure space and $Y_1, Y_2 \in \mathcal{A}$ such that $Y_j^{\complement}$ are sets of measure zero. Let $f : Y_1 \to \mathbb{R}$ be a measurable function such that for some $g : Y_2 \to \mathbb{R}$ we have $f = g$ μ-a.e. Prove that g is measurable.

9. Solve Exercise 3.20, i.e. prove that the completion $\tilde{\lambda}^{(n)}$ of the n-dimensional Borel-Lebesgue measure $\lambda^{(n)}$ is translation invariant.

10. Let $(\Omega, \mathcal{A}, \mu)$ be a measure space and assume that there exists a sequence of measurable sets $(B_k)_{k \in \mathbb{N}}$ such that $\Omega = \bigcup_{k \in \mathbb{N}} B_k$ and $\mu(B_k) < \infty$, i.e. the measure μ is σ-finite. A set $N \in \mathcal{A}$ has **measure zero locally** if $\mu(N \cap C) = 0$ for every $C \in \mathcal{A}, \mu(C) < \infty$. Prove that in the case of a σ-finite measure μ every set that has measure zero locally is a set of measure zero.

11. Prove that a bijective continuous function $g : \mathbb{R} \to \mathbb{R}$ maps Borel sets onto Borel sets.
Hint: prove that $\{A \subset \mathbb{R} | g(A) \in \mathcal{B}^{(1)}\}$ is a σ-field.

12. a) For an interval $I \subset \mathbb{R}$ with non-empty interior let $\gamma : I \to \mathbb{R}^n$ be a C^1-curve. Prove that for every interval $\tilde{I} \subset I$ with non-empty interior we have $\mathcal{H}_\alpha(\gamma(\tilde{I})) = 0$ for $\alpha > 1$.

b) Let $G \subset \mathbb{R}^2$ be an open set and $f : G \to \mathbb{R}^3$ a C^1-mapping. Suppose that $\overline{B_1(0)} \subset G$ and consider the parametric surface $f : B_1(0) \to \mathbb{R}^3$. Prove that $\mathcal{H}_\beta(f(B_1(0))) = 0$ for $\beta > 2$.

c) The following theorem is essentially due to G. Peano and it had enormous influence in the development of many branches of mathematics:

There exists a mapping $P : [0,1] \to [0,1] \times [0,1]$ which is continuous and surjective and satisfies the Hölder condition

$$|P(t) - P(s)| \le L|t - s|^{\frac{1}{2}}.$$

Such a curve is called a **Peano curve** or a **space filling curve**. We refer to [84] where a detailed proof and discussion is provided.
Prove that it is not possible to have a continuous surjective mapping $R : [0,1] \to [0,1] \times [0,1]$ which is Hölder continuous for some exponent $s > \frac{1}{2}$.

13.   a) Let $T : \mathbb{R}^n \to \mathbb{R}^n$ be a symmetric positive definite linear mapping, where T is called positive definite if its matrix has this property. Prove that the Gram determinant of T is given by $g_T = \det T$.

b) Given a parametric C^1-surface $f : G \to \mathbb{R}^3$, $x = (u,v) \mapsto \begin{pmatrix} f_1(x) \\ f_2(x) \\ f_3(x) \end{pmatrix}$, where $G \subset \mathbb{R}^2$ is an open set. We denote by $g_f(x)$ the Gram determinant of $d_x f$ where $d_x f : \mathbb{R}^2 \to \mathbb{R}^3$ is the differential of f at $x \in G$. Find the expression of $g_f(x)$ in terms of $E(x) := \langle f_u(x), f_u(x) \rangle$, $F(x) := \langle f_u(x), f_v(x) \rangle$ and $G(x) := \langle f_v(x), f_v(x) \rangle$.

4 Measurable Mappings

We have already introduced measurable mappings $f : \Omega \to \Omega'$, where $(\Omega, \mathcal{A})$ and $(\Omega', \mathcal{A}')$ are two measurable spaces. These are mappings having the property that the pre-image of $\mathcal{A}'$-measurable sets are $\mathcal{A}$-measurable, i.e. $f^{-1}(A') \in \mathcal{A}$ for $A' \in \mathcal{A}'$, see Definition 1.15. In this chapter we want to investigate measurable mappings further, but first let us recollect what is already known to us:

- the composition of measurable mappings is measurable (Remark 1.16.C);

- f is measurable if and only if $f^{-1}(E') \in \mathcal{A}$ for a generator E' of $\mathcal{A}'$ (Lemma 1.17);

- for measurable spaces $(\Omega_j, \mathcal{A}_j)$, $j \in I$, and mappings $f_j : \Omega \to \Omega_j$ exists a smallest σ-field in Ω such that all f_j become measurable (Example 1.14 and Remark 1.16.A);

- if $f : \Omega \to \Omega'$ is measurable and μ is a measure on $(\Omega, \mathcal{A})$ then the image $f(\mu)$ is a well defined measure on $(\Omega', \mathcal{A}')$ (Theorem 3.7);

- forming image measures is transitive (Corollary 3.9).

We can derive more results if $(\Omega, \mathcal{A})$ and/or $(\Omega', \mathcal{A}')$ have additional structures respected by certain mappings. For example if $(\Omega, \mathcal{O})$ and $(\Omega', \mathcal{O}')$ are two topological spaces and $\mathcal{A} = \mathcal{B}(\Omega)$, $\mathcal{A}' = \mathcal{B}'(\Omega')$ are the Borel σ-fields generated by the topologies, then every continuous mapping $f : \Omega \to \Omega'$ is measurable, see Corollary 1.18 the proof of which extends without change to the general situation.

In the following we want to discuss results when we have on Ω' (and maybe on Ω) additional algebraic and topological structures. For example when $\Omega' = \mathbb{R}^n$ and $\mathcal{A}' = \mathcal{B}^{(n)}$ then we can define $f \pm g$ and αf, and we may ask whether $f \pm g$ or αf are measurable if f and g are. The function $f + g$ can be written as composition. We define add : $\mathbb{R}^n \times \mathbb{R}^n \to \mathbb{R}^n$ by $\text{add}(x, y) = x + y$, and $x \mapsto f(x) + g(x)$ is obtained as $x \mapsto (f(x), g(x)) \mapsto \text{add}(f(x), g(x)) = f(x) + g(x)$. If we can prove that $x \mapsto (f(x), g(x))$ and $(x, y) \mapsto \text{add}(x, y)$ are measurable, we can deduce that $x \mapsto f(x) + g(x)$ is measurable. Of general use is

Lemma 4.1. *Let $(\Omega, \mathcal{A})$ and $(\Omega_j, \mathcal{A}_j)$, $j = 1, 2$, be measurable spaces. A mapping $f : \Omega \to \Omega_1 \times \Omega_2$ is $\mathcal{A}/\mathcal{A}_1 \otimes \mathcal{A}_2$-measurable if and only if the mappings $\mathrm{pr}_j \circ f : \Omega \to \Omega_j$ are $\mathcal{A}/\mathcal{A}_j$-measurable, $j = 1, 2$.*

Proof. Since by the construction of $\mathcal{A}_1 \otimes \mathcal{A}_2$ the projections are measurable, the measurability of f implies that of $\mathrm{pr}_j \circ f$. Conversely, since $\mathcal{A}_1 \times \mathcal{A}_2$ is a generator of $\mathcal{A}_1 \otimes \mathcal{A}_2$ and since for $A_1 \times A_2 \in \mathcal{A}_1 \times \mathcal{A}_2$ it follows that

$$\begin{aligned} f^{-1}(A_1 \times A_2) &= f^{-1}\left(\mathrm{pr}_1^{-1}(A_1) \cap \mathrm{pr}_2^{-1}(A_2)\right) \\ &= f^{-1}\left(\mathrm{pr}_1^{-1}(A_1)\right) \cap f^{-1}\left(\mathrm{pr}_2^{-1}(A_2)\right) \\ &= (\mathrm{pr}_1 \circ f)^{-1}(A_1) \cap (\mathrm{pr}_2 \circ f)^{-1}(A_2) \in \mathcal{A} \end{aligned}$$

the measurability of $\mathrm{pr}_j \circ f$ implies the measurability of f. $\qquad\square$

In our motivating example it remains to study "add" but that the mapping $(x, y) \mapsto x + y$ is in $\mathbb{R}^n$ continuous, hence measurable, is well known to us. We want to handle this problem in a wider context for example we want to consider mappings with values in $GL(n; \mathbb{R})$ considered as a group, or mappings defined on some measurable space mapping into the space of all continuous, real-valued functions defined on a compact space and this space we equip with the supremum norm. We need as preparation some additional background from topology. Since the examples most interesting to us are those where the topologies are induced by a metric, here we only discuss the case of metric spaces. We refer to Appendix I where we discuss the more general case.

Definition 4.2. *Let (X, d_X) and (Y, d_Y) be two metric spaces and define for $(x_1, y_1), (x_2, y_2) \in X \times Y$*

$$d_{X \times Y}\left((x_1, y_1), (x_2, y_2)\right) = d_X(x_1, x_2) + d_Y(y_1, y_2). \tag{4.1}$$

*Then $(X \times Y, d_{X \times Y})$ is called the **product metric space** of (X, d_X) and (Y, d_Y) and $d_{X \times Y}$ the **product metric**.*

Clearly, this definition extends to the case of N metric spaces (X_j, d_{X_j}), $j = 1, \ldots, N$, in the obvious way. Given two metrics d_1 and d_2 on a set X, we call d_1 and d_2 **equivalent** if for $0 < \gamma_1 \leq \gamma_2$ and all $x, y \in X$ the following holds

$$\gamma_1 d_1(x, y) \leq d_2(x, y) \leq \gamma_2 d_1(x, y). \tag{4.2}$$

For the corresponding open balls $B_\rho^{d_j}(x) := \{y \in X \mid d_j(x, y) < \rho\}$ this implies

$$B_{\frac{\rho}{\gamma_2}}^{d_1}(x) \subset B_\rho^{d_2}(x) \subset B_{\frac{\rho}{\gamma_1}}^{d_1}(x)$$

and hence equivalent metrics induce the same open sets in X, which yields that the corresponding Borel σ-fields are the same.

Exercise 4.3. *Prove that by $d_{X \times Y}$ a metric on $X \times Y$ is given which is for $1 \leq p < \infty$ equivalent to each of the metrics $d_{X \times Y, p}$ where*

$$d_{X \times Y, p}((x_1, y_1), (x_2, y_2)) = (d_X^p(x_1, x_2) + d_Y^p(y_1, y_2))^{\frac{1}{p}}.$$

Show further that the projections $\mathrm{pr}_X : X \times Y \to X$, $\mathrm{pr}_X(x, y) = x$, and $\mathrm{pr}_Y : X \times Y \to Y$, $\mathrm{pr}_Y(x, y) = y$, are continuous. In addition give a formulation of these results in the case of N, $N \in \mathbb{N}$, metric spaces (X_j, d_{X_j}), $1 \leq j \leq N$.

In the following, when discussing metric spaces, we always consider them equipped with their metric topology, i.e. the topology induced by the metric, and the corresponding Borel σ-field, i.e. the σ-field generated by the open sets. On finite products of metric spaces we consider the topology induced by the product metric and the corresponding Borel σ-field which coincides with the product of the underlying σ-fields of each factor, see Problem 2.

Definition 4.4. A. *Let $(G, \cdot)$ be a group and suppose that on G a metric d_G is given. We call G (or more precisely $(G, \circ, d_G)$) a **metric group** if the two mappings $\mathrm{comp} : G \times G \to G$, $\mathrm{comp}(h_1, h_2) = h_1 \circ h_2$, and $\mathrm{inv} : G \to G$, $\mathrm{inv}(h) = h^{-1}$, are continuous.*
B. *Let $(V, +, \cdot)$ be an $\mathbb{R}^n$-vector space and let d_V be a metric on V. We call V a **metric vector space** if $+ : V \times V \to V$, $(x, y) \mapsto x + y$, and $\cdot : \mathbb{R} \times V \to V$, $(\alpha, x) \mapsto \alpha x$, are continuous, where on $\mathbb{R}$ we choose the topology induced by the absolute value.*

Remark 4.5. In Appendix I we discuss the product topology of two topological spaces. With this at hand we can immediately introduce the notion of a topological group and that of a topological vector space. If $(G, \circ)$ is a group and on G a topology $\mathcal{O}_G$ is given, then we call G (or $(G, \circ, \mathcal{O}_G)$) a **topological group** if the mappings comp and inv (as defined previously) are continuous. We call $(V, +, \cdot)$ a **topological vector space** if on V a topology $\mathcal{O}_V$ is given such that $+ : V \times V \to V$ and $\cdot : \mathbb{R} \times V \to V$ are continuous where on $\mathbb{R}$ we choose as usual the Euclidean topology induced by the absolute value.

Example 4.6. A. The vector space $\mathbb{R}^n$ equipped with the Euclidean metric is a metric vector space since we know that adding and multiplying with scalars is continuous.

B. The general linear group $GL(n; \mathbb{R})$ is a metric group. We can identify $GL(n; \mathbb{R})$ with an open subset of $\mathbb{R}^{n^2}$, namely $GL(n; \mathbb{R}) = \det^{-1}(\mathbb{R} \backslash \{0\})$, and the group operations can be decomposed into a finite sequence of adding, subtracting, multiplying and dividing by a non-zero number. Since the projections $\mathrm{pr}_j : \mathbb{R}^{n^2} \to \mathbb{R}$ are continuous, all these operations are continuous too.

C. The groups $O(n)$ and $SO(n)$ are metric groups. The continuity of the group operations is inherit from $GL(n; \mathbb{R})$.

D. Every normed vector space is a metric vector space. This example includes infinite dimensional vector spaces such as $(C_b(K), \|.\|_\infty)$ where K is a metric space, or even a topological space.

Corollary 4.7. A. *Let $(G, \circ)$ be a metric group. Then the mappings* comp $: G \times G \to G$, $\mathrm{comp}(h_1, h_2) = h_1 \circ h_2$ *and* inv $: G \to G$, $\mathrm{inv}(h) = h^{-1}$, *are measurable.*

***B.** Let $(V, +, \cdot)$ be a metric $\mathbb{R}$-vector space. Then the vector space operations are measurable.*

Corollary 4.8. A. *Let $(\Omega, \mathcal{A})$ be a measurable space and $(G, \circ)$ a metric group. Further let $u, v : \Omega \to G$ be two measurable mappings. Then the mappings $u \circ v : \Omega \to G$, $(u \circ v)(\omega) := u(\omega) \circ v(\omega)$, and $u^{-1} : \Omega \to G$, $u^{-1}(\omega) := (u(\omega))^{-1}$, are measurable.*

***B.** Let $(\Omega, \mathcal{A})$ be a measurable space and $(V, +, \cdot)$ a metric vector space. Let $f, h : \Omega \to V$ be two measurable mappings and $\alpha, \beta \in \mathbb{R}$. Then the mapping $\alpha f + \beta h : \Omega \to V$, $\omega \mapsto \alpha f(\omega) + \beta h(\omega)$, is measurable too.*

Turning to real-valued functions $f, h : \Omega \to \mathbb{R}$ we can also consider the function $f \cdot h : \Omega \to \mathbb{R}$, $\omega \mapsto f(\omega)h(\omega)$. Since multiplication in $\mathbb{R}$ is continuous we deduce that if f and h are measurable then $f \cdot h$ is measurable. (See also Problem 3.)

The summary of all these considerations is simple: consider mappings from a measurable space $(\Omega, \mathcal{A})$ into a set H which carries an algebraic structure (group, vector space, etc) and a metric (or even a topology) such that the algebraic operations in H are continuous. Define the corresponding algebraic operations pointwisely for mappings $f, g : \Omega \to H$, $(f \circ g)(\omega) = f(\omega) \circ g(\omega)$, etc. Then these mappings are measurable when H is equipped with the Borel σ-field.

Measures are certain mappings defined on a σ-field with values in $[0, \infty]$. When trying to define an integral with respect to a measure it makes sense to allow functions which may attain the values $+\infty$ or $-\infty$ too. The measurability of such functions will be investigated next. We have already introduced the arithmetic for $\overline{\mathbb{R}} = [-\infty, +\infty]$, see (1.29) - (1.32). Now we want to find a σ-field on $\overline{\mathbb{R}}$ extending $\mathcal{B}^{(1)}$. There are two equivalent routes. We can first consider $\overline{\mathbb{R}}$ as a certain compactification of $\mathbb{R}$ and then we can work with the corresponding Borel σ-field. This will be discussed in Appendix I. Another way is to define $\overline{\mathcal{B}}^{(1)}$ directly by

$$\overline{\mathcal{B}}^{(1)} := \left\{ \tilde{A} = A \cup S \,\middle|\, A \in \mathcal{B}^{(1)} \text{ and } S \in \{\emptyset, \{-\infty\}, \{+\infty\}, \{-\infty, +\infty\}\} \right\}. \quad (4.3)$$

It is easy to see that $\overline{\mathcal{B}}^{(1)}$ is indeed a σ-field and that the trace σ-field $\overline{\mathcal{B}}^{(1)}_{\mathbb{R}} = \mathbb{R} \cap \overline{\mathcal{B}}^{(1)}$ is the Borel σ-field $\mathcal{B}^{(1)}$, see Problem 7.

Lemma 4.9. *Each of the following families is a generator of* $\overline{\mathcal{B}}^{(1)}$:

$$\{[a, \infty] \,|\, a \in \mathbb{R}\}, \quad \{[a, \infty] \,|\, a \in \mathbb{Q}\}, \quad (4.4)$$
$$\{(b, \infty] \,|\, b \in \mathbb{R}\}, \quad \{(b, \infty] \,|\, b \in \mathbb{Q}\}, \quad (4.5)$$
$$\{[-\infty, c) \,|\, c \in \mathbb{R}\}, \quad \{[-\infty, c) \,|\, c \in \mathbb{Q}\}, \quad (4.6)$$
$$\{[-\infty, d] \,|\, d \in \mathbb{R}\}, \quad \{[-\infty, d] \,|\, d \in \mathbb{Q}\}. \quad (4.7)$$

Proof. We prove that $\sigma\left(\{[a, \infty] \,|\, a \in \mathbb{R}\}\right) = \overline{\mathcal{B}}^{(1)}$, the other cases can be proved in a similar way. In particular, switching from $\mathbb{R}$ to $\mathbb{Q}$ relies on arguments analogous to those employed in Chapter 1. Since $[a, \infty] = [a, \infty) \cup \{+\infty\}$ and $[a, \infty) \in \mathcal{B}^{(1)}$ it follows that $[a, \infty] \in \overline{\mathcal{B}}^{(1)}$, i.e. $\sigma\left(\{[a, \infty] \,|\, a \in \mathbb{R}\}\right) \subset \overline{\mathcal{B}}^{(1)}$. For $b, c \in \mathbb{R}$ we have $[b, c) = [b, \infty] \setminus [c, \infty] \in \sigma\left(\{[a, \infty] \,|\, a \in \mathbb{R}\}\right)$ and since the right open intervals $[b, c)$ generate $\mathcal{B}^{(1)}$ we find in addition $\mathcal{B}^{(1)} \subset \sigma\left(\{[a, \infty] \,|\, a \in \mathbb{R}\}\right) \subset \overline{\mathcal{B}}^{(1)}$. Finally we observe that $\{+\infty\} = \bigcap_{k \in \mathbb{N}} [k, \infty]$ and $\{-\infty\} = \bigcap_{k \in \mathbb{N}} [-k, \infty]^{\complement}$ which yields $\{-\infty\}, \{+\infty\} \in \sigma\left(\{[a, \infty] \,|\, a \in \mathbb{R}\}\right)$ as well as

$$A, \ A \cup \{+\infty\}, \ A \cup \{-\infty\}, \ A \cup \{-\infty, \infty\} \in \sigma\left(\{[a, \infty] \,|\, a \in \mathbb{R}\}\right)$$

for all $A \in \mathcal{B}^{(1)}$, i.e. $\overline{\mathcal{B}}^{(1)} = \sigma\left(\{[a, \infty] \,|\, a \in \mathbb{R}\}\right)$. $\qquad \square$

Remark 4.10. Note that in Chapter 1 we have already seen that $\mathcal{B}^{(1)}$ is generated by any of the systems $\{(c, \infty) \,|\, c \in \mathbb{K}\}$, $\{[c, \infty) \,|\, c \in \mathbb{K}\}$ and $\{(-\infty, c] \,|\, c \in \mathbb{K}\}$ where $\mathbb{K} \in \{\mathbb{Q}, \mathbb{R}\}$.

Definition 4.11. *For any set $G \neq \emptyset$ we call $f : G \to \overline{\mathbb{R}}$ a **numerical function**.*

Thus saying that a certain numerical function is real-valued means to emphasise that it takes only elements of $\mathbb{R}$ as values, i.e. it does not attain the values $-\infty$ or $+\infty$.

Combining Lemma 1.17 with Lemma 4.9 we find that for a measurable space $(\Omega, \mathcal{A})$ the numerical function $f : \Omega \to \overline{\mathbb{R}}$ is $\mathcal{A} \backslash \overline{\mathcal{B}}^{(1)}$-measurable if and only if one of the following equivalent conditions hold for all $c \in \mathbb{K}$, $\mathbb{K} \in \{\mathbb{Q}, \mathbb{R}\}$,

$$f^{-1}([c, \infty]) = \{\omega \in \Omega \,|\, f(\omega) \geq c\} \in \mathcal{A}, \qquad (4.8)$$
$$f^{-1}((c, \infty]) = \{\omega \in \Omega \,|\, f(\omega) > c\} \in \mathcal{A}, \qquad (4.9)$$
$$f^{-1}([-\infty, c)) = \{\omega \in \Omega \,|\, f(\omega) < c\} \in \mathcal{A}, \qquad (4.10)$$
$$f^{-1}([-\infty, c]) = \{\omega \in \Omega \,|\, f(\omega) \leq c\} \in \mathcal{A}. \qquad (4.11)$$

Note that the equivalence for (4.8) to (4.11) follows already from

$$\{\omega \in \Omega \,|\, f(\omega) > c\} = \bigcup_{k \in \mathbb{N}} \left\{\omega \in \Omega \,|\, f(\omega) \geq c + \frac{1}{k}\right\},$$

$$\{\omega \in \Omega \,|\, f(\omega) \leq c\} = \{\omega \in \Omega \,|\, f(\omega) > c\}^{\complement},$$

$$\{\omega \in \Omega \,|\, f(\omega) < c\} = \bigcup_{k \in \mathbb{N}} \left\{\omega \in \Omega \,|\, f(\omega) \leq c - \frac{1}{k}\right\}$$

and

$$\{\omega \in \Omega \,|\, f(\omega) \geq c\} = \{\omega \in \Omega \,|\, f(\omega) < c\}^{\complement}.$$

We make our life easier by introducing the notation

$$\{f \leq g\} := \{\omega \in \Omega \,|\, f(\omega) \leq g(\omega)\} \qquad (4.12)$$

with $\{f = g\}$, $\{f \neq g\}$, $\{f < g\}$, $\{f \geq g\}$ and $\{f > g\}$ being analogously defined and for example for $\{f \geq h\} \cap \{f \leq g\}$ we write $\{h \leq f \leq g\}$. Since

$$\{f < g\} = \bigcup_{r \in \mathbb{Q}} (\{f < r\} \cap \{r < g\}),$$

$$\{f \leq g\} = \{f > g\}^{\complement},$$

$$\{f = g\} = \{f \leq g\} \cap \{g \leq f\},$$

and

$$\{f \neq g\} = \{f = g\}^{\complement},$$

all these sets are measurable for measurable functions $f, g : \Omega \to \overline{\mathbb{R}}$.

Constant functions are measurable and since for $\alpha, \beta \in \mathbb{R}$, $\beta \neq 0$, we have $\{\alpha + \beta f \geq c\} = \{f \geq \frac{1}{\beta}(c - \alpha)\}$, $\beta > 0$, and $\{\alpha + \beta f \geq c\} = \{f \leq \frac{1}{\beta}(c - \alpha)\}$, $\beta < 0$, we deduce that $\alpha + \beta f$ is measurable if f is, clearly the case $\beta = 0$ is trivial. Consequently for two numerical functions f and g such that $f + g$ is defined we have $\{f + g \geq c\} = \{f \geq c - g\}$ and hence $f + g$ is measurable. Moreover, whenever f^2 is defined we find $\{f^2 \geq c\} = \{f \geq \sqrt{c}\} \cup \{f \leq -\sqrt{c}\}$ for $c \geq 0$ and $\{f^2 \geq c\} = \Omega$ for $c < 0$, hence f^2 is measurable if f is and from $f \cdot g = \frac{1}{4}(f + g)^2 - \frac{1}{4}(f - g)^2$ we deduce that if f and g are measurable and $f \cdot g$ is defined, then $f \cdot g$ is measurable.

Note that since we have on $\overline{\mathbb{R}}$ no vector space or algebra structure we cannot use our previous arguments to deduce that $f \pm g$, αf or $f \cdot g$ are measurable if f and g are.

We know that the pointwise limit of a sequence of continuous functions need not be continuous. For a sequence of measurable functions the situation is different.

Theorem 4.12. *Let $(\Omega, \mathcal{A})$ be a measurable space and for $k \in \mathbb{N}$ let $f_k : \Omega \to \overline{\mathbb{R}}$ be a measurable numerical function. Suppose that for all $\omega \in \Omega$ the limit $f(\omega) := \lim_{k \to \infty} f_k(\omega)$ exists (in $\overline{\mathbb{R}}$). Then the function $f : \Omega \to \overline{\mathbb{R}}$ is $\mathcal{A}/\overline{\mathcal{B}}^{(1)}$-measurable. In particular if f is real-valued, i.e. $f : \Omega \to \mathbb{R}$, then f is $\mathcal{A}/\mathcal{B}^{(1)}$-measurable.*

Proof. The last remark is obvious. Since for $\omega \in \Omega$

$$f(\omega) = \liminf_{k \to \infty} f_k(\omega) = \limsup_{k \to \infty} f_k(\omega)$$

holds, it is sufficient to prove that $\liminf f_k$ or $\limsup f_k$ are measurable. Moreover, by definition, compare Definition I.19.19,

$$\limsup_{k \to \infty} f_k = \inf_{k \in \mathbb{N}} \left(\sup_{l \geq k} f_l \right)$$

and

$$\liminf_{k \to \infty} f_k = \sup_{k \in \mathbb{N}} \left(\inf_{l \geq k} f_l \right).$$

Thus we need to prove that $\sup_{k \in \mathbb{N}} f_k$ and $\inf_{k \in \mathbb{N}} f_k$ are measurable as functions from $\Omega \to \overline{\mathbb{R}}$. However for $c \in \mathbb{R}$ we have

$$\left\{ \sup_{k \in \mathbb{N}} f_k \leq c \right\} = \bigcap_{k \in \mathbb{N}} \{ f_k \leq c \}$$

implying already the measurability of $\sup f_k$, but $\inf f_k$ is given by $\inf_{k \in \mathbb{N}} f_k = -\sup_{k \in \mathbb{N}}(-f_k)$ and the theorem follows. $\qquad\square$

For reference purpose we note as corollary to the proof of Theorem 4.12.

Corollary 4.13. *For a sequence $f_k : \Omega \to \overline{\mathbb{R}}$ of measurable functions the functions $\sup_{k \in \mathbb{N}} f_k$, $\inf_{k \in \mathbb{N}} f_k$, $\limsup_{k \to \infty} f_k$ and $\liminf_{k \to \infty} f_k$ are measurable.*

We want to introduce a further helpful notation: For $a, b \in \overline{\mathbb{R}}$ we write

$$a \wedge b := \min(a, b) \tag{4.13}$$

and

$$a \vee b := \max(a, b) \tag{4.14}$$

with the extensions $a_1 \wedge \cdots \wedge a_N = \min\{a_1, \ldots, a_N\}$ and $a_1 \vee \cdots \vee a_N = \max\{a_1, \ldots, a_N\}$. For numerical functions $f, g : G \to \overline{\mathbb{R}}$ we define further $f \wedge g$ and $f \vee g$ pointwisely as $(f \wedge g)(\omega) := f(\omega) \wedge g(\omega)$ and $(f \vee g)(\omega) := f(\omega) \vee g(\omega)$.

Corollary 4.14. *For a finite number of measurable functions $f_k : \Omega \to \overline{\mathbb{R}}$, $1 \leq k \leq N$, the functions $f_1 \wedge \cdots \wedge f_N$ and $f_1 \vee \cdots \vee f_N$ are measurable.*

Proof. We apply Corollary 4.13 to the sequence $(f_k)_{k \in \mathbb{N}}$, $f_l = f_N$ for $l \geq N$. $\qquad\square$

Since $f^+ = f \vee 0$, $f^- = -(f \wedge 0)$ and $f = f^+ - f^-$ as well as $|f| = f^+ + f^-$ it follows that together with f, f^+, f^- and $|f|$ are also measurable. In fact, f is measurable if and only if f^+ and f^- are measurable.

Finally we want to investigate a special class of measurable functions, namely step functions, or as they are often called in the context of Lebesgue's theory simple functions or elementary functions.

Let $(\Omega, \mathcal{A})$ be a measurable space. We start with the observation that for the characteristic function χ_A, $\chi_A(\omega) = \begin{cases} 1, & \omega \in A \\ 0, & \omega \notin A \end{cases}$, of a set $A \subset \Omega$ we have

$$\{\chi_A > c\} = \begin{cases} \emptyset, & c \geq 1 \\ A, & 1 > c \geq 0, \\ \Omega, & 0 > c \end{cases} \tag{4.15}$$

and therefore χ_A is measurable if and only if $A \in \mathcal{A}$.

Definition 4.15. *Let $(\Omega, \mathcal{A})$ be a measurable space. We call a non-negative, real-valued measurable function $u : \Omega \to [0, \infty)$ a **simple function** if it attains only finitely many values. The set of all simple functions on Ω we denote by $S(\Omega)$.*

Let $u : \Omega \to \mathbb{R}$ be a simple function and $\{\gamma_1, \ldots, \gamma_N\} \subset [0, \infty)$ its range. Then there exist measurable sets $C_1, \ldots, C_N \in \mathcal{A}$ such that $u|_{C_j} = \gamma_j$ and assuming $\gamma_k \neq \gamma_l$ for $k \neq l$ it follows that $C_k \cap C_l = \emptyset$. Hence we can write

$$u = \sum_{j=1}^{N} \gamma_j \chi_{C_j}. \tag{4.16}$$

Indeed, for every partition of Ω into measurable sets $B_1, \ldots, B_M \in \mathcal{A}$ such that $u|_{B_j}$ is constant we can write

$$u = \sum_{k=1}^{M} \beta_k \chi_{B_k}, \quad \beta_k \in \{\gamma_1, \ldots, \gamma_N\}. \tag{4.17}$$

In this general case however we cannot assume that $\beta_k \neq \beta_l$ for $k \neq l$.

Remark 4.16. As mentioned before, simple functions are also called **elementary functions**. Note that step functions in the sense of Definition I.25.2 and Definition II.18.5 are simple functions if they attain only non-negative values. But even on an interval a simple function is not necessarily a step function since the sets C_j in (4.16) (or B_k in (4.17)) are not necessarily intervals (or hyper-cubes in the case of $\mathbb{R}^n$).

It turns out that it is not practical to always assume we have a representation of the form (4.16), i.e. $\gamma_k \neq \gamma_l$ for $k \neq l$. For example if we want to have a representation of the sum of two simple functions as a simple function we need to use a representation as (4.17).

Definition 4.17. *Let $u \in S(\Omega)$. For any finite partition $(A_j)_{j=1,\ldots,N}$ of Ω into measurable sets $A_j \in \mathcal{A}$ we call*

$$u = \sum_{j=1}^{N} \alpha_j \chi_{A_j} \qquad (4.18)$$

*a **normal representation** of u.*

Let $u, v \in S(\Omega)$ and $\alpha \in [0, \infty)$. Then the following functions belong to $S(\Omega)$ too:

$$u^+, \; u^-, \; |u|, \; \alpha u, \; u + v, \; u \cdot v, \; u \wedge v, \; u \vee v. \qquad (4.19)$$

Clearly each of these functions attains only finitely many values and they are non-negative. Moreover by our previous considerations these functions are all measurable.

Let $u, v \in S(\Omega)$ and assume that they have the normal representation

$$u = \sum_{j=1}^{N} \alpha_j \chi_{A_j} \text{ and } v = \sum_{k=1}^{M} \beta_k \chi_{B_k}. \qquad (4.20)$$

Introducing

$$C_{jk} := A_j \cap B_k, \quad j = 1, \ldots, N \text{ and } k = 1, \ldots, M \qquad (4.21)$$

we obtain a partition of Ω, i.e. $C_{jk} \cap C_{nm} = \emptyset$ for $(j, k) \neq (n, m)$ and

$$\Omega = \bigcup_{j=1}^{N} \bigcup_{k=1}^{M} C_{jk}. \qquad (4.22)$$

In addition we have

$$A_j = \bigcup_{k=1}^{M} C_{jk} \text{ and } B_k = \bigcup_{j=1}^{N} C_{jk} \tag{4.23}$$

as well as

$$u = \sum_{j=1}^{N} \sum_{k=1}^{M} \xi_{jk} \chi_{C_{jk}} \text{ and } v = \sum_{j=1}^{N} \sum_{k=1}^{M} \eta_{jk} \chi_{C_{jk}} \tag{4.24}$$

where $\xi_{jk} = \alpha_j$ and $\eta_{jk} = \beta_k$. The advantage of the representations (4.24) is that we can now easily reduce operations for u and v to operations on the numbers ξ_{jk} and η_{jk}. For example we find

$$u + v = \sum_{j=1}^{N} \sum_{k=1}^{M} (\xi_{jk} + \eta_{jk}) \chi_{C_{jk}},$$

and since $\chi_{C_{jk}} \cdot \chi_{C_{nm}} = 0$ for $(j, k) \neq (n, m)$ we have

$$(u \cdot v) = \left(\sum_{j=1}^{N} \sum_{k=1}^{M} \xi_{jk} \chi_{C_{jk}} \right) \left(\sum_{n=1}^{N} \sum_{m=1}^{M} \eta_{nm} \chi_{C_{nm}} \right)$$
$$= \sum_{r=1}^{N} \sum_{s=1}^{M} \xi_{rs} \eta_{rs} \chi_{C_{rs}}.$$

Finally we want to prove that measurable functions are pointwise limits of simple functions.

Theorem 4.18. *A non-negative numerical function $f : \Omega \to [0, \infty]$ is measurable if and only if it is the increasing limit of simple functions $u_k \in S(\Omega)$ such that $u_k \leq u_{k+1}$ and $\lim_{k \to \infty} u_k(\omega) = f(\omega)$ for all $\omega \in \Omega$.*

Proof. Clearly, for every increasing sequence of simple functions the limit is non-negative and measurable. Conversely let $f : \Omega \to [0, \infty]$ be a measurable function. For $k \in \mathbb{N}$ we define the sets

$$A_{jk} := \left\{ j2^{-k} \leq f < (j+1)2^{-k} \right\}, \quad j = 0, \ldots, k2^k - 1$$

and

$$A_{k2^k, k} = \{ k \leq f \}$$

which are measurable and in addition for every $k \in \mathbb{N}$ they form a partition of Ω. Hence

$$u_k := \sum_{j=1}^{k2^k} j2^{-k} \chi_{A_{jk}} \tag{4.25}$$

is a simple function in normal representation and by construction we have

$$|f(\omega) - u_k(\omega)| \leq 2^{-k} \text{ for } \omega \in \{f < k\}.$$

Since $u_{k+1}|_{A_{jk}}(\omega)$ can attain only the values $(2j)2^{-(k+1)}$ and $(2j+1)2^{-(k+1)}$ for $j = 0, \ldots, k2^{-k} - 1$, it follows that $u_k \leq u_{k+1}$. If $f(\omega) = +\infty$ we have $u_k(\omega) = k$ for all $k \in \mathbb{N}$, and for $f(\omega) < \infty$ we find $u_k(\omega) \leq f(\omega) \leq u_k(\omega) + 2^{-k}$ for all $k > f(\omega)$. Hence $\sup_{k \in \mathbb{N}} u_k(\omega) = f(\omega)$, $\omega \in \Omega$, and since the sequence is monotone increasing, i.e. $u_k(\omega) \leq u_{k+1}(\omega)$ for all $\omega \in \Omega$ and $k \in \mathbb{N}$, the supremum is indeed a limit. $\qquad \square$

If $f : \Omega \to [-\infty, \infty]$ is any numerical measurable function then we can decompose f according to $f = f^+ - f^-$. Moreover for f^+ exists a sequence $(u_k)_{k \in \mathbb{N}}$, $u_k \in S(\Omega)$, converging pointwisely to f^+, and for f^- exists a sequence $(v_k)_{k \in \mathbb{N}}$, $v_k \in S(\Omega)$, converging pointwisely to f^-. Consequently the sequence $w_k := u_k - v_k$ converges pointwisely to f and further we have $|w_k| = u_k + v_k \leq f^+ + f^- = |f|$. Thus we have proved

Corollary 4.19. *A numerical function $f : \Omega \to \overline{\mathbb{R}}$ defined on a measurable space $(\Omega, \mathcal{A})$ is measurable if and only if there exists a sequence $(w_k)_{k \in \mathbb{N}}$, $w_k \in S(\Omega) - S(\Omega)$, such that $\lim_{k \to \infty} w_k(\omega) = f(\omega)$ holds for all $\omega \in \Omega$.*

Recall that $w \in S(\Omega) - S(\Omega)$ means that $w = u - v$, $u, v \in S(\Omega)$. There is a different way to phrase Corollary 4.19. We introduce $\mathrm{span}\,(S(\Omega))$, the span of $S(\Omega)$, and we denote by $\overline{E}(\Omega)$ the set of all numerical measurable functions $f : \Omega \to [-\infty, \infty]$. Then Corollary 4.19 states that $\mathrm{span}\,(S(\Omega))$ is sequentially dense in $\overline{E}(\Omega)$ with respect to pointwise convergence.

When we have to emphasise the σ-field $\mathcal{A}$, we will write $\overline{E}(\Omega, \mathcal{A})$ instead of $\overline{E}(\Omega)$. Moreover, by $E(\Omega)$ (or $E(\Omega, \mathcal{A})$) we denote all real-valued measurable functions $f : \Omega \to \mathbb{R}$.

Problems

In the case of mappings between subsets on $\mathbb{R}^k$ and $\mathbb{R}^m$ we always consider the corresponding Borel σ-field except in some cases where we explicitly give the σ-field of interest.

1.   a) For arbitrary sets $A, B, A_k \subset \Omega$, $k \in \mathbb{N}$, prove the following:

     i) $A \subset B$ implies $\chi_A \le \chi_B$;

     ii) $\chi_{A^c} = 1 - \chi_A$;

     iii) $\chi_{\bigcup_{k \in \mathbb{N}} A_k} = \sup_{k \in \mathbb{N}} \chi_{A_k}$;

     iv) $\chi_{\bigcap_{k \in \mathbb{N}} A_k} = \inf_{k \in \mathbb{N}} \chi_{A_k}$.

     b) Let $(\Omega, \mathcal{A})$ be a measurable space. Suppose that $\sup_{k \in \mathbb{N}} \chi_{A_k}$ is measurable, $A_k \subset \Omega$, $k \in \mathbb{N}$. Deduce that $\bigcup_{k \in \mathbb{N}} A_k$ is measurable. Does this imply that all of the sets A_k are measurable, i.e. elements of $\mathcal{A}$?

2.   a) Let (X_j, d_j), $j = 1, \ldots, N$ be metric spaces and (X, d) their product space. Denote by $\mathcal{B}(X_j)$ and $\mathcal{B}(X)$ the corresponding Borel σ-fields and prove that $\otimes_{j=1}^{N} \mathcal{B}(X_j) = \mathcal{B}(X)$.

     b) Solve Exercise 4.3.

3. Prove that for M measurable functions $g_j : \mathbb{R}^n \to \mathbb{R}$ the product $g := \prod_{j=1}^{M} g_j : \mathbb{R}^n \to \mathbb{R}$ is measurable too.

4.   a) Let $K \subset \mathbb{R}^n$ be a convex set and $h : K \to \mathbb{R}$ a concave function. Is h a $\mathcal{B}^{(n)}/\mathcal{B}^{(1)}$-measurable function?

     b) Let $g : (a, b) \to \mathbb{R}$ be differentiable. Prove that $g' : (a, b) \to \mathbb{R}$ is measurable. (We do not assume that $g \in C^1((a, b))$, but only that g is differentiable.)

5.   a) For a measurable function $f : [a, \infty) \to \mathbb{R}$ consider the function $g : \mathbb{R}^n \to \mathbb{R}$, $g(x) = f(||x||)$. Is f measurable?

     b) Let $f : \mathbb{R}^n \to \mathbb{R}$ be a measurable function and $\sigma \in S_n$ where S_n denotes the symmetric group. Is the function $\sigma(f) : \mathbb{R}^n \to \mathbb{R}$, $\sigma(f)(x) = f(x_{\sigma(1)}, \ldots, x_{\sigma(n)})$ measurable?

6. Let $f : \Omega \to \mathbb{R}$ be a measurable function on a measurable space $(\Omega, \mathcal{A})$. Prove the existence of a sequence of functions $\varphi_k \in S(\Omega) - S(\Omega)$ such that $|\varphi_k(\omega)| \le |\varphi_{k+1}(\omega)|$ and $\lim_{k \to \infty} \varphi_k(\omega) = f(\omega)$, $\omega \in \Omega$.

7. Prove that $\overline{\mathcal{B}}_{\mathbb{R}}^{(1)} = \mathbb{R} \cap \overline{\mathcal{B}} = \mathcal{B}^{(1)}$.

8. We call $u : \mathbb{R} \to \mathbb{R}$ of Baire class zero if u is continuous. If $u : \mathbb{R} \to \mathbb{R}$ is the pointwise limit of functions of Baire class zero but not of Baire class zero itself we call u of Baire class 1. In general $u : \mathbb{R} \to \mathbb{R}$ is of **Baire class k** $\in \mathbb{N}_0$ if u is the pointwise limit of functions of Baire class $k - 1$ but not of Baire class $k - 1$ itself. Prove that if u is of Baire class k it is measurable. For u defined by

$$u(x) = \begin{cases} 1, & |x| < 1, \\ \frac{1}{2}, & |x| = 1, \\ 0, & |x| > 1, \end{cases}$$

show that u is of Baire class 1.

Hint: consider the sequence $u_k : \mathbb{R} \to \mathbb{R}$, $u_k(x) = \frac{1}{1 + x^{2k}}$.

5 Integration with Respect to a Measure — The Lebesgue Integral

Let $(\Omega, \mathcal{A}, \mu)$ be a measure space and denote by $S(\Omega)$ the set of all simple functions $u : \Omega \to \mathbb{R}$. Suppose that $\{\gamma_1, \ldots, \gamma_K\} \subset [0, \infty)$ is the range of u and $\gamma_j \neq \gamma_k$ for $j \neq k$. In this case the sets $C_l := \{u = \gamma_l\}$, $l = 1, \ldots, K$, are mutually disjoint and we can define the $(\mu-)$**integral** of u by

$$\int u \, d\mu := \sum_{l=1}^{K} \gamma_l \mu(C_l) \in [0, \infty]. \tag{5.1}$$

Lemma 5.1. *For $u \in S(\Omega)$ with normal representation $u = \sum_{k=1}^{M} \beta_k \chi_{B_k}$ we have*

$$\sum_{k=1}^{M} \beta_k \mu(B_k) = \int u \, d\mu. \tag{5.2}$$

Proof. We show that for two normal representations $u = \sum_{k=1}^{M} \beta_k \chi_{B_k}$ and $u = \sum_{j=1}^{N} \alpha_j \chi_{A_j}$ the following holds

$$\sum_{j=1}^{N} \alpha_j \mu(A_j) = \sum_{k=1}^{M} \beta_k \mu(B_k). \tag{5.3}$$

With $C_{jk} = A_j \cap B_k$, $\xi_{jk} = \alpha_j$ and $\eta_{jk} = \beta_k$ we find as in (4.24) that

$$\sum_{j=1}^{N} \sum_{k=1}^{M} \xi_{jk} \chi_{C_{jk}} = \sum_{k=1}^{M} \sum_{j=1}^{N} \eta_{jk} \chi_{C_{jk}}$$

and therefore

$$\sum_{j=1}^{N} \sum_{k=1}^{M} \xi_{jk} \mu(C_{jk}) = \sum_{k=1}^{M} \sum_{j=1}^{N} \eta_{jk} \mu(C_{jk}),$$

but

$$\sum_{j=1}^{N} \sum_{k=1}^{M} \xi_{jk} \mu(C_{jk}) = \sum_{j=1}^{N} \alpha_j \mu \left(\bigcup_{k=1}^{M} C_{jk} \right) = \sum_{j=1}^{N} \alpha_j \mu(A_j)$$

and

$$\sum_{k=1}^{M} \sum_{j=1}^{N} \eta_{jk} \mu(C_{jk}) = \sum_{k=1}^{M} \beta_k \mu \left(\bigcup_{j=1}^{N} C_{jk} \right) = \sum_{k=1}^{M} \beta_k \mu(B_k).$$

$\square$

Before we investigate the μ-integral further, let us compare the situation with the Riemann integral where we started to define the integral on a hyper-rectangle $K \subset \mathbb{R}^n$ for step functions $f : K \to \mathbb{R}$. In this case we first chose a partition of K into hyper-rectangles K_α and then we considered functions being constant on K_α, say $f|_{K_\alpha} = c_\alpha$. The volume or measure of K_α was taken as the classical volume as defined for hyper-rectangles in $\mathbb{R}^n$. The last point was the serious and difficult restriction: only for hyper-rectangles we know the volume from classical geometry and consequently only for a partition of K into hyper-rectangles K_α we can form

$$\int_K f(x)\, dx := \sum_{\alpha \in J} c_\alpha \mathrm{vol}_n(K_\alpha). \tag{5.4}$$

Now we start with a measurable function attaining (as a step function) only finitely many values, from this we obtain the partition of the domain into measurable sets and hence we can form (5.1). This allows us *a priori* to consider as the domain of the function any measurable set in a measurable space $(\Omega, \mathcal{A})$ and we are neither restricted to hyper-rectangles nor to $\mathbb{R}^n$. Of course we have to pay a price: we have to introduce measure spaces and measurable mappings. In the case of the Riemann integral we just do this after having introduced an integral for functions we can approximate with step functions. It turns out that for $n \geq 2$ this is a rather involved and less satisfactory approach, although it looks to be the more natural one.

Proposition 5.2. *Let $(\Omega, \mathcal{A}, \mu)$ be a measure space. For $u, v \in S(\Omega)$, $\alpha \geq 0$ and $A \in \mathcal{A}$ the following hold*

$$\int \chi_A \, d\mu = \mu(A); \tag{5.5}$$

$$\int (\alpha u)\, d\mu = \alpha \int u \, d\mu; \tag{5.6}$$

$$\int (u + v)\, d\mu = \int u \, d\mu + \int v \, d\mu; \tag{5.7}$$

$$u \leq v \; \text{implies} \; \int u \, d\mu \leq \int v \, d\mu. \tag{5.8}$$

Proof. Of course, (5.5) and (5.6) are consequences of the definition. For (5.7) we start with normal representations of u and v, i.e. $u = \sum_{j=1}^N \alpha_j \chi_{A_j}$ and

$v = \sum_{k=1}^{M} \beta_k \chi_{B_k}$, and we switch to the common partition $C_{jk} := A_j \cap B_k, 1 \leq j \leq N, 1 \leq k \leq M$, as in (4.24). Hence we have

$$u = \sum_{j=1}^{N} \sum_{k=1}^{M} \xi_{jk} \chi_{C_{jk}} \quad \text{and} \quad v = \sum_{k=1}^{M} \sum_{j=1}^{N} \eta_{jk} \chi_{C_{jk}}$$

which yields

$$u + v = \sum_{j=1}^{N} \sum_{k=1}^{M} (\xi_{jk} + \eta_{jk}) \chi_{C_{jk}}$$

and (5.7) follows from

$$\sum_{j=1}^{N} \sum_{k=1}^{M} (\xi_{jk} + \eta_{jk}) \mu(C_{jk}) = \sum_{j=1}^{N} \sum_{k=1}^{M} \xi_{jk} \mu(C_{jk}) + \sum_{k=1}^{M} \sum_{j=1}^{N} \eta_{jk} \mu(C_{jk})$$

$$= \sum_{j=1}^{N} \alpha_j \mu(A_j) + \sum_{k=1}^{M} \beta_k \mu(B_k).$$

Moreover, since $u \leq v$ implies $\xi_{jk} \leq \eta_{jk}$ for $1 \leq j \leq N$, $1 \leq k \leq M$, we deduce (5.8) from

$$\sum_{j=1}^{N} \alpha_j \mu(A_j) = \sum_{j=1}^{N} \sum_{k=1}^{M} \xi_{jk} \mu(C_{jk}) \leq \sum_{k=1}^{M} \sum_{j=1}^{N} \eta_{jk} \mu(C_{jk}) = \sum_{k=1}^{M} \beta_k \mu(B_k).$$

$\square$

Exercise 5.3. *Let $u = \sum_{j=1}^{N} \alpha_j \chi_{A_j}$ for a partition of Ω into measurable sets but do not assume that $\alpha_j \neq \alpha_k$ for $j \neq k$. Prove that $\sum_{j=1}^{N} \alpha_j \mu(A_j) = \int u \, d\mu$ still holds.*

Example 5.4. In Example I.25.12.A we have seen that the Dirichlet function $\chi_{\mathbb{Q} \cap [0,1]} : [0,1] \to \mathbb{R}$ is not Riemann integrable. Since $\mathbb{Q} \cap [0,1]$ is a measurable set in the trace σ-field $\mathcal{B}^{(1)}([0,1]) = [0,1] \cap \mathcal{B}^{(1)}$, it follows that $\chi_{\mathbb{Q} \cap [0,1]}$ is an element of $S([0,1])$, hence its $\lambda^{(1)}$-integral is well defined. The set $\mathbb{Q}$ is countable and therefore $\lambda^{(1)}(\mathbb{Q}) = 0$ implying $\int \chi_{\mathbb{Q} \cap [0,1]} \, d\lambda^{(1)}_{|[0,1]} = 0$ where $\lambda^{(1)}_{|[0,1]}$ is the Lebesgue measure on $\mathbb{R}$ restricted to $[0,1]$. The example already shows that not all $\lambda^{(1)}$-integrable functions are Riemann integrable.

Let $u \in S(\Omega)$ with $\mathrm{ran}(u) = \{\alpha_1, \ldots, \alpha_N\} \subset [0, \infty)$. The function $\omega \mapsto u^p(\omega)$ is for every $p > 0$ well defined, has range $\{\alpha_1^p, \ldots, \alpha_N^p\}$ and since $\{u \geq \alpha\} = \Omega$ for $\alpha \leq 0$ and $\{u^p \geq \alpha\} = \{u \geq \alpha^{\frac{1}{p}}\}$ for $\alpha > 0$, it follows that u^p is measurable, hence $u^p \in S(\Omega)$. For $p > 1$ we define as usual the conjugate exponent q by $\frac{1}{p} + \frac{1}{q} = 1$. Consider $u, v \in S(\Omega)$ with the normal representation

$$u = \sum_{j=1}^{N} \alpha_j \chi_{A_j} \text{ and } v = \sum_{j=1}^{N} \beta_j \chi_{A_j}.$$

Since $\alpha_j, \beta_j \geq 0$ and $u \cdot v$, u^p, $v^q \in S(\Omega)$ we deduce using Hölder's inequality (for finite sums)

$$\int |u \cdot v| \, d\mu = \sum_{j=1}^{N} |\alpha_j \cdot \beta_j| \mu(A_j)$$

$$= \sum_{j=1}^{N} |\alpha_j| \mu(A_j)^{\frac{1}{p}} |\beta_j| \mu(A_j)^{\frac{1}{q}}$$

$$\leq \left(\sum_{j=1}^{N} |\alpha_j|^p \mu(A_j) \right)^{\frac{1}{p}} \left(\sum_{j=1}^{N} |\beta_j|^q \mu(A_j) \right)^{\frac{1}{q}}$$

$$= \left(\int |u|^p \, d\mu \right)^{\frac{1}{p}} \left(\int |v|^q \, d\mu \right)^{\frac{1}{q}}$$

and we have proved

Lemma 5.5. *For $u, v \in S(\Omega)$ and $\frac{1}{p} + \frac{1}{q} = 1$, $p > 1$, **Hölder's inequality***

$$\int |u \cdot v| \, d\mu \leq \left(\int |u|^p \, d\mu \right)^{\frac{1}{p}} \left(\int |v|^q \, d\mu \right)^{\frac{1}{q}} \tag{5.9}$$

holds.

Corollary 5.6. *For $u, v \in S(\Omega)$ and $p \geq 1$ **Minkowski's inequality** holds, i.e.*

$$\left(\int |u + v|^p \, d\mu \right)^{\frac{1}{p}} \leq \left(\int |u|^p \, d\mu \right)^{\frac{1}{p}} + \left(\int |v|^p \, d\mu \right)^{\frac{1}{p}}. \tag{5.10}$$

Proof. First we note that we do not need the absolute value in (5.10) since $u, v \geq 0$. Further note that for $p = 1$ we even have by (5.7) equality in (5.10). For $p > 1$ it follows

$$\int |u + v|^p \, d\mu = \int (u + v)(u + v)^{p-1} \, d\mu$$
$$= \int u(u + v)^{p-1} \, d\mu + \int v(u + v)^{p-1} \, d\mu,$$

and now Hölder's inequality yields with $\frac{1}{p} + \frac{1}{q} = 1$

$$\int |u + v|^p \, d\mu \leq \left(\int u^p \, d\mu \right)^{\frac{1}{p}} \left(\int |u + v|^{q(p-1)} \, d\mu \right)^{\frac{1}{q}}$$
$$+ \left(\int |v|^p \, d\mu \right)^{\frac{1}{p}} \left(\int |u + v|^{q(p-1)} \, d\mu \right)^{\frac{1}{q}}.$$

However $q(p - 1) = p$ and therefore we find

$$\int |u + v|^p \, d\mu \leq \left(\left(\int u^p \, d\mu \right)^{\frac{1}{p}} + \left(\int v^p \, d\mu \right)^{\frac{1}{p}} \right) \left(\int |u + v|^p \, d\mu \right)^{\frac{1}{q}}$$

or

$$\left(\int |u + v|^p \, d\mu \right)^{\frac{1}{p}} \leq \left(\int u^p \, d\mu \right)^{\frac{1}{p}} + \left(\int v^p \, d\mu \right)^{\frac{1}{p}}.$$

$\square$

Clearly, for $u, v \in S(\Omega)$ with $\int u \, d\mu < \infty$ and $\int v \, d\mu < \infty$ we can define

$$\int (u - v) \, d\mu := \int u \, d\mu - \int v \, d\mu \tag{5.11}$$

and for $\alpha < 0$

$$\int (\alpha u) \, d\mu := - \int (-\alpha u) \, d\mu. \tag{5.12}$$

But we prefer to investigate this "linear" extension in a larger frame after having studied the behaviour of the integral on $S(\Omega)$ under monotone increasing limits.

Theorem 5.7. *Let $(\Omega, \mathcal{A}, \mu)$ be a measurable space and $(u_l)_{l\in\mathbb{N}}$, $u_l \in S(\Omega)$, a monotone increasing sequence, i.e. $u_l \leq u_{l+1}$. For $u \in S(\Omega)$ we have*

$$u \leq \sup_{l\in\mathbb{N}} u_l \ \ implies \ \ \int u \, d\mu \leq \sup_{l\in\mathbb{N}} \int u_l \, d\mu. \tag{5.13}$$

Proof. Let $u = \sum_{k=1}^{N} \alpha_k \chi_{A_k}$ be a normal representation of u and $\alpha \in (0,1)$. It follows that $B_l := \{u_l \geq \alpha u\} \in \mathcal{A}$ and $u_l \geq \alpha u \chi_{B_l}$, implying for $l \in \mathbb{N}$

$$\int u_l d\mu \geq \alpha \int u \chi_{B_l} \, d\mu.$$

Since $(u_l)_{l\in\mathbb{N}}$ is monotone increasing and $u \leq \sup_{l\in\mathbb{N}} u_l$ we find $B_l \subset B_{l+1}$ and $\bigcup_{l\in\mathbb{N}} B_l = \Omega$ as well as $A_k \cap B_l \subset A_k \cap B_{l+1}$ and $\bigcup_{l\in\mathbb{N}}(A_k \cap B_l) = A_k$, $k = 1, \ldots, N$. We know that μ is continuous from below, Theorem 2.10.A, and therefore we find

$$\sup_{l\in\mathbb{N}} \int u_l \, d\mu \geq \sup_{l\in\mathbb{N}} \alpha \int u \chi_{B_l} \, d\mu$$

$$= \alpha \lim_{l\to\infty} \int u \chi_{B_l} \, d\mu = \alpha \int u \, d\mu.$$

Since $\alpha \in (0,1)$ was arbitrary (5.13) follows. □

Corollary 5.8. *For two monotone increasing sequences $(u_n)_{n\in\mathbb{N}}$ and $(v_m)_{m\in\mathbb{N}}$ in $S(\Omega)$ it follows that*

$$\sup_{n\in\mathbb{N}} u_n = \sup_{m\in\mathbb{N}} v_m \ \ implies \ \ \sup_{n\in\mathbb{N}} \int u_n \, d\mu = \sup_{m\in\mathbb{N}} \int v_m \, d\mu. \tag{5.14}$$

Proof. For $k, l \in \mathbb{N}$ we have $v_l \leq \sup_{n\in\mathbb{N}} u_n$ and $u_k \leq \sup_{m\in\mathbb{N}} v_m$ and Theorem 5.7 now implies

$$\int v_l \, d\mu \leq \sup_{n\in\mathbb{N}} \int u_n \, d\mu \ \ and \ \ \int u_k \, d\mu \leq \sup_{m\in\mathbb{N}} \int v_m \, d\mu.$$

Now taking the supremum with respect to l and k, respectively, gives (5.14). □

With Theorem 4.18 in mind we can now give

Definition 5.9. *Let $(\Omega, \mathcal{A}, \mu)$ be a measure space and $f : \Omega \to [0, \infty]$ a non-negative, numerical measurable function. The **integral** of f over Ω with respect to μ is defined by*

$$\int f \, \mathrm{d}\mu = \sup_{k \in \mathbb{N}} \int u_k \, \mathrm{d}\mu \qquad (5.15)$$

where $(u_k)_{k \in \mathbb{N}}$ is any monotone increasing sequence of functions in $S(\Omega)$ such that $u = \sup_{k \in \mathbb{N}} u_k = \lim_{k \to \infty} u_k \le \infty$.

Remark 5.10. A. By Corollary 5.8 the definition is independent of the choice of the approximating sequence.
B. Sometimes we call $\int f \, \mathrm{d}\mu$ the μ-**integral** if we want to emphasise the measure.

From Proposition 5.2 we deduce (see Problem 5)

Corollary 5.11. *For $f, g : \Omega \to [0, \infty]$ measurable and $\alpha \ge 0$, $A \in \mathcal{A}$ the following hold*

$$\int \chi_A \, \mathrm{d}\mu = \mu(A); \qquad (5.16)$$

$$\int (f + g) \, \mathrm{d}\mu = \int f \, \mathrm{d}\mu + \int g \, \mathrm{d}\mu; \qquad (5.17)$$

$$\int (\alpha f) \, \mathrm{d}\mu = \alpha \int f \, \mathrm{d}\mu; \qquad (5.18)$$

$$f \le g \ \text{implies} \ \int f \, \mathrm{d}\mu \le \int g \, \mathrm{d}\mu. \qquad (5.19)$$

Let $u, v : \Omega \to [0, \infty]$ be measurable functions such that $u = \sup_{k \in \mathbb{N}} u_k = \lim_{k \to \infty} u_k$ and $v = \sup_{l \in \mathbb{N}} v_l = \lim_{l \to \infty} v_l$ for monotone increasing sequences $(u_k)_{k \in \mathbb{N}}$ and $(v_l)_{l \in \mathbb{N}}$ of simple functions on Ω. It follows that $u \cdot v = \lim_{k \to \infty} u_k \cdot \lim_{k \to \infty} v_k = \lim_{k \to \infty} (u_k v_k)$ and for $w_k := u_k v_k$ we find $w_k = u_k v_k \le u_{k+1} v_{k+1} = w_{k+1}$, hence $(u_k \cdot v_k)_{k \in \mathbb{N}}$ is an increasing sequence of simple functions approximating $u \cdot v$. Moreover we have $u^p = |u|^p = \sup_{k \in \mathbb{N}} |u_k|^p$ and $v^p = |v|^p = \sup_{l \in \mathbb{N}} |v_l|^p$. Consequently we can extend **Hölder's inequality** and **Minkowski's inequality** to non-negative, numerical measurable functions.

Corollary 5.12. *For $u, v : \Omega \to [0, \infty]$ measurable and $p > 1$, $\frac{1}{p} + \frac{1}{q} = 1$ we have*

$$\int |uv| \, \mathrm{d}\mu \leq \left(\int |u|^p \, \mathrm{d}\mu \right)^{\frac{1}{p}} \left(\int |v|^q \, \mathrm{d}\mu \right)^{\frac{1}{q}} \qquad (5.20)$$

and for $p \geq 1$

$$\left(\int |u + v|^p \, \mathrm{d}\mu \right)^{\frac{1}{p}} \leq \left(\int |u|^p \, \mathrm{d}\mu \right)^{\frac{1}{p}} + \left(\int |v|^p \, \mathrm{d}\mu \right)^{\frac{1}{p}}. \qquad (5.21)$$

The following result is due to B. Levi and gives an extremely powerful tool when studying the interchangeability of limits and integrals. It is either called **Theorem of Beppo Levi** or **Monotone Convergence Theorem**.

Theorem 5.13. *Let $(\Omega, \mathcal{A}, \mu)$ be a measure space and $(f_k)_{k \in \mathbb{N}}$, $f_k : \Omega \to [0, \infty]$, an increasing sequence of measurable functions. Then $\sup_{k \in \mathbb{N}} f_k$ is a numerical, non-negative measurable function, hence μ-integrable, and we have*

$$\int \sup_{k \in \mathbb{N}} f_k \, \mathrm{d}\mu = \sup_{k \in \mathbb{N}} \int f_k \, \mathrm{d}\mu. \qquad (5.22)$$

Proof. Set $f := \sup_{k \in \mathbb{N}} f_k$. We prove the existence of a sequence of simple functions $(v_n)_{n \in \mathbb{N}}$ which is monotone increasing and satisfies $\sup_{n \in \mathbb{N}} v_n = f$ as well as $v_n \leq f_n$. For such a sequence we obtain from (5.15)

$$\int f \, \mathrm{d}\mu = \int \sup_{n \in \mathbb{N}} v_n \, \mathrm{d}\mu = \sup_{n \in \mathbb{N}} \int v_n \, \mathrm{d}\mu$$

and on the other hand we have $\int v_k \, \mathrm{d}\mu \leq \int f_n \, \mathrm{d}\mu$ implying

$$\int f \, \mathrm{d}\mu = \sup_{n \in \mathbb{N}} \int v_n \, \mathrm{d}\mu \leq \sup_{n \in \mathbb{N}} \int f_n \, \mathrm{d}\mu,$$

but $f_n \leq f$ further implies that $\int f_n \, \mathrm{d}\mu \leq \int f \, \mathrm{d}\mu$, i.e. we get $\sup_{n \in \mathbb{N}} \int f_n \, \mathrm{d}\mu \leq \int f \, \mathrm{d}\mu$ and (5.22) follows.

Now we construct the sequence $(v_n)_{n \in \mathbb{N}}$, $v_n \in S(\Omega)$. For f_k there exists a monotone increasing sequence $(u_{mk})_{m \in \mathbb{N}}$ of simple functions such that $\sup_{m \in \mathbb{N}} u_{mk} = f_k$. Clearly we have $v_m := v_{m1} \vee \cdots \vee u_{mm} \in S(\Omega)$ and further $v_m \leq v_{m+1}$ by construction. Since $(f_k)_{k \in \mathbb{N}}$ is increasing it follows that

$v_m \leq f_m$ for all $m \in \mathbb{N}$, implying $\sup_{m \in \mathbb{N}} v_m \leq f$. Moreover, for $m \geq k$ we find $u_{mk} \leq v_m$ and therefore

$$\sup_{m \in \mathbb{N}} u_{mk} = f_k \leq \sup_{m \in \mathbb{N}} v_m \leq f,$$

which yields $\sup_{m \in \mathbb{N}} v_m = \sup_{k \in \mathbb{N}} f_k = f$. Hence, $(v_m)_{m \in \mathbb{N}}$ has all the desired properties. $\square$

Remark 5.14. The name "monotone convergence theorem" becomes more clear when we give it the following formulation: let $(f_k)_{k \in \mathbb{N}}$ be a monotone increasing sequence of non-negative, numerical functions on Ω converging pointwisely to f, i.e. $\lim_{k \to \infty} f_k = \sup_{k \in \mathbb{N}} f_k = f$. Then we have

$$\lim_{k \to \infty} \int f_k \, d\mu = \int \lim_{k \to \infty} f_k \, d\mu. \tag{5.23}$$

Corollary 5.15. *Let $(f_k)_{k \in \mathbb{N}}$ be any sequence of non-negative, numerical functions on the measure space $(\Omega, \mathcal{A}, \mu)$. Then the numerical function $\sum_{k=1}^{\infty} f_k = \lim_{N \to \infty} \sum_{k=1}^{N} f_k$ is measurable and*

$$\int \sum_{k=1}^{\infty} f_k \, d\mu = \sum_{k=1}^{\infty} \int f_k \, d\mu. \tag{5.24}$$

Proof. We only have to apply Theorem 5.13 to the sequence of partial sums $g_N := \sum_{k=1}^{N} f_k$. $\square$

Remark 5.16. It is important to emphasise that (5.22), (5.23) or (5.24) are statements for non-negative numerical functions, hence the value $+\infty$ is not excluded and convergence refers to convergence in $[0, \infty]$. Of different nature is the question whether we have convergence in $\mathbb{R}$ provided all functions f_k are real-valued.

Example 5.17. Let $(\Omega, \mathcal{A})$ be any measurable space and for $\omega \in \Omega$ let ϵ_ω be the Dirac measure at ω, i.e. $\epsilon_\omega(A) = 1$ for $\omega \in A$ and $\epsilon_\omega(A) = 0$ for $\omega \notin A$. For any non-negative measurable numerical functions $f : \Omega \to [0, \infty]$ we have

$$\int f \, d\epsilon_\omega = f(\omega). \tag{5.25}$$

Indeed, for $f \in S(\Omega)$ with normal representation $f = \sum_{k=1}^{N} \alpha_k \chi_{A_k}$ it follows

$$\int f \, d\epsilon_\omega = \sum_{k=1}^{N} \alpha_k \epsilon_\omega(A_k) = \alpha_{k_0} = f(\omega)$$

for some $\alpha_{k_0} \in \{\alpha_1, \ldots, \alpha_N\}$ and $\omega \in A_{k_0}$. Now, if $f : \Omega \to [0, \infty]$ is measurable and $(f_k)_{k \in \mathbb{N}}$, $f_k \in S(\Omega)$, is an increasing sequence converging to f we find

$$\int f \, d\epsilon_\omega = \int \lim_{k \to \infty} f_k \, d\epsilon_\omega = \lim_{k \to \infty} \int f_k \, d\epsilon_\omega = \lim_{k \to \infty} f_k(\omega) = f(\omega).$$

Example 5.18. Consider the measurable space $(\mathbb{N}, \mathcal{P}(\mathbb{N}))$. Since a measure is σ-additive every measure on $\mathcal{P}(\mathbb{N})$ is determined by the sequence $\alpha_k := \mu(\{k\}) \in [0, \infty]$, $k \in \mathbb{N}$. A non-negative, numerical measurable function $f : \mathbb{N} \to [0, \infty]$ can be described in the following way. Set $f_k := f(k)\chi_{\{k\}}$. This is a measurable function and for $f(k) < \infty$ it is a simple function, i.e. $f(k) < \infty$ implies $f_k \in S(\Omega)$. Moreover $f = \lim_{N \to \infty} \sum_{k=1}^{N} f_k = \sum_{k=1}^{\infty} f_k$. Therefore we find with Example 5.17 that

$$\int f \, d\mu = \int \sum_{k=1}^{\infty} f_k \, d\mu = \sum_{k=1}^{\infty} f(k) \, \alpha_k = \sum_{k=1}^{\infty} f(k)\mu(\{k\}),$$

i.e.

$$\int f \, d\mu = \sum_{k=1}^{\infty} f(k)\mu(\{k\}). \tag{5.26}$$

Example 5.19. A. For $N \in \mathbb{N}$ and $p \in [0, 1]$, $q := 1 - p$, consider on $(\mathbb{R}, \mathcal{B}^{(1)})$ the **Bernoulli distribution** $B(n, p) := \beta_N^p := \sum_{k=0}^{N} \binom{N}{k} p^k q^{N-k} \epsilon_k$. For the non-negative measurable function $f(x) = \chi_{[0,\infty)}(x) \cdot x$ we find

$$\int_{\mathbb{R}} f \, d\beta_N^p = \sum_{k=0}^{N} \binom{N}{k} p^k q^{N-k} \int f \, d\epsilon_k$$

$$= \sum_{k=0}^{N} \binom{N}{k} p^k q^{N-k} f(k) = \sum_{k=1}^{N} \binom{N}{k} p^k q^{N-k} k$$

$$= Np(p+q)^{N-1} = Np.$$

B. For the **Poisson distribution** with parameter $\gamma > 0$, i.e. for the measure $\pi_\gamma := \sum_{k=0}^\infty e^{-\gamma}\frac{\gamma^k}{k!}\epsilon_k$ on $(\mathbb{R}, \mathcal{B}^{(1)})$, we find for f as in part A

$$\int f\,\mathrm{d}\pi_\gamma = \sum_{k=1}^\infty e^{-\gamma}\frac{\gamma^k}{k!}k = \gamma.$$

(It would be a beneficial exercise to complete the remaining details in these two calculations.)

Let $[a, b]$, $a < b$ be an interval and $a = t_0 < t_1 < \ldots < t_N = b$ be a partition of $[a, b]$. Further let $\varphi : [a, b] \to \mathbb{R}$ be a bounded, non-negative step function with values $\varphi|_{(t_{k-1}, t_k)} = c_k \geq 0$. This function is also a simple function when using on $[a, b] \in [a, b] \cap \mathcal{B}^{(1)}$ the partition into measurable sets (t_{k-1}, t_k), $k = 1, \ldots, N$, and $\{t_k\}$, $k = 0, \ldots, N$. Since with respect to the measure $\lambda^{(1)}_{[a,b]}$ the sets $\{t\}$ are negligible, i.e. $\lambda^{(1)}_{[a,b]}(\{t_k\}) = 0$, it follows that

$$\int \varphi\,\mathrm{d}\lambda^{(1)}_{[a,b]} = \sum_{k=1}^N c_k\lambda^{(1)}_{[a,b]}((t_{k-1}, t_k)) + \sum_{k=0}^N \varphi(t_k)\lambda^{(1)}_{[a,b]}(\{t_k\})$$

$$= \sum_{k=1}^N c_k\lambda^{(1)}_{[a,b]}((t_{k-1}, t_k)) = \int_a^b \varphi(x)\,\mathrm{d}x$$

where $\lambda^{(1)}_{[a,b]}$ denotes the restriction of $\lambda^{(1)}$ to $[a, b] \cap \mathcal{B}^{(1)}$ and as usual $\int_a^b \varphi(x)\,\mathrm{d}x$ denotes the Riemann integral of ψ. Thus we conclude that the Riemann integral of a non-negative step function coincides with its $\lambda^{(1)}_{[a,b]}$-integral (Lebesgue integral) of the corresponding simple function. But we already know that there are simple functions which are not step functions and which are not Riemann integrable.

In the following we extend the integral to a class of real-valued functions which are not necessarily non-negative anymore. Before this it is maybe appropriate to acknowledge the influence of H. Bauer [11] on our presentation in parts of this in the following chapters. This is partly due to the fact that more than 25 years ago the first named author was much occupied with a critical proof -reading of a first draft of [11].

Definition 5.20. *Let* $(\Omega, \mathcal{A}, \mu)$ *be a measure space and* $f : \Omega \to \overline{\mathbb{R}}$ *be a numerical, measurable function. We call* f *($\mu-$)**integrable** *if the integrals*

$\int f^+ \, d\mu$ and $\int f^- \, d\mu$ are real numbers, i.e. finite. The value

$$\int f \, d\mu := \int f^+ \, d\mu - \int f^- \, d\mu \qquad (5.27)$$

is called the (μ-)*integral* of f.

Remark 5.21. A. For a measurable function $f : \Omega \to [0, \infty]$ we have $f^+ = f$ and $f^- = 0$, hence (5.27) is consistent with our previous definitions of a (μ-)integral for non-negative numerical and measurable functions. **B.** The following notations are sometimes useful:

$$\int f(\omega) \, d\mu(\omega) := \int f(\omega) \, \mu(d\omega) := \int f \, d\mu. \qquad (5.28)$$

Our first result provides us with equivalent criteria for integrability.

Theorem 5.22. *Necessary and sufficient for the integrability of a measurable numerical function $f : \Omega \to \overline{\mathbb{R}}$ is any of the following conditions:*

i) *f^+ and f^- have a finite integral;*

ii) *$f = u - v$ for two integrable functions $u \geq 0$ and $v \geq 0$ with finite integral;*

iii) *there exists an integrable function g with finite integral such that $|f| \leq g$;*

iv) *$|f|$ has a finite integral.*

Proof. Statement i) is just the definition of the integrability of f and with $u = f^+$ and $v = f^-$ it implies ii). Suppose now that ii) holds. Since $u + v$ is integrable (with finite integral) and $f = u - v \leq u \leq u + v$ and $-f = v - u \leq v \leq u + v$ we deduce iii) from ii) with $g := u + v$. Further we note that (5.19) entails that iii) implies iv). Finally, if $|f|$ has a finite integral, then $f^+ \leq |f|$ and $f^- \leq |f|$ have finite integrals too and hence iv) yields i). $\qquad \Box$

Theorem 5.23. *The set of all μ-integrable functions $f : \Omega \to \overline{\mathbb{R}}$ form an $\mathbb{R}$-vector space and the integral is a positivity preserving linear form on this* *vector space, i.e. for integrable functions f and g and for $\alpha, \beta \in \mathbb{R}$ the function $\alpha f + \beta g$ is integrable too and the following hold:*

$$\int (\alpha f + \beta g)\, \mathrm{d}\mu = \alpha \int f\, \mathrm{d}\mu + \beta \int g\, \mathrm{d}\mu; \tag{5.29}$$

and

$$f \geq 0 \ \text{and} \ \int f\, \mathrm{d}\mu \geq 0. \tag{5.30}$$

Proof. Since $f = f^+ - f^-$ and $g = g^+ - g^-$ we find $f + g = (f^+ + g^+) - (f^- + g^-)$ implying the integrability of $f + g$. Further, for $\alpha \geq 0$ we have $(\alpha f)^+ = \alpha f^+$ and $(\alpha f)^- = \alpha f^-$ whereas for $\alpha \leq 0$ we have $(\alpha f)^+ = |\alpha| f^-$ and $(\alpha f)^- = |\alpha| f^+$, hence $\alpha f = \alpha f^+ - \alpha f^-$, $\alpha \geq 0$, or $\alpha f = |\alpha| f^- - |\alpha| f^+$ for $\alpha \leq 0$. Thus αf is integrable and therefore we deduce that $\alpha f + \beta g$ is integrable. From $f + g = (f + g)^+ - (f + g)^- = f^+ + g^+ - (f^- + g^-)$ we deduce

$$(f + g)^+ + f^- + g^- = f^+ + g^+ + (f + g)^-$$

and using the additivity of the integral for non-negative functions we obtain

$$\int (f + g)^+\, \mathrm{d}\mu - \int (f + g)^-\, \mathrm{d}\mu = \int f^+\, \mathrm{d}\mu - \int f^-\, \mathrm{d}\mu + \int g^+\, \mathrm{d}\mu - \int g^-\, \mathrm{d}\mu,$$

i.e. the additivity of the integral. Further, for $\alpha \geq 0$ we find

$$\int \alpha f\, \mathrm{d}\mu = \int (\alpha f)^+\, \mathrm{d}\mu - \int (\alpha f)^-\, \mathrm{d}\mu$$

$$= \alpha \int f^+\, \mathrm{d}\mu - \alpha \int f^-\, \mathrm{d}\mu = \alpha \int f\, \mathrm{d}\mu,$$

and for $\alpha \leq 0$ we get

$$\int \alpha f\, \mathrm{d}\mu = \int (\alpha f)^+\, \mathrm{d}\mu - \int (\alpha f)^-\, \mathrm{d}\mu$$

$$= |\alpha| \int f^-\, \mathrm{d}\mu - |\alpha| \int f^+\, \mathrm{d}\mu = -|\alpha| \left(\int f^+\, \mathrm{d}\mu - \int f^-\, \mathrm{d}\mu \right)$$

$$= \alpha \int f\, \mathrm{d}\mu$$

and (5.29) is proved. Finally, (5.30) follows from Remark 5.21.A and (5.19).

$\square$

Corollary 5.24. *For two integrable functions $f, g : \Omega \to \mathbb{R}$ the functions $f \wedge g$ and $f \vee g$ are integrable too. Furthermore $f \leq g$ implies*

$$\int f \, d\mu \leq \int g \, d\mu \tag{5.31}$$

and for all f we have

$$\left| \int f \, d\mu \right| \leq \int |f| \, d\mu. \tag{5.32}$$

*In addition **Hölder's inequality** (5.20) and **Minkowski's inequality** (5.21) hold for integrable functions f and g, i.e. for $p > 1$ and $\frac{1}{p} + \frac{1}{q} = 1$ we have*

$$\int |fg| \, d\mu \leq \left(\int |f|^p \, d\mu \right)^{\frac{1}{p}} \left(\int |g|^q \, d\mu \right)^{\frac{1}{q}} \tag{5.33}$$

and for $p \geq 1$

$$\left(\int |f + g|^p \, d\mu \right)^{\frac{1}{p}} \leq \left(\int |f|^p \, d\mu \right)^{\frac{1}{p}} + \left(\int |g|^p \, d\mu \right)^{\frac{1}{p}}. \tag{5.34}$$

Note that no assumption is made about the finiteness of the integrals in (5.33) and (5.34).

Proof. Since $f \wedge g = \frac{1}{2}(f + g - |f - g|)$ and $f \vee g = \frac{1}{2}(f + g + |f - g|)$ the integrability of $f \vee g$ and $f \wedge g$ follows immediately. Moreover, (5.31) follows from (5.30) and (5.29) since $0 \leq g - f$. In order to see (5.32) we note that

$$\left| \int f \, d\mu \right| = \left| \int f^+ \, d\mu - \int f^- \, d\mu \right| \leq \int f^+ \, d\mu + \int f^- \, d\mu = \int |f| \, d\mu.$$

Since $|fg| = |f||g|$ Hölder's inequality follows from (5.20). Noting that $|f + g| \leq |f| + |g|$ we derive from (5.21)

$$\left(\int |f + g|^p \, d\mu \right)^{\frac{1}{p}} \leq \left(\int (|f| + |g|)^p \, d\mu \right)^{\frac{1}{p}}$$

and now we may apply (5.21). $\qquad \square$

Example 5.25. For $(\mathbb{N}, \mathcal{P}(\mathbb{N}), \mu)$, $\mu(k) = \alpha_k \geq 0$, and a function $f : \mathbb{N} \to \overline{\mathbb{R}}$ the integrability with respect to μ is equivalent to the condition

$$\sum_{k=1}^{\infty} |f(k)|\alpha_k < \infty,$$

and the μ-integral is given by

$$\int f \, d\mu = \sum_{k=1}^{\infty} f(k) \, \alpha_k.$$

Let $(\Omega, \mathcal{A}, \mu)$ be a measure space and $A \in \mathcal{A}$. For a measurable function $f : \Omega \to \overline{\mathbb{R}}$ we may consider the restriction $f|_A : A \to \overline{\mathbb{R}}$ and we may ask whether f_A is integrable. There are two possibilities to approach the question: we may consider $f\chi_A : \Omega \to \overline{\mathbb{R}}$ which coincides on A with $f|_A$ and is zero in the complement of A, or we may consider the measure space $(A, A \cap \mathcal{A}, \mu|_A)$ where $\mu|_A$ denotes the restriction of μ to $A \cap \mathcal{A}$. Of course we expect that both approaches lead to the same result. We start with

Definition 5.26. *Let $(\Omega, \mathcal{A}, \mu)$ be a measure space and $f : \Omega \to \overline{\mathbb{R}}$ be a measurable function which is either non-negative or integrable. For $A \in \mathcal{A}$ we define the (μ-)integral of f over A as*

$$\int_A f \, d\mu := \int \chi_A f \, d\mu. \tag{5.35}$$

This definition is of course consistent with notation

$$\int_\Omega f \, d\mu = \int f \, d\mu. \tag{5.36}$$

The next lemma relates the two approaches described above.

Lemma 5.27. *For $A \in \mathcal{A}$ and every measurable function $f : \Omega \to \overline{\mathbb{R}}$ which is either non-negative or integrable we have*

$$\int_A f \, d\mu = \int f|_A \, d\mu|_A, \tag{5.37}$$

where the integral on the right hand side is an integral in the measure space $(A, A \cap \mathcal{A}, \mu|_A)$.

Proof. First consider the case $f \geq 0$. Since $A \in \mathcal{A}$ it follows that $f_A : A \to \overline{\mathbb{R}}$ is measurable (with $\mathcal{A}$ replaced by $A \cap \mathcal{A}$). For f we can find simple functions $(u_k)_{k \in \mathbb{N}}$, $u_k \in S(\Omega)$, such that $u_k \leq u_{k+1}$ and $\lim_{k \to \infty} u_k = \chi_A f$, moreover $u_k|_A \in S(A)$, $u_k|_A \leq u_{k+1}|_A$ and $\lim_{k \to \infty} u_k|_A = f|_A$. Consequently we have

$$\int_A f \, d\mu = \sup_{k \in \mathbb{N}} \int u_k \, d\mu \quad \text{and} \quad \int f|_A \, d\mu|_A = \sup_{k \in \mathbb{N}} \int u_k|_A \, d\mu|_A. \tag{5.38}$$

99

Since $0 \le u_k \le \chi_A f$ we deduce that $u_k|_{A^{\complement}} = 0$, hence $u_k \chi_A = u_k|_A$ and (5.38) implies the lemma for $f \ge 0$. In the case where $f : \Omega \to \overline{\mathbb{R}}$ and is integrable, we decompose f into positive and negative parts and apply the statement just proved for $f \ge 0$. $\qquad\square$

A trivial consequence is

Corollary 5.28. *The function* $f : A \to \overline{\mathbb{R}}$, $A \in \mathcal{A}$, *is* μ-*integrable if and only if* f_A *defined by*

$$f_A(\omega) := \begin{cases} f(\omega), & \omega \in A \\ 0, & \omega \in A^{\complement} \end{cases}$$

is μ-*integrable and in this case* $\int_A f \, d\mu = \int f_A \, d\mu$ *holds.*

The following properties follow in a straightforward way from properties of the integral and those of characteristic functions. For a measure space $(\Omega, \mathcal{A}, \mu)$, sets $A, B \in \mathcal{A}$ and measurable functions $f, g : \Omega \to \overline{\mathbb{R}}$ such that they are either non-negative or μ-integrable we have

$$\int_{A \cup B} f \, d\mu + \int_{A \cap B} f \, d\mu = \int_A f \, d\mu + \int_B f \, d\mu, \qquad (5.39)$$

in particular for $A \cap B = \emptyset$ we have

$$\int_{A \cup B} f \, d\mu = \int_A f \, d\mu + \int_B f \, d\mu. \qquad (5.40)$$

Moreover we have that

$$f|_A \le g|_A \text{ implies } \int_A f \, d\mu \le \int_A g \, d\mu. \qquad (5.41)$$

Lemma 5.29. *Let* $(\Omega, \mathcal{A}, \mu)$ *be a measurable space and* $N \in \mathcal{A}$ *a set of measure zero, i.e.* $\mu(N) = 0$. *Further let* $f : \Omega \to \overline{\mathbb{R}}$ *be a measurable numerical function. Then* f *is integrable over* N *and* $\int_N f \, d\mu = 0$. *Moreover, if* f *is integrable then* $\mu\left(\{|f| = \infty\}\right) = 0$.

Proof. For $k \in \mathbb{N}$ the function $u_k := k\chi_N$ is a simple function and $\int u_k \, d\mu = k\mu(N) = 0$. Moreover the sequence $(u_k)_{k \in \mathbb{N}}$ is increasing and $\lim_{k \to \infty} u_k = \sup_{k \in \mathbb{N}} u_k$ implying that $\int \sup_{k \in \mathbb{N}} u_k \, d\mu = 0$. Since $\chi_N f \le \sup_{k \in \mathbb{N}} u_k$ holds for a non-negative, measurable numerical function $f : \Omega \to \overline{\mathbb{R}}$, we conclude that $\int_N f \, d\mu = 0$. The general case now follows from the decomposition $f = f^+ - f^-$ and the definition of the integral. Now suppose that f is integrable, implying $\int f^+ \, d\mu < \infty$ and $\int f^- \, d\mu < \infty$. Hence we need to prove that for an integrable, non-negative numerical function f we have $\mu(\{f = \infty\}) = 0$. If $\mu(\{f = \infty\}) > 0$ then we have

$$+\infty = +\infty\mu(\{f = \infty\}) = \int (+\infty)\chi_{\{f=\infty\}} \, d\mu \leq \int f \, d\mu < \infty$$

which is a contradiction. □

Example 5.30. A. For $f \in \mathbb{R}^n \to \mathbb{R}$ the **support** supp f is defined by supp $f = \overline{\{x \in \mathbb{R}^n \mid f(x) \neq 0\}}$, see (II.14.12). As a closed set supp f is Borel measurable and it follows that f is $\lambda^{(n)}$-integrable if and only if $f|_{\text{supp} f}$ is $\lambda^{(n)}$-integrable over supp f and the following holds

$$\int f \, d\mu = \int_{\text{supp} f} f \, d\mu.$$

B. Let $f : \mathbb{R}^n \to \mathbb{R}$ be a continuous function with compact support supp f. In this case $|f|$ is bounded by $M := \max_{x \in \text{supp} f} |f(x)|$ and of course we have $|f|_{\text{supp} f} \leq M$. On the compact set supp f the constant function $x \mapsto M$ is a simple function, hence it is integrable and by Theorem 5.22.iii) we deduce that $f|_{\text{supp} f}$ is integrable and

$$\int_{\text{supp} f} f \, d\lambda^{(n)} = \int (\chi_{\text{supp} f}) f \, d\lambda^{(n)}. \tag{5.42}$$

Thus all continuous functions $f : \mathbb{R}^n \to \mathbb{R}$ with compact support are $\lambda^{(n)}$-integrable as they are $\lambda^{(n)}$-integrable over every measurable subset of their support.

C. Let $K \subset \mathbb{R}^n$ be a compact set and $h : K \to \mathbb{R}$ a bounded measurable function, $|h(x)| \leq M$ for all $x \in K$. We extend h to $H : \mathbb{R}^n \to \mathbb{R}$ by

$$H(x) := \begin{cases} h(x), & x \in K \\ 0, & x \in K^{\complement} \end{cases}. \text{ Since } x \mapsto M \text{ is } \lambda^{(n)}_{|K}\text{-integrable it follows that } h$$

is $\lambda^{(n)}_{|K}$-integrable and

$$\int h \, d\lambda^{(n)}_{|K} = \int H \, d\lambda^{(n)} = \int H\chi_K \, d\lambda^{(n)}.$$

In particular continuous functions defined on a compact set $K \subset \mathbb{R}^n$ are $\lambda^{(n)}$-integrable over K and of its Borel subsets.

Let us compare these results with the situation we have encountered in Chapter II.20 when introducing the Riemann integral for subsets of $\mathbb{R}^n$ not being a hyper-rectangle. The starting point was a bounded function $f : G \to \mathbb{R}$ defined on a bounded Jordan measurable set $G \subset \mathbb{R}^n$ and with f we considered its canonical extension $\tilde{f}_G(x) = \begin{cases} f(x), & x \in G \\ 0, & x \notin G \end{cases}$ to $\mathbb{R}^n$, see Definition II.20.1 and note that H in Example 5.30.C is nothing but the canonical extension of h. We then defined the Riemann integral of f over G by

$$\int_G f(x)\, \mathrm{d}x := \int_K \tilde{f}_G|_K(x)\, \mathrm{d}x \tag{5.43}$$

where K is any compact cell in $\mathbb{R}^n$ such that $\overline{G} \subset \mathring{K}$, see (II.20.2). The crucial point was the Jordan measurability of G which is essentially a property of ∂G, i.e. it depends on the topology. Now we can construct or define integrals over measurable subsets independent of any topological conditions. Of course, if the σ-field under consideration is related to a topology, e.g. if it is a Borel σ-field, then measurable sets will have in general relations to the elements of the topology, however usually in a less restricted sense as Jordan measurable sets do have. In Chapter 8 we will discuss cases where the Riemann approach to define volume integrals over a Jordan measurable subset of $\mathbb{R}^n$ coincides with the $\lambda^{(n)}$-integral.

In Definition 3.8 we have introduced the image measure: Let $(\Omega, \mathcal{A})$ and $(\tilde{\Omega}, \tilde{\mathcal{A}})$ be two measurable spaces and μ a measure on $(\Omega, \mathcal{A})$. Further let $T : \Omega \to \tilde{\Omega}$ be a measurable mapping. Then the **image measure** $T(\mu)$ was defined as a measure on $(\tilde{\Omega}, \tilde{\mathcal{A}})$ by

$$T(\mu)(\tilde{A}) = \mu(T^{-1}(\tilde{A})), \quad \tilde{A} \in \tilde{\mathcal{A}}. \tag{5.44}$$

We now want to study integration with respect to an image measure.

Theorem 5.31. *Let $(\Omega, \mathcal{A}, \mu)$ be a measure space and $(\tilde{\Omega}, \tilde{\mathcal{A}})$ a measurable space. Further let $T : \Omega \to \tilde{\Omega}$ be a measurable mapping and $\tilde{f} : \tilde{\Omega} \to [0, \infty]$ a numerical, non-negative, measurable function. In this situation we have*

$$\int_{\tilde{\Omega}} \tilde{f}\, \mathrm{d}T(\mu) = \int_{\Omega} (\tilde{f} \circ T)\, \mathrm{d}\mu. \tag{5.45}$$

Proof. First we note that since $\tilde{f} \geq 0$ the integral on the left hand side is defined and the measurability of T implies that $\tilde{f} \circ T$ is measurable and of course $\tilde{f} \circ T \geq 0$, hence the integral on the right hand side is also defined. Now let $\tilde{f}$ be a simple function on $\tilde{\Omega}$, i.e. $\tilde{f} \in S(\Omega)$, which yields that

$$\tilde{f} = \sum_{k=1}^{N} \alpha_k \chi_{\tilde{A}_k}$$

for $\alpha_k \geq 0$ and mutually disjoint sets $\tilde{A}_k \in \tilde{\mathcal{A}}$, $k = 1, \ldots, N$. The sets $A_k := T^{-1}(\tilde{A}_k)$ belong to $\mathcal{A}$ and further we find

$$\tilde{f} \circ T = \sum_{k=1}^{N} \alpha_k \chi_{A_k}.$$

Taking into account that $T(\mu)(\tilde{A}_k) = \mu(A_k)$ we deduce

$$\sum_{k=1}^{N} \alpha_k T(\mu)(\tilde{A}_k) = \sum_{k=1}^{N} \alpha_k \mu(A_k), \tag{5.46}$$

i.e. (5.45) holds for simple functions. The general result follows now by approximation. For $\tilde{f} \geq 0$ we can find a sequence of simple functions $(\tilde{u}_k)_{k \in \mathbb{N}}$, $\tilde{u}_k \leq \tilde{u}_{k+1}$ and $\lim_{k \to \infty} \tilde{u}_k = f$. The sequence $(\tilde{u}_k \circ T)_{k \in \mathbb{N}}$ is now a sequence in $S(\Omega)$, i.e. $\tilde{u}_k \circ T \in S(\Omega)$, $\tilde{u}_k \circ T \leq \tilde{u}_{k+1} \circ T$ and $\lim_{k \to \infty} \tilde{u}_k \circ T = \tilde{f} \circ T$. Thus we may pass in (5.45) to the limit and (5.44) follows. $\qquad\square$

For $\tilde{f} : \tilde{\Omega} \to \overline{\mathbb{R}}$ and T as in Theorem 5.31 we have

$$(\tilde{f} \circ T)^+ = \tilde{f}^+ \circ T \text{ and } (\tilde{f} \circ T)^- = \tilde{f}^- \circ T \tag{5.47}$$

and therefore we get immediately

Corollary 5.32. *In the situation of Theorem 5.31 let $\tilde{f} : \tilde{\Omega} \to \overline{\mathbb{R}}$ be a numerical measurable function. If $\tilde{f}$ is $T(\mu)$-integrable then $\tilde{f} \circ T$ is μ-integrable and if $\tilde{f} \circ T$ is μ-integrable then $\tilde{f}$ is $T(\mu)$-integrable and in each case*

$$\int_{\tilde{\Omega}} \tilde{f} \, dT(\mu) = \int_{\Omega} \tilde{f} \circ T \, d\mu \tag{5.48}$$

holds.

Corollary 5.33. *If in the situation of Theorem 5.31 the mapping $T : \Omega \to \tilde{\Omega}$ is bijective with a measurable inverse then for every numerical function $\tilde{f} : \tilde{\Omega} \to \overline{\mathbb{R}}$ the $T(\mu)$-integrability of $\tilde{f}$ is equivalent to the μ-integrability of $\tilde{f} \circ T$.*

Proof. The key point is that if $\tilde{f} \circ T$ is integrable then it is measurable and hence $\tilde{f} = \tilde{f} \circ T \circ T^{-1}$ is measurable too. $\qquad\square$

The result of Theorem 5.31 and its corollaries is usually referred to as **integration with respect to the image measure**. Our first application of Theorem 5.31 relates to symmetry or invariance of a measure. Let $(\Omega, \mathcal{A}, \mu)$ be a measure space, $T : \Omega \to \Omega$ a measurable mapping. Suppose that $T(\mu) = \mu$ and $T(\Omega) = \Omega$, and let $f : \Omega \to \overline{\mathbb{R}}$ be μ-integrable (or $f \geq 0$). From (5.48) we deduce that

$$\int f \, d\mu = \int f \, dT(\mu) = \int f \circ T \, d\mu \tag{5.49}$$

or

$$\int_\Omega f(x)\, \mu(dx) = \int_\Omega f(Tx)\mu(dx). \tag{5.50}$$

In particular if μ is invariant with respect to a group G of measurable mappings $T : \Omega \to \Omega$, i.e. $T(\mu) = \mu$ for all $T \in G$, then (5.50) holds for $T \in G$ and we say that the **integral is invariant under G or the action of G**.

Example 5.34. By Theorem 3.2 the Lebesgue-Borel measure $\lambda^{(n)}$ is translation invariant, i.e. $\lambda^{(n)}(x_0 + B) = \lambda^{(n)}(B)$ for all $B \in \mathcal{B}^{(n)}$. With $\tau_{x_0}(x) = x_0 + x$ we may rewrite this equality as $\tau_{x_0}(\lambda^{(n)})(B) = \lambda^{(n)}(\tau_{x_0}^{-1}(B)) = \lambda^{(n)}(-x_0 + B)$ and therefore we find by Corollary 5.32 for an appropriate integrable function $f : \mathbb{R}^n \to \overline{\mathbb{R}}$ (or $f \geq 0$) that

$$\int_{\mathbb{R}^n} f(x \mp x_0)\lambda^{(n)}(d\mu) = \int_{\mathbb{R}^n} (f \circ \tau_{\mp x_0})(x)\lambda^{(n)}(dx)$$

$$= \int_{\mathbb{R}^n} f(x)\tau_{\mp x_0}(\lambda^{(n)})(dx) = \int_{\mathbb{R}^n} f(x)\lambda^{(n)}(dx).$$

Example 5.35. According to Theorem 3.6 the Lebesgue-Borel measure $\lambda^{(n)}$ is invariant under the action of $O(n)$, i.e. $T(\lambda^{(n)}) = \lambda^{(n)}$ for $T \in O(n)$. Consequently we find for suitable functions $f : \mathbb{R}^n \to \mathbb{R}$ (or $f : \mathbb{R}^n \to \overline{\mathbb{R}}$) that

$$\int_{\mathbb{R}^n} f(Tx)\lambda^{(n)}(dx) = \int_{\mathbb{R}^n} f(x)\lambda^{(n)}(dx), \quad T \in O(n).$$

As a particular case we mention the reflection at the origin $S : \mathbb{R}^n \to \mathbb{R}^n$, $S(x) = -x$, which yields for all suitable functions f

$$\int_{\mathbb{R}^n} f(-x) \lambda^{(n)}(\mathrm{d}x) = \int_{\mathbb{R}^n} f(x) \lambda^{(n)}(\mathrm{d}x).$$

Example 5.36. Consider the measure space $(\mathbb{Z}^n, \mathcal{P}(\mathbb{Z}^n), \mu_{\mathbb{Z}^n})$, $\mu_{\mathbb{Z}^n}(A) = \sum_{k \in \mathbb{Z}^n} \epsilon_k(A)$, as introduced in Example 3.5. We have seen that $\mu_{\mathbb{Z}^n}$ is invariant under the translation τ_{l_0}, $\tau_{l_0} : \mathbb{Z}^n \to \mathbb{Z}^n$, $\tau_{l_0}(k) = k + l_0$, $l_0 \in \mathbb{Z}^n$. For $f : \mathbb{Z}^n \to [0, \infty)$ we find

$$\int_{\mathbb{Z}^n} f \, \mathrm{d}\mu_{\mathbb{Z}^n} = \sum_{k \in \mathbb{Z}^n} f(k),$$

and Theorem 5.31 yields

$$\int_{\mathbb{Z}^n} f \, \mathrm{d}\mu_{\mathbb{Z}^n} = \int_{\mathbb{Z}^n} f \, \mathrm{d}\tau_{l_0}(\mu_{\mathbb{Z}^n}) = \int_{\mathbb{Z}^n} f \circ \tau_{l_0} \, \mathrm{d}\mu_{\mathbb{Z}^n},$$

or

$$\sum_{k \in \mathbb{Z}^n} f(k) = \sum_{k \in \mathbb{Z}^n} f(k + l_0) \quad \text{for all } l_0 \in \mathbb{Z}^n,$$

which is of course evident.

Next we want to see how we can use Theorem 5.31 and its corollaries in probability theory. Let $(\Omega, \mathcal{A}, P)$ be a probability space and $(\tilde{\Omega}, \tilde{\mathcal{A}})$ a measurable space. A measurable mapping $X : \Omega \to \tilde{\Omega}$ is called a **random variable**, thus we may speak of $\mathbb{R}^n$-valued random variables, real-valued random variables, numerical random variables etc. The image measure of P under X is called the **distribution** of P under X and it is denoted by P_X. Let X be a numerical random variable which we assume either to be non-negative or P-integrable. We call

$$E(X) = \int X \, \mathrm{d}P \tag{5.51}$$

the expectation of X. Since $P(\Omega) = 1$ we have

$$E(X) = \frac{1}{P(X)} \int X \, \mathrm{d}P,$$

i.e. $E(X)$ has the interpretation of a mean with respect to P. Now let $f : \overline{\mathbb{R}} \to \mathbb{R}$ be a measurable function integrable with respect to P_X (or $f \geq 0$). Theorem 5.31 or Corollary 5.32 yields

$$E(f \circ X) = \int f \, dP_X. \tag{5.52}$$

In particular for the identity $\mathrm{id}(x) = x$ we find

$$E(X) = \int x P_X(\mathrm{d}x). \tag{5.53}$$

Thus knowing the distribution P_X of X allows us already to calculate its expectation and for this we need not to have any knowledge of Ω, $\mathcal{A}$ or even P.

In light of these definitions and observations we can give Example 5.19 a probabilistic interpretation:

If a random variable X is Bernoulli-distributed according to β_N^P, then its expectation is

$$E(X) = Np,$$

if a random variable Y is Poisson-distributed with $\gamma > 0$, then its expectation is

$$E(Y) = \gamma.$$

After we have introduced an integral with respect to a measure, we want to clarify the common façon de parler. The integration theory discussed in this chapter is called **Lebesgue-integration** (theory) and hence the μ-integral for a measure μ is also called the integral with respect to μ in the sense of Lebesgue and often one speaks just about the Lebesgue integral with respect to the measure μ. At the same time it is common to understand as **Lebesgue integral** the integral constructed in this chapter for the Lebesgue or the Lebesgue-Borel measure $\lambda^{(n)}$. In practice this will not cause any confusion.

In the next chapter we will use the results on integration with respect to the image measure to derive a version of the change of variables formula or the transformation theorem we have discussed in Chapter II.21.

Problems

1. For $\Omega = \{1, \ldots, N\}$ find $S(\Omega)$ and for a measure μ on $\mathcal{P}(\Omega)$ find for $u \in S(\Omega)$ the integral $\int u \, d\mu$.
   (As usual, the σ-field in Ω is the power set $\mathcal{P}(\Omega)$.)

2. Let $(\Omega, \mathcal{A}, \mu)$ be a measure space and $u = \sum_{j=1}^{N} \alpha_j \chi_{A_j}$ where $(A_j)_{j=1,\ldots,N}$ is a partition of Ω into measurable sets, but it is not assumed that $\alpha_j \neq \alpha_k$ for $j \neq k$. Prove that

$$\sum_{j=1}^{N} \alpha_j \mu(A_j) = \int u \, d\mu$$

   still holds.

3. For $0 < r_1 < r_2 < \cdots < r_N$ define the measurable sets $A_1 := B_{r_1}(0)$, $A_k := B_{r_k}(0) \setminus B_{r_{k-1}}(0)$, $k = 2, \ldots, N$. Consider the function $u = \sum_{k=1}^{N} \gamma_k \chi_{A_k}$, $\gamma_1 := \frac{1}{r_1}, \gamma_k = \frac{1}{r_k - r_{k-1}}, k = 2, \ldots, N$. Using the fact that $\lambda^{(2)}(B_r(0)) = \pi r^2$ find $\int u \, d\lambda^{(2)}$.

4. On $[0, 1)$ consider the sequence of elementary functions

$$u_k = \sum_{l=1}^{k} \frac{l-1}{k} \chi_{\left[\frac{l-1}{k}, \frac{l}{k}\right)}.$$

   Prove that $u_k \in S([0, 1))$, and find $\sup_{k \in \mathbb{N}} \int u_k \, d\lambda_{[0,1)}^{(1)}$. Is the sequence $(u_k)_{k \in \mathbb{N}}$ increasing?

5. Give details of the proof of (5.17).

6. Let $\rho : \mathbb{R} \to \mathbb{R}$ be a measurable function such that $\rho > 0$ and $\int \rho \, d\lambda^{(1)} < \infty$. Define on $\mathcal{B}^{(1)}$ the mapping $\nu := \rho\lambda^{(1)}$ by $\nu(A) := \int_A \rho(x)\lambda^{(1)}(dx)$ and prove that ν is a measure on $\mathcal{B}^{(1)}$. Let $g : \mathbb{R} \to \mathbb{R}$ be a bounded measurable function. Prove that g is ν-integrable.

7. For $g(x) = x^2 \chi_{[0,\infty)}(x)$ find the following integrals:
   i) $\int_{\mathbb{R}} g \, d\beta_N^p$, where β_N^p is a Bernoulli distribution;
   ii) $\int_{\mathbb{R}} g \, d\pi_\gamma$, where π_γ is a Poisson distribution.

8. Show that $h : \mathbb{R} \to \mathbb{R}$ defined by $h(x) := \begin{cases} \frac{\sin x}{x}, & x \neq 0 \\ 1, & x = 0 \end{cases}$ is a continuous function vanishing at infinity, i.e. $\lim_{|x| \to \infty} h(x) = 0$. With the help of Theorem 5.22 prove that h is not integrable over $\mathbb{R}$ with respect to $\lambda^{(1)}$.

Hint: use the decomposition $\int_0^\infty \left| \frac{\sin x}{x} \right| dx = \sum_{k=0}^\infty \int_{k\pi}^{(k+1)\pi} \left| \frac{\sin x}{x} \right| \lambda^{(1)}(dx)$ and now use integration with respect to the image measure to find

$$\int_{k\pi}^{(k+1)\pi} \left| \frac{\sin x}{x} \right| \lambda^{(1)}(dx) = \int_0^\pi \left| \frac{\sin x}{x + k\pi} \right| \lambda^{(1)}(dx).$$

9. Denote by $W_1^{(n)} = [-1, 1]^n \subset \mathbb{R}^n$ the unit cube on $\mathbb{R}^n$. Let $T \in GL(n; \mathbb{R})$ and $g : W_1^{(n)} \to \mathbb{R}$ be a measurable function. Prove that

$$\int_{W_1^{(n)}} g \, d\lambda^{(n)} = \frac{1}{|\det T|} \int_{T(W_1^{(n)})} (g \circ T^{-1}) d\lambda^{(n)}.$$

10. Consider the mappings $U(\varphi) := \begin{pmatrix} \cos \varphi & -\sin \varphi \\ \sin \varphi & \cos \varphi \end{pmatrix}$ for $\varphi = 0, \frac{\pi}{2}, \pi, \frac{3\pi}{2}$.

Show that $U(\varphi)W_1^{(2)} = W_1^{(2)}$ and deduce that

$$\int_{W_1^{(2)}} f(U(\varphi)x)\lambda^{(n)}(dx) = \int_{W_1^{(2)}} f(x)\lambda^{(n)}(dx)$$

for all continuous functions $f : W_1^{(2)} \to \mathbb{R}$.

6 The Radon-Nikodym Theorem and the Transformation Theorem

In the last chapter we discussed integration with respect to an image measure, compare for example with the statement of Corollary 5.32, and we looked at measures being invariant under transformations and the behaviour of corresponding integrals. One purpose of this chapter is to study the behaviour of integrals over Borel sets $G \subset \mathbb{R}^n$ with respect to the measure $\lambda^{(n)}$ under suitable transformations. We start with

Example 6.1. For $T \in GL(n; \mathbb{R})$ we know from Theorem 3.11 that

$$T(\lambda^{(n)}) = \frac{1}{|\det T|} \lambda^{(n)}. \tag{6.1}$$

Now let $f : \mathbb{R}^n \to \mathbb{R}$ be an integrable function. According to Corollary 5.32 we have

$$\int f \frac{1}{|\det T|} \, d\lambda^{(n)} = \int f \, dT(\lambda^{(n)}) = \int f \circ T \, d\lambda^{(n)}$$

or

$$\int f \, d\lambda^{(n)} = \int (f \circ T) |\det T| \, d\lambda^{(n)} = \int f \circ T |\det J_T| \, d\lambda^{(n)}, \tag{6.2}$$

where J_T denotes the Jacobi matrix (the differential) of T which is of course T itself.

Our aim is to extend (6.2) to a diffeomorphism $\varphi : G \to H$, $G, H \subset \mathbb{R}^n$ open, with T replaced by φ and $|\det J_T|$ by $|\det J_\varphi|$. We start with the following observation:

For a homeomorphism $\varphi : G \to H$ between two open sets $G, H \subset \mathbb{R}^n$ the mapping $\Phi_\varphi : C(H) \to C(G)$, $\Phi_\varphi(h) = h \circ \varphi$, is an algebra homeomorphism, i.e. Φ_φ is linear and in addition we have $\Phi_\varphi(h_1 h_2) = \Phi_\varphi(h_1) \Phi_\varphi(h_2)$. The compatibility with the algebraic structure is obvious, see also Problem 1. Moreover we have with $\Phi_{\varphi^{-1}} : C(G) \to C(H)$, $\Phi_{\varphi^{-1}}(g) = g \circ \varphi^{-1}$, that $\Phi_\varphi \circ \Phi_{\varphi^{-1}} = \mathrm{id}_{C(G)}$ and $\Phi_{\varphi^{-1}} \circ \Phi_\varphi = \mathrm{id}_{C(H)}$ where as usual id_A denotes the identity on A. Note that in general we have for open sets $F, G, H \subset \mathbb{R}^n$ and homeomorphisms $\varphi : F \to G$ and $\psi : G \to H$ the relation $\Phi_{\psi \circ \varphi} = \Phi_\varphi \circ \Phi_\psi$.

Thus for integrable $g \in C(\mathbb{R}^n)$, $g = f \circ T$ with $f \in C(\mathbb{R}^n)$ integrable and $T \in GL(n; \mathbb{R}^n)$, we have

$$\int f \, d\lambda^{(n)} = \int g |\det T| \, d\lambda^{(n)} \tag{6.3}$$

and we want to consider the right hand side of (6.3) as an integral of g with respect to the measure $|\det T| \lambda^{(n)}$. More generally for a measurable non-negative function $\rho : \mathbb{R}^n \to \mathbb{R}$ we may investigate $\mu := \rho \lambda^{(n)}$ on $\mathbb{R}^n$, or even better on $\mathcal{B}^{(n)}$. Indeed we have the following more general result:

Proposition 6.2. *Let $(\Omega, \mathcal{A})$ be a measurable space and $\rho : \Omega \to [0, \infty]$ be a measurable function. For a measure μ we can define a new measure ν on $\mathcal{A}$ by*

$$\nu(A) := \int_A \rho \, d\mu = \int \chi_A \rho \, d\mu. \tag{6.4}$$

Proof. Since $\chi_\emptyset = 0$ it follows that $\nu(\emptyset) = 0$ and since $\rho \geq 0$ by assumption we find $\nu(A) \geq 0$ for all $A \in \mathcal{A}$. Now let $(A_k)_{k \in \mathbb{N}}$, $A_k \in \mathcal{A}$, be a sequence of mutually disjoint sets with union $A = \bigcup_{k \in \mathbb{N}} A_k$. Since $\chi_A \rho = \sum_{k=1}^{\infty} \chi_{A_k} \rho$ Corollary 5.15 yields

$$\nu(A) = \int \chi_A \rho \, d\mu = \int \sum_{k=1}^{\infty} \chi_{A_k} \rho \, d\mu = \sum_{k=1}^{\infty} \int \chi_{A_k} \rho \, d\mu = \sum_{k=1}^{\infty} \nu(A_k).$$

$\square$

Definition 6.3. *In the situation of Proposition 6.2 we call ρ the **density** of ν with respect to μ and we write*

$$\nu = \rho \mu. \tag{6.5}$$

Example 6.4. By $|\det T| \lambda^{(n)}$, $T \in GL(n; \mathbb{R})$, a measure is given on $\mathcal{B}^{(n)}$ having density $|\det T|$ with respect to the Lebesgue-Borel measure $\lambda^{(n)}$.

We know that the Lebesgue-Borel measure $\lambda^{(n)}$ is determined on the set $\mathcal{I}_{r,n} = \mathcal{I}_r^n$, see Chapter 2, and hence on the n-dimensional figures $\mathcal{F}^{(n)}$. Moreover, for $f : \mathbb{R}^n \to [0, \infty]$ measurable, a pre-measure is given on $\mathcal{F}^{(n)}$ defined on $\mathcal{I}_{r,n}$ by

$$\nu(Q) := \int \chi_Q f \, d\lambda^{(n)}. \tag{6.6}$$

This pre-measure has a unique extension to $\mathcal{B}^{(n)}$ implying that inequalities such as $\nu(Q) \le \lambda^{(n)}(Q)$ or $\lambda^{(n)}(Q) \le \nu(Q)$ for all $Q \in \mathcal{I}_{r,n}$ (hence all $A \in \mathcal{F}^{(n)}$) extend to $\mathcal{B}^{(n)}$.

Here are two simple consequences of Definition 6.3.

Corollary 6.5. *If* $\nu = \rho\mu$ *and* $\kappa = \sigma\nu$ *then* $\kappa = (\rho\sigma)\mu$.

Proposition 6.6. *Let* $(\Omega, \mathcal{A})$ *be a measurable space and* $\rho : \Omega \to [0, \infty]$ *be a measurable function. For a measure* μ *on* $\mathcal{A}$ *consider the measure* ν *having density* ρ *with respect to* μ, *i.e.* $\nu = \rho\mu$. *For every measurable function* $f : \Omega \to [0, \infty]$ *we have*

$$\int f \, d\nu = \int f\rho \, d\mu. \tag{6.7}$$

Moreover, for every measurable function $f : \Omega \to \overline{\mathbb{R}}$ *the* ν-*integrability is equivalent to the* μ-*integrability of* $f\rho$ *and in the integrable case* (6.7) *holds.*

Proof. For a simple function $u = \sum_{k=1}^{N} \alpha_k \chi_{A_k}$ we find

$$\int u \, d\nu = \sum_{k=1}^{N} \alpha_k \nu(A_k) = \sum_{k=1}^{N} \alpha_k \int \chi_{A_k} \rho \, d\mu = \int u\rho \, d\mu, \tag{6.8}$$

and for an arbitrary non-negative measurable function $f : \Omega \to [0, \infty]$ we can find an increasing sequence of simple functions $(u_l)_{l \in \mathbb{N}}$ converging to f. Passing for such a sequence in (6.8) to the limit $l \to \infty$ gives the first statement. The second statement is derived from the first one by noting that

$$\int f^\pm \, d\nu = \int f^\pm \rho \, d\mu = \int (f\rho)^\pm \, d\mu.$$

$\square$

We return to our original problem and to (6.1). Taking advantage of the fact that $T \in GL(n; \mathbb{R})$ has an image which is again measurable we deduce from

$$T(\lambda^{(n)})(B) = \lambda^{(n)}(T^{-1}(B)) = \frac{1}{|\det T|} \lambda^{(n)}(B)$$

with $A = T^{-1}(B)$ that

$$\lambda^{(n)}(A) = \frac{1}{|\det T|} \lambda^{(n)}(T(A))$$

or

$$\lambda^{(n)}(T(A)) = |\det T|\lambda^{(n)}(A) \tag{6.9}$$

for all $A \in \mathcal{B}^{(n)}$. We want to find the analogue to (6.9) when T is replaced by a diffeomorphism.

Let $U, V \subset \mathbb{R}^n$ be open sets and $\varphi : U \to V$ a C^1-diffeomorphism. We denote, as usual, the differential of φ by $d\varphi$ which we represent at $x_0 \in U$ by its Jacobi matrix $J_\varphi(x_0)$. We claim

Theorem 6.7. *Let $\varphi : U \to V$ be a C^1-diffeomorphism between two open sets $U, V \subset \mathbb{R}^n$ and $A \subset \overline{A} \subset U$ a Borel set. Assume that*

$$\int_A |\det J_\varphi(x)|\lambda_U^{(n)}(dx) < \infty$$

exists. Then $\varphi(A) \in \mathcal{B}^{(n)}$ and the following holds

$$\lambda_V^{(n)}(\varphi(A)) = \int_A |\det J_\varphi(x)|\lambda_U^{(n)}(dx). \tag{6.10}$$

Note that for $\varphi = T \in GL(n; \mathbb{R})$ we recover (6.9) from (6.10) since in this case $\det J_\varphi(x) = \det T$ for all $x \in \mathbb{R}^n$.

Proof of Theorem 6.7. We will prove this theorem in several steps. First let $\varphi : U \to V$ be a Lipschitz continuous mapping, i.e.

$$\|\varphi(x) - \varphi(y)\|_\infty \le L\|x - y\|_\infty$$

for all $x, y \in U$. Further let $Q \subset \overline{Q} \subset U$ be a cube, $\mathring{Q} \ne \emptyset$, which can be open, half-open or closed. We denote the centre of Q by $c = (c_1, \ldots, c_n)$ and we assume that Q has side length $2d$, $d > 0$. Thus we have

$$\overline{Q} = \underset{k=1}{\overset{n}{\times}}[c_k - d, c_k + d].$$

Since $\overline{Q} = \{y \in \mathbb{R}^n \,|\, \|c - y\|_\infty \le d\}$ we find for the image of $\overline{Q}$ under $\varphi = (\varphi_1, \ldots, \varphi_n)$

$$\varphi(\overline{Q}) \subset \underset{k=1}{\overset{n}{\times}}[\varphi_k(c) - Ld, \varphi_k(c) + Ld],$$

112

and since $\lambda^{(n)}(\mathring{Q}) = \lambda^{(n)}(Q) = \lambda^{(n)}(\overline{Q})$ we obtain

$$\lambda^{(n)}(\varphi(Q)) \leq L^n \lambda^{(n)}(Q). \tag{6.11}$$

With the help of (6.11) we now prove for $Q \in \mathcal{I}_{r,n}$, $\overline{Q} \subset U$,

$$\lambda^{(n)}(\varphi(Q)) \leq \int_Q |\det J_\varphi(x)| \lambda^{(n)}(dx). \tag{6.12}$$

Let $Q \in \mathcal{I}_{r,n}$, $\overline{Q} \subset U$. On the compact set $\overline{Q}$ the mapping J_φ is uniformly continuous and therefore for every $\epsilon > 0$ there exists $\delta > 0$ such that

$$\sup_{\substack{x,x' \in \overline{Q} \\ \|x-x'\|_\infty \leq \delta}} \|J_\varphi(x) - J_\varphi(x')\|_\infty < \frac{\epsilon}{L}, \tag{6.13}$$

where $L := \sup_{x \in \overline{Q}} \|(J_\varphi(x))^{-1}\|_\infty$. Here $\|\cdot\|_\infty$ denotes the maximum norm in $\mathbb{R}^m$, i.e. $\|z\|_\infty = \max\{|z_l| \,|\, 1 \leq l \leq m\}$. Note that for a matrix A we calculate $\|A\|_\infty$ by identifying A with an element in $\mathbb{R}^{n^2}$. We also note that

$$\sup_{x \in \overline{Q}} \|(J_\varphi(x))^{-1}\|_\infty \leq \sup_{y \in \varphi(\overline{Q})} \|(J_{\varphi^{-1}})(y)\|_\infty. \tag{6.14}$$

We now choose a partition of Q, $Q = \bigcup_{l=1}^N Q_l$ into cubes $Q_l \in \mathcal{I}_{r,n}$ such that the side length of each Q_l is less that $\delta > 0$, δ as in (6.13). The continuity of J_φ implies that for every $l = 1, \ldots, N$ there exists $x_l \in \overline{Q}_l$ such that

$$|\det J_\varphi(x_l)| = \inf_{x \in \overline{Q}_l} |\det J_\varphi(x)|.$$

With $T_l := J_\varphi(x_l) \in GL(n; \mathbb{R})$ we now find using the chain rule that

$$J_{T_l^{-1} \circ \varphi}(x) = T_l^{-1} \circ J_\varphi(x) = \mathrm{id}_n + T_l^{-1} \circ (J_\varphi(x) - J_\varphi(x_l)).$$

From (6.13) and (6.14) we deduce for all $l = 1, \ldots, N$

$$\sup_{x \in \overline{Q}_l} \left\| J_{T_l^{-1} \circ \varphi}(x) \right\|_\infty \leq 1 + L\frac{\epsilon}{L} = 1 + \epsilon, \tag{6.15}$$

where we used that $\overline{Q}_l$ has side length less than δ. Estimate (6.15) implies the Lipschitz continuity of the mapping $T_l^{-1} \circ \varphi$ on the convex set $\overline{Q}_l$. Now it follows with (6.9)

$$\begin{aligned}
\lambda^{(n)}(\varphi(Q_l)) &= \lambda^{(n)}\left((T_l \circ T_l^{-1} \circ \varphi)(Q_l)\right) \\
&= |\det T_l| \lambda^{(n)}\left((T_l^{-1} \circ \varphi)(Q_l)\right) \\
&\leq |\det T_l|(1+\epsilon)^n \lambda^{(n)}(Q_l),
\end{aligned}$$

where the last step follows from (6.11). Using the fact that the cubes Q_l, $l = 1, \ldots, N$, form a partition of Q and that by construction for $l = 1, \ldots, N$ we have $|\det T_l| \leq |\det J_\varphi(x)|$ for all $x \in Q_l$, we obtain

$$\lambda^{(n)}(\varphi(Q)) \leq \sum_{l=1}^{N} \lambda^{(n)}(\varphi(Q_l))$$

$$\leq (1 + \epsilon)^n \sum_{l=1}^{N} |\det T_l| \, \lambda^{(n)}(Q_l)$$

$$\leq (1 + \epsilon)^n \sum_{l=1}^{N} \int_{Q_l} |\det J_\varphi(x)| \, \lambda^{(n)}(dx)$$

$$= (1 + \epsilon)^n \int_{Q} |\det J_\varphi(x)| \, \lambda^{(n)}(dx).$$

If we now pass to the limit as ϵ tends to 0 we arrive at (6.12). As mentioned before, we can extend (6.12) to all $A \in \mathcal{B}^{(n)}$, $A \subset U$, i.e.

$$\lambda^{(n)}(\varphi(A)) \leq \int_{A} |\det J_\varphi(x)| \lambda^{(n)}(dx) \qquad (6.16)$$

holds for these Borel sets $A \in \mathcal{B}^{(n)}$. Note that we allowed ourselves some imprecise notation, we should have always written $\lambda_U^{(n)} = \lambda^{(n)}|_U$ and $\lambda_V^{(n)} = \lambda^{(n)}|_V$, i.e. (6.16) must be read as

$$\lambda_V^{(n)}(\varphi(A)) \leq \int_{A} |\det J_\varphi(x)| \lambda_U^{(n)}(dx) \qquad (6.17)$$

and for the following it is helpful to be more strict with the notation. Since $\varphi : U \to V$ is a diffeomorphism we further have

$$\lambda_V^{(n)}(\varphi(A)) = \varphi^{-1}(\lambda_V^{(n)})(A).$$

For $A \in U \cap \mathcal{B}^{(n)}$ we have for some $B \in V \cap \mathcal{B}^{(n)}$ that $A = \varphi^{-1}(B)$ and (6.17) yields

$$\int_{V} \chi_B(y) \lambda_V^{(n)}(dy) = \lambda_V^{(n)}(B)$$

$$\leq \int_{U} \chi_{\varphi^{-1}(B)}(x) |\det J_\varphi(x)| \lambda_U^{(n)}(dx)$$

$$= \int_{U} \chi_B(\varphi(x)) |\det J_\varphi(x)| \lambda_U^{(n)}(dx),$$

i.e. we have

$$\int_V \chi_B(y)\lambda_V^{(n)}(dy) \leq \int_U \chi_B(\varphi(x))|\det J_\varphi(x)|\lambda_U^{(n)}(dx) \qquad (6.18)$$

for all $B \in V \cap \mathcal{B}^{(n)}$ which extends to all non-negative, measurable functions $u : V \to [0, \infty]$:

$$\int_V u(y)\lambda_V^{(n)}(dy) \leq \int_U u(\varphi(x))|\det J_\varphi(x)|\lambda_U^{(n)}(dx). \qquad (6.19)$$

With (6.19) at our disposal, we can switch from $\varphi : U \to V$ to $\varphi^{-1} : V \to U$ and to the function $v : U \to \mathbb{R}$ defined for some $A = \varphi^{-1}(B)$ by $v(x) := (\chi_{\varphi(A)} \circ \varphi)(x)|\det J_\varphi(x)|$ replacing u. We obtain from (6.19)

$$\int_U \left(\chi_{\varphi(A)} \circ \varphi\right)(x)|\det J_\varphi(x)|\lambda_U^{(n)}(dx)$$

$$\leq \int_V \left((\chi_{\varphi(A)} \circ \varphi)|\det J_\varphi|\right)(\varphi^{-1}(y))|\det J_{\varphi^{-1}}(y)|\lambda_V^{(n)}(dy)$$

$$= \int_V \chi_{\varphi(A)}(y)|\det(J_\varphi(\varphi^{-1}(y)) \circ J_{\varphi^{-1}}(y))|\lambda_V^{(n)}(dy)$$

$$= \int_V \chi_{\varphi(A)}(y)\lambda_V^{(n)}(dy) = \lambda_V^{(n)}(\varphi(A)).$$

With (6.16) we now get

$$\lambda_V^{(n)}(\varphi(A)) = \int_A |\det J_\varphi(x)|\lambda_U^{(n)}(dx),$$

i.e. (6.10) and the theorem is proved. □

Since (6.10) is an equality of measures we can extend it (as we have done before with inequality (6.16)) to an equality for integrals. This is the famous and very important **transformation theorem for integrals** which was first noted by C. G. J. Jacobi.

Theorem 6.8. *Let $U, V \subset \mathbb{R}^n$ be open sets and $\varphi : U \to V$ be a C^1-diffeomorphism. Further let $f : V \to \overline{\mathbb{R}}$ be a numerical function. This function is integrable with respect to $\lambda_V^{(n)}$ if and only if the function $(f \circ \varphi) \cdot$*

$|\det J_\varphi| : U \to \overline{\mathbb{R}}$ *is integrable with respect to* $\lambda_U^{(n)}$ *and in the case where these functions are integrable we find*

$$\int_V f(y)\lambda_V^{(n)}(\mathrm{d}y) = \int_U f(\varphi(x))|\det J_\varphi(x)|\lambda_U^{(n)}(\mathrm{d}x), \qquad (6.20)$$

or compare with (II.21.5), we have

$$\int_{\varphi(U)} f(y)\lambda^{(n)}(\mathrm{d}y) = \int_U (f \circ \varphi)(x)|\det J_\varphi(x)|\lambda^{(n)}(\mathrm{d}x). \qquad (6.21)$$

Remark 6.9. A. All the proofs of Theorem 6.8 we know follow essentially the same idea, unfortunately some authors of textbooks leave serious gaps. In our presentation we have combined the discussions of R. Schilling [75] and [76].

B. The reader of Volume II might remember that for the Riemann integral (in $\mathbb{R}^n$) we only discussed the transformation theorem, Theorem II.21.8, but we postponed the proof to this part of our Course. At this stage we still need to identify in certain situations the Riemann integral with the Lebesgue integral in order to claim that Theorem 6.8 implies Theorem II.21.8 in these cases. Once this is done we can, in admissible situations, transfer the calculations made in the Riemannian context to Lebesgue integrals.

C. We have already seen that sets of measure zero do not contribute to integrals. In the next chapter we will use this observation in a more precise sense to give an extension of Theorem 6.8.

So far we have used the idea of a measure with density only in one case, namely to prove Theorem 6.8. It turns out that this notion has many more far reaching consequences and applications.

Let $(\Omega, \mathcal{A}, \mu)$ be a measure space and $\rho : \Omega \to [0, \infty]$ a measurable function with finite, non-zero integral, i.e. $0 < \int \rho\, \mathrm{d}\mu < \infty$. With $\|\rho\|_{L^1} := \int \rho\, \mathrm{d}\mu$ it follows that

$$\nu := \frac{1}{\|\rho\|_{L^1}}\rho\mu$$

is a probability measure and $(\Omega, \mathcal{A}, \nu)$ is a probability space. The most interesting case is when $(\Omega, \mathcal{A}, \mu) = (\mathbb{R}^n, \mathcal{B}^{(n)}, \lambda^{(n)})$. Suppose that $(\tilde{\Omega}, \tilde{\mathcal{A}}, P)$ is a probability space and $X : \tilde{\Omega} \to \mathbb{R}^n$ is a random variable with distribution P_X which is by definition a probability measure on $\mathbb{R}^n$ (or better $\mathcal{B}^{(n)}$). If P_X

has a density $p : \mathbb{R}^n \to [0, \infty]$ with respect to $\lambda^{(n)}$, i.e. $P_X = p\lambda^{(n)}$, we call p the **probability density** or just the **density** of X with respect to $\lambda^{(n)}$. When p is the probability density of X we find for the expectation of X

$$E(X) = \int_{\tilde{\Omega}} XP(\mathrm{d}\omega) = \int_{\mathbb{R}^n} xp(x)\lambda^{(n)}(\mathrm{d}x). \qquad (6.22)$$

Hence the calculation of $E(X)$ is reduced to the evaluation of an integral in $\mathbb{R}^n$ with respect to the Lebesgue-Borel measure (and we will see that often we can do this by using the corresponding Riemann integral). Interesting probability densities are for example for $n = 1$

$$g_{\alpha,\sigma}(x) = (2\pi\sigma^2)^{-\frac{1}{2}} e^{-\frac{(x-\alpha)^2}{2\sigma^2}} \qquad (6.23)$$

which is the one-dimensional **Gauss** or **normal distribution** with mean α and standard deviation $\sigma > 0$, or the one-dimensional **Cauchy distribution**

$$c_\alpha(x) = \frac{\alpha}{\pi} \frac{1}{\alpha^2 + x^2} \qquad (6.24)$$

and we will see in Problem 6 of Chapter 7 that a random variable which is Cauchy distributed does not have a well defined expectation since the integral $\int_{\mathbb{R}} \frac{x}{\alpha^2 + x^2} \lambda^{(1)}(\mathrm{d}x)$ does not exist.

However, recalling the Bernoulli or the Poisson distribution, compare with Example 1.25 or Example 5.19.A and B, we note that not all real-valued random variables admit a density with respect to $\lambda^{(1)}$. The natural question which arises is when for two measures μ and ν on a measurable space $(\Omega, \mathcal{A})$ one measure has a density with respect to the other. A first observation yields the following necessary condition:

Lemma 6.10. *Suppose that $\nu = f\mu$ on the measurable space $(\Omega, \mathcal{A})$ with a non-negative, numerical measurable density $f : \Omega \to [0, \infty]$. If $A \in \mathcal{A}$ is a set of measure zero with respect to μ, then A is also a set of measure zero with respect to ν, i.e. $\mu(A) = 0$ implies $\nu(A) = 0$.*

Proof. Since $\nu(A) = \int_A f \, \mathrm{d}\mu$ the result follows from Lemma 5.29. $\qquad \square$

Definition 6.11. *We call a measure ν on the measurable space $(\Omega, \mathcal{A})$ **absolutely continuous** with respect to the measure μ on $(\Omega, \mathcal{A})$ if for every set $A \in \mathcal{A}$ with $\mu(A) = 0$ it follows that $\nu(A) = 0$. If ν is absolutely continuous with respect to μ we write*

$$\nu \ll \mu. \qquad (6.25)$$

For a finite measure ν on $(\Omega, \mathcal{A})$ we can prove

Theorem 6.12. *A finite measure ν on $(\Omega, \mathcal{A})$ is absolutely continuous with respect to a measure μ on $(\Omega, \mathcal{A})$ if and only if for every $\epsilon > 0$ there exists $\delta > 0$ such that for every $A \in \mathcal{A}$ with $\mu(A) \leq \delta$ it follows that $\nu(A) \leq \epsilon$.*

Proof. If $A \in \mathcal{A}$ and $\mu(A) = 0$ the assumptions made above imply that $\nu(A) \leq \epsilon$ for every $\epsilon > 0$, hence $\nu(A) = 0$, and ν is absolutely continuous with respect to μ. Now assume that $\nu \ll \mu$. Suppose that there exists $\epsilon > 0$ such that for all $\delta > 0$ there exists $A_\delta \in \mathcal{A}$ such that $\mu(A_\delta) \leq \delta$ and $\nu(A_\delta) \geq \epsilon$. We may choose for $\delta = 2^{-k-1}$ a set $A_k \in \mathcal{A}$ such that

$$\mu(A_k) \leq 2^{-k-1} \text{ and } \nu(A_k) \geq \epsilon.$$

For $B_k := \bigcup_{l=k}^{\infty} A_l \in \mathcal{A}$ we deduce

$$\mu(B_k) \leq \sum_{l=k}^{\infty} 2^{-l-1} = 2^{-k} \text{ and } \nu(B_k) \geq \nu(A_k) \geq \epsilon.$$

Since the sequence $(B_k)_{k \in \mathbb{N}}$ is decreasing, i.e. $B_{k+1} \subset B_k$, and ν is a finite measure it follows by Theorem 2.10.B that

$$\nu\left(\bigcap_{k=1}^{\infty} B_k\right) = \lim_{k \to \infty} \nu(B_k) \geq \epsilon.$$

On the other hand we know for every k_0 that $\mu\left(\bigcap_{k=1}^{\infty} B_k\right) \leq \mu(B_{k_0}) \leq 2^{-k_0}$ which yields that $\mu\left(\bigcap_{k=1}^{\infty} B_k\right) = 0$ and this is a contradiction. $\square$

We will also need

Lemma 6.13. *Let $(\Omega, \mathcal{A})$ be a measurable space and let μ and ν be two finite measures on $\mathcal{A}$. For $\tau : \mathcal{A} \to \mathbb{R}$, $\tau := \mu - \nu$, exists $\Omega_0 \in \mathcal{A}$ such that*

$$\tau(\Omega_0) \geq \tau(\Omega) \tag{6.26}$$

and

$$\tau(A) \geq 0 \text{ for all } A \in \Omega_0 \cap \mathcal{A}. \tag{6.27}$$

Proof. (See [11]) We start by proving that for every $n \in \mathbb{N}$ there exists $\Omega_n \in \mathcal{A}$ such that

$$\tau(\Omega_n) \geq \tau(\Omega) \tag{6.28}$$

and

$$\tau(A) > -\frac{1}{n} \text{ for all } A \in \Omega_n \cap \mathcal{A}. \tag{6.29}$$

If $\tau(\Omega) = 0$ we choose $\Omega_n = \emptyset$, so we assume $\tau(\Omega) > 0$. If $\tau(\Omega) > 0$ and $\tau(A) > -\frac{1}{n}$ for all $A \in \mathcal{A}$, we may choose $\Omega_n := \Omega$. Thus we assume the existence of $A_1 \in \mathcal{A}$ such that $\tau(A_1) \leq -\frac{1}{n}$. It follows that

$$\tau(A_1^{\mathsf{C}}) = \tau(\Omega) - \tau(A_1) \geq \tau(\Omega) + \frac{1}{n} > \tau(\Omega) > 0.$$

If $\tau(A) > -\frac{1}{n}$ for all $A \in A_1^{\mathsf{C}} \cap \mathcal{A}$ we choose $\Omega_n := A_1^{\mathsf{C}}$. Otherwise we can find $A_2 \in A_1^{\mathsf{C}} \cap \mathcal{A}$ such that $\tau(A_2) \leq -\frac{1}{n}$, and since $A_1 \cap A_2 = \emptyset$ we have

$$\tau\left((A_1 \cup A_2)^{\mathsf{C}}\right) \geq \tau(\Omega) + \frac{2}{n} > \tau(\Omega) > 0.$$

We can iterate this process. If it stops after N steps we take $\Omega_N := \left(\bigcup_{j=1}^{N} A_j\right)^{\mathsf{C}}$ where the sets A_j are mutually disjoint. Suppose that this process does not stop. Then we can find a sequence $(A_k)_{k \in \mathbb{N}}$ of mutually disjoint sets $A_k \in \mathcal{A}$ such that for all $N \in \mathbb{N}$

$$\tau\left(\Omega \setminus \bigcup_{k=1}^{N} A_k\right) > \tau(\Omega) > 0 \text{ and } \tau(A_n) \leq -\frac{1}{n}.$$

The measures μ and ν are finite additive and hence

$$\tau\left(\bigcup_{k=1}^{N} A_k\right) = \sum_{k=1}^{N} \tau(A_k) \leq -\frac{N}{n},$$

implying the divergence of $\sum_{k=1}^{\infty} \tau(A_k)$. However the σ-additivity of μ and ν yields

$$\sum_{k=1}^{\infty} \tau(A_k) = \mu\left(\bigcup_{k=1}^{\infty} A_k\right) - \nu\left(\bigcup_{k=1}^{\infty} A_k\right)$$

which is finite since μ and ν are finite measures. Hence, given $n \in \mathbb{N}$ there exists mutually disjoint sets $A_1, \ldots, A_N \in \mathcal{A}$ such that with $\Omega_n := \left(\bigcup_{k=1}^{N} A_k\right)^{\mathsf{C}}$

condition (6.28) and (6.29) hold. We may apply (6.28), (6.29) to Ω_n, if necessary, we may assume $\Omega_{n+1} \subset \Omega_n$. For $\Omega_0 := \bigcup_{n \in \mathbb{N}} \Omega_n$ we obtain from (6.28) now (6.26) and since in addition we have $\tau(A) \geq -\frac{1}{n}$ for all $A \in \Omega_0 \cap \mathcal{A}$ implying (6.27). $\qquad\square$

With the help of Lemma 6.13 we can give a first answer to the question when ν has a density with respect to μ, the **Radon-Nikodym theorem** for finite measures.

Theorem 6.14. *For two finite measures μ and ν on the measurable space $(\Omega, \mathcal{A})$ the absolute continuity of ν with respect to μ is equivalent to the existence of a density f for ν with respect to μ, i.e. $\nu \ll \mu$ if and only if $\nu = f\mu$.*

Proof. We already know that the existence of a density always, not only for finite measures, implies the absolute continuity. So we need to prove that the absolute continuity of ν implies the existence of a density. Consider the set

$$G := \left\{ g : \Omega \to [0, \infty] \,\Big|\, g \text{ is measurable}, \int_A g \, d\mu \leq \nu(A) \text{ for all } A \in \mathcal{A} \right\}.$$

This set is not empty since $g(\omega) = 0$ for all $\omega \in \Omega$ belongs to G. For $g, h \in G$ we find further

$$\int_A (g \vee h) \, d\mu = \int_{A \cap \{g \geq h\}} g \, d\mu + \int_{A \cap \{g \geq h\}^C} h \, d\mu$$
$$\leq \nu(A \cap \{g \geq h\}) + \nu\left(A \cap \{g \geq h\}^C\right) = \nu(A),$$

i.e. $g, h \in G$ implies $g \vee h \in G$. Since ν is a finite measure it follows that

$$\gamma := \sup_{g \in G} \int g \, d\mu \leq \nu(\Omega) < \infty.$$

Let $(\tilde{g}_k)_{k \in \mathbb{N}}$, $g_k \in G$, be a sequence such that $\lim_{k \to \infty} \int \tilde{g}_k \, d\mu = \gamma$. The sequence $g_N := \tilde{g}_1 \vee \cdots \vee \tilde{g}_N$ is increasing, $\tilde{g}_N \leq g_N$ and $\lim_{k \to \infty} \int g_k \, d\mu = \gamma$. The monotone convergence theorem yields that $f := \sup_{k \in \mathbb{N}} g_k$ belongs to G and $\int f \, d\mu = \gamma$. Our claim is that $\nu = f\mu$. By construction we have $f\mu \leq \nu$, i.e. $\int_A f \, d\mu \leq \nu(A)$ for all $A \in \mathcal{A}$, and therefore the measure

$$\tau := \nu - f\mu$$

is absolutely continuous with respect to μ. If $\tau(A) = 0$ for all A we are done. Suppose that $\tau(\Omega) > 0$ which also implies by absolute continuity that $\mu(\Omega) > 0$. For

$$\beta := \frac{1}{2}\frac{\tau(\Omega)}{\mu(\Omega)} > 0$$

we have $\tau(\Omega) = 2\beta\mu(\Omega) > \beta\mu(\Omega)$ and applying Lemma 6.13 to $\beta\mu$ instead of ν and τ as defined above we deduce the existence of $\Omega_0 \in \mathcal{A}$ such that

$$\tau(\Omega_0) - \beta\mu(\Omega_0) \geq \tau(\Omega) - \beta\mu(\Omega) > 0$$

and $\tau(A) \geq \beta\mu(A)$ for all $A \in \Omega_0 \cap \mathcal{A}$. The function $f_0 := f + \beta\chi_{\Omega_0}$ is measurable and non-negative implying that

$$\int_A f_0 \, d\mu = \int_A f \, d\mu + \beta\mu(A \cap \Omega_0) \leq \int_A f \, d\mu + \tau(A \cap \Omega_0)$$

$$\leq \int_A f \, d\mu + \tau(A) = \nu(A)$$

for all $A \in \mathcal{A}$. This estimate yields that $f_0 \in G$ and that

$$\int f_0 \, d\mu = \int f \, d\mu + \beta\mu(\Omega_0) = \gamma + \beta\mu(\Omega_0) > \gamma$$

since $\tau(\Omega_0) > \beta\mu(\Omega_0)$ and $\mu(\Omega_0) > 0$ due to the absolute continuity of τ with respect to μ. Hence we have constructed a contradiction to the assumption $\tau(\Omega) > 0$ implying the theorem. $\qquad\square$

The following example shows that we cannot expect the Radon-Nikodym theorem to hold for general, not necessarily finite measures.

Example 6.15. Let Ω be a non-denumerable set and $\mathcal{A} := \{A \in P(\Omega) \,|\, A$ or $A^{\complement}$ is denumerable$\}$ the σ-field from Example 1.3.A. Further let ν be the measure on $\mathcal{A}$ given by $\nu(A) = 0$ for A denumerable and $\nu(A) = +\infty$ for A non-denumerable. In addition we consider the measure μ on $\mathcal{A}$ defined for a finite set A by $\mu(A) = \#(A)$, i.e. for a finite set μ gives the number of its elements, and $\mu(A) = +\infty$ for any infinite set. Since $\mu(A) = 0$ if and only if $A = \emptyset$ it follows that ν is absolutely continuous with respect to μ. However, ν cannot have a density with respect to μ. Indeed, if $\nu = f\mu$ with an $\mathcal{A}$-measurable function $f : \Omega \to [0, \infty]$, then for every $\omega \in \Omega$ we find

$$0 = \nu(\{\omega\}) = \int_{\{\omega\}} f \, d\mu = f(\omega)\mu(\{\omega\}) = f(\omega)$$

implying $f = 0$, i.e. $\nu(A) = 0$ for all $A \in \mathcal{A}$ which is a contradiction.

As it turns out we can rescue the Radon-Nikodym theorem for σ-finite measures. Recall that by Definition 2.12.B a measure μ is σ-finite on $(\Omega, \mathcal{A})$ if there exists a sequence $(A_k)_{k \in \mathbb{N}}$ of measurable sets $A_k \in \mathcal{A}$ with finite measure $\mu(A_k) < \infty$ such that $\Omega = \bigcup_{k=1}^{\infty} A_k$. The following lemma links σ-finite measures to the existence of certain integrable functions.

Lemma 6.16. *Let $(\Omega, \mathcal{A}, \mu)$ be a measure space. The measure μ is σ-finite if there exists a μ-integrable function $h : \Omega \to \mathbb{R}$ such that $0 < h(\omega) < \infty$ for all $\omega \in \Omega$. Conversely, if on a measure space $(\Omega, \mathcal{A}, \mu)$ such a function h exists, then μ is σ-finite.*

Proof. Let $(A_k)_{k \in \mathbb{N}}$ be a sequence of measurable sets $A_k \in \mathcal{A}$ such that $A_k \subset A_{k+1}$, $\Omega = \bigcup_{k \in \mathbb{N}} A_k$ and $\mu(A_k) < \infty$. Note that by Problem 5 to Chapter 2 such a sequence exists. We now choose $c_k \in (0, 2^{-k})$ such that $c_k \mu(A_k) < 2^{-k}$, $k \in \mathbb{N}$. It follows that $h := \sum_{k=1}^{\infty} c_k \chi_{A_k}$ is a strictly positive measurable function with $h(\omega) \leq 1$ for all $\omega \in \Omega$ and

$$\int h \, d\mu \leq \sum_{k=1}^{\infty} c_k \int \chi_{A_k} \, d\mu = \sum_{k=1}^{\infty} c_k \mu(A_k) \leq 1.$$

Conversely, suppose that on $(\Omega, \mathcal{A}, \mu)$ we can find a measurable function h such that $0 < h(\omega) < \infty$ for all $\omega \in \Omega$ and $\int h \, d\mu < \infty$. With $A_k := \{h \geq \frac{1}{k}\} \in \mathcal{A}$ we get $\chi_{A_k} \leq kh$ and consequently $\mu(A_k) \leq k \int h \, d\mu < \infty$ and $\bigcup_{k \in \mathbb{N}} A_k = \Omega$. $\square$

Theorem 6.17 (Radon-Nikodym). *Let μ be a σ-finite measure on $(\Omega, \mathcal{A})$ and ν a measure. The existence of a density f for ν with respect to μ is equivalent to the absolute continuity of ν with respect to μ.*

Proof. We first consider the case where μ is finite, i.e. $\mu(\Omega) < \infty$, but ν is not finite, i.e. $\nu(\Omega) = +\infty$. We will prove that we can decompose Ω into a sequence $(A_k)_{k \in \mathbb{N} \cup \{\infty\}}$ of measurable sets $A_k \in \mathcal{A}$ such that $\nu(A_k) < \infty$ for $k \in \mathbb{N}$ and if $A \in A_\infty \cap \mathcal{A}$ then $\nu(A) < \infty$ implies $\nu(A) = 0$. For this let

$$\mathcal{A}(\nu) := \{A \in \mathcal{A} \,|\, \nu(A) < \infty\} \quad \text{and} \quad a := \sup \{\mu(A) \,|\, A \in \mathcal{A}(\nu)\}.$$

Since μ is finite $a < \infty$. We choose now an increasing sequence $\tilde{A}_k \in \mathcal{A}$ such that $a = \sup_{k \in \mathbb{N}} \mu(\tilde{A}_k)$. For $A_0 := \bigcup_{k \in \mathbb{N}} \tilde{A}_k$ it follows that $\mu(A_0) = a$ and on

$A_0 \cap \mathcal{A}$ the measure ν is σ-finite. Hence, see Problem 5 of Chapter 2, we can represent A_0 as disjoint union $A_0 = \bigcup_{k=1}^{\infty} A_k$, $A_k \in \mathcal{A}$, and $\nu(A_k) < \infty$. We set $A_\infty := \Omega \backslash A_0$ and claim that for $A \in A_\infty \cap \mathcal{A}$ and $\nu(A) < \infty$ it follows that $\mu(A) = 0$. The following holds

$$a = \sup_{k \in \mathbb{N}} \mu(\tilde{A}_k) = \sup\left\{ \mu(\tilde{A}) \mid \tilde{A} \in \mathcal{A}(\nu) \right\}$$

$$\geq \sup_{k \in \mathbb{N}} \mu(A \cup \tilde{A}_k)$$

since $A \cup \tilde{A}_k \in \mathcal{A}(\nu)$. Thus, since $A \cap \tilde{A}_k = \emptyset$ we have

$$a \geq \sup_{k \in \mathbb{N}} \mu(A \cup \tilde{A}_k) = \sup(\mu(A) + \mu(\tilde{A}_k))$$

$$= \mu(A) + a,$$

and since $0 \leq a < \infty$ it follows $\mu(A) = 0$. On $A_k \cap \mathcal{A}$ the measure ν is finite too and according to Theorem 6.14 we can find a measurable function $f_k : A_k \to [0, \infty]$ such that $\nu|_{A_k} = f_k \mu|_{A_k}$. We define $f : \Omega \to [0, \infty]$ by

$$f(\omega) := \begin{cases} f_k(\omega), & \omega \in A_k, \ k \in \mathbb{N} \\ +\infty, & \omega \in A_\infty. \end{cases}$$

Clearly, f is measurable and $\nu = f\mu$, since for $B \in \mathcal{A}$ we find the decomposition of $B = \bigcup_{k \in \mathbb{N}} (A_k \cap B) \cup (A_\infty \cap B)$ into mutually disjoint sets.

Finally we drop the assumption that μ is finite but assume that μ is σ-finite. By Lemma 6.16 we can find a strictly positive function $h : \Omega \to \mathbb{R}$ with finite μ-integral, i.e. $0 < h < \infty$ and $\int h \, d\mu < \infty$. Thus the measure $h\mu$ is on $\mathcal{A}$ a finite measure and hence ν admits a density f with respect to $h\mu$, hence it follows that

$$\nu = f(h\mu) = (fh)\mu.$$

The converse statement, i.e. that the existence of a density implies the absolute continuity has already been discussed in the proof of Theorem 6.14.

□

Remark 6.18. Our proof is one of the standard proofs and we combined arguments from [11] and [13]. For a proof using martingale theory we refer to [75] or [56]. The latter proof is related to a "Hilbert space proof", see [21], which is due to J. v. Neumann.

A natural question is to which extent a density in $\nu = f\mu$ is uniquely determined. From $\nu = f_1\mu$ and $\nu = f_2\mu$ we can deduce

$$\int_A (f_2 - f_1)\,\mathrm{d}\mu = 0 \text{ for all } A \in \mathcal{A},$$

which however in general does not imply $f_1 = f_2$. If A is a set of measure zero with respect to μ, i.e. $\mu(A) = 0$, then $\int_A g\,\mathrm{d}\mu = 0$ for all measurable numerical functions $g : \Omega \to [0,\infty]$. How to handle properties on sets of measure zero will be studied in the next chapter.

Problems

Although the transformation theorem is one of the central topics of this chapter we do not include problems related to the application of the transformation theorem. There are two reasons for this. Firstly, in Chapter II.21 (and further chapters in Volume II) we have already discussed many applications. Secondly, for interesting applications we need more tools such as Fubini's theorem, Theorem 9.17, or we need criteria to identify the Lebesgue integral with the Riemann integral. Later on in this Course we will encounter again and again the transformation theorem and its applications.

1.   a) Let X and Y be two topological spaces and $\varphi : X \to Y$ a homeomorphism. Denote by $C(X)$ the space of all continuous real-valued functions from X to $\mathbb{R}$. Prove that $C(X)$ and $C(Y)$ are isomorphic as algebras.

   b) Let G and H be open sets in $\mathbb{R}^n$ and $\psi : G \to H$ a diffeomorphism of class C^k, $k \in \mathbb{N}_0$. Consider the mapping $\Psi : C^k(H) \to C^k(G)$, $\Psi(u) = u \circ \psi$, and prove that Ψ is a vector space isomorphism.

2. Prove that if $g : \mathbb{R}^n \to \mathbb{R}^n$ is Lipschitz continuous with respect to the norm $\|\cdot\|_p, 1 \le p \le \infty$, i.e. $\|g(x) - g(y)\|_p \le \gamma_{p,p}\|x - y\|_p$ for all $x, y \in \mathbb{R}^n$, then g is also Lipschitz continuous from $(\mathbb{R}^n, \|\cdot\|_q)$ to $(\mathbb{R}^n, \|\cdot\|_r)$, $1 \le q, r \le \infty$, i.e. we have $\|g(x) - g(y)\|_r \le \gamma_{q,r}\|x - y\|_q$ for all $x, y \in \mathbb{R}^n$.

3.   a) Let $G \subset \mathbb{R}^n$ be a measurable set with $\lambda^{(n)}(G) > 0$ and consider the measure $\nu := \chi_G \lambda^{(n)}$. Find all Borel measurable functions $f : \mathbb{R}^n \to \mathbb{R}$ which are integrable with respect to ν.

b) For $g_s(x) := (1+|x|^2)^{-\frac{s}{2}}$, $s \in \mathbb{R}$, find a power growth condition for a continuous function $u : \mathbb{R} \to \mathbb{R}$ which implies that u is integrable with respect to $\nu = g_s \lambda^{(n)}$, i.e. find $r \in \mathbb{R}$ such that $|u(x)| \leq c_u(1 + |x|^2)^{\frac{r}{2}}$ implies that u integrable with respect to ν.

4.   a) Let $\mu_j, j = 1, \ldots, N$ be measures on the measurable space $(\Omega, \mathcal{A})$. Find a measure μ on $(\Omega, \mathcal{A})$ such that $\mu_j \ll \mu$ for all $j = 1, \ldots, N$.

b) Let μ, ν and π be measures on $(\Omega, \mathcal{A})$. Suppose that $\pi \ll \nu$ and $\nu \ll \mu$. Show that π is absolutely continuous with respect to μ.

5. Let $g, h : \mathbb{R}^n \to [0, \infty)$ be continuous functions such that for all $x \in \mathbb{R}^n$ we have $0 < c_0 \leq \frac{g(x)}{h(x)} \leq c_1$. Consider two measures $\nu_1 := g\lambda^{(n)}$ and $\nu_2 := h\lambda^{(n)}$. Prove that $\nu_2 \ll \nu_1$ and $\nu_1 \ll \nu_2$.

6. Let $\nu_k = g_k\mu, k \in \mathbb{N}$, and $\nu = g\mu$ be absolutely continuous measures with respect to μ where μ is a measure on $(\Omega, \mathcal{A})$. Suppose that $\mu(\mathcal{A}) < \infty$ and $\lim_{k \to \infty} \|g_k - g\|_\infty = 0$. Prove for every $A \in \mathcal{A}$ that $\lim_{k \to \infty} \nu_k(A) = \nu(A)$.

7.   a) Is the Dirac measure $\epsilon_{x_0}, x_0 \in \mathbb{R}$, absolutely continuous with respect to the Borel-Lebesgue measure $\lambda^{(1)}$? Is $\lambda^{(1)}$ absolutely continuous with respect to ϵ_{x_0}?

b) Let $\mu = \sum_{k \in \mathbb{N}} a_k \epsilon_k$ and $\nu = \sum_{k \in \mathbb{N}} b_k \epsilon_k$, $a_k, b_k \leq 0$, be two measures on $\mathcal{B}^{(1)}$. Give conditions on the sequences $(a_k)_{k \in \mathbb{N}}$ and $(b_k)_{k \in \mathbb{N}}$ which will imply that ν is absolutely continuous with respect to μ.

8. Let X be a real-valued random variable on the probability space $(\Omega, \mathcal{A}, P)$ which is Cauchy distributed. Prove that $E(X)$ is not defined.

7 Almost Everywhere Statements, Convergence Theorems

Let $(\Omega, \mathcal{A}, \mu)$ be a measure space and $(u_k)_{k \in \mathbb{N}}$, $u_k : \Omega \to \mathbb{R}$, be a sequence of measurable functions. We want to discuss several different types of convergence of the sequence $(u_k)_{k \in \mathbb{N}}$ to a function $u : \Omega \to \mathbb{R}$. As we will see later in this chapter we can often replace $\mathbb{R}$ by $\mathbb{C}$, $\mathbb{R}^n$ or even a normed vector space $(V, \|.\|)$, and sometimes we may replace $\mathbb{R}$ by $\overline{\mathbb{R}}$. Independent of $\mathcal{A}$ and μ we have of course the notion of **pointwise convergence**: a sequence of functions $u_k : \Omega \to \mathbb{R}$ converges pointwisely to $u : \Omega \to \mathbb{R}$ if for every $\omega \in \Omega$ and every $\epsilon > 0$ there exists $N = N(\epsilon, \omega) \in \mathbb{N}$ such that $k \geq N(\epsilon, \omega)$ implies $|u_k(\omega) - u(\omega)| < \epsilon$. In this definition we can obviously replace $(\mathbb{R}, |.|)$ by a normed $\mathbb{R}$-vector space $(V, \|.\|)$, i.e. $u_k : \Omega \to V$ converges pointwisely to $u : \Omega \to V$ if for every $\omega \in \Omega$ and every $\epsilon > 0$ there exists $N = N(\epsilon, \omega)$ such that $k \geq N(\epsilon, \omega)$ implies $\|u_k(\omega) - u(\omega)\| < \epsilon$.

We recall that Theorem 4.12 states that the pointwise limit of measurable functions $u_k : \Omega \to \mathbb{R}$ is measurable. The following example shows that interchanging pointwise limits with integrals is in general a problem.

Example 7.1. On $\left((0, 1), \mathcal{B}^{(1)}((0, 1)), \lambda^{(1)}_{|(0,1)} \right)$ we consider the sequence $u_k :$ $(0, 1) \to \mathbb{R}$, $u_k(x) := k\chi_{(1 - \frac{1}{k}, 1)}(x)$. Each function u_k is a simple function since $(1 - \frac{1}{k}, 1)$ and its complement in $(0, 1)$ belongs to $\mathcal{B}^{(1)}((0, 1))$ $(= (0, 1) \cap \mathcal{B}^{(1)})$. The integral of u_k is

$$\int u_k(x)\, \lambda^{(1)}_{(0,1)}(\mathrm{d}x) = k\lambda^{(1)}\left(\left(1 - \frac{1}{k}, 1 \right) \right) = k \cdot \frac{1}{k} = 1$$

for all $k \in \mathbb{N}$ and therefore $\lim_{k \to \infty} \int u_k \, \mathrm{d}\lambda^{(1)}_{(0,1)} = 1$. We claim that on $(0, 1)$ the sequence $(u_k)_{k \in \mathbb{N}}$ converges pointwisely to zero. Given $x \in (0, 1)$ we can find $N = N(x) \in \mathbb{N}$ such that $k \geq N$ implies $x < 1 - \frac{1}{k}$ and hence $u_k(x) = 0$ for $k \geq N(x)$. Hence $\lim_{k \to \infty} u_k(x) = 0$ for all $x \in (0, 1)$. Thus we have constructed a sequence $(u_k)_{k \in \mathbb{N}}$ of measurable functions converging pointwisely to the measurable function u, $u(x) = 0$ for all $x \in (0, 1)$, but while the sequence of corresponding integrals converges to 1 the limit function has integral 0.

In the case of continuous functions and the Riemann integral we know that uniform convergence yields the desired result:

$$\int \left(\lim_{k \to \infty} u_k(x) \right) dx = \lim_{k \to \infty} \int u_k(x)\, dx.$$

However for an arbitrary measure space $(\Omega, \mathcal{A}, \mu)$ the notion of continuity is not defined. Thus a natural question is to look for criteria as to when we can interchange limits of converging sequences of measurable functions with corresponding (converging) sequences of integrals. A key notion needed for this is that of almost everywhere convergence.

Definition 7.2. *Let $(\Omega, \mathcal{A}, \mu)$ be a measure space and $(u_k)_{k \in \mathbb{N}}$, $u_k : \Omega \to \mathbb{R}$ (or $\overline{\mathbb{R}}$) be a sequence of functions. We say that this sequence **converges** μ-**almost everywhere** $(\mu$-**a.e.**$)$ to the function $u : \Omega \to \mathbb{R}$ (or $\overline{\mathbb{R}}$) if there exists a set $\mathcal{N} \in \mathcal{A}$ of measure zero, i.e. $\mu(\mathcal{N}) = 0$, such that for every $\omega \in \Omega \backslash \mathcal{N}$ the sequence $(u_k(\omega))_{k \in \mathbb{N}}$ converges in $\mathbb{R}$ (or $\overline{\mathbb{R}}$) to $u(\omega)$.*

Remark 7.3. A. We can not change Definition 7.2 and require that the set of all $\omega \in \Omega$ for which $(u_k(\omega))_{k \in \mathbb{N}}$ does not converge to $u(\omega)$ is a set of μ-measure zero since this set is not necessarily measurable. However in the case where $(\Omega, \mathcal{A}, \mu)$ is a complete measure space, then this set will be a subset of a μ-null set and hence measurable.
B. Clearly, if the sequence $(u_k)_{k \in \mathbb{N}}$ converges pointwisely to u, then u is measurable provided all u_k are measurable and the convergence also holds almost everywhere.

Consider the sequence $u_k : [0, 1] \to \mathbb{R}$, $u_k(t) = t^k$, $k \in \mathbb{N}$, on the measure space $\left([0, 1], [0, 1] \cap \mathcal{B}^{(1)}, \lambda^{(1)}_{[0,1]} \right)$. This sequence converges $\lambda^{(1)}_{[0,1]}$-a.e. to the function $u : [0, 1] \to \mathbb{R}$, $u(t) = 0$ for all $t \in [0, 1]$. Indeed, for $t \in [0, 1)$ we know that $\lim_{k \to \infty} t^k = 0$ and $\{1\} \subset [0, 1]$ has measure zero, i.e. $\lambda^{(1)}_{[0,1]}(\{1\}) = 0$. We note further that $(u_k)_{k \in \mathbb{N}}$ converges pointwisely to $v : [0, 1] \to \mathbb{R}$,
$$v(t) = \begin{cases} 0, & t \in [0, 1) \\ 1, & t = 1 \end{cases}.$$
Hence $(u_k)_{k \in \mathbb{N}}$ converges also $\lambda^{(1)}_{[0,1]}$-a.e. to v and $v \neq u$. Thus a μ-almost everywhere limit is in general not uniquely determined. For dealing with this and related non-uniqueness results we need

Definition 7.4. *Let $(\Omega, \mathcal{A}, \mu)$ be a measure space and $P(\omega)$ a statement we can make for every $\omega \in \Omega$. We say that $P(\omega)$ holds μ-**almost everywhere** $(\mu$-**a.e.**$)$ if there exists a set $\mathcal{N} \in \mathcal{A}$, $\mu(\mathcal{N}) = 0$, such that $P(\omega)$ holds for all $\omega \in \Omega \backslash \mathcal{N}$.*

Thus, if $P(\omega)$ is the statement: "$(u_k(\omega))_{k\in\mathbb{N}}$ converges to $u(\omega)$" then Definition 7.4 gives Definition 7.2. Statements such as $f \geq g$ μ-a.e. are now well defined.

In particular we can consider μ-almost everywhere equality of functions $u, v :$ $\Omega \to Y$ where $Y \neq \emptyset$ is an arbitrary set.

Definition 7.5. *Let* $(\Omega, \mathcal{A}, \mu)$ *be a measure space and* $Y \neq \emptyset$ *be a set. We call* $u, v : \Omega \to Y$ μ-**equivalent** *and write* $u \sim_\mu v$, *if there exists a set* $\mathcal{N} \in \mathcal{A}$, $\mu(\mathcal{N}) = 0$, *such that* $u(\omega) = v(\omega)$ *for all* $\omega \in \Omega\backslash\mathcal{N}$.

Corollary 7.6. *The relation* $\sim_\mu$ *is an equivalence relation on the set of all mappings* $u : \Omega \to Y$.

Proof. Clearly $u \sim_\mu u$, we just take $\mathcal{N} = \emptyset$, and if $u \sim_\mu v$, i.e. $u(\omega) = v(\omega)$ for all $\omega \in \Omega\backslash\mathcal{N}$, $\mu(\mathcal{N}) = 0$, we have with the same set $\mathcal{N}$ that $v \sim_\mu u$. Finally, if $u \sim_\mu v$ and $v \sim_\mu z$ there exists sets $\mathcal{N}_1, \mathcal{N}_2 \in \mathcal{A}$, $\mu(\mathcal{N}_1) = \mu(\mathcal{N}_2) = 0$ such that $u(\omega) = v(\omega)$ for $\omega \in \Omega\backslash\mathcal{N}_1$ and $v(\omega) = z(\omega)$ for $\omega \in \Omega\backslash\mathcal{N}_2$. Hence for $\omega \in \Omega\backslash(\mathcal{N}_1 \cup \mathcal{N}_2)$ we have $u(\omega) = z(\omega)$, and since $\mu(\mathcal{N}_1 \cup \mathcal{N}_2) = 0$ it follows that $u \sim_\mu z$. $\square$

By $[u]_\mu$ we denote the equivalence class of $u : \Omega \to Y$ with respect to $\sim_\mu$. We can now reword the second statement of Lemma 5.29: every numerical integrable function is μ-a.e. finite.

Proposition 7.7. *Suppose that the sequence* $u_k : \Omega \to \mathbb{R}$, $k \in \mathbb{N}$, *converges* μ-*almost everywhere to* $u : \Omega \to \mathbb{R}$. *Then every further* μ-a.e. *limit* v *of* $(u_k)_{k\in\mathbb{N}}$ *belongs to* $[u]_\mu$.

Proof. Let $\mathcal{N}_1, \mathcal{N}_2 \in \mathcal{A}$, $\mu(\mathcal{N}_1) = \mu(\mathcal{N}_2) = 0$, and for $\omega \in \Omega\backslash\mathcal{N}_1$ we have $\lim_{k\to\infty} u_k(\omega) = u(\omega)$ and for $\omega \in \Omega\backslash\mathcal{N}_2$ we have $\lim_{k\to\infty} u_k(\omega) = v(\omega)$. This implies for $\omega \in \Omega\backslash(\mathcal{N}_1\cup\mathcal{N}_2)$ that $u(\omega) = v(\omega)$ holds and since $\mu(\mathcal{N}_1\cup\mathcal{N}_2) = 0$ we have $u \sim_\mu v$ or $v \in [u]_\mu$. $\square$

Theorem 7.8. *If for a measurable function* $f : \Omega \to [0, \infty]$ *on* $(\Omega, \mathcal{A}, \mu)$ *the integral is zero then* f *is* μ-*almost everywhere the zero function, i.e.*

$$\int f \, d\mu = 0, \quad f \geq 0, \quad \text{implies } f = 0 \ \mu - a.e. \tag{7.1}$$

Conversely, $f = 0$ μ-a.e. *yields* $\int f \, d\mu = 0$.

Proof. Since f is measurable it follows that $\{f > 0\} \in \mathcal{A}$ and (7.1) is proved if we can show that $\mu(\{f > 0\}) = 0$. Since $\{f \geq \frac{1}{k}\} \subset \{f \geq \frac{1}{k+1}\}$ and $\bigcup_{k \in \mathbb{N}}\{f \geq \frac{1}{k}\} = \{f > 0\}$ we need to prove that $\lim_{k \to \infty} \mu\{f > \frac{1}{k}\} = 0$. Observing that $f \geq \frac{1}{k}\chi_{\{f \geq \frac{1}{k}\}}$ we find

$$0 = \int f \, d\mu \geq \int \frac{1}{k}\chi_{\{f \geq \frac{1}{k}\}} \, d\mu = \frac{1}{k}\mu\left\{f \geq \frac{1}{k}\right\} \geq 0,$$

i.e. $\mu(\{f \geq \frac{1}{k}\}) = 0$ implying $f = 0$ μ-a.e. The converse statement is just a consequence of Lemma 5.29. $\qquad\square$

From Theorem 7.8 we can derive a few helpful consequences.

Corollary 7.9. A. *If two measurable functions $f, g : \Omega \to [0, \infty]$ are μ-a.e. equal then they have the same integral.*
B. *If two measurable functions $f, g : \Omega \to \overline{\mathbb{R}}$ are μ-a.e. equal and f is μ-integrable then g is μ-integrable and both integrals are equal.*
C. *If $f, g : \Omega \to \overline{\mathbb{R}}$ are measurable functions, $|f| \leq g$ μ-a.e. and g is μ-integrable, then f is μ-integrable too.*

Proof. **A.** Since $\mu(\{f \neq g\}) = 0$, by Lemma 5.29 $\int_{\{f \neq g\}} f \, d\mu = \int_{\{f \neq g\}} g \, d\mu = 0$ and we find

$$\int f \, d\mu = \int_{\{f \neq g\}} f \, d\mu + \int_{\{f = g\}} f \, d\mu = \int_{\{f = g\}} f \, d\mu$$

$$= \int_{\{f = g\}} g \, d\mu = \int_{\{f = g\}} g \, d\mu + \int_{\{f \neq g\}} g \, d\mu = \int g \, d\mu.$$

B. By assumption we have $f^+ = g^+$ μ-a.e. and $f^- = g^-$ μ-a.e., hence by part A we find $\int f^+ \, d\mu = \int g^+ \, d\mu$ and $\int f^- \, d\mu = \int g^- \, d\mu$ and the integrability of f implies now the integrability of g as well as $\int f \, d\mu = \int g \, d\mu$.
C. The function $g \vee |f|$ is measurable and μ-a.e. we have $g \vee |f| = g$. It follows that $g \vee |f|$ is integrable and from Theorem 5.22.iii) we deduce that f is integrable. $\qquad\square$

Now we may return to the Radon-Nikodym theorem and discuss the uniqueness of densities. We do this in two steps combining the presentation in [13] and [11].

Theorem 7.10. *Let $(\Omega, \mathcal{A})$ be a measurable space and μ a measure on $\mathcal{A}$.* **A.** *If two non-negative measurable functions $f, g : \Omega \to [0, \infty]$ are μ-a.e. equal, then the measures $f\mu$ and $g\mu$ are equal on $\mathcal{A}$.* **B.** *If f or g is integrable and the measures $f\mu$ and $g\mu$ are equal on $\mathcal{A}$ then $f = g$ μ-a.e.*

Proof. **A.** For $A \in \mathcal{A}$ it follows that $f\chi_A = g\chi_A$ μ-a.e. and therefore

$$(f\mu)(A) = \int_A f \, d\mu = \int_A g \, d\mu = (g\mu)(A).$$

B. Suppose that f is integrable and $f\mu = g\mu$. We have to show that $\{f \neq g\}$ is contained in a set of measure zero. By the second statement in Lemma 5.29 there exist sets $\mathcal{N}_f$ and $\mathcal{N}_g$, $\mu(\mathcal{N}_f) = \mu(\mathcal{N}_g) = 0$ and $f|_{\Omega \setminus \mathcal{N}_f} < \infty$ as well as $g|_{\Omega \setminus \mathcal{N}_g} < \infty$. For $n \in \mathbb{N}$ consider the set

$$\Omega_n := \left\{ \omega \in \Omega \setminus (\mathcal{N}_f \cup \mathcal{N}_g) \,\middle|\, |f(\omega) - g(\omega)| \geq \frac{1}{n} \right\}.$$

We claim that $\mu(\Omega_n) = 0$. Suppose $\mu(\Omega_n) > 0$. Then either $\Omega_{n,1} := \{\omega \in \Omega \setminus (\mathcal{N}_f \cup \mathcal{N}_g) \mid f(\omega) - g(\omega) \geq \frac{1}{n}\}$ or $\Omega_{n,2} := \{\omega \in \Omega \setminus (\mathcal{N}_f \cup \mathcal{N}_g) \mid f(\omega) - g(\omega) \leq -\frac{1}{n}\}$ must have positive measure. For $\Omega_{n,1}$ this would imply

$$(f\mu)(\Omega_{n,1}) = \int_{\Omega_{n,1}} f \, d\mu \geq \int_{\Omega_{n,1}} (g + \epsilon\chi_{\Omega_{n,1}}) \, d\mu$$
$$= (g\mu)(\Omega_{n,1}) + \epsilon\mu(\Omega_{n,1}) > (g\mu)(\Omega_{n,1}),$$

and analogously for $\Omega_{n,2}$ we would obtain $(f\mu)(\Omega_{n,2}) < (g\mu)(\Omega_{n,2})$, both statements contradict $f\mu = g\mu$. Thus $\mu(\Omega_n) = 0$ and since $\{f \neq g\} \subset \mathcal{N}_f \cup \mathcal{N}_g \cup \bigcup_{n=1}^{\infty} \Omega_n$ it follows that $f = g$ μ-a.e. $\square$

Theorem 7.11. *In the situation of Theorem 6.17 the density of f of ν with respect to μ is μ-a.e. unique.*

Proof. Since μ is σ-finite there exists a strictly positive integrable function h on $(\Omega, \mathcal{A}, \mu)$. The measure $h\nu = h(f\mu) = f(h\mu)$ has the density f with respect to the finite measure $h\mu$, hence by Theorem 7.10 we have that f is a $h\mu$-a.e. uniquely determined function. Since μ and $h\mu$ have the same sets of measure zero, recall that $h > 0$, it follows that f is μ-a.e. uniquely determined. $\square$

We now return to discuss convergence results for which we give

Definition 7.12. *Let $(\Omega, \mathcal{A}, \mu)$ be a measurable space and $(u_k)_{k\in\mathbb{N}}$, $u_k : \Omega \to \mathbb{R}$ (or $\overline{\mathbb{R}}$) a sequence of measurable functions, and let $u : \Omega \to \mathbb{R}$ (or $\overline{\mathbb{R}}$) be a further measurable function.* **A.** *We say that $(u_k)_{k\in\mathbb{N}}$* **converges in the pth mean**, *$1 \le p < \infty$, to u if*

$$\lim_{k\to\infty} \int |u_k - u|^p \, d\mu = 0. \tag{7.2}$$

B. *The sequence $(u_k)_{k\in\mathbb{N}}$* **converges in measure** *to u if for every set $A \in \mathcal{A}$ with finite measure $\mu(A) < \infty$ the following holds for every $\epsilon > 0$ that*

$$\lim_{k\to\infty} \mu\left(\{|u_k - u| > \epsilon\} \cap A\right) = 0. \tag{7.3}$$

Remark 7.13. A. Convergence in the pth mean for $p = 1$ is called **convergence in the mean** and for $p = 2$ it is called **convergence in the quadratic mean**.
B. In the case where $(\Omega, \mathcal{A}, \mu)$ is a finite measure space we can replace (7.3) by

$$\lim_{k\to\infty} \mu\left(\{|u_k - u| > \epsilon\}\right) = 0. \tag{7.4}$$

C. If $(\Omega, \mathcal{A}, P)$ is a probability space convergence in measure is called **stochastic convergence**.

Theorem 7.14. *Let $(\Omega, \mathcal{A}, \mu)$ be a σ-finite measure space and suppose that the sequence $(u_k)_{k\in\mathbb{N}}$ of measurable mappings converges in measure to the measurable function u. Any further limit in measure v of $(u_k)_{k\in\mathbb{N}}$ belongs to $[u]_\mu$, i.e. $u = v$ μ-a.e. Conversely, every measurable function $z = u$ μ-a.e. is a further limit in measure of $(u_k)_{k\in\mathbb{N}}$.*

Proof. The triangle inequality yields

$$\left\{|u - v| \ge \epsilon\right\} \subset \left\{|u_k - u| \ge \frac{\epsilon}{2}\right\} \cup \left\{|u_k - v| \ge \frac{\epsilon}{2}\right\}$$

implying for $A \in \mathcal{A}$, $\mu(A) < \infty$

$$\mu\left(\left\{|u - v| \ge \epsilon\right\} \cap A\right) \le \mu\left(\left\{|u_k - u| \ge \frac{\epsilon}{2}\right\} \cap A\right) + \mu\left(\left\{|u_k - v| \ge \frac{\epsilon}{2}\right\} \cap A\right)$$

and for $k \to \infty$ we obtain

$$\mu\left(\left\{|u - v| \ge \epsilon\right\} \cap A\right) = 0.$$

Since $\{u \neq v\} \cap A = \bigcup_{n \in \mathbb{N}} \{|u - v| \geq \frac{1}{n}\} \cap A$ we deduce that $u|_A = v|_A$ μ-a.e.
Now we let A run through an increasing sequence $(A_l)_{l \in \mathbb{N}}$ of measurable sets
$A_l \subset A_{l+1}$, $\mu(A_l) < \infty$, and $\bigcup_{l \in \mathbb{N}} A_l = \Omega$, to find that $u = v$ μ-a.e.
The converse statement is almost trivial since there exists a set $\mathcal{N}_z$, $\mu(\mathcal{N}_z) = 0$, such that on $\mathcal{N}_z^C$ we have $u = z$ and therefore $\{|u_k - u| \geq \epsilon\} \cap A$ and $\{|u_k - z| \geq \epsilon\} \cap A$ differ only by a set of measure zero. $\qquad \square$

In order to relate convergence in the pth mean to convergence in measure we prove the following, often very useful estimate:

Lemma 7.15 (Chebyshev-Markov inequality). *For every measurable function $f : \Omega \to \overline{\mathbb{R}}$ defined on a measure space $(\Omega, \mathcal{A}, \mu)$ the following holds*

$$\mu\left(\{|f| \geq \alpha\}\right) \leq \frac{1}{\alpha^p} \int |f|^p \, d\mu \qquad (7.5)$$

for every $\alpha > 0$ and $p > 1$.

Proof. Since $\{|f| \geq \alpha\} \in \mathcal{A}$ we find

$$\alpha^p \mu\left(\{|f| \geq \alpha\}\right) = \int_{\{|f| \geq \alpha\}} \alpha^p \, d\mu \leq \int_{\{|f| \geq \alpha\}} |f|^p \, d\mu \leq \int |f|^p \, d\mu.$$

$\qquad \square$

Corollary 7.16. *If the sequence $(u_k)_{k \in \mathbb{N}}$ of measurable functions converges in the pth mean to u then it converges in measure to u.*

Proof. For every $A \in \mathcal{A}$ we have $\mu\left(\{|u_k - u| \geq \epsilon\} \cap A\right) \leq \mu(\{|u_k - u| \geq \epsilon\})$ and the Chebyshev-Markov inequality yields

$$\mu\left(\{|u_k - u| \geq \epsilon\} \cap A\right) \leq \frac{1}{\alpha^p} \int |u_k - u|^p \, d\mu$$

hence the result follows. $\qquad \square$

We postpone the discussion of the converses of Theorem 7.14 and Corollary 7.16 as we do with the discussion of the relation of μ-a.e. convergence to convergence in measure or convergence in the pth means and we first study in more detail the latter notion.

Let $(u_k)_{k \in \mathbb{N}}$ be a sequence of functions converging in the pth mean, $p \geq 1$, to u and v. Using Minkowski's inequality (5.34) we find

$$\left(\int |u - v|^p \, d\mu \right)^{\frac{1}{p}} \leq \left(\int |u - u_k|^p \, d\mu \right)^{\frac{1}{p}} + \left(\int |u_k - v|^p \, d\mu \right)^{\frac{1}{p}}$$

implying that $|u - v| = 0$ μ-a.e., hence $u = v$ $\mu - a.e.$ and we have proved

Corollary 7.17. *Limits in the pth mean are μ-a.e. uniquely determined.*

Definition 7.18. *Let $(\Omega, \mathcal{A}, \mu)$ be a measure space and $p \geq 1$. By $\mathcal{L}^p(\Omega)$ or $\mathcal{L}^p(\Omega; \mu)$ we denote the set of all measurable functions $u : \Omega \to \mathbb{R}$ for which*

$$N_p(u) := \left(\int |u|^p \, d\mu \right)^{\frac{1}{p}} \tag{7.6}$$

*is finite. Elements of $\mathcal{L}^p(\Omega)$ are called **p-fold integrable**.*

Again we deduce from Lemma 5.29 that if for $u : \Omega \to \overline{\mathbb{R}}$ and $p \geq 1$ it follows that $N_p(u) < \infty$ then u is finite μ-a.e. However, by assumption elements of $\mathcal{L}^p(\Omega)$ are always real-valued, i.e. have on Ω finite values.

Theorem 7.19. *With the natural algebraic operations $\mathcal{L}^p(\Omega)$ is an $\mathbb{R}$-vector space. Furthermore, with $u, v \in \mathcal{L}^p(\Omega)$ it follows that $u \vee v$ and $u \wedge v$ belong to $\mathcal{L}^p(\Omega)$ too. In addition $u \in \mathcal{L}^p(\Omega)$ if and only if $u^+, u^- \in \mathcal{L}^p(\Omega)$.*

Proof. Note that elements in $\mathcal{L}^p(\Omega)$ are real-valued and hence $\alpha u + \beta v$ is for all $\alpha, \beta \in \mathbb{R}$ and $u, v \in \mathcal{L}^p(\Omega)$ defined. Since for $\alpha \in \mathbb{R}$ we have $N_p(\alpha u) = |\alpha| N_p(u)$ and Minkowski's inequality implies $N_p(u + v) \leq N_p(u) + N_p(v)$ it is clear that $\mathcal{L}^p(\Omega)$ is an $\mathbb{R}$-vector space. Moreover, $u \in \mathcal{L}^p(\Omega)$ implies $|u| \in \mathcal{L}^p(\Omega)$ since $N_p(u) = N_p(|u|)$ and from $u \vee v = \frac{1}{2}(u + v + |u - v|)$ and $u \wedge v = \frac{1}{2}(u + v - |u - v|)$ it follows that $u \vee v, u \wedge v \in \mathcal{L}^p(\Omega)$ for $u, v \in \mathcal{L}^p(\Omega)$. In particular we find $u^+, u^- \in \mathcal{L}^p(\Omega)$ for $u \in \mathcal{L}^p(\Omega)$. Conversely, if $u^+, u^- \in \mathcal{L}^p(\Omega)$ then $u = u^+ - u^- \in \mathcal{L}^p(\Omega)$. $\square$

As a corollary to Hölder's inequality (5.33) we obtain

Corollary 7.20. *Let $p > 1$ and $\frac{1}{p} + \frac{1}{q} = 1$. If $u \in \mathcal{L}^p(\Omega)$ and $v \in \mathcal{L}^q(\Omega)$ then $u \cdot v \in \mathcal{L}^1(\Omega)$. In particular if $u, v \in \mathcal{L}^2(\Omega)$ then $u \cdot v \in \mathcal{L}^1(\Omega)$.*

Exercise 7.21. *Prove that on $\mathcal{L}^2(\Omega)$ by*

$$\langle u, v \rangle := \int uv \, \mathrm{d}\mu \tag{7.7}$$

a symmetric, non-negative bilinear form is defined and $\langle u, u \rangle = 0$ yields $u = 0$ μ-a.e.

Corollary 7.22. *Let $(\Omega, \mathcal{A}, \mu)$ be a finite measure space and $1 \leq q \leq p$. If $u \in \mathcal{L}^p(\Omega)$ then $u \in \mathcal{L}^q(\Omega)$ and we have*

$$N_q(u) \leq \mu(\Omega)^{\frac{p-q}{pq}} N_p(u). \tag{7.8}$$

Proof. We may take $\frac{1}{r} = \frac{p-q}{p}$ to find $\frac{1}{r} + \frac{q}{p} = 1$ and now Hölder's inequality applied to $|u|^{\frac{p}{q}}$ and $1 = 1^{\frac{p}{p-q}}$ gives

$$\int |u|^q \, \mathrm{d}\mu = \int |u|^q \cdot 1^{\frac{p}{p-q}} \, \mathrm{d}\mu \leq \left(\int 1 \, \mathrm{d}\mu \right)^{\frac{p-q}{p}} \left(\int |u|^p \, \mathrm{d}\mu \right)^{\frac{q}{p}}$$

which implies $u \in \mathcal{L}^q(\Omega)$ and (7.8). $\qquad\qquad\square$

In Problem 6 we will see that in Corollary 7.22 we cannot in general remove the condition, μ to be a finite measure.

We want to add to the scale $\mathcal{L}^p(\Omega)$, $1 \leq p < \infty$, a space which corresponds to bounded functions.

Definition 7.23. *Let $(\Omega, \mathcal{A}, \mu)$ be a measure space and $f : \Omega \to \overline{\mathbb{R}}$ a measurable function. We call f **essentially bounded** if for some $M_f \geq 0$ we have $|f| \leq M_f$ μ-a.e. and*

$$N_\infty(f) := \inf \left\{ M \geq 0 \,\middle|\, \mu\left(\{|f| > M\}\right) = 0 \right\}$$

*is called the **essential supremum** of f. The set of all real-valued essentially bounded functions $f : \Omega \to \mathbb{R}$ is denoted by $\mathcal{L}^\infty(\Omega)$ or $\mathcal{L}^\infty(\Omega; \mu)$.*

The triangle inequality implies that $\mathcal{L}^\infty(\Omega)$ is a vector space over $\mathbb{R}$, and in Problem 8 we will see that $N_\infty(\cdot)$ is a semi-norm.
Moreover, for $u \in \mathcal{L}^p(\Omega)$, $1 \leq p \leq \infty$, and $v \in \mathcal{L}^\infty(\Omega)$ it follows that $u \cdot v \in \mathcal{L}^p(\Omega)$, in particular $\mathcal{L}^\infty(\Omega)$ is an algebra. Indeed we have

$$\int |u \cdot v|^p \, \mathrm{d}\mu \leq \int |u|^p N_\infty^p(v) \, \mathrm{d}\mu = N_\infty^p(v) \int |u|^p \, \mathrm{d}\mu,$$

or

$$N_p(u \cdot v) \le N_\infty(v)N_p(u). \tag{7.9}$$

On $\mathcal{L}^p(\Omega)$, $1 \le p \le \infty$, the mapping or functional $N_p : \mathcal{L}^p(\Omega) \to \mathbb{R}$ satisfies

$$N_p(u) \ge 0, \tag{7.10}$$
$$N_p(\alpha u) = |\alpha| N_p(u), \tag{7.11}$$
$$N_p(u + v) \le N_p(u) + N_p(v), \tag{7.12}$$

however $N_p(u) = 0$ does not imply $u = 0$, but $u = 0$ μ-a.e. Hence N_p is on $\mathcal{L}^p(\Omega)$ a semi-norm in the sense of Definition II.20.11.

Definition 7.24. *Let V be an $\mathbb{R}$-vector space and $\rho : V \to [0, \infty)$ be a semi-norm on V. We call a sequence $(x_k)_{k \in \mathbb{N}}$, $x_k \in V$, **convergent** to x **with respect to the semi-norm** ρ if $\lim_{k \to \infty} \rho(x_k - x) = 0$, i.e. for every $\epsilon > 0$ there exists $K \in \mathbb{N}$ such that $k \ge K$ implies $\rho(x_k - x) < \epsilon$.*

Proposition 7.25. *If $(x_k)_{k \in \mathbb{N}}$ converges to x with respect to ρ then $(x_k)_{k \in \mathbb{N}}$ is bounded in the sense that $\sup_{k \in \mathbb{N}} \rho(x_k) \le M < \infty$. For two limits x and y of $(x_k)_{k \in \mathbb{N}}$ with respect to ρ it follows that $\rho(x - y) = 0$. Given two sequences $(x_k)_{k \in \mathbb{N}}$ and $(y_k)_{k \in \mathbb{N}}$ converging with respect to ρ to x and y respectively then the sequence $(x_k + y_k)_{k \in \mathbb{N}}$ converges with respect to ρ to $x + y$ and for $\alpha \in \mathbb{R}$ the sequence $(\alpha x_k)_{k \in \mathbb{N}}$ converges with respect to ρ to αx.*

Proof. We can essentially rely on the proofs for convergence with respect to a metric or a norm. Given $\epsilon = 1$ the convergence of $(x_k)_{k \in \mathbb{N}}$ to x implies for some $K \in \mathbb{N}$ that $\rho(x_k) \le \rho(x_k - x) + \rho(x) \le 1 + \rho(x)$ for $k \ge K$ and hence $\sup_{k \in \mathbb{N}} \rho(x_k) \le \max_{1 \le k < K} \{\rho(x_1), \dots, \rho(x_k), 1 + \rho(x)\}$. Further, $\lim_{k \to \infty} \rho(x_k - x) = \lim_{k \to \infty} \rho(x_k - y) = 0$ yields for every $\epsilon \ge 0$ the existence of $K \in \mathbb{N}$ such that $\rho(x - y) \le \rho(x_k - x) + \rho(x_k - y) < \epsilon$ if $k > K$, hence $\rho(x - y) = 0$. Again we employ the triangle inequality (7.12) for ρ to find

$$\rho(x_k + y_k - (x + y)) \le \rho(x_k - x) + \rho(y_k - y)$$

implying $\lim_{k \to \infty} \rho(x_k + y_k - (x + y)) = 0$. Finally we note that

$$\rho(\alpha x_k - \alpha x) = |\alpha| \rho(x_k - x)$$

implying that $\lim_{k \to \infty} \rho(\alpha x_k - \alpha x) = 0$. $\qquad \square$

136

Corollary 7.26. *If $(u_k)_{k\in\mathbb{N}}$ converges to u with respect to N_p, $1 \le p \le \infty$, then $\sup_{k\in\mathbb{N}} N_p(u_k) \le M < \infty$ and two limits of $(u_k)_{k\in\mathbb{N}}$ with respect to N_p are μ-a.e. equal, i.e. $\lim_{k\to\infty}(u_k - u) = 0$ and $\lim_{k\to\infty} N_p(u_k - v) = 0$ implies $u = v$ μ-a.e. Further, for $\lim_{k\to\infty} N_p(u_k - u) = 0$ and $\lim_{k\to\infty} N_p(v_k - v) = 0$ it follows for all $\alpha, \beta \in \mathbb{R}$ that $\lim_{k\to\infty} N_p((\alpha u_k + \beta v_k) - (\alpha u + \beta v)) = 0$, i.e. if $(u_k)_{k\in\mathbb{N}}$ converges to u and $(v_k)_{k\in\mathbb{N}}$ converges to v with respect to N_p then $(\alpha u_k + \beta v_k)_{k\in\mathbb{N}}$ converges to $\alpha u + \beta v$ with respect to N_p.*

Furthermore we have

Proposition 7.27. *Let $(u_k)_{k\in\mathbb{N}}$ converge to u with respect to N_p, $1 \le p \le \infty$, and let $(v_k)_{k\in\mathbb{N}}$ converge to v in $\mathcal{L}^\infty(\Omega)$. Then $(v_k u_k)_{k\in\mathbb{N}}$ converges in $\mathcal{L}^p(\Omega)$ to vu. Moreover if $1 < p < \infty$, $\frac{1}{p} + \frac{1}{q} = 1$, and $(\omega_k)_{k\in\mathbb{N}}$ converges to ω in $\mathcal{L}^q(\Omega)$ then $(u_k \omega_k)_{k\in\mathbb{N}}$ converges to $u\omega$ in $\mathcal{L}^1(\Omega)$.*

Proof. For the first statement let M_u be a bound for $(N_p(u_k))_{k\in\mathbb{N}}$. It follows by (7.9) that

$$
\begin{aligned}
N_p(u_k v_k - uv) &\le N_p(u_k(v_k - v)) + N_p((u_k - u)v) \\
&\le N_p(u_k)N_\infty(v_k - v) + N_p(u_k - u)N_\infty(v) \\
&\le M_u N_\infty(v_k - v) + N_\infty(v)N_p(u_k - u)
\end{aligned}
$$

implying that $\lim_{k\to\infty} N_p(u_k v_k - uv) = 0$. By Hölder's inequality we find $u_k \omega_k, u\omega \in \mathcal{L}^1(\Omega)$ and further since $N_q(\omega_k) \le M_\omega$ for all $k \in \mathbb{N}$ we get

$$
\begin{aligned}
N_1(u_k \omega_k - u\omega) &= \int |u_k \omega_k - u\omega| \, \mathrm{d}\mu \\
&\le \int |u_k \omega_k - u\omega_k| \, \mathrm{d}\mu + \int |u\omega_k - u\omega| \, \mathrm{d}\mu \\
&\le \left(\int |u_k - u|^p \, \mathrm{d}\mu\right)^{\frac{1}{p}} \left(\int |\omega_k|^q \, \mathrm{d}\mu\right)^{\frac{1}{q}} + \left(\int |u|^p \, \mathrm{d}\mu\right)^{\frac{1}{p}} \left(\int |\omega_k - \omega|^q \, \mathrm{d}\mu\right)^{\frac{1}{q}} \\
&\le M_\omega N_p(u_k - u) + N_p(u)N_q(\omega_k - \omega)
\end{aligned}
$$

implying $\lim_{k\to\infty} N_1(u_k \omega_k - u\omega) = 0$. $\qquad\square$

A further useful result is

Lemma 7.28. *Let* $(\Omega, \mathcal{A}, \mu)$ *be a measure space and* $(u_k)_{k \in \mathbb{N}}$ *a sequence of non-negative, numerical measurable functions. For* $1 \leq p < \infty$ *the following holds*

$$N_p \left(\sum_{k=1}^{\infty} u_k \right) \leq \sum_{k=1}^{\infty} N_p(u_k) \tag{7.13}$$

or

$$\left(\int \left| \sum_{k=1}^{\infty} u_k \right|^p d\mu \right)^{\frac{1}{p}} \leq \sum_{k=1}^{\infty} \left(\int |u_k|^p d\mu \right)^{\frac{1}{p}}. \tag{7.14}$$

Proof. For the partial sum $\sum_{k=1}^{N} u_k$ we find using the triangle inequality

$$N_p \left(\sum_{k=1}^{N} u_k \right) \leq \sum_{k=1}^{N} N_p(u_k) \leq \sum_{k=1}^{\infty} N_p(u_k). \tag{7.15}$$

Moreover the sequences $\left(\sum_{k=1}^{N} u_k \right)_{N \in \mathbb{N}}$ and $\left(\sum_{k=1}^{N} |u_k|^p \right)_{N \in \mathbb{N}}$ are monotone increasing with $\sup_{N \in \mathbb{N}} \sum_{k=1}^{N} u_k = \sum_{k=1}^{\infty} u_k$ and $\sup_{N \in \mathbb{N}} \sum_{k=1}^{N} |u_k|^p = \sum_{k=1}^{\infty} |u_k|^p$. The monotone convergence theorem, Theorem 5.13, now gives

$$N_p \left(\sum_{k=1}^{\infty} u_k \right) = \sup_{N \in \mathbb{N}} N_p \left(\sum_{k=1}^{N} u_k \right)$$

and with (7.15) the lemma follows. $\square$

Since the triangle inequality holds for a semi-norm ρ on a vector space V we can deduce with the standard arguments,

$$|\rho(x) - \rho(y)| \leq \rho(x - y), \tag{7.16}$$

compare with Lemma II.1.4. For $p = 1$ this yields with $A \in \mathcal{A}$ that

$$\left| \int_A f \, d\mu - \int_A g \, d\mu \right| \leq \int_A |f - g| \, d\mu \leq N_1(f - g) \tag{7.17}$$

for all $f, g \in \mathcal{L}^1(\Omega)$, and for $1 < p < \infty$ and $f, g \in \mathcal{L}^p(\Omega)$, $A \in \mathcal{A}$ we find

$$\left| \left(\int_A |f|^p d\mu \right)^{\frac{1}{p}} - \left(\int_A |g|^p d\mu \right)^{\frac{1}{p}} \right| = |N_p(\chi_A f) - N_p(\chi_A g)|$$

$$\leq N_p(\chi_A(f - g)) \leq N_p(f - g),$$

i.e.

$$|N_p(\chi_A f) - N_p(\chi_A g)| \le N_p(f - g). \tag{7.18}$$

From (7.17) and (7.18) we derive

Proposition 7.29. *If $(f_k)_{k \in \mathbb{N}}$, $f_k \in \mathcal{L}^1(\Omega)$, converges in the mean to $f \in \mathcal{L}^1(\Omega)$ then for all $A \in \mathcal{A}$ we have*

$$\lim_{k \to \infty} \int_A f_k \, d\mu = \int_A f \, d\mu. \tag{7.19}$$

For a sequence $(g_k)_{k \in \mathbb{N}}$, $g_k \in \mathcal{L}^p(\Omega)$, converging in the pth mean to $g \in \mathcal{L}^p(\Omega)$, $1 \le p < \infty$, we have for all $A \in \mathcal{A}$

$$\lim_{k \to \infty} \int_A |g_k|^p \, d\mu = \int_A |g|^p \, d\mu. \tag{7.20}$$

To proceed further we need a result which is of great use in many situations:

Lemma 7.30 (P. Fatou). *For a sequence $(f_k)_{k \in \mathbb{N}}$ of non-negative measurable numerical functions $f_k : \Omega \to [0, \infty]$ on $(\Omega, \mathcal{A}, \mu)$ we always have*

$$\int \liminf_{k \to \infty} f_k \, d\mu \le \liminf_{k \to \infty} \int f_k \, d\mu. \tag{7.21}$$

Proof. Recall that $\liminf_{k \to \infty} f_k = \sup_{k \in \mathbb{N}} \inf_{l \ge k} f_l$ and that both $\inf_{l \ge k} f_l \ge 0$ and $f := \liminf_{k \to \infty} f_k \ge 0$ are measurable. Moreover, the sequence $(\inf_{l \ge k} f_l)_{k \in \mathbb{N}}$ is increasing and therefore by the monotone convergence theorem, Theorem 5.13, it follows

$$\int f \, d\mu = \sup_{k \in \mathbb{N}} \int \left(\inf_{l \ge k} f_l \right) d\mu = \lim_{k \to \infty} \int \left(\inf_{l \ge k} f_l \right) d\mu.$$

Since $\inf_{l \ge k} f_l \le f_k$ we deduce

$$\int \left(\inf_{l \ge k} f_l \right) d\mu \le \inf_{l \ge k} \int f_l \, d\mu$$

or

$$\int \left(\liminf_{k \to \infty} f_k \right) d\mu \le \liminf_{k \to \infty} \int f_k \, d\mu.$$

$$\square$$

The following corollary to Fatou's lemma is in particular in probability theory of great help.

Corollary 7.31. *For a sequence $(A_k)_{k\in\mathbb{N}}$, $A_k \in \mathcal{A}$, the following holds*

$$\mu\left(\bigcup_{k=1}^{\infty}\bigcap_{l=k}^{\infty} A_l\right) \leq \liminf_{k\to\infty} \mu(A_k) \tag{7.22}$$

and if μ is a finite measure we have

$$\mu\left(\bigcap_{k=1}^{\infty}\bigcup_{l=k}^{\infty} A_l\right) \geq \limsup_{k\to\infty} \mu(A_k). \tag{7.23}$$

Proof. For a proof of (7.22) we refer to Problem 9.c). We arrive at (7.23) when applying (7.22) to the sequence $(A_k^{\complement})_{k\in\mathbb{N}}$:

$$\mu(\Omega) - \mu\left(\bigcap_{k=1}^{\infty}\bigcup_{l=k}^{\infty} A_l\right) = \mu\left(\left(\bigcap_{k=1}^{\infty}\bigcup_{l=k}^{\infty} A_l\right)^{\complement}\right)$$

$$= \mu\left(\bigcup_{k=1}^{\infty}\bigcap_{l=k}^{\infty} A_l^{\complement}\right) \leq \liminf_{k\to\infty} \mu(A_k^{\complement})$$

$$= \mu(\Omega) - \limsup_{k\to\infty} \mu(A_k)$$

or

$$\mu\left(\bigcap_{k=1}^{\infty}\bigcup_{l=k}^{\infty} A_l\right) \geq \limsup_{k\to\infty} \mu(A_k).$$

$\square$

With the help of Fatou's lemma we can prove the following convergence theorem due to F. Riesz:

Theorem 7.32 (F. Riesz). *Let $(u_k)_{k\in\mathbb{N}}$, $u_k \in \mathcal{L}^p(\Omega)$, be a sequence converging μ-a.e. to $u \in \mathcal{L}^p(\Omega)$, $1 \leq p < \infty$. If*

$$\lim_{k\to\infty} \int |u_k|^p \,\mathrm{d}\mu = \int |u|^p \,\mathrm{d}\mu \tag{7.24}$$

then $(u_k)_{k\in\mathbb{N}}$ converges to u in the pth mean, i.e.

$$\lim_{k\to\infty} \left(\int |u_k - u|^p \,\mathrm{d}\mu\right)^{\frac{1}{p}} = 0. \tag{7.25}$$

Proof. For $a, b \geq 0$ and $1 \leq p < \infty$ we find

$$(a + b)^p \leq (2(a \vee b))^p = 2^p(a^p \vee b^p) \leq 2^p(a^p + b^p)$$

and since $|a - b| \leq a + b$ we have

$$|a - b|^p \leq 2^p(a^p + b^p).$$

Given $(u_k)_{k \in \mathbb{N}}$ as in the assumption, we deduce that

$$v_k := 2^p \left(|u_k|^p + |u|^p \right) - |u_k - u|^p \in \mathcal{L}^1(\Omega) \tag{7.26}$$

is a sequence of non-negative functions converging μ-a.e. to $2^{p+1}|u|^p$, hence $\liminf_{k \to \infty} v_k = 2^{p+1}|u|^p$. Applying Fatou's lemma to $(v_k)_{k \in \mathbb{N}}$ and taking into account (7.24) we get

$$2^{p+1} \int |u|^p \, d\mu = \int \left(\liminf_{k \to \infty} v_k \right) d\mu \leq \liminf_{k \to \infty} \int v_k \, d\mu$$

$$= 2^{p+1} \int |u|^p \, d\mu - \limsup_{k \to \infty} \int |u_k - u|^p \, d\mu,$$

or

$$0 \leq \limsup_{k \to \infty} \int |u_k - u|^p \, d\mu \leq 0.$$

$\square$

By far the most important convergence result which in particular shows the advantage the Lebesgue integral has over the Riemann integral is the dominated convergence theorem due to H. Lebesgue.

Theorem 7.33 (Dominated Convergence Theorem). *Let $(\Omega, \mathcal{A}, \mu)$ be a measure space and $(u_k)_{k \in \mathbb{N}}$, $u_k : \Omega \to \mathbb{R}$, be a sequence of integrable functions. Suppose that $(u_k)_{k \in \mathbb{N}}$ converges μ-a.e. to $u : \Omega \to \mathbb{R}$ and that for an integrable function $g : \Omega \to [0, \infty]$ we have $|u_k| \leq g$ μ-a.e. Then the function u is integrable, μ-a.e. bounded by g, and*

$$\lim_{k \to \infty} \int u_k \, d\mu = \int \lim_{k \to \infty} u_k \, d\mu = \int u \, d\mu. \tag{7.27}$$

Proof. Since the μ-a.e. convergence of $(u_k)_{k \in \mathbb{N}}$ to u implies $|u| \leq g$ μ-a.e. and since g is integrable it follows from Theorem 5.22 that u is integrable. We first assume that u_k, u and g are real-valued which implies that $u_k, u, g \in \mathcal{L}^1(\Omega)$. In this situation it is sufficient to prove that $N_1(|u_k - u|)$ converges to 0 in order to arrive at (7.27). For this set

$$v_k := |u_k - u|$$

and observe that

$$0 \leq v_k \leq |u_k| + |u| \leq 2g,$$

hence v_k is integrable. Applying the Lemma of Fatou to the sequence $(|u| + g - v_k)_{k \in \mathbb{N}}$ we get

$$\int \left(\liminf_{k \to \infty} (|u| + g - v_k) \right) d\mu \leq \liminf_{k \to \infty} \int (|u| + g - v_k) \, d\mu$$

$$= \int (|u| + g) \, d\mu - \limsup_{k \to \infty} \int v_k \, d\mu.$$

Since $(|u|+g-v_k)_{k \in \mathbb{N}}$ converges μ-a.e. to $|u|+g$ it follows that $\liminf_{k \to \infty}(|u| + g - v_k) = |u| + g$ μ-a.e. which yields

$$\int \left(\liminf_{k \to \infty} (|u| + g - v_k) \right) d\mu = \int (|u| + g) \, d\mu$$

and therefore

$$\limsup_{k \to \infty} \int v_k \, d\mu \leq 0.$$

But $v_k \geq 0$ μ-a.e. and we get $\lim_{k \to \infty} \int |u_k - u| \, d\mu = 0$ and hence (7.27). Now we turn to the general case. The integrability of u_k, u and g implies the existence of a set $\mathcal{N}_1 \in \mathcal{A}$ of measure zero such that in $\mathcal{N}_1^C$ we have

$$|u_k(\omega)| \leq g(\omega) < \infty \text{ and } |u(\omega)| < \infty.$$

Further, from the μ-a.e. convergence of $(u_k)_{k \in \mathbb{N}}$ to u follows the existence of $\mathcal{N}_2 \in \mathcal{A}$, $\mu(\mathcal{N}_2) = 0$, such that for $\omega \in \mathcal{N}_2^C$ we have $\lim_{k \to \infty} u_k(\omega) = u(\omega)$. Let $\mathcal{N} := \mathcal{N}_1 \cup \mathcal{N}_2$. The first part of the proof yields

$$\lim_{k \to \infty} \int (\chi_{\mathcal{N}^C} u_k) \, d\mu = \int (\chi_{\mathcal{N}^C} u) \, d\mu.$$

However $\mathcal{N}$ is a set of measure zero and therefore we have

$$\int (\chi_{\mathcal{N}^C} u_k) \, d\mu = \int u_k \, d\mu \text{ and } \int (\chi_{\mathcal{N}^C} u) \, d\mu = \int u \, d\mu$$

and the theorem follows. $\qquad \square$

Corollary 7.34. *For $1 \leq p < \infty$ let $(u_k)_{k \in \mathbb{N}}$ be a sequence in $\mathcal{L}^p(\Omega)$ converging μ-a.e. to u. Further assume that $|u_k| \leq g$ holds for some $g \in \mathcal{L}^p(\Omega)$. It follows that $u \in \mathcal{L}^p(\Omega)$ and $\lim_{k \to \infty} N_p(u_k - u) = 0$.*

Proof. As in the proof of Theorem 7.33 we deduce $u_k, u, g \in \mathcal{L}^p(\Omega)$ and replacing v_k by $w_k := |u_k - u|^p$ we have only to note that

$$0 \leq w_k \leq (|u_k| + |u|)^p \leq (|g| + |u|)^p$$

and with $h := (|g| + |u|)^p$ replacing $|u| + g$ we can now follow the proof of Theorem 7.33 unchanged. $\qquad \square$

For every semi-norm ρ on a vector space V the notion of a Cauchy sequence makes sense: $(x_k)_{k \in \mathbb{N}}$, $x_k \in V$, is called a Cauchy sequence with respect to ρ if for every $\epsilon > 0$ there exists $N = N(\epsilon) \in \mathbb{N}$ such that $k, l \geq N$ implies $\rho(x_k - x_l) < \epsilon$. From the triangle inequality which holds for ρ we deduce as usual that if $(x_k)_{k \in \mathbb{N}}$ converges with respect to ρ then it must be also a Cauchy sequence with respect to ρ. The converse is not necessarily true as we already know for norms (or metrics).

The following result, often called the **Fischer-Riesz theorem** claims the convergence of Cauchy sequences in $\mathcal{L}^p(\Omega)$, $1 \leq p < \infty$.

Theorem 7.35. *Every Cauchy sequence $(u_k)_{k \in \mathbb{N}}$ in $\mathcal{L}^p(\Omega)$, $1 \leq p < \infty$, converges in the pth mean to some $u \in \mathcal{L}^p(\Omega)$.*

Proof. The sequence $(u_k)_{k \in \mathbb{N}}$ admits a subsequence $(u_{k_l})_{l \in \mathbb{N}}$ such that for all $l \in \mathbb{N}$

$$N_p(u_{k_{l+1}} - u_{k_l}) < 2^{-l}$$

holds. With

$$v_l := u_{k_{l+1}} - u_{k_l} \text{ and } v := \sum_{l-1}^{\infty} |v_l| \geq 0,$$

we get by Lemma 7.28

$$N_p(v) \leq \sum_{l=1}^{\infty} N_p(v_l) \leq \sum_{l=1}^{\infty} 2^{-l} = 1.$$

Since v is measurable it is p-fold integrable, hence it is μ-a.e. real-valued, i.e. $\sum_{l=1}^{\infty} v_l$ converges μ-a.e. absolutely. Since for the Nth partial sum of

$\sum_{l=1}^{\infty} v_l$ we find $\sum_{l=1}^{N} v_l = u_{k_{N+1}} - u_{k_1}$ it follows that $(u_{k_l})_{l\in\mathbb{N}}$ converges μ-a.e. Moreover we have

$$|u_{k_{l+1}}| = |v_1 + \cdots + v_l + u_{k_1}| \leq v + |u_{k_1}|$$

and $v + |u_{k_1}| \in \mathcal{L}^p(\Omega)$. Hence we may apply the dominated convergence theorem in form of Corollary 7.34 to $(u_{k_l})_{l\in\mathbb{N}}$ with $u = \sum_{l=1}^{\infty} v_l$, i.e. $u = \sum_{l=1}^{\infty} v_l \in \mathcal{L}^p(\Omega)$ and

$$\lim_{l\to\infty} N_p(u_{k_l} - u) = 0. \tag{7.28}$$

We claim that (7.28) also holds for the sequence $(u_k)_{k\in\mathbb{N}}$ which indeed follows from a standard argument. Since $(u_k)_{k\in\mathbb{N}}$ is a Cauchy sequence with respect to N_p, given $\epsilon > 0$ there exists $K_1(\epsilon) \in \mathbb{N}$ such that $k, l \geq K_1(\epsilon)$ implies $N_p(u_k - u_l) < \frac{\epsilon}{2}$. Further, since $(u_{k_m})_{m\in\mathbb{N}}$ converges in the pth mean to u there exists m such that $m \geq K_1(\epsilon)$ implies $N_p(u_{k_m} - u) < \frac{\epsilon}{2}$. Now the triangle inequality for N_p yields for $k \geq K_1(\epsilon)$

$$N_p(u_k - u) \leq N_p(u_k - u_{k_m}) + N_p(u_{k_m} - u) < \epsilon.$$

$\square$

For a sequence $(u_k)_{k\in\mathbb{N}}$ of measurable functions on a measure space $(\Omega, \mathcal{A}, \mu)$ we now have the following possible modes of convergence:

- convergence in measure

- μ-a.e. convergence

- pointwise convergence

- convergence in the pth mean

- uniform convergence.

So far we know the following relations

uniform convergence $\Rightarrow$ pointwise convergence $\Rightarrow$ μ-a.e. convergence
convergence in the pth mean $\Rightarrow$ convergence in measure

and none of these implications can be reversed.
In the case where $(\Omega, \mathcal{A}, \mu)$ is a finite measure space, Corollary 7.22 implies

144

that for $1 \leq q < p$ convergence in the pth mean implies convergence in the qth mean. Moreover from

$$N_p(u) = \left(\int |u|^p \, d\mu \right)^{\frac{1}{p}} \leq \mu(\Omega)^{\frac{1}{p}} \sup_{\omega \in \Omega} |u(\omega)| = \mu(\Omega)^{\frac{1}{p}} \|u\|_\infty$$

which holds for a finite measure space $(\Omega, \mathcal{A}, \mu)$ and bounded measurable functions we deduce that for a finite measure space uniform convergence implies convergence in the pth mean for all $p \geq 1$.

Theorem 7.36. *Let $(u_k)_{k \in \mathbb{N}}$ be a sequence of real-valued measurable functions on the same measure space $(\Omega, \mathcal{A}, \mu)$. **A.** If $(u_k)_{k \in \mathbb{N}}$ converges μ-a.e. to a real-valued measurable function u on Ω, then $(u_k)_{k \in \mathbb{N}}$ converges also in measure to u. **B.** If $(u_k)_{k \in \mathbb{N}}$ converges in measure to u then for every set A of finite measure the sequence $(u_k|_A)_{k \in \mathbb{N}}$ has a subsequence converging μ_A-a.e. to u.*

Proof. **A.** Let $A \in \mathcal{A}$, $\mu(A) < \infty$, $\epsilon > 0$, and define the following sets:

$$B := \{\omega \in A \mid (u_k(\omega))_{k \in \mathbb{N}} \text{ does not converge to } u(\omega)\};$$
$$B_k(\epsilon) := \{\omega \in A \mid |u_k(\omega) - u(\omega)| \geq \epsilon\};$$
$$R_n(\epsilon) := \bigcup_{k=n}^{\infty} B_k(\epsilon);$$
$$M(\epsilon) := \bigcap_{n \in \mathbb{N}} R_n(\epsilon).$$

Since the functions u_k and u are measurable all these sets belong to $\mathcal{A}$. First we claim $M \subset B$. For $\omega_0 \notin B$ it follows that $\lim_{k \to \infty} u_k(\omega_0) = u(\omega_0)$. Hence for every $\eta > 0$ there exists $k_0 \in \mathbb{N}$ such that $k \geq k_0$ implies that $|u_k(\omega_0) - u(\omega_0)| < \eta$, i.e. $\omega_0 \notin B_k$ for $k \geq k_0$ and therefore $\omega_0 \notin M$. Now $\mu|_A$ is a finite measure and $R_n(\epsilon) \supset R_{n+1}(\epsilon)$ therefore by Theorem 2.10, we get $\mu|_A(R_n(\epsilon)) \to \mu|_A(M)$, but $B_n(\epsilon) \subset R_n(\epsilon)$, which yields $\mu|_A(B_n(\epsilon)) \to 0$ implying $\lim_{k \to \infty} \mu(\{\omega \in \Omega \mid |u_k - u| > \epsilon\} \cap A) = 0$, i.e. $u_k \to u$ in measure. **B.** Suppose that $(u_k)_{k \in \mathbb{N}}$ converges in measure to u and let $A \in \mathcal{A}$, $\mu(A) < \infty$. Switching to $(A, A \cap \mathcal{A}, \mu|_A)$ we may assume that $(\Omega, \mathcal{A}, \mu)$ is a finite measure space. For $\alpha > 0$ and $k, l \in \mathbb{N}$ the triangle inequality yields

$$\{|u_k - u_l| \geq \alpha\} \subset \left\{|u_k - u| \geq \frac{\alpha}{2}\right\} \cup \left\{|u_l - u| \geq \frac{\alpha}{2}\right\}.$$

Since $(u_k)_{k\in\mathbb{N}}$ converges in measure to u, for a sequence $(\eta_j)_{j\in\mathbb{N}}$, $\eta_j > 0$, such that $\sum_{j=1}^{\infty} \eta_j < \infty$ we can find n_j such that for $k \geq n_j$ we have

$$\mu\left(\{|u_k - u_{n_j}| \geq \eta_j\}\right) < \eta_j$$

and $n_j < n_{j+1}$ for all $j \in \mathbb{N}$. The sets

$$B_j := \{\omega \in \Omega \mid |u_{n_{j+1}} - u_{n_j}| \geq \eta_j\}$$

are measurable and

$$\sum_{j=1}^{\infty} \mu(B_j) \leq \sum_{j=1}^{\infty} \eta_j < \infty,$$

implying that $\lim_{N\to\infty} \sum_{j=N}^{\infty} \mu(B_j) = 0$. For $B = \bigcap_{N=1}^{\infty} \bigcup_{j=N}^{\infty} B_j$ it follows further that $B \subset \bigcap_{j=N}^{\infty} B_j$ and hence $\mu(B) \leq \sum_{j=N}^{\infty} \mu(B_j)$ for all N, i.e. $\mu(B) = 0$. For $\omega \in B^{\complement}$ the inequality

$$|u_{n_{j+1}}(\omega) - u_{n_j}(\omega)| \geq \eta_j$$

can hold only for finitely many j. Further since $\sum_{j=1}^{\infty} \eta_j$ converges the series $\sum_{j=1}^{\infty}(u_{n_{j+1}}(\omega) - u_{n_j}(\omega))$ converges μ-a.e. to a real-valued measurable function $v : \Omega \to \mathbb{R}$. By part A v is also a limit in measure of $(u_{n_j})_{j\in\mathbb{N}}$ and therefore we have $u = v$ μ-a.e. Note we have proved that $(u_k)_{k\in\mathbb{N}}$ has a subsequence $(u_{n_j})_{j\in\mathbb{N}}$ converging μ-a.e. on $A \subset \Omega$, $\mu(A) < \infty$, to u. $\square$

Corollary 7.37. *If $(u_k)_{k\in\mathbb{N}}$, $u_k \in \mathcal{L}^p(\Omega)$, converges in the pth mean to $u \in \mathcal{L}^p(\Omega)$, then for every measurable set $A \in \mathcal{A}$, $\mu(A) < \infty$, $(u_k)_{k\in\mathbb{N}}$ admits a subsequence converging on A μ-a.e. to u.*

Finally we want to discuss an interesting relation between μ-a.e. convergence and uniform convergence.

Theorem 7.38 (D. F. Egorov). *Let $(\Omega, \mathcal{A}, \mu)$ be a finite measure space and $(u_k)_{k\in\mathbb{N}}$ a sequence of real-valued measurable functions converging μ-a.e. to the real-valued (measurable) function u. For every $\delta > 0$ there exists $A_\delta \in \mathcal{A}$ such that*

$$\mu(A_\delta) > \mu(\Omega) - \delta, \quad i.e. \quad \mu(A_\delta^{\complement}) < \delta \tag{7.29}$$

and

$$(u_k|_{A_\delta})_{k\in\mathbb{N}} \text{ converges uniformly to } u|_{A_\delta}. \tag{7.30}$$

Proof. (Following [50]) We consider the measurable sets

$$A_{m,n} := \bigcap_{k \geq n} \left\{ \omega \in \Omega \,\Big|\, |u_k(\omega) - u(\omega)| < \frac{1}{m} \right\},$$

i.e. for $m, n \in \mathbb{N}$ fixed $A_{m,n}$ is the set of all $\omega \in \Omega$ where $|u_k(\omega) - u(\omega)| < \frac{1}{m}$ for all $k \geq n$. From the definition it follows that $A_{m,n} \subset A_{m,n+1}$ and we set

$$A_m := \bigcup_{n \in \mathbb{N}} A_{m,n}.$$

The continuity of μ, see Theorem 2.10, implies that for every $m \in \mathbb{N}$ and $\delta > 0$ there exists $n_0(m) \in \mathbb{N}$ such that

$$\mu\left(A_m \backslash A_{m,n_0(m)} \right) < \frac{\delta}{2^m}.$$

We claim that

$$A_\delta := \bigcap_{m \in \mathbb{N}} A_{m,n_0(m)}$$

satisfies (7.29) and (7.30). First we note that for $m \in \mathbb{N}$ and $\omega \in A_\delta$ given, we have for $k > n_0(m)$ that $|u_k(\omega) - u(\omega)| < \frac{1}{m}$ which is however the uniform convergence of $(u_k|_{A_\delta})_{k \in \mathbb{N}}$ to $u|_{A_\delta}$. It remains to prove (7.29). For $\omega_0 \in \Omega \backslash A_m$ we can find k sufficiently large such that $|u_k(\omega_0) - u(\omega_0)| \geq \frac{1}{m}$, i.e. at ω_0 $(u_k)_{k \in \mathbb{N}}$ does not converges to u, but since this sequence converges μ-a.e. to u we have $\mu(\Omega \backslash A_m) = 0$ for every $m \in \mathbb{N}$. This now implies

$$\mu\left(\Omega \backslash A_{m,n_0(m)} \right) = \mu\left(A_m \backslash A_{m,n_0(m)} \right) < \frac{\delta}{2^m}$$

and further

$$\mu\left(\Omega \backslash A_\delta \right) = \mu\left(\Omega \backslash \bigcap_{m \in \mathbb{N}} A_{m,n_0(m)} \right) = \mu\left(\bigcup_{m \in \mathbb{N}} \left(\Omega \backslash A_{m,n_0(m)} \right) \right)$$

$$\leq \sum_{m=1}^{\infty} \mu\left(\Omega \backslash A_{m,n_0(m)} \right) \leq \sum_{m=1}^{\infty} \frac{\delta}{2^m} = \delta$$

and the theorem is proved. □

We end this chapter by an obvious extension to the monotone convergence theorem, see [75].

147

Theorem 7.39 (General monotone convergence theorem). *Let* $(\Omega, \mathcal{A}, \mu)$ *be a measure space and* $(u_k)_{k \in \mathbb{N}}$, $u_k \in \mathcal{L}^1(\Omega)$, *an increasing sequence. If* $\sup_{k \in \mathbb{N}} \int u_k \, d\mu$ *is finite then* $u := \sup_{k \in \mathbb{N}} u_k$ *belongs to* $\mathcal{L}^1(\Omega)$ *and the following holds*

$$\sup_{k \in \mathbb{N}} \int u_k \, d\mu = \int \sup_{k \in \mathbb{N}} u_k \, d\mu. \tag{7.31}$$

Proof. We may apply Theorem 5.13 to the sequence $(u_k - u_1)_{k \in \mathbb{N}}$ which is non-negative, and we obtain

$$0 \leq \sup_{k \in \mathbb{N}} \int (u_k - u_1) \, d\mu = \int \sup_{k \in \mathbb{N}} (u_k - u_1) \, d\mu. \tag{7.32}$$

Since $\sup_{k \in \mathbb{N}} \int u_k \, d\mu$ is finite it follows that $u = u_1 + (u - u_1)$ belongs to $\mathcal{L}^1(\Omega)$ and now (7.32) implies (7.31). $\qquad \square$

Remark 7.40. A. Note that $u := \sup_{k \in \mathbb{N}} u_k$ implies

$$\sup_{k \in \mathbb{N}} \int u_k \, d\mu = \int u \, d\mu < \infty,$$

and hence $\sup_{k \in \mathbb{N}} \int u_k \, d\mu$ is finite if and only if $u \in \mathcal{L}^1(\Omega)$.
B. Changing the sign we can reformulate Theorem 7.39 for decreasing sequences: $(v_k)_{k \in \mathbb{N}}$, $v_k \in \mathcal{L}^1(\Omega)$, decreasing and $\inf_{k \in \mathbb{N}} \int v_k \, d\mu > -\infty$ implies

$$\inf_{k \in \mathbb{N}} \int v_k \, d\mu = \int \inf_{k \in \mathbb{N}} v_k \, d\mu$$

and $\inf_{k \in \mathbb{N}} \int v_k \, d\mu > -\infty$ if and only if $\inf_{k \in \mathbb{N}} v_k \in \mathcal{L}^1(\Omega)$.

Problems

1.   a) On the measurable space $(\Omega, \mathcal{P}(\Omega))$, $\Omega = \{1, \ldots, N\}$, consider the measure $\mu = \frac{1}{N} \sum_{k=1}^{N} \epsilon_k$. Let $(u_j)_{j \in \mathbb{N}}$ be a sequence of functions $u_j : \Omega \to \mathbb{R}$ which converges μ-almost everywhere to $u : \Omega \to \mathbb{R}$. Prove that the convergence of $(u_j)_{j \in \mathbb{N}}$ is in fact pointwise.

   b) Now consider on $(\mathbb{R}, \mathcal{B}^{(1)})$ the measure $\nu = \frac{1}{N} \sum_{k=1}^{N} \epsilon_k$ and the sequence of functions $(v_j)_{j \in \mathbb{N}}$, $v_j = \chi_{[\frac{1}{4} - \frac{1}{j}, \frac{1}{2} + \frac{1}{j}]}$ for $j \geq 5$, and $v_j = j$ for $j = 1, 2, 3, 4$. Prove that $(v_j)_{j \in \mathbb{N}}$ converges pointwisely to $\chi_{[\frac{1}{4}, \frac{1}{2}]}$

148

and consequently ν-a.e. Prove further that $(v_j)_{j\in\mathbb{N}}$ converges also ν-almost everywhere to the function $\tilde{v} = 0$. Now consider the $\lambda^{(1)}$-almost everywhere limit of $(v_j)_{j\in\mathbb{N}}$. Is this $\lambda^{(1)}$-almost everywhere limit equal to $\tilde{v}$?

2. On the measure space $(\mathbb{R}, \mathcal{B}^{(1)}, \sum_{k=1}^{\infty})$ consider $N_p(u), p \geq 1$, defined as in (7.6). Let $u_l, u : \mathbb{R} \to \mathbb{R}, l \in \mathbb{N}$, be measurable functions such that $\lim_{l\to\infty} N_p(u_l - u) = 0$. Prove that this implies for all $k \in \mathbb{N}$ the convergence of $(u_l(k))_{l\in\mathbb{N}}$ to $u(k)$.

3. Let $f : \mathbb{R} \to \mathbb{R}$ be a Lipschitz continuous function, i.e. $|f(x) - f(y)| \leq L|x - y|$ with $L \geq 0$. Further let $(u_k)_{k\in\mathbb{N}}$ be a sequence of measurable functions $u_k : \mathbb{R} \to \mathbb{R}$ converging $\lambda^{(1)}$-almost everywhere to the measurable function $u : \mathbb{R} \to \mathbb{R}$. Prove that $(f(u_k))_{k\in\mathbb{N}}$ converges $\lambda^{(1)}$-almost everywhere to $f(u)$.

4.   a) Let $g : \mathbb{R}^n \to \mathbb{R}, g \geq 0$ and $g \in \mathcal{L}^q(\mathbb{R}^n)$. Consider on $\mathcal{B}^{(n)}$ the measure $\nu = g\lambda^{(n)}$. For $f \in \mathcal{L}^p(\mathbb{R}^n), \frac{1}{p} + \frac{1}{q} = 1$, prove

$$\nu(\{|f| \geq \alpha\}) \leq \frac{1}{\alpha}||g||_{\mathcal{L}^q}||f||_{\mathcal{L}^p}.$$

b) With g, q and p as above let $(f_k)_{k\in\mathbb{N}}, f_k \in L^p(\mathbb{R}^n)$, be a sequence for which $\lim_{k\to\infty} \int_{\mathbb{R}^n} |f_k - h| g \, d\lambda^{(n)} = 0$ holds for some $h \in L^p(\mathbb{R}^n)$. Prove that $(f_k)_{k\in\mathbb{N}}$ converges in ν-measure to h.

5. For $k \in \mathbb{N}$ we define on $[0, 1)$ the functions $f_j^{(k)}, 1 \leq j \leq k$, by

$$f_j^{(k)}(x) := \begin{cases} 1, & x \in [\frac{j-1}{k}, \frac{j}{k}) \\ 0, & x \in [0, 1) \setminus [\frac{j-1}{k}, \frac{j}{k}). \end{cases}$$

These functions form a sequence $(g_l)_{l\in\mathbb{N}}$ defined by

$$g_1 := f_1^{(1)}, g_2 := f_1^{(2)}, g_3 := f_2^{(2)}, g_4 := f_1^{(3)}, g_5 := f_2^{(3)}, \ldots$$

Prove that the sequence $(g_k)_{k\in\mathbb{N}}$ converges in $\lambda^{(1)}_{|[0,1)}$-measure to zero but not $\lambda^{(1)}_{|[0,1)}$-almost everywhere.

(This is the standard counter example for a sequence converging in measure but not almost everywhere, see for example J.P.Natanson [59].)

6. Prove that the function $g : \mathbb{R} \to \mathbb{R}$, $g(x) = \frac{1}{1+|x|}$, belongs to $\mathcal{L}^2(\mathbb{R})$ but not to $\mathcal{L}^1(\mathbb{R})$. Deduce that estimate (7.8) in Corollary 7.22 in general cannot hold for an unbounded set.

7.　　a) Solve Exercise 7.21.

　　b) Prove that by $p_1(u) := \sum_{x \in [0,1]} |u'(x)|$ a semi-norm is defined on $C_b^1([0,1]) = \{u \in C_b([0,1]) | u' \in C_b([0,1])\}$ which is not a norm. For $u_k, u \in C_b^1([0,1])$, $k \in \mathbb{N}$, suppose that $\lim_{k \to \infty} p_1(u_k - u) = 0$. Show that for every $c \in \mathbb{R}$ we also have $\lim_{k \to \infty} p_1(u_k - (u + c)) = 0$.

　　c) Prove that on $C_b^1([0,1])$ a further semi-norm is given by

$$ q_1(u) := \left(\int_{[0,1]} |u'(x)|^2 \lambda^{(1)}(dx) \right)^{\frac{1}{2}}. $$

Characterise the functions $u \in C_b^1([0,1])$ for which $q_1(u) = 0$ holds. Suppose that $(u_k)_{k \in \mathbb{N}}$, $u_k \in C_b^1([0,1])$, and that for $v \in \mathcal{L}^2([0,1])$ we have

$$ \lim_{k \to \infty} \left(\int_{[0,1]} |u_k'(x) - v(x)|^2 \lambda^{(1)}(dx) \right)^{\frac{1}{2}} = 0 $$

and assume that $\lim_{k \to \infty} p_1(u_k - u) = 0$, $u \in C_b^1([0,1])$. What can we state about v?

8. Prove that by $N_\infty(\cdot)$ a semi-norm is given on $\mathcal{L}^\infty(\Omega)$.

9.　　a) Verify Fatou's Lemma for the sequence $u_k : \mathbb{R} \to \mathbb{R}$,

$$ u_k(x) = \begin{cases} k^2, & 0 < x < \frac{1}{k^2} \\ 0, & \text{otherwise.} \end{cases} $$

　　b) Let $(\Omega, \mathcal{A}, \mu)$ be a measure space and $v_k, v : \Omega \to \mathbb{R}$, $k \in \mathbb{N}$, are non-negative functions such that $v_k(x) \leq v(x)$ and $v_k \to v$ μ-almost everywhere. Use Fatou's Lemma to prove that $\lim_{k \to \infty} \int v_k \, d\mu = \int v \, d\mu$.

　　c) Prove (7.22).

10. Let $(\Omega, \mathcal{A}, \mu)$ be a finite measure space and $u : \Omega \to \mathbb{R}$ an essentially bounded, measurable function. Prove that

$$\lim_{p \to \infty} \left(\int_\Omega |u(\omega)|^p \mu(d\omega) \right)^{\frac{1}{p}} = ||u||_\infty.$$

11. Let $(\Omega, \mathcal{A}, \mu)$ be a finite measure space and let $(f_k)_{k \in \mathbb{N}}$ be a sequence of measurable functions converging on Ω in measure to the measurable function f. Show that a subsequence of $(f_k)_{k \in \mathbb{N}}$ converges μ-almost everywhere to f.

12. Use the result of Problem 10 to prove the following variant of the dominated convergence theorem: let $(\Omega, \mathcal{A}, \mu)$ be a finite measure space and $(f_k)_{k \in \mathbb{N}}$ be a sequence of measurable functions on Ω converging in measure to the measurable function f. If there exists an integrable function g such that $|f_k(\omega)| \le g(\omega)$ for all $k \in \mathbb{N}$ and $\omega \in \Omega$ then

$$\lim_{k \to \infty} \int_\Omega f_k d\mu = \int f \, d\mu.$$

13. Let $(\Omega, \mathcal{A}, \mu)$ be a finite measure space and $(g_k)_{k \in \mathbb{N}}, g_k \in \mathcal{L}^p(\Omega)$, a sequence of functions such that $\lim_{k \to \infty} N_p(g_k - g) = 0$ for some $g \in \mathcal{L}^p(\Omega)$. Further let $(h_k)_{k \in \mathbb{N}}$ be a sequence of measurable functions satisfying $|h_k(\omega)| \le c < \infty$ for all $k \in \mathbb{N}$ and $\omega \in \Omega$ and which converges μ-almost everywhere to the measurable function h. Prove that

$$\lim_{k \to \infty} N_p(g_k h_k - gh) = 0.$$

8 Applications of the Convergence Theorems and More

In this chapter we collect various applications of the convergence results proved in the last chapter. Some are just useful tools, others are important statements in their own right. We will investigate the relations between the Lebesgue and the Riemann integral for domains in $\mathbb{R}^n$. Finally we also discuss how we can pass from the spaces $\mathcal{L}^p(\Omega)$ to the Banach spaces $L^p(\Omega)$ and therefore prepare some topics needed in functional analysis.

The first topic we want to deal with is parameter dependent integrals. Already when discussing the Γ-function in I.28 we encountered integrals depending on a parameter, and in Part 4 we devoted in the context of the Riemann integral two chapters, II.17 and II.22 to these questions. We will now see that when using the Lebesgue integral certain results become clearer and more easy. We start with

Theorem 8.1. *Let $(\Omega, \mathcal{A}, \mu)$ be a measure space and (X, d) be a metric space. Further let $u : X \times \Omega \to \mathbb{R}$ be a function satisfying*

i) for every $x \in X$ the function $u(x, \cdot) : \Omega \to \mathbb{R}$ is μ-integrable;

ii) for every $\omega \in \Omega$ the function $u(\cdot, \omega) : X \to \mathbb{R}$ is continuous at $x_0 \in X$;

iii) there exists a μ-integrable function $h : \Omega \to \mathbb{R}$, $h \geq 0$, such that

$$|u(x, \omega)| \leq h(\omega)$$

for all $(x, \omega) \in X \times \Omega$.

Under these conditions the function $g : X \to \mathbb{R}$ defined by

$$g(x) := \int u(x, \omega)\mu(d\omega) \tag{8.1}$$

is continuous at $x_0 \in X$.

Proof. Let $(x_k)_{k \in \mathbb{N}}$, $x_k \in X$, be any sequence converging in (X, d) to x_0. Define the sequence $u_k : \Omega \to \mathbb{R}$ by $u_k(\omega) := u(x_k, \omega)$. By assumption it follows that $|u_k| \leq h$ for all $k \in \mathbb{N}$ and since u is continuous at x_0 the sequence

$(u_k)_{k\in\mathbb{N}}$ converges pointwisely to $u(x_0, \cdot)$. From the dominated convergence theorem, Theorem 7.33, we now deduce that

$$\lim_{k\to\infty} g(x_k) = \lim_{k\to\infty} \int u_k(\omega)\mu(\mathrm{d}\omega)$$

$$= \lim_{k\to\infty} \int u(x_k, \omega)\mu(\mathrm{d}\omega)$$

$$= \int \lim_{k\to\infty} u(x_k, \omega)\mu(\mathrm{d}\omega)$$

$$= \int u(x_0, \omega)\,\mathrm{d}\mu = g(x_0),$$

i.e. the continuity of g at x_0. $\qquad\square$

Corollary 8.2. *If in the situation of Theorem 8.1 the function $u(\cdot, \omega) : X \to \mathbb{R}$ is for every $\omega \in \Omega$ continuous, then the function g defined by (8.1) is continuous on X.*

Note that we can apply Theorem 8.1 and its corollary in particular to the situation where $X \subset \mathbb{R}^m$ and $d(x, y) = \|x - y\|$, and $(\Omega, \mathcal{A}, \mu) = \left(Y, Y \cap \mathcal{B}^{(n)}, \lambda^{(n)}_{|Y}\right)$ where $Y \in \mathcal{B}^{(n)}$. For example if $X \subset \mathbb{R}^m$ and $Y \subset \mathbb{R}^n$ are two open (or closed) sets and $u : X \times Y \to \mathbb{R}$ is continuous, thus $x \mapsto u(x, y)$ is continuous and $y \mapsto u(x, y)$ is continuous, hence measurable, then the integrability of $u(x, \cdot)$ with respect to $\lambda^{(n)}_{|Y}$ for all $x \in X$ and the existence of an $\lambda^{(n)}_{|Y}$-integrable function $h : Y \to \mathbb{R}$ such that $|u(x, y)| \leq h(y)$ for all $(x, y) \in X \times Y$, entails the continuity of $\int_Y u(x, y)\lambda^{(n)}(\mathrm{d}y)$.

Example 8.3. Let $u \in \mathcal{L}^1(\mathbb{R}^n)$. The **cosine-transform** of u is defined for $\xi \in \mathbb{R}^n$ by

$$\tilde{u}(\xi) := (2\pi)^{-\frac{n}{2}} \int \cos\langle x, \xi\rangle u(x)\, \lambda^{(n)}(\mathrm{d}x) \tag{8.2}$$

and $\tilde{u} : \mathbb{R}^n \to \mathbb{R}$ is a continuous function. Indeed, we consider the function $(x, \xi) \mapsto \cos\langle x, \xi\rangle u(x)$ on $\mathbb{R}^n \times \mathbb{R}^n$. For every $\xi \in \mathbb{R}^n$ fixed this function is $\lambda^{(n)}$-integrable as product of a bounded measurable function with an integrable function, and for $x \in \mathbb{R}^n$ fixed this function is continuous since $\xi \mapsto \cos\langle x, \xi\rangle$ is continuous. Further we have uniformly in ξ the estimate

$$|\cos\langle x, \xi\rangle u(x)| \leq |u(x)|$$

and $|u| \in \mathcal{L}^1(\mathbb{R}^n)$. Hence by Theorem 8.1 the function $\tilde{u}$ is continuous on $\mathbb{R}^n$.

We now turn to the differentiability of parameter dependent integrals. Once we understand the structure of the argument for functions $u : I \times \Omega \to \mathbb{R}$ where $I \subset \mathbb{R}$, $\mathring{I} \neq \emptyset$, is an interval, extending to functions $v : X \times \Omega \to \mathbb{R}$, $X \subset \mathbb{R}^m$ open, is almost obvious. Therefore we concentrate on the first mentioned case.

Theorem 8.4. *Let $(\Omega, \mathcal{A}, \mu)$ be a measure space and $I \subset \mathbb{R}$, $\mathring{I} \neq \emptyset$, an interval. For the function $u : I \times \Omega \to \mathbb{R}$ we assume*

   i) for all $x \in I$ the function $u(x, \cdot) : \Omega \to \mathbb{R}$ is μ-integrable;

   ii) for all $\omega \in \Omega$ the function $u(\cdot, \omega) : I \to \mathbb{R}$ is differentiable with derivative $\frac{\partial u}{\partial x}(\cdot, \omega)$;

   iii) for a μ-integrable function $h : \Omega \to [0, \infty)$ we have on $I \times \Omega$

$$\left| \frac{\partial u}{\partial x}(x, \omega) \right| \leq h(\omega). \tag{8.3}$$

Then the function

$$g(x) := \int u(x, \omega) \mu(d\omega) \tag{8.4}$$

is on I differentiable, $\omega \mapsto \frac{\partial u}{\partial x}(x, \omega)$ is for all $x \in I$ integrable and

$$g'(x) = \int \frac{\partial u}{\partial x}(x, \omega) \mu(d\omega) \tag{8.5}$$

holds.

Proof. Let $(x_k)_{k \in \mathbb{N}}$, $x_k \in I$, $x_k \neq x_0$, be an arbitrary sequence converging to $x_0 \in I$ and define on Ω the functions

$$u_k(\omega) := \frac{u(x_k, \omega) - u(x_0, \omega)}{x_k - x_0}.$$

These are μ-integrable functions and $\lim_{k \to \infty} u_k(\omega) = \frac{\partial u(x_0, \omega)}{\partial x}$ for all $\omega \in \Omega$. By the mean value theorem we can find $\xi = \xi(x, x_0, \omega)$, $x \wedge x_0 \leq \xi \leq x \vee x_0$, such that

$$\frac{u_k(x, \omega) - u(x_0, \omega)}{x - x_0} = \frac{\partial u}{\partial x}(\xi, \omega).$$

For x_k condition (8.3) implies with $\xi = \xi_k$ as above

$$|u_k(\omega)| = \left|\frac{\partial u}{\partial x}(\xi_k, \omega)\right| \leq h(\omega).$$

Again we can apply the dominated convergence theorem to derive that $\frac{\partial u}{\partial x}(x_0, \cdot)$ is integrable and that

$$\lim_{k\to\infty} \int \frac{u(x_k, \omega) - u(x_0, \omega)}{x_k - x_0} \mu(d\omega) = \lim_{k\to\infty} \int u_k(\omega)\mu(d\omega)$$
$$= \int \lim_{k\to\infty} u_k(\omega)\mu(d\omega) = \int \frac{\partial u}{\partial x}(x_0, \omega)\mu(d\omega).$$

$\square$

Remark 8.5. Of course, if we substitute in Theorem 8.4 the function u by $\frac{\partial^m u(x,\omega)}{\partial x^m}$ and now assume the conditions $i) - iii)$ modified accordingly, we can obtain statements for higher order derivatives, see Problem 2.

The following corollary is a version of Theorem 8.4 for functions $u : U \times \Omega \to \mathbb{R}$, $U \subset \mathbb{R}^m$ open, and again it extends to higher order partial derivatives.

Corollary 8.6. *Let $(\Omega, \mathcal{A}, \mu)$ be a measure space and $U \subset \mathbb{R}^m$ an open set. Suppose that $u : U \times \Omega \to \mathbb{R}$ satisfies*

i) for all $x \in U$ the function $u(x, \cdot)$ is μ-integrable;

ii) for all $\omega \in \Omega$ the partial derivative $\frac{\partial}{\partial x_j}u(x, \omega)$ exists, $1 \leq j \leq m$;

iii) for a μ-integrable function $h : \Omega \to [0, \infty)$ the following holds on $U \times \Omega$

$$\left|\frac{\partial u}{\partial x_j}(x, \omega)\right| \leq h(\omega). \tag{8.6}$$

In this case the function $v : U \to \mathbb{R}$,

$$v(x) := \int u(x, \omega)\mu(d\omega)$$

has in U the partial derivative $\frac{\partial v}{\partial x_j}$, the function $\frac{\partial u}{\partial x_j}(x, \cdot)$ is for all $x \in U$ μ-integrable and we have

$$\frac{\partial v}{\partial x_j}(x) = \int \frac{\partial u}{\partial x_j}(x, \omega)\mu(d\omega). \tag{8.7}$$

Proof. We can employ the proof of Theorem 8.4 when fixing all variables $x_1, \ldots, x_{j-1}, x_{j+1}, \ldots, x_m$. □

Of course, in Theorem 8.4 and its corollary we may take for $(\Omega, \mathcal{A}, \mu)$ the measure space $(V, V \cap \mathcal{B}^{(n)}, \lambda^{(n)}_{|_V})$ for $V \in \mathcal{B}^{(n)}$ and hence we can handle situations such as

$$\partial_x^\alpha \int_V u(x, y) \lambda^{(n)}_{|_V}(dy).$$

Example 8.7. Let $u \in \mathcal{L}^1(\mathbb{R}^n)$ be a function such that the function $x \mapsto |x|u(x)$ belongs to $\mathcal{L}^1(\mathbb{R}^n)$ too. In this case the cosine-transform $\tilde{u}$ has all first order partial derivatives and they are bounded. We consider the function defined on $\mathbb{R}^n \times \mathbb{R}^n$ by $(x, \xi) \mapsto \cos\langle x, \xi \rangle u(x)$. This function is with respect to x for all ξ integrable and since for $1 \leq j \leq n$

$$\frac{\partial}{\partial \xi_j}(\cos\langle x_l, \xi \rangle u(x)) = -x_j \sin\langle x, \xi \rangle u(x)$$

we have the estimate

$$\left| \frac{\partial}{\partial \xi_j}(\cos\langle x, \xi \rangle u(x)) \right| \leq |x| \, |u(x)|$$

and by assumption $x \mapsto |x||u(x)|$ belongs to $\mathcal{L}^1(\mathbb{R}^n)$, i.e. is integrable. It follows that

$$\frac{\partial}{\partial \xi_j}\tilde{u}(\xi) = \int (-x_j \sin\langle x, \xi \rangle u(x)) \lambda^{(n)}(dx)$$

and further

$$\left| \frac{\partial}{\partial \xi_j}\tilde{u}(\xi) \right| \leq \int |x||u(x)|\lambda^{(n)}(dx) = M_u < \infty.$$

Our next result is Jensen's inequality which we have encountered already in several versions in Volume I and II. It is in particular of great use in probability theory and a more general variant than ours can be found in [75].

In Theorem I.23.5 we have proved that a convex function $f : I \to \mathbb{R}, I \subset \mathbb{R}$ an interval with end points $a < b$, is differentiable from the right and the proof yields

$$\frac{f(x_0) - f(x)}{x_0 - x} \leq f'_+(x_0), \tag{8.8}$$

where $f'_+(x_0)$ is the derivative from the right of f at $x_0 \in I$. We can rewrite (8.8) as

$$f(x) \geq f'_+(x_0)(x - x_0) + f(x_0). \tag{8.9}$$

From (8.9) combined with Proposition I.23.10 we deduce, since $x_0 \in I$, that

$$f(x) = \sup_{y \in I}\{f'_+(y)(x - y) + f(y)\}. \tag{8.10}$$

In particular, for $u : \Omega \to \mathbb{R}$ we get

$$f(u(\omega)) \geq f'_+(y)(u(\omega) - y) + f(y) \tag{8.11}$$

for every $y \in I$. With these preparations we can prove

Theorem 8.8 (Jensen's inequality). *Let $(\Omega, \mathcal{A}, \mu)$ be a measure space and $A \in \mathcal{A}$ a set of finite measure $\mu(A) > 0$. Let $f : I \to \mathbb{R}$ be a convex function and $u : A \to \mathbb{R}$ be a $\mu|_A$-integrable function such that $u(A) \subset I$. Then the following holds*

$$f\left(\frac{1}{\mu(A)} \int_A u \, d\mu\right) \leq \frac{1}{\mu(A)} \int_A (f \circ u) \, d\mu. \tag{8.12}$$

Proof. We take in (8.11) $y = \frac{1}{\mu(A)} \int_A u \, d\mu \in I$ and we deduce

$$f\left(\frac{1}{\mu(A)} \int_A u \, d\mu\right) + f'_+(y)\left(u(\omega) - \frac{1}{\mu(A)} \int_A u \, d\mu\right) \leq f(u(\omega)).$$

Integrating over A we find

$$\mu(A)f\left(\frac{1}{\mu(A)} \int_A u \, d\mu\right) + f'_+(y)\left(\int u \, d\mu - \int u \, d\mu\right) \leq \int (f \circ u) \, d\mu,$$

and dividing by $\mu(A)$ we obtain (8.12). $\qquad\square$

Corollary 8.9. *Let $(\Omega, \mathcal{A}, P)$ be a probability space and $X : \Omega \to \mathbb{R}$ a random variable with range contained in the interval $I \subset \mathbb{R}$. For every convex function $f : I \to \mathbb{R}$ we have*

$$f(E(X)) \leq E(f \circ X). \tag{8.13}$$

Example 8.10. For a finite measure space $(\Omega, \mathcal{A}, \mu)$ we have for an integrable function $u : \Omega \to \mathbb{R}$ with $\int u \, d\mu = 1$ the estimate

$$\int (\ln u) \, d\mu \leq \mu(\Omega) \ln \frac{1}{\mu(\Omega)}.$$

Since $(-\ln)''(x) = \frac{1}{x^2}$ the function $-\ln : (0, \infty) \to \mathbb{R}$ is convex and now (8.12) yields

$$-\ln \left(\frac{1}{\mu(\Omega)} \right) = -\ln \left(\frac{1}{\mu(\Omega)} \int u \, d\mu \right) \leq \frac{1}{\mu(\Omega)} \int (-\ln u) \, d\mu$$

or

$$\int \ln u \, d\mu \leq \mu(\Omega) \ln \frac{1}{\mu(\Omega)}.$$

We now want to compare the Riemann (Darboux-Jordan) intergration theory with the Lebesgue theory. We do this in three steps. First for functions defined on an interval $I \subset \mathbb{R}$, then for functions defined on a non-degenerate cell or hyper-rectangle and then we turn to Jordan measurable sets. Our results will allow us to use in many cases the Riemann theory to evaluate integrals, while on the other hand they will provide us with a proof of the transformation theorem or the change of variable formula, Theorem II.21.8, for a large class of domains. When speaking in the following about Lebesgue integrable functions on subsets of $\mathbb{R}^n$ we always consider the Borel σ-field $\mathcal{B}^{(n)}$ and the Lebesgue-Borel measure $\lambda^{(n)}$ restricted to this subset. Discussing integrability over a subset $A \subset \mathbb{R}^n$ we always assume $A \in \mathcal{B}^{(n)}$ and for simplicity we also write $\lambda^{(n)}$ for $\lambda^{(n)}_{|A}$. We also adopt the notation of Volume I and II when dealing with the Riemann integral, see in particular Chapter I.25.

Theorem 8.11. *Let $a < b$ and $u : [a, b] \to \mathbb{R}$ be a Riemann integrable function. Then u is Lebesgue integrable over $[a, b]$ and the Riemann and the Lebesgue integral of u coincide, i.e.*

$$\int_a^b u(x) \, dx = \int_{[a,b]} u \, d\lambda^{(1)}. \tag{8.14}$$

Proof. Let $u : [a, b] \to \mathbb{R}$ be Riemann integrable, in particular u is bounded. By decomposing $u = u^+ - u^-$ we may assume that $u \geq 0$. Take a sequence

$(Z_k)_{k\in\mathbb{N}}$ of partition of $[a,b]$ with $\mathrm{mesh}(Z_k)$ tending to zero for $k \to \infty$ and for an interval $[x_j^{(k)}, x_{j+1}^{(k)}]$, $x_j^{(k)}$, $x_{j+1}^{(k)} \in Z_k$, let

$$\xi_j^{(k)} = \sup_{x\in\left(x_j^{(k)},x_{j+1}^{(k)}\right)} u(x) \quad \text{and} \quad \eta_j^{(k)} := \inf_{x\in\left(x_j^{(k)},x_{j+1}^{(k)}\right)} u(x).$$

We assume further that $Z_k \subset Z_{k+1}$, i.e. Z_{k+1} is a refinement of Z_k. It follows that the Riemann sums $S(u, Z_k, \xi^{(k)})$ and $S(u, Z_k, \eta^{(k)})$ converge to the Riemann integral $\int_a^b u(x)\, dx$. We can interpret $S(u, Z_k, \xi^{(k)})$ and $S(u, Z_k, \eta^{(k)})$ as Lebesgue integrals of the corresponding simple functions v_k and ω_k where $v_k|_{[x_j^{(k)},x_{j+1}^{(k)}]} = \xi_j^{(k)}$ and $\omega_k|_{[x_j^{(k)},x_{j+1}^{(k)}]} = \eta_j^{(k)}$. The sequence $(v_k)_{k\in\mathbb{N}}$ is decreasing whereas the sequence $(\omega_k)_{k\in\mathbb{N}}$ is increasing and we have $\sup_{k\in\mathbb{N}} \omega_k \leq u \leq \inf_{k\in\mathbb{N}} v_k$. The general monotone convergence theorem, Theorem 7.39, now yields

$$\int_a^b u(x)\, dx = \lim_{k\to\infty} S(u, Z_k, \eta^{(k)}) = \lim_{k\to\infty} \int \omega_k \, d\lambda^{(1)}$$
$$= \int \sup_{k\in\mathbb{N}} \omega_k \, d\lambda^{(1)},$$

as well as

$$\int_a^b u(x)\, dx = \lim_{k\to\infty} S(u, Z_k, \xi^{(k)}) = \lim_{k\to\infty} \int v_k \, d\lambda^{(1)}$$
$$= \int \inf_{k\in\mathbb{N}} v_k \, d\lambda^{(1)},$$

which implies $\sup_{k\in\mathbb{N}} \omega_k, \inf_{k\in\mathbb{N}} v_k \in \mathcal{L}^1(\Omega)$ and

$$0 \geq \int \left(\inf_{k\in\mathbb{N}} v_k - \sup_{k\in\mathbb{N}} \omega_k \right) d\lambda^{(1)} = 0,$$

i.e. $\sup_{k\in\mathbb{N}} \omega_k = \inf_{k\in\mathbb{N}} v_k = u$ $\lambda^{(1)}$-a.e., i.e. $u \in \mathcal{L}^1(\Omega)$ and $\int_a^b u(x)\, dx = \int_{[a,b]} u \, d\lambda^{(1)}$. $\square$

An analysis of the proof of Theorem 8.11 immediately gives

Corollary 8.12. *Let $K \subset \mathbb{R}^n$ be a non-degenerate compact cell (a compact hyper-rectangle with $\mathrm{vol}_n(K) > 0$) and $u : K \to \mathbb{R}$ a measurable Riemann*

integrable function. Then u is Lebesgue integrable and the Riemann and the
Lebesgue integral of u over K coincide, i.e.

$$\int_K u(x)\,\mathrm{d}x = \int_K u\,\mathrm{d}\lambda^{(n)}. \tag{8.15}$$

Using the set-additivity of both the Riemann and the Lebesgue integral we
deduce that if $G = \bigcup_{j=1}^{N} K_j$ is the union of finitely many compact non-
degenerate cells $K_j \subset \mathbb{R}^n$ then a measurable Riemann integrable function
$u : G \to \mathbb{R}$ is also Lebesgue integrable and the two integrals of u over G
coincide.

In Theorem II.19.25 we have seen that a bounded set $G \subset \mathbb{R}^n$ is Jordan
measurable if ∂G has Lebesgue measure zero and we have mentioned that this
condition indeed characterises Jordan measurable sets. Every compact set
and every (bounded) open set in $\mathbb{R}^n$ is Borel measurable. Thus once we know
that such a set is a Jordan measurable set, using the definition of the Riemann
integral, i.e. Definition II.20.2, we can apply Corollary 8.12 to identify the
corresponding Riemann and Lebesgue integral for a measurable and Riemann
integrable function. A discussion when a given Borel measurable set in $\mathbb{R}^n$ is
already Jordan measurable is provided in G. Folland [27]. The key problem
is, when is it possible to "approximate" $G \subset \mathbb{R}^n$ by open cubes from the
interior and by the compact cubes from the exterior (meaning the union
of these cubes contains G) in such a way that the difference between the
measures of the approximating sets tend to zero? See Figure 8.1 below.

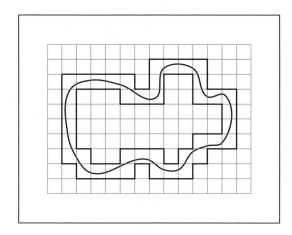

Figure 8.1

In Theorem II.20.4 we have proved that for a bounded function $f : G \to \mathbb{R}$ on a bounded Jordan measurable set the Riemann integrability follows if the set of discontinuities has Jordan content zero. There are more general results, for example as stated in Theorem I.32.17 a bounded function $u : [a, b] \to \mathbb{R}$ is Riemann integrable if and only if its set of discontinuity is a Lebesgue null set. We now want to provide a proof of Theorem I.32.17 which was not possible with the tools at our disposal in Volume I.

Proof of Theorem I.32.17. As in the proof of Theorem 8.11 let $(Z_k)_{k\in\mathbb{N}}$, $Z_k \subset Z_{k+1}$, be a sequence of partitions of $[a, b]$ with $\mathrm{mesh}(Z_k)$ tending to 0 for $k \to \infty$ and with $\xi_j^{(k)} := \sup_{x\in[x_j^{(k)}, x_{j+1}^{(k)})} u(x)$, $\eta_j^{(k)} := \inf_{x\in[x_j^{(k)}, x_{j+1}^{(k)})} u(x)$ denoting the sequences $(v_k)_{k\in\mathbb{N}}$ and $(w_k)_{k\in\mathbb{N}}$ of simple functions corresponding to $S(u, Z_k, \xi)$ and $S(u, Z_k, \eta)$, respectively. Note that $\bigcup_{k\in\mathbb{N}} Z_k$ is a countable set, hence it has $\lambda^{(1)}$-measure zero. The Riemann integrability of f implies that for $\epsilon > 0$ and $x \in [a, b]$ there exists $N = N(\epsilon, x) \in \mathbb{N}$ and $t_{l-1}, t_l \in Z_{N(\epsilon,x)}$ such that $x \in (t_{l-1}, t_l)$ implies for $k \geq N(\epsilon, x)$ that

$$\left| w_k(x) - \sup_{k\in\mathbb{N}} w_k(x) \right| + \left| v_k(x) - \inf_{k\in\mathbb{N}} v_k(x) \right| \leq \epsilon.$$

For $x, y \in (t_{l-1}, t_l)$ we find

$$|u(x) - u(y)| \leq \sup_{t\in[t_{l-1}, t_l]} u(t) - \inf_{t\in[t_{l-1}, t_l]} u(t)$$

$$= v_{N(\epsilon,x)}(x) - w_{N(\epsilon,x)}(x)$$

$$\leq \epsilon + \left| \sup_{k\in\mathbb{N}} w_k(x) - \inf_{k\in\mathbb{N}} v_k(x) \right|.$$

We know from the proof of Theorem 8.11 that $\sup_{k\in\mathbb{N}} w_k = \inf_{k\in\mathbb{N}} v_k = u$ $\lambda^{(1)}$-a.e. The inequality just proven shows that $\{x \in [a, b] \mid u \text{ is discontinuous at } x\}$ is contained in $\{x \in [a, b] \mid |\sup_{k\in\mathbb{N}} w_k(x) - \inf_{k\in\mathbb{N}} v_k(x)| \neq 0\} \cup \bigcup_{k\in\mathbb{N}} Z_k$, hence it is contained in a set of measure zero, i.e. a null set. On the other hand, we can deduce from the same inequality that

$$\left\{ x \in [a, b] \mid \sup_{k\in\mathbb{N}} w_k(x) = \inf_{k\in\mathbb{N}} v_k(x) \right\} \subset \{x \in [a, b] \mid u \text{ is continuous at } x\} \cup \bigcup_{k\in\mathbb{N}} Z_k.$$

Thus, if the set of discontinuities of u is a null set, then the measurable set $\{x \in [a, b] \mid \sup_{k \in \mathbb{N}} w_k(x) = \inf_{k \in \mathbb{N}} v_k(x)\}$ is a set of measure zero and since

$$\int_* u = \lim_{k \to \infty} S\left(u, Z_k, \eta^{(k)}\right) = \lim_{k \to \infty} \int w_k \, d\lambda^{(1)} = \int \sup_{k \in \mathbb{N}} w_k \, d\lambda^{(1)}$$

and

$$\int^* u = \lim_{k \to \infty} S\left(u, Z_k, \xi^{(k)}\right) = \lim_{k \to \infty} \int v_k \, d\lambda^{(1)} = \int \inf_{k \in \mathbb{N}} v_k \, d\lambda^{(1)}$$

this implies that $\int_* u = \int^* u$, i.e. the Riemann integrability of u. $\qquad\square$

Remark 8.13. In the proofs of Theorem 8.11 and Theorem I.32.17 we follow [76].

In Chapter I.28 and Chapter II.22 we discussed improper Riemann integrals which have a lot of important applications, we just want to mention the Γ-function or the Fourier transform which we will discuss in more detail in Part 8. Here we want to investigate the relations between improper Riemann integrals and the Lebesgue integral for $n = 1$, the higher dimensional case does not give a further insight. In Chapter I.28 we have discussed different types of improper Riemann integrals depending on whether we are working with bounded or unbounded domains and/or bounded or unbounded functions. For simplicity we have chosen here to treat in more detail the case of the half line $[0, \infty)$ which covers for example (almost) the Γ-function. The corresponding results for the other cases we will state without proof. We also extend the definition of improper Riemann integrals slightly. While in Chapter I.28 and II.22 we worked with continuous functions only, we now assume that the functions under consideration are on every compact subinterval of their domain Riemann integrable. None of the results about the existence of improper Riemann integrals will change after this alteration.

Theorem 8.14. *Let $f : [0, \infty) \to \mathbb{R}$ be a measurable function and suppose that f is improper Riemann integrable in the sense that $f|_{[0,R]}$ is for every $R > 0$ Riemann integrable and that*

$$\int_0^\infty f(x) \, dx := \lim_{R \to \infty} \int_0^R f(x) \, dx \tag{8.16}$$

exists. The function f is Lebesgue integrable on $[0, \infty)$ and the improper Riemann integral on $[0, \infty)$ coincides with the Lebesgue integral of f over

$[0, \infty)$ *if and only if* $|f|$ *is improper Riemann integrable which means in particular that*

$$\lim_{R \to \infty} \int_0^R |f(x)| \, dx = \int_0^\infty |f(x)| \, dx \qquad (8.17)$$

exists. In this case we also have

$$\int_0^\infty |f(x)| \, dx = \int_{[0,\infty)} |f| \, d\lambda^{(1)}.$$

Proof. If f is Riemann integrable over $[0, R]$, $R > 0$, then f^+ and f^- are also Riemann integrable over $[0, R]$ and both functions are measurable by assumption. By Theorem 8.11 both, f^+ and f^-, are Lebesgue integrable over $[0, R]$, $R > 0$, and in each case the Lebesgue integral is equal to the Riemann integral, i.e.

$$\int_0^R f^+(x) \, dx = \int \chi_{[0,R]} f^+ \, d\lambda^{(1)} \qquad (8.18)$$

and

$$\int_0^R f^-(x) \, dx = \int \chi_{[0,R]} f^- \, d\lambda^{(1)}. \qquad (8.19)$$

For every sequence $(R_N)_{N \in \mathbb{N}}$, $0 < R_N < R_{N+1}$, the sequence $\left(\chi_{[0,R_N]} f^+\right)_{N \in \mathbb{N}}$ and $\left(\chi_{[0,R_N]} f^-\right)_{N \in \mathbb{N}}$ are monotone increasing and if (8.16) and (8.17) hold the monotone convergence theorem applied to (8.18) and (8.19) yields the existence of $\int \chi_{[0,\infty)} f^+ \, d\lambda^{(1)}$ and $\int \chi_{[0,\infty)} f^- \, d\lambda^{(1)}$, hence $f\chi_{[0,\infty)}$ is Lebesgue integrable and we have

$$\int_0^\infty f(x) \, dx = \int \chi_{[0,\infty)} f \, d\lambda^{(1)} = \int_{[0,\infty)} f \, d\lambda^{(1)}.$$

Conversely, if $f \in \mathcal{L}^1([0, \infty))$ then the functions f^+, f^- as well as $g\chi_{[0,R]}$, $g \in \{f, f^+, f^-\}$, are elements of $\mathcal{L}^1([0, \infty))$ and since f is by assumption improper Riemann integrable all these functions are Riemann integrals over $[0, R]$, $R > 0$, and their Lebesgue and Riemann integrable over $[0, R]$, $R > 0$, coincide. For every increasing sequence $(R_j)_{j \in \mathbb{N}}$, $R_j > R_{j-1} > 0$, the monotone convergence theorem gives

$$\lim_{j \to \infty} \int \chi_{[0,R_j]} f^+ \, d\lambda^{(1)} = \int \chi_{[0,\infty)} f^+ \, d\lambda^{(1)}$$

164

and

$$\lim_{j\to\infty} \int \chi_{[0,R_j]} f^- \, d\lambda^{(1)} = \int \chi_{[0,\infty)} f^- \, d\lambda^{(1)},$$

which implies the existence of the limits (8.18) and (8.19), and hence (8.17).

$\square$

Example 8.15. In Example I.28.19.B we have seen that the improper Riemann integral $\int_0^\infty \frac{\sin x}{x} \, dx$ exists. However the function $g : [0, \infty) \to \mathbb{R}$, $g(x) = \frac{\sin x}{x}$, $x \neq 0$, $g(0) = 1$, is not Lebesgue integrable over $[0, \infty)$ as we have seen already in Problem 8 of Chapter 5.

Remark 8.16. With the same type of arguments we can prove Theorem 8.14 for any type of improper Riemann integrals, compare with Definition I.28.1 and I.28.6. In the language of Definition I.28.14 we can state that an improper Riemann integral is a Lebesgue integral if it is absolutely convergent. In particular in the situation of Theorem I.28.17 an improper Riemann integral is a Lebesgue integral.

Example 8.17. Since for $x > 0$ the function $t \mapsto t^{x-1} e^{-t}$ is non-negative, Theorem 8.14 (or Remark 8.16) implies that the Γ-function $\Gamma(x) = \int_0^\infty t^{x-1} e^{-t} \, dt$ can be looked at as a parameter dependent Lebesgue integral.

Finally we want to return to the fact that for a measure space $(\Omega, \mathcal{A}, \mu)$ by $N_p(f) = \left(\int |f|^p \, d\mu \right)^{\frac{1}{p}}$ a semi-norm is given on $\mathcal{L}^p(\Omega)$ which is in general not a norm. This fact has caused a number of problems in uniqueness statements, and moreover, since the theory of normed spaces, especially that of complete normed spaces, i.e. Banach spaces, gives much better tools to handle concrete problems such as solving equations, the question arises as to whether we can turn $\mathcal{L}^p(\Omega)$ into a Banach space.

The starting point is that for every semi-norm ρ on an $\mathbb{R}$-vector space V an equivalence relation $\sim_\rho$ is given by $x \sim_\rho y$ if $\rho(x - y) = 0$. Indeed, since $\rho(x-x) = \rho(0) = 0$, $\rho(x-y) = \rho(y-x)$ as well as $\rho(x-z) \leq \rho(x-y)+\rho(y-z)$, the fact that $\sim_\rho$ is an equivalence relation is trivial. As an equivalence relation $\sim_\rho$ induces on V a partition into equivalence classes which we denote by $[x] = \{y \in V \mid x \sim_\rho y\}$. As usual we denote by $V/\sim_\rho$ the set of all equivalence classes induced by $\sim_\rho$. With the operations

$$[x] + [y] := [x + y], \quad x, y \in V, \tag{8.20}$$

165

and

$$\lambda[x] := [\lambda x], \quad x \in V, \ \lambda \in \mathbb{R}, \tag{8.21}$$

we can turn $V/\sim_\rho$ into a vector space. In particular the zero element $0 \in V/\sim_\rho$ is $[0_V] = \{y \in V \mid 0_V \sim_\rho y\}$ where for the moment 0_V is the neutral element with respect to addition in V. Moreover we define on $V/\sim_\rho$ the mapping

$$\|.\|_\rho : V/\sim_\rho \to \mathbb{R}, \quad \|[x]\|_\rho := \rho(x). \tag{8.22}$$

First we note that for $x_1, x_2 \in [x]$ it follows that

$$\rho(x_1 - x_2) \le \rho(x_1 - x) + \rho(x - x_2) = 0,$$

and therefore

$$\|[x_1]\|_\rho = \rho(x_1) \le \rho(x_1 - x_2) + \rho(x_2) = \|[x_2]\|_\rho$$

as well as

$$\|[x_2]\|_\rho = \rho(x_2) \le \rho(x_2 - x_1) + \rho(x_1) = \|[x_1]\|_\rho$$

i.e. $\|[x_1]\|_\rho = \|[x_2]\|_\rho$ for all $x_1, x_2 \in [x]$ implying that $\|[x]\|_\rho$ is independent of the representative and $\|.\|_\rho$ is on $V/\sim_\rho$ well defined.

Lemma 8.18. *For every semi-norm ρ on the $\mathbb{R}$-vector space V a norm is given on $V/\sim_\rho$ by $\|.\|_\rho$.*

Proof. Clearly, $\|[x]\|_\rho \ge 0$ and $\|[x]\|_\rho = 0$ implies $\rho(x) = 0$, i.e. $x \in [0_V]$ or $[x] = 0 \in V/\sim_\rho$ is the zero element. For $\lambda \in \mathbb{R}$ and $[x] \in V/\sim_\rho$ we have with $x_1 \in [x]$

$$\|\lambda[x]\|_\rho = \|[\lambda x]\|_\rho = \rho(\lambda x_1) = |\lambda|\rho(x_1) = |\lambda|\|[x]\|_\rho,$$

and for $[x], [y] \in V/\sim_\rho$ with $x_1 \in [x]$ and $y_1 \in [y]$ it follows that

$$\|[x] + [y]\|_\rho = \|[x + y]\|_\rho = \rho(x_1 + y_1)$$
$$\le \rho(x_1) + \rho(y_1) = \|[x]\|_\rho + \|[y]\|_\rho.$$

$\square$

Thus $\left(V/\sim_\rho, \|\cdot\|_\rho\right)$ is a normed space.

Since for a measure space $(\Omega, \mathcal{A}, \mu)$ the term $N_p(u) = \left(\int |u|^p \, d\mu\right)^{\frac{1}{p}}$ is for $1 \leq p < \infty$ a semi-norm on $\mathcal{L}^p(\Omega)$ it follows that

$$L^p(\Omega) := \mathcal{L}^p(\Omega) / \sim_{N_p} \tag{8.23}$$

is a normed space. We know that $N_p(u - v) = 0$ if and only if $u = v$ μ-a.e., $1 \leq p < \infty$, and hence we have

Proposition 8.19. *The space* $L^p(\Omega) = \mathcal{L}^p(\Omega) / \sim_\mu$ *is a normed space with norm*

$$\|[u]\|_{L^p} := \|[u]\|_p := N_p(u). \tag{8.24}$$

We now want to lift the Fischer-Riesz theorem, Theorem 7.35, to $L^p(\Omega)$.

Theorem 8.20. *The normed space* $(L^p(\Omega), \|.\|_{L^p})$, $1 \leq p < \infty$, *is a Banach space where we have written* $\|u\|_{L^p}$ *for* $\|[u]\|_{L^p}$.

Proof. Let $([u_k])_{k \in \mathbb{N}}$ be a Cauchy sequence in $L^p(\Omega)$. For $v_k \in [u_k]$ it follows that $(v_k)_{k \in \mathbb{N}}$ is a Cauchy sequence in $\mathcal{L}^p(\Omega)$ and this property is independent of the chosen representatives $v_k \in [u_k]$. Hence by Theorem 7.35, $(v_k)_{k \in \mathbb{N}}$ determines μ-a.e. an element $u \in \mathcal{L}^p(\Omega)$ satisfying $N_p(v_k - u) \to 0$ as $k \to \infty$. We claim that

$$\lim_{k \to \infty} \|[u_k] - [u]\|_{L^p} = 0. \tag{8.25}$$

We note that $[u_k] - [u] = [u_k - u]$ and consequently

$$\|[u_k] - [u]\|_{L^p} = \|[u_k - u]\|_{L^p} = N_p(v_k - u)$$

and these equalities are independent from the representatives. Since $\lim_{k \to \infty} N_p(v_k - u) = 0$ it follows that (8.25) holds, i.e. in $(L^p(\Omega), \|.\|_{L^p})$ every Cauchy sequence has a limit and hence this is a Banach space. $\quad\square$

Theorem 8.21. *Let* $(\Omega, \mathcal{A}, \mu)$ *be a measure space. The span of the set of all simple functions* $u \in \mathcal{S}(\Omega)$ *with* $\mu(\{x \in \Omega \mid u(x) \neq 0\}) < \infty$ *is dense in* $L^p(\Omega)$, $1 \leq p < \infty$.

Proof. Decompose $f \in L^p(\Omega)$ according to $f = f^+ - f^-$. For f^+ we can find a sequence $(u_k)_{k \in \mathbb{N}}$ of simple functions such that $0 \leq u_k \leq u_{k+1} \leq f^+$ and $\lim_{k \to \infty} u_k = f$ μ-a.e. This implies first of all that $\mu(\{x \in \Omega \mid u_k(x) \neq 0\}) < \infty$ and secondly $u_k \in L^p(\Omega)$. Moreover we have $|f^+ - u_k|^p \leq |f|^p$ and the dominated convergence theorem gives the result for f^+ which implies the general result. $\quad\square$

In Problem 11 we will discuss the space $L^\infty(\Omega) := \mathcal{L}^\infty(\Omega)/\sim_\mu$.

The Banach spaces $(L^p(\Omega), \|.\|_{L^p})$ are from the structural point of view nice objects. However we must be very careful when working in $L^p(\Omega)$ since elements of $L^p(\Omega)$ are not any more functions but they are equivalence classes of functions. The common approach is to pretend that elements of $L^p(\Omega)$ are functions, i.e. to work with representatives, and to check that results are independent of the choice of the representatives. This needs some caution and experience. For example for $u \in \mathcal{L}^p(\Omega)$ which is a pointwisely defined function $u : \Omega \to \mathbb{R}$ the statement

$$u(x_0) = \max_{x \in \Omega} u(x)$$

makes sense. But for $[u] \in L^p(\Omega)$ such a statement does not make sense since $[u](x_0)$ is not defined independently of the representatives - it is not defined at all! Still we will later on follow the custom to write $u \in L^p(\Omega)$ when we should write $[u] \in L^p(\Omega)$.

We leave the spaces $L^p(\Omega)$ here but return to them in Part 8 in this volume and in particular in Part 12 where we will discuss in more detail the properties of $L^p(\Omega)$ as Banach spaces.

Problems

Where appropriate, results on the relationships between (improper) Riemann and Lebesgue integrals should be used to evaluate certain integrals below.

1.   a) For $x \geq 0$ consider the integral

$$\varphi(x) := \int_{\mathbb{R}} \frac{e^{-x(1+y^2)}}{1+y^2} \lambda^{(1)}(dy)$$

and prove that $\varphi : [0, \infty) \to \mathbb{R}$ is continuous with $\varphi(0) = \pi$.

   b) Show that the function

$$\psi(\xi) := \int_{(0,\infty)} e^{-\xi x} \left(\frac{\sin x}{x}\right)^k \lambda^{(1)}(dx), k \geq 2,$$

is well defined and continuous on $[0, \infty)$.

2. Give precise conditions on u and μ which will imply for $m \in \mathbb{N}$ that

$$\frac{\partial^m g(x)}{\partial x^m} = \int \frac{\partial u(x, \omega)}{\partial x^m} \mu(d\omega)$$

where $g(x) = \int u(x, \omega) \mu(d\omega)$.

3. For the function in Problem 1 a) show

$$\varphi'(x) = -\sqrt{\frac{\pi}{x}} e^{-x}, x > 0.$$

4.   a) Prove that the function $g : \mathbb{R} \to \mathbb{R}, g(x) = e^{-\frac{x^2}{2}}$, solves the differential equation

$$(*) \qquad g'(x) + xg(x) = 0.$$

   b) Define $\tilde{g}(\xi) = \frac{1}{\sqrt{2\pi}} \int_{\mathbb{R}} \cos(\xi x) e^{-\frac{x^2}{2}} \lambda^{(1)}(dx)$ and prove that $\tilde{g}$ also satisfies $(*)$.

   c) Use the fact that two differentiable solutions h_1 and h_2 of $(*)$ differ only by a multiplicative constant, i.e. $h_1 = ch_2$, to conclude that $\tilde{g} = g$.
Hint: use the fact that $\int_{\mathbb{R}} e^{-x^2} dx = \sqrt{\pi}$, see (I.28.29) or Theorem I.30.14.

5. Let μ be a bounded measure on $\mathcal{B}^{(1)}$.

   a) Show that

$$\tilde{\mu}(y) := \int_{\mathbb{R}} \cos(yx) \, \mu(dx)$$

is a continuous function $\tilde{\mu} : \mathbb{R} \to \mathbb{R}$.

   b) Find conditions on μ in terms of integrability properties of the functions $x \mapsto |x|^k$, $k \in \mathbb{N}$, which will imply that $\tilde{\mu}$ is an N-times continuously differentiable function.

6. Formulate Jensen's inequality for the probability measure $\frac{1}{N} \sum_{k=1}^{N} \epsilon_k$ on $\mathcal{B}^{(1)}$.

7. For an integrable function $u : [0,1] \to \mathbb{R}$ such that $|u(x)| \le 1$ show that

$$\int_0^1 (1 - u^2(x))^{\frac{1}{2}} dx \le \left(1 - \left(\int_0^1 u(x)dx\right)^2\right)^{\frac{1}{2}}.$$

8.   a) Let $u, w : [a,b] \to \mathbb{R}$ be integrable functions and suppose that $w(x) \ge 0$ but $\int_a^b w(x)\lambda^{(1)}(dx) > 0$. Further assume that $m \le u \le M$. For a convex function $\varphi : [m, M] \to \mathbb{R}$ prove

$$\varphi\left(\frac{1}{\int_{[a,b]} w(x)\lambda^{(1)}(dx)} \int_{[a,b]} u(x)w(x)\lambda^{(1)}(dx)\right)$$

$$\le \frac{1}{\int_{[a,b]} w(x)\lambda^{(1)}(dx)} \int_a^b \varphi(u(x))w(x)\lambda^{(1)}(dx).$$

Hint: prove that $\mu(A) := \int_A w(x)\lambda^{(1)}(dx)$ is a bounded measure on $[a, b] \cap \mathcal{B}^{(1)}$.

b) For a continuous function $u : [0,1] \to \mathbb{R}$ prove that

$$\int_0^1 u(x) \, dx \le \ln\left(\int_0^1 e^{u(x)} dx\right).$$

9. Prove that the Fresnel integrals

$$\int_0^\infty \cos(x^2)dx \quad \text{and} \quad \int_0^\infty \sin(x^2)dx$$

exist as improper Riemann integrals but not as Lebesgue integrals. **Hint:** use the transformation $x^2 = y$ and investigate the transformed integrals. Now try to find the value of $\int_0^\infty \sin(x^2)dx$ using methods from real analysis. We will also evaluate these two integrals later by using complex variable methods, see Chapter 21.

10. Show that for every $\alpha > 0$ the improper Riemann integral $\int_0^\infty e^{-x^\alpha} dx$ are Lebesgue integrals.

11. Discuss in detail the definition of $L^\infty(\Omega) := \mathcal{L}^\infty(\Omega)/\sim_\mu$ as a Banach space with norm $\|u\|_\infty = \text{esssup}|u(x)|$.

12.   a) Let $1 < q$ and $q \le r \le \infty$. Prove that $f \in L^q(\mathbb{R}^n)$ admits a decomposition $f = g + h$, $g \in L^1(\mathbb{R}^n)$ and $h \in L^r(\mathbb{R}^n)$.

   b) For $1 < q < r \le \infty$ show that $L^1(\mathbb{R}^n) \cap L^r(\mathbb{R}^n) \subset L^q(\mathbb{R}^n)$ and for $u \in L^1(\mathbb{R}^n) \cap L^r(\mathbb{R}^n)$ we have

$$\|u\|_{L^q} \le \|u\|_{L^1}^{\lambda} \|u\|_{L^r}^{1-\lambda}, \quad \lambda = \frac{r - q}{q(r - 1)}.$$

13. This problem aims to remind us that we can also apply the convergence results in L^p spaces. The result itself is often used in the calculus of variations. Use Fatou's Lemma to prove for a sequence $(f_k)_{k \in \mathbb{N}}$, $f_k \in L^1(\mathbb{R}^n)$, $g \le f_k$ for some $g \in L^1(\mathbb{R}^n)$, that $f_k \to f$ $\lambda^{(n)}$-almost everywhere implies $\int f d\lambda^{(n)} \le \liminf_{k \to \infty} \int f_k d\lambda^{(n)}$.

9 Integration on Product Spaces and Applications

Before introducing the Riemann integral for certain domains in $\mathbb{R}^n$ we studied in Chapter II.17 iterated integrals and then, in Chapter II.18 their relation to integrals over a hyper-rectangle in $\mathbb{R}^n$, see Theorem II.18.19. We can interpret a hyper-rectangle as a product of intervals and an interesting question is whether we can find a natural measure on the product of two or finitely many measure spaces such that the integral with respect to this "product measure" will be for certain functions an iterated integral. In Chapter 1 we discussed product σ-fields and their generators and we will use the notations introduced there, see especially Definition 1.19 and Theorem 1.20.

As a first step we will now construct a product measure. Let $(\Omega_j, \mathcal{A}_j, \mu_j)$, $j = 1, \ldots, N$, be measure spaces and let $\mathcal{A} := \bigotimes_{j=1}^{N} \mathcal{A}_j$ be the corresponding product σ-field in $\Omega := \bigtimes_{j=1}^{N} \Omega_j$. Further let $\mathcal{E}_j$ be a generator of $\mathcal{A}_j$, $j = 1, \ldots, N$. We are seeking for a measure μ on $\mathcal{A}$ such that for all $E^{(j)} \in \mathcal{E}_j$ the following holds

$$\mu\left(E^{(1)} \times \cdots \times E^{(N)}\right) = \mu_1(E^{(1)}) \cdot \ldots \cdot \mu_N(E^{(N)}). \tag{9.1}$$

Having in mind the corresponding situation for a hyper-rectangle in $\mathbb{R}^n$, $K := \bigtimes_{j=1}^{n} [a_j, b_j]$, $a_j < b_j$, where we find

$$\lambda^{(n)}(K) = \prod_{j=1}^{n} (b_j - a_j) = \prod_{j=1}^{n} \lambda^{(1)}([a_j, b_j]), \tag{9.2}$$

it is natural to ask a condition such as (9.1) for a product measure. We want to establish the existence and uniqueness of a product measure μ satisfying (9.1) and it turns out that the uniqueness is quite easy to show.

Theorem 9.1. *For $j = 1, \ldots, N$ let $(\Omega_j, \mathcal{A}_j, \mu_j)$ be measure spaces and let $\mathcal{E}_j$ be a generator of $\mathcal{A}_j$, i.e. $\sigma(\mathcal{E}_j) = \mathcal{A}_j$. We assume that $\mathcal{E}_j$ is stable under finite intersections, i.e. $E_1^{(j)}, E_2^{(j)} \in \mathcal{E}_j$ implies that $E_1^{(j)} \cap E_2^{(j)} \in \mathcal{E}_j$, and we also require the existence of sequences $\left(E_k^{(j)}\right)_{k \in \mathbb{N}}$, $E_k^{(j)} \in \mathcal{E}_k$, $E_k^{(j)} \subset E_{k+1}^{(j)}$ and $\bigcup_{k \in \mathbb{N}} E_k^{(j)} = \Omega_j$, such that $\mu(E_k^{(j)}) < \infty$, i.e. each measure space $(\Omega_j, \mathcal{A}_j, \mu_j)$ is σ-finite. Then $\mathcal{A}$ admits at most one measure μ such that (9.1) is satisfied.*

Proof. The σ-field $\mathcal{A}$ is generated by $\mathcal{E} := \{E^{(1)} \times \cdots \times E^{(N)} \mid E^{(j)} \in \mathcal{E}_j,\ 1 \leq j \leq N\}$, and $\mathcal{E}$ is also stable under finite intersections. For the sequence $(E_k)_{k\in\mathbb{N}}$, $E_k := E_k^{(1)} \times \cdots \times E_k^{(N)} \in \mathcal{E}$, we find that $E_k \subset E_{k+1}$ and $\bigcup_{k\in\mathbb{N}} E_k = \Omega_1 \times \cdots \times \Omega_N = \Omega$. Furthermore, we have

$$\mu(E_k) = \mu_1\left(E_k^{(1)}\right) \cdot \ldots \cdot \mu_N\left(E_k^{(N)}\right) < \infty.$$

Now we apply Theorem 2.20 to obtain the uniqueness of a measure μ on $\mathcal{A}$ satisfying (9.1). $\qquad\square$

Remark 9.2. From Theorem 9.1 we may already deduce that (9.2) uniquely determines $\lambda^{(n)}$ on $\mathcal{B}^{(n)}$ given $\lambda^{(1)}$ on $\mathcal{B}^{(1)}$.

Now we turn to the existence problem and we urge the reader to first recollect the definition of a ring and an algebra in a set $\Omega \neq \emptyset$, see Definition 2.3. Let $(\Omega_1, \mathcal{A}_1, \mu_1)$ and $(\Omega_2, \mathcal{A}_2, \mu_2)$ be two σ-finite measure spaces. We are searching for a measure τ on $\mathcal{A} := \mathcal{A}_1 \otimes \mathcal{A}_2$ with the property that for $A \times B \in \mathcal{A}_1 \times \mathcal{A}_2$

$$\tau(A \times B) = \mu_1(A)\mu_2(B) \tag{9.3}$$

holds. The system $\mathcal{A}_1 \times \mathcal{A}_2 \subset \mathcal{P}(\Omega_1 \times \Omega_2)$ is in some sense too small and it does not have the structure to apply general existence results such as Theorem 2.24 to τ. Therefore we introduce

$$\mathcal{A}_{1,2} := \left\{ \bigcup_{j=1}^{N} A_j \times B_j \,\middle|\, A_j \in \mathcal{A}_1,\ B_j \in \mathcal{A}_2,\ A_j \times B_j \cap A_l \times B_l = \emptyset \text{ for } j \neq l,\ N \in \mathbb{N} \right\}.$$

$$\tag{9.4}$$

Clearly $\emptyset$ and $\Omega_1 \times \Omega_2$ belong to $\mathcal{A}_{1,2}$. Furthermore we find for $A_1 \times B_1$, $A_2 \times B_2 \in \mathcal{A}_1 \times \mathcal{A}_2$ that

$$(A_1 \times B_1) \cap (A_2 \times B_2) = (A_1 \cap A_2) \times (B_1 \cap B_2)$$

and

$$(A_1 \times B_1)^{\complement} = (\Omega_1 \times B_1^{\complement}) \cup (A_1^{\complement} \times \Omega_2)$$

implying that $\mathcal{A}_{1,2}$ is a ring, in fact an algebra and $\sigma(\mathcal{A}_{1,2}) = \mathcal{A}$. For $A \times B \in \mathcal{A}_1 \times \mathcal{A}_2$ we set as expected

$$\tau(A \times B) := \mu_1(A)\mu_2(B),$$

174

where we now use the convention $0 \cdot \infty = \infty \cdot 0 = 0$.

Consider $A \times B = \bigcup_{k \in \mathbb{N}} A_k \times B_k$, $A_k \in \mathcal{A}_1$, $B_k \in \mathcal{A}_2$, where we suppose that the sets $(A_k \times B_k)_{k \in \mathbb{N}}$ are mutually disjoint. For the characteristic functions we find

$$\chi_A(\omega_1)\chi_B(\omega_2) = \chi_{A \times B}(\omega_1, \omega_2) = \sum_{k=1}^{\infty} \chi_{A_k \times B_k}(\omega_1, \omega_2) = \sum_{k=1}^{\infty} \chi_{A_k}(\omega_1)\chi_{B_k}(\omega_2)$$

and the monotone convergence theorem in form of Corollary 5.15 yields

$$\mu_1(A)\chi_B(\omega_2) = \int_{\Omega_1} \chi_A(\omega_1)\chi_B(\omega_2)\mu_1(d\omega_1)$$

$$= \sum_{k=1}^{\infty} \int_{\Omega_1} \chi_{A_k}(\omega_1)\chi_{B_k}(\omega_2)\mu_1(d\omega_1) = \sum_{k=1}^{\infty} \mu_1(A_k)\chi_{B_k}(\omega_2)$$

and further

$$\mu_1(A)\mu_2(B) = \sum_{k=1}^{\infty} \mu_1(A_k) \int_{\Omega_2} \chi_{B_k}(\omega_2)\mu_2(d\omega_2) = \sum_{k=1}^{\infty} \mu_1(A_k)\mu_2(B_k). \quad (9.5)$$

This allows us to define τ on $\mathcal{A}_{1,2}$ by

$$\tau(C) = \sum_{j=1}^{N} \mu_1(A_j)\mu_2(B_j), \quad C = \bigcup_{j=1}^{N} A_j \times B_j, \quad (9.6)$$

where the sets $(A_j \times B_j)_{j=1,\ldots,N}$ are mutually disjoint. Indeed the above calculation shows that (9.5) is independent of the representation of C: if $C = \bigcup_{j=1}^{N} A_j \times B_j = \bigcup_{l=1}^{M} D_l \times F_l$ then we can switch to the joint refinement $((A_j \cap D_l) \times (B_j \cap F_l))_{j=1,\ldots,N, l=1,\ldots,M}$ and find

$$C = \bigcup_{j=1}^{N} \bigcup_{l=1}^{M} ((A_j \cap D_l) \times (B_j \cap F_l)). \quad (9.7)$$

The calculation leading to (9.5) also yields the σ-additivity of τ on $\mathcal{A}_{1,2}$: let $C_k \in \mathcal{A}_{1,2}$, $C_k = \bigcup_{l=1}^{N_k} A_{k,l} \times B_{k,l}$, such that $C_k \cap C_j = \emptyset$ for $k \neq j$, and consider $C := \bigcup_{k \in \mathbb{N}} C_k \in \mathcal{A}$, i.e. $C = \bigcup_{i=1}^{M} D_i \times F_i$ with mutually disjoint sets $D_i \times F_i$, $i \in M$. Since $C = \bigcup_{k \in \mathbb{N}} \bigcup_{l=1}^{N_k} A_{k,l} \times B_{k,l} = \bigcup_{i=1}^{M} D_i \times F_i$ we can conclude that

$$
\tau(C) = \sum_{i=1}^{M} \mu_1(D_i)\mu_2(F_i) = \sum_{k=1}^{\infty} \sum_{l=1}^{N_k} \mu_1(A_{k,l})\mu_2(B_{k,l})
$$
$$
= \sum_{k=1}^{\infty} \tau(C_k).
$$

Finally we remark that $\emptyset = (\Omega_1 \times \Omega_2)^{\complement} \in \mathcal{A}_{1,2}$, i.e. $\tau(\emptyset) = 0$ and hence we have proved that τ is a pre-measure on the ring $\mathcal{A}_{1,2}$. Using Theorem 2.24 we arrive finally at

Theorem 9.3. *Let $(\Omega_1, \mathcal{A}_1, \mu_1)$ and $(\Omega_2, \mathcal{A}_2, \mu_2)$ be two σ-finite measure spaces. On the product $(\Omega_1 \times \Omega_2, \mathcal{A}_1 \otimes \mathcal{A}_2)$ there exists a unique σ-finite measure $\mu_1 \otimes \mu_2$ such that for all sets $A \times B \in \mathcal{A}_1 \times \mathcal{A}_2$ we have*

$$
\mu(A \times B) = \mu_1(A)\mu_2(B). \tag{9.8}
$$

Remark 9.4. If in the situation of Theorem 9.3 both measure spaces are finite, then $\mu_1 \otimes \mu_2$ is a finite measure on $\mathcal{A}_1 \otimes \mathcal{A}_2$. In addition for two probability spaces $(\Omega_1, \mathcal{A}_1, P_1)$ and $(\Omega_2, \mathcal{A}_2, P_2)$ the measure $P_1 \otimes P_2$ is a probability measure too.

The proof of Theorem 9.3 can be extended to the case of finitely many σ-finite measure spaces, i.e. we have

Corollary 9.5. *Let $(\Omega_j, \mathcal{A}_j, \mu_j)$, $j = 1, \ldots, N$, be a finite family of σ-finite measure spaces. On the product space $(\Omega, \mathcal{A})$, where $\Omega = \bigtimes_{j=1}^{N} \Omega_j$ and $\mathcal{A} = \bigotimes_{j=1}^{N} \mathcal{A}_j$, there exists a unique measure denoted by $\mu_1 \otimes \cdots \otimes \mu_N$ such that for all $A_j \in \mathcal{A}_j$ the following holds*

$$
(\mu_1 \otimes \cdots \otimes \mu_N)(A_1 \times \cdots \times A_N) = \mu_1(A_1) \cdot \ldots \cdot \mu_N(A_N). \tag{9.9}
$$

Definition 9.6. *The measure $\mu := \mu_1 \otimes \cdots \otimes \mu_N$ in Corollary 9.5 is called the **product measure** of the measures μ_j, $j = 1, \ldots, N$, and $(\Omega, \mathcal{A}, \mu)$ is called the **product (measure) space** of the measure spaces $(\Omega_j, \mathcal{A}_j, \mu_j)$.*

The calculation leading to (9.5) allows us to immediately state

$$\int f(\omega_1)g(\omega_2)(\mu_1 \otimes \mu_2)(\mathrm{d}(\omega_1, \omega_2)) = \int f(\omega_1)\mu_1(\mathrm{d}\omega_1) \int g(\omega_2)\mu_2(\mathrm{d}\omega_2)$$

$$= \int \left(\int f(\omega_1)\mu_1(\mathrm{d}\omega_1) \right) g(\omega_2)\mu_2(\mathrm{d}\omega_2)$$

$$= \int \left(\int f(\omega_1)g(\omega_2)\mu_1(\mathrm{d}\omega_1) \right) \mu_2(\mathrm{d}\omega_2)$$

$$= \int \left(\int f(\omega_1)g(\omega_2)\mu_2(\mathrm{d}\omega_2) \right) \mu_1(\mathrm{d}\omega_1)$$

for integrable functions $f : \Omega_1 \to \mathbb{R}$ and $g : \Omega_2 \to \mathbb{R}$. Our next aim is to prove such a result for a function $h : \Omega_1 \times \Omega_2 \to \mathbb{R}$ which is integrable with respect to $\mu_1 \otimes \mu_2$, i.e. we want to find conditions when

$$\int h(\omega_1, \omega_2)(\mu_1 \otimes \mu_2)(\mathrm{d}(\omega_1, \omega_2)) = \int \left(\int h(\omega_1, \omega_2)\mu_1(\mathrm{d}\omega_1) \right) \mu_2(\mathrm{d}\omega_2)$$

$$= \int \left(\int h(\omega_1, \omega_2)\mu_2(\mathrm{d}\omega_2) \right) \mu_1(\mathrm{d}\omega_1)$$

holds. In other words we want to prove that under certain conditions on h the integral with respect to a product measure is equal to an iterated integral and iterated integrals are independent of the order of integration. This needs some preparation.

Definition 9.7. *Let X and Y be two non-empty sets and $S \subset X \times Y$. We define the **x-section** $S_x \subset Y$ and the **y-section** $S^y \subset X$ of S by*

$$S_x := \left\{ y \in Y \,\middle|\, (x, y) \in S \right\}, \quad x \in X \tag{9.10}$$

and

$$S^y := \left\{ x \in X \,\middle|\, (x, y) \in S \right\}, \quad y \in Y. \tag{9.11}$$

Lemma 9.8. *Given two measurable spaces $(\Omega_1, \mathcal{A}_1)$ and $(\Omega_2, \mathcal{A}_2)$ with product σ-field $\mathcal{A} = \mathcal{A}_1 \otimes \mathcal{A}_2$. For every $S \in \mathcal{A}_1 \otimes \mathcal{A}_2$ and all $\omega_1 \in \Omega_1$, $\omega_2 \in \Omega_2$ the ω_1-sections and the ω_2-sections are measurable sets, i.e.*

$$S_{\omega_1} \in \mathcal{A}_2 \quad and \quad S^{\omega_2} \in \mathcal{A}_1. \tag{9.12}$$

Proof. We prove that $\mathcal{E} := \{S \in \mathcal{A}_1 \otimes \mathcal{A}_2 \,|\, S_{\omega_1} \in \mathcal{A}_2, \omega_1 \in \Omega_1\}$ is a σ-field and $\mathcal{A} \subset \mathcal{E}$. This of course implies the result for the ω_1-sections and the case of the ω_2-sections goes analogously. First we note that for $S = A \times B$, $A \in \mathcal{A}_1$, $B \in \mathcal{A}_2$, it follows for $\omega_1 \in \Omega_1$ that $S_{\omega_1} \in \{\emptyset, B\} \subset \mathcal{A}_2$. Since $\sigma(\mathcal{A}_1 \times \mathcal{A}_2) = \mathcal{A}_1 \otimes \mathcal{A}_2$ it follows that $\mathcal{A}_1 \otimes \mathcal{A}_2 \subset \mathcal{E}$, and in particular we have $\Omega_1 \times \Omega_2 \in \mathcal{E}$. Moreover, if $S \in \mathcal{E}$ we find for all $\omega_1 \in \Omega_1$ that

$$((\Omega_1 \times \Omega_2)\backslash S)_{\omega_1} = \Omega_2 \backslash S_{\omega_1} \in \mathcal{A}_2.$$

Now let $(S_k)_{k\in\mathbb{N}}$ be a sequence of mutually disjoint elements in $\mathcal{E}$. Since $\mathcal{A}_2$ is a σ-field it follows that $\left(\bigcup_{k\in\mathbb{N}} S_k\right)_{\omega_1} = \bigcup_{k\in\mathbb{N}} (S_k)_{\omega_1}$ and hence $\bigcup_{k\in\mathbb{N}} S_k \in \mathcal{E}$ implying that $\mathcal{E}$ is a σ-field and $\mathcal{A}_1 \otimes \mathcal{A}_2 \subset \mathcal{E}$, and hence the lemma is shown. $\qquad\square$

Analogously to the ω_1- and ω_2-sections of sets we now define sections for mappings.

Definition 9.9. *Let $f : X \times Y \to Z$ be a mapping. The x-section $f_x : Y \to Z$ and the y-section $f^y : X \to Z$ of f are defined by*

$$f_x(y) = f^y(x) = f(x, y). \tag{9.13}$$

Example 9.10. For arbitrary sets $X, Y \neq \emptyset$ we find for the characteristic functions of $S \subset X \times Y$, $S_x \subset Y$ and $S^y \subset X$

$$(\chi_S)_x = \chi_{S_x} \quad \text{and} \quad (\chi_S)^y = \chi_{S^y}. \tag{9.14}$$

Corollary 9.11. *Let $(\Omega_1, \mathcal{A}_1)$, $(\Omega_2, \mathcal{A}_2)$ and $(\Omega_3, \mathcal{A}_3)$ be three measurable spaces and let $\mathcal{A} = \mathcal{A}_1 \otimes \mathcal{A}_2$. If $f := \Omega_1 \times \Omega_2 \to \Omega_3$ is $\mathcal{A} - \mathcal{A}_3$-measurable then for every $\omega_1 \in \Omega$, the ω_1-section f_{ω_1} of f is $\mathcal{A}_2 - \mathcal{A}_3$-measurable and for every $\omega_2 \in \Omega_2$ the ω_2-section f^{ω_2} of f is $\mathcal{A}_1 - \mathcal{A}_3$-measurable.*

Proof. First we note that for $A \subset \Omega_3$

$$(f_{\omega_1})^{-1}(A) = \{\omega_2 \in \Omega_2 \,|\, (\omega_1, \omega_2) \in f^{-1}(A)\} = (f^{-1}(A))_{\omega_1}$$

and

$$(f^{\omega_2})^{-1}(A) = (f^{-1}(A))^{\omega_2},$$

and now the measurability of f and Lemma 9.8 imply the result. $\qquad\square$

Often we have used Dynkin systems to prove that a certain family of subsets of a given set form a σ-field. Sometimes a different way is more convenient, namely a monotone class argument.

Definition 9.12. *Let $\Omega \neq \emptyset$ be a set and $\mathcal{C} \subset \mathcal{P}(\Omega)$. We call $\mathcal{C}$ a **monotone class** if it is closed under countable increasing unions and countable decreasing intersections, i.e. whenever for $A_k, B_k \in \mathcal{C}$ such that $A_k \subset A_{k+1}$ and $B_k \supset B_{k+1}$ it follows that*

$$\bigcup_{k \in \mathbb{N}} A_k \in \mathcal{C} \quad and \quad \bigcap_{k \in \mathbb{N}} B_k \in \mathcal{C}.$$

Remark 9.13. Every σ-field is a monotone class as is the intersection of any family of monotone classes again a monotone class. Hence for $\mathcal{E} \subset \mathcal{P}(\Omega)$ we can look at all intersections of monotone classes containing $\mathcal{E}$ which gives the monotone class $\mathcal{C}(\mathcal{E})$ generated by $\mathcal{E}$, see also Problem 5.

Lemma 9.14 (Monotone class lemma). *For every algebra $\mathcal{A}$ in $\Omega \neq \emptyset$ we have*

$$\sigma(\mathcal{A}) = \mathcal{C}(\mathcal{A}). \tag{9.15}$$

Proof. From Remark 9.13 it follows that $\mathcal{C}(\mathcal{A}) \subset \sigma(\mathcal{A})$. We will prove that $\mathcal{C}(\mathcal{A})$ is a σ-field if $\mathcal{A}$ is an algebra which will imply $\sigma(\mathcal{A}) \subset \mathcal{C}(\mathcal{A})$, i.e. (9.15). For $A \in \mathcal{C}(\mathcal{A})$ define

$$\mathcal{C}(A) := \left\{ B \in \mathcal{C}(\mathcal{A}) \,\middle|\, A \backslash B, \, B \backslash A, \, A \cap B \in \mathcal{C}(\mathcal{A}) \right\}.$$

Since $\mathcal{A}$ is an algebra it follows that $\emptyset$ and A belong to $\mathcal{C}(A)$ and $A \in \mathcal{C}(B)$ if and only if $B \in \mathcal{C}(A)$. Moreover $\mathcal{C}(A)$ is itself a monotone class which is proved by inspection. For $A \in \mathcal{A}$ it follows that $B \in \mathcal{C}(A)$ for all $B \in \mathcal{A}$ since $\mathcal{A}$ is an algebra, implying that $\mathcal{A} \subset \mathcal{C}(A)$, hence $\mathcal{C}(\mathcal{A}) \subset \mathcal{C}(A)$ for all $A \in \mathcal{A}$. Further $B \in \mathcal{C}(\mathcal{A})$ yields $B \in \mathcal{C}(A)$ for all $A \in \mathcal{A}$, i.e. $A \in \mathcal{C}(B)$ for all $A \in \mathcal{A}$, which gives $\mathcal{A} \subset \mathcal{C}(B)$ and therefore $\mathcal{C}(\mathcal{A}) \subset \mathcal{C}(B)$. Thus, $A, B \in \mathcal{C}(\mathcal{A})$ yields $A \backslash B \in \mathcal{C}(\mathcal{A})$ and $A \cap B \in \mathcal{C}(\mathcal{A})$, and since $\Omega \in \mathcal{A}$ we have $\Omega \in \mathcal{C}(\mathcal{A})$. Thus we have proved that $\mathcal{C}(A)$ is an algebra. For a sequence $(A_k)_{k \in \mathbb{N}}$ of sets $A_k \in \mathcal{C}(\mathcal{A})$ we can first use the fact that $\mathcal{C}(\mathcal{A})$ is an algebra to deduce that $\bigcup_{k=1}^{N} A_k \in \mathcal{C}(\mathcal{A})$ and now, since $\mathcal{C}(\mathcal{A})$ is closed under countable increasing unions we derive that $\bigcup_{k \in \mathbb{N}} A_k \in \mathcal{C}(\mathcal{A})$, i.e. $\mathcal{C}(\mathcal{A})$ is a σ-field and the lemma is proved. $\qquad\square$

The following result already links the product measure with integrals of sections.

Theorem 9.15. *Let $(\Omega_1, \mathcal{A}_1, \mu_1)$ and $(\Omega_2, \mathcal{A}_2, \mu_2)$ be two σ-finite measure spaces and $S \in \mathcal{A} = \mathcal{A}_1 \otimes \mathcal{A}_2$. For all $\omega_1 \in \Omega_1$ and all $\omega_2 \in \Omega_2$ the mappings $\omega_1 \mapsto \mu_2(S_{\omega_1})$ and $\omega_2 \mapsto \mu_1(S^{\omega_2})$ are measurable on $(\Omega_1, \mathcal{A}_1)$ and $(\Omega_2, \mathcal{A}_2)$, respectively, and the following holds*

$$(\mu_1 \otimes \mu_2)(S) = \int_{\Omega_1} \mu_2(S_{\omega_1})\mu_1(\mathrm{d}\omega_1) = \int_{\Omega_2} \mu_1(S^{\omega_2})\mu_2(\mathrm{d}\omega_2). \tag{9.16}$$

Proof. We prove the theorem first under the additional assumption that both measures are finite. We denote by $\mathcal{C}$ all sets in $\mathcal{A} = \mathcal{A}_1 \otimes \mathcal{A}_2$ for which the theorem holds. First we note that for $S = A \times B \in \mathcal{A}_1 \times \mathcal{A}_2$ it follows that

$$\mu_2(S_{\omega_1}) = \chi_A(\omega_1)\mu_2(B) \quad \text{and} \quad \mu_1(S^{\omega_2}) = \mu_1(A)\chi_B(\omega_2)$$

implying that $S \in \mathcal{C}$. Since μ_1 and μ_2 are additive, we deduce that finite unions of mutually disjoint "rectangles" $S_k = A_k \times B_k \in \mathcal{A}_1 \times \mathcal{A}_2$ also belong to $\mathcal{C}$ as does $(A \times B)^{\complement} = (\Omega_1 \times B^{\complement}) \cup (A^{\complement} \times \Omega_2)$ for $A \times B \in \mathcal{A}_1 \times \mathcal{A}_2$. Now let $(S_k)_{k\in\mathbb{N}}$ be a sequence in $\mathcal{C}$, $S_k \subset S_{k+1}$ and $S = \bigcup_{k\in\mathbb{N}} S_k$, and consider the sequence $u_k : \Omega_2 \to \mathbb{R}$ defined by $u_k(\omega_2) = \mu_1((S_k)^{\omega_2})$. Each function u_k is measurable since $S_k \in \mathcal{C}$ and further $u_k \leq u_{k+1}$ with $\lim_{k\to\infty} u_k(\omega_2) = \mu_1(S^{\omega_2})$. It follows that S^{ω_2} is measurable and further, the monotone convergence theorem yields

$$\int_{\Omega_2} \mu_1(S^{\omega_2})\mu_2(\mathrm{d}\omega_2) = \lim_{k\to\infty} \int_{\Omega_2} \mu_1((S_k)^{\omega_2})\mu_2(\mathrm{d}\omega)$$
$$= \lim_{k\to\infty} (\mu_1 \otimes \mu_2)(S_k) = (\mu_1 \otimes \mu_2)(S).$$

Analogously we see that $(\mu_1 \otimes \mu_2)(S) = \int_{\Omega_1} \mu_2(S_{\omega_1})\mu_1(\mathrm{d}\omega_1)$ and hence $S = \bigcup_{k\in\mathbb{N}} S_k \in \mathcal{C}$. Now let $(Q_k)_{k\in\mathbb{N}}$, $Q_k \in \mathcal{C}$, be a decreasing sequence $Q_k \supset Q_{k+1}$. Since μ_1 and μ_2 are finite measures, as assumed for the moment, we deduce that by $v_k : \Omega_2 \to \mathbb{R}$, $v_k(\omega_2) = \mu_1(Q_k)^{\omega_2}$, a decreasing sequence of functions $v_k \in \mathcal{L}^1(\Omega_2)$ is defined, where the integrability follows from $\mu_1(Q_k)^{\omega_2} \leq \mu_1(Q_1)^{\omega_1} \leq \mu_1(\Omega_1) < \infty$. Applying the dominated convergence theorem to this sequence we obtain that $Q := \bigcap_{k\in\mathbb{N}} Q_k \in \mathcal{C}$. Thus $\mathcal{C}$ is a monotone class containing $\mathcal{A}_1 \times \mathcal{A}_2$, and by Lemma 9.14 it follows that $\mathcal{C}(\mathcal{A}_1 \times \mathcal{A}_2) = \mathcal{A} = \mathcal{A}_1 \otimes \mathcal{A}_2$.

Now we assume that both μ_1 and μ_2 are σ-finite and let $A_k \in \mathcal{A}_1$, $B_k \in \mathcal{A}_2$ be increasing sequences such that $\bigcup_{k\in\mathbb{N}} A_k = \Omega_1$ and $\bigcup_{k\in\mathbb{N}} B_k = \Omega_2$ and $\mu_1(A_k) < \infty$, $\mu_2(B_k) < \infty$. For $S \subset \mathcal{A} = \mathcal{A}_1 \otimes \mathcal{A}_2$, we can apply the results of the theorem as well as the arguments of its proof to the sets $S \cap (A_k \times B_k)$ which yields

$$(\mu_1 \otimes \mu_2)(S \cap (A_k \times B_k))$$

$$= \int_{\Omega_1} \chi_{A_k}(\omega_1)\mu_2(S_{\omega_1} \cap B_k)\mu(d\omega_1) = \int_{\Omega_2} \chi_{B_k}(\omega_2)\mu_1(S^{\omega_2} \cap A_k)\mu_2(d\omega_2),$$

and an application of the monotone convergence theorem finally induces the result. $\qquad\square$

The following two theorems are cornerstones in the integration theory following Lebesgue's approach and they have many important consequences.

Theorem 9.16 (L. Tonelli). *Let $(\Omega_1, \mathcal{A}_1, \mu_1)$ and $(\Omega_2, \mathcal{A}_2, \mu_2)$ be two σ-finite measure spaces and $f : \Omega_1 \times \Omega_2 \to [0, \infty]$ an $\mathcal{A}_1 \otimes \mathcal{A}_2$-measurable function. For every $\omega_1 \in \Omega_1$ and $\omega_2 \in \Omega_2$ the functions*

$$\omega_1 \mapsto \int f_{\omega_1}\, d\mu_2 \quad and \quad \omega_2 \mapsto \int f^{\omega_2}\, d\mu_1$$

are $\mathcal{A}_1$- and $\mathcal{A}_2$-measurable, respectively. Moreover we have

$$\int f\, d(\mu_1 \otimes \mu_2) = \int \left(\int f^{\omega_2}\, d\mu_1 \right) \mu_2(d\omega_2) \qquad (9.17)$$

$$= \int \left(\int f_{\omega_1}\, d\mu_2 \right) \mu_1(d\omega_1).$$

Proof. For characteristic functions of sets $S \in \mathcal{A}_1 \otimes \mathcal{A}_2$ this is just the statement of Theorem 9.15 and it can be extended immediately to simple functions by linearity (with non-negative scalars). Since f is measurable it is a pointwise limit of an increasing sequence $(u_k)_{k\in\mathbb{N}}$ of simple functions u_k. From the monotone convergence theorem we deduce first that $(u_k)_{\omega_1}$ increases to $\omega_1 \mapsto \int f_{\omega_1}\, d\mu_2$ and $(u_k)^{\omega_2}$ increases to $\omega_2 \mapsto \int f^{\omega_2}\, d\mu_1$ implying in particular that these functions are measurable. A further application of the monotone convergence theorem now yields

$$\int \left(\int f_{\omega_1}\, d\mu_2 \right) \mu_1(d\omega_1) = \lim_{k\to\infty} \int \left(\int u_k\, d\mu_2 \right)_{\omega_1} \mu_1(d\omega_1)$$

$$= \lim_{k\to\infty} \int u_k\, d(\mu_1 \otimes \mu_2) = \int f\, d(\mu_1 \otimes \mu_2)$$

and

$$\int \left(\int f^{\omega_2} \, d\mu_1 \right) \mu_2(d\omega_2) = \lim_{k \to \infty} \int \left(\int u_k \, d\mu_1 \right)^{\omega_2} \mu_2(d\omega_2)$$

$$= \lim_{k \to \infty} \int u_k \, d(\mu_1 \otimes \mu_2) = \int f \, d(\mu_1 \otimes \mu_2).$$

$\square$

Replacing in Tonelli's theorem f by a function integrable with respect to $\mu_1 \otimes \mu_2$ we obtain

Theorem 9.17 (G. Fubini). *Let $(\Omega_1, \mathcal{A}_1, \mu_1)$ and $(\Omega_2, \mathcal{A}_2, \mu_2)$ be two σ-finite measure spaces and $f : \Omega_1 \times \Omega_2 \to \overline{\mathbb{R}}$ be an $\mathcal{A}$-measurable function, $\mathcal{A} = \mathcal{A}_1 \otimes \mathcal{A}_2$. Consider the following three integrals*

$$\int |f| \, d(\mu_1 \otimes \mu_2), \quad \int \left(\int |f| \, d\mu_1 \right) d\mu_2, \quad \int \left(\int |f| \, d\mu_2 \right) d\mu_1. \quad (9.18)$$

If one of these integrals is finite then all three are finite. In this case $f \in \mathcal{L}^1(\Omega_1 \times \Omega_2)$ and further we have

$$\omega_1 \mapsto f(\omega_1, \omega_2) \text{ is } \mu_2\text{-a.e. an element of } \mathcal{L}^1(\Omega_1) \quad (9.19)$$
$$\omega_2 \mapsto f(\omega_1, \omega_2) \text{ is } \mu_1\text{-a.e. an element of } \mathcal{L}^1(\Omega_2) \quad (9.20)$$
$$\omega_1 \mapsto \int f(\omega_1, \omega_2) \mu_2(d\omega_2) \text{ is an element of } \mathcal{L}^1(\Omega_1) \quad (9.21)$$
$$\omega_2 \mapsto \int f(\omega_1, \omega_2) \mu_1(d\omega_1) \text{ is an element of } \mathcal{L}^1(\Omega_2). \quad (9.22)$$

Moreover (9.17) holds.

Proof. From Tonelli's theorem we deduce that if one of the integrals in (9.18) is finite then they all are finite which in turn implies that $f \in \mathcal{L}^1(\Omega_1 \times \Omega_2)$. The theorem of Tonelli also entails the measurability of the functions $\omega_1 \mapsto f^+(\omega_1, \omega_2)$, $\omega_1 \mapsto f^-(\omega_1, \omega_2)$ as well as of $\omega_2 \mapsto \int f^+(\omega_1, \omega_2)\mu_1(d\omega_1)$ and $\omega_2 \mapsto \int f^-(\omega_1, \omega_2)\mu_1(d\omega_1)$. Since $f^+ \leq |f|$ and $f^- \leq |f|$ we deduce that

$$\int f^+(\omega_1, \omega_2)\mu_1(d\omega_1) \leq \int |f(\omega_1, \omega_2)|\mu_1(d\omega_1) \quad (9.23)$$

and

$$\int f^-(\omega_1, \omega_2)\mu_1(d\omega_1) \leq \int |f(\omega_1, \omega_2)|\mu_1(d\omega_1) \quad (9.24)$$

and the μ_2-integrability of the right hand side of (9.23) and (9.24) imply the μ_2-integrability of the left hand sides as well as the μ_2-a.e. integrability of $\omega_1 \mapsto f^+(\omega_1, \omega_2)$ and $\omega_1 \mapsto f^-(\omega_1, \omega_2)$ which yields (9.19) and (9.21). The statements (9.20) and (9.22) follow completely analogously. Finally, since (9.17) holds for f^+ and f^- we conclude that it holds for $f = f^+ - f^-$. $\quad\square$

Before we pay attention to some theoretical applications of Fubini's theorem, some remarks are in order with respect to its role to evaluate integrals. So far we have no workable method to find the value of an integral, except in some very special cases. In one dimension when handling integrals with respect to the Lebesgue-Borel measure $\lambda^{(1)}$ our position is better: whenever we can identify the Lebesgue integral with the Riemann integral and the fundamental theorem is applicable to the corresponding Riemann integral, then we can take advantage of the theory developed in Volume I to evaluate integrals. Of course this applies to iterated integrals over hyper-rectangles in $\mathbb{R}^n$. However, when combining the transformation theorem, Theorem 6.8, with Fubini's theorem the situation improves dramatically: if the domain of integration is the (almost everywhere) image of a hyper-rectangle, then we can transform the integral to an integral over this hyper-rectangle (with respect to $\lambda^{(n)}$), which in turn by an application of Fubini's theorem becomes an iterated integral, and now we may use the tools available for one-dimensional Riemann integrals.

Our first application of Fubini's theorem, or better of Tonelli's theorem, is a reduction of an integral with respect to a measure μ to a one-dimension integral with respect to the Lebesgue measure $\lambda^{(1)}$. We start with a detailed example.

Consider the rectangle $K := [a, b] \times [c, d] \subset \mathbb{R}^2$, $a < b$, $c < d$. Its Lebesgue measure is of course $\lambda^{(2)}(K) = (b-a)(d-c)$. Since

$$\lambda^{(2)}(K) = \int_K 1 \, d\lambda^{(2)} = \int_{\mathbb{R}^2} \chi_K(z)\lambda^{(2)}(dz), \quad z = (x, y)$$

and since $\chi_K(z) = \chi_K(x, y) = \chi_{[a,b]}(x)\chi_{[c,d]}(y)$ we find by Fubini's or Tonelli's theorem

$$\lambda^{(2)}(K) = \int_{\mathbb{R}} \left(\int_{\mathbb{R}} \chi_K(x,y) \lambda^{(1)}(dy) \right) \lambda^{(1)}(dx) = \int_{\mathbb{R}} \left(\int_{\mathbb{R}} \chi_K(x,y) \lambda^{(1)}(dx) \right) \lambda^{(1)}(dy)$$

$$= \int_{\mathbb{R}} \chi_{[a,b]}(x) \lambda^{(1)}(dx) \int_{\mathbb{R}} \chi_{[c,d]}(y) \lambda^{(1)}(dy) = \int_a^b 1 \lambda^{(1)}(dx) \int_c^d 1 \lambda^{(2)}(dy).$$

For a compact interval $\emptyset \neq \mathring{I} \subset I = [a,b]$ we may now consider a non-negative continuous function $f : I \to [0,\infty)$ and we interpret the integral $\int_a^b f(x)\,dx = \int_I f\,d\lambda^{(1)}$ as area of the set $K \subset \mathbb{R}^2$ bounded by $I \times \{0\} \cup \Gamma(f) \cup \{a\} \times [0, f(a)] \cup \{b\} \times [0, f(b)]$, i.e.

$$K = \{(x,y) \in \mathbb{R}^2 \mid x \in [a,b],\ 0 \leq y \leq f(x)\} = \{(x,y) \in [a,b] \times [0,\infty) \mid y \leq f(x)\},$$

see Figure 9.1.

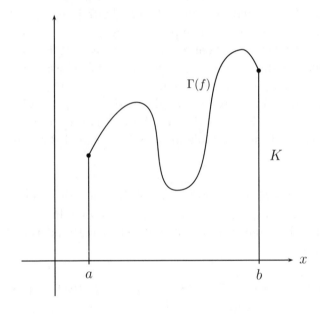

Figure 9.1

On the other hand we know that

$$\lambda^{(2)}(K) = \int_{\mathbb{R}^2} \chi_K(z) \lambda^{(2)}(dz) = \int_K 1\,d\lambda^{(2)}, \quad z = (x,y),$$

and we observe further that

$$\chi_K(x,y) = 1 \text{ if and only if } x \in [a,b] \text{ and } f(x) \geq y$$

as well as

$$\chi_{(0,f(x)]}(y) = 1 \text{ if and only if } x \in [a,b] \text{ and } f(x) \geq y.$$

Therefore we find using the equivalence of the Lebesgue to the Riemann integral in our situation and Tonelli's theorem

$$
\begin{aligned}
\int_a^b f(x)\,dx &= \int_{[a,b]} f\,d\lambda^{(1)} = \int_{[a,b]} \left(\int_0^{f(x)} 1\,dy \right) \lambda^{(1)}(dx) \\
&= \int_{[a,b]} \left(\int_{(0,f(x)]} 1\,\lambda^{(1)}(dy) \right) \lambda^{(1)}(dx) \\
&= \int_{[a,b]} \left(\int_{[0,\infty)} \chi_{(0,f(x)]}(y)\lambda^{(1)}(dy) \right) \lambda^{(1)}(dx) \\
&= \int_{[0,\infty)} \left(\int_{[a,b]} \chi_K(x,y)\lambda^{(1)}(dx) \right) \lambda^{(1)}(dy) \\
&= \int_{[0,\infty)} \lambda^{(1)}\left(\{x \in [a,b] \mid f(x) \geq y\}\right) \lambda^{(1)}(dy) \\
&= \int_0^\infty \lambda^{(1)}\left(\{f \geq y\}\right) \lambda^{(1)}(dy),
\end{aligned}
$$

thus we have derived the formula

$$\int_a^b f(x)\,dx = \int_0^\infty \lambda^{(1)}(\{f \geq y\})\lambda^{(1)}(dy). \tag{9.25}$$

Note that if $f(x) = d$ for all $x \in [a,b]$ then it follows that

$$\{x \in [a,b] \mid f(x) = d \geq y\} = \begin{cases} \emptyset, & y > d \\ [a,b], & y \leq d \end{cases}$$

and consequently we have

$$\lambda^{(1)}(\{f = d \geq y\}) = \begin{cases} 0, & y > d \\ b - a, & 0 < y \leq d \end{cases}$$

implying

$$
\begin{aligned}
\int_0^\infty \lambda^{(1)}(\{f \geq y\})\lambda^{(1)}(dy) &= \int_{[0,d]} (b-a)\lambda^{(1)}(dy) + \int_{(d,\infty)} 0\lambda^{(1)}(dy) \\
&= (b-a)d = (b-a)(d-c), \quad c = 0,
\end{aligned}
$$

185

i.e. we recover as expected $\lambda^{(1)}(K)$ for $K = [a,b] \times [0,d]$, and for general $[a,b] \times [c,d]$ we can use for example the invariance of the integral under translation. To illustrate the calculation we have a look at the set K from Figure 9.2, see below.

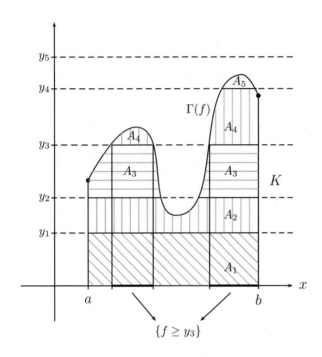

Figure 9.2

$$\{f \geq y_5\} = \mathrm{pr}_1(\emptyset) = \emptyset$$
$$\{f \geq y_4\} = \mathrm{pr}_1(A_5)$$
$$\{f \geq y_3\} = \mathrm{pr}_1(A_4 \cup A_5)$$
$$\{f \geq y_2\} = \mathrm{pr}_1(A_3 \cup A_4 \cup A_5)$$
$$\{f \geq y_1\} = \mathrm{pr}_1(A_2 \cup A_3 \cup A_4 \cup A_5).$$

The important observation is now that the core result derived above holds for every σ-finite measure space and every non-negative measurable function.

Definition 9.18. *Let $(\Omega, \mathcal{A}, \mu)$ be a finite σ-finite measure space and let $f : \Omega \to [0, \infty)$ a measurable function. We call the function defined on $[0, \infty)$ by*

$$y \mapsto \mu(\{f \geq y\}) \tag{9.26}$$

186

the (measure theoretical) **distribution function** of f with respect to μ.

Remark 9.19. A. The function $y \mapsto \mu(\{f \geq y\})$ is obviously a numeric function which is decreasing and the fact that μ is continuous implies the left-continuity of the distribution functions. **B.** In probability theory the function $y \mapsto \mu(\{X \leq y\})$ is of greater importance and it is called the distribution function of the random variable X. We will call it **probabilistic distribution function** to make a difference with the measure theoretical distribution function.

Theorem 9.20. For a σ-finite measure space $(\Omega, \mathcal{A}, \mu)$ and a non-negative measurable function $f : \Omega \to [0, \infty)$ we have

$$\int f \, d\mu = \int_{(0,\infty)} \mu(\{f \geq y\}) \lambda^{(1)}(dy). \tag{9.27}$$

Proof. We consider on $\Omega \times [0, \infty)$ the measurable function $F : \Omega \times [0, \infty) \to \mathbb{R}^2$, $(x, y) \mapsto F(x, y) = (f(x), y)$. It follows that the set $A := \{(x, y) \in \Omega \times [0, \infty) \mid f(x) \geq y\}$ is measurable and

$$\chi_A(x, y) = \chi_{(0, f(x))}(y),$$

by the same argument used before. Now it follows again by using Tonelli's theorem

$$\int_\Omega f \, d\mu = \int_\Omega \left(\int_0^{f(x)} (1 \, dy) \right) \mu(dx) = \int_\Omega \left(\int_{(0,\infty)} \chi_{(0, f(x))}(y) \lambda^{(1)}(dy) \right) \mu(dx)$$

$$= \int_{\Omega \times (0,\infty)} \chi_A(x, y)(\mu \otimes \lambda^{(1)})(d(x, y)) = \int_{(0,\infty)} \left(\int_\Omega \chi_A(x, y) \mu(dx) \right) \lambda^{(1)}(dy)$$

$$= \int_0^\infty \mu(\{f \geq y\}) \lambda^{(1)}(dy).$$

$\square$

Example 9.21. The function $g : \mathbb{R}^n \to \mathbb{R}$, $g(x) = e^{-\|x\|^2}$, is Lebesgue integrable as well as improper Riemann integrable and by Example II.22.12 we know that

$$\int_{\mathbb{R}^n} e^{-\|x\|^2} \, dx = \pi^{\frac{n}{2}}.$$

According to Theorem 9.20 we have

$$\int_{\mathbb{R}^n} e^{-\|x\|^2} \, dx = \int_0^\infty \lambda^{(n)}(\{g \geq y\}) \lambda^{(1)}(dy).$$

187

We first note that

$$\lambda^{(n)}(\{g \geq y\}) = \lambda^{(n)}\left(\{x \in \mathbb{R}^n \mid g(x) \geq y\}\right)$$
$$= \lambda^{(n)}\left(\{x \in \mathbb{R}^n \mid e^{-\|x\|^2} \geq y\}\right)$$
$$= \lambda^{(n)}\left(\{x \in \mathbb{R}^n \mid \|x\|^2 \leq -\ln y\}\right),$$

hence

$$\int_{\mathbb{R}^n} e^{-\|x\|^2} \, dx = \int_0^\infty \lambda^{(n)}\left(\{x \in \mathbb{R}^n \mid \|x\|^2 \leq -\ln y\}\right) \lambda^{(1)}(dy)$$

and with $-\ln y = r$, i.e. $\lambda^{(1)}(dy) = e^{-r}\lambda^{(1)}(dr)$ we find

$$\int_{\mathbb{R}^n} e^{-\|x\|^2} \, dx = \int_0^\infty \lambda^{(n)}\left(\{x \in \mathbb{R}^n \mid \|x\|^2 \leq r\}\right) e^{-r}\lambda^{(1)}(dr)$$
$$= \int_0^\infty \lambda^{(n)}\left(B_{\sqrt{r}}(0)\right) e^{-r}\lambda^{(1)}(dr).$$

We know by Proposition II.21.13, in particular (II.21.10), that

$$\lambda^{(n)}\left(B_{\sqrt{r}}(0)\right) = \mathrm{vol}_n\left(B_{\sqrt{r}}(0)\right) = \frac{\pi^{\frac{n}{2}} r^{\frac{n}{2}}}{\Gamma(\frac{n}{2}+1)}.$$

This implies now

$$\int_{\mathbb{R}^n} e^{-\|x\|^2} \, dx = \frac{\pi^{\frac{n}{2}}}{\Gamma(\frac{n}{2}+1)} \int_0^\infty r^{\frac{n}{2}} e^{-r} \, dr = \pi^{\frac{n}{2}}$$

since $\Gamma(t) = \int_0^\infty y^{t-1} e^{-y} \, dy$, i.e. $\Gamma(\frac{n}{2}+1) = \int_0^\infty y^{\frac{n}{2}} e^{-y} \, dy$. While this looks just like a further method to evaluate the Gauss integral given the volume of balls in $\mathbb{R}^n$, the formula

$$\int_{\mathbb{R}^n} e^{-\|x\|^2} \, dx = \int_0^\infty \lambda^{(n)}\left(B_{\sqrt{r}}(0)\right) e^{-r} \, dr \qquad (9.28)$$

is in itself of interest. It states that we can express the Gauss integral entirely with the help of the volumes of balls. We give two extensions of this observation.

188

Example 9.22. A. Let $g : [0, \infty) \to [0, \infty)$ be a bijective, continuous function with continuous derivative and suppose that $\int_{\mathbb{R}^n} g(\|x\|) \, dx$ is a finite improper Riemann integral which then is equal to the Lebesgue integral of g. Then the following holds

$$\int_{\mathbb{R}^n} g(\|x\|) \, dx = \int_0^\infty \lambda^{(n)}(\{g(\|x\|) \geq y\}) \, dy$$
$$= \int_0^\infty \lambda^{(n)} \left(B_r^{\complement}(0) \right) g'(r) \, dr.$$

B. Let $f : [0, \infty) \to [0, \infty)$ be a monotone increasing function and $d(x, y) = \psi^{\frac{1}{2}}(x - y)$ a metric on $\mathbb{R}^n$ which generates on $\mathbb{R}^n$ the Euclidean topology. Here ψ is assumed to be a continuous function such that $e^{-\psi}$ is integrable. In this case we find

$$\int_{\mathbb{R}^n} e^{-d^2(x,z)} \, dx = \int_{\mathbb{R}^n} e^{-\psi(x-z)} \, dx = \int_{\mathbb{R}^n} e^{-\psi(x)} \, dx,$$

and further

$$\int_{\mathbb{R}^n} e^{-\psi(x)} \, dx = \int_0^\infty \lambda^{(n)} \left(\{e^{-\psi} \geq y\} \right) dy$$
$$= \int_0^\infty \lambda^{(n)} \left(\{\psi \leq -\ln y\} \right) dy$$
$$= \int_0^\infty \lambda^{(n)} \left(\{\psi \leq r\} \right) e^{-r} \, dr$$
$$= \int_0^\infty \lambda^{(n)} \left(B_{\sqrt{r}}^{\psi}(0) \right) e^{-r} \, dr,$$

where $B_R^{\psi}(0)$ is the ball with centre $0 \in \mathbb{R}^n$ and radius $R > 0$ in the metric space $(\mathbb{R}^n, d) = (\mathbb{R}^n, \psi^{\frac{1}{2}})$. The formula

$$\int_{\mathbb{R}^n} e^{-\psi(x)} \, dx = \int_0^\infty \lambda^{(n)} \left(B_{\sqrt{r}}^{\psi}(0) \right) e^{-r} \, dr \tag{9.29}$$

is of some interest when discussing transition functions of certain Lévy processes, see [40].

The examples given above have in principle already provided the proof to

Corollary 9.23. *Let $\varphi : [0, \infty) \to [0, \infty)$ be a continuously differentiable function such that $\varphi(0) = 0$ and $\varphi'(0) > 0$. Further let $f : \mathbb{R}^n \to [0, \infty)$ be a continuous function such that $\int_{\mathbb{R}^n} (\varphi \circ f)(x)\, dx$ exists. Then we have*

$$\int_{\mathbb{R}^n} (\varphi \circ f)(x)\, dx = \int_0^\infty \lambda^{(n)}(\{f \geq y\})\varphi'(y)\, dy.$$

Proof. We apply Theorem 9.20 and use the substitution $y = \varphi(s)$:

$$\int_{\mathbb{R}^n} (\varphi \circ f)(x)\, dx = \int_0^\infty \lambda^{(n)}(\{\varphi \circ f \geq y\})\, dy$$

$$= \int_0^\infty \lambda^{(n)}(\{\varphi \circ f \geq \varphi(s)\})\varphi'(s)\, ds$$

$$= \int_0^\infty \lambda^{(n)}(\{f \geq s\})\varphi'(s)\, ds.$$

$\square$

In Problem 10 we will give a further extension of Corollary 9.23.

Example 9.24. Applying Corollary 9.23 to the function $\varphi(y) = y^p$, $p > 1$, we find

$$\|u\|_{L^p}^p = \int_{\mathbb{R}^n} |u(x)|^p\, dx = \int_0^\infty \lambda^{(n)}(\{|u| \geq s\})ps^{p-1}\, ds$$

and in Problem 10 we see that this formula holds for more general measure spaces $(\Omega, \mathcal{A}, \mu)$.

Our final application in this chapter is known as **Minkowski's integral inequality**. We follow closely the presentation in K. Kuttler [52]

Theorem 9.25. *Let $(\Omega_1, \mathcal{A}_1, \mu)$ and $(\Omega_2, \mathcal{A}_2, \nu)$ be two σ-finite measure spaces and $f : \Omega_1 \times \Omega_2 \to \overline{\mathbb{R}}$ an $\mathcal{A}_1 \otimes \mathcal{A}_2$-measurable function. Further let $p \geq 1$. Then*

$$\left(\int_{\Omega_2} \left(\int_{\Omega_1} |f(\omega_1, \omega_2)| \mu(d\omega_1) \right)^p \nu(d\omega_2) \right)^{\frac{1}{p}} \leq \int_{\Omega_1} \left(\int_{\Omega_2} |f(\omega_1, \omega_2)|^p \nu(d\omega_2) \right)^{\frac{1}{p}} \mu(d\omega_1).$$

$$(9.30)$$

Proof. Let $(A_k)_{k \in \mathbb{N}}$ be a sequence of sets $A_k \in \mathcal{A}_1$ such that $\mu(A_k) < \infty$, $A_k \subset A_{k+1}$ and $\bigcup_{k \in \mathbb{N}} A_k = \Omega_1$, and let $(B_l)_{l \in \mathbb{N}}$ be a sequence of sets $B_l \in \mathcal{A}_2$

190

such that $\nu(B_l) < \infty$, $B_l \subset B_{l+1}$ and $\bigcup_{l\in\mathbb{N}} B_l = \Omega_2$. Further, for $m \in \mathbb{N}$ we define

$$f_m(x,y) := \begin{cases} f(x,y) & \text{if } |f(x,y)| \le m \\ 0 & \text{if } |f(x,y)| > m. \end{cases}$$

For $k, l, m \in \mathbb{N}$ it follows that

$$\left(\int_{B_l} \left(\int_{A_l} |f_m(\omega_1,\omega_2)| \mu(d\omega_1) \right)^p \nu(d\omega_2) \right)^{\frac{1}{p}} < \infty$$

and further the function $F_{m,k}$ defined on Ω_2 by

$$F_{m,k}(\omega_2) := \int_{A_k} |f_m(\omega_1,\omega_2)| \mu(d\omega_1)$$

is measurable. By Fubini's theorem we find

$$\int_{B_l} \left(\int_{A_k} |f_m(\omega_1,\omega_2)| \mu(d\omega_1) \right)^p \nu(d\omega_2) = \int_{B_l} \left(F_{m,k}^{p-1}(\omega_2) \int_{A_k} |f_m(\omega_1,\omega_2)| \mu(d\omega_1) \right) \nu(d\omega_2)$$

$$= \int_{A_k} \left(\int_{B_l} F_{m,k}^{p-1}(\omega_2) |f_m(\omega_1,\omega_2)| \nu(d\omega_2) \right) \mu(d\omega_1). \tag{9.31}$$

We now apply Hölder's inequality to (9.31) to find with $\frac{1}{p} + \frac{1}{q} = 1$ that

$$\int_{B_l} \left(\int_{A_k} |f_m(\omega_1,\omega_2)| \mu(d\omega_1) \right)^p \nu(d\omega_2)$$

$$\le \int_{A_k} \left(\int_{B_l} F_{m,k}^p(\omega_2) \nu(d\omega_2) \right)^{\frac{1}{q}} \left(\int_{B_l} |f_m(\omega_1,\omega_2)|^p \nu(d\omega_2) \right)^{\frac{1}{p}} \mu(d\omega_1)$$

$$= \left(\int_{B_l} F_{m,k}^p(\omega_2) \nu(d\omega_2) \right)^{\frac{1}{q}} \int_{A_k} \left(\int_{B_l} |f_m(\omega_1,\omega_2)|^p \nu(d\omega_2) \right)^{\frac{1}{p}} \mu(d\omega_1)$$

$$= \left(\int_{B_l} \left(\int_{A_k} |f_m(\omega_1,\omega_2)| \mu(d\omega_1) \right)^p \nu(d\omega_2) \right)^{\frac{1}{q}} \int_{A_k} \left(\int_{B_l} |f_m(\omega_1,\omega_2)|^p \nu(d\omega_2) \right)^{\frac{1}{p}} \mu(d\omega_1)$$

which yields

$$\left(\int_{B_l} \left(\int_{A_k} |f_m(\omega_1,\omega_2)| \mu(d\omega_1) \right)^p \nu(d\omega_2) \right)^{\frac{1}{p}} \le \int_{A_k} \left(\int_{B_l} |f_m(\omega_1,\omega_2)|^p \nu(d\omega_2) \right)^{\frac{1}{p}} \mu(d\omega_1).$$

$$\tag{9.32}$$

Now we use the monotone convergence theorem: first we let $m \to \infty$ and note that by construction $f_m \le f_{m+1}$, hence we arrive by (9.32) with f_m replaced by f. Next we let $k \to \infty$ and we obtain (9.32) with f_m replaced by f and A_k replaced by Ω_1. Finally we let $l \to \infty$ and eventually we get (9.32) with f_m replaced by f, A_k replaced by Ω_1 and B_l replaced by Ω_2, which is however nothing but (9.30). □

Problems

1. Let $(\Omega_j, \mathcal{A}_j), j = 1, 2$, be two σ-finite measurable spaces and $A_j \in \mathcal{A}_j$. Find a natural generator of the trace σ-field $(A_1 \times A_2) \cap A_1 \otimes A_2$.

2. Show that the Dirac measure $\epsilon_x, x \in \mathbb{R}^n$, defined on $(\mathbb{R}^n, \mathcal{B}^{(n)})$ is the product measure of n Dirac measures defined on $(\mathbb{R}, \mathcal{B}^{(1)})$.

3.   a) Let $(\Omega_j, \mathcal{A}_j, \mu_j)$, $j = 1, \ldots, N$, be σ-finite measure spaces and let $\nu_j := g_j \mu_j$ a further measure on $\mathcal{A}_j$ with non-negative, μ_j-integrable density. Prove that the product measure $\nu_1 \otimes \cdots \otimes \nu_N$ is a finite measure on $\mathcal{A}_1 \otimes \cdots \otimes \mathcal{A}_N$ which has a density with respect to $\mu_1 \otimes \cdots \otimes \mu_N$.

   b) Let $(\Omega_j, \mathcal{A}_j, \mu_j)$, $j = 1, 2$, be two σ-finite measure spaces and denote by $\mathcal{N}_2$ the sets of measure zero in $(\Omega_2, \mathcal{A}_2, \mu_2)$. Prove that for every $N \in \mathcal{N}_2$ and every $A \in \mathcal{A}_1$ the set $A \times N$ belongs to $\mathcal{N} \subset \mathcal{A}_1 \otimes \mathcal{A}_2$, where $B \in \mathcal{N}$ if and only if $(\mu_1 \otimes \mu_2)(B) = 0$, and that $(\mu_1 \otimes \mu_2)(A \times N) = 0$.

4.   a) For the ellipsoid $\mathcal{E} := \left\{ (x, y) \in \mathbb{R}^2 \middle| \frac{x^2}{4} + \frac{y^2}{36} \le 1 \right\}$ find the x-sections and the y-sections.

   b) For the triangle $S = ABC \subset \mathbb{R}^2$ with vertices $A = (0, 0)$, $B = (1, 0)$, $C = (1, 1)$ find the x-section and the y-section. Now verify (9.16) for $S = ABC$, $(\Omega_1, \mathcal{A}_1, \mu_1) = (\Omega_2, \mathcal{A}_2, \mu_2) = (\mathbb{R}, \mathcal{B}^{(1)}, \lambda^{(1)})$. (It may be taken for granted that $S = ABC \in \mathcal{B}^{(2)}$.)

5. Prove that the intersection of an arbitrary family of monotone classes $C_j \subset \mathcal{P}(\Omega), j \in I, I \neq \emptyset$, is again a monotone class.

6. Let $f : \mathbb{R}^n \to [0, \infty)$ be a function and define

$$\Gamma_s(f) := \{ (x, y) \in \mathbb{R}^n \times [0, \infty) | 0 \le y \le f(x) \}.$$

Prove that $\Gamma_s(f)$ belongs to $\mathcal{B}^{(n+1)}$ if and only if f is $\mathcal{B}^{(n)}/\mathcal{B}^{(1)}$-measurable and that we have

$$\lambda^{(n+1)}(\Gamma_s(f)) = \int_{\mathbb{R}^n} f \, d\lambda^{(n)}.$$

Give an interpretation of this result.

7.* Let $K \subset \mathbb{R}^3$ be a closed set contained in the bounded cube $Q \subset \mathbb{R}^3$. For $\rho > 0$ the **Marcinkiewicz integral** is defined by $M_{K,Q,\rho} : \mathbb{R}^3 \to \mathbb{R}$,

$$M_{K,Q,\rho}(x) := \int_Q \frac{(\operatorname{dist}(K,y))^\rho}{||x - y||^{3+\rho}} \lambda^{(3)}(dy).$$

Use Tonelli's theorem to prove that $M_{K,Q,\rho}$ is $\lambda^{(3)}$-almost everywhere finite in K and that $x \mapsto M_{K,Q,\rho}(x)$ is integrable over K. Moreover prove the estimate

$$\int_K M_{K,Q,\rho}(x)\lambda^{(3)}(dx) \leq \frac{4\pi}{\rho}\lambda^{(3)}(Q \setminus K).$$

Hint: use the fact that $x \mapsto \operatorname{dist}(Q,x)$ is Lipschitz continuous and try to prove $\int_K \frac{\lambda^{(3)}(dx)}{|x-y|^{3+\rho}} \leq 4\pi \int_{\operatorname{dist}(Q,y)} \frac{dr}{r^{1+\rho}}$.

8. Construct an example of a function $f : \mathbb{R}^2 \to \mathbb{R}$ such that

$$\int_{\mathbb{R}} \left(\int_{\mathbb{R}} f(x,y)\lambda^{(1)}(dx) \right) \lambda^{(1)}(dy) = \int_{\mathbb{R}} \left(\int_{\mathbb{R}} f(x,y)\lambda^{(1)}(dy) \right) \lambda^{(1)}(dx)$$

but f is not $\lambda^{(2)}$-integrable.
Hint: try $f(x,y) = \chi_{(-1,1)\times(-1,1)}(x,y)\frac{xy}{(x^2+y^2)^2}$ and use polar coordinates.

9.* In the situation of Example 9.22.B assume that for the measure of $B_\rho^\psi(0)$ the following holds for $c > 1$ and $\rho > 0$

$$\lambda^{(n)}(B_{c\rho}^\psi(0)) \leq \gamma_0(c)\lambda^{(n)}(B_\rho^\psi(0))$$

and $\gamma_0(c) \leq \gamma_0(1)c^\alpha$ for some $\alpha \geq 0$. Prove that there exists two constants $\kappa_0 > 0$ and $\kappa_1 > 0$ such that

$$\kappa_0\lambda^{(n)}\left(B_{\frac{1}{\sqrt{t}}}^\psi(0)\right) \leq \int_{\mathbb{R}^n} e^{-t\psi(x)}\lambda^{(n)}(dx) \leq \kappa_1\lambda^{(n)}\left(B_{\frac{1}{\sqrt{t}}}^\psi(0)\right).$$

Hint: first derive the equality

$$\int_{\mathbb{R}^n} e^{-t\psi(x)} \lambda^{(n)}(dx) = t \int_0^\infty \lambda^{(n)}(B_{\sqrt{\rho}}^\psi(0))e^{-t\rho}d\rho.$$

10. Let $(\Omega, \mathcal{A}, \mu)$ be a σ-finite measure space and $\varphi : [0, \infty) \to [0, \infty)$ an increasing C^1-function with $\varphi(0) = 0$. Prove that

$$\int_\Omega \varphi \circ u \, d\mu = \int_0^\infty \mu(\{u \geq y\})\varphi'(y)dy$$

holds for every measurable function $u : \Omega \to [0, \infty)$.

11. Let $k \in L^2([a, b] \times [a, b])$ and $u \in L^2([a, b]), a < b$. Use Minkowski's integral inequality to show that

$$\left(\int_{[a,b]} \left| \int_{[a,b]} k(x, y)u(y)\lambda^{(1)}(dy) \right|^2 \lambda^{(1)}(dx) \right)^{\frac{1}{2}} \leq \|k\|_{L^2} \|u\|_{L^2}.$$

Now conclude that by $K_{op} : L^2([a, b]) \to L^2([a, b])$ a linear mapping is defined which satisfies $\|K_{op}u\|_{L^2} \leq \|k\|_{L^2} \|u\|_{L^2}$.
(Note that $\|u\|_{L^2}$ is the L^2-norm in $L^2([a, b]), \|k\|_{L^2}$ is the L^2-norm in $L^2([a, b] \times [a, b])$.)

10 Convolutions of Functions and Measures

In Problem 10 of Chapter II.22 we briefly discussed the convolution for continuous functions with compact support, i.e. the function

$$(u * v)(x) := \int_{\mathbb{R}^n} u(x - y)v(y)\, dy, \tag{10.1}$$

and in the lengthy Problem 11 of Chapter II.22 we investigated properties of the Friedrichs mollifier which we can interpret as a convolution operator. Convolutions of functions and measures are encountered in many situations, for example as solutions of partial differential equations with constant coefficients or as operators related to expectations of certain stochastic processes. In this chapter we will study convolutions rather systematically and by this provide tools that will be often employed in the following volumes.

In order to handle integrals such as the one in (10.1) we need the additive group $(\mathbb{R}^n, +)$ and the invariance of the Lebesgue-Borel $\lambda^{(n)}$ under translations. In later volumes we will see that many results of this chapter extend to so called locally compact Abelian groups on which a unique (up to a normalization) translation invariant measure, the Haar measure exists. In this chapter we will only work in $\mathbb{R}^n$ and as long as we are dealing with the spaces $\mathcal{L}^p(\mathbb{R}^n)$ or $L^p(\mathbb{R}^n)$ (or some of their subspaces) the underlying measure is the Lebesgue-Borel measure $\lambda^{(n)}$. However, in the second part of this chapter we discuss the convolution $\mu * \nu$ of certain measures μ and ν on $\mathcal{B}^{(n)}$. Our first aim is to make sense of integrals such as the one in (10.1) for elements in $\mathcal{L}^p(\mathbb{R}^n)$ and even in $L^p(\mathbb{R}^n)$. Recall that elements in $L^p(\mathbb{R}^n)$ are not pointwisely defined functions but equivalence classes of elements from $\mathcal{L}^p(\mathbb{R}^n)$ under the equivalence relation "$\sim_{\lambda^{(n)}}$", i.e. $u \sim_{\lambda^{(n)}} v$ if $u = v$ $\lambda^{(n)}$-a.e. Thus for $[u] \in L^1(\mathbb{R}^n)$ an expression such as $[u](x - y)$ is undefined and $[u(x - \cdot)]$ needs an explanation. In order to simplify notation, in the following we write a.e. for $\lambda^{(n)}$-a.e. and if no confusion can arise, we also write u for the equivalence class generated by u.

Let $u : \mathbb{R}^n \to \mathbb{R}$ be a measurable function and let $\delta : \mathbb{R}^n \times \mathbb{R}^n \to \mathbb{R}^n$ be defined by $\delta(x, y) := x - y$. The mapping δ is continuous, hence measurable, and therefore the function $\omega := u \circ \delta : \mathbb{R}^n \times \mathbb{R}^n \to \mathbb{R}$, $(u \circ \delta)(x, y) = u(\delta(x, y)) = u(x - y)$ is measurable implying that for each $x \in \mathbb{R}^n$ fixed the function $y \mapsto u(x - y)$ is also measurable. Thus for two measurable functions

$u, v : \mathbb{R}^n \to \mathbb{R}$ the function $(x, y) \mapsto u(x - y)v(y)$ is measurable as a function from $\mathbb{R}^n \times \mathbb{R}^n$ to $\mathbb{R}$. The function on $(x, y) \mapsto |u(x - y)v(y)|$ is a non-negative measurable function defined on $\mathbb{R}^n \times \mathbb{R}^n$ and by Tonelli's theorem we find

$$\int_{\mathbb{R}^n \times \mathbb{R}^n} |u(x - y)v(y)| \, \lambda^{(n)} \otimes \lambda^{(n)}(\mathrm{d}x, \mathrm{d}y) = \int_{\mathbb{R}^n} |u(x - y)| \left(\int_{\mathbb{R}^n} |v(y)| \lambda^{(n)}(\mathrm{d}y) \right) \lambda^{(n)}(\mathrm{d}x)$$

$$= \int_{\mathbb{R}^n} |v(y)| \lambda^{(n)}(\mathrm{d}y) \int_{\mathbb{R}^n} |u(x - y)| \lambda^{(n)}(\mathrm{d}x) = \int_{\mathbb{R}^n} |v(y)| \lambda^{(n)}(\mathrm{d}y) \int_{\mathbb{R}^n} |u(x)| \lambda^{(n)}(\mathrm{d}x)$$

where in the last step we used that $\lambda^{(n)}$ is translation invariant. If we add as an assumption that both u and v are integrable, we then deduce by Fubini's theorem that

$$\int_{\mathbb{R}^n \times \mathbb{R}^n} |u(x - y)v(y)| \, \lambda^{(n)} \otimes \lambda^{(n)}(\mathrm{d}x, \mathrm{d}y) = \|v\|_{L^1} \|u\|_{L^1} < \infty, \qquad (10.2)$$

and further that for almost all $x \in \mathbb{R}^n$ the integral

$$\int_{\mathbb{R}^n} |u(x - y)v(y)| \, \lambda^{(n)}(\mathrm{d}y) \qquad (10.3)$$

is finite. Thus we can define for those $x \in \mathbb{R}^n$ the function

$$(u * v)(x) := \int_{\mathbb{R}^n} u(x - y)v(y) \, \lambda^{(n)}(\mathrm{d}y) \qquad (10.4)$$

which determines an equivalence class with respect to "$\sim_{\lambda^{(n)}}$". This equivalence class we denote again by $u * v$ and from (10.2) we deduce $u * v \in L^1(\mathbb{R}^n)$ and

$$\|u * v\|_{L^1} \le \|u\|_{L^1} \|v\|_{L^1}. \qquad (10.5)$$

Moreover, when replacing in (10.2) both u and v by functions $\tilde{u}$ and $\tilde{v}$ being a.e. equal to u and v, respectively, nothing will change in our arguments. In particular (10.4) will be a.e. equal to $\tilde{u} * \tilde{v}$. Thus we have proved

Theorem 10.1. *Let $u, v \in L^1(\mathbb{R}^n)$. Then there exists a unique element $u * v \in L^1(\mathbb{R}^n)$ satisfying (10.5) and $\lambda^{(n)}$-a.e. this element $u * v$ is equal to (10.4) where in (10.4) we may take any representative of u and v, respectively.*

Definition 10.2. *The element $u * v$ from Theorem 10.1 is called the **convolution** of u and v.*

Remark 10.3. Once we have given a precise meaning of $u * v$ for $u, v \in L^1(\mathbb{R}^n)$, we turn to the common façon de parler and call $u * v$ the convolution of the functions u and v and we will also consider $u * v$ from time to time as a function.

Corollary 10.4. *For $u, v, w \in L^1(\mathbb{R}^n)$ the following hold:*

$$u * v = v * u; \tag{10.6}$$

$$(u * v) * w = u * (v * w). \tag{10.7}$$

Proof. The first result follows from the change of variable $z = x - y$:

$$(u * v)(x) = \int u(x - y)v(y)\,\lambda^{(n)}(dy) = \int u(z)v(x - z)\lambda^{(n)}(dz) = (v * u)(z).$$

To see (10.7) we use (10.6) and Fubini's theorem, and of course the fact that by Theorem 10.1 we have $u * v \in L^1(\mathbb{R}^n)$:

$$((u * v) * w)(x) = \left(\left(\int u(y)v(\cdot - y)\,\lambda^{(n)}(dy) \right) * w \right)(x)$$

$$= \int \left(\int u(y)v(x - z - y)w(z)\,\lambda^{(n)}(dy) \right)\lambda^{(n)}(dz)$$

$$= \int \left(\int u(y)v(x - y - z)w(z)\lambda^{(n)}(dz) \right)\lambda^{(n)}(dy)$$

$$= \left(u * \int w(z)v(\cdot - z)\,\lambda^{(n)}(dz) \right)(x)$$

$$= (u * (v * w))(x).$$

$\square$

Remark 10.5. From Theorem 10.1 and Corollary 10.4 we deduce that $* : L^1(\mathbb{R}^n) \times L^1(\mathbb{R}^n) \to L^1(\mathbb{R}^n)$ is an associative and commutative operation which we can interpret as multiplication, i.e. we may speak about the convolution product, which turns the vector space $L^1(\mathbb{R}^n)$ into an algebra, also see Problem 2.

Before we study the convolution for different L^p-spaces we want to change our point of view for understanding terms such as $u(\cdot - x)$ for $x \in \mathbb{R}^n$ and $u \in L^1(\mathbb{R}^n)$. Recall that $C_0(\mathbb{R}^n)$ denotes the vector space of all continuous functions $u : \mathbb{R}^n \to \mathbb{R}$ with compact support $\operatorname{supp} u$, $C_\infty(\mathbb{R}^n)$ is the vector space of all continuous functions vanishing at infinity, and $C_b(\mathbb{R}^n)$ is the space of all bounded continuous functions. We have the inclusions $C_0(\mathbb{R}^n) \subset C_\infty(\mathbb{R}^n) \subset C_b(\mathbb{R}^n)$ and $\|u\|_\infty := \sup_{x \in \mathbb{R}^n} |u(x)|$ is a norm on $C_b(\mathbb{R}^n)$, hence it is also a norm on $C_0(\mathbb{R}^n)$ and $C_\infty(\mathbb{R}^n)$. Let $X \in \{C_0(\mathbb{R}^n), C_\infty(\mathbb{R}^n), C_b(\mathbb{R}^n)\}$, $x \in \mathbb{R}^n$, and $A \in O(n)$. For every $u \in X$ it follows that v defined by $v(y) := (T_{A,x}u)(y) = u(Ay + x)$ is again an element in X, see Problem 1. In particular $u \in X$ implies $u(x - \cdot) \in X$. Moreover the mapping $T_{A,x} : X \to X$, $u \mapsto T_{A,x}u$ is linear and $\|T_{A,x}u\|_\infty = \|u\|_\infty$. Now let $1 \le p < \infty$ and $u \in C_0(\mathbb{R}^n) \subset L^p(\mathbb{R}^n)$. It follows that

$$\|T_{A,x}u\|_{L^p} = \left(\int_{\mathbb{R}^n} |(T_{A,x}u)(y)|^p \, \mathrm{d}y \right)^{\frac{1}{p}} = \left(\int_{\mathbb{R}^n} |u(Ay + x)|^p \, \mathrm{d}y \right)^{\frac{1}{p}}$$

and the substitution $z := Ay + x$ yields

$$\left(\int_{\mathbb{R}^n} |u(Ay + x)|^p \, \mathrm{d}y \right)^{\frac{1}{p}} = \left(\int_{\mathbb{R}^n} |u(z)|^p \, \mathrm{d}z \right)^{\frac{1}{p}} = \|u\|_{L^p},$$

where we used that the absolute value of the Jacobi determinant of the underlying transformation is 1. Thus we have for all $u \in C_0(\mathbb{R}^n)$ and all $A \in O(n)$, $x \in \mathbb{R}^n$,

$$\|T_{A,x}u\|_{L^p} = \|u\|_{L^p}, \tag{10.8}$$

and (10.8) just reflects the invariance of the Lebesgue-Borel measure $\lambda^{(n)}$ under translations and $O(n)$. We know by Theorem 8.20 that $L^p(\mathbb{R}^n)$, $1 \le p < \infty$, is a Banach space, hence continuity for a mapping $T : L^p(\mathbb{R}^n) \to L^p(\mathbb{R}^n)$ is well defined. In particular if with some constant c_T the following holds for all $u, v \in L^p(\mathbb{R}^n)$

$$\|Tu - Tv\|_{L^p} \le c_T \|u - v\|_{L^p} \tag{10.9}$$

then T is even Lipschitz continuous. If T is linear then it follows that $Tu - Tv = T(u - v)$ and the estimate $\|Tw\|_{L^p} \le c_T \|w\|_{L^p}$ for all $w \in L^p(\mathbb{R}^n)$ implies (10.9). Thus (10.8) means that $T_{A,x}|_{C_0(\mathbb{R}^n)}$ is a continuous linear mapping from $C_0(\mathbb{R}^n)$ to $L^p(\mathbb{R}^n)$ when we choose on $C_0(\mathbb{R}^n)$ and on $L^p(\mathbb{R}^n)$ the norm $\|.\|_{L^p}$.

Proposition 10.6. Let $Z \subset L^p(\mathbb{R}^n)$ be a linear subspace which is dense in $L^p(\mathbb{R}^n)$. If $T : Z \to L^p(\mathbb{R}^n)$ is a linear operator satisfying

$$\|Tw\|_{L^p} \le c_T \|w\|_{L^p} \tag{10.10}$$

for all $w \in Z$, then T has a unique extension $\tilde{T} : L^p(\mathbb{R}^n) \to L^p(\mathbb{R}^n)$ which is linear and satisfies (10.10) for all $u \in L^p(\mathbb{R}^n)$. In particular $\tilde{T}$ is continuous.

Remark 10.7. Recall that a set $Z \subset X$ is dense in the metric space (X, d) if for every $x \in X$ and every $\epsilon > 0$ there exists $z_\epsilon \in Z$ such that $d(z_\epsilon, x) < \epsilon$, or equivalently, for every $x \in X$ there exists a sequence $(z_k)_{k\in\mathbb{N}}$, $z_k \in Z$, such that $\lim_{k\to\infty} d(z_k, x) = 0$, see also Definition II.3.4.

Proof of Proposition 10.6. Let $u \in L^p(\mathbb{R}^n)$ and $(u_k)_{k\in\mathbb{N}}$ a sequence in Z converging in $L^p(\mathbb{R}^n)$ to u. From (10.10) we deduce that

$$\|Tu_k - Tu_l\|_{L^p} \leq c_T \|u_k - u_l\|_{L^p},$$

and therefore $(Tu_k)_{k\in\mathbb{N}}$ is a Cauchy sequence in $L^p(\mathbb{R}^n)$, hence it has a limit $v := \lim_{k\to\infty} Tu_k$. If $(\tilde{u}_k)_{k\in\mathbb{N}}$ a further sequence in Z converging in $L^p(\mathbb{R}^n)$ to u we find

$$\|Tu_k - T\tilde{u}_k\|_{L^p} = \|T(u_k - \tilde{u}_k)\|_{L^p}$$
$$\leq c_T \|u_k - \tilde{u}_k\|_{L^p} \leq c_T \|u_k - u\|_{L^p} + c_T \|\tilde{u}_k - u\|_{L^p},$$

i.e. $v := \lim_{k\to\infty} Tu_k = \lim_{k\to\infty} T\tilde{u}_k$ is independent of the approximating sequence. Therefore

$$\tilde{T}u := \lim_{k\to\infty} Tu_k \quad (L^p\text{-limit}) \tag{10.11}$$

is well defined for all $u \in L^p(\mathbb{R}^n)$ and for $u \in Z$ we may choose as approximating sequence $u_k = u$ for all $k \in \mathbb{N}$ to find that $\tilde{T}u = Tu$ for $u \in Z$, thus $\tilde{T}|_Z = T$, i.e. $\tilde{T}$ is an extension of T. For $u, v \in L^p(\mathbb{R}^n)$ and $\alpha, \beta \in \mathbb{R}$ we choose sequences $(u_k)_{k\in\mathbb{N}}$, $(v_k)_{k\in\mathbb{N}}$, $u_k, v_k \in Z$, such that in $L^p(\mathbb{R}^n)$ we have $\lim_{k\to\infty} u_k = u$ and $\lim_{k\to\infty} v_k = v$. Thus $\lim_{k\to\infty}(\alpha u_k + \beta v_k) = \alpha u + \beta v$ and we find

$$\tilde{T}(\alpha u + \beta v) = \lim_{k\to\infty} T(\alpha u_k + \beta v_k)$$
$$= \alpha \lim_{k\to\infty} Tu_k + \beta \lim_{k\to\infty} Tv_k = \alpha \tilde{T}u + \beta \tilde{T}v,$$

i.e. $\tilde{T}$ is linear on $L^p(\mathbb{R}^n)$. Moreover, for $(u_k)_{k\in\mathbb{N}}$ as above we have

$$\|\tilde{T}u\|_{L^p} \leq \|\tilde{T}(u - u_k)\|_{L^p} + \|Tu_k\|_{L^p}$$
$$\leq \|\tilde{T}(u - u_k)\|_{L^p} + c_T \|u_k\|_{L^p}.$$

Since $\big|\|u_k\|_{L^p} - \|u\|_{L^p}\big| \leq \|u_k - u\|_{L^p}$ and since $\lim_{k\to\infty} \|\tilde{T}(u - u_k)\|_{L^p} = 0$, we arrive at $\|\tilde{T}u\|_{L^p} \leq c_T\|u\|_{L^p}$. Finally suppose that $\tilde{T}$ and $\tilde{S}$ are two continuous and linear extensions of T. For $u \in L^p(\mathbb{R}^n)$ with $u = \lim_{k\to\infty} u_k$, $u_k \in Z$, we find

$$\|\tilde{S}u - \tilde{T}u\|_{L^p} = \left\|\lim_{k\to\infty} \tilde{S}u_k - \lim_{k\to\infty} \tilde{T}u_k\right\|_{L^p}$$
$$= \lim_{k\to\infty} \|Tu_k - Tu_k\|_{L^p} = 0,$$

i.e. $\tilde{S} = \tilde{T}$. $\qquad\qquad\qquad\qquad\qquad\qquad\qquad\qquad\qquad\qquad\square$

Remark 10.8. In the following, when no confusion may arise, we will denote the linear continuous extension of T again by T.

Our next goal is to prove that $C_0(\mathbb{R}^n) \subset L^p(\mathbb{R}^n)$ is dense. As a first preparation we need the following metric variant of the **Lemma of Urysohn**.

Lemma 10.9. *Let $\emptyset \neq K \subset V \subset \mathbb{R}^n$ where K is compact and V is open. For any open set U such that $K \subset U \subset V$ and $\overline{U}$ is compact the function $h : \mathbb{R}^n \to \mathbb{R}$ defined by*

$$h(x) := \frac{\operatorname{dist}\left(x, U^\complement\right)}{\operatorname{dist}(x, K) + \operatorname{dist}\left(x, U^\complement\right)} \qquad (10.12)$$

is continuous with support $\operatorname{supp} h \subset \overline{U}$, $h|_K = 1$, $h|_{U^\complement} = 0$ and $0 \leq h(x) \leq 1$. Thus $h \in C_0(\mathbb{R}^n)$ and $\chi_K \leq h \leq \chi_U \leq \chi_V$.

Proof. For $x \in K$ it follows that $\operatorname{dist}\left(x, U^\complement\right) > 0$ and for $x \in K^\complement$ it follows that $\operatorname{dist}(x, K) > 0$, i.e. h is well defined on $\mathbb{R}^n$. By Example II.3.28 the mappings $x \mapsto \operatorname{dist}\left(x, U^\complement\right)$ and $x \mapsto \operatorname{dist}(x, K)$ are continuous, so h is continuous. Further, for $x \in K$ we have $h(x) = \frac{\operatorname{dist}\left(x, U^\complement\right)}{\operatorname{dist}\left(x, U^\complement\right)} = 1$ and for $x \in U^\complement$ we have $\operatorname{dist}\left(x, U^\complement\right) = 0$, i.e. $h(x) = 0$. Clearly $0 \leq h(x) \leq 1$ and since $\overline{U}$ is compact, $\operatorname{supp} h \subset \overline{U}$ is compact too. $\qquad\square$

We will also need the next theorem, a proof of which we will discuss in Appendix II. Indeed it is worth discussing this result in a wider context as we will do in Part 12 when proving the Riész representation theorem.

Theorem 10.10. *Let $A \in \mathcal{B}^{(n)}$ with $\lambda^{(n)}(A) < \infty$. Then for every $\eta > 0$ we can find a compact set K_η and an open set U_η such that $K_\eta \subset A \subset U_\eta$, $\lambda^{(n)}(U_\eta) < \infty$ and $\lambda^{(n)}(U_\eta \backslash K_\eta) < \eta$.*

With these preparations we can prove

Theorem 10.11. *For $1 \leq p < \infty$ the set $C_0(\mathbb{R}^n)$ is a dense linear subspace of $L^p(\mathbb{R}^n)$.*

Proof. We want to approximate elements in a linear space and we have achieved our aim if we can approximate elements whose linear combinations contain for every element in $L^p(\mathbb{R}^n)$ a sequence converging in $L^p(\mathbb{R}^n)$ to the given element. Since $u = u^+ - u^-$ we first can reduce the problem to approximate non-negative elements in $L^p(\mathbb{R}^n)$. Since these are measurable functions (or better, every representant is a $\lambda^{(n)}$-a.e. non negative function) we can find an increasing sequence of simple functions $(u_k)_{k\in\mathbb{N}}$ such that $0 \leq u_k \leq u_{k+1} \leq u$ and $\lim_{k\to\infty} u_k = u$ $\lambda^{(n)}$-a.e. The dominated convergence theorem applied to $|u_k - u|^p$ now yields that $\lim_{k\to\infty} \|u_k - u\|_{L^p} = 0$. Thus, if we can approximate in the norm $\|.\|_{L^p}$ every characteristic function χ_A of a Borel measurable set A with $\lambda^{(n)}(A) < \infty$ by continuous functions with compact support, the result will follow. Now we use Theorem 10.10: given $A \in \mathcal{B}^{(n)}$ with $\lambda^{(n)}(A) < \infty$ we can find for $\epsilon > 0$ a compact set K and an open set U such that $\lambda^{(n)}(U) < \infty$ and

$$\lambda^{(n)}(U \backslash K) = \int \chi_{U \backslash K} \, d\lambda^{(n)} < \left(\frac{\epsilon}{2}\right)^p,$$

or

$$\|\chi_U - \chi_K\|_{L^p} < \frac{\epsilon}{2}.$$

Further, by Lemma 10.9 we find $h \in C_0(\mathbb{R}^n)$ such that $\chi_K \leq h \leq \chi_U$ or $0 \leq \chi_U - h \leq \chi_U - \chi_K$ which yields

$$\|\chi_U - h\|_{L^p} < \frac{\epsilon}{2}.$$

Noting that $\chi_K \leq \chi_A \leq \chi_U$, we find also that

$$\|\chi_U - \chi_A\|_{L^p} < \frac{\epsilon}{2}$$

which eventually yields

$$\|\chi_A - h\|_{L^p} \leq \|\chi_A - \chi_U\|_{L^p} + \|\chi_U - h\|_{L^p} < \epsilon.$$

$\square$

Combining Theorem 10.11 with Proposition 10.6 we arrive at

Corollary 10.12. *The operator $T_{A,x} : C_0(\mathbb{R}^n) \to L^p(\mathbb{R}^n)$ defined for $A \in O(n)$ and $x \in \mathbb{R}^n$ by $(T_{A,x}u)(y) = u(Ay + x)$ has a continuous extension to a linear continuous operator from $L^p(\mathbb{R}^n)$ into itself which satisfies (10.8). We will denote this extension again by $T_{A,x}$.*

When no confusion may arise we even write for $u \in L^p(\mathbb{R}^n)$ and $T_{A,x}$ instead of $T_{A,x}u$ just $u(A \cdot +x)$ or even $u(Ay + x)$. In this context it is worth noting that if $u = v$ $\lambda^{(n)}$-a.e. and $u = w$ $\lambda^{(n)}$-a.e. with two continuous functions v and w, then $v = w$ as continuous functions, i.e. pointwisely. Indeed, suppose $v(x_0) \neq w(x_0)$ for some $x_0 \in \mathbb{R}^n$. It follows that $|v(w_0) - w(x_0)| > 0$ and since $x \mapsto |v(x) - w(x)|$ is a continuous function there exists some $\rho > 0$ such that $|v(x) - w(x)| > 0$ for all $x \in B_\rho(x_0)$. It follows that $v \neq w$ on a set of positive measure, a contradiction to the fact that $v = w$ a.e. which follows from our assumption. Hence $u \in L^p(\mathbb{R}^n)$ can have at most one continuous representative.

So far we have proved that $u * v$ is defined for $u, v \in L^1(\mathbb{R}^n)$ and it is now our aim to study $u * v$ for u and v belonging to different spaces. The first result is the classical **Young's inequality**

Theorem 10.13. *Let $u \in L^1(\mathbb{R}^n)$ and $v \in L^p(\mathbb{R}^n)$, $1 \leq p \leq \infty$. Then $(u * v)(x)$ exists $\lambda^{(n)}$-a.e., $(u * v)$ is an element of $L^p(\mathbb{R}^n)$ and we have*

$$\|u * v\|_{L^p} \leq \|u\|_{L^1} \|v\|_{L^p}. \tag{10.13}$$

Proof. First we prove (10.13) for all $v \in C_0(\mathbb{R}^n) \subset L^p(\mathbb{R}^n)$, $1 \leq p < \infty$. For this we note that $(x, y) \mapsto u(y)v(x - y)$ is $\lambda^{(n)} \otimes \lambda^{(n)}$-measurable and furthermore, since $u * v = v * u$ for $u, v \in L^1(\mathbb{R}^n)$, see Corollary 10.4, we find using Minkowski's integral inequaltiy, Theorem 9.25,

$$\|u * v\|_{L^p} = \left(\int \left| \int u(y)v(x - y)\,dy \right|^p dx \right)^{\frac{1}{p}} \leq \left(\int \left(\int |u(y)v(x - y)|\,dy \right)^p dx \right)^{\frac{1}{p}}$$

$$\leq \int \left(\int |u(y)|^p |v(x - y)|^p\,dx \right)^{\frac{1}{p}} dy = \int |u(y)| \left(\int |v(x - y)|^p\,dx \right)^{\frac{1}{p}} dy$$

$$= \int |u(y)| \, \|v\|_{L^p}\,dy = \|u\|_{L^1} \|v\|_{L^p},$$

where we used once more that the Lebesgue-Borel measure is translation invariant. If now $v \in L^p(\mathbb{R}^n)$, we choose a sequence $(v_k)_{k \in \mathbb{N}}$, $v_k \in C_0(\mathbb{R}^n)$, converging in $L^p(\mathbb{R}^n)$ to v and (10.13) already proved for $u \in L^1(\mathbb{R}^n)$ and $v_k \in C_0(\mathbb{R}^n)$ follows now by passing to the limit $v_k \to v$. Since we now know $u * v \in L^p(\mathbb{R}^n)$ it follows that $u * v$ must be $\lambda^{(n)}$-a.e. finite and further by Fubini's theorem we have $\lambda^{(n)}$-a.e. that

$$(u * v)(x) = \int u(x - y)v(y)\, dy,$$

recall that $u(x - y)$ stands for $(T_{-\mathrm{id},0} \circ T_{\mathrm{id},-x})u$.

The case $p = \infty$ is of course much simpler since in this case we have

$$\|u * v\|_\infty = \left\| \int u(y)v(\cdot - y)\, dy \right\|_\infty \leq \|v\|_\infty \int |u(y)|\, dy.$$

$\square$

Proposition 10.14. *Let $1 \leq p < \infty$ and $\frac{1}{p} + \frac{1}{q} = 1$. If $u \in L^p(\mathbb{R}^n)$ and $v \in L^q(\mathbb{R}^n)$ then $u * v$ is defined and belongs to $C_b(\mathbb{R}^n)$. Moreover we have*

$$\|u * v\|_\infty \leq \|u\|_{L^p} \|v\|_{L^q}. \tag{10.14}$$

Proof. Interpreting $v(x - \cdot) \in L^q(\mathbb{R}^n)$ as mentioned above we find

$$\left| \int_{\mathbb{R}^n} u(y)v(x - y)\, dy \right| \leq \int_{\mathbb{R}^n} |u(y)|\, |v(x - y)|\, dy$$

$$\leq \left(\int_{\mathbb{R}^n} |u(y)|^p\, dy \right)^{\frac{1}{p}} \left(\int_{\mathbb{R}^n} |v(x - y)|^q\, dy \right)^{\frac{1}{q}}$$

$$= \|u\|_{L^p} \|v\|_{L^q}$$

where we used once again the translation invariance of $\lambda^{(n)}$. This estimate already implies that $u*v$ is almost everywhere defined and a bounded function satisfying (10.14). From the definition of $u * v$ it is obvious that $(u, v) \mapsto u * v$ is a bilinear mapping, say from $L^p(\mathbb{R}^n) \times L^q(\mathbb{R}^n)$ to $L^\infty(\mathbb{R}^n)$. Further, for $u, v \in C_0(\mathbb{R}^n)$ it follows that $u * v$ is continuous and bounded. Given $u \in L^p(\mathbb{R}^n)$ and $v \in L^q(\mathbb{R}^n)$ we can find a sequence $(u_k)_{k \in \mathbb{R}^n}$, $u_k \in C_0(\mathbb{R}^n)$, converging in $L^p(\mathbb{R}^n)$ to u, and we can find a sequence $(v_k)_{k\in\mathbb{N}}$, $v_k \in C_0(\mathbb{R}^n)$, converging in $L^q(\mathbb{R}^n)$ to v. It follows that

$$\|(u*v) - (u_k * v_k)\|_\infty \le \|u*(v - v_k)\|_\infty + \|(u - u_k)*v_k\|_\infty$$
$$\le \|u\|_{L^p}\|v - v_k\|_{L^q} + \|u - u_k\|_{L^p}\|v_k\|_{L^q}.$$

As a convergent sequence in $L^q(\mathbb{R}^n)$ the sequence $(v_k)_{k\in\mathbb{N}}$ is in $L^q(\mathbb{R}^n)$ bounded and we find

$$\|(u*v) - (u_k * v_k)\|_\infty \le M\left(\|v - v_k\|_{L^q} + \|u - u_k\|_{L^p}\right)$$

which means that $u*v$ is with respect to the sup-norm limit of a sequence of continuous functions, hence it is continuous. $\qquad\square$

For a pointwisely defined function $f : \mathbb{R}^n \to \mathbb{R}$ (or $f : G \to \mathbb{R}$, $G \subset \mathbb{R}^n$) we can define the support f as

$$\operatorname{supp} f := \overline{\{x \in \mathbb{R}^n \mid f(x) \ne 0\}}, \qquad (10.15)$$

compare with (II.14.12). This is a definition that we cannot transfer to the equivalence class $[f]$ of f with respect to "$\sim_{\lambda^{(n)}}$". But there are other possibilities to characterise $\operatorname{supp} f$, say for measurable functions which are integrable over every compact set $K \subset \mathbb{R}^n$. Take $x \notin \operatorname{supp} f$. Since $\{x\}$ is compact and $\operatorname{supp} f$ is closed, $\operatorname{dist}(x, \operatorname{supp} f) = \delta > 0$. Let $\varphi \in C_0(\mathbb{R}^n)$, $\operatorname{supp}\varphi \subset B_\delta(0)$. It follows that $\int_{\mathbb{R}^n} f\varphi \, d\lambda^{(n)} = 0$. In fact for every $\varphi \in C_0(\mathbb{R}^n)$ with $\operatorname{supp}\varphi \subset (\operatorname{supp} f)^\complement$ it follows that $\int_{\mathbb{R}^n} f\varphi \, d\lambda^{(n)} = 0$. Given f we may consider the set U_f defined by

$$U_f := \left\{x \in \mathbb{R}^n \;\middle|\; \text{for some } \rho > 0 \text{ we have for all } \varphi \in C_0(B_\rho(x)) \text{ that } \int_{\mathbb{R}^n} f\varphi \, d\lambda^{(n)} = 0\right\}.$$

This is an open set since for $x \in U_f$ and $y \in B_{\frac{\rho}{2}}(x)$ it follows for all $\varphi \in C_0\left(B_{\frac{\rho}{2}}(y)\right)$ that $\varphi \in C_0(B_\rho(x))$ and hence $\int f\varphi \, d\lambda^{(n)} = 0$, i.e. $B_{\frac{\rho}{2}}(x) \subset U_f$. Now for $f \in L^p(\mathbb{R}^n)$ the set U_f is independent of the representative and we define $\operatorname{supp} f := U_f^\complement$. For a continuous function f we recover the old definition. We already know for f continuous that $(\operatorname{supp} f)^\complement \subset U_f$, i.e. $U_f^\complement \subset \operatorname{supp} f$. Now let $x \in \operatorname{supp} f$. For every $\epsilon > 0$ we can find y such that $\|x - y\| < \epsilon$ and $f(y) \ne 0$. Suppose $f(y) > 0$, the case $f(y) < 0$ goes analogously. Due to the continuity of f we can find $\eta > 0$ such that $f(z) > 0$ for all $z \in B_\eta(y)$. However in this case we cannot have $\int_{\mathbb{R}^n} f\varphi \, d\lambda^{(n)} = 0$ for all $\varphi \in C_0(\mathbb{R}^n)$, $\text{supp}\,\varphi \subset B_{\frac{\eta}{2}}(y)$, hence $x \in U_f^\complement$, i.e. $\text{supp}\, f \subset U_f^\complement$.

Proposition 10.15. *Let $u \in L^p(\mathbb{R}^n)$ and $v \in L^q(\mathbb{R}^n)$ such that $u * v$ is defined and belong to some space $L^r(\mathbb{R}^n)$. For the support of $u * v$ we have*

$$\text{supp}(u * v) \subset \overline{\text{supp}\, u + \text{supp}\, v}, \tag{10.16}$$

i.e. the support of $u * v$ is contained in the closure of $\text{supp}\, u + \text{supp}\, v$.

Proof. Recall that for $A, B \subset \mathbb{R}^n$ the set $A + B$ is defined as $A + B = \{z \in \mathbb{R}^n \mid z = x + y, \, x \in A, \, y \in B\}$. When both u and v are continuous we may argue pointwisely: if $x \notin \overline{\text{supp}\, u + \text{supp}\, v}$ then for any $y \in \text{supp}\, u$ we find $x - y \notin \text{supp}\, v$ and therefore $u(y)v(x - y) = 0$ implying that $u * v = 0$, i.e. $\text{supp}\, u * v \subset \overline{\text{supp}\, u + \text{supp}\, v}$. Now for the general case let $\varphi \in C_0(\mathbb{R}^n)$. By our assumptions the integral $\int \varphi(u * v) \, d\lambda^{(n)}$ exists and we have

$$\int_{\mathbb{R}^n} \varphi(x) \left(\int_{\mathbb{R}^n} u(y)v(x - y) \, dy \right) dx = \int_{\text{supp}\, v} \left(\int_{\text{supp}\, u} \varphi(z + y)u(y)v(z) \, dy \right) dz.$$

Thus, if $\text{supp}\,\varphi \subset \overline{\text{supp}\, u + \text{supp}\, v}^\complement$ then the integral vanishes and consequently we have $\text{supp}(u * v) \subset \overline{\text{supp}\, u + \text{supp}\, v}$. $\square$

Using the Friedrichs mollifier which we have already introduced in Problem 11 of Chapter II.22 we will prove now that $C_0^\infty(\mathbb{R}^n)$ is dense in $L^p(\mathbb{R}^n)$, $1 \leq p < \infty$. With

$$a^{-1} := \int_{\mathbb{R}^n} \exp\left((\|x\|^2 - 1)^{-1} \right) dx$$

we define $j : \mathbb{R}^n \to \mathbb{R}$ by

$$j(x) := \begin{cases} a \exp\left((\|x\|^2 - 1)^{-1} \right), & \|x\| < 1 \\ 0, & \|x\| \geq 1. \end{cases} \tag{10.17}$$

It follows that $j \in C_0^\infty(\mathbb{R}^n)$, $\int j(x) \, dx = 1$, $j(x) \geq 0$, and $\text{supp}\, j \subset \overline{B_1(0)}$. For $\epsilon > 0$ we introduce further $j_\epsilon(x) := \epsilon^{-n} j\left(\frac{x}{\epsilon} \right)$ and we know that $j_\epsilon \in C_0^\infty(\mathbb{R}^n)$, $\text{supp}\, j_\epsilon \subset \overline{B_\epsilon(0)}$, $\int j_\epsilon(x) \, dx = 1$ and $j_\epsilon \geq 0$. For $u \in L^p(\mathbb{R}^n)$ we can now define the convolution operator (the **Friedrichs mollifier**)

$$J_\epsilon(u)(x) := (j_\epsilon * u)(x) = \int_{\mathbb{R}^n} j_\epsilon(x - y)u(y) \, dy. \tag{10.18}$$

From Theorem 8.4 we deduce that $J_\epsilon(u) \in C^\infty(\mathbb{R}^n)$, whereas Theorem 10.13 yields $J_\epsilon(u) \in L^p(\mathbb{R}^n)$, i.e. $J_\epsilon(u) \in C^\infty(\mathbb{R}^n) \cap L^p(\mathbb{R}^n)$. Since $C_0^\infty(\mathbb{R}^n) \subset \bigcap_{r \geq 1} L^r(\mathbb{R}^n)$, using Hölder's inequality, $\frac{1}{p} + \frac{1}{q} = 1$, we find

$$
|J_\epsilon(u)(x)|^p = \left| \int_{\mathbb{R}^n} u(x-y) j_\epsilon(y) \, \mathrm{d}y \right|^p
$$

$$
= \left| \int_{\mathbb{R}^n} \left(u(x-y) j_\epsilon(y)^{\frac{1}{p}} \right) j_\epsilon(y)^{\frac{1}{q}} \, \mathrm{d}y \right|^p
$$

$$
\leq \int_{\mathbb{R}^n} |u(x-y)|^p j_\epsilon(y) \, \mathrm{d}y \left(\int_{\mathbb{R}^n} j_\epsilon(y) \, \mathrm{d}y \right)^{\frac{p}{q}}
$$

$$
= \int_{\mathbb{R}^n} |u(x-y)|^p j_\epsilon(y) \, \mathrm{d}y.
$$

Integration with respect to x yields further

$$
\int_{\mathbb{R}^n} |J_\epsilon(u)(x)|^p \, \mathrm{d}x \leq \int_{\mathbb{R}^n} \left(\int_{\mathbb{R}^n} |u(x-y)|^p j_\epsilon(y) \, \mathrm{d}y \right) \mathrm{d}x
$$

$$
= \int_{\mathbb{R}^n} j_\epsilon(y) \left(\int_{\mathbb{R}^n} |u(x-y)|^p \, \mathrm{d}x \right) \mathrm{d}y = \|u\|_{L^p}^p,
$$

thus we arrive at

$$
\|J_\epsilon(u)\|_{L^p} \leq \|u\|_{L^p}. \tag{10.19}
$$

Note that for $p = 1$ there is no need to use Hölder's inequality to derive (10.19).

Lemma 10.16. *If $u \in C_0(\mathbb{R}^n)$, then $J_\epsilon(u)$ converges as $\epsilon \to 0$ in $L^p(\mathbb{R}^n)$ to u.*

Proof. For $\epsilon > 0$ and $x \in \mathbb{R}^n$ we find by the same arguments as above that

$$
|J_\epsilon(u)(x) - u(x)|^p = \left| \int_{\mathbb{R}^n} j_\epsilon(y) \left(u(x-y) - u(x) \right) \mathrm{d}y \right|^p
$$

$$
\leq \int_{\mathbb{R}^n} j_\epsilon(y) |u(x-y) - u(x)|^p \, \mathrm{d}y.
$$

Since by assumption supp u is compact we can find $R > 0$ such that $u|_{B_R^c(0)} = 0$ and therefore, for $\|y\| \leq R$ it follows that

$$
\int_{\|x\| \geq 2R} |u(x-y)|^p \, \mathrm{d}x \leq \int_{\|x\| \geq R} |u(x)|^p \, \mathrm{d}x = 0.
$$

In addition, since u is continuous, for $\eta > 0$ and $\epsilon = \epsilon(R, \eta)$ sufficiently small we have

$$\sup_{\|y\| \leq \epsilon} \int_{\|x\| \leq 2R} |u(x-y) - u(x)|^p \, \mathrm{d}x < \eta.$$

For these $\epsilon > 0$ we get

$$\int_{\mathbb{R}^n} |J_\epsilon(u)(x) - u(x)|^p \, \mathrm{d}x \leq \int_{\mathbb{R}^n} j_\epsilon(y) \left(\int_{\mathbb{R}^n} |u(x-y) - u(x)|^p \, \mathrm{d}x \right) \mathrm{d}y$$

$$\leq \sup_{\|y\| \leq \epsilon} \int_{\mathbb{R}^n} |u(x-y) - u(x)|^p \, \mathrm{d}x$$

$$\leq \sup_{\|y\| \leq \epsilon} \left(\int_{\|x\| \geq 2R} |u(x-y) - u(x)|^p \, \mathrm{d}x + \int_{\|x\| \leq 2R} |u(x-y) - u(x)|^p \, \mathrm{d}x \right)$$

$$= \sup_{\|y\| \leq \epsilon} \int_{\|x\| \leq 2R} |u(x-y) - u(x)|^p \, \mathrm{d}x < \eta,$$

which implies that $\lim_{\epsilon > 0} \|J_\epsilon(u) - u\|_{L^p} = 0$. $\square$

Theorem 10.17. *The space $C_0^\infty(\mathbb{R}^n)$ of all arbitrarily often differentiable functions with compact support is dense in $L^p(\mathbb{R}^n)$, $1 \leq p < \infty$.*

Proof. By Theorem 10.11 we know that $C_0(\mathbb{R}^n)$ is dense in $L^p(\mathbb{R}^n)$, $1 \leq p < \infty$. Hence, given $\eta > 0$ we can find for $u \in L^p(\mathbb{R}^n)$ a function $v \in C_0(\mathbb{R}^n)$ such that $\|u - v\|_{L^p} < \frac{\eta}{2}$. Further, by Lemma 10.16 we can find $\epsilon > 0$ such that $\|v - J_\epsilon(v)\|_{L^p} < \frac{\eta}{2}$ and $J_\epsilon(v) \in C^\infty(\mathbb{R}^n)$ with supp $J_\epsilon(v) \subset$ supp $v + \overline{B}_\epsilon(0)$ by Proposition 10.15. Thus $J_\epsilon(v) \in C_0^\infty(\mathbb{R}^n)$ and the triangle inequality gives

$$\|u - J_\epsilon(v)\|_{L^p} \leq \|u - v\|_{L^p} + \|v - J_\epsilon(v)\|_{L^p} < \eta.$$

$\square$

Let $g : \mathbb{R}^n \to \mathbb{R}$ be a non-negative integrable function. On $\mathcal{B}^{(n)}$ we define the measure $\mu := g\lambda^{(n)}$ and for $u \in L^p(\mathbb{R}^n)$, $1 \leq p < \infty$, we may consider the convolution

$$\int_{\mathbb{R}^n} u(x-y)g(y)\lambda^{(n)}(\mathrm{d}y) = \int_{\mathbb{R}^n} u(x-y)\mu(\mathrm{d}y). \qquad (10.20)$$

Thus we can take (10.20) to extend the definition of the convolution of two functions to define the convolution of a measure with a suitable function.

Suppose that $u \geq 0$ is integrable and let $B \in \mathcal{B}^{(n)}$. It follows with $\mu = g\lambda^{(n)}$ and $\nu = u\lambda^{(n)}$ that

$$\int_B \left(\int_{\mathbb{R}^n} u(x-y)g(y)\lambda^{(n)}(\mathrm{d}y) \right) \lambda^{(n)}(\mathrm{d}x)$$

$$= \int_{\mathbb{R}^n} \left(\int_{\mathbb{R}^n} \chi_B(x)u(x-y)g(y)\lambda^{(n)}(\mathrm{d}y) \right) \lambda^{(n)}(\mathrm{d}x)$$

$$= \int_{\mathbb{R}^n} \left(\int_{\mathbb{R}^n} \chi_B(x)u(x-y)\lambda^{(n)}(\mathrm{d}x) \right) g(y)\lambda^{(n)}(\mathrm{d}y)$$

$$= \int_{\mathbb{R}^n} \left(\int_{\mathbb{R}^n} \chi_B(y+z)u(z)\lambda^{(n)}(\mathrm{d}z) \right) g(y)\lambda^{(n)}(\mathrm{d}y)$$

$$= \int_{\mathbb{R}^n} \left(\int_{\mathbb{R}^n} \chi_B(x+y)\nu(\mathrm{d}x) \right) \mu(\mathrm{d}y) = \int_{\mathbb{R}^n} \left(\int_{\mathbb{R}^n} \chi_B(x+y)\mu(\mathrm{d}y) \right) \nu(\mathrm{d}x).$$

The mapping

$$B \mapsto \int_{\mathbb{R}^n} \left(\int_{\mathbb{R}^n} \chi_B(x+y)\nu(\mathrm{d}x) \right) \mu(\mathrm{d}y)$$

is however a measure on $\mathcal{B}^{(n)}$, hence

$$(\mu * \nu)(B) := \int_{\mathbb{R}^n} \left(\int_{\mathbb{R}^n} \chi_B(x+y)\nu(\mathrm{d}x) \right) \mu(\mathrm{d}y) \tag{10.21}$$

defines a measure on $\mathcal{B}^{(n)}$ called the convolution of $\mu = g\lambda^{(n)}$ and $\nu = u\lambda^{(n)}$. We have already seen that $\mu * \nu = \nu * \mu$ holds.

Definition 10.18. *Let μ and ν be two finite measures on $\mathcal{B}^{(n)}$. Their **convolution** $\mu * \nu$ is the measure defined on $\mathcal{B}^{(n)}$ by*

$$(\mu * \nu)(B) = \int_{\mathbb{R}^n} \left(\int_{\mathbb{R}^n} \chi_B(x+y)\nu(\mathrm{d}x) \right) \mu(\mathrm{d}y). \tag{10.22}$$

Lemma 10.19. *For finite measures μ and ν on $\mathcal{B}^{(n)}$ the following hold*

$$\mu * \nu = \nu * \mu; \tag{10.23}$$

and

$$(\mu * \nu)(\mathbb{R}^n) = \mu(\mathbb{R}^n)\nu(\mathbb{R}^n). \tag{10.24}$$

In particular the convolution of two probability measures is a probability measure. Moreover, if $\mu = g\lambda^{(n)}$ and $\nu = h\lambda^{(n)}$ with non-negative integrable functions g and h, then $\mu * \nu$ has also a density with respect to $\lambda^{(n)}$ given by $g * h$, i.e.

$$\mu * \nu = (g * h)\lambda^{(n)}. \tag{10.25}$$

Proof. The commutativity relation (10.23) follows by applying Fubini's theorem to the defining equation (10.22) and choosing $B = \mathbb{R}^n$ in (10.22) we find immediately

$$(\mu * \nu)(\mathbb{R}^n) = \int_{\mathbb{R}^n} \left(\int_{\mathbb{R}^n} 1\nu(dx) \right) \mu(dy) = \int_{\mathbb{R}^n} \nu(\mathbb{R}^n)\mu(dy) = \nu(\mathbb{R}^n)\mu(\mathbb{R}^n).$$

The final statement was already discussed:

$$\begin{aligned}
(\mu * \nu)(B) &= \int_{\mathbb{R}^n} \left(\int_{\mathbb{R}^n} \chi_B(x + y)\nu(dx) \right) \mu(dy) \\
&= \int_B \left(\int_{\mathbb{R}^n} h(x - y)g(y)\lambda^{(n)}(dy) \right) \lambda^{(n)}(dx) \\
&= \left((g * h)\lambda^{(n)} \right)(B).
\end{aligned}$$

$\square$

We want to change our point of view and give a different interpretation of $\mu * \nu$ starting with (10.22). For a simple function $f : \mathbb{R}^n \to [0, \infty)$, $f = \sum_{k=1}^N \alpha_k \chi_{A_k}$, we find

$$\int f \, d(\mu * \nu) = \int_{\mathbb{R}^n} \left(\int_{\mathbb{R}^n} f(x + y)\mu(dy) \right) \nu(dx). \tag{10.26}$$

Further, if u is a $(\mu*\nu)$-integrable non-negative function on $\mathbb{R}^n$ we may choose a sequence $(f_k)_{k\in\mathbb{N}}$ of non-negative measurable functions such that $f_k \leq f_{k+1}$ and $\lim_{k\to\infty} f_k(x) = u(x)$ and it follows that (10.26) holds for u, i.e. for all non-negative $(\mu*\nu)$-integrable functions. By Fubini's theorem it now follows that $f(x + \cdot)$ is μ- as well as ν-integrable. Next we apply our results on integration with respect to an image measure to the product measure $\mu \otimes \nu$ and the mapping $\mathrm{add}_2 : \mathbb{R}^n \times \mathbb{R}^n \to \mathbb{R}^n$, $\mathrm{add}_2(x, y) = x + y$, to find

$$\begin{aligned}
\int_{\mathbb{R}^n \times \mathbb{R}^n} (f \circ \mathrm{add}_2)(x, y)(\mu \otimes \nu)(dy, dx) &= \int_{\mathbb{R}^n} \left(\int_{\mathbb{R}^n} f(x + y)\mu(dy) \right) \nu(dx) \\
&= \int_{\mathbb{R}^n} f(z)(\mu * \nu)(dz),
\end{aligned}$$

i.e. we have proved

Lemma 10.20. *The convolution $\mu * \nu$ is the image of the measure $\mu \otimes \nu$ under the mapping add_2.*

With Lemma 10.20 in mind we may introduce for finite measures $\mu_1, \ldots, \mu_N$ on $\mathcal{B}^{(n)}$ the measure $\mu_1 * \cdots * \mu_N$ by

$$\mu_1 * \cdots * \mu_N := \mathrm{add}_N(\mu_1 \otimes \cdots \otimes \mu_N) \tag{10.27}$$

where $\mathrm{add}_N : \mathbb{R}^n \times \cdots \times \mathbb{R}^n \to \mathbb{R}^n$, $(x_1, \ldots, x_N) \mapsto x_1 + \cdots + x_N$. Let $B_{N+1} : \mathbb{R}^{n(N+1)} \to \mathbb{R}^{2n}$, $B_{N+1}(x_1, \ldots, x_{N+1}) = (x_1 + \cdots + x_N, x_{N+1})$ and $D_{N+1} : \mathbb{R}^{n(N+1)} \to \mathbb{R}^{2n}$, $D_{N+1}(x_1, \ldots, x_{N+1}) = (x_1, x_2 + \cdots + x_{N+1})$. Both mappings are continuous, hence measurable, and we have

$$B_{N+1}(\mu_1 \otimes \cdots \otimes \mu_{N+1}) = \mathrm{add}_N(\mu_1 \otimes \cdots \otimes \mu_N) \otimes \mu_{N+1}$$

as well as

$$D_{N+1}(\mu_1 \otimes \cdots \otimes \mu_{N+1}) = \mu_1 \otimes \mathrm{add}_N(\mu_2 \otimes \cdots \otimes \mu_{N+1}).$$

Moreover we have $\mathrm{add}_{N+1} = \mathrm{add}_2 \circ B_{N+1} = \mathrm{add}_2 \circ D_{N+1}$, implying

$$\begin{aligned}
\mu_1 * \cdots * \mu_{N+1} &= \mathrm{add}_{N+1}(\mu_1 \otimes \cdots \otimes \mu_{N+1}) \\
&= \mathrm{add}_2((\mu_1 * \cdots * \mu_N) \otimes \mu_{N+1}) \\
&= (\mu_1 * \cdots * \mu_N) * \mu_{N+1}
\end{aligned}$$

as well as

$$\begin{aligned}
\mu_1 * \cdots * \mu_{N+1} &= \mathrm{add}_{N+1}(\mu_1 \otimes \cdots \otimes \mu_{N+1}) \\
&= \mathrm{add}_2(\mu_1 \otimes (\mu_2 * \cdots * \mu_{N+1})) \\
&= \mu_1 * (\mu_2 * \cdots * \mu_{N+1}).
\end{aligned}$$

In particular we have that iterating our initial definition is consistent with (10.27) and convolution of finite measures is associative, i.e.

$$(\mu_1 * \mu_2) * \mu_3 = \mu_1 * (\mu_2 * \mu_3). \tag{10.28}$$

Example 10.21. A. A trivial consequence of the properties of integrals is that for finite measures μ_1, μ_2 and μ_3 on $\mathcal{B}^{(n)}$ and for $\alpha \geq 0$ we have

$$\mu_1 * (\mu_2 + \mu_3) = \mu_1 * \mu_2 + \mu_1 * \mu_3 \tag{10.29}$$

and

$$(\alpha\mu_1) * \mu_2 = \mu_1 * (\alpha\mu_2) = \alpha(\mu_1 * \mu_2). \tag{10.30}$$

B. For a finite measure μ on $\mathcal{B}^{(n)}$ and the translation $T_{x_0}(y) = x_0 + y$ we always find that

$$\mu(B - x_0) = \mu(T_{-x_0}(B)) = \mu\left(\{y \in \mathbb{R}^n \mid y \in -x_0 + B\}\right)$$
$$= \mu\left(\{y \in \mathbb{R}^n \mid y + x_0 \in B\}\right) = T_{x_0}(\mu)(B)$$

and therefore we find for the Dirac measure at x_0

$$(\mu * \epsilon_{x_0})(B) = (\epsilon_{x_0} * \mu)(B) = \int \mu(B - x)\epsilon_{x_0}(dx)$$
$$= \mu(B - x_0) = T_{x_0}(\mu)(B),$$

i.e. we have

$$\mu * \epsilon_{x_0} = \epsilon_{x_0} * \mu = T_{x_0}(\mu). \tag{10.31}$$

In particular we find

$$\epsilon_{x_1} * \epsilon_{x_2} = \epsilon_{x_1 + x_2} \tag{10.32}$$

and

$$\mu * \epsilon_0 = \mu = \epsilon_0 * \mu, \tag{10.33}$$

i.e. in the set of all finite measures on $\mathcal{B}^{(n)}$ with convolution as operation ϵ_0 behaves as a unit element.

C. Consider for $t, s > 0$ the Poisson distributions

$$\pi_t = \sum_{k=0}^{\infty} e^{-t} \frac{t^k}{k!} \epsilon_k \quad \text{and} \quad \pi_s = \sum_{l=0}^{\infty} e^{-s} \frac{s^l}{l!} \epsilon_l.$$

These are probability measures on $\mathcal{B}^{(1)}$ and therefore their convolution $\pi_t * \pi_s$ is well defined and we find using Part A and B and the binomial theorem

$$\pi_t * \pi_s = \sum_{k=0}^{\infty} \sum_{l=0}^{\infty} e^{-(t+s)} \frac{t^k}{k!} \frac{s^l}{l!} \epsilon_k * \epsilon_l$$
$$= e^{-(t+s)} \sum_{m=0}^{\infty} \left(\sum_{k=1}^{m} \frac{t^m}{k!} \frac{s^{m-k}}{(m-k)!} \right) \epsilon_m$$
$$= e^{-(t+s)} \sum_{m=0}^{\infty} \frac{(t+s)^m}{m!} \epsilon_m = \pi_{t+s}.$$

Once we have mastered the basic properties of the Fourier transform in Part 8, we will understand this example in a much broader context.

We will encounter convolutions of functions or of measures, and later on of distributions, in many places in the rest of these treatises. Often they appear as "linear operators" in our considerations.

Example 10.22. With j_ϵ, $\epsilon > 0$, and j as in (10.17) we can define a linear mapping or operator $J_\epsilon : L^p(\mathbb{R}^n) \to L^p(\mathbb{R}^n)$, $1 \le p < \infty$ satisfying the estimate (10.19), hence J_ϵ is continuous with respect to the norm $\|.\|_{L^p}$, $1 \le p < \infty$.

More generally let $k : \mathbb{R}^n \to \mathbb{R}$ be an element of $L^1(\mathbb{R}^n)$ and define on $L^p(\mathbb{R}^n)$, $1 \le p < \infty$, the term

$$K_{\mathrm{op}}u := k * u, \tag{10.34}$$

or

$$(K_{\mathrm{op}}u)(x) = \int k(x - y)u(y)\,\mathrm{d}y. \tag{10.35}$$

Young's inequality, i.e. Theorem 10.13, yields that

$$\|K_{\mathrm{op}}u\|_{L^p} \le \|k\|_{L^1}\|u\|_{L^p},$$

which implies that $K_{\mathrm{op}} : L^p(\mathbb{R}^n) \to L^p(\mathbb{R}^n)$ is a linear and (Lipschitz) continuous operator. Operators of this type are called **convolution operators** and k or $\tilde{k}(x, y) := k(x - y)$ are called the **kernel** of the operator K_{op}, sometimes k is called the **convolution kernel** whereas $\tilde{k}$ is called the kernel.

We claim that every convolution operator K_{op} is translation invariant, i.e. for $x_0 \in \mathbb{R}^n$ we have

$$T_{x_0}(K_{\mathrm{op}}u) = K_{\mathrm{op}}(T_{x_0}u) \tag{10.36}$$

or

$$T_{x_0} \circ K_{\mathrm{op}} = K_{\mathrm{op}} \circ T_{x_0}, \tag{10.37}$$

i.e. for the **commutator** $[K_{\mathrm{op}}, T_{x_0}] = K_{\mathrm{op}} \circ T_{x_0} - T_{x_0} \circ K_{\mathrm{op}}$ it follows that $[K_{\mathrm{op}}, T_{x_0}] = 0$. Indeed we have for $u \in L^p(\mathbb{R}^n)$

$$T_{x_0}(K_{\mathrm{op}}u)(x) = (K_{\mathrm{op}}u)(x_0 + x) = \int_{\mathbb{R}^n} k(x + x_0 - y)u(y)\,\mathrm{d}y$$

$$= \int_{\mathbb{R}^n} k(x - z)u(x_0 + z)\,\mathrm{d}z = \int_{\mathbb{R}^n} k(x - z)(T_{x_0}u)(z)\,\mathrm{d}z$$

$$= K_{\mathrm{op}}(T_{x_0}u)(x).$$

We will see much later that all translation invariant operators (satisfying some type of continuity condition) are of convolution type, but $L^p(\mathbb{R}^n)$ must be replaced by certain spaces of generalised functions.

Problems

1. Let $X \in \{C_0(\mathbb{R}^n), C_\infty(\mathbb{R}^n), C_b(\mathbb{R}^n), C^k(\mathbb{R}^n)\}$ and define for $u \in X$ the function $v(y) := (T_{A,x}u)(y) = u(Ay + x)$, where $A \in O(n)$ and $x \in \mathbb{R}^n$. Prove that $v \in X$.

2. Prove that $L^1(\mathbb{R}^n, +, *)$ is a commutative algebra over $\mathbb{R}$.

3.   a) Let $a \in C_b(\mathbb{R}^n)$ and define on $C_0(\mathbb{R}^n)$ the linear operator $A : C_0(\mathbb{R}^n) \to L^p(\mathbb{R}^n)$, $1 \le p < \infty$, by $Au := au$. Prove that A has a continuous extension $\tilde{A} : L^p(\mathbb{R}^n) \to L^p(\mathbb{R}^n)$ satisfying $\|\tilde{A}u\|_{L^p} \le \|a\|_\infty \|u\|_{L^p}$.
    Hint: Theorem 10.17 may be used.

     b) Prove that the operator $\frac{d}{dx} : C^1([0,1]) \to L^2([0,1])$, $u \mapsto \frac{du}{dx}$, does not have a linear continuous extension satisfying $\left\|\frac{du}{dx}\right\|_{L^2} \le c\|u\|_{L^2}$.
    Hint: investigate the functions $u_k(x) = \sin 2\pi kx$.

4. Recall the definition of the Schwartz space $\mathcal{S}(\mathbb{R}^n)$, see Problem 3 of Chapter II.20: $\mathcal{S}(\mathbb{R}^n) = \{u \in C^\infty(\mathbb{R}^n) | p_{\alpha,\beta}(u) < \infty \text{ for all } \alpha, \beta \in \mathbb{N}_0^n\}$ where $p_{\alpha,\beta}(u) = \sup_{x \in \mathbb{R}^n} |x^\beta \partial^\alpha u(x)|$. Prove that for $u, v \in \mathcal{S}(\mathbb{R}^n)$ it follows that $u * v \in \mathcal{S}(\mathbb{R}^n)$.
    Hint: first prove that for $u \in \mathcal{S}(\mathbb{R}^n)$ and all $m_1, m_2 \in \mathbb{N}_0$ we have $\sup_{x \in \mathbb{R}^n}(1 + \|x\|^2)^{\frac{m_2}{2}} \sum_{|\alpha| \le m_1} |\partial^\alpha u(x)| < \infty$. Further recall Peetre's inequality (I.23.14).

5. Prove the following C^∞-version of Urysohn's lemma: let $K \subset G \subset \mathbb{R}^n$ where K is compact and G is an open bounded set. Then there exists a function $\varphi \in C_0^\infty(\mathbb{R}^n)$ such that $\operatorname{supp}\varphi \subset G$, $0 \le \varphi \le 1$ and $\varphi|_K = 1$.

6.* (**C^∞-partition of unity**) Let $K \subset \mathbb{R}^n$ be a compact set and $G_j \subset \mathbb{R}^n$, $j = 1, \ldots, N$, be bounded open sets such that $K \subset \bigcup_{j=1}^N G_j$. Then there exist functions $\varphi_j \in C_0^\infty(\mathbb{R}^n)$, $\operatorname{supp}\varphi_j \subset G_j$, such that $\sum_{j=1}^N \varphi_j(x) = 1$ for all $x \in K$.

7. For $t > 0$ and $a \geq 0$ define on $(\mathbb{R}, \mathcal{B}^{(1)})$ the measures $\mu_t := e^{-at}\epsilon_0$ and prove that $\mu_t * \mu_s = \mu_{t+s}$ for all $t, s > 0$.

8. Let $(\mu_t)_{t>0}$ be a family of probability measures on $\mathbb{R}^n$ having the property that $\mu_t * \mu_s = \mu_{t+s}$ for all $t, s > 0$. See Example 10.21.C or Problem 7 for examples. For $u \in C_\infty(\mathbb{R}^n)$ and $t > 0$ define

$$T_t u(x) := \int u(x - y)\mu_t(dy),$$

and take for granted that $T_t u \in C_\infty(\mathbb{R}^n)$ for all $t > 0$, a proof of this will be provided in Part 8. Thus $T_t : C_\infty(\mathbb{R}^n) \to C_\infty(\mathbb{R}^n)$ by assumption. Prove that T_t is a linear operator satisfying $\|T_t u\|_\infty \leq \|u\|_\infty$ for all $t > 0$ and $u \in C_\infty(\mathbb{R}^n)$. Furthermore, prove for $t, s > 0$ that $T_{s+t} = T_s \circ T_t$, i.e. $T_{s+t}u = T_s(T_t u)$.

9. For $0 < \alpha < 1$ the function $k_\alpha(x) = \chi_{[-1,1]}|x|^{-\alpha}$, $x \neq 0$, $k_\alpha(0) = 0$, belongs to $L^1(\mathbb{R})$. (Why is this the case?) Define on $C_0(\mathbb{R})$ the integral operator K_{op} by
$$K_{op}u := k_\alpha * u$$
(such an operator is called a **convolution operator**). Prove that for $1 \leq p < \infty$ the operator K_{op} has a continuous extension from $L^p(\mathbb{R})$ to $L^p(\mathbb{R})$.

Now switch to $\mathbb{R}^n$ and give conditions on $\gamma > 0$ such that with $k_{\gamma,n}(\|x\|)$ $:= \chi_{\overline{B_1(0)}}(x)\|x\|^{-\gamma}$, $\gamma > 0$, $k_{\gamma,n}(0) = 0$, the operator

$$K_{op}^{(n)}u(x) := \int k_{\gamma,n}(\|x - y\|)u(y)\lambda^{(n)}(dy)$$

originally defined on $C_0(\mathbb{R}^n)$ extends to a continuous linear operator from $L^p(\mathbb{R}^n)$ to $L^p(\mathbb{R}^n)$.

11 Differentiation Revisited

Let $f : [a, b] \to \mathbb{R}$ be a Borel measurable function. We want to investigate questions such as: for which points $x \in [a, b]$ is f differentiable, or if f is integrable, when is $x \mapsto \int_{[a,x]} f(t)\lambda^{(1)}(dt)$ differentiable, and does the fundamental theorem hold?

Before we start with our studies we want to provide a result seemingly unrelated to our problems, but as it will turn out, it is a key technical ingredient for deriving central results.

Definition 11.1. *Let $A \subset \mathbb{R}$ be a Borel measurable set. A family $\mathcal{V} := (I_j)_{j \in J}$, $J \neq \emptyset$, of closed intervals $I_j \subset \mathbb{R}$, $\mathring{I}_j \neq \emptyset$, is called a **Vitali covering** of A if for each $x \in A$ and $\epsilon > 0$ there exists an interval $I = I_{x,\epsilon} \in \mathcal{V}$ such that $x \in I$ and $\lambda^{(1)}(I) < \epsilon$.*

Theorem 11.2 (Vitali's covering theorem). *Let $A \subset \mathbb{R}$ be a Borel measurable set and $\mathcal{V}$ a Vitali covering of A. Then we can find a denumerable family $(I_{j_k})_{k \in \mathbb{N}}$ of mutually disjoint intervals $I_{j_k} \in \mathcal{V}$ such that*

$$\lambda^{(1)}\left(A \cap \left(\bigcup_{k \in \mathbb{N}} I_{j_k}\right)^{\complement}\right) = 0. \tag{11.1}$$

Moreover, if $\lambda^{(1)}(A) < \infty$ then for every $\epsilon > 0$ we can find a finite family of mutually disjoint intervals $I_{j_1}, \ldots, I_{j_N} \in \mathcal{V}$ such that

$$\lambda^{(1)}\left(A \cap \left(\bigcup_{k=1}^{N} I_{j_k}\right)^{\complement}\right) < \epsilon. \tag{11.2}$$

Proof. (Following [35]) Suppose first that $\lambda^{(1)}(A) < \infty$ and take $U \subset \mathbb{R}$ open such that $A \subset U$ and $\lambda^{(1)}(U) < \infty$, recall Theorem 10.10. Now consider $\mathcal{V}_0 := \{I \subset \mathcal{V} \mid I \subset U\}$ which is a further Vitali covering of A. Pick any $I_1 \in \mathcal{V}_0$. In the case that $A \subset I_1$ it follows that $A \cap I_1^{\complement} = \emptyset$, i.e. $\lambda^{(1)}(A \cap I_1^{\complement}) = 0$ and the assertion is proved. Assume that $A \cap I_1^{\complement} \neq \emptyset$, i.e. A is not contained in I_1. We now set

$$A_1 := I_1, \quad U_1 := U \cap A_1^{\complement}$$

215

and we note that A_1 is closed, U_1 is open and $U_1 \cap A \neq \emptyset$. We choose now $I_2 \in \mathcal{V}_0$ such that $I_2 \subset U_1$, in particular $I_1 \cap I_2 = \emptyset$, and such that $\lambda^{(1)}(I_2) > \frac{1}{2}\delta_1$ where

$$\delta_1 = \sup\left\{\lambda^{(1)}(I) \,\middle|\, I \in \mathcal{V}_0,\, I \subset U_1\right\} \leq \lambda^{(1)}(U) < \infty.$$

If $A \subset I_1 \cup I_2$ we are done. Otherwise we continue this construction of intervals I_j. Thus suppose that we have already selected mutually disjoint intervals $I_1, \ldots, I_N$, $I_j \in \mathcal{V}_0$ and $A \cap \left(\bigcup_{k=1}^{N} I_k\right)^{\mathsf{c}} \neq \emptyset$. We set

$$A_N := \bigcup_{k=1}^{N} I_k, \quad U_N := U \cap A_N^{\mathsf{c}}$$

and we note again that A_N is closed and U_N is open as well as $U_N \cap A \neq \emptyset$. Let

$$\delta_N := \sup\left\{\lambda^{(1)}(I) \,\middle|\, I \in \mathcal{V}_0,\, I \subset U_N\right\} \tag{11.3}$$

and choose $I_{N+1} \in \mathcal{V}_0$ such that $I_{N+1} \subset U_N$, $\lambda^{(1)}(I_{N+1}) > \frac{1}{2}\delta_N$. If this process stops for some N_0, i.e. $A \subset \bigcup_{k=1}^{N_0} I_k$ then the theorem is already proved. Otherwise we obtain an infinite sequence $(I_k)_{k \in \mathbb{N}}$ of mutually disjoint closed intervals $I_k \in \mathcal{V}_0$. We need to prove in this case with $D := \bigcup_{k \in \mathbb{N}} I_k$ that $\lambda^{(1)}\left(A \cap D^{\mathsf{c}}\right) = 0$. For $k \in \mathbb{N}$ we denote by J_k the closed interval with the same midpoint as I_k and satisfying

$$\lambda^{(1)}(J_k) = 5\lambda^{(1)}(I_k).$$

By our assumptions we have

$$\lambda^{(1)}\left(\bigcup_{k=1}^{\infty} J_k\right) \leq 5\sum_{k=1}^{\infty} \lambda^{(1)}(I_k)$$
$$= 5\lambda^{(1)}(D) \leq 5\lambda^{(1)}(U) < \infty.$$

From Corollary 2.13 we deduce that

$$\lim_{M \to \infty} \lambda^{(1)}\left(\bigcup_{k=M}^{\infty} J_k\right) = 0. \tag{11.4}$$

We claim that $A \cap D^{\mathsf{c}} \subset \bigcup_{k=M}^{\infty} J_k$ holds for every $M \in \mathbb{N}$ which implies by (11.3) that $\lambda^{(1)}\left(A \cap D^{\mathsf{c}}\right) = 0$, i.e. (11.2). We fix $M \in \mathbb{N}$ and pick

$x \in A \cap D^{\complement}$. It follows that $x \in A \cap A_M^{\complement} \subset U_M$. Hence we can find $I \in \mathcal{V}_0$ such that $x \in I \subset U_M$. Clearly $\delta_M < 2\lambda^{(1)}(I_{M+1})$ and by (11.4) we have $\lim_{k \to \infty} \lambda^{(1)}(I_k) = 0$, i.e. for some $k_0 \in \mathbb{N}$ we have $\delta_{k_0} < \lambda^{(1)}(I)$. On the other hand, by (11.4) there exists $l_0 \in \mathbb{N}$ such that I is not a subset of U_{l_0}, and therefore there is a smallest $M_0 \in \mathbb{N}$ with this property. Therefore we must have $M < M_0$, and consequently we find

$$I \cap A_{M_0} \neq \emptyset \quad \text{and} \quad I \cap A_{M_0-1} = \emptyset,$$

hence

$$I \cap I_{M_0} \neq \emptyset. \tag{11.5}$$

Since $I \subset U_{M_0-1}$ we get

$$\lambda^{(1)}(I) \leq \delta_{M_0-1} < 2\lambda^{(1)}(I_{M_0}). \tag{11.6}$$

Using that $\lambda^{(1)}(I_{M_0}) = 5\lambda^{(1)}(I_{M_0})$, (11.5) and (11.6) we deduce that

$$I \subset J_{M_0} \subset \bigcup_{k=M}^{\infty} J_k,$$

i.e. $x \in \bigcup_{k=M}^{\infty} J_k$, which eventually yields $\lambda^{(1)}\left(A \cap D^{\complement}\right) = 0$. Given $\epsilon > 0$ we can find $M \in \mathbb{N}$ such that

$$\sum_{k=M+1}^{\infty} \lambda^{(1)}(I_k) < \epsilon$$

which implies by

$$A \cap A_M^{\complement} \subset \left(A \cap D^{\complement}\right) \cup \left(\bigcup_{k=M+1}^{\infty} I_k\right)$$

that

$$\lambda^{(1)}\left(A \cap A_M^{\complement}\right) \leq \lambda^{(1)}\left(\bigcup_{k=M+1}^{\infty} I_k\right) = \sum_{k=M+1}^{\infty} \lambda^{(1)}(I_k) < \epsilon,$$

i.e. the theorem is proved for $\lambda^{(1)}(A) < \infty$.

Now let $\lambda^{(1)}(A) = \infty$. For $m \in \mathbb{Z}$ we define $A^{(m)} := A \cap (m, m+1)$ and $\mathcal{V}_m := \{I \in \mathcal{V} \mid I \subset (m, m+1)\}$ which is a Vitali covering of $A^{(m)}$ and $\lambda^{(1)}(A^{(m)}) \leq 1$. Thus we can apply the result of the first part to find a denumerable family $\mathcal{I}_m \subset \mathcal{V}_m$ of mutually disjoint intervals such that with $D_m = \bigcup\{K \mid K \in \mathcal{I}_m\}$ we have $\lambda^{(1)}\left(A^{(m)} \cap D_m^{\complement}\right) = 0$, $m \in \mathbb{Z}$. The family $\mathcal{I} = \bigcup_{k \in \mathbb{Z}} \mathcal{I}_k$ is denumerable and consists of mutually disjoint elements of $\mathcal{V}$. Moreover we have with $D_\infty = \bigcup_{m \in \mathbb{Z}} D_m$ that

$$A \cap D_\infty^{\complement} \subset \mathbb{Z} \cup \bigcup_{m \in \mathbb{Z}} \left(A^{(m)} \cap D_m^{\complement}\right)$$

implying

$$\lambda^{(1)}\left(A \cap D_\infty^{\complement}\right) \leq \lambda^{(1)}(\mathbb{Z}) + \sum_{m \in \mathbb{Z}} \lambda^{(1)}\left(A^{(m)} \cap D_m^{\complement}\right) = 0.$$

$\square$

Before we now turn to differentiability questions we recollect some results from Chapter I.32. Let $a < b$ and $f : [a, b] \to \mathbb{R}$ be a bounded increasing function. We know that f can have at most countably many jumps on $[a, b]$. By

$$[f](x_0) := f(x_0+) - f(x_0-) \tag{11.7}$$

we denote the **jump** of f at x_0 where

$$f(x_0+) = \lim_{\substack{x \to x_0 \\ x > x_0}} f(x) \quad \text{and} \quad f(x_0-) = \lim_{\substack{x \to x_0 \\ x < x_0}} f(x). \tag{11.8}$$

The **jump function** s_f of f is defined as

$$s_f(x) = \begin{cases} 0, & x = a \\ (f(a+) - f(a)) + \sum_{y < x} [f]y + (f(x) - f(x-)), & 0 < x \leq b. \end{cases} \tag{11.9}$$

By Theorem I.32.7 the function

$$\varphi_f(x) := f(x) - s_f(x) \tag{11.10}$$

is an increasing continuous function.

For a bounded function $g : [a, b] \to \mathbb{R}$ we define, see Definition I.32.8,

$$V_Z(g) = \sum_{k=0}^{N-1} |g(x_{k+1}) - g(x_k)| \tag{11.11}$$

where $Z(x_0, x_1, \ldots, x_{N-1}, x_N)$, $a = x_0 < x_1 < \cdots < x_{N-1} < x_N = b$ is a partition of $[a, b]$. Now the **variation** or **total variation** of g is given by

$$V(g) := \sup_Z V_Z(g) \tag{11.12}$$

where the supremum is taken over all partitions of $[a, b]$. According to Definition I.32.8, we call g of **bounded variation** if $V(g) < \infty$. Functions of bounded variation are bounded, and monotone as well as Lipschitz continuous functions are of bounded variation. However, there are continuous functions which are not of bounded variation, see Example I.32.13. By Theorem I.32.14 the set $BV([a, b])$ of all functions on $[a, b]$ with bounded variation is an algebra and by Corollary I.32.16 we know that $g \in BV([a, b])$ if and only if with two increasing monotone functions h_1 and h_2 we have $g = h_1 - h_2$.

The next class of functions we have introduced, see Problem 5 of Chapter I.32, are absolutely continuous functions. Recall that $h : [a, b] \to \mathbb{R}$ is called **absolutely continuous** on $[a, b]$ if for every $\epsilon > 0$ there exists $\delta > 0$ such that for all $m \in \mathbb{N}$ and any choice of mutually disjoint open intervals $(a_j, b_j) \subset [a, b]$, $j = 1, \ldots, m$, the estimate $\sum_{j=1}^m (b_j - a_j) < \delta$ implies $\sum_{j=1}^m |h(b_j) - h(a_j)| < \epsilon$. Any Lipschitz continuous function is absolutely continuous and every absolutely continuous function is of bounded variation.

Definition 11.3. *By $AC([a, b])$ we denote the set of all absolutely continuous functions on $[a, b]$.*

From Problem 6 of Chapter I.32 we know that $AC([a, b])$ is an algebra.

We have already seen relation between BV- and AC-functions and integrals:

$$f \in C([a, b]) \quad \text{implies} \quad F(\cdot) = \int_a^\bullet f(t) \, dt \in BV([a, b]) \quad \text{and} \quad V(f) = \int_a^b f(t) \, dt, \tag{11.13}$$

$$f \in C([a, b]) \cup BV([a, b]) \quad \text{implies} \quad F(\cdot) = \int_a^\bullet f(t) \, dt \in AC([a, b]). \tag{11.14}$$

Thus there are indicators that $BV([a, b])$ and $AC([a, b])$ are related to differentiability properties and the fundamental theorem.

Let $f : [a, b] \to \mathbb{R}$ be a function. The **Dini derivatives** or **Dini numbers** of f at $x \in [a, b]$ are defined as

$$D_+ f(x) = \liminf_{h \to 0+} \frac{f(x+h) - f(x)}{h}, \qquad (11.15)$$

$$D_- f(x) = \liminf_{h \to 0-} \frac{f(x+h) - f(x)}{h}, \qquad (11.16)$$

$$D^+ f(x) = \limsup_{h \to 0+} \frac{f(x+h) - f(x)}{h}, \qquad (11.17)$$

$$D^- f(x) = \limsup_{h \to 0-} \frac{f(x+h) - f(x)}{h}. \qquad (11.18)$$

Clearly, if f is differentiable at x then all four Dini derivatives are equal to $f'(x)$. For $x = a$ we cannot define $D_- f(a)$ and $D^- f(a)$ while for $x = b$ we cannot define $D_+ f(b)$ and $D^+ f(b)$. We call $D_+ f(x)$ the **lower right derivative** of f at x, $D^+ f(x)$ the **upper right derivative** of f at x, and further $D_- f(x)$ is the **lower left derivative** of f at x whereas $D^- f(x)$ is the **upper left derivative** of f at x. Note that the values $+\infty$ and $-\infty$ are not excluded.

Definition 11.4. *We say that $f : [a, b] \to \mathbb{R}$ has **right derivative** at $x \in [a, b]$ or is **differentiable from the right** at x if $D^+ f(x) = D_+ f(x)$, and in this case we write*

$$f'_+(x) := D^+ f(x) = D_+ f(x). \qquad (11.19)$$

*If $D^- f(x) = D_- f(x)$ we call f **differentiable from the left** or we say that f has a **left derivative** at x, and we write*

$$f'_-(x) := D^- f(x) = D_- f(x). \qquad (11.20)$$

We also note that in (11.15) - (11.18) we can switch to sequences, e.g.

$$D_+ f(x) = \liminf_{n \to \infty} \frac{f(x + h_n) - f(x)}{h_n}$$

where $h_n > 0$ and $\lim_{n \to \infty} h_n = 0$. This implies immediately that for f measurable the functions $D_+ f, D^+ f, D_- f$ and $D^- f$ are measurable. Thus for $0 < p < q$, $p, q \in \mathbb{Q}$ the sets

$$A_{p,q}(f) := \left\{ x \in [a, b] \,\middle|\, D_+ f(x) < p < q < D^+ f(x) \right\} \qquad (11.21)$$

and

$$A(f) := \{x \in [a, b] \mid D_+ f(x) < D^+ f(x)\} = \bigcup_{\substack{p, q \in \mathbb{Q} \\ 0 < p < q}} A_{p,q}(f) \tag{11.22}$$

are measurable.

Lemma 11.5. *For a monotone increasing function* $f : [a, b] \to \mathbb{R}$ *we have* $\lambda^{(1)}(A(f)) = 0$.

Proof. According to (11.22) it is sufficient to prove that for $0 < p < q$, $p, q \in \mathbb{Q}$, we have $\lambda^{(1)}(A_{p,q}(f)) = 0$. Suppose the existence of $0 < p < q$, $p, q \in \mathbb{Q}$, such that $\lambda^{(1)}(A_{p,q}(f)) = \alpha > 0$. Choose $\epsilon > 0$ such that

$$0 < \epsilon < \frac{\alpha(q - p)}{p + 2q}$$

and pick an open set $U \subset \mathbb{R}$ such that $\lambda^{(1)}(U) < \alpha + \epsilon$. For $x \in A_{p,q}(f)$ we can find $\delta > 0$ such that $[x, x + \delta] \subset U \cap [a, b]$ and

$$f(x + \delta) - f(x) < p\delta. \tag{11.23}$$

If we let x vary over all points in $A_{p,q}(f)$ and allow all $\delta > 0$ with the above property we obtain a Vitali covering $\mathcal{V}$ of $A_{p,q}(f)$ consisting of closed intervals. By Theorem 11.2 there exists a finite subfamily of mutually disjoint intervals $[x_1, x_1 + \delta_1], \ldots, [x_N, x_N + \delta_N]$ belonging to $\mathcal{V}$, such that

$$\lambda^{(1)} \left(A_{p,q}(f) \cap \left(\bigcup_{k=1}^{N} [x_k, x_k + \delta_k] \right)^{\complement} \right) < \epsilon.$$

For the open set $V := \bigcup_{k=1}^{N} (x_k, x_k + \delta_k)$ we also find

$$\lambda^{(1)} \left(A_{p,q}(f) \cap V^{\complement} \right) < \epsilon \tag{11.24}$$

and since $V \subset U$ it follows that

$$\sum_{k=1}^{N} \delta_k = \lambda^{(1)}(V) \leq \lambda^{(1)}(U) < \alpha + \epsilon.$$

Now (11.23) yields

$$\sum_{k=1}^{N} (f(x_k + \delta_k) - f(x_k)) \leq p \sum_{k=1}^{N} \delta_k < p(\alpha + \epsilon). \tag{11.25}$$

Moreover, for $y \in A_{p,q}(f) \cap V$ there exists $\eta > 0$ such that $[y, y + \eta] \subset V$ and

$$f(y + \eta) - f(y) > q\eta, \tag{11.26}$$

and we get a Vitali covering of $A_{p,q}(f) \cap V$ by closed intervals as collection of all these closed intervals for $y \in A_{p,q}(f) \cap V$ and $\eta > 0$ such that (11.26) holds. A further application of Vitali's covering theorem allows us to pick a finite family $[y_1, y_1 + \eta_1], \ldots, [y_M, y_M + \eta_M]$ of mutually disjoint intervals of this covering such that

$$\lambda^{(1)} \left(A_{p,q}(f) \cap V \cap \left(\bigcup_{j=1}^{M} [y_j, y_j + \eta_j] \right)^C \right) < \epsilon. \tag{11.27}$$

Combining (11.24) with (11.27) yields

$$\alpha = \lambda^{(1)}(A_{p,q}(f)) \leq \lambda^{(1)} \left(A_{p,q}(f) \cap V^C \right) + \lambda^{(1)}(A_{p,q}(f) \cap V) \tag{11.28}$$

$$< \epsilon + \left(\epsilon + \sum_{j=1}^{M} \eta_j \right),$$

which yields with (11.26)

$$q(\alpha - 2\epsilon) < q \sum_{j=1}^{M} \eta_j < \sum_{j=1}^{M} (f(y_j + \eta_j) - f(y_j)). \tag{11.29}$$

By construction we have $\bigcup_{j=1}^{M} [y_j, y_j + \eta_j] \subset \bigcup_{k=1}^{N} [x_k, x_k + \delta_k]$, and since f is increasing we deduce

$$\sum_{j=1}^{M} (f(y_j + \eta_j) - f(y_j)) \leq \sum_{k=1}^{N} (f(x_k + \delta_k) - f(x_k)). \tag{11.30}$$

Now (11.29), (11.30) and (11.25) lead to

$$q(\alpha - 2\epsilon) < p(\alpha + \epsilon)$$

222

or
$$\frac{\alpha(q-p)}{p+2q} < \epsilon$$

which contradicts our choice of $\epsilon > 0$. Hence $\lambda^{(1)}(A(f)) = 0$. □

Theorem 11.6. *Let $[a, b]$, $a < b$, be a closed interval and $f : [a, b] \to \mathbb{R}$ a monotone function. Then there exists a set $A \subset [a, b]$, $\lambda^{(1)}(A) = 0$, such that f has for all $x \in A^{\complement}$ a finite derivative, i.e. f is $\lambda^{(1)}$-a.e. differentiable on $[a, b]$.*

Proof. Since for a decreasing function f the function $-f$ is increasing, we only need to consider the case of an increasing function. According to Lemma 11.5 the function f is $\lambda^{(1)}$-a.e. differentiable from the right, i.e. $f'_+(x)$ exists $\lambda^{(1)}$-a.e. Similarly we can prove that $f'_-(x)$ exists $\lambda^{(1)}$-a.e. We claim that $f'_+(x) \neq f'_-(x)$ only for at most denumerable many points $x \in (a, b)$. For this let

$$A_1 := \left\{ x \in (a, b) \,\middle|\, f'_+(x), \, f'_-(x) \text{ exists and } f'_+(x) < f'_-(x) \right\}$$

and

$$A_2 := \left\{ x \in (a, b) \,\middle|\, f'_+(x), \, f'_-(x) \text{ exists and } f'_+(x) > f'_-(x) \right\}.$$

For $x \in A_1$ we choose $r(x) \in \mathbb{Q}$ such that $f'_+(x) < r(x) < f'_-(x)$. Further we pick $p(x), q(x) \in \mathbb{Q}$ such that $a < p(x) < q(x) < b$ and

$$\frac{f(y) - f(x)}{y - x} > r(x) \quad \text{for} \quad p(x) < y < x \tag{11.31}$$

and

$$\frac{f(y) - f(x)}{y - x} < r(x) \quad \text{for} \quad x < y < q(x) \tag{11.32}$$

hold. These two inequalities imply for $y \neq x$ and $p(x) < y < q(x)$ that

$$f(y) - f(x) < r(x)(y - x). \tag{11.33}$$

We define $g : A_1 \to \mathbb{Q}^3$, $g(x) = (r(x), p(x), q(x))$, and we are going to show that g is an injective mapping, hence A_1 must be a denumerable set. Let $x, y \in A_1$, $x \neq y$, and suppose that $g(x) = g(y)$. It follows that $(p(y), q(y)) = (p(x), q(x))$ (as open intervals) and $x, y \in (p(y), q(y)) = (p(x), q(x))$. From (11.33) we deduce now that

$$f(y) - f(x) < r(x)(y - x) \quad \text{and} \quad f(x) - f(y) < r(y)(x - y). \tag{11.34}$$

223

By assumption we have $r(x) = r(y)$, and hence by adding the two inequalities from (11.34) we obtain $0 < 0$, i.e. a contradiction. Thus we have to conclude that g is injective and therefore $f'_+(x) = f'_-(x)$ $\lambda^{(1)}$-a.e. The case of points belonging to A_2 goes analogously.

Finally we prove that the set $B = \{x \in (a, b) \mid f'(x) = \infty\}$ has measure zero, i.e. $\lambda^{(1)}(B) = 0$. We follow the ideas of the proof of Lemma 11.5. Let $R > 0$. For $x \in B$ there exists $\delta > 0$ such that $[x, x + \delta] \subset (a, b)$ and

$$f(x + \delta) - f(x) > R\delta. \tag{11.35}$$

The family of all these closed intervals is a Vitali covering of B. By Theorem 11.2 there exists a countable subfamily of mutually disjoint intervals $([x_k, x_k + \delta_k])_{k \in \mathbb{N}}$ of this covering such that

$$\lambda^{(1)}\left(B \cap \left(\bigcup_{k \in \mathbb{N}} [x_k, x_k + \delta]\right)^c\right) = 0,$$

which together with (11.35) implies

$$R\lambda^{(1)}(B) \leq R\sum_{k=1}^{\infty} \delta_k < \sum_{k=1}^{\infty} (f(x_k + \delta_k) - f(x_k)) \leq f(b) - f(a),$$
$$R\lambda^{(1)}(B) < f(b) - f(a)$$

for all $R > 0$, i.e. $\lambda^{(1)}(B) = 0$, and the theorem is proved. $\square$

Remark 11.7. A. For proving Lemma 11.5 and Theorem 11.6 we have adopted the discussion in E. Hewitt and K. Stromberg [35], originally Theorem 11.6 was proved by H. Lebesgue.

B. Note that the proof of Theorem 11.6 yields the following interesting result: let $f : (a, b) \to \mathbb{R}$ be any function, then the set of all points $x \in (a, b)$ for which $f'_+(x)$ and $f'_-(x)$ exist but are not equal is at most countable.

Corollary 11.8. *Every function $f : [a, b] \to \mathbb{R}$ of bounded variation, and hence every absolutely continuous function $f : [a, b] \to \mathbb{R}$ is $\lambda^{(1)}$-a.e. differentiable with finite derivative. Moreover the derivative f' of $f \in BV([a, b])$ belongs to $L^1([a, b])$ and if f is increasing the following holds*

$$\int_a^b f'(x)\,\mathrm{d}x \leq f(b) - f(a). \tag{11.36}$$

In addition, if f is increasing then $f' \geq 0$ a.e. and if f is decreasing then $f' \leq 0$.

Proof. The first part of the corollary is trivial in light of Theorem 11.6 since $f = f_1 - f_2$ and f_1, f_2 are increasing functions. In order to prove the remaining part it is sufficient that for $f : [a, b] \to \mathbb{R}$ increasing it follows $f' \in L^1([a, b])$ and $f' \geq 0$ $\lambda^{(1)}$-a.e. Consider the sequence $(f_k)_{k \in \mathbb{N}}$ defined by

$$0 \leq f_k(x) := \frac{f\left(x + \frac{1}{k}\right) - f(x)}{\frac{1}{k}} = k\left(f\left(x + \frac{1}{k}\right) - f(x)\right).$$

By Theorem 11.6 we know that $\lambda^{(1)}$-a.e. we have $\lim_{k \to \infty} f_k(x) = f'(x)$, hence $f' \geq 0$ $\lambda^{(1)}$-a.e. Using Fatou's lemma we find further

$$0 \leq \int_a^b f'(x)\, dx \leq \liminf_{k \to \infty} \int f_k(x)\, dx$$

$$\leq \liminf_{k \to \infty} \left(k \int_b^{b + \frac{1}{k}} f(x)\, dx - k \int_a^{a + \frac{1}{k}} f(x)\, dx \right) \leq f(b) - f(a).$$

This proves that $f' \in L^1([a, b])$ for f increasing as well as (11.36). Since a general function belonging to $BV([a, b])$ is the difference between two increasing functions the corollary follows.   □

Consider now $f \in L^1([a, b])$ and note that $f|_{[a,x]} \in L^1([a, x])$ for every $x \in [a, b]$. Therefore we can define a function $F : [a, b] \to \mathbb{R}$ by

$$F(x) := \int_a^x f(t)\, dt. \tag{11.37}$$

Note that the value $F(x)$ is pointwisely defined, i.e. if $f_1 = f_2$ $\lambda^{(1)}$-a.e. we have $\int_a^x f_1(t)\, dt = \int_a^x f_2(t)\, dt$. Of course we would like to obtain a result such as $F'(x) = f(x)$ $\lambda^{(1)}$-a.e. In preparation for a proof of such a statement we show the following proposition which we can interpret as the **absolute continuity of the integral**.

Proposition 11.9. *Let $(\Omega, \mathcal{A}, \mu)$ be a measure space and $f \in L^1(\Omega)$. Then for every $\epsilon > 0$ there exists $\delta > 0$ such that $A \in \mathcal{A}$ and $\mu(A) < \delta$ implies $\int_A |f|\, d\mu < \epsilon$.*

Proof. If $\|f\|_\infty < \infty$, i.e. f is $\lambda^{(1)}$-a.e. bounded then we have for $\delta := \frac{\epsilon}{\|f\|_\infty}$ (the case $\|f\|_\infty = 0$ is trivial) the estimate

$$\int_A |f|\,\mathrm{d}\mu \le \|f\|_\infty \mu(A) < \epsilon,$$

i.e. the assertion holds. Now let $f \in L^1(\Omega)$ and define

$$g_k(\omega) := |f(\omega)| \wedge k \le |f(\omega)|.$$

The sequence $(g_k)_{k\in\mathbb{N}}$ consists of measurable functions and $\|g_k\|_\infty \le k$. Moreover $g_k \to |f|$ holds pointwisely. By the dominated convergence theorem we conclude that

$$\int |f|\,\mathrm{d}\mu = \lim_{k\to\infty} \int |g_k|\,\mathrm{d}\mu,$$

or for every $\epsilon > 0$ exists $N \in \mathbb{N}$ such that $k \ge N$ implies that

$$\int (|f(\omega)| - g_k(\omega)|)\,\mu(\mathrm{d}\omega) < \frac{\epsilon}{2}.$$

For $0 < \delta < \frac{\epsilon}{2N}$ we find for $A \in \mathcal{A}$ and $\mu(A) < \delta$ that

$$\left| \int_A f\,\mathrm{d}\mu \right| \le \int_A |f|\,\mathrm{d}\mu = \int_A (|f| - g_N)\,\mathrm{d}\mu + \int_A g_N\,\mathrm{d}\mu$$

$$\le \int_A (|f| - g_N)\,\mathrm{d}\mu + N\mu(A) < \epsilon.$$

$\square$

In particular for $\Omega = [a, b]$, $\mathcal{A} = \mathcal{B}^{(1)}([a, b])$ and $\mu = \lambda^{(1)}$ (or more precisely $\mu = \lambda^{(1)}_{[a,b]}$) we can read the statement of Proposition 11.9 as follows: for every $\epsilon > 0$ and any choice of mutually disjoint intervals $(a_1, b_1), \ldots, (a_M, b_M) \subset [a, b]$ such that $\sum_{j=1}^{M}(b_j - a_j) < \delta$ we have with $A = \bigcup_{j=1}^{M}(a_j, b_j)$

$$\left| \int_A f\,\mathrm{d}\lambda^{(1)} \right| \le \int_A |f|\,\mathrm{d}\lambda^{(1)} = \sum_{j=1}^{M} \int_{a_j}^{b_j} |f(x)|\lambda^{(1)}(\mathrm{d}x) < \epsilon.$$

Corollary 11.10. *For $f \in L^1([a, b])$ the function F defined by (11.37) is absolutely continuous and hence $\lambda^{(1)}$-a.e. differentiable.*

Proof. Since for $a < x < y < b$ we have

$$F(y) - F(x) = \int_a^y f(t)\,dt - \int_a^x f(t)\,dt = \int_x^y f(t)\,dt$$

and we find for a finite number of mutually disjoint intervals $(x_1, y_1), \ldots,$ $(x_M, y_M) \subset [a, b]$ that

$$\sum_{j=1}^M (F(y_j) - F(x_j)) = \sum_{j=1}^M \int_{x_j}^{y_j} f(t)\,dt.$$

Hence by Proposition 11.9, for $\epsilon > 0$ there exists $\delta > 0$ such that whenever $\sum_{j=1}^M (y_j - x_j) < \delta$ it follows that

$$\sum_{j=1}^M |F(y_j) - F(x_j)| \leq \sum_{j=1}^M \int_{x_j}^{y_j} |f(t)|\,dt < \epsilon,$$

i.e. F is absolutely continuous, hence $\lambda^{(1)}$-a.e. differentiable. $\qquad\square$

Proposition 11.11. *If* $f \in L^1([a, b])$ *and* $F(x) := \int_a^x f(t)\,dt = 0$ *for all* $x \in [a, b]$ *then* $f = 0$ $\lambda^{(1)}$*-a.e.*

Proof. Suppose that for a set $A \in \mathcal{B}([a, b])$, $\lambda^{(1)}(A) > 0$, we have $f|_A > 0$ (the case $f|_A < 0$ goes analogously). By Theorem 10.10 we can find a compact set $K \subset A$ such that $\lambda^{(1)}(K) > 0$. The set $U := (a, b) \setminus K$ is open, hence by Theorem I.19.27 we have $U = \bigcup_{k=1}^M (a_k, b_k)$ where the open intervals (a_k, b_k) are mutually disjoint and $M \in \mathbb{N} \cup \{\infty\}$. We handle the case $M = \infty$, the other cases are then trivial. Since

$$0 = F(b) = \int_a^b f(t)\,dt = \int_K f(t)\,dt + \int_U f(t)\,dt$$

we find

$$\int_U f(t)\,dt < 0,$$

hence

$$0 \neq \int_U f(t)\,dt = \sum_{k=1}^\infty \int_{a_k}^{b_k} f(t)\,dt$$

and therefore for some k_0 we must have $\int_{a_{k_0}}^{b_{k_0}} f(t)\,dt \neq 0$. However we have

$$\int_{a_{k_0}}^{b_{k_0}} f(t)\,dt = \int_a^{b_{k_0}} f(t)\,dt - \int_a^{a_{k_0}} f(t)\,dt = F(b_{k_0}) - F(a_{k_0}) = 0,$$

which is a contradiction. $\qquad\square$

Proposition 11.12. *For a bounded measurable function $f : [a,b] \to \mathbb{R}$ the derivative of F as defined by (11.37) is f, i.e. $F' = f$, $\lambda^{(1)}$-a.e.*

Proof. Since F is absolutely continuous it is $\lambda^{(1)}$-a.e. differentiable, see Corollary 11.10. Suppose that $|f(x)| \leq M$ and consider the sequence g_k of functions defined by

$$g_k(x) = \frac{F\left(x + \frac{1}{k}\right) - F(x)}{\frac{1}{k}} = k\left(F\left(x + \frac{1}{k}\right) - F(x)\right)$$

where we agree to set $F(y) = F(b)$ for $y \geq b$. It follows that

$$g_k(x) = k \int_x^{x + \frac{1}{k}} f(t)\, \mathrm{d}t$$

implying that $|g_k(x)| \leq M$. Furthermore $(g_k(x))_{k \in \mathbb{N}}$ converges $\lambda^{(1)}$-a.e. to $F'(x)$ and the dominated convergence theorem yields that for any $c \in [a,b]$ we have

$$\int_a^c F'(x)\, \mathrm{d}x = \lim_{k \to \infty} \int_a^c g_k(x)\, \mathrm{d}x$$

$$= \lim_{k \to \infty} k \int_a^c \left(F\left(x + \frac{1}{k}\right) - F(x)\right) \mathrm{d}x$$

$$= \lim_{k \to \infty} k \left(\int_c^{c + \frac{1}{k}} F(x)\, \mathrm{d}x - \int_a^{a + \frac{1}{k}} F(x)\, \mathrm{d}x\right)$$

$$= F(c) - F(a)$$

where for the last step we use the fact that F is absolutely continuous, hence continuous and the mean-value theorem for Riemann integrals applies. Thus we arrive at

$$\int_a^c F'(x)\, \mathrm{d}x = F(c) - F(a) = \int_a^c f(x)\, \mathrm{d}x$$

for all $c \in [a,b]$ and by Proposition 11.11 it follows that $F' = f$ $\lambda^{(1)}$-a.e. $\square$

With the help of Proposition 11.12 we eventually can show

Theorem 11.13. *For $f \in L^1([a,b])$ and F defined as in (11.37) we have*

$$F' = f \;\lambda^{(1)}\text{-a.e.} \tag{11.38}$$

Proof. The basic idea is to approximate f by bounded functions. Since we can decompose f into positive and negative parts, we may assume that $f \geq 0$. We define the approximating sequence of functions $h_k \in L^1([a,b])$ by

$$h_k(x) := f(x) \wedge k. \tag{11.39}$$

Clearly $f - h_k \geq 0$ and $|h_k(x)| \leq k$. The function $H_k : [a,b] \to \mathbb{R}$

$$H_k(x) := \int_a^x (f(t) - h_k(t))\, dt$$

is increasing, $\lambda^{(1)}$-a.e. differentiable and $H_k'(x) \geq 0$ $\lambda^{(1)}$-a.e. By Proposition 11.12 we have

$$\frac{d}{dx} \int_a^x h_k(t)\, dt = h_k(x)\ \lambda^{(1)}\text{-a.e.}$$

and therefore $\lambda^{(1)}$-a.e.

$$F'(x) = H_k'(x) + \frac{d}{dx} \int_a^x h_k(x)\, dx \geq h_k(x).$$

Since k was arbitrary we find that $F'(x) \geq f(x)$ $\lambda^{(1)}$-a.e. It follows that

$$\int_a^b F'(x)\, dx \geq \int_a^b f(x)\, dx = F(b) - F(a),$$

and (11.36) in Corollary 11.8 yields now

$$\int_a^b F'(x)\, dx = F(b) - F(a) = \int_a^b f(x)\, dx$$

as well as

$$\int_a^b (F'(x) - f(x))\, dx = 0.$$

Since $F'(x) \geq f(x)$ $\lambda^{(1)}$-a.e. it follows now that $F'(x) = f(x)$ $\lambda^{(1)}$-a.e. and the theorem is proved. $\qquad\square$

Corollary 11.14. *For $f \in L^1([a,b])$ and F_a defined by*

$$F_a(x) = F(a) + \int_a^x f(t)\, dt$$

it follows that $F' = f$ $\lambda^{(1)}$-a.e.

Recall that according to Definition I.12.6 a differentiable function $F : [a, b] \to \mathbb{R}$ is a **primitive** of $f : [a, b] \to \mathbb{R}$ if $F' = f$. The fundamental theorem of calculus, Theorem I.12.7, states that if f is continuous then it admits a primitive F and

$$\int_a^b f(t)\,dt = F(b) - F(a).$$

We will now show that a function F is a primitive if and only if it is absolutely continuous. For this we need

Lemma 11.15. *An absolutely continuous function $f : [a, b] \to \mathbb{R}$ with derivative $f' = 0$ $\lambda^{(1)}$-a.e. is constant.*

Proof. We want to prove that $f(a) = f(c)$ for every $c \in [a, b]$. By assumption we can find a measurable set $A \subset (a, c)$ such that $f'(x) = 0$ for all $x \in A$. Let $\epsilon > 0$. For every $x \in A$ there exists y such that $[x, y] \subset [a, c]$ and $|f(y) - f(x)| < \frac{\epsilon}{2(c-a)}(y - x)$ since $f'(x) = 0$ for $x \in A$. We choose now $\delta > 0$ such that δ is determined by the absolute continuity of f with ϵ replaced by $\frac{\epsilon}{2}$. This δ we now use in Vitali's covering theorem, Theorem 11.2 as ϵ, to get that we can find mutually disjoint intervals $[x_1, y_1], \ldots, [x_M, y_M] \subset [a, c]$, $x_k < x_{k+1}$, $k = 1, \ldots, M - 1$, such that

$$|f(y_k) - f(x_k)| < \frac{\epsilon}{2(c - a)}(y_k - x_k) \tag{11.40}$$

and

$$\lambda^{(1)} \left(A \backslash \left(\bigcup_{k=1}^{M} [x_k, y_k] \right)^{\complement} \right) < \delta. \tag{11.41}$$

With $y_0 = a$ and $x_{M+1} = c$ it follows from (11.41) that

$$\sum_{k=0}^{M} |x_{k+1} - y_k| < \delta$$

and further using again the absolute continuity of f and (11.40)

$$|f(c) - f(a)| = \left| \sum_{k=0}^{M} (f(x_{k+1}) - f(y_k)) + \sum_{k=1}^{M} (f(y_k) - f(x_k)) \right|$$

$$< \frac{\epsilon}{2} + \frac{\epsilon}{2(c - a)} \sum_{k=1}^{M} (y_k - x_k) \leq \frac{\epsilon}{2} + \frac{\epsilon}{2(c - a)}(c - a) = \epsilon.$$

Since $\epsilon > 0$ was arbitrary it follows that $f(c) = f(a)$. $\qquad\square$

Now we can prove the extension of the **fundamental theorem of calculus**:

Theorem 11.16. *Let $f \in L^1([a,b])$ then $F(x) := \int_a^x f(t)\,dt$ is absolutely continuous on $[a,b]$ and $F' = f$ holds in $L^1([a,b])$ as well as*

$$F(x) - F(a) = \int_a^x f(t)\,dt. \tag{11.42}$$

Conversely, if $F : [a,b] \to \mathbb{R}$ is an absolutely continuous function then there exists $f \in L^1([a,b])$ such that $F(x) = \int_a^x f(t)\,dt$ for all $x \in [a,b]$.

Proof. We know already from Corollary 11.10 and Theorem 11.13 that for $f \in L^1([a,b])$ the function F is absolutely continuous and that $F' = f$ holds. Since $F(a) = 0$ we also have (11.42). Now let F be absolutely continuous, hence $F \in BV([a,b])$ and with two increasing functions F_1 and F_2 we have $F(x) = F_1(x) - F_2(x)$. It follows that F' exists $\lambda^{(1)}$-a.e. and

$$|F'(x)| \leq F_1'(x) + F_2'(x),$$

and therefore

$$\int_a^b |F'(x)|\,dx \leq F_1(b) + F_2(b) - F_1(a) - F_2(a)$$

by (11.36) in Corollary 11.8. This shows already that $F' \in L^1([a,b])$. We consider now the two absolutely continuous functions

$$H(x) := \int_a^x F'(t)\,dt \quad \text{and} \quad g := F - H.$$

By Theorem 11.13 we have $g' = F' - H' = 0$ $\lambda^{(1)}$-a.e., hence by Lemma 11.15 the function g is constant implying that

$$F(x) = \int_a^x F'(t)\,dt + F(a)$$

and the second part of the theorem is proved with $f := F'$. $\square$

The fundamental theorem allows us to generalise the **integration by parts** formula:

Theorem 11.17. *Let $f, g \in L^1([a,b])$ and denote by F and G the absolutely continuous functions*

$$F(x) = \int_a^x f(t)\,dt \quad and \quad G(x) = \int_a^x g(t)\,dt.$$

The following formula

$$\int_a^b f(t)G(t)\,dt = F \cdot G|_a^b - \int_a^b F(t)g(t)\,dt \tag{11.43}$$

holds.

Proof. We also know that $F \cdot G$ is absolutely continuous, hence $\lambda^{(1)}$-a.e. differentiable and

$$\int_a^b (F \cdot G)'(t)\,dt = F(b)G(b) - F(a)G(a) = F \cdot G|_a^b,$$

while on the other hand by Leibniz's rule

$$\int_a^b (F \cdot G)'(t)\,dt = \int_a^b (fG)(t)\,dt + \int_a^b (Fg)(t)\,dt,$$

implying (11.43). $\qquad\square$

Corollary 11.18. *For two absolutely continuous functions $f, g : [a,b] \to \mathbb{R}$ we have*

$$\int_a^b f(t)g'(t)\,dt = fg|_a^b - \int_a^b f'(t)g(t)\,dt \tag{11.44}$$

Remark 11.19. Our presentation of this part of "differentiation theory" is much influenced by [35], [70], also see [14].

In Chapter 6 we introduced measures ν having a density with respect to given measure μ and where led to the notion of absolutely continuous measures, see Definition 6.11: ν is absolutely continuous with respect to μ if $\mu(A) = 0$ implies that $\nu(A) = 0$. If $\nu = f\lambda^{(1)}$, $f \in L^1([a,b])$, i.e. ν is absolutely continuous with respect to $\lambda^{(1)}$, then $F(\cdot) = \int_a^\cdot f(t)dt = \int_a^\cdot 1\,d\nu$ is absolutely continuous and hence both notions are related.

The property that $f : A \to \mathbb{R}$, $A \in \mathcal{B}^{(1)}$, maps sets of measure zero onto sets of measure zero is often called the **Lusin property**. The relation of the

Lusin property to other properties of f such as being of bounded variation, Lipschitz continuous etc. is discussed in detail in [6].

Without proof we state the **Banach-Zaretzky theorem**, see [14]:

Theorem 11.20. *A function $F : [a, b] \to \mathbb{R}$ of bounded variation is absolutely continuous if and only if for every $A \in \mathcal{B}^{(1)}([a, b])$ it follows that $\lambda^{(1)}(A) = 0$ implies $\lambda^{(1)}(F(A)) = 0$.*

We now want to study two examples.

Example 11.21. In Example I.32.13 we have seen that the function $f : [0, 1] \to \mathbb{R}$ defined by

$$f(x) = \begin{cases} x \sin \frac{1}{x}, & x \in (0, 1] \\ 0, & x = 0 \end{cases}$$

is not of bounded variation, hence not absolutely continuous. We extend this example in the following direction: For $\alpha, \beta > 0$ define

$$g_{\alpha,\beta}(x) := \begin{cases} x^{\alpha} \sin \frac{1}{x^{\beta}}, & x \in (0, 1] \\ 0, & x = 0. \end{cases}$$

We claim that if $0 < \beta < \alpha$ then $g_{\alpha,\beta}$ is absolutely continuous while for $0 < \alpha \leq \beta$ it is not. For $0 < \beta < \alpha$ we find for $x > 0$ that

$$g'_{\alpha,\beta}(x) = \alpha x^{\alpha-1} \sin \frac{1}{x^{\beta}} - \beta x^{\alpha-\beta-1} \cos \frac{1}{x^{\beta}}$$

and this function belongs to $L^1([0, 1])$, thus we have

$$g_{\alpha,\beta}(x) = \int_0^x \left(\alpha y^{\alpha-1} \sin \frac{1}{y^{\beta}} - \beta y^{\alpha-\beta-1} \cos \frac{1}{y^{\beta}} \right) dy,$$

i.e. $g_{\alpha,\beta}$ is for $0 < \beta < \alpha$ absolutely continuous. Of course, for $0 < \alpha \leq \beta$ the expression of $g'_{\alpha,\beta}$ does not change, but now $g'_{\alpha,\beta}$ does not belong any more to $L^1([0, 1])$ and hence by Corollary 11.8 the function $g_{\alpha,\beta}$ is not of bounded variation.

Example 11.22. Let $F : [0,1] \to \mathbb{R}$ be the Cantor-Lebesgue function as introduced in Definition 3.42. By Lemma 3.43 we know that F is Hölder continuous with exponent $\alpha = \frac{\ln 2}{\ln 3}$, furthermore $F' = 0$ $\lambda^{(1)}$-a.e. by its construction, recall that F is constant on the intervals forming $C^{\complement}$ and $\lambda^{(1)}(C) = 0$. Thus F cannot be absolutely continuous, since otherwise we would have $F(x) = \int_0^x F'(t)\,dt = 0$ for all $x \in [0,1]$. This example shows that neither continuous monotone functions nor Hölder continuous functions need to be absolutely continuous.

Let $f : [a,b] \to \mathbb{R}$ be a continuous function and $x_0 \in (a,b)$. It follows for h sufficiently small that

$$\frac{1}{2h} \int_{x_0-h}^{x_0+h} (f(t) - f(x_0))dt = \frac{1}{2h} \int_{x_0-h}^{x_0+h} f(t)\,dt - f(x_0)$$

$$= \frac{1}{2h}(F(x_0 + h) - F(x_0 - h)) - f(x_0)$$

where $F(x) = \int_a^x f(t)\,dt$. Thus we find

$$\lim_{h \to 0} \frac{1}{2h} \int_{x_0-h}^{x_0+h} (f(t) - f(x_0))\,dt = 0 \tag{11.45}$$

or

$$\lim_{h \to 0} \frac{1}{2h} \int_{x_0-h}^{x_0+h} f(t)\,dt = f(x_0). \tag{11.46}$$

This result has a far reaching extension to $\mathbb{R}^n$ known as Lebesgue's differentiation theorem which we are going to discuss next. In doing so we will also introduce a new, rather powerful tool, namely the Hardy-Littlewood maximal function. As preparation we need a further covering result, sometimes called **Wiener's covering lemma**.

Lemma 11.23. *For every collection $\mathcal{B}_0 = \{B_1, \ldots, B_N\}$ of open balls $B_j \subset \mathbb{R}^n$ we can find a subcollection $B_{j_1}, \cdots, B_{j_M}$ of $\mathcal{B}_0$ such that*

$$\lambda^{(n)}\left(\bigcup_{k=1}^{N} B_k\right) \leq 3^n \sum_{l=1}^{M} \lambda^{(n)}(B_{j_l}). \tag{11.47}$$

Proof. Let $B_{j_1} = B_{r_1}(x_1) \in \mathcal{B}_0$ be a ball with largest radius and consider now $\mathcal{B}_1 := \{B_k \in \mathcal{B}_0 \,|\, B_k \cap B_{j_1} = \emptyset\}$. Since B_{j_1} has the largest radius, by the

234

triangle inequality it follows that

$$\bigcup_{B\in\mathcal{B}_0\setminus\mathcal{B}_1} B \subset B_{3r_1}(x_1).$$

We now apply the same procedure to $\mathcal{B}_1$ and thus obtain an open ball $B_{j_2} = B_{r_2}(x_2) \in \mathcal{B}_1$ and we introduce $\mathcal{B}_2 := \{B_k \in \mathcal{B}_1 \mid B_k \cap B_{r_2} = \emptyset\}$. After $M \leq N$ steps we have a subcollection of mutually disjoint balls $B_{j_1}, \ldots, B_{j_M} \in \mathcal{B}_0$ which covers $\bigcup_{k=1}^{N} B_k$ and therefore we have

$$\lambda^{(n)}\left(\bigcup_{k=1}^{N} B_k\right) \leq \lambda^{(n)}\left(\bigcup_{l=1}^{M} B_{3r_l}(x_l)\right)$$

$$\leq \sum_{l=1}^{M} \lambda^{(n)}\left(B_{3r_l}(x_l)\right) = 3^n \sum_{l=1}^{M} \lambda^{(n)}(B_{j_l}).$$

$\square$

Definition 11.24. A. *We call a measurable function* $u : \mathbb{R}^n \to \mathbb{R}$ **locally integrable** *with respect to* $\lambda^{(n)}$ *if* $\int_K |u(x)|\lambda^{(n)}(\mathrm{d}x) < \infty$ *for every compact set* $K \subset \mathbb{R}^n$. *The space of all locally integrable functions* u *we denote by* $L^1_{\mathrm{loc}}(\mathbb{R}^n)$. *(As usual* $L^1_{\mathrm{loc}}(\mathbb{R}^n)$ *consists of equivalence classes induced by the equivalence relation "$\sim_{\lambda^{(n)}}$").*
B. *For* $u \in L^1_{\mathrm{loc}}(\mathbb{R}^n)$ *we define its* **average** *over* $B_r(x) \subset \mathbb{R}^n$ *by*

$$(A_r u)(x) := \frac{1}{\lambda^{(n)}(B_r(x))} \int_{B_r(x)} u(y)\,\mathrm{d}y. \tag{11.48}$$

Lemma 11.25. *For* $u \in L^1_{\mathrm{loc}}(\mathbb{R}^n)$ *the mapping* $(r, x) \mapsto (A_r u)(x)$ *is continuous from* $(0, \infty) \times \mathbb{R}^n$ *to* $\mathbb{R}$.

Proof. Recall that $\lambda^{(n)}(B_r(x)) = \omega_n r^n$, $\omega_n = \dfrac{\pi^{\frac{n}{2}}}{\Gamma(\frac{n}{2}+1)}$, see (II.21.11), and that $\partial B_r(x)$ is a set of measure zero for $\lambda^{(n)}$. In addition we have on $\mathbb{R}^n \setminus \partial B_{r_0}(x_0)$ pointwisely

$$\lim_{(r,x)\to(r_0,x_0)} \chi_{B_r(x)}(y) = \chi_{B_{r_0}(x_0)}(y),$$

i.e. $\lim_{(r,x)\to(r_0,x_0)} \chi_{B_r(x)} = \chi_{B_{r_0}(x_0)}$ $\lambda^{(n)}$-a.e. Furthermore, for $r < r_0 + \frac{1}{2}$ and $\|x_0 - x\| < \frac{1}{2}$ the triangle inequality yields $\left|\chi_{B_r(x)}(y)\right| \leq \chi_{B_{r_0+1}(x_0)}(y)$. Now the dominated convergence theorem implies

$$\lim_{(r,x)\to(r_0,x_0)} \int_{B_r(x)} u(y)\, \mathrm{d}y = \lim_{(r,x)\to(r_0,x_0)} \int_{\mathbb{R}^n} \chi_{B_r(x)}(y)u(y)\, \mathrm{d}y$$

$$= \int_{\mathbb{R}^n} \chi_{B_{r_0}(x_0)}(y)u(y)\, \mathrm{d}y = \int_{B_{r_0}(x_0)} u(y)\, \mathrm{d}y,$$

and consequently

$$\lim_{(r,x)\to(r_0,x_0)} (A_r u)(x) = (A_{r_0} u)(x_0),$$

proving the lemma. $\qquad\square$

Definition 11.26. *For $f \in L^1_{\text{loc}}(\mathbb{R}^n)$ the **Hardy-Littlewood maximal function**, in short the **maximal function**, of f is defined by*

$$\mathcal{M}(f)(x) := \sup_{r>0}(A_r|f|)(x) = \sup_{r>0} \frac{1}{\lambda^{(n)}(B_r(x))} \int_{B_r(x)} |f(y)|\, \mathrm{d}y. \qquad (11.49)$$

Lemma 11.27. *The Hardy-Littlewood maximal function is measurable on $\mathbb{R}^n$.*

Proof. From Lemma 11.25 we know that for $a > 0$ the pre-image of (a, ∞) under $A_r|f| : \mathbb{R}^n \to \mathbb{R}$, i.e. $(A_r|f|)^{-1}(a, \infty)$, is open. Since $\mathcal{M}(f)^{-1}(a, \infty) = \bigcup_{r>0}(A_r|f|)^{-1}(a, \infty)$ is open it follows that $\mathcal{M}(f)$ is measurable. $\qquad\square$

The following theorem, often called **maximal** or **Hardy-Littlewood maximal theorem**, is of great importance and has many applications in real analysis.

Theorem 11.28. *The estimate*

$$\lambda^{(n)}\left(\{x \in \mathbb{R}^n \mid \mathcal{M}(f)(x) > \alpha\}\right) \leq \frac{3^n}{\alpha} \int_{\mathbb{R}^n} |f(t)|\, \mathrm{d}t = \frac{3^n}{\alpha}\|f\|_{L^1} \qquad (11.50)$$

holds for all $\alpha > 0$ and $f \in L^1(\mathbb{R}^n)$.

Proof. Let $A_\alpha := \{x \in \mathbb{R}^n \mid \mathcal{M}(f)(x) > \alpha\}$. From the definition of $\mathcal{M}(f)$ for $x \in A_\alpha$ we can find a ball $B_x := B_{r(x)}(x)$ such that

$$\frac{1}{\lambda^{(n)}(B_x)} \int_{B_x} |f(y)|\, \mathrm{d}y > \alpha.$$

For a compact set $K \subset A_\alpha$ the family $\{B_x \mid x \in K\}$ is an open covering of K, hence we can cover K by finitely many of such balls, i.e. $K \subset \bigcup_{k=1}^{M} B_k$, $B_k \in \{B_x \mid x \in K\}$. Applying Lemma 11.23 we can find a further subcovering of mutually disjoint balls $B_{k_1}, \ldots, B_{k_N} \in \{B_1, \ldots, B_M\}$ such that

$$\lambda^{(n)} \left(\bigcup_{k=1}^{M} B_k \right) \leq 3^n \sum_{j=1}^{N} \lambda^{(n)}(B_{k_j}). \tag{11.51}$$

For each B_{k_j} we have

$$\lambda^{(n)}(B_{k_j}) < \frac{1}{\alpha} \int_{B_{k_j}} |f(y)| \, dy. \tag{11.52}$$

Now it follows that

$$\lambda^{(n)}(K) \leq \lambda^{(n)} \left(\bigcup_{k=1}^{M} B_k \right) \leq 3^n \sum_{j=1}^{N} \lambda^{(n)}(B_{k_j})$$

$$\leq \frac{3^n}{\alpha} \sum_{j=1}^{N} \int_{B_{k_j}} |f(y)| \, dy = \frac{3^n}{\alpha} \int_{\bigcup_{j=1}^{N} B_{k_j}} |f(y)| \, dy$$

$$\leq \frac{3^n}{\alpha} \int_{\mathbb{R}^n} |f(y)| \, dy,$$

i.e. for every compact set $K \subset A_\alpha$

$$\lambda^{(n)}(K) \leq \frac{3^n}{\alpha} \|f\|_{L^1}$$

holds, implying (11.50). $\qquad\square$

Eventually we can show **Lebesgue's differentiation theorem:**

Theorem 11.29. *For every $f \in L^1(\mathbb{R}^n)$ the limit*

$$\lim_{\substack{\lambda^{(n)}(B) \to 0 \\ x \in B}} \frac{1}{\lambda^{(n)}(B)} \int_B f(y) \, dy = f(x) \tag{11.53}$$

exists $\lambda^{(n)}$-a.e. where $B \subset \mathbb{R}^n$ denotes an open ball $B \subset \mathbb{R}^n$ containing x.

237

Proof. Suppose that we can prove for every $\alpha > 0$ that the set

$$A_\alpha := \left\{ x \in \mathbb{R}^n \mid \limsup_{\substack{\lambda^{(n)}(B) \to 0 \\ x \in B}} \left| \frac{1}{\lambda^{(n)}(B)} \int_B f(y)\, dy - f(x) \right| > 2\alpha \right\}$$

has measure zero, i.e. $\lambda^{(n)}(A_\alpha) = 0$. It follows that the set $A = \bigcup_{n \in \mathbb{N}} A_{\frac{1}{n}}$ also has measure zero and (11.53) holds for $x \in A^\complement$, i.e. $\lambda^{(n)}$-a.e. Let $\alpha > 0$ be fixed and let $\epsilon > 0$. By Theorem 10.11 we can find $g \in C_0(\mathbb{R}^n)$ such that $\|f - g\|_{L^1} < \epsilon$. For g we have

$$\lim_{\substack{\lambda^{(n)}(B) \to 0 \\ x \in B}} \frac{1}{\lambda^{(n)}(B)} \int_B g(y)\, dy = g(x).$$

Indeed, this we can see as follows: first we note that

$$\frac{1}{\lambda^{(n)}(B)} \int_B g(y)\, dy - g(x) = \frac{1}{\lambda^{(n)}(B)} \int_B (g(y) - g(x))\, dy.$$

Since g is continuous, given $\eta > 0$ there exists $\delta > 0$ such that $\|x - y\| < \delta$ implies $|g(x) - g(y)| < \eta$. For every ball $B \subset \mathbb{R}^n$ with $x \in B$ and diameter δ we find now

$$\left| \frac{1}{\lambda^{(n)}(B)} \int_B g(y)\, dy - g(x) \right| \leq \frac{1}{\lambda^{(n)}(B)} \int_B |g(y) - g(x)|\, dy < \eta.$$

We return to f and write

$$\frac{1}{\lambda^{(n)}(B)} \int_B f(y)\, dy - f(x) = \frac{1}{\lambda^{(n)}(B)} \int_B (f(y) - f(x))\, dy$$

$$= \frac{1}{\lambda^{(n)}(B)} \int_B (f(y) - g(y))\, dy + \frac{1}{\lambda^{(n)}(B)} \int_B (g(y) - g(x))\, dy + g(x) - f(x).$$

Since for $\lambda^{(n)}(B) \to 0$, $x \in B$, the second term tends to 0 we get

$$\limsup_{\substack{\lambda^{(n)}(B) \to 0 \\ x \in B}} \left| \frac{1}{\lambda^{(n)}(B)} \int_B f(y)\, dy - f(x) \right|$$

$$\leq \limsup_{\substack{\lambda^{(n)}(B) \to 0 \\ x \in B}} \frac{1}{\lambda^{(n)}(B)} \int_B |f(y) - g(y)|\, dy + |g(x) - f(x)|$$

$$\leq M(f - g)(x) + |g(x) - f(x)|.$$

We observe that

$$A_\alpha \subset \{x \in \mathbb{R}^n \,|\, \mathcal{M}(f-g)(x) > \alpha\} \cup \{x \in \mathbb{R}^n \,|\, |f(x) - g(x)| > \alpha\}.$$

By the Chebyshev-Markov inequality, Lemma 7.15, we find

$$\lambda^{(n)}\left(\{x \in \mathbb{R}^n \,|\, |f(x) - g(x)| > \alpha\}\right) \leq \frac{1}{\alpha}\|f-g\|_{L^1},$$

and the maximal theorem yields

$$\lambda^{(n)}\left(\{x \in \mathbb{R}^n \,|\, \mathcal{M}(f-g)(x) > \alpha\}\right) \leq \frac{3^n}{\alpha}\|f-g\|_{L^1},$$

or

$$\lambda^{(n)}(A_\alpha) \leq \left(\frac{3^n}{\alpha} + \frac{1}{\alpha}\right)\|f-g\|_{L^1} \leq \frac{3^n + 1}{\alpha}\epsilon$$

by our choice of g. Since $\epsilon > 0$ was arbitrary the theorem follows.   □

Remark 11.30. A. Theorem 11.29 extends to $f \in L^1_{\text{loc}}$, see for example [27] or the remark [84]. Our proof is a combination of the arguments in [27], [84] and [14].
B. In [27] it is also discussed that we can replace the balls in Theorem 11.29 by more general families of sets shrinking to x.
C. For $f \in L^1_{\text{loc}}(\mathbb{R}^n)$ the **Lebesgue set** of f, $\Lambda(f)$, is defined by

$$\Lambda(f) := \left\{x \in \mathbb{R}^n \,\Big|\, \lim_{r \to 0} \frac{1}{\lambda^{(n)}(B_r(x))} \int_{B_r(x)} |f(y) - f(x)|\, dy = 0\right\}, \quad (11.54)$$

and in [27] it is proved that $\lambda^{(n)}\left(\Lambda(f)^{\complement}\right) = 0$ for $f \in L^1_{\text{loc}}(\mathbb{R}^n)$, i.e. we can strengthen Theorem 11.29 to

$$\lim_{r \to 0} \frac{1}{\lambda^{(n)}(B_r(0))} \int_{B_r(x)} |f(y) - f(x)|\, dy = 0, \quad \lambda^{(n)}\text{-a.e.}$$

for $f \in L^1_{\text{loc}}(\mathbb{R}^n)$.

Problems

1. Let $f, g : [a, b] \to \mathbb{R}$ be two absolutely continuous functions and $g(x) \neq 0$ for all $x \in [a, b]$. Prove that $\frac{f}{g}$ is also an absolutely continuous function.

2. a) Let $f : [a, b] \to \mathbb{R}$ be an absolutely continuous function with $f([a, b]) = [c, d]$. Further let $\varphi : [c, d] \to \mathbb{R}$ be a Lipschitz continuous function. Prove that $\varphi \circ f : [a, b] \to \mathbb{R}$ is an absolutely continuous function.

b) Let $f, g : [a, b] \to \mathbb{R}$ be absolutely continuous and g in addition a monotone function. Show that $f \circ g$ is also an absolutely continuous function.

3. Let $f : [a, b] \to \mathbb{R}$ be an absolutely continuous function and $N \subset [a, b]$ be a set of measure zero. Show that $f(N)$ is of measure zero too.

4. Let $g \in L^p([a, b])$, $1 \le p < \infty$, and extend g to be 0 in $[a, b]^{\complement}$. For $h > 0$ define the function

$$g_h(x) := \frac{1}{2h} \int_{x-h}^{x+h} g(t)dt.$$

Prove that g_h is a continuous function and $\|g_h\|_{L^p} \le \|g\|_{L^p}$.

5. We call a non-constant continuous function $s : [a, b] \to \mathbb{R}$ which is of bounded variation a **singular function** if $s' = 0$ $\lambda^{(1)}|_{[a,b]}$-almost everywhere.

a) Prove that if $f : [a, b] \to \mathbb{R}$ is a continuous function with bounded variation then f admits a unique decomposition $f = \varphi + s$ where φ is absolutely continuous with $\varphi(a) = f(a)$, and a function s which is either identically zero or a singular function.

b) For a monotone function f with decomposition $f = \varphi + s$ as in part a) show that φ and s are monotone too.

6. Consider $u : [0, 1] \to \mathbb{R}$, $u(x) = \begin{cases} x^2 \cos \frac{\pi}{x^2}, & x \in (0, 1] \\ 0, & x = 0 \end{cases}$, and prove that u has a (finite) derivative $u'(x)$ everywhere, but $u' \notin L^1([0, 1])$.

7.* Let $g : (a, b) \to \mathbb{R}$ be a convex function. Prove that then $g'(x)$ exists for all $x \in (a, b) \setminus A$, where $A \subset (a, b)$ is at most a countable set. Furthermore g' is an increasing function.
Hint: recall Theorem I.23.5.

8.* For $f \in L^1_{\mathrm{loc}}(\mathbb{R})$ denote by $\mathcal{M}(f)$ its Hardy-Littlewood maximal function. For $1 < p < \infty$ prove that $f \in L^p(\mathbb{R})$ implies $\mathcal{M}(f) \in L^p(\mathbb{R})$ and that $\|\mathcal{M}(f)\|_{L^p} \leq c\|f\|_{L^p}$.

Hint: Example 9.24 may be used to state for $\mathcal{M}(f)$ that

$$\|\mathcal{M}(f)\|_{L^p}^p = p \int_0^\infty \lambda^{(1)}(\{\mathcal{M}(f) > s\})s^{p-1}ds.$$

9.* In this problem we construct a continuous function which is at no point differentiable. The first example of such a function is due to Weierstrass, his example was modified by many authors. We have chosen the function discussed in [45]. Let $g : \mathbb{R} \to \mathbb{R}$ be the periodic extension from $x \mapsto |x|$ defined on $[-1, 1]$, i.e. $g(x) = g(x + 2)$ and $g|_{[-1,1]} = |x|$. Prove that the **Weierstrass function** $x \mapsto f(x) := \sum_{k=0}^\infty \left(\frac{3}{4}\right)^k g(4^k x)$ is continuous on $\mathbb{R}$ and at no point $x \in \mathbb{R}$ differentiable.

12 Selected Topics

In this chapter we want to discuss three topics with some of their applications: the theorem of Sard; the theorem of Lusin; and Kolmogorov's version of the Arzela-Ascoli theorem for L^p-spaces. In some discussions we will need auxilliary results (mainly from point set topology) which we will quote but not prove. The main idea of this chapter is to present some useful results and to indicate that the theory developed so far has far reaching consequences in quite different, and maybe unexpected fields of mathematics.

We start with a version of Sard's theorem. In Chapter II.9 we introduced the notion of a critical point of a differentiable function $f : G \to \mathbb{R}$, $G \subset \mathbb{R}^n$, as a point $x_0 \in G$ for which $\operatorname{grad} f(x_0) = 0$. For differentiable mappings $f : G \to \mathbb{R}^n$, $G \subset \mathbb{R}^n$, we can extend this definition to points $x_0 \in G$ where $\det J_f(x_0) = 0$, i.e. $x_0 \in G$ is called a critical point if the Jacobi determinant of f at x_0 vanishes. From here it is an easy step to

Definition 12.1. *Let $G \subset \mathbb{R}^n$ be an open set and $f : G \to \mathbb{R}^m$ a differentiable mapping. We call $x_0 \in G$ a **critical point** of f if $d_{x_0} f = J_f(x_0)$ has no maximal rank. The set of all critical points of f we denote by $\operatorname{crit}(f)$. The set $f(\operatorname{crit}(f))$ is the set of **critical values** and its complement in $f(G)$ is the set of all **regular values**.*

The version of **Sard's theorem** we are going to prove is

Theorem 12.2. *For a C^1-mapping $f : G \to \mathbb{R}^n$, $G \subset \mathbb{R}^n$ open, the set $f(\operatorname{crit}(f))$ has Lebesgue measure zero, i.e. $f(\operatorname{crit}(f))$ is a Borel set and $\lambda^{(n)}(f(\operatorname{crit}(f))) = 0$.*

Remark 12.3. In Volume VI we will extend the definition of critical points even to differentiable mappings between differentiable manifolds, and if these manifolds have a countable base, Sard's theorem still holds, compare with [58]. For C^∞-mappings $f : G \to \mathbb{R}^m$, $G \subset \mathbb{R}^n$ open, a complete proof of Sard's theorem is given in [57] stating that the measure of the set $f(\{x \in G \,|\, \text{rank of } J_f(x) < m\})$ is zero. A different extension of Sard's theorem is given in [78] and it reads as follows: for $G \subset \mathbb{R}^n$ and $f : G \to \mathbb{R}^n$ continuously differentiable the set $f(A)$ is measurable if $A \subset G$ is measurable and we have

$$\lambda^{(n)}(f(A)) \leq \int_A |\det J_f(x)| \, dx. \tag{12.1}$$

Clearly (12.1) implies Theorem 12.2.

For the proof of Theorem 12.2 we need an easy extension of Theorem II.9.1: For $f : G \to \mathbb{R}^n$, $G \subset \mathbb{R}^n$ convex and f continuously differentiable

$$f(x) - f(y) - (\mathrm{d}f(y))(x-y) = \int_0^1 (\mathrm{d}f(y + t(x-y)) - \mathrm{d}f(y))\,(x-y)\,\mathrm{d}t$$
(12.2)

holds for all $x, y \in G$.

In addition the proof requires

Lemma 12.4. *Let $G \subset \mathbb{R}^n$ be an open set. Then G is the union of countably many compact cubes.*

We do not prove this result which can be derived from general results for separable, locally compact and σ-compact spaces, compare with [65], otherwise we would have to derive a more lengthy concrete proof for the special case of $G \subset \mathbb{R}^n$.

Proof of Theorem 12.2. (Compare with [19] or [74]). As stated in Lemma 12.4, G is a countable union of compact cubes Q_j, $j \in \mathbb{N}$. Once we can prove the result for a compact cube $Q \subset G$, the result will follow since

$$f(\mathrm{crit}(f)) = \bigcup_{j \in \mathbb{N}} f(\mathrm{crit}(f|_{Q_j})).$$

Let $Q \subset G$ a compact cube with side length $r > 0$. Since by assumption $\mathrm{d}f(x)$ is continuous on G it is bounded and uniformly continuous on Q. Thus $\|\mathrm{d}f(x)\| \leq M$ and for every $\epsilon > 0$ there exists $l \in \mathbb{N}$ such that with $\delta := \frac{\sqrt{n}r}{l}$ the following holds

$$\|x - y\| < \delta \quad \text{implies} \quad \|\mathrm{d}f(x) - \mathrm{d}f(y)\| < \epsilon.$$

Note that for x and $f(x)$ the norm $\|x\|$ and $\|f(x)\|$ is the norm in $\mathbb{R}^n$ whereas for $\mathrm{d}f(x)$ the norm $\|\mathrm{d}f(x)\|$ is the norm (matrix norm) in $\mathbb{R}^{n^2}$. For $x, y \in Q$ such that $\|x - y\| < \delta$ we now find

$$\|f(x) - f(y) - (\mathrm{d}f(y))(x-y)\| \leq \int_0^1 \|\mathrm{d}f(y + t(x-y)) - \mathrm{d}f(y)\| \, \|x - y\| \, \mathrm{d}y$$
$$\leq \epsilon \|x - y\|.$$

Let $N = l^n$ and divide Q into N cubes Q_j, $j = 1, \ldots, N$, with diameter δ, i.e. side length $\frac{\delta}{\sqrt{n}} = \frac{r}{l}$. For $x, y \in Q_j$ we have

$$f(x) = f(y) + (\mathrm{d}f(y))(x - y) + R(x, y), \quad \|R(x, y)\| \leq \epsilon\delta.$$

Now let $x_j \in Q_j \cap \mathrm{crit}(f)$ and set

$$D := (\mathrm{d}f)(x_j)$$

and

$$g(y) := f(x_j + y) - f(x_j), \quad y \in Q_j - x_j,$$

which gives

$$g(y) = Dy + \tilde{R}(y), \quad \|\tilde{R}(y)\| = \|R(x_j + y, x_j)\| \leq \epsilon\delta.$$

We note that $\det(J_f(x_j)) = \det(\mathrm{d}f(x_j)) = 0$ implies that $D(Q_j - x_j)$ is contained in an $(n - 1)$-dimensional subspace of $\mathbb{R}^n$, say the subspace characterised by $\langle z, b_1 \rangle = 0$ for all $z \in D(Q_j - x_j)$ with some $b_1 \in \mathbb{R}^n$, $\|b_1\| = 1$. Let $\{b_1, \ldots, b_n\}$ be an orthogonal basis of $\mathbb{R}^n$ obtained by extending $\{b_1\}$. It follows that

$$g(y) = \sum_{k=1}^{n} \langle g(y), b_k \rangle b_k$$

and

$$\begin{aligned}|\langle g(y), b_1 \rangle| &\leq |\langle Dy, b_1 \rangle| + |\langle R(x_j + y, x_j), b_1 \rangle| \\ &\leq \|R(x_j + y, x_j)\| \leq \epsilon\delta,\end{aligned} \tag{12.3}$$

whereas for $2 \leq k \leq n$ we find

$$\begin{aligned}|\langle g(y), b_k \rangle| &\leq \|D\| \, \|y\| + \|R(x_j + y, x_j)\| \\ &\leq \|D\|\delta + \epsilon\delta.\end{aligned} \tag{12.4}$$

Using the definition of g we deduce that $f(Q_j)$ is contained in some cube W_j with centre $f(x_j)$ such that

$$\begin{aligned}\lambda^{(n)}(W_j) &= (2(\|D\|\delta + \epsilon\delta))^{n-1} 2\epsilon\delta \\ &= 2^n (\|D\| + \epsilon)^{n-1} \delta^n \epsilon.\end{aligned}$$

Consequently we have

$$f\left(\mathrm{crit}f|_Q\right) \subset \bigcup_{j=1}^{N} W_j$$

and

$$\sum_{j=1}^{N} \lambda^{(n)}(W_j) \leq \sum_{j=1}^{N} (2(\|\mathrm{d}f(x_j)\| + \epsilon))^{n-1} 2\delta^n \epsilon$$

$$\leq N2^n(M + \epsilon)^{n-1}(\sqrt{n}r)^n \epsilon,$$

implying that $\bigcup_{j=1}^{N} W_j$ is a set of measure zero, and hence $f\left(\mathrm{crit}f|_Q\right)$ is a null set, hence a Lebesgue set of measure zero. Since $f(\mathrm{crit}(f))$ is the union of countably many compact set it is also a Borel set. $\qquad\square$

Corollary 12.5. *Let $G \subset \mathbb{R}^n$ be an open set and $f : G \to \mathbb{R}^m$ a continuously differentiable mapping. If $n < m$ then $f(G)$ is a Borel set and $\lambda^{(m)}(f(G)) = 0$.*

Proof. Let $\mathrm{pr} : \mathbb{R}^m \to \mathbb{R}^n$ be the canonical projection and consider the mapping $f \circ \mathrm{pr} : \mathrm{pr}^{-1}(G) \to \mathbb{R}^m$. Since $\mathrm{d}(f \circ \mathrm{pr}) = \mathrm{d}f \circ \mathrm{d}\,\mathrm{pr}$ the Jacobi determinant of $f \circ \mathrm{pr}$ is identically zero, however $(f \circ \mathrm{pr})(\mathrm{pr}^{-1}(G)) = f(G)$. $\qquad\square$

Sard's theorem has many applications in particular when discussing the critical or regular values of mappings between differentiable manifolds, but it is also possible to give a proof of the Brouwer fixed point theorem based on Sard's theorem. Of particular importance are applications in the degree theory of mappings, compare for example with [78] or [19].

The next theorem we want to discuss is Lusin's theorem. Although it holds for a much more general situation, see Appendix II, we state and prove it here for the case $\mathbb{R}^n$ and the Lebesgue measure. For the proof we need once again a result relating measurable sets to closed sets, namely

Lemma 12.6. *Let $A \in \mathcal{B}^{(n)}$. For every $\epsilon > 0$ we can find a closed set $C \subset A$ such that $\lambda^{(n)}(A\backslash C) \leq \epsilon$.*

Of course, this lemma is closely related to Theorem 10.10 and we will discuss it in Appendix II.

Theorem 12.7 (Lusin). *For $A \in \mathcal{B}^{(n)}$, $\lambda^{(n)}(A) < \infty$, let $f : A \to \mathbb{R}$ be a measurable function. Given $\epsilon > 0$ we can find a closed set $C_\epsilon \subset A$ such that $\lambda^{(n)}(A \backslash C_\epsilon) < \epsilon$ and $f|_{C_\epsilon}$ is continuous.*

Proof. (Adopted from [84]). Since f is measurable it is the $\lambda^{(n)}$-a.e. limit of simple functions $f_k : A \to \mathbb{R}$. For each f_k we can find a measurable set $A_k \subset A$ such that $\lambda^{(n)}(A_k) < \frac{1}{2^k}$ and $f|_{A_k^c}$ is continuous, see Problem 5. Further by Egorov's theorem, Theorem 7.38, there exists a measurable set $B_{\frac{\epsilon}{3}}$ such that $B_{\frac{\epsilon}{3}} \subset A$, $\lambda^{(n)}\left(A \backslash B_{\frac{\epsilon}{3}}\right) < \frac{\epsilon}{3}$ and on $B_{\frac{\epsilon}{3}}$ the sequence $(f_k)_{k \in \mathbb{N}}$ converges uniformly to f. For $\epsilon > 0$ we can always find $N \in \mathbb{N}$ such that $\sum_{k=N}^{\infty} 2^{-k} \leq \frac{\epsilon}{3}$ and with this N we define

$$\tilde{C}_\epsilon := B_{\frac{\epsilon}{3}} \backslash \bigcup_{k \geq N} A_k.$$

Since $f_k|_{\tilde{C}_\epsilon}$ is continuous and $\left(f_k|_{\tilde{C}_\epsilon}\right)_{k \in \mathbb{N}}$ converges on $\tilde{C}_\epsilon$ uniformly to f, it follows that $f|_{\tilde{C}_\epsilon}$ is continuous. Clearly $\tilde{C}_\epsilon$ is measurable and therefore we can find by Lemma 12.6 a closed set C_ϵ such that $C_\epsilon \subset \tilde{C}_\epsilon$ and $\lambda^{(n)}\left(\tilde{C}_\epsilon \backslash C_\epsilon\right) < \frac{\epsilon}{3}$ implying that $\lambda^{(n)}(A \backslash C_\epsilon) < \epsilon$ and the theorem is proved. $\square$

Remark 12.8. A. As is emphasised in [84], and rarely in other books, the statement of Lusin's theorem is that $f|_{C_\epsilon}$ is continuous, but we cannot expect that $f : A \to \mathbb{R}$ (as a function defined on A) is continuous at all points $C_\epsilon \subset A$. We also want to note that in general we have no possibility to identify C_ϵ.
B. In [94] the equivalence of measurability and the property stated in Lusin's theorem is discussed.
C. For justifiable reasons some authors call Lusin's theorem the Vitali-Lusin theorem.

We want to use Lusin's theorem to characterise a new, rather important type of convergence in $L^p(G)$, $G \in \mathcal{B}^{(n)}$, $\lambda^{(n)}(G) < \infty$. First we give

Definition 12.9. *Let $G \subset \mathcal{B}^{(n)}$ be any measurable set and $1 < p < \infty$. We call a sequence $(g_k)_{k \in \mathbb{N}}$, $g_k \in L^p(G)$, **weakly convergent** to $g \in L^p(G)$ if for all $\varphi \in L^q(G)$, $\frac{1}{p} + \frac{1}{q} = 1$, we have*

$$\lim_{k \to \infty} \int_G g_k \varphi \, d\lambda^{(n)} = \int_G g\varphi \, d\lambda^{(n)}. \tag{12.5}$$

Corollary 12.10. *Every sequence* $(g_k)_{k\in\mathbb{N}}$ *converging in* $L^p(G)$ *to* g, *i.e.* $\lim_{k\to\infty}\|g_k - g\|_{L^p} = 0$, *converges weakly to* g. *Moreover the weak limit of a sequence is uniquely determined.*

Proof. From Hölder's inequality we deduce

$$\left|\int_G g_k\varphi\,\mathrm{d}\lambda^{(n)} - \int_G g\varphi\,\mathrm{d}\lambda^{(n)}\right| \leq \|g_k - g\|_{L^p}\|\varphi\|_{L^q}$$

implying that $\lim_{k\to\infty}\|g_k - g\|_{L^p}$ entails weak convergence. Furthermore for two weak limits g and h of $(g_k)_{k\in\mathbb{N}}$ we find

$$\left|\int_G (g-h)\varphi\,\mathrm{d}\lambda^{(n)}\right| \leq \lim_{k\to\infty}\left|\int_G (g-g_k)\varphi\,\mathrm{d}\lambda^{(n)}\right| + \lim_{k\to\infty}\left|\int_G (g_k-h)\varphi\,\mathrm{d}\lambda^{(n)}\right| = 0,$$

i.e. for all $\varphi \in L^q(G)$ the following holds

$$\int_G (g-h)\varphi\,\mathrm{d}\lambda^{(n)} = 0.$$

Let $\tilde{g}$ be any representant of g and $\tilde{h}$ of h. Since $(p-1)q = p$ the function

$$\tilde{\varphi}_{\tilde{g},\tilde{h}}(x) = \begin{cases} \frac{\tilde{g}-\tilde{h}}{|\tilde{g}-\tilde{h}|}|\tilde{g}-\tilde{h}|^{p-1}, & \tilde{g} \neq \tilde{h} \\ 0, & \tilde{g} = \tilde{h} \end{cases}$$

is measurable and $|\tilde{\varphi}_{\tilde{g},\tilde{h}}|^q$ is integrable. It follows that

$$0 = \int_G (\tilde{g}-\tilde{h})\tilde{\varphi}_{\tilde{g},\tilde{h}}\,\mathrm{d}\lambda^{(n)} = \int_G |\tilde{g}-\tilde{h}|^p\,\mathrm{d}\lambda^{(n)}$$

implying that $\tilde{g} = \tilde{h}$ $\lambda^{(n)}$-a.e., hence $g = h$ in $L^p(G)$. $\qquad\square$

Example 12.11. In $L^2([0, 2\pi])$ consider the function $g_k(x) = \sin kx$. Since for $k \neq l$

$$\|g_k - g_l\|_{L^2}^2 = \int_0^{2\pi} g_k^2\,\mathrm{d}\lambda^{(1)} + \int_0^{2\pi} g_l^2\,\mathrm{d}\lambda^{(1)} - 2\int_0^{2\pi} g_k g_l\,\mathrm{d}\lambda^{(1)}$$

$$= \int_0^{2\pi} (\sin kx)^2\,\mathrm{d}x + \int_0^{2\pi} (\sin lx)^2\,\mathrm{d}x - 2\int_0^{2\pi} \sin kx \sin lx\,\mathrm{d}x$$

$$= 2\pi,$$

the sequence $(g_k)_{k\in\mathbb{N}}$ is not a Cauchy sequence in $L^2([0, 2\pi])$, hence it is not convergent. On the other hand, for $\varphi = \chi_{[a,b]}$, $[a, b] \subset [0, 2\pi]$, we have as $k \to \infty$

$$\int_0^{2\pi} (\sin kx)\varphi(x)\,\mathrm{d}x = \frac{1}{k}(\cos ak - \cos bk) \to 0$$

and this extends to all linear combinations $\sum_{j=1}^N \gamma_j \chi_{[a_j, b_j]}$, $[a_j, b_j] \subset [0, 2\pi]$, $\gamma_j \in \mathbb{R}$, and hence by Theorem 8.21 to all $\varphi \in L^2([0, 2\pi])$ implying that the sequence $(g_k)_{k\in\mathbb{N}}$ converges weakly in $L^2([0, 2\pi])$ to 0, but it does not converge in $L^2([0, 2\pi])$.

Remark 12.12. A. In order to have a simpler façon de parler, a sequence $(g_k)_{k\in\mathbb{N}}$ converging in $L^p(G)$ to g is also called **strongly convergent** or **norm convergent**.
B. In light of Corollary 12.10 we know that if $(g_k)_{k\in\mathbb{N}}$, $g_k \in L^p(G)$, has a weak limit $g \in L^p(G)$, then the only candidate for a strong limit of $(g_k)_{k\in\mathbb{N}}$ is g.

With the help of Egorov's and Lusin's theorem we can prove

Proposition 12.13. *Let $G \in \mathcal{B}^{(n)}$, $\lambda^{(n)}(G) < \infty$, and let $(g_k)_{k\in\mathbb{N}}$, $g_k \in L^p(G)$, $1 < p < \infty$, be a sequence of functions converging weakly in $L^p(G)$ to g. In addition assume that $(g_k)_{k\in\mathbb{N}}$ converges also $\lambda^{(n)}$-a.e. to h. It follows that $g = h\ \lambda^{(n)}$-a.e.*

Proof. Combining the statements of Egorov's and Lusin's theorem, for $\epsilon > 0$ we can find a closed set $G_\epsilon \subset G$ such that $\lambda^{(n)}(G \backslash G_\epsilon) < \epsilon$ and $(g_k)_{k\in\mathbb{N}}$ converges on G_ϵ uniformly to h and $g|_{G_\epsilon}$ is continuous and bounded. This implies also that $(h - g)\chi_{G_\epsilon} \in L^\infty(G)$, i.e. $(h - g)\chi_{G_\epsilon} \in L^q(G)$ for all $q \geq 1$. The uniform convergence of $(g_k)_{k\in\mathbb{N}}$ on G_ϵ to h yields

$$\lim_{k\to\infty} \int_{G_\epsilon} g_k(h - g)\,\mathrm{d}\lambda^{(n)} = \int_{G_\epsilon} h(h - g)\,\mathrm{d}\lambda^{(n)},$$

while the weak convergence of $(g_k)_{k\in\mathbb{N}}$ to g implies

$$\lim_{k\to\infty} \int_{G_\epsilon} g_k(h - g)\,\mathrm{d}\lambda^{(n)} = \lim_{k\to\infty} \int_G g_k(h - g)\chi_{G_\epsilon}\,\mathrm{d}\lambda^{(n)}$$
$$= \int_{G_\epsilon} g(h - g)\,\mathrm{d}\lambda^{(n)}.$$

Subtracting these lines gives for every $\epsilon > 0$

$$0 = \int_{G_\epsilon} (h - g)^2 \, d\lambda^{(n)}$$

which in turn implies $h = g$ $\lambda^{(n)}$-a.e. $\qquad\qquad\qquad\qquad$ □

Combining Proposition 12.13 with Theorem 7.35 we get

Corollary 12.14. *Let $G \in \mathcal{B}^{(n)}$, $\lambda^{(n)}(G) < \infty$. If a sequence $(g_k)_{k\in\mathbb{N}}$, $g_k \in L^p(G)$, $1 < p < \infty$, converges weakly to $g \in L^p(G)$ and in measure to h then $g = h$ $\lambda^{(n)}$-a.e.*

Results such as in Proposition 12.13 or its corollary are important tools in the analysis of (non-linear) partial differential equations when combined with certain compactness results. Our next aim is to characterise pre-compact sets in the spaces $L^p(G)$, $G \in \mathcal{B}^{(n)}$. The space $L^p(\mathbb{R}^n)$, $1 \leq p < \infty$, is a Banach space, hence a complete metric space and compact subsets of $L^p(\mathbb{R}^n)$ are defined. We are longing to find a characterisation of all **relatively compact sets** $K \subset L^p(\mathbb{R}^n)$, i.e. sets whose closures are compact, compare with Definition II.3.15.A. An important tool to check relative compactness is the notion of totally bounded sets (see Definition II.3.15.B): a subset Y of a metric space (X, d) is called **totally bounded** if for every $\epsilon > 0$ there exists a finite covering $(U_j)_{j=1,\dots,M}$, $M \in \mathbb{N}$, of Y with open sets $U_j \subset X$ all having diameter $\operatorname{diam} U_j < \epsilon$. We can reformulate this definition in the following way: for every $\epsilon > 0$ exist finitely many points $x_1, \dots, x_M$ such that $Y \subset \bigcup_{j=1}^{M} B_\epsilon(x_j)$. The set $\{x_1, \dots, x_M\}$ is often called an ϵ-**net** for Y. For every ϵ-net for Y we have for all $x \in Y$

$$d(x, \{x_1, \dots, x_M\}) = \inf \{d(x, x_j) \,|\, j = 1, \dots, M\} < \epsilon. \qquad (12.6)$$

Now we can state Theorem II.3.18 in the following way

Theorem 12.15. *For a subset Y of a metric space (X, d) the following are equivalent*

    i) Y is compact;

    ii) every infinite sequence $(x_k)_{k\in\mathbb{N}}$, $x_k \in Y$, has at least one accumulation point $x \in Y$;

iii) Y is complete (as metric subspace) and totally bounded;

iv) Y is complete and for every $\epsilon > 0$ there exists an ϵ-net for Y.

The metric space we are interested in is $L^p(\mathbb{R}^n)$, $1 \leq p < \infty$, which is a Banach space, hence complete, but it is an infinite dimensional space. We have already seen a compactness result for an infinte dimensional Banach space, namely the Arzela-Ascoli theorem, Theorem II.14.25. We will use the Arzela-Ascoli in our proof for a compactness result for $L^p(\mathbb{R}^n)$. Recall that equi-continuity, Definition II.14.24, was one condition a relatively compact subset of $C(Y)$, Y a compact metric space, must satisfy. For L^p-spaces we need a similar condition and we start by investigating the continuity of the translation operator T_h in $L^p(\mathbb{R}^n)$ as well as the **averaging operator**

$$(M_R u)(x) := \frac{1}{\lambda^{(n)}(B_R(0))} \int_{B_R(0)} (T_h u)(x)\, dh, \qquad (12.7)$$

$R > 0$ and $u \in L^p(\mathbb{R}^n)$. In Chapter 10 we have seen that for $u \in L^p(\mathbb{R}^n)$ the translation operator $(T_h u)(x) = u(x + h)$, $h \in \mathbb{R}^n$, is well defined and hence it makes sense to consider the linear mapping $T_h : L^p(\mathbb{R}^n) \to L^p(\mathbb{R}^n)$. The translation invariance of $\lambda^{(n)}$ yields of course

$$\|T_h u\|_{L^p}^p = \int_{\mathbb{R}^n} |u(x + h)|^p \lambda^{(n)}(dx) = \int_{\mathbb{R}^n} |u(x)|^p \lambda^{(n)}(dx) = \|u\|_{L^p}^p. \quad (12.8)$$

Theorem 12.16. *Let $1 \leq p < \infty$ and T_h as well as M_R be defined as above.*
A. *The operator T_h maps to $L^p(\mathbb{R}^n)$ boundedly into itself and is continuous with respect to h as $h \to 0$ in the sense that*

$$\lim_{h \to 0} \|T_h u - u\|_{L^p} = 0. \qquad (12.9)$$

B. *For $R > 0$ fixed the operator M_R satisfies for all $u \in L^p(\mathbb{R}^n)$, $1 \leq p < \infty$, the estimate*

$$\|M_R u - u\|_{L^p} \leq \sup_{\|h\| \leq R} \|T_h u - u\|_{L^p} \qquad (12.10)$$

as well as

$$\|M_R u\|_\infty \leq \left(\frac{1}{\lambda^{(n)}(B_R(0))} \|u\|_{L^p}^p \right)^{\frac{1}{p}}. \qquad (12.11)$$

Proof. **A.** For $\epsilon > 0$ we can find $\varphi \in C_0(\mathbb{R}^n)$ such that $\|u - \varphi\|_{L^p} < \frac{\epsilon}{3}$, recall $C_0(\mathbb{R}^n)$ is dense in $L^p(\mathbb{R}^n)$. Suppose that $\operatorname{supp} \varphi \subset \overline{B_R(0)}$. The function φ is uniformly continuous and therefore we can find δ, $0 < \delta < 1$, such that $\|h\| < \delta$ implies

$$|\varphi(x + h) - \varphi(x)| < \frac{\epsilon}{3}\left(\frac{1}{\lambda^{(n)}(B_R(0))}\right)^{\frac{1}{p}}.$$

For $h \in \mathbb{R}^n$ such that $\|h\| < \delta$ we find

$$\int_{\mathbb{R}^n} |\varphi(x + h) - \varphi(x)|^p \, dx = \int_{\overline{B_R(0)}} |\varphi(x + h) - \varphi(x)|^p \, dx < \left(\frac{\epsilon}{3}\right)^p,$$

i.e. $\|T_h\varphi - \varphi\|_{L^p} < \frac{\epsilon}{3}$. Furthermore, by the translation invariance of the Lebesgue measure $\lambda^{(n)}$ we have $\|T_h u - T_h\varphi\|_{L^p} = \|u - \varphi\|_{L^p}$ for all $h \in \mathbb{R}^n$ and thus we arrive at

$$\|T_h u - u\|_{L^p} \leq \|T_h u - T_h\varphi\|_{L^p} + \|T_h\varphi - \varphi\|_{L^p} + \|\varphi - u\|_{L^p} < \epsilon.$$

B. For $u \in L^p(\mathbb{R}^n)$ and $R > 0$ fixed we have with $V(R) := \lambda^{(n)}(B_R(0))$ and $\frac{1}{q} + \frac{1}{p} = 1$ that

$$\|M_R u - u\|_{L^p} = \left(\int_{\mathbb{R}^n} \left|\frac{1}{V(R)}\int_{\overline{B_R(0)}} ((T_h u)(x) - u(x)) \, dh\right|^p dx\right)^{\frac{1}{p}}$$

$$= \left(\int_{\mathbb{R}^n} \left|\frac{1}{V(R)}\int_{\overline{B_R(0)}} (u(x + h) - u(x)) \, dh\right|^p dx\right)^{\frac{1}{p}}$$

$$\leq \frac{1}{V(R)}\left(\int_{\mathbb{R}^n} \left|\int_{\overline{B_R(0)}} 1 \cdot (u(x + h) - u(x)) \, dh\right|^p dx\right)^{\frac{1}{p}}$$

$$\leq \frac{1}{V(R)}\left(\int_{\mathbb{R}^n} \left(\int_{\overline{B_R(0)}} 1^q \, dh\right)^{\frac{p}{q}} \left(\int_{\overline{B_R(0)}} |u(x + h) - u(x)|^p \, dh\right) dx\right)^{\frac{1}{p}},$$

where in the last step we used Hölder's inequality. Thus we arrive by using

252

Fubini's theorem at

$$\|M_R u - u\|_{L^p} \le V(R)^{\frac{1}{q}-1} \left(\int_{\mathbb{R}^n} \left(\int_{B_R(0)} |u(x+h) - u(x)|^p \, dh \right) dx \right)^{\frac{1}{p}}$$

$$= V(R)^{-\frac{1}{p}} \left(\int_{B_R(0)} \left| \int_{\mathbb{R}^n} |u(x+h) - u(x)|^p \, dx \right| dh \right)^{\frac{1}{p}}$$

$$= \frac{1}{\lambda^{(n)}(B_R(0))^{\frac{1}{p}}} \lambda^{(n)}(B_R(0))^{\frac{1}{p}} \sup_{\|h\| \le R} \|T_h u - u\|_{L^p}$$

$$= \sup_{\|h\| \le R} \|T_h u - u\|_{L^p},$$

which proves (12.10). In order to derive (12.11) we can use parts of the calculations made above to find

$$|(M_R u)(x)| \le \frac{1}{V(R)} \int_{B_R(0)} |u(x+h)| \, dh$$

$$\le \frac{1}{V(R)} (V(R))^{\frac{1}{q}} \left(\int_{B_R(0)} |u(x+h)|^p \, dh \right)^{\frac{1}{p}}$$

$$\le V(R)^{\frac{1}{q}-1} \left(\int_{\mathbb{R}^n} |u(x+h)|^p \, dh \right)^{\frac{1}{p}}$$

$$= \left(\frac{1}{V(R)} \|u\|_{L^p}^p \right)^{\frac{1}{p}}.$$

Note that the proof of (12.10) and (12.11) for $p = 1$ follows when we agree to consider in this case $\int |uv| \, d\lambda^{(n)} \le \|u\|_\infty \|v\|_{L^1}$, $u \in L^\infty(\mathbb{R}^n)$ and $v \in L^1(\mathbb{R}^n)$ also as Hölder's inequality with $p = 1$ and $q = \infty$. $\square$

We can now prove the **Kolmogorov-Riesz theorem** which characterises relatively compact sets in $L^p(\mathbb{R}^n)$, $1 \le p < \infty$.

Theorem 12.17. *A subset* $K \subset L^p(\mathbb{R}^n)$, $1 \le p < \infty$, *is relatively compact in* $L^p(\mathbb{R}^n)$ *if and only if the following three conditions hold:*

i) K *is bounded in* $L^p(\mathbb{R}^n)$, *i.e.*

$$\sup_{u \in K} \|u\|_{L^p} \le M < \infty; \tag{12.12}$$

*ii) K is **equi-continuous** in $L^p(\mathbb{R}^n)$, i.e.*

$$\lim_{h\to 0}\sup_{u\in K} \|T_h u - u\|_{L^p} = 0; \tag{12.13}$$

iii)

$$\lim_{R\to\infty}\sup_{u\in K} \left(\int_{B_R^{\complement}(0)} |u(x)|^p \lambda^{(n)}(\mathrm{d}x)\right)^{\frac{1}{p}} = 0. \tag{12.14}$$

Proof. Assume first that K is relatively compact. Since $\overline{K}$ is compact in $L^p(\mathbb{R}^n)$ it must be bounded, hence $K \subset \overline{K}$ must be bounded. Furthermore K is totally bounded, i.e. for $\epsilon > 0$ we can find an $\frac{\epsilon}{2}$-net $\{u_1, \ldots, u_N\}$ for K such that for every $u \in K$ there exists $j \in \{1, \ldots, N\}$ such that $\|u - u_j\|_{L^p} < \frac{\epsilon}{2}$. Now let $\varphi_1, \ldots, \varphi_N \in C_0(\mathbb{R}^n)$ be such that $\|u_j - \varphi_j\|_{L^p} < \frac{\epsilon}{2}$, since $C_0(\mathbb{R}^n)$ is dense in $L^p(\mathbb{R}^n)$, $1 \le p < \infty$, such functions exists. We may also assume that $\operatorname{supp}\varphi_j \subset \overline{B_R(0)}$ for some $R > 0$ and all $j = 1, \ldots, N$. For $u \in K$ it follows that

$$\left(\int_{B_R^{\complement}(0)} |u(x)|^p \, \mathrm{d}x\right)^{\frac{1}{p}} \le \left(\int_{B_R^{\complement}(0)} |u(x) - \varphi_j(x)|^p \, \mathrm{d}x\right)^{\frac{1}{p}} + \left(\int_{B_R^{\complement}(0)} |\varphi_j(x)|^p \, \mathrm{d}x\right)^{\frac{1}{p}}$$

$$= \left(\int_{B_R^{\complement}(0)} |u(x) - \varphi_j(x)|^p \, \mathrm{d}x\right)^{\frac{1}{p}}$$

$$\le \|u - \varphi_j\|_{L^p} \le \|u - u_j\|_{L^p} + \|u_j - \varphi_j\|_{L^p} < \epsilon,$$

implying that

$$\lim_{R\to\infty}\sup_{u\in K} \left(\int_{B_R^{\complement}(0)} |u(x)|^p \mathrm{d}x\right)^{\frac{1}{p}} = 0,$$

i.e. *iii)* holds. We now prove *ii)*. For this we will use Theorem 12.16.A. Note that we cannot just apply this theorem to every single function $u \in K$ since we long for uniformity with respect to $u \in K$ in (12.13). Given $\epsilon > 0$ let $\{u_1, \ldots, u_N\}$ be now an $\frac{\epsilon}{3}$-net of K. It follows that

$$\|T_h u - u\|_{L^p} \le \|T_h u - T_h u_j\|_{L^p} + \|T_h u_j - u_j\|_{L^p} + \|u_j - u\|_{L^p}$$

$$\le 2\|u - u_j\|_{L^p} + \|T_h u_j - u_j\|_{L^p},$$

where we used that $\|T_h v\|_{L^p} = \|v\|_{L^p}$ for every $v \in L^p(\mathbb{R}^n)$ and $T_h u - T_h u_j = T_h(u - u_j)$. In light of Theorem 12.16.A we may choose $\delta > 0$ such that $\|h\| < \delta$ implies for all $j = 1, \ldots, N$ that $\|T_h u_j - u_j\|_{L^p} < \frac{\epsilon}{3}$ and we arrive at

$$\sup_{u \in K} \|T_h u - u\|_{L^p} < \epsilon$$

for all $\|h\| < \delta$, i.e. $ii)$ is shown.

We are now going to prove the more important converse statement. Condition $ii)$ requires $\lim_{h \to 0} \|T_h u - u\|_{L^p} = 0$ uniformly for $u \in K$, hence by (12.10) we have $\lim_{h \to 0} \|M_R u - u\|_{L^p} = 0$ uniformly for $u \in K$. Fix $R > 0$ and pick $x_1, x_2 \in \mathbb{R}^n$. It follows with $V(R) := \lambda^{(n)}(B_r(0))$ that

$$\begin{aligned}
|M_R u(x_1) - M_R u(x_2)| &\leq \frac{1}{V(R)} \int_{B_R(0)} |u(x_1 + h) - u(x_2 + h)| \, dh \\
&\leq \frac{1}{V(R)^{\frac{1}{p}}} \left(\int_{B_R(0)} |u(x_1 + h) - u(x_2 + h)|^p \, dh \right)^{\frac{1}{p}} \\
&\leq \frac{1}{V(R)^{\frac{1}{p}}} \left(\int_{\mathbb{R}^n} |u(x_1 + h) - u(x_2 + h)|^p \, dh \right)^{\frac{1}{p}} \\
&= \frac{1}{V(R)^{\frac{1}{p}}} \|T_{x_1} u - T_{x_2} u\|_{L^p} \\
&\leq \frac{1}{\lambda^{(n)}(B_R(0))^{\frac{1}{p}}} \left(\|T_{x_1} u - u\|_{L^p} + \|T_{x_2} u - u\|_{L^p} \right),
\end{aligned}$$

and now (12.13) implies the equi-continuity of the function $M_R u$ for $R > 0$ fixed. In particular M_R is continuous. Moreover (12.11) yields that the family $\{M_R u \mid u \in K\}$, $R > 0$ fixed, is bounded with respect to the sup-norm $\|\cdot\|_\infty$. We consider the set

$$D_R := \left\{ v = \chi_{\overline{B_R(0)}} \cdot M_R u \mid u \in K \right\}.$$

This set can be considered as a subset of $C\left(\overline{B_R(0)}\right)$ and in the Banach space $\left(C\left(\overline{B_R(0)}\right), \|\cdot\|_\infty\right)$ this set D_R is bounded and equi-continuous. Since $\overline{B_R(0)}$ is compact, by Arzela-Ascoli theorem, Theorem II.14.25, the set D_R is relatively compact in $C(\overline{B_R(0)})$. Now we prove that K is relatively compact in $L^p(\mathbb{R}^n)$, $1 \leq p < \infty$. Given $\epsilon > 0$. By $iii)$ we can find $R_0 > 0$ such that

$$\left(\int_{B_{R_0}^{\complement}(0)} |u(x)|^p \, dx\right)^{\frac{1}{p}} < \frac{\epsilon}{3}, \tag{12.15}$$

and further we can find $R_1 > 0$ such that for all $u \in K$

$$\|M_{R_1} u - u\|_{L^p} < \frac{\epsilon}{3} \tag{12.16}$$

holds. Moreover, in $C\left(\overline{B_{R_0}(0)}\right)$ the set D_{R_0} is relative compact. Hence there exist functions $u_1, \ldots, u_M \in K$ such that for every $u \in K$ there exists $j \in \{1, \ldots, M\}$ such that for all $x \in \overline{B_{R_0}(0)}$ we have

$$|M_{R_1} u(x) - M_{R_1} u_j(x)| \leq \frac{\epsilon}{3} \frac{1}{V(R_0)^{\frac{1}{p}}}.$$

For $j = 1, \ldots, M$ we define $g_j := \chi_{\overline{B_{R_0}(0)}} M_{R_1} u_j$ and we claim that $\{g_1, \ldots, g_M\}$ is an ϵ-net for K. Let $u \in K$ and pick u_j, $1 \leq j \leq M$, as above. Then it follows

$$\|u - g_j\|_{L^p} = \left(\int_{B_{R_0}^{\complement}(0)} |u(x)|^p \, dx + \int_{B_{R_0}(0)} |u(x) - M_{R_1} u_j(x)|^p \, dx\right)^{\frac{1}{p}}$$

$$\leq \left(\int_{B_{R_0}^{\complement}(0)} |u(x)|^p \, dx \right.$$

$$\left. + \left(\left(\int_{B_{R_0}(0)} |u(x) - M_{R_1} u(x)|^p \, dx\right)^{\frac{1}{p}} + \left(\int_{B_{R_0}(0)} |M_{R_1} u(x) - M_{R_1} u_j(x)|^p \, dx\right)^{\frac{1}{p}}\right)^p\right)^{\frac{1}{p}}$$

$$\leq \left(\left(\frac{\epsilon}{3}\right)^p + \left(\|u - M_{R_1} u\|_{L^p} + \left(\left(\frac{\epsilon}{3}\right)^p \frac{1}{V(R_0)} \int_{B_{R_0}(0)} 1 \, dx\right)^{\frac{1}{p}}\right)^p\right)^{\frac{1}{p}}$$

$$\leq \left(\left(\frac{\epsilon}{3}\right)^p + \left(\frac{2\epsilon}{3}\right)^p\right)^{\frac{1}{p}} = \frac{(1 + 2^p)^{\frac{1}{p}}}{3} \epsilon < \epsilon,$$

i.e. the set $\{g_1, \ldots, g_M\}$ is indeed an ϵ-net for K and the theorem is proved.
□

Remark 12.18. Our proof of Theorem 12.17 is a modification and an adaption of the proof given in [90] for the case $p = 2$.

Example 12.19. Consider on $L^2(\mathbb{R})$ the sequence of functions $u_k(x) := \chi_{[0,2\pi]}(x) \sin kx$. From Example 12.11 we deduce that $\|u_k - u_l\| = 2\pi$, hence no subsequence of $(u_k)_{k \in \mathbb{N}}$ will ever be a Cauchy sequence, and hence $(u_k)_{k \in \mathbb{N}}$ has no accumulation point, i.e. $\{u_k \,|\, k \in \mathbb{N}\}$ is not relatively compact in $L^2(\mathbb{R})$. With the arguments of Example 12.11 it still follows that $(u_k)_{k \in \mathbb{N}}$ converges weakly in $L^2(\mathbb{R})$ to 0. The calculations of Example 12.11 also yields that $\|u_k\|_{L^2} = \sqrt{2\pi}$, i.e. the set $\{u_k \,|\, k \in \mathbb{N}\}$ is uniformly bounded in $L^2(\mathbb{R})$ and since $u_k|_{[0,2\pi]^\complement} = 0$, condition (12.14) in Theorem 12.17 holds. However the equi-continuity of $\{u_k \,|\, k \in \mathbb{N}\}$ fails to hold since

$$\int_{\mathbb{R}} |u_k(x+h) - u_k(x)|^2 \, dx = \int_0^{2\pi} |\sin k(x+h) - \sin kx|^2 \, dx$$

$$= 2\pi - 2\pi \cos hk$$

and therefore $\limsup_{h \to 0} (\sup_{k \in \mathbb{N}} \|T_h u_k - u_k\|_{L^2}) = 2\sqrt{\pi}$.

Problems

1. Let $G \subset \mathbb{R}^n$ be a bounded open set, $y_0 \in \mathbb{R}^n$, and $f : \overline{G} \to \mathbb{R}^n$ a C^1-mapping. Consider the equation $f(x) = y_0$ and suppose that no critical point of f solves this equation. Prove that then $f(x) = y_0$ can have at most finitely many solutions in G.

2. Let $h := [0,1] \to \mathbb{R}$ be a continuously differentiable function and denote by $\text{Ex}(h)$ the set of all local extreme values of h, i.e. $\text{Ex}(h) = \{h(x) | x \in [0,1], h \text{ has a local extreme value at } x\}$. Clearly we have $\text{Ex}(h) \subset h(\text{crit}(h|_{(0,1)})) \cup \{h(0), h(1)\}$. Prove the existence of a function $h \in C^1([0,1])$ such that $\text{Ex}(h)$ is finite but $h(\text{crit}(h|_{(0,1)}))$ is non-denumerable.
   Hint: consider for the Cantor set C the function $g : \mathbb{R} \to \mathbb{R}$, $g(x) := \text{dist}(x, C)$ and then investigate $h(x) := \int_0^x g(t) dt$, $x \in [0,1]$.

3. Why is the trace of every parametrized C^1-surface $f : G \to \mathbb{R}^3$, $G \subset \mathbb{R}^2$ open, a set of $\lambda^{(3)}$-measure zero?

4. Let $A \in \mathcal{B}^{(n)}$ be a bounded set and $s : A \to \mathbb{R}$ a simple function. Prove that for every $\epsilon > 0$ there exists a closed set $C_\epsilon \subset A$ such that $\lambda^{(n)}(A \setminus C_\epsilon) \leq \epsilon$ and $s|_{C_\epsilon}$ is continuous.

5. Let $f : A \to \mathbb{R}$, $A \in \mathcal{B}^{(n)}$. Suppose that f has the following property (compare with Lusin's theorem, Theorem 12.7): for every $\epsilon > 0$ there exists a closed set $C_\epsilon \subset A$ such that $\lambda^{(n)}(A \setminus C_\epsilon) < \epsilon$ and $f|_{C_\epsilon}$ is continuous. Prove that f is a measurable function.

6. Prove that if a subset Y of a metric space (X, d) has the property that for every $\delta > 0$ there exist closed balls $\overline{B_\delta(x_1)}, \ldots, \overline{B_\delta(x_N)}$ such that $Y \subset \bigcup_{j=1}^{N} \overline{B_\delta(x_j)}$ then Y admits for every $\epsilon > 0$ a finite ϵ-net.

7.   a) Prove that $(a, b) \subset \mathbb{R}$ has for every $\epsilon > 0$ a finite ϵ-net.

   b) Consider the Banach space $(C([0, 1]), \|\cdot\|_\infty)$ and its subset $Y := \{u \in C([0, 1]) \mid 0 = u(0) \le u(t) \le u(1) = 1\}$. Prove that $Y \subset C([0, 1])$ is bounded and that for $\epsilon = \frac{1}{2}$ the set Y admits a finite $\frac{1}{2}$-net.
Hint: use Problem 6 and consider the function $u_0 \in C([0, 1])$, $u_0(t) = \frac{1}{2}$ for all $t \in [0, 1]$. Further show that for $0 < \epsilon < \frac{1}{2}$ the set Y does not have an ϵ-net.
Hint: suppose that for $0 < \epsilon < \frac{1}{2}$ the set $\{u_1, \ldots, u_m\}$ is a finite ϵ-net for Y. Investigate the function $u : [0, 1] \to \mathbb{R}$ defined by $u(0) = 0$, $u(1) = 1$

$$u\left(\frac{j}{m+1}\right) := \begin{cases} 1, & u_j\left(\frac{j}{m+1}\right) \le \frac{1}{2} \\ 0, & u_j\left(\frac{j}{m+1}\right) > \frac{1}{2} \end{cases},$$

for $j = 1, \ldots, m$, and for $t \in \left(\frac{j}{m+1}, \frac{j+1}{m+1}\right)$, $j = 0, \ldots, m$, we define u as the linear function connecting $\left(j, u\left(\frac{j}{m+1}\right)\right)$ and $\left(j + 1, u\left(\frac{j+1}{m+1}\right)\right)$. This problem is taken from [7].

Part 7: Complex-valued Functions of a Complex Variable

13 The Complex Numbers as a Complete Field

Although we assume that the reader is familiar with the complex numbers $\mathbb{C}$ from studies about algebra or geometry, we want to discuss in some detail the algebraic properties of $\mathbb{C}$ as well as the fact that $\mathbb{C}$ equipped with the modulus or absolute value is metric in a complete metric space. In fact $\mathbb{C}$ is also a complete algebraic field in the sense that every polynomial factorises into linear factors, but for this we will give a proof much later in Chapter 23 and our proof will be analytic not algebraic. This chapter also serves to fix notations and to make a start with getting used to the traditional façon de parler in the theory of complex functions of a complex variable.

A complex number $z \in \mathbb{C}$ is written as

$$z = x + iy, \ x, y \in \mathbb{R} \tag{13.1}$$

where

$$x = \operatorname{Re} z \text{ and } y = \operatorname{Im} z \tag{13.2}$$

are the **real** and the **imaginary part** of z, respectively. Thus we can identify the complex numbers with pairs (x, y) of real numbers x, y, hence we can identify $\mathbb{C}$ with $\mathbb{R}^2$ using the bijective mapping

$$j : \mathbb{C} \mapsto \mathbb{R}^2, z = x + iy \mapsto j(z) = (x, y). \tag{13.3}$$

This identification will soon become very useful when discussing the topology of $\mathbb{C}$. However to handle the algebraic structure of $\mathbb{C}$ another identification is more useful, namely

$$h : \mathbb{C} \to H := \left\{ \begin{pmatrix} x & -y \\ y & x \end{pmatrix} \middle| x, y \in \mathbb{R} \right\} \subset M(2, \mathbb{R}) \tag{13.4}$$

$$z = x + iy \mapsto h(z) = \begin{pmatrix} x & -y \\ y & x \end{pmatrix}.$$

Let us study the algebraic structure of H with respect to the standard matrix operations. It is an easy exercise to see that H with the standard matrix addition forms an Abelian group where $\begin{pmatrix} -x & y \\ -y & -x \end{pmatrix}$ is inverse to $\begin{pmatrix} x & -y \\ y & x \end{pmatrix}$ and the neutral element with respect to addition is of course $\begin{pmatrix} 0 & 0 \\ 0 & 0 \end{pmatrix}$. Moreover,

$H \setminus \left\{ \begin{pmatrix} 0 & 0 \\ 0 & 0 \end{pmatrix} \right\}$ is also an Abelian group with respect to matrix multiplication where $\frac{1}{x^2+y^2} \begin{pmatrix} x & y \\ -y & x \end{pmatrix}$ is inverse to $\begin{pmatrix} x & -y \\ y & x \end{pmatrix}$ and $\begin{pmatrix} 1 & 0 \\ 0 & 1 \end{pmatrix}$ is the neutral element with respect to multiplication. Since matrix multiplication is in general not commutative it is worth noting for two elements of H that

$$\begin{pmatrix} x & -y \\ y & x \end{pmatrix} \begin{pmatrix} u & -v \\ v & u \end{pmatrix} = \begin{pmatrix} xu - yv & -xv - yu \\ xv + yu & xu - yv \end{pmatrix} = \begin{pmatrix} u & -v \\ v & u \end{pmatrix} \begin{pmatrix} x & -y \\ y & x \end{pmatrix},$$

i.e. restricted to H the multiplication of matrices is commutative. We also have the law of distributivity, i.e.

$$\begin{pmatrix} x & -y \\ y & x \end{pmatrix} \left(\begin{pmatrix} a & -b \\ b & a \end{pmatrix} + \begin{pmatrix} c & -d \\ d & c \end{pmatrix} \right)$$

$$= \left(\begin{pmatrix} x & -y \\ y & x \end{pmatrix} \begin{pmatrix} a & -b \\ b & a \end{pmatrix} \right) + \left(\begin{pmatrix} x & -y \\ y & x \end{pmatrix} \begin{pmatrix} c & -d \\ d & c \end{pmatrix} \right),$$

which is easily verified by inspection. Thus H with these operations is a commutative field.

Using $h^{-1} \left(\begin{pmatrix} x & -y \\ y & x \end{pmatrix} \right) = x + iy = z$, we now find with $z_k = x_k + iy_k = h^{-1} \left(\begin{pmatrix} x_k & -y_k \\ y_k & x_k \end{pmatrix} \right)$ that

$$z_1 + z_2 = (x_1 + x_2) + i(y_1 + y_2) \tag{13.5}$$

or

$$\operatorname{Re}(z_1 + z_2) = \operatorname{Re} z_1 + \operatorname{Re} z_2 \text{ and } \operatorname{Im}(z_1 + z_2) = \operatorname{Im} z_1 + \operatorname{Im} z_2, \tag{13.6}$$

as well as

$$z_1 + z_2 = z_2 + z_1, \quad z_1 \cdot z_2 = z_2 \cdot z_1, \tag{13.7}$$

$$(z_1 + z_2) + z_3 = z_1 + (z_2 + z_3), \quad (z_1 \cdot z_2)z_3 = z_1 \cdot (z_2 \cdot z_3), \tag{13.8}$$

and
$$z_1(z_2 + z_3) = z_1 z_2 + z_1 z_3. \tag{13.9}$$

The neutral element with respect to addition is given by

$$0 := 0 + i\,0 = h^{-1}\left(\begin{pmatrix} 0 & 0 \\ 0 & 0 \end{pmatrix}\right) \tag{13.10}$$

and the neutral element with respect to multiplication is given by

$$1 := 1 + i\,0 = h^{-1}\left(\begin{pmatrix} 1 & 0 \\ 0 & 1 \end{pmatrix}\right). \tag{13.11}$$

We will not introduce different symbols for the zero element (neutral element with respect to addition) in $\mathbb{R}, \mathbb{C}$ or H and usually we also take 0 as the symbol for the zero element in $\mathbb{R}^2$. Similarly 1 denotes the neutral element with respect to multiplication in $\mathbb{R}$ and $\mathbb{C}$, however the corresponding element in H (or $M_2(\mathbb{R})$) we often denote id or id_H.

For $z_1 \cdot z_2$ we find

$$z_1 \cdot z_2 = x_1 x_2 - y_1 y_2 + i(x_2 y_1 + x_1 y_2), \tag{13.12}$$

i.e.

$$\mathrm{Re}\, z_1 \cdot z_2 = x_1 x_2 - y_1 y_2, \, \mathrm{Im}\, z_1 \cdot z_2 = x_2 y_1 + x_1 y_2. \tag{13.13}$$

For $z_1 \neq 0$ it follows that

$$z_1^{-1} = \frac{1}{z_1} = \frac{x_1}{x_1^2 + y_1^2} + i\left(\frac{-y_1}{x_1^2 + y_1^2}\right) = \frac{x_1}{x_1^2 + y_1^2} - i\frac{y_1}{x_1^2 + y_1^2} \tag{13.14}$$

and we have

$$\frac{z_2}{z_1} = z_2 \cdot z_1^{-1} = z_1^{-1} \cdot z_2. \tag{13.15}$$

The most interesting relations are of course

$$i = 0 + i.1 = h^{-1}\left(\begin{pmatrix} 0 & -1 \\ 1 & 0 \end{pmatrix}\right) \tag{13.16}$$

implying

$$i^2 = -1 \tag{13.17}$$

where we adopted the notation $z_1^k = z_1 \cdot \ldots \ldots \cdot z_k$, for $k \in \mathbb{Z}$, with the convention $z_1^0 = 1$.

By $\operatorname{Re} H := \left\{ \begin{pmatrix} x & 0 \\ 0 & x \end{pmatrix} \middle| x \in \mathbb{R} \right\}$ a subfield of H is given and we can identify $\operatorname{Re} H$ via h^{-1} with $\mathbb{R}$, hence we can identify $\mathbb{R}$ with a subfield of $\mathbb{C}$ and we simply consider $\mathbb{R}$ as a subset, i.e. $\mathbb{R} \subset \mathbb{C}$, with the interpretation that $z \in \mathbb{R}$ means $z = x + i0$ or z is identified with $(x, 0)$, or as we started with $\begin{pmatrix} x & 0 \\ 0 & x \end{pmatrix}$.

When comparing or identifying $\mathbb{C}$ with $\mathbb{R}^2$, i.e. using the mapping j, we can use the additive group operation on $\mathbb{C}$ and the multiplication with scalars from $\mathbb{R} \subset \mathbb{C}$ to consider $\mathbb{C}$ as a two dimensional $\mathbb{R}$-vector space, but we do not have a multiplication on $\mathbb{R}^2$.

From now on, as in the beginning, we write $z = x + iy$ for a complex number, and for $z_1 \cdot z_2$ we write simply $z_1 z_2$. We will rarely make explicit use of the identification of $\mathbb{C}$ by H, and the identification of $\mathbb{C}$ by $\mathbb{R}^2$ will be used and is in fact of importance when dealing with the geometry and the topology of $\mathbb{C}$, but again, in the algebraic context it is rarely used.

On $\mathbb{C}$ we have a further algebraic operation, we may consider the **conjugate complex number**

$$\bar{z} := \overline{x + iy} = x - iy, \tag{13.18}$$

i.e.

$$\operatorname{Re} \bar{z} = \operatorname{Re} z, \ \operatorname{Im} \bar{z} = -\operatorname{Im} z. \tag{13.19}$$

Obviously we have

$$\overline{(\bar{z})} = z, \tag{13.20}$$

i.e. the mapping $z \mapsto \bar{z}$ is an **involution** and we can look at $\mathbb{C}$ as an algebra with involution. The following rules follow from the definition and are easily verified by inspection

$$\overline{z_1 + z_2} = \bar{z}_1 + \bar{z}_2, \tag{13.21}$$

$$\overline{z_1 \cdot z_2} = \bar{z}_1 \cdot \bar{z}_2, \tag{13.22}$$

264

$$\overline{\left(\frac{z_1}{z_2}\right)} = \frac{\bar{z}_1}{\bar{z}_2}, \ z_2 \neq 0, \tag{13.23}$$

$$z \cdot \bar{z} = (\text{Re } z)^2 + (\text{Im } z)^2 = x^2 + y^2, \tag{13.24}$$

$$\text{Re } z = \frac{z + \bar{z}}{2}, \ \text{Im } z = \frac{z - \bar{z}}{2i}, \tag{13.25}$$

see also Problem 3.

Note that for $z \neq 0$ we have

$$\frac{z \cdot \bar{z}}{z \cdot \bar{z}} = z \cdot \frac{\bar{z}}{z \cdot \bar{z}} = 1,$$

i.e.

$$z^{-1} = \frac{\bar{z}}{z \cdot \bar{z}} = \frac{x}{x^2 + y^2} - i\frac{y}{x^2 + y^2}, \tag{13.26}$$

where the last equality is of course (13.14).

It is important to become familiar in manipulating complex numbers. In particular, as we will suggest in some problems, i.e. Problems 1 and 2, to get experience with handling fractions of complex numbers in an efficient way.

Definition 13.1. *The **modulus** or **absolute value** $|z|$ of $z = x + iy \in \mathbb{C}$ is defined by*

$$|z| := \sqrt{z\bar{z}} = \sqrt{(\text{Re } z)^2 + (\text{Im } z)^2} = \sqrt{x^2 + y^2}. \tag{13.27}$$

Note that (13.27) implies

$$|\text{Re } z| \leq |z| \text{ and } |\text{Im } z| \leq |z| \tag{13.28}$$

and

$$|z|^2 = z\bar{z} \geq 0. \tag{13.29}$$

Furthermore we have

$$|z_1 \cdot z_2| = |z_1||z_2| \tag{13.30}$$

which follows from $(z_1 z_2)\overline{(z_1 z_2)} = (z_1 \bar{z}_1)(z_2 \bar{z}_2)$, and for $z_2 \neq 0$ we find

$$\left|\frac{z_1}{z_2}\right| = \frac{|z_1|}{|z_2|}. \tag{13.31}$$

For $z \in \mathbb{R}$, i.e. Im $z = 0$, we have of course

$$|z| = \sqrt{x^2} = |x|$$

which justifies the name absolute value.

Identifying $\mathbb{C}$ with $\mathbb{R}^2$, i.e. $z = x + iy$ with (x, y), we note that $|z|$ is nothing but the Euclidean norm $||(x, y)||_2 = \sqrt{(x^2 + y^2)}$. Consequently we have

$$|z| \geq 0 \text{ and } |z| = 0 \quad \text{if and only if } z = 0 \tag{13.32}$$

and the triangle inequality

$$|z_1 + z_2| \leq |z_1| + |z_2| \tag{13.33}$$

holds. Moreover, the $\mathbb{R}$-positive homogeneity of the Euclidean norm extends to multiplication with complex numbers, i.e. the equality

$$||\lambda(x, y)||_2 = |\lambda| ||(x, y)||_2 \text{ or } |\lambda z| = |\lambda| |z|$$

for $\lambda \in \mathbb{R}$ and $(x, y) \in \mathbb{R}^2$, $z = x + iy$, extends to

$$|z_1 \cdot z_2| = |z_1| |z_2|, \quad z_1, z_2 \in \mathbb{C}. \tag{13.34}$$

Definition 13.2. *Let V be a vector space over the field $\mathbb{C}$. A **norm** on V is a mapping $||.|| : V \to \mathbb{R}$ satisfying*

i) $||w|| \geq 0$ and $||w|| = 0$ if and only if $w = 0, w \in V$,

ii) $||zw|| = |z| ||w||, z \in \mathbb{C}, w \in V$,

iii) $||w_1 + w_2|| \leq ||w_1|| + ||w_2||$.

Whenever $||.||$ is a norm on V the **converse triangle inequality** holds

$$|||w_1|| - ||w_2||| \leq ||w_1 - w_2||. \tag{13.35}$$

Moreover by
$$d(w_1, w_2) := ||w_1 - w_2|| \tag{13.36}$$

a metric d is given on V, see Problem 6. Thus the modulus $|.|$ is a norm on $\mathbb{C}$ (which is a of course a $\mathbb{C}$-vector space of complex dimension 1) and hence it induces a metric d on $\mathbb{C}$. For $z_1, z_2 \in \mathbb{C}$ we find

$$d(z_1, z_2) = |z_1 - z_2| = \sqrt{\text{Re}(z_1 - z_2)^2 + \text{Im}(z_1 - z_2)^2}$$

or with $z_1 = (x_1, y_1)$ and $z_2 = (x_2, y_2)$ we have

$$d(z_1, z_2) = \sqrt{(x_1 - x_2)^2 + (y_1 - y_2)^2}, \tag{13.37}$$

i.e. d is just the Euclidean metric of $\mathbb{R}^2$.

We can now consider convergence in the metric space $(\mathbb{C}, d)$ which must correspond to convergence in the Euclidean space $\mathbb{R}^2$. Since we fix on $\mathbb{C}$ the metric d, we will write just $\mathbb{C}$ instead of $(\mathbb{C}, d)$ when meaning the metric space $(\mathbb{C}, d)$.

In order to get a better understanding of how to handle complex numbers, the following considerations are much more detailed than needed given our identification of the metric space $\mathbb{C}$ with the metric space $\mathbb{R}^2$. We start with

Definition 13.3. *Let $(z_n)_{n \in \mathbb{N}}$, $z_n \in \mathbb{C}$, be a sequence of complex numbers and $z \in \mathbb{C}$. We call z the **limit** of $(z_n)_{n \in \mathbb{N}}$ and we write*

$$\lim_{n \to \infty} z_n = z \tag{13.38}$$

if for every $\epsilon > 0$ there exists $N = N(\epsilon)$ such that

$$n \geq N(\epsilon) \text{ implies } |z_n - z| = d(z_n, z) < \epsilon.$$

Since in metric space every limit is unique, z is uniquely determined. Moreover, we call a sequence $(z_n)_{n \in \mathbb{N}}$ **bounded** if $|z_n| \leq M$ for all $n \in \mathbb{N}$ and some $M \geq 0$, and it follows that every convergent sequence is bounded.

Lemma 13.4. *Let $(z_n)_{n \in \mathbb{N}}$, $z_n = x_n + iy_n \in \mathbb{C}$, and $z = x + iy$ be the limit of $(z_n)_{n \in \mathbb{N}}$. Then we have*

$$\lim_{n \to \infty} x_n = x \text{ and } \lim_{n \to \infty} y_n = y. \tag{13.39}$$

Proof. First we recall (13.28), i.e. for every complex number $w \in \mathbb{C}$ the following holds
$$|\text{Re } w| \leq \sqrt{(\text{Re } w)^2 + (\text{Im } w)^2} = |w|$$
and
$$|\text{Im } w| \leq \sqrt{(\text{Re } w)^2 + (\text{Im } w)^2} = |w|.$$

Now we find

$$|x_n - x| \leq |z_n - z| \tag{13.40}$$

as well as

$$|y_n - y| \leq |z_n - z|. \tag{13.41}$$

Hence, given $\epsilon > 0$ and chosen $N(\epsilon)$ such that $n \geq N(\epsilon)$ implies $|z_n - z| < \epsilon$ it follows also for $n \geq N(\epsilon)$ that $|x_n - x| < \epsilon$ and $|y_n - y| < \epsilon$, i.e.

$$\lim_{n \to \infty} x_n = x \text{ and } \lim_{n \to \infty} y_n = y. \tag{13.42}$$

$\square$

Suppose now that two sequences $(x_n)_{n \in \mathbb{N}}, (y_n)_{n \in \mathbb{N}}$ of real numbers $x_n, y_n \in \mathbb{R}$ are given and that with $x, y \in \mathbb{R}$ we have in $\mathbb{R}$ that $\lim_{n \to \infty} x_n = x$ and $\lim_{n \to \infty} y_n = y$. For the sequence $(z_n)_{n \in \mathbb{N}}$ of complex numbers $z_n := x_n + iy_n$ we find

$$|z_n - z|^2 = |x_n - x|^2 + |y_n - y|^2$$

or

$$|z_n - z| \leq (|x_n - x| + |y_n - y|).$$

Thus, given $\epsilon > 0$ we can find $N = N(\epsilon)$ such that $n \geq N(\epsilon)$ implies $|x_n - x| < \frac{\epsilon}{2}$ as well as $|y_n - y| < \frac{\epsilon}{2}$ and consequently $|z_n - z| < \epsilon$, i.e. $\lim_{n \to \infty} z_n = z$.
Hence, we have proved.

Proposition 13.5. *A sequence of complex numbers* $(z_n)_{n \in \mathbb{N}}$ *converges to a complex number* z *if and only if the sequence* $(\operatorname{Re} z_n)_{n \in \mathbb{N}}$ *converges to* $\operatorname{Re} z$ *and the sequence* $(\operatorname{Im} z_n)_{n \in \mathbb{N}}$ *to* $\operatorname{Im} z$.

From here it is easy to derive the basic statements for the algebra of limits in $\mathbb{C}$.

Proposition 13.6. *Let* $(z_n)_{n \in \mathbb{N}}$ *and* $(w_n)_{n \in \mathbb{N}}$ *be two sequences of complex numbers* $z_n = x_n + iy_n \in \mathbb{C}$ *and* $w_n = u_n + iv_n \in \mathbb{C}$. *Suppose that* $(z_n)_{n \in \mathbb{N}}$ *converges to* z *and that* $(w_n)_{n \in \mathbb{N}}$ *converges to* w. *Then we have*

$$\lim_{n \to \infty} (z_n \pm w_n) = z \pm w, \tag{13.43}$$

$$\lim_{n \to \infty} z_n \cdot w_n = z \cdot w, \tag{13.44}$$

and if $w \neq 0$ *it follows that*

$$\lim_{n \to \infty} \left| \frac{z_n}{w_n} \right| = \frac{z}{w}. \tag{13.45}$$

Proof. We will reduce each statement to a statement involving only the sequences of the real and imaginary parts and then we may apply our results to sequences of real numbers. Of course we have

$$z_n \pm w_n = (x_n \pm u_n) + i(y_n \pm v_n) \tag{13.46}$$

and since $(x_n \pm u_n)_{n \in \mathbb{N}}$ converges to $x \pm u$, and $(y_n \pm v_n)_{n \in \mathbb{N}}$ converges to $y \pm v$, the first statement follows.
Now

$$z_n \cdot w_n = x_n \cdot u_n - y_n v_n + i(u_n y_n + x_n v_n)$$

and since

$$\lim_{n \to \infty} (x_n u_n - y_n v_n) = xu - yv = \operatorname{Re}(z \cdot w)$$

and

$$\lim_{n \to \infty} (y_n u_n + x_n v_n) = yu + xv = \operatorname{Im}(z \cdot w)$$

the second statement follows.
Finally we observe for $w_n \neq 0$

$$\frac{z_n}{w_n} = \frac{z_n \bar{w}_n}{|w_n|^2}$$

$$= \frac{x_n u_n + y_n v_n + i(y_n u_n - x_n v_n)}{u_n^2 + v_n^2}$$

$$= \frac{x_n u_n + y_n v_n}{u_n^2 + v_n^2} + i\left(\frac{y_n u_n - x_n v_n}{u_n^2 + v_n^2}\right).$$

Since $|w|^2 = u^2 + v^2 \neq 0$ and $\lim_{n \to \infty}(u_n^2 + v_n^2) = u^2 + v^2$ we deduce first that $w_n \neq 0$ for n sufficiently large, hence $u_n^2 + v_n^2 \neq 0$ for n sufficiently large, and now it follows that

$$\lim_{n \to \infty} \frac{z_n}{w_n} = \frac{z}{w}.$$

$\square$

Corollary 13.7. *Let* $(z_n)_{n \in \mathbb{N}}$, $z_n = x_n + iy_n$, *be a sequence of complex numbers converging to* $z = x + iy \in \mathbb{C}$. *Then we have*

$$\lim_{n \to \infty} \bar{z}_n = \bar{z} \tag{13.47}$$

and

$$\lim_{n \to \infty} |z_n| = |z|. \tag{13.48}$$

Proof. We need only to note that $\bar{z}_n = x_n - iy_n$ and that $|z_n| = \sqrt{x_n^2 + y_n^2}$ and that the square root is on $\mathbb{R}_+$ a continuous function. $\square$

As in $\mathbb{R}$ or in any metric space we can introduce the notion of a Cauchy sequence of complex numbers.

Definition 13.8. *Let $(z_n)_{n \in \mathbb{N}}$ be a sequence of complex numbers $z_n \in \mathbb{C}$. We call $(z_n)_{n \in \mathbb{N}}$ a **Cauchy sequence** if for every $\epsilon > 0$ there exists $N = N(\epsilon) \in \mathbb{N}$ such that $n, m \geq N(\epsilon)$ implies $|z_n - z_m| < \epsilon$.*

Corollary 13.9. *Let $(z_n)_{n \in \mathbb{N}}$ be a sequence of complex numbers $z_n = x_n + iy_n$. The sequence $(z_n)_{n \in \mathbb{N}}$ is a **Cauchy sequence** if and only if the sequences $(x_n)_{n \in \mathbb{N}}$ and $(y_n)_{n \in \mathbb{N}}$ are Cauchy sequences.*

Proof. We just have to note the inequalities

$$|\text{Re}z| \leq |z|, \ |\text{Im}z| \leq |z|, \ |z| \leq (|\text{Re}z| + |\text{Im}z|)$$

to deduce that $|z_n - z_m| < \epsilon$ implies

$$|x_n - x_m| < \epsilon \text{ and } |y_n - y_m| < \epsilon$$

as well as $|x_n - x_m| < \frac{\epsilon}{2}$ and $|y_n - y_m| < \frac{\epsilon}{2}$ will imply

$$|z_n - z_m| < \epsilon.$$

$\square$

Thus $(z_n)_{n \in \mathbb{N}}$ is a Cauchy sequence in $\mathbb{C}$ if and only if its real parts $(x_n)_{n \in \mathbb{N}}$ and its imaginary parts $(y_n)_{n \in \mathbb{N}}$ form a Cauchy sequence in $\mathbb{R}$. However, in $\mathbb{R}$ every Cauchy sequence has a limit, i.e. $\mathbb{R}$ is complete. With Proposition 13.5 we therefore arrive at

Theorem 13.10. *The completeness of $\mathbb{R}$ implies the completeness of $\mathbb{C}$, i.e. every Cauchy sequence $(z_n)_{n \in \mathbb{N}}$ in $\mathbb{C}$ has a limit $z \in \mathbb{C}$.*

As in the case of the real numbers we can introduce series of complex numbers. Let $(z_n)_{n \in \mathbb{N}}$ be a sequence of complex numbers $z_n = x_n + iy_n$. We can construct a new sequence $(S_n)_{n \in \mathbb{N}}$ by

$$S_n := \sum_{k=1}^{n} z_k = \sum_{k=1}^{n} x_k + i \sum_{k=1}^{n} y_k. \tag{13.49}$$

Thus

$$\operatorname{Re} S_n = \sum_{k=1}^{n} \operatorname{Re} z_k \text{ and } \operatorname{Im} S_n = \sum_{k=1}^{n} \operatorname{Im} z_k.$$

Again we call S_n the nth **partial sum** of the (infinite) series $\sum_{k=1}^{\infty} z_k$ with the understanding that $\sum_{k=1}^{\infty} z_k$ should be read as the symbol of $(S_n)_{n \in \mathbb{N}}$. However, in the case where

$$\lim_{n \to \infty} S_n = S \in \mathbb{C}$$

exists, then $\sum_{k=1}^{\infty} z_k$ also means this limit, i.e. the complex number S and we write

$$\sum_{k=1}^{\infty} z_k = S.$$

Further, as in the case of series of real numbers we can consider series such as $\sum_{k=m}^{\infty} z_k$, $m \in \mathbb{Z}$ fixed. Operating with complex series of the type $\sum_{k=m}^{\infty} z_k$ is similar to the case of series of real numbers.

Clearly, $\sum_{k=1}^{\infty} z_k$ converges if and only if $\sum_{k=1}^{\infty} \operatorname{Re} z_k$ and $\sum_{k=1}^{\infty} \operatorname{Im} z_k$ converge and we have

$$\sum_{k=1}^{\infty} z_k = \sum_{k=1}^{\infty} \operatorname{Re} z_k + i \sum_{k=1}^{\infty} \operatorname{Im} z_k.$$

We call the series $\sum_{k=1}^{\infty} z_k$ **absolutely convergent** if the series $\sum_{k=1}^{\infty} |z_k|$ converges.

Now it is easy to transfer certain results for series of real numbers to series of complex numbers. We start with the **Cauchy criterion** for (complex) series, for the real case see Theorem I.18.1.

Theorem 13.11. *Given a sequence $(z_n)_{n \in \mathbb{N}}$ of complex numbers. The series $\sum_{k=1}^{\infty} z_k$ converges if and only if for every $\epsilon > 0$ there exists $N = N(\epsilon) \in \mathbb{N}$ such that $n \geq m > N$ implies*

$$\left| \sum_{k=m}^{n} z_k \right| < \epsilon.$$

Proof. Let $S_p := \sum_{k=1}^{p} z_k$ be the pth partial sum. It follows that

$$S_n - S_{m-1} = \sum_{k=m}^{n} z_k,$$

271

and the criterion is nothing but the Cauchy criterion for the sequence of partial sums. □

As in the case of series of real numbers, see Theorem I.18.2, we have

Theorem 13.12. *If the series $\sum_{k=1}^{\infty} z_k$ converges, then $\lim_{n \to \infty} z_n = 0$.*

Proof. If $\sum_{k=1}^{\infty} z_k$ converges then by Theorem 13.11, for any $\epsilon > 0$ it follows that $|\sum_{k=m}^{n} z_k| < \epsilon$ for $n \geq m \geq N$. Taking $n = m$ we find $|z_n| < \epsilon$ for $n \geq N$, i.e. $\lim_{n \to \infty} z_n = 0$. □

Of great importance is that we can extend the result about the limit of the geometric series from $\mathbb{R}$ to $\mathbb{C}$:

Theorem 13.13. *For $z \in \mathbb{C}, |z| < 1$, the geometric series $\sum_{k=0}^{\infty} z^k$ converges to $\frac{1}{1-z}$, i.e.*

$$\sum_{k=0}^{\infty} z^k = \frac{1}{1-z}, \quad |z| < 1. \tag{13.50}$$

Moreover, the geometric series converges absolutely.

Proof. As in the real case we find compare with Theorem I. 16.4

$$\sum_{k=0}^{n} z^k = \frac{1 - z^{n+1}}{1 - z}. \tag{13.51}$$

Now we note that

$$\lim_{n \to \infty} |z^{n+1}| = |z| \lim_{n \to \infty} |z|^n = 0$$

since $|z| < 1$. This implies however that $\lim_{n \to \infty} z^{n+1} = 0$ and passing in (13.51) to the limit gives the convergence. Further we note that

$$\sum_{k=0}^{\infty} |z^k| = \sum_{k=0}^{\infty} |z|^k = \frac{1}{1 - |z|}, \quad |z| < 1,$$

which implies the absolute convergence. □

More generally, as in the real case, see Theorem I.18.11, we have

Theorem 13.14. *Any absolutely convergent series of complex numbers is convergent.*

Proof. Suppose that $\sum_{k=1}^{\infty} z_k$ converges absolutely. According to the Cauchy criterion applied to the series $\sum_{k=1}^{\infty} |z_k|$, for $\epsilon > 0$ there exists $N(\epsilon) \in \mathbb{N}$ such that $n \geq m \geq N(\epsilon)$ implies

$$\sum_{k=m}^{n} |z_k| < \epsilon.$$

Now the triangle inequality yields for $n \geq m \geq N(\epsilon)$

$$|\sum_{k=m}^{n} z_k| \leq \sum_{k=m}^{n} |z_k| < \epsilon,$$

i.e. the Cauchy criterion holds also for $\sum_{k=1}^{\infty} z_k$ implying convergence.   □

Theorem 13.15 (Comparison test). *Let $\sum_{k=1}^{\infty} c_k$ be a convergent series of non-negative real numbers $c_k \geq 0$. Further let $(z_k)_{k \in \mathbb{N}}$, $z_k \in \mathbb{C}$, be a series of complex numbers such that $|z_k| \leq c_k$ for $k \in \mathbb{N}$(or $k \geq N_0$). Then the series $\sum_{k=1}^{\infty} z_k$ converges absolutely.*

Proof. Given $\epsilon > 0$ there exists $N(\epsilon) \in \mathbb{N}$ such that

$$|\sum_{k=m}^{n} c_k| = \sum_{k=m}^{n} c_k < \epsilon \text{ for } n \geq m \geq N(\epsilon).$$

Therefore we find

$$\sum_{k=m}^{n} |z_k| \leq \sum_{k=m}^{n} c_k < \epsilon \text{ for } n \geq m \geq N(\epsilon)$$

which proves the result.   □

Finally we prove the **ratio test** for the complex series, see Theorem I.18.14 for the real case.

Theorem 13.16. *Let $\sum_{k=0}^{\infty} z_k$ be a series of complex numbers such that $z_n \neq 0$ for all $n \geq N_0$. Suppose that there exists $\nu, 0 < \nu < 1$, such that*

$$|\frac{z_{k+1}}{z_k}| \leq \nu \text{ for all } k \geq N_0.$$

Then the series $\sum_{k=0}^{\infty} z_k$ converges absolutely.

Proof. The convergence of the series $\sum_{k=0}^{\infty} z_k$ does not depend on the first N_0 terms. Now

$$\left|\frac{z_{k+1}}{z_k}\right| \leq \nu \ for \ k \geq N_0$$

implies $|z_{N_0+n}| \leq |z_{N_0}|\nu^n$. Since $0 < \nu < 1$ the series $\sum_{n=N_0}^{\infty} \nu^n$ converges, and by Theorem 13.15 the result follows. □

Problems

1. Simplify the following expressions:

   a) $\frac{3+5i}{2i-7} + \frac{4i}{3+i}$;

   b) $\frac{(2i)^8 - 128}{(1+2i)(1-2i)}$;

   c) $\frac{\frac{1-i}{2+3i}}{\frac{2+4i}{6-2i}}$;

   d) $\left(-\frac{1}{2} + \frac{\sqrt{3}}{2}i\right)^2$.

2. Verify:

   a) $\frac{(a+ib)^2 - 2iab}{i(a-b)} = -i(a+b), \quad a, b \in \mathbb{R}, a \neq b$;

   b) For $p, q \in \mathbb{R}, \frac{p^2}{4} < q$, the numbers $z_1 = -\frac{p}{2} + i\sqrt{q - \frac{p^2}{4}}$ and $z_2 = -\frac{p}{2} - i\sqrt{q - \frac{p^2}{4}}$ are solutions to the equation $z^2 + pz + q = 0$.

3. Verify the following for complex numbers z_1, z_2 and $z = x + iy$:

   a) $\overline{z_1 \cdot z_2} = \overline{z_1} \cdot \overline{z_2}$;

   b) $z \cdot \bar{z} = x^2 + y^2$;

   c) $\operatorname{Re} z = \frac{z + \bar{z}}{2}$ and $\operatorname{Im} z = \frac{z - \bar{z}}{2i}$.

4. Formulate and indicate the proof of the binomial theorem for two complex numbers.

5. Prove the following inequalities:

   a) $||z| - |w|| \leq |z - w| \wedge |z + w|$, recall that $a \wedge b = \min\{a, b\}$;

   b) $\frac{|x| + |y|}{\sqrt{2}} \leq |z| \leq |x| + |y|$, for $z = x + iy$.

6.   a) Prove that if V is a vector space over $\mathbb{C}$ and $\|\cdot\|$ is a norm on V then $d(f, g) := \|f - g\|$, $f, g \in V$, is a metric on V.

b) For $z = (z_1, \ldots, z_n) \in \mathbb{C}^n$ define $\|z\| := \left(\sum_{j=1}^{n} |z_j|^2\right)^{\frac{1}{2}}$ and prove that this gives a norm on $\mathbb{C}^n$.

7. Find the following limits:

a) $\lim\limits_{n\to\infty} \dfrac{3n^2 - 5in}{n^2 i}$;

b) $\lim\limits_{k\to\infty} \dfrac{(k - ik)^2(2k + 5i)}{k^2 + (3 + ik)^3}$;

c) $\lim\limits_{N\to\infty} \sum\limits_{k=4}^{N} z^k$, $|z| < 1$, $N \geq 4$.

8. Prove the convergence of the following power series for the range of $z \in \mathbb{C}$ as indicated. You may use any criterion for convergence we have discussed.

a) $\sum\limits_{k=0}^{\infty} \dfrac{z^k}{k!}$, $z \in \mathbb{C}$;

b) $\sum\limits_{k=0}^{\infty} (-1)^k \dfrac{z^{2k}}{(2k)!}$, $z \in \mathbb{C}$;

c) $\sum\limits_{k=1}^{\infty} (-1)^{k-1} \dfrac{z^{2k-1}}{(2k-1)!}$, $z \in \mathbb{C}$;

d) $\sum\limits_{k=1}^{\infty} (-1)^{k-1} \dfrac{z^k}{k}$, $|z| < 1$;

e) $\sum\limits_{k=1}^{\infty} (-1)^{k-1} \dfrac{z^{2k-1}}{2k-1}$, $|z| < 1$.

9. Let $\sum_{k=0}^{\infty} a_k$ and $\sum_{k=0}^{\infty} b_k$ be two absolutely convergent series of complex numbers. Define their **Cauchy product** as $\sum_{n=0}^{\infty} c_n$ where $c_n := \sum_{k=0}^{n} a_{n-k} b_k$. Prove that the Cauchy product converges absolutely and find its value.
Hint: compare with Theorem I.29.21.

10. Define for $z \in \mathbb{C}$ the number $e^z := \sum_{k=0}^{\infty} \dfrac{z^k}{k!}$ which is justified by Problem 8 a). For $z_1, z_2 \in \mathbb{C}$ prove with the help of Problem 4 and Problem 9 the functional equation of the exponential function: $e^{z_1 + z_2} = e^{z_1} e^{z_2}$.

11. Let $(a_k)_{k \in \mathbb{N}}$ be a sequence of complex numbers $a_k \neq 0$ and set $P_N := \prod_{k=1}^{N} a_k$. In the case where $P := \lim_{N \to \infty} P_N$ exists, we write $P = \prod_{k=1}^{\infty} a_k$ and call P the **infinite product** of the sequence $(a_k)_{k \in \mathbb{N}}$. Prove:

a) If $\prod_{k=1}^{\infty} a_k$ converges then $\lim_{k \to \infty} a_k = 1$;

b) If the product $\prod_{k=1}^{\infty} a_k$ converges then for every $\epsilon > 0$ there exists $N = N(\epsilon) \in \mathbb{N}$ such that $n > m > N(\epsilon)$ implies

$$\left| \prod_{k=m+1}^{n} a_k - 1 \right| < \epsilon.$$

(Compare with Proposition I.30.7.)

12. Let $(a_k)_{k \in \mathbb{N}}$, $a_k \neq -1$, be a sequence of complex numbers. We say that the infinite product $\prod_{k=1}^{\infty} (1 + a_k)$ converges absolutely if the product $\prod_{k=1}^{\infty} (1 + |a_k|)$ converges.

a) Prove that if $\prod_{k=1}^{\infty} (1 + a_k)$ converges absolutely, then it converges.

b) Prove that $\prod_{k=1}^{\infty} (1 + a_k)$ converges absolutely if and only if the series $\sum_{k=1}^{\infty} |a_k|$ converges, i.e. if and only if the series $\sum_{k=1}^{\infty} a_k$ converges absolutely.

Hint: compare with Proposition I.30.9 and Proposition I.30.10.

14 A Short Digression: Complex-valued Mappings

Part 7 is essentially devoted to functions defined on some subset $D \subset \mathbb{C}$ with values in $\mathbb{C}$. The fact that $\mathbb{C}$ is a field allows us to consider difference quotients $\frac{f(z)-f(z_0)}{z-z_0}$ and hence complex differentiation which is significantly different from differentiating functions $g : G \to \mathbb{R}^2, G \subset \mathbb{R}^2$.

However, before turning to this theory, in this chapter we want to discuss mappings $f : X \to \mathbb{C}$ where X might be just a non-empty set or a subset of $\mathbb{R}^n$, $n \geq 1$, or a metric space, etc. We want to see how the additional algebraic structure of $\mathbb{C}$ gives new results or allows to reformulate results for mappings $f : X \to \mathbb{R}^2$. Clearly, the set of all mappings $f : X \to \mathbb{C}, X \neq \emptyset$, equipped with the natural operations

$$(f \pm g)(x) = f(x) + g(x) \qquad \text{(addition of complex numbers)}$$
$$(\lambda f)(x) = \lambda f(x) \qquad (\lambda \in \mathbb{C})$$
$$(f \cdot g)(x) = f(x)g(x) \qquad \text{(multiplication of complex numbers)},$$

is an algebra over $\mathbb{C}$ which allows the involution $f \mapsto \overline{f}$. Since for $x \in X$ a complex number is given by $f(x)$ we can introduce its real and imaginary part $u(x)$ and $v(x)$, respectively, and we write

$$f(x) = u(x) + i\,v(x) \tag{14.1}$$

where $u, v : X \to \mathbb{R}$ are real-valued mappings. For two mappings $f_k : X \to \mathbb{C}$, $f_k = u_k + i\,v_k$, we find

$$\text{Re}\,(f_1 + f_2) = u_1 + u_2 \ \text{ and } \ \text{Im}\,(f_1 + f_2) = v_1 + v_2. \tag{14.2}$$

Moreover, the modulus of $f = u + iv$ is given at $x \in X$ by

$$|f(x)| = (u^2(x) + v^2(x))^{1/2}. \tag{14.3}$$

For a sequence $(f_k)_{k \in \mathbb{N}}$, $f_k : X \to \mathbb{C}$, we can define **pointwise convergence** to a function $f : X \to \mathbb{C}$ by

$$\lim_{k \to \infty} f_k(x) = f(x) \ \text{ for all } x \in X, \tag{14.4}$$

and **uniform convergence** on X to f by

$$\lim_{k \to \infty} \left(\sup_{x \in X} |f_k(x) - f(x)| \right) = \lim_{k \to \infty} ||f_k - f||_\infty = 0, \tag{14.5}$$

where we have introduced the sup-norm of $f : X \to \mathbb{C}$ as

$$||f||_\infty := \sup_{x \in X} |f(x)|. \tag{14.6}$$

In Problem 1 we will discuss norms and scalar products on vector spaces over $\mathbb{C}$ and see that $||f||_\infty$ is indeed a norm on the vector space of all mappings $f : X \to \mathbb{C}$ satisfying that $||f||_\infty < \infty$.

Depending on additional structures on X we may carry over further properties and ideas from the real case to the complex case. The first case we want to discuss is the case where on X a σ-field and a measure is given. Instead of writing X we now deal with a measure space $(\Omega, \mathscr{A}, \mu)$ and look at complex-valued mappings $f : \Omega \to \mathbb{C}$. If not stated otherwise we consider on $\mathbb{C}$ the Borel σ-field $\mathcal{B}^{(2)}$ induced when identifying $\mathbb{C}$ with the Euclidean space $\mathbb{R}^2$. It follows immediately that $f : \Omega \to \mathbb{C}, f = u + iv$, is $\mathcal{A}/\mathcal{B}^{(2)}$ measurable if and only if u and v are $\mathcal{A}/\mathcal{B}^{(1)}$ measurable. For this we only need to note that the projections $\mathrm{pr}_1 : \mathbb{C} \to \mathbb{R}, z \mapsto \mathrm{Re}\, z$, $\mathrm{pr}_2 : \mathbb{C} \to \mathbb{R}, z \to \mathrm{Im}\, z$, are measurable since after identification of $\mathbb{C}$ with $\mathbb{R}^2$ they coincide with the projections $\mathrm{pr}_k : \mathbb{R}^2 \to \mathbb{R}, k = 1, 2$. In the case where f, hence u and v, are measurable it is natural to extend integrability, more precisely μ-integrability, of f by

$$\int f(w)\, \mu(dw) := \int u(w)\, \mu(dw) + i \int v(w)\, \mu(dw), \tag{14.7}$$

which we define for μ-integrable function u and v. Note that $\int f\, d\mu$ is now a complex number. All μ-integrable complex-valued functions $f : \Omega \to \mathbb{C}$ form a vector space and f is μ-integrable if and only if $|f|$ is μ-integrable. We can further define the (semi-)norms

$$||f||_{L^p} := \left(\int |f(w)|^p \mu(dw) \right)^{1/p}, \tag{14.8}$$

note that

$$||f||_{L^p} = \left(\int (|u(w)|^2 + |v(w)|^2)^{p/2} \mu(dw) \right)^{1/p}$$

where $f = u + iv$. This leads to the complex vector spaces $\mathcal{L}^p(\Omega\,;\mathbb{C})$ and $L^p(\Omega\,;\mathbb{C})$, and if no confusion may arise we often will write just as before $\mathcal{L}^p(\Omega)$ and $L^p(\Omega)$, respectively. As we will see in Problem 2, typical inequalities such as Hölder's inequality or the Minkowski's inequality carry over to the case of complex-valued mappings.

Moreover, on $L^2(\Omega; \mathbb{C})$ a scalar product is given by

$$< f, g >= \int f(w) \, \overline{g(w)} \, \mu(dw). \tag{14.9}$$

Indeed, we can also prove that for $1 \leq p < \infty$ the spaces $L^p(\Omega; \mathbb{C})$ are Banach spaces, i.e. complete with respect to the norm $||.||_{L^p}$.

Next we assume that X carries some topology $\mathcal{O}_X$. In this case continuity of $f : X \to \mathbb{C}$ is well defined when on $\mathbb{C}$ we have as usual chosen the 2 - dimensional Euclidean topology induced by the identification of $\mathbb{C}$ with $\mathbb{R}^2$. As expected we find that $f = u + iv$ is continuous if and only if u and v are continuous. In the following assume that $(X, \mathcal{O}_X)$ is a Hausdorff space so that compactness is well defined. If $K \subset X$ is compact and $f : X \to \mathbb{C}$ is continuous then $f|_K$ is bounded, i.e. $|f(x)| \leq M$ for all $x \in K$ and some $M < \infty$. Further, if X is even a metric space and $K \subset X$ compact, then $f|_K$ is uniformly continuous. If X is a Hausdorff space and $f_k : X \to \mathbb{C}$ is a sequence of functions we call $(f_k)_{k \in \mathbb{N}}$ **locally uniform** or **uniformly on compact sets convergent** to $f : X \to \mathbb{C}$ if for every compact set $K \subset X$ the following holds

$$\lim_{k \to \infty} \sup_{X \in K} |f_k(x) - f(x)| = 0. \tag{14.10}$$

Locally uniform convergence will become a very important notion later on when studying complex-differentiable mappings $f : D \to \mathbb{C}, D \subset \mathbb{C}$.

In general we denote by $C(X; \mathbb{C})$ the set of all continuous functions $f : X \to \mathbb{C}$ which is of course a vector space over $\mathbb{C}$, and as before, if no confusion may arise we just write $C(X)$. The space $C_b(X; \mathbb{C})$ is the subspace of all bounded continuous function on X and equipped with the sup-norm $||.||_\infty$ it is again a Banach space. For a Hausdorff space $(X, \mathcal{O}_X)$ we denote by $C_0(X; \mathbb{C})$ the space of all continuous functions $f : X \to \mathbb{C}$ with compact support, i.e. $\operatorname{supp} f := \{x \in X | f(x) \neq 0\}$ is compact for $f \in C_0(X; \mathbb{C})$.

In the topological space $(X, \mathcal{O}_X)$ we can always consider the Borel σ - field $\mathcal{B}(X) := \sigma(\mathcal{O}_X)$. Now the properties of $\mathcal{O}_X$ will determine the relations of the spaces $L^p(X; \mathbb{C}), C(X; \mathbb{C}), C_b(X; \mathbb{C})$ or $C_0(X; \mathbb{C})$ and we do not want to discuss these details now, some will be discussed in Part 12, i.e. Volume IV. In the special case of $\mathbb{R}^n$ equipped with the Euclidean topology we can also consider the space $C_\infty(\mathbb{R}^n; \mathbb{C})$ of all continuous complex-valued funtions $f : \mathbb{R}^n \to \mathbb{C}$ vanishing at infinity, i.e. functions for which we have

$$\lim_{||x|| \to \infty} |f(x)| = 0.$$

Finally, if $G \subset \mathbb{R}^n$, $\overset{\circ}{G} \neq \emptyset$, we can consider functions $f : G \to \mathbb{C}$ with real partial derivatives, i.e. for $\alpha \in \mathbb{N}_0^n$ we may look at

$$D^\alpha f(x) = D^\alpha u(x) + i D^\alpha v(x). \tag{14.11}$$

Thus spaces such as $C^m(G; \mathbb{C})$ or $C_b^m(G; \mathbb{C})$ as well as $C_0^m(G; \mathbb{C}), C^\infty(G; \mathbb{C})$ or $C_0^\infty(G; \mathbb{C})$ are all well defined. But even in the case $n = 2$ we make and have to make a strict distinction between functions having some real partial derivatives at a point (or in G) and functions admitting a complex derivative as we will introduce it soon.

Once more, if $H(X; \mathbb{C})$ is a certain space of functions $f : X \to \mathbb{C}$ and no confusion may arise, we will write simply $H(X)$.

Problems

1.   a) For $z, w \in \mathbb{C}^n$ define the unitary scalar product by $\langle z, w \rangle := \sum_{k=1}^n z_k \overline{w_k}$ and prove the Cauchy-Schwarz inequality

$$|\langle z, w \rangle| \leq \|z\| \|w\|$$

where $\|z\| = \left(\sum_{k=1}^n |z_k|^2 \right)^{\frac{1}{2}}$.

   b) Show that for a compact set $K \subset \mathbb{R}^n$ norms are given on $C(K; \mathbb{C})$ by $\|u\|_\infty := \sup_{x \in K} |u(x)|$ and $\|u\|_{L^2} := \left(\int_K |u(x)|^2 dx \right)^{\frac{1}{2}}$.

2. Let $f, g \in C(K; \mathbb{C})$ where $K \subset \mathbb{R}^n$ is a compact set. Sketch a proof of the Hölder and the Minkowski inequalities, i.e.

$$\int_K |f(x)g(x)| dx \leq \|f\|_{L^p} \|g\|_{L^q}, \quad 1 < p < \infty \text{ and } \frac{1}{p} + \frac{1}{q} = 1,$$

and

$$\|f + g\|_{L^p} \leq \|f\|_{L^p} + \|g\|_{L^p}, 1 \leq p < \infty.$$

280

3. Let $(\Omega, \mathcal{A})$ be a measurable space and $f : \Omega \to \mathbb{C}$ a function. Prove that f is measurable if and only if $\mathrm{Re}\, f$ and $\mathrm{Im}\, f$ are measurable.

4. Let $\emptyset \neq G \subset \mathbb{R}^n$ be an open set and $f : G \to \mathbb{C}$, $f = u + iv$, a function. Define differentiability of f at x_0 as follows: there exists $a \in \mathbb{C}^n$ and φ_{x_0} defined in a neighbourhood of x_0 such that $\lim\limits_{x \to x_0} \dfrac{\varphi_{x_0}(x)}{\|x - x_0\|} = 0$ and $f(x) = f(x_0) = \langle a, (x - x_0)\rangle + \varphi_{x_0}(x)$ in the domain of φ_{x_0}.

a) Prove that $a_k := \frac{\partial f}{\partial x_k}(x_0), 1 \le k \le n$.

b) Prove that $f : G \to \mathbb{C}$ is differentiable at $x_0 \in G$ if and only if $F : G \to \mathbb{R}^2, F(x) = \begin{pmatrix} u(x) \\ x(x) \end{pmatrix}$ is differentiable at x_0.

15 Complex Numbers and Geometry

The preparations of Chapter 13 allow us to start immediately with the investigation of complex-valued functions defined on a subset $G \subset \mathbb{C}$ and nowadays many textbooks on this topic follow the same approach. However complex analysis is also a rather geometric theory and we prefer to collect early in a more or less coherent manner interpretations and results linking geometry and algebra in $\mathbb{C}$.

This is a more classical approach and we refer to the classical text of C. Carathéodory [16] or the book of E. Peschl [64].

Complex numbers can be identified as points in the plane, see Figure 15.1, and by its very definition, the addition of two complex numbers z_1 and z_2 is vector addition in $\mathbb{R}^2$, see Figure 15.2. Moreover, mapping z onto $\bar{z}$ is just a reflection in the real axis, see Figure 15.3.

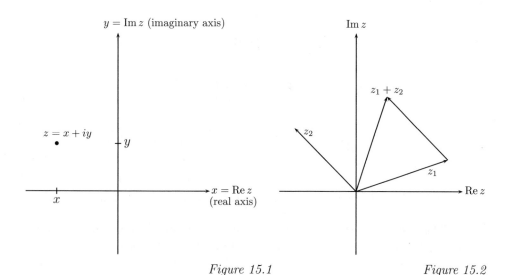

Figure 15.1   Figure 15.2

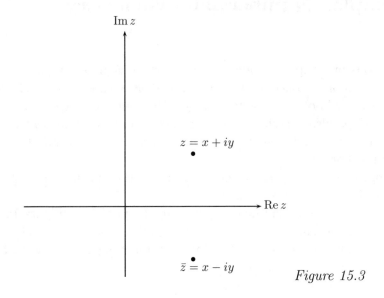

Figure 15.3

The multiplication of two complex numbers $z_1, z_2 \in \mathbb{C}, z_j \neq 0$, can be interpreted in the following way: we write z_1 as a matrix, i.e. $h(z_1) = \begin{pmatrix} x_1 & -y_1 \\ y_1 & x_1 \end{pmatrix}$, and z_2 as a matrix $h(z_2) = \begin{pmatrix} x_2 & -y_2 \\ y_2 & x_2 \end{pmatrix}$, and then $z_1 \cdot z_2$ corresponds to

$$h(z_1 \cdot z_2) = \begin{pmatrix} x_1 & -y_1 \\ y_1 & x_1 \end{pmatrix} \begin{pmatrix} x_2 & -y_2 \\ y_2 & x_2 \end{pmatrix}$$

$$= (x_1^2 + y_1^2)^{1/2} \begin{pmatrix} \frac{x_1}{(x_1^2+y_1^2)^{1/2}} & \frac{-y_1}{(x_1^2+y_1^2)^{1/2}} \\ \frac{y_1}{(x_1^2+y_1^2)^{1/2}} & \frac{x_1}{(x_1^2+y_1^2)^{1/2}} \end{pmatrix} \begin{pmatrix} x_2 & -y_2 \\ y_2 & x_2 \end{pmatrix}$$

The matrix $U = \begin{pmatrix} \frac{x_1}{(x_1^2+y_1^2)^{1/2}} & \frac{-y_1}{(x_1^2+y_1^2)^{1/2}} \\ \frac{y_1}{(x_1^2+y_1^2)^{1/2}} & \frac{x_1}{(x_1^2+y_1^2)^{1/2}} \end{pmatrix}$ has determinant 1 and therefore $U \in SO(2)$, i.e. it corresponds to a rotation. Thus it looks as if complex multiplication is related to a rotation followed by a dilation. In order to come to a firm statement we introduce polar coordinates. We have already encountered polar coordinates (r, φ) in the plane $\mathbb{R}^2$, see Chapter II.12, in particular Example II.12.3. We want to treat polar coordinates in $\mathbb{C}$ and we will make use of Euler's relation $e^{i\varphi} = \cos\varphi + i\sin\varphi, \varphi \in \mathbb{R}$, which we

284

already needed in Chapter II.17, also see Problem 6 of Chapter II.17. Using the ratio test we have seen in Problem 8 of Chapter 13 that for every $z \in \mathbb{C}$ the series

$$\exp z := \sum_{k=0}^{\infty} \frac{z^k}{k!}, \tag{15.1}$$

$$\cos z := \sum_{l=0}^{\infty} (-1)^l \frac{z^{2l}}{(2l)!} \tag{15.2}$$

and

$$\sin z := \sum_{m=1}^{\infty} (-1)^{m-1} \frac{z^{2m-1}}{(2m-1)!} \tag{15.3}$$

converge for all $z \in \mathbb{C}$ and for $z = i\varphi, \varphi \in \mathbb{R}$, we find using $i^2 = -1$ that

$$e^{i\varphi} = \sum_{k=0}^{\infty} \frac{(i\varphi)^k}{k!}$$

$$= \sum_{l=0}^{\infty} \frac{(i\varphi)^{2l}}{(2l)!} + \sum_{m=1}^{\infty} \frac{(i\varphi)^{2m-1}}{(2m-1)!}$$

$$= \sum_{l=0}^{\infty} \frac{i^{2l}\varphi^{2l}}{(2l)!} + \sum_{l=0}^{\infty} \frac{i^{2m-1}\varphi^{2m-1}}{(2m-1)!}$$

$$= \sum_{l=0}^{\infty} (-1)^l \frac{\varphi^{2l}}{(2l)!} + i \sum_{l=0}^{\infty} (-1)^{m-1} \frac{\varphi^{2m-1}}{(2m-1)!}$$

$$= \cos \varphi + i \sin \varphi.$$

With **polar coordinates**

$$r = |z| = (x^2 + y^2)^{1/2} \text{ and } \varphi = \arctan\frac{y}{x}, \tag{15.4}$$

where some care is needed in defining φ, compare with Chapter II.12, and see also Figure 15.4 below, we get for $z \neq 0$

$$z = r (\cos \varphi + i \sin \varphi) = re^{i\varphi}, \tag{15.5}$$

where φ is only determined modulo 2π, see Figure 15.4.

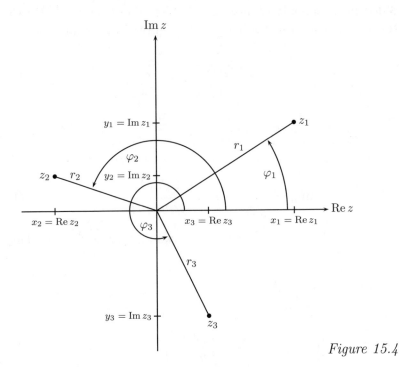

Figure 15.4

The function tan is π -periodic, not 2π - periodic and it has the natural real domain $\left(\frac{-\pi}{2}, \frac{\pi}{2}\right)$. In order to use arg $z = \arctan\frac{y}{x}$ for all $z = x + iy$ we use the following (justifiable) convention

$$\arctan\frac{y}{x} := \begin{cases} \arctan\frac{y}{x}, & x > 0 \text{ and } y \geq 0 \\ \frac{\pi}{2}, & x = 0 \text{ and } y > 0 \\ \frac{\pi}{2} + \arctan\frac{y}{|x|}, & x < 0 \text{ and } y > 0 \\ \pi + \arctan\frac{|y|}{|x|}, & x < 0 \text{ and } y \leq 0 \\ \frac{3\pi}{2}, & x = 0 \text{ and } y < 0 \\ \frac{3\pi}{2} + \arctan\frac{|y|}{x}, & x > 0 \text{ and } y < 0. \end{cases}$$

We have already mentioned that $|z|$ is called the **modulus** of z and arg $z := \varphi$ is called the **argument** of z. In order to have arg z uniquely defined we reduce the range of φ to a half-open interval of length 2π. Traditionally in complex analysis the choice is $-\pi < \varphi \leq \pi$, the modern choice which we will

286

follow is $0 \leq \varphi < 2\pi$.

So far we may summarize our consideration in

Corollary 15.1. *Every complex number $z \in \mathbb{C}\backslash\{0\}$ has a unique representation*

$$z = re^{i\varphi} \qquad (15.6)$$

*with $r = |z| > 0$ and $\varphi = \arg z \in [0, 2\pi)$. This is called the **polar representation** of z.*

Now we find for complex multiplication of $z_1, z_2 \in \mathbb{C}\backslash\{0\}$

$$z_1 \cdot z_2 = r_1 \, e^{i\varphi_1} \, r_2 \, e^{i\varphi_2} = r_1 \, r_2 \, e^{i(\varphi_1 + \varphi_2)}$$
$$= |z_1||z_2|e^{i \, \arg(z_1 z_2)}$$
$$= |z_1||z_2|(\cos(\arg(z_1 \cdot z_2)) + i \, \sin(\arg(z_1 \cdot z_2)),$$

where $\arg(z_1 \cdot z_2) \equiv (\varphi_1 + \varphi_2) \mod 2\pi$, $\arg(z_1 \cdot z_2) \in [0, 2\pi)$. This also leads to

Corollary 15.2. *The inverse of $z \in \mathbb{C}\backslash\{0\}$ with respect to multiplication is given by*

$$z^{-1} = \frac{1}{z} = \frac{1}{r} e^{-i\varphi}.$$

The reader may recall from algebra or number theory that for integers $k, l, m \in \mathbb{Z}$ the meaning of $k \equiv l \mod m$ (read k is congruent to l modulus m) is that $k - l$ is divisible by m. This notation is extended to real numbers in the sense that $a \equiv b \mod c$ means that $\frac{a-b}{c} = k$ for some $k \in \mathbb{Z}$. Thus

$$\arg(z_1 \cdot z_2) \equiv (\varphi_1 + \varphi_2) \mod 2\pi \qquad (15.7)$$

means that

$$\arg(z_1 \cdot z_2) = \varphi_1 + \varphi_2 + 2\pi k, k \in \mathbb{Z}, \ 0 \leq \arg(z_1 \cdot z_2) < 2\pi. \qquad (15.8)$$

Applying now the polar coordinate representation to the matrix representation we find with $z_j = r_j \, e^{i\varphi_j}$ that

$$h(z_1 \cdot z_2) = h\left(r_1 e^{i\varphi_1} r_2 \, e^{i\varphi_2}\right) = r_1 \begin{pmatrix} \cos\varphi_1 & -\sin\varphi_1 \\ \sin\varphi_1 & \cos\varphi_1 \end{pmatrix} h(z_2),$$

and our geometric interpretation becomes more clear: multiplying z_2 by z_1 corresponds to a rotation by $U(\arg z_1) = U(\varphi_1) = \begin{pmatrix} \cos \varphi_1 & -\sin \varphi_1 \\ \sin \varphi_1 & \cos \varphi_2 \end{pmatrix}$ follo- wed by a dilation of size r_1.

This observation leads to some interesting geometric consequences. Consider a subset $A \subset \mathbb{C}$. For $w \in \mathbb{C}$, $|w| = 1$, we can study the new set $wA := \{\zeta = wz | z \in A\}$. This set is obtained from A by a rotation around the origin with an angle $\arg w$. Hence invariance of A under a certain rotation can be expresses by the equality $A = wA$. We will see further cases where geometry in the plane is best described algebraically with the help of complex numbers.

Using the law of the exponential function, compare with Problem 10 of Chapter 13, we now find the **formula of de Moivre**

Lemma 15.3. *For $z = re^{i\varphi}$ the following holds*

$$z_n = r^n e^{in\varphi} = r^n(\cos n\varphi + i \sin n\varphi). \tag{15.9}$$

This lemma allows us to determine the $\mathbf{n^{th}}$ **roots of unity**, $n \in \mathbb{N}$, i.e. all complex numbers z with $z^n = 1$ or $z^n - 1 = 0$. We take for granted the fundamental theorem of algebra stating that a polynomial $p(z) = \sum_{k=0}^{N} a_k z^k, a_k \in \mathbb{C}, z \in \mathbb{C}$, of degree N, i.e. $a_N \neq 0$, has N complex roots (taking multiplicity into account).

An analytic proof of the fundamental theorem of algebra will also be given later in Chapter 23, Theorem 23.15. Now, if $z = re^{i\varphi}$ satisfies $z^n = 1$ it follows that $r^n e^{in\varphi} = 1 = 1e^{i0}$. Thus we must have $r = 1$ and $\varphi = \varphi_k = \frac{2\pi}{n}k$, $k = 0, \dots, n - 1$. Indeed for $r = 1$ and $\varphi = \frac{2\pi}{n}k$, $k = 0, \dots, n - 1$, we have

$$z^n = (1 e^{\frac{2\pi}{n} \varphi_k i})^n = e^{2\pi k i} = 1.$$

In Figure 15.5 and Figure 15.6 we have shown the positions on the unit circle S^1 of the 5th and the 12th roots of unity, respectively. These figures also demonstrate nicely that the associated polygons are invariant under certain rotations which can be expressed by showing that multiplying by the corresponding root of unity does not change the set.

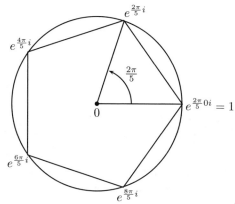

Figure 15.5

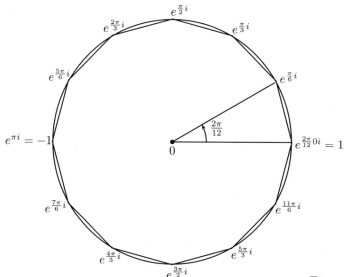

Figure 15.6

Thus the number of the nth roots of unity is exactly n and they have the representation $z_k = e^{\frac{2\pi}{n}ki}$, $k = 0, \dots, n-1$. Of course for other values $k \in \mathbb{Z}$ we still have $z_k^n = 1$, but z_k can not anymore be considered as given in polar coordinates with argument in $[0, 2\pi)$. Indeed if $z_m = e^{\frac{2\pi}{n}mi}$, $m \in \mathbb{Z}\backslash\{0, 1, \dots, n-1\}$, we have $z_m^n = 1$ and we can find a unique $k_m \in \{0, 1, \dots, n-1\}$ and $l_m \in \mathbb{Z}$

such that

$$\frac{m - k_m}{l_m} = n \text{ or } m = k_m \bmod n.$$

Consider

$$E(n) := \{e^{\frac{2\pi}{n} k i} | k = 0, 1, ..., n - 1\}$$

and

$$C(n) := \{e^{\frac{2\pi}{n} k i} | k \in \mathbb{Z}\}.$$

Both sets are subsets of S^1, and S^1 with complex multiplication is a subgroup of $\mathbb{C}\backslash\{0\}$. It follows that both $E(n)$ and $C(n)$ are subgroups of S^1 which as point sets are equal. However we can identify $C(n)$ with the infinite group $n\mathbb{Z} = \{n\,k | k \in \mathbb{Z}\}$ with the group operation being addition in $\mathbb{Z}$, whereas $E(n)$ we can identify with the finite group, the group $\mathbb{Z}/n$ formed by the system of representation $\{0, 1, ..., n-1\}$ and the corresponding group operation and $E(n)$ is considered as the group of nth unit roots.

We urge the student to get used to the idea that with groups we can associate geometric objects which are invariant under the action or operation of the group under consideration. First let us give

Definition 15.4. *Let $X \neq \emptyset$ be a set and G a group.*

A. *We say that G **operates** or **acts** on X if for every $g \in G$ there exists a bijective mapping $T_g : X \to X$. We call the mapping $g \mapsto T_g$ from G into $M(X)$ a **representation** of G if $T_{g^{-1}} = T_g^{-1}$ and $T_{g_1 g_2} = T_{g_1} \cdot T_{g_2}$. As usual $M(X)$ denotes the set of all mappings of X into itself.*

B. *A set $Y \subset X$ is called **invariant** under T_g if $T_g Y = Y$.*

C. *If $R : G \to M(X)$ is a representation of G and $Y \subset X$ is invariant under all $T_g := R(g)$ then we call Y invariant under the representation R of G and if no confusion may arise we say that Y is invariant under G.*

Example 15.5. Let $X = \mathbb{C}$ and $G = \mathbb{C}\backslash\{0\}$ the multiplicative group of $\mathbb{C}$. For every $w \in G$ we define $T_w : \mathbb{C} \to \mathbb{C}$ by $T_w z = w \cdot z$. Note that the inverse of T_w is given by $T_w^{-1} = T_{w^{-1}}$ and that $T_{w_2} \cdot T_{w_1}$ as well as $T_1 = \text{id}_{\mathbb{C}}$. We may consider S^1 as a subgroup of G and it follows for $w \in S^1$ and $z \in S^1$ that we have $T_w z = w z = S^1$, i.e. S^1 as a subgroup of $\mathbb{C}\backslash\{0\}$ leaves S^1 as subset of $\mathbb{C}$ invariant.

Example 15.6. Consider the regular n-gon $P_n \subset \mathbb{C}$ with vertices $z_k = e^{\frac{2\pi}{n}ki}$, $k = 0, \ldots, n - 1$, as subset of $\mathbb{C}$. Thus the vertices of P_n are just the elements of $E(n)$, for $n = 5$ and $n = 12$ compare with Figure 15.5 and Figure 15.6, respectively. Since multiplication with an nth root of unity corresponds to a rotation around the origin by an angle $\varphi_k \in \{0, \frac{2\pi}{n}, \ldots, \frac{2\pi}{n}(n-1)\}$, it follows that $P_n \subset \mathbb{C}$ is invariant under the action of the representation of $E(n)$ as multiplication operators.

We now want to describe some simple geometric objects (subsets) in the plane with the help of complex numbers. Let us start with the following observation. We can identify $\mathbb{C}$ with $\mathbb{R}^2$, i.e. $z = x + iy$ is identified with (x, y). If we can characterise a geometric object in the plane $\mathbb{R}^2$ by a relation $h(x, y) = 0$ or an inequality $g(x, y) \geq 0$ with some functions $h, g : \mathbb{R}^2 \to \mathbb{R}$, then we may switch to complex numbers using the relations

$$x = \operatorname{Re} z = \frac{z + \bar{z}}{2} \text{ and } y = \operatorname{Im} z = \frac{z - \bar{z}}{2i}.$$

Thus $h(x, y) = 0$ becomes $h\left(\frac{z+\bar{z}}{2}, \frac{z-\bar{z}}{2i}\right) = 0$ and similarly $g(x, y) \geq 0$ becomes $g\left(\frac{z+\bar{z}}{2}, \frac{z-\bar{z}}{2i}\right) \geq 0$. If we now can find H and G such that

$$H(z, \bar{z}) = h\left(\frac{z + \bar{z}}{2}, \frac{z - \bar{z}}{2i}\right)$$

and

$$G(z, \bar{z}) = g\left(\frac{z + \bar{z}}{2}, \frac{z - \bar{z}}{2i}\right).$$

We can give a description of the corresponding geometric object with the help of complex numbers. However, in concrete situations direct (geometric) considerations are often easier.

A straight line in $\mathbb{C}$ we can give in its parametric form by

$$z = z_1 + t(z_2 - z_1), \ t \in \mathbb{R}, \tag{15.10}$$

which corresponds to

$$\begin{pmatrix} x \\ y \end{pmatrix} = \begin{pmatrix} x_1 \\ y_1 \end{pmatrix} + t \begin{pmatrix} x_2 - x_1 \\ y_2 - y_1 \end{pmatrix} \tag{15.11}$$

and which passes through z_1, i.e. $\begin{pmatrix} x_1 \\ y_1 \end{pmatrix}$, and z_2, i.e. $\begin{pmatrix} x_2 \\ y_2 \end{pmatrix}$.

With $c := \frac{z_2 - z_1}{|z_2 - z_1|} \in S^1 \subset \mathbb{C}$ we find

$$\frac{z - z_1}{|z_2 - z_1|} = c\,t, \tag{15.12}$$

where c, interpreted as vector in $\mathbb{R}^2$, gives the direction of the line. If we rewrite (15.12) as

$$\frac{z - z_1}{|z_2 - z_1|}\,\bar{c} = t, \tag{15.13}$$

note $c\bar{c} = 1$, we observe that the left hand side in (15.13) must be a real number for all z on the straight line given by (15.10). Thus for points on this line

$$\operatorname{Im}\left(\bar{c}\,(z - z_1)\right) = 0 \tag{15.14}$$

must hold and it is easy to derive from (15.14) that $\{z \in \mathbb{C} \,|\, \operatorname{Im}\left(\bar{c}\,(z - z_1)\right) = 0\}$, $|c| = 1$, must be a straight line passing through z_1 and $z_2 = c + z_1$. In Problem 8 we will also find the line orthogonal to the line (15.10). The normal form of (15.10) is given by

$$\bar{d}\,z + d\bar{z} - 2\,p = 0, \quad d = i\,c, \quad p = \operatorname{Re}\left(d z_1\right). \tag{15.15}$$

Next consider for $a, b \in \mathbb{R}$ and $c \in \mathbb{C}$ the equation

$$a\,z\bar{z} + \bar{c}\,z + c\bar{z} + b = 0 \tag{15.16}$$

under the constraint

$$D := |c|^2 - ab > 0. \tag{15.17}$$

For $a = 0$ it follows from (15.15) that (15.16) gives a straight line. In the case where $a \neq 0$ we can rewrite (15.16) as

$$\left(z + \frac{c}{a}\right)\left(\bar{z} + \frac{\bar{c}}{a}\right) = \frac{|c|^2 - ab}{a^2} = \frac{D}{a^2},$$

or with $R^2 = \frac{D}{a^2}$ it follows from (15.16) that

$$\left|z + \frac{c}{a}\right| = R \tag{15.18}$$

292

must hold, whereas for $D < 0$ and $a \neq 0$ no $z \in \mathbb{C}$ can solve (15.16).
For $\varrho > 0$ the set $\{z \in \mathbb{C} | |z - z_0| = \varrho\}$ is however a circle with centre z_0 and radius ϱ: with $z = x + iy$ and $z_0 = x_0 + iy_0$ we have indeed

$$|z - z_0|^2 = (x - x_0)^2 + (y - y_0)^2 = \varrho^2. \tag{15.19}$$

Thus (15.18) gives a circle with centre $z_0 = -\frac{c}{a}$ and radius R. In polar coordinates we can rewrite the equation $|z - z_0| = \varrho$ as $z = z_0 + \varrho e^{i\varphi}, \varphi \in [0, 2\pi)$.

Note that when we choose in (15.18) the value $c = 0$ we have a circle with the origin 0 as centre. For every $w \in \mathbb{C}$, $|w| = 1$, we have of course $|wz| = |z| = R$ if z lies on the circle $|z| = R$. This nicely demonstrates our geometric interpretation of the multiplication with a complex number of modulus 1: a circle with centre 0 is of course invariant under all rotations around the origin.

We can identify the **open disc** with centre z_0 and radius $\varrho > 0$ with

$$D_\varrho(z_0) = \{z \in \mathbb{C} | |z - z_0| < \varrho\} \tag{15.20}$$

for which we more often write as before $B_\varrho(z_0) = B_\varrho((x, y)), z = x + iy$, see Figure 15.7. The closed disc with centre z_0 and radius ϱ is of course $\overline{D_\varrho(z_0)} = \{z \in \mathbb{C} | |z - z_0| \leq \varrho\} = \overline{B_\varrho(z_0)}$. However we also use $\mathbb{D}$ for $B_1(0)$. We also note that the inequality

$$\operatorname{Im} z > 0 \tag{15.21}$$

characterises the open **upper half plane**, see Figure 15.8

$$\mathbb{H} := \{z \in \mathbb{C} | \operatorname{Im} z > 0\} = \{(x, y) \in \mathbb{R}^2 | y > 0\}, \tag{15.22}$$

and the closed upper half plane $\overline{\mathbb{H}}$ as well as the open and closed lower half planes $\mathbb{H}_-$ and $\overline{\mathbb{H}}_-$ are defined analogously.

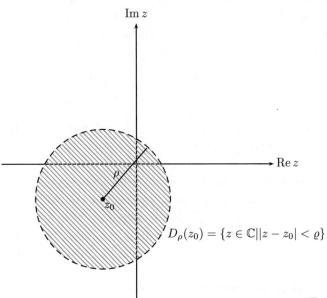

$$D_\rho(z_0) = \{z \in \mathbb{C} \,|\, |z - z_0| < \varrho\}$$

Figure 15.7

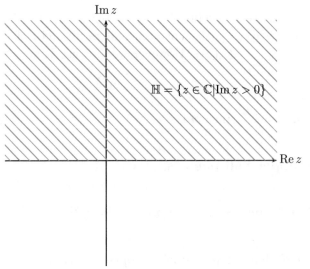

$$\mathbb{H} = \{z \in \mathbb{C} \,|\, \mathrm{Im}\, z > 0\}$$

Figure 15.8

On $\mathbb{R}$ we have a natural order structure and limit points $+\infty$ and $-\infty$ were introduced. In the plane this needs a different approach: there are too many directions leading to "infinity". We overcome this problem by mapping $S^2 \backslash \{N\}$, where $S^2 \subset \mathbb{R}^3$ is the unit sphere and $N = (0,0,1)$ is the "north pole", bijectively onto the complex plane $\mathbb{C}$. Then we extend this mapping to S^2 and consider the image of N as "point at infinity" for $\mathbb{C}$.
We consider the sphere $S^2 \subset \mathbb{R}^3$ with centre $(0,0,0)$ and radius 1 in the ξ, η, ζ coordinate system, $S^2 = \{(\xi, \eta, \zeta) \in \mathbb{R}^3 \,|\, \xi^2 + \eta^2 + \zeta^2 = 1\}$, and we consider the $\xi - \eta$ - plane also as $z = x + iy$ plane on which we want to project S^2, see Figure 15.9.

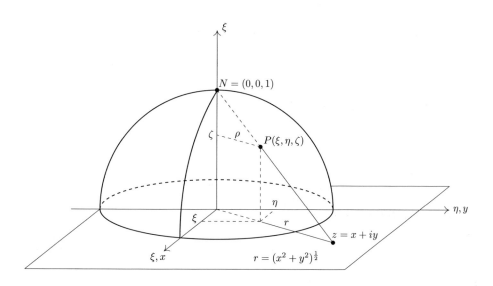

Figure 15.9

The point $z \in \mathbb{C}, z = x + iy$, is connected with the north pole of the sphere by a straight line. This line hits the sphere in the point $P = (\xi, \eta, \zeta)$. Now we want to determine $z = x + iy$ as function of ξ, η, ζ and then we want to determine the inverse mapping. Using Figure 15.9 we get the following set of equations

$$\xi^2 + \eta^2 + \zeta^2 = 1 \tag{15.23}$$

and

$$\frac{x}{\xi} = \frac{y}{\eta} = \frac{r}{\rho} = \frac{1}{1-\zeta},$$

(15.24)

implying

$$x = \frac{\xi}{1-\zeta}, \quad y = \frac{\eta}{1-\zeta},$$

(15.25)

and therefore

$$z = \frac{\xi + iy}{1-\zeta} \quad \text{and} \quad \bar{z} = \frac{\xi - i\eta}{1-\zeta}.$$

(15.26)

Furthermore we find

$$z\bar{z} = \frac{\xi^2 + \eta^2}{(1-\zeta)^2} = \frac{1-\zeta^2}{(1-\zeta)^2} = \frac{1+\zeta}{1-\zeta},$$

$$z\bar{z} - 1 = \frac{1+\zeta}{1-\zeta} - \frac{1-\zeta}{1-\zeta} = \frac{2\zeta}{1-\zeta},$$

$$z\bar{z} + 1 = \frac{2}{1-\zeta},$$

and therefore (15.25) gives

$$\xi = 2x\frac{1-\zeta}{2} = \frac{2x}{z\bar{z}+1} = \frac{z+\bar{z}}{z\bar{z}+1}$$

$$\eta = \frac{2y}{z\bar{z}+1} = i\frac{(\bar{z}-z)}{z\bar{z}+1}.$$

Thus we have the mapping and its inverse:

$$z = \frac{\xi + i\eta}{1-\zeta}$$

(15.27)

and

$$\xi = \frac{z+\bar{z}}{z\bar{z}+1}, \quad \eta = \frac{i(\bar{z}-z)}{z\bar{z}+1}, \quad \zeta = \frac{z\bar{z}-1}{z\bar{z}+1}.$$

(15.28)

The mapping $S : S^2 \backslash \{N\} \to \mathbb{C}, (\xi, \eta, \zeta) \mapsto S(\xi, \eta, \zeta) = z = \frac{\xi+i\eta}{1-\zeta}$, is called the **stereographic projection**.

Our problem is to extend S to S^2, so the question is to which point shall we map N? We introduce the "ideal" point $\infty, \infty \notin \mathbb{C}$, hence we extend $\mathbb{C}$ to $\hat{\mathbb{C}} := \mathbb{C} \cup \{\infty\}$, and now we extend S to a mapping $\tilde{S} : S^2 \to \hat{\mathbb{C}} = \mathbb{C} \cup \{\infty\}$ where

$$\tilde{S}(\xi, \eta, \zeta) = \begin{cases} S(\xi, \eta, \zeta), & (\xi, \eta, \zeta) \neq N = (0, 0, 1) \\ \infty, & (\xi, \eta, \zeta) = N = (0, 0, 1). \end{cases}$$

We call $\hat{\mathbb{C}}$ the closed or **extended complex plane** or the **Riemann sphere**. First we extend arithmetic operations to $\hat{\mathbb{C}}$

$$z + \infty = \infty \text{ for } z \in \mathbb{C}, \tag{15.29}$$

$$\frac{z}{\infty} = 0 \text{ for } z \in \mathbb{C}, \tag{15.30}$$

$$z \cdot \infty = \infty \text{ for } z \in \mathbb{C}\backslash\{0\}, \tag{15.31}$$

$$\frac{z}{0} = \infty \text{ for } z \in \mathbb{C}, \tag{15.32}$$

however we do not define $\infty + \infty$, $0 \cdot \infty$, $\frac{\infty}{\infty}$ or $\frac{0}{0}$.

Let $(z_n)_{n \in \mathbb{N}}$ be a sequence of complex numbers. If for every $R > 0$ there exists $N(R) \in \mathbb{N}$ such that $n \geq N(R)$ implies $|z_n| \geq R$ we say that $(z_n)_{n \geq N}$ converges to ∞ and we write

$$\lim_{n \to z_n} z_n = \infty. \tag{15.33}$$

In Appendix I we will give a further description of the topology on $\hat{\mathbb{C}}$.

Problems

1. Find the polar representations of:

   a) $z = 4 + 4\sqrt{3}i$;

   b) $z = -1 + i$;

   c) $z = -\sqrt{24} - \sqrt{8}i$.

2.   a) Express $\cos(5\varphi)$ as a polynomial in $\cos\varphi$, i.e. find $N \in \mathbb{N}$ and $a_k \in \mathbb{C}$ such that $\cos(5\varphi) = \sum_{k=0}^{N} a_k \cos^k \varphi$.

   b) Use Euler's formula to verify

   $$8\cos^4(2\varphi) - 3 = \cos(8\varphi) + 4\cos(4\varphi).$$

   c) Find the value of $\left(\dfrac{1 + \frac{1}{3}\sqrt{3}i}{1 - \frac{1}{3}\sqrt{3}i} \right)^{12}$.

3. a) Let $z_1, \ldots, z_n$ be the roots of the polynomial $p(z) = \sum_{k=0}^{n} a_k z^k$. Prove that $\sum_{k=1}^{n} z_k = -\frac{a_{n-1}}{a_n}$ and $\prod_{k=1}^{n} z_k = (-1)^n \frac{a_0}{a_n}$. (For $n = 2$ these formulae are known as the **formulae of Viéta**.)

    b) Show that the sum of all n^{th} roots of unity is equal to 0 provided $n \geq 2$.

4. a) Sketch the positions of all 8th roots of unity.

    b) Prove that if n divides m then $E(n)$ is a subgroup of $E(m)$.

5. Show that the annulus $A_{r,R} = \{z \in \mathbb{C} | 0 < r < |z| < R\}$ is invariant under the action of S^1. (Compare with Example 15.5.)

6. Find the subset $\mathcal{E} \subset \mathbb{C}$ determined by the equation $|z-3|+|z+3| = 10$.

7. Find the non-parametric equation of the line passing through $z_1 = -2 - i$ and $z_2 = 3 + 5i$.

8. Let $z = z_1 + t(z_2 - z_1)$, $t \in \mathbb{R}$, be a straight line L in $\mathbb{C}$ given in parametric form. Find a non-parametric equation for the line passing through z_1 and orthogonal to L.

9. Find the image of a line $L \subset \mathbb{C}$ passing through $z_0 = 0$ on the Riemann sphere.

10. Let $n \in \mathbb{N}$ and $z_n := z_0^n$, $|z_0| > 1$. Prove that $\lim_{n \to \infty} z_n = \infty$. Compare this result with the analogous result in the real case.

16 Complex-Valued Functions of a Complex Variable

In Chapter 14 we discussed complex-valued functions defined on an arbitrary non-empty set. When specifying this set to a subset $D \subset \mathbb{C}$ we can of course apply the earlier results. We denote points in D by $z = x + iy$ and it follows that every $f : D \to \mathbb{C}$ admits a decomposition

$$f(z) = \operatorname{Re} f(z) + i \operatorname{Im} f(z). \tag{16.1}$$

It is very helpful to consider $\operatorname{Re} f(z)$ and $\operatorname{Im} f(z)$ also as real-valued functions of the two variables (x, y) and we write

$$f(z) = u(x, y) + iv(x, y). \tag{16.2}$$

Since D equipped with the metric $d(z_1 - z_2) = |z_1 - z_2|$ is a metric space as is $\mathbb{C}$, continuity of $f : D \to \mathbb{C}$ is well defined. Using the results of Chapter II. 2 we know:

A function $f : D \to \mathbb{C}, D \subset \mathbb{C}$, is **continuous at** $z_0 \in D$ if for every $\epsilon > 0$ there exists $\delta > 0$ such that for all $z \in D$ with $0 < |z - z_0| < \delta$ it follows that $|f(z) - f(z_0)| < \epsilon$. If D contains a sequence $(z_n)_{n \in \mathbb{N}}, z_n \neq z_0$, then this is equivalent to the fact that for every sequence $(z_n)_{n \in \mathbb{N}}, z_n \in D$, converging to z_0 it follows that $(f(z_n))_{n \in \mathbb{N}}$ converges to $f(z_0)$, i.e.

$$\lim_{n \to \infty} z_n = z_0 \quad \text{implies} \quad \lim_{n \to \infty} f(z_n) = f(z_0)(= f(\lim_{n \to \infty} z_n)).$$

If $f : D \to \mathbb{C}$ is continuous for all $z \in D$ we call f **continuous in D**. The continuous functions in D form a vector space, in fact an algebra, over $\mathbb{C}$ which we denote by $C(D, \mathbb{C})$ or if no confusion may arise simply by $C(D)$. We have already noted in Chapter 14 that f is continuous (at z_0 or in D) if and only if u and v are continuous (at (x_0, y_0) or in D considered as subset of $\mathbb{R}^2$). If $D \subset \mathbb{C}$ is compact then a continuous function $f : D \to \mathbb{C}$ is uniformly continuous. Since on $\mathbb{C}$ we do not have a natural order structure the notion of a monotone function does not make sense and we call $f : D \to \mathbb{C}$ **bounded** if $|f(z)| \leq M$ for all $z \in D$. Continuous functions on compact sets are bounded and for some $z_1, z_2 \in D$ the following hold:

$$|f(z_1)| = \sup_{z \in D} |f(z)| \quad \text{and} \quad |f(z_2)| = \inf_{z \in D} |f(z)|. \tag{16.3}$$

This follows from the fact that real-valued functions which are continuous on a compact set are bounded and attain their supremum and infimum and the observation that

$$|f(z)| = (u^2(x,y) + v^2(x,y))^{1/2}.$$

If $f : D \to \mathbb{C}$ is continuous and $g : f(D) \to \mathbb{C}$ is continuous too, then $g \circ f : D \to \mathbb{C}$ is continuous. Using explicitly the fact that $\mathbb{C}$ is a field for two functions $f, g : D \to \mathbb{C}$ we can define on $\{z \in D \,|\, g(z) \neq 0\}$ the quotient $\frac{f}{g} : D \backslash \{z \in \mathbb{C} \,|\, g(z) = 0\} \to \mathbb{C}$ and if both, f and g are continuous, then $\frac{f}{g}$ is continuous too.

So far we can deduce that every **polynomial**

$$p(z) := \sum_{k=0}^{N} a_k \, z^k, \quad N \in \mathbb{N} \cup \{0\}, \quad a_k \in \mathbb{C}, \tag{16.4}$$

is continuous on $\mathbb{C}$ and hence on every subset $D \subset \mathbb{C}$. As usual we call the largest $k_0 \in \{0, 1, ..., N\}$ with $a_{k_0} \neq 0$ the **degree** of p and usually, if we write $p(z) = \sum_{k=0}^{N} a_k \, z^k$ we assume that p is of degree N.

Moreover we know that **rational functions** are continuous on their natural domain, i.e. if with two polynomials p and q we have

$$r(z) := \frac{p(z)}{q(z)} \tag{16.5}$$

then $r : \mathbb{C} \backslash \{z \in \mathbb{C} \,|\, q(z) = 0\} \to \mathbb{C}$ is continuous. Note that r might have a continuous extension to a larger set:

The rational function $r(z) = \frac{(z-z_0)^2}{z-z_0}$ is defined and continuous on $\mathbb{C} \backslash \{z_0\}$ but it admits a continuous extension (or continuation) to $\mathbb{C}$ by $h(z) = z - z_0$. We will discuss such problems for so called meromorphic function later on in Chapters 22 and 24.

We next want to discuss whether it is possible to extend elementary transcendental functions defined on subsets of $\mathbb{R}$, for example $\exp, \cos, \arctan$, etc., to subsets of $\mathbb{C}$. Having in mind that we have derived power series expansions for these functions, namely their Taylor series, we may try to extend the domain of convergence of such a series (which is an interval in $\mathbb{R}$) to some complex domain. In fact we have already seen that for example the exponential series $\sum_{k=0}^{\infty} \frac{z^k}{k!}$ converges for all $z \in \mathbb{C}$ and uniformly on compact sets, hence we can extend $\exp : \mathbb{R} \to \mathbb{R}$ to $\exp : \mathbb{C} \to \mathbb{C}$ using the power series. Thus we were led to investigate power series in $\mathbb{C}$. We start with recollecting some results on uniform convergence.

Definition 16.1. *A sequence $f_k : D \to \mathbb{C}$, $D \subset \mathbb{C}$, of complex-valued functions* **converges uniformly** *to $f : D \to \mathbb{C}$ if for every $\epsilon > 0$ there exists $N = N(\epsilon) \in \mathbb{N}$ such that $k \geq N(\epsilon)$ implies*

$$\sup_{z \in D} | f_k(z) - f(z)| < \epsilon. \tag{16.6}$$

Remark 16.2. A. We know that uniform convergence is convergence with respect to the supremum norm $||f||_\infty := \sup_{z \in D} |f(z)|$.
B. By Theorem II.2.32 the uniform limit of a sequence of continuous functions is continuous.
C. We call a series $\sum\limits_{k=0}^{\infty} f_k$ of functions $f_k : D \to \mathbb{C}$ **absolutely convergent** if the series $\sum\limits_{k=0}^{\infty} |f_k(z)|$ converges for all $z \in D$.

Next we prove a complex version of the Weierstrass M-test, compare with Theorem I.29.1.

Theorem 16.3. *Let $(f_k)_{k \in \mathbb{N}_0}$, $f_k : D \to \mathbb{C}$, $D \subset \mathbb{C}$, be a sequence of functions and suppose that $\sum\limits_{k=0}^{\infty} ||f_k||_\infty < \infty$. Then the series $\sum\limits_{k=0}^{\infty} f_k$ converges absolutely and uniformly on D to a function $f : D \to \mathbb{C}$. If all functions f_k are continuous then f is continuous too.*

Proof. First we show that $\sum\limits_{k=0}^{\infty} f_k(z)$, $z \in D$, converges pointwisely to some function $f : D \to \mathbb{C}$. Since $|f_k(z)| \leq ||f_k||_\infty$ the series $\sum\limits_{k=0}^{\infty} f_k(z)$ converges absolutely by the comparison test, Theorem 13.15.
Thus for $z \in D$ we can define

$$f(z) := \sum_{k=0}^{\infty} f_k(z) \tag{16.7}$$

which gives a function $f : D \to \mathbb{C}$. Now we prove that this convergence is uniformly, i.e. the sequence of partial sums $(S^N)_{N \in \mathbb{N}_0}$, $S^N(z) := \sum\limits_{k=0}^{N} f_k(z)$, converges uniformly to f. Given $\epsilon > 0$ the convergence of $\sum_{k=0}^{\infty} ||f_k||_\infty$ implies the existence of $N_0(\epsilon) \in \mathbb{N}$ such that $N \geq N_0(\epsilon)$ implies $\sum_{k=N+1}^{\infty} ||f_k||_\infty < \epsilon$. Now it follows for $N \geq N_0(\epsilon)$ that

$$||S^N - f||_\infty = \sup_{z \in D} |S^N(z) - f(z)|$$

$$= \sup_{z \in D} |\sum_{k=N+1}^{\infty} f_k(z)| \leq \sum_{k=N+1}^{\infty} \sup_{z \in D} |f_k(z)|$$

$$= \sum_{k=N+1}^{\infty} ||f_k||_\infty < \epsilon,$$

i.e. the uniform convergence of $(S^N)_{N \in \mathbb{N}_0}$ to f. It follows further from Remark 16.2.A that the continuity of all $f_k, k \in \mathbb{N}_0$, implies the continuity of f. □

Using the arithmetic in $\mathbb{C}$ we find immediately by first looking at partial sums

Corollary 16.4. *Let* $(f_k)_{k \in \mathbb{N}_0}$ *and* $(g_k)_{k \in \mathbb{N}_0}$ *be two sequences of functions* $f_k, g_k : D \to \mathbb{C}, D \subset \mathbb{C}$. *If the series* $\sum_{k=0}^{\infty} f_k$ *and* $\sum_{k=0}^{\infty} g_k$ *converge pointwise (absolutely, uniformly) to* f *and* g *respectively, then for* $a, b \in \mathbb{C}$ *we have*

$$\sum_{k=0}^{\infty} (a\, f_k + b\, g_k) = af + bg \tag{16.8}$$

and

$$\sum_{k=0}^{\infty} \overline{f_k} = \overline{f} \tag{16.9}$$

where the convergence is pointwise (absolutely, uniformly).

We now turn to power series and the reader may compare our discussion with Chapter I.29.

Definition 16.5. *Let $(a_k)_{k \in \mathbb{N}_0}$ be a sequence of complex numbers and let $c \in \mathbb{C}$. We call*

$$T^c_{(a_k)}(z) := \sum_{k=0}^{\infty} a_k (z - c)^k, z \in \mathbb{C}, \tag{16.10}$$

the (formal) **power series** *associated with the sequence $(a_k)_{k \in \mathbb{N}_0}$ and centre c. The numbers a_k are called the* **coefficients** *of the power series.*

Of course, the most urgent questions is: for which $z \in \mathbb{C}$ does $T^c_{(a_k)}(z)$ converge?

Theorem 16.6 (N.H.Abel). *Let $(a_k)_{k \in \mathbb{N}_0}$ be a sequence of complex numbers and $c \in \mathbb{C}$. If $T^c_{(a_k)}(z)$ converges for some $z_1 \neq c$, then it converges for all $z \in \mathbb{C}$ such that $|z - c| \leq \varrho < |z_1 - c|$, i.e. $T^c_{(a_k)}(z)$ converges in $\overline{B_\varrho(z)}$. In addition, on $\overline{B_\varrho(z)}$ the convergence is uniform and the same statement holds for $\tilde{T}^c_{(a_k)}(z) := \sum_{k=1}^{\infty} k \, a_k (z - c)^k$. In particular on $\overline{B_\varrho(z)}$ the functions $z \mapsto T^c_{(a_k)}(z)$ and $z \mapsto \tilde{T}^c_{(a_k)}(z)$ are continuous.*

Proof. We set $f_k(z) := a_k (z - c)^k$, and formally we find with $f(z) := T^c_{(a_k)}(z)$ that $f = \sum_{k=0}^{\infty} f_k$. Since $\sum_{k=0}^{\infty} f_k(z_1)$ converges by assumption there exists $M \geq 0$ such that $|f_k(z_n)| \leq M$ for all $k \in \mathbb{N}_0$. For $z \in [c - \varrho, c + \varrho]$ it follows with $0 < \varrho < |z_1 - c|$ that

$$|f_k(z)| = |a_k(z - c)^k| = \left| a_k(z_1 - c)^k \left| \frac{z - c}{z_1 - c} \right|^k \right| \leq M \vartheta^k,$$

where $\vartheta := \frac{\varrho}{z_1 - c} < 1$. Thus we have

$$\|f_k\|_{\infty, \overline{B_\varrho(c)}} = \sup_{z \in B_\varrho(c)} |f_k(z)| \leq M \vartheta^k$$

implying that

$$\sum_{k=0}^{\infty} \|f_k\|_{\infty, \overline{B_\vartheta(c)}} \leq M \frac{1}{1 - \vartheta}.$$

Now Theorem 16.3 implies that the series $\sum_{k=0}^{\infty} f_k$ converges on $\overline{B_\varrho(c)}$ absolutely and uniformly.

In order to prove the second statement we define $g_k(z) := k\, a_k(z-c)^{k-1}$ and $g := \sum\limits_{k=1}^{\infty} g_k$. As before we can derive

$$\|g_k\|_{\infty, \overline{B_\varrho(c)}} \leq k\, M\, \vartheta^{k-1}$$

and the ratio test, Theorem 13.16, implies the convergence of $\sum\limits_{k=1}^{\infty} k\, M\, \vartheta^{k-1}$. Note that $\frac{(k+1)M\vartheta^k}{kM\vartheta^{k-1}} = \frac{k+1}{k}\vartheta < 1$ for k large since $\vartheta < 1$ and

$$\lim_{k\to\infty} \frac{k+1}{k} = 1.$$

Theorem 16.3 gives the convergence result for the series $\sum\limits_{k=1}^{\infty} k\, a_k\,(z-c)^{k-1}$. $\square$

Remark 16.7. Since Theorem 16.6 gives uniform and absolute convergence for all $\overline{B_\varrho(c)}, \varrho < |z_1 - c|$, we obtain pointwise absolute convergence in the open disc $B_{\varrho_0}(c), \varrho_0 = |z_1 - c|$, i.e.

$$B_{\varrho_0}(c) = \bigcup_{\varrho < \varrho_0} \overline{B_\varrho(c)}.$$

In general we do not know convergence results of $T^c_{(a_k)}(z)$ for $z \in \partial B_{\varrho_0}(c) = \{z \in \mathbb{C} | |z - \mathbb{C}| = |z_1 - c|\}$.

Definition 16.8. Let $T^c_{(a_k)}(z)$ be a formal power series. We call the largest $\varrho_0 > 0$ such that $T^c_{(a_k)}(z)$ converges for all $z \in B_{\varrho_0}(c)$ the **radius of convergence** of $T^c_{(a_k)}(z)$.

Of interest is

Corollary 16.9. If the radius of convergence $\varrho_0 > 0$ of a power series $T^c_{(a_k)}(z) = \sum\limits_{k=0}^{\infty} a_k(z - c)^k$ is finite, then for $z \in \mathbb{C}$ such that $|z - c| > \varrho_0$, i.e. $z \in \overline{B_\varrho(c)}^C$, we have

$$\sup_{k\in\mathbb{N}_0} |a_k||z - c|^k = \infty. \tag{16.11}$$

Proof. Suppose that $z_1 \in B_{\varrho_0}(c)^{\complement}$ and $|a_k (z_1 - c)^k| \leq M < \infty$ for all $k \in \mathbb{N}_0$. The proof of Theorem 16.6 yields that the power series $T^c_{(a_k)}(z)$ must converge for all $z \in \mathbb{C}$ such that $|z - c| < |z_1 - c|$. However for $z \in \mathbb{C}$ in the annulus $\varrho_0 < |z - c| < |z_1 - c|$ we get a contradiction to the maximality of ϱ_0. $\quad\square$

The next result tells us how we can (in principle) determine ϱ_0.

Theorem 16.10 (Cauchy and Hadamard). *The radius convergence ϱ_0 of the power series $\sum_{k=0}^{\infty} a_k (z - c)^k$ is given by*

$$\varrho_0 = \liminf_{k \to \infty} \frac{1}{\sqrt[k]{|a_k|}} = \frac{1}{\limsup_{k \to \infty} \sqrt[k]{|a_k|}}. \tag{16.12}$$

Proof. We know that $\sum_{k=0}^{\infty} a_k (z - c)^k$ converges for $|z - c| = r < \varrho_0$ which implies that $\lim_{k \to \infty} |a_k| r^k = 0$. Thus there exists $N_0 \in \mathbb{N}$ such that $|a_k| r^k < 1$ for $k \geq N_0$ or $r < \frac{1}{\sqrt[k]{|a_k|}}$ for $k \geq N_0$. This implies of course that $r \leq \liminf_{k \to \infty} \frac{1}{\sqrt[k]{|a_k|}}$, but $r < \varrho_0$ was arbitrary, hence

$$\varrho_0 \leq \liminf_{k \to \infty} \frac{1}{\sqrt[k]{|a_k|}}.$$

When $\varrho_0 = \infty$ it follows, of course, that $\liminf_{k \to \infty} \frac{1}{\sqrt[k]{|a_k|}} = \infty = \varrho_0$. If $\varrho_0 < \infty$ and $|z - c| = r > \varrho_0$ then the sequence $(|a_k| r^k)_{k \in \mathbb{N}_0}$ is by Corollary 16.9 unbounded. Therefore we have $|a_k| r^k > 1$, i.e. $r > \frac{1}{\sqrt[k]{|a_k|}}$, for an infinite number of values of $k \in \mathbb{N}_0$. This implies $r \geq \liminf_{k \to \infty} \frac{1}{\sqrt[k]{|a_k|}}$ and again since r was arbitrary but larger than ϱ_0 we arrive at

$$\varrho_0 \geq \liminf_{k \to \infty} \frac{1}{\sqrt[k]{|a_k|}},$$

and the result follows. $\quad\square$

For power series, using the Cauchy-Hadamard result, we can obtain an improved ratio test, see R.Remmert [67].

Proposition 16.11. *Let* $(a_k)_{k \in \mathbb{N}_0}$, $a_k \in \mathbb{C}$ *and* $a_k \neq 0$ *for all* $k \in \mathbb{N}_0$, *and* $c \in \mathbb{C}$. *Suppose that* $\varrho_0 > 0$ *is the radius of convergence of the power series* $\sum\limits_{k=0}^{\infty} a_k (z - c)^k$. *Then*

$$\liminf_{k \to \infty} \frac{|a_k|}{|a_{k+1}|} \leq \varrho \leq \limsup_{k \to \infty} \frac{|a_k|}{|a_{k+1}|} \tag{16.13}$$

holds and in particular in the case that $\lim\limits_{k \to \infty} \frac{|a_k|}{|a_{k+1}|} = A$ *exists we have* $\varrho_0 = A$.

Proof. Let $0 < \sigma < \liminf_{k \to \infty} \frac{|a_k|}{|a_{k+1}|}$. It follows the existence of $l \in \mathbb{N}$ such that $\left| \frac{a_k}{a_{k+1}} \right| > \sigma$, i.e. $|a_{k+1}| \sigma < |a_k|$, for all $k \geq l$. With $\alpha := |a_l| \sigma^l$ it follows that $|a_{l+m}| \sigma^{l+m} \leq \alpha$ for all $m \geq 0$. Hence $(|a_k| \sigma^k)_{k \in \mathbb{N}}$ is a bounded sequence and Corollary 16.9 implies that $\sigma < \varrho_0$. Now assume that $0 < \limsup_{k \to \infty} \frac{|a_k|}{|a_{k+1}|} < \tau < \infty$.

In this case we can find $m \in \mathbb{N}$ such that $\left| \frac{a_n}{a_{n+1}} \right| > \tau$ or $|a_{n+1}| \tau > |a_n|$ for $n \geq m$. This yields for $0 < \beta := |a_m| \tau^m$ that $|a_{n+m}| \tau^{n+m} \geq \beta$ for all $n \geq 0$ and therefore the sequence $(|a_n| \tau^n)_{n \in \mathbb{N}}$ cannot converge to 0, implying that $\sum\limits_{k=0}^{\infty} |a_k| \tau^k$ cannot converge, hence $\tau \geq \varrho_0$ and the proposition is proved. $\square$

Finally we can link the convergence of power series with real coefficients and a real variable to that of the power series with the same coefficients but a complex variable.

Lemma 16.12. *Let* $(a_k)_{k \in \mathbb{N}_0}$, $a_k \in \mathbb{R}$, *and* $c \in \mathbb{R}$. *Suppose that the power series* $\sum\limits_{k=0}^{\infty} a_k (x - c)^k$ *converges for all* $x \in (c - \varrho_0, c - \varrho_0)$, $\varrho_0 > 0$. *Then the power series* $\sum\limits_{k=0}^{\infty} a_k (z - c)^k$ *converges for all* $z \in \mathbb{C}$ *such that* $|z - c| < \varrho_0$, *i.e.* $z \in D_{\varrho_0}(c) = B_{\varrho_0}(c) \subset \mathbb{C}$.

Proof. Let $0 < r < \varrho_0$. We know that $\sum\limits_{k=0}^{\infty} |a_k| r^k$ converges, hence for all $z \in \mathbb{C}$, $|z - c| \leq r$ it follows the convergence of $\sum\limits_{k=0}^{\infty} |a_k| |z - c|^k$ implying the convergence of $\sum\limits_{k=0}^{\infty} a_k (z - c)^k$ for $|z - c| \leq r$. Since $r < \varrho_0$ was arbitrary we have proved that $\sum\limits_{k=0}^{\infty} a_k (z - c)^k$ converges for all $z \in \mathbb{C}, |z - c| < \varrho_0$. $\square$

Corollary 16.13. *Let $I \subset \mathbb{R}$ be an interval and $f : I \to \mathbb{R}$ a function which is arbitrary often differentiable in $\mathring{I}$. Suppose that for some $c \in \mathring{I}$ the function f has the convergent Taylor expansion*

$$f(x) = \sum_{k=0}^{\infty} \frac{f^{(k)}(c)}{k!}(x - c)^k$$

in the interval $(c - \varrho_0, c + \varrho_0) \subset \mathring{I}$, $\varrho_0 > 0$. Then the power series

$$\sum_{k=0}^{\infty} \frac{f^{(k)}(c)}{k!}(z - c)^k$$

converges for all $z \in \mathbb{C}$, $|z - c| < \varrho_0$, to a continuous function extending f to the open disc with centre c and radius ϱ_0.

Thus we obtain the following continuous complex-valued functions by extending the corresponding Taylor series of the known real-valued function:

$$\exp(z) = \sum_{k=0}^{\infty} \frac{z^k}{k!}, \qquad\qquad |z| < \infty; \qquad (16.14)$$

$$\cos(z) = \sum_{k=0}^{\infty} \frac{(-1)^k}{(2k)!} z^{2k}, \qquad\qquad |z| < \infty; \qquad (16.15)$$

$$\sin(z) = \sum_{k=1}^{\infty} \frac{(-1)^{k+1}}{(2k-1)!} z^{2k-1}, \qquad\qquad |z| < \infty; \qquad (16.16)$$

$$\arctan(z) = \sum_{k=0}^{\infty} \frac{(-1)^k}{2k+1} z^{2k+1}, \qquad\qquad |z| < 1; \qquad (16.17)$$

$$\tilde{\ln}(1+z) = \sum_{k=1}^{\infty} \frac{(-1)^{k+1}}{k} z^k, \qquad\qquad |z| < 1; \qquad (16.18)$$

$$\frac{1}{1-z} = \sum_{k=0}^{\infty} z^k, \qquad\qquad |z| < 1; \qquad (16.19)$$

$$J_l = \sum_{k=0}^{\infty} \frac{(-1)^k}{2^{2k+l} k!(k+l)!} z^{2n+l}, \qquad\qquad |z| < \infty, \qquad (16.20)$$

where J_l is the Bessel function of order l, see Problem 7 to Chapter I.29. We will discuss some more examples below and in the problems. Some of these and other complex-valued functions we want to study now in more detail. The following three examples demonstrate a very important point: extending functions from subsets of $\mathbb{R}$ to subsets of $\mathbb{C}$ may dramatically change their properties.

Example 16.14. The function $g_{2N+1} : \mathbb{R} \to \mathbb{R}$, $g(x) = x^{2N+1}$ is for all $N \in \mathbb{N}$ a strictly monotone increasing bijective function with inverse $g_{2N+1}^{-1}(y) = \operatorname{sgn}(y)|y|^{\frac{1}{2N+1}}$. The complex extension h_{2N+1} of g_{2N+1} defined by $h_{2N+1} : \mathbb{C} \to \mathbb{C}$, $h_{2N+1}(z) = z^{2N+1}$ is for no $N \in \mathbb{N}$ injective. In fact for every real number $R > 0$ we find $2N + 1$ distinct pre-images of R namely

$$z_k = R^{\frac{1}{2N+1}} e^{\frac{2\pi i}{2N+1}k}, \quad k = 0, ..., 2N,$$

or interpreting $h_{2N+1}^{-1}(R)$ as the pre-image of $\{R\} \subset \mathbb{C}$

$$h_{2N+1}^{-1}(R) = R^{\frac{1}{2N+1}} E(2N+1)$$

where $E(2N+1)$ is again the group of $(2N+1)^{\text{th}}$ roots of unity. (Note that for $G \subset \mathbb{C}$ and $a \in \mathbb{C}$ we write $G = \{z = ag \mid g \in G\}$).
Indeed, the equation $h_{2N+1}(z) = z^{2N+1} = R$ is equivalent to

$$|z|^{2N+1} e^{(2N+1)i\varphi} = Re^{i0}$$

which yields $|z| = R^{\frac{1}{2N+1}}$ and $\varphi \in E(2N+1)$.

Example 16.15. The complex exponential function $\exp(z) = \sum_{k=0}^{\infty} \frac{z^k}{k!}$ is a periodic function with period $\tau = 2\pi i$. To see this we first note that $\exp(z_1 + z_2) = \exp(z_1)\exp(z_2)$ also holds for all $z_1, z_2 \in \mathbb{C}$, see Problem 10 of Chapter 13. Now it follows for $z = x + iy$ that

$$\begin{aligned}
\exp(z + 2\pi i) &= \exp(x + i(y + 2\pi)) \\
&= e^x(\cos(y + 2\pi) + i\sin(y + 2\pi)) \\
&= e^x(\cos y + i\sin y) \\
&= \exp(x + iy) = \exp(z).
\end{aligned}$$

Thus the strictly monotone, hence injective function $\exp : \mathbb{R} \to \mathbb{R}$ is extended to a periodic function $\exp : \mathbb{C} \to \mathbb{C}$.

Example 16.16. Using the power series for $\exp, \cos$ and $\sin$ we find

$$\cos z = \frac{\exp(iz) + \exp(-iz)}{2} \tag{16.21}$$

$$\sin z = \frac{\exp(iz) - \exp(-iz)}{2i}. \tag{16.22}$$

For $z = iy$, i.e. a purely imaginary number, we find

$$\cos(iy) = \frac{e^{-y} + e^y}{2} \quad \text{and} \quad \sin(iy) = \frac{e^{-y} - e^y}{2i},$$

implying that $|\cos(iy)|$ and $|\sin(iy)|$ are unbounded. While $\cos$ and $\sin$ as functions from $\mathbb{R}$ to $\mathbb{R}$ are bounded functions, their complex extensions to $\mathbb{C}$ are not.

We want to extend the binomial formula

$$(1+z)^n = \sum_{k=0}^{n} \binom{n}{k} z^k \tag{16.23}$$

to non-integer values. For this we define for $a \in \mathbb{C}$ and $n \in \mathbb{N}$ the **binomial coefficients**

$$\binom{a}{n} := \frac{a(a-1) \cdot \ldots \cdot (a-n+1)}{n!} = \prod_{k=1}^{n} \frac{a-k+1}{k}, \qquad \binom{a}{0} := 1. \tag{16.24}$$

An easy calculation shows that

$$\binom{a}{n+1} = \frac{a-n}{n+1} \binom{a}{n}, \qquad a \in \mathbb{C} \text{ and } n \in \mathbb{N}_0. \tag{16.25}$$

Definition 16.17. *For $a \in \mathbb{C}$ we define the **binomial series** by*

$$B_a(z) := \sum_{k=0}^{\infty} \binom{a}{k} z^k. \tag{16.26}$$

Clearly, for $a \in \mathbb{N}$ formula (16.26) reduces to (16.23). For $a \in \mathbb{C} \backslash \mathbb{N}$ we have

Proposition 16.18. *For all $a \in \mathbb{C} \backslash \mathbb{N}$ the binomial series has radius of convergence 1.*

Proof. For $a \in \mathbb{C}\backslash\mathbb{N}$ we always have $\binom{a}{k} \neq 0$ and with (16.25) we find

$$\binom{a}{k} \Big/ \binom{a}{k+1} = \frac{k+1}{a-k} = -\frac{1+\frac{1}{k}}{1-\frac{a}{k}}, \quad k \in \mathbb{N},$$

implying $\lim\limits_{k \to \infty} \left| \binom{a}{k} \Big/ \binom{a}{k+1} \right| = 1$ and therefore by Theorem 16.11 it follows that the radius of convergence ϱ_0 is equal to 1. □

In Problem 8 we will use the formula

$$\binom{a-1}{k} + \binom{a-1}{k-1} = \binom{a}{k}, \quad k \geq 1, \tag{16.27}$$

to derive

$$(1+z) B_{a-1}(z) = B_a(z). \tag{16.28}$$

For the next example we need some notational preparation. For a real number $\alpha \in \mathbb{R}$ the **Pochhammer symbols** are defined by

$$(\alpha)_0 := 1, (\alpha)_n = \alpha \, (\alpha+1) \cdot \ldots \cdot (\alpha+n-1), \quad n \in \mathbb{N}. \tag{16.29}$$

It follows that $(1)_n = n!$, and further $(-k)_n = 0$ for $k \in \mathbb{N}$ and $k \leq n-1$.

Theorem 16.19. *Let $\alpha, \beta \in \mathbb{R}$ and $\gamma \in \mathbb{R}\backslash(-\mathbb{N}_0)$. The series*

$$_2F_1(\alpha, \beta; \gamma; z) := \sum_{k=0}^{\infty} \frac{(\alpha)_k \, (\beta)_k}{(\gamma)_k \, k!} z^k \tag{16.30}$$

converges for $|z| < 1$.

Proof. With $a_k := \frac{(\alpha)_k \, (\beta)_k}{(\gamma)_k \, k!}$ we find

$$\left| \frac{a_{k+1} z^{k+1}}{a_k z^k} \right| = \frac{|(\alpha)_{k+1} (\beta)_{k+1} (\gamma)_k k!|}{|(\alpha)_k (\beta)_k (\gamma)_{k+1}| (k+1)!} |z|$$

$$= \frac{|(\alpha+k)(\beta+k)|}{|(\gamma+k)(k+1)|} |z|.$$

Since

$$\lim_{k \to \infty} \frac{|(\alpha+k)(\beta+k)|}{|(\gamma+k)(k+1)|} = 1$$

the result follows. □

Definition 16.20. *The series* $_2F_1(\alpha, \beta; \gamma; z)$ *is called the* **hypergeometric series** *associated with* α, β, γ.

Example 16.21. The following hold:

$$_2F_1(1, 1; 1; z) = \sum_{k=0}^{\infty} z^k; \tag{16.31}$$

$$_2F_1(\alpha, 1; 1; z) = (1 - z)^{-\alpha}; \tag{16.32}$$

$$_2F_1(-\alpha, 1; 1; -z) = (1 + z)^{\alpha}; \tag{16.33}$$

$$z\,_2F_1(1, 1; 2; -z) = \ln(1 + z); \tag{16.34}$$

$$z\,_2F_1\left(\frac{1}{2}, 1; \frac{3}{2}; z^2\right) = \frac{1}{2}\ln\left(\frac{1 + z}{1 - z}\right); \tag{16.35}$$

$$z\,_2F_1\left(\frac{1}{2}, 1; \frac{3}{2}; -z^2\right) = \arctan z. \tag{16.36}$$

Thus we can identify many well-known functions as (modified) hypergeometric series.

There are many questions to answer on the behaviour of the complex extension of functions such as $\exp, \cos, \sin, \ln$ etc. In particular, since $\exp$ is periodic on $\mathbb{C}$ we can not expect $\ln$ to be the inverse, thus the relation $\ln \circ \exp$ and $\exp \circ \ln$ are in $\mathbb{C}$ at first glance meaningless. In order to investigate these functions it is helpful to introduce and discuss complex derivatives of complex-valued functions of a complex variable. This we will do in the next chapter. In the remaining part of this chapter we want to return to more geometric considerations by discussing the mapping properties of so called linear transformations or Möbius transformation.

Definition 16.22. *Let* $a, b, c, d \in \mathbb{C}$ *such that*

$$\det\begin{pmatrix} a & b \\ c & d \end{pmatrix} = ad - bc \neq 0. \tag{16.37}$$

We call the expression

$$w = w(z) := \frac{az + b}{cz + d}, \qquad z \in \mathbb{C}, \tag{16.38}$$

the **Möbius transformation** *or* **linear transformation** *associated with* a, b, c, d.

If $c = 0$ then it follows from (16.37) that $a \neq 0$ and $d \neq 0$. In particular, in the case where w is defined for all $z \in \mathbb{C}$. If $c \neq 0$ then w is defined for all $z \in \mathbb{C} \backslash \{-\frac{d}{c}\}$.

Now, if $c = 0$, then (16.38) yields

$$w = \frac{a}{d} z + \frac{b}{d} \quad \text{and} \quad z = \frac{d}{a} w - \frac{b}{d}, \tag{16.39}$$

i.e. by (16.38) a bijective mapping from $\mathbb{C}$ into $\mathbb{C}$ is defined. The classical and still used façon de parler is that for $c = 0$ a one-to-one mapping of the z-plane onto the w-plane is defined by (16.38).

In the case that $c \neq 0$ and $z \neq -\frac{d}{c}$ we find by a simple calculation

$$w = \frac{az + b}{cz + d} \quad \text{and} \quad z = \frac{-dw + b}{cw - a}, \tag{16.40}$$

and w attains all values of the complex plane except $w_0 = \frac{a}{c}$. For $z_0 \in \mathbb{C}$ we call $\mathbb{C} \backslash \{z_0\}$ the **complex plane punctured at z_0**. Our considerations so far yield

Proposition 16.23. *The Möbius transformation $w = \frac{az+b}{cz+d}$ maps the punctured plane $\mathbb{C} \backslash \{-\frac{d}{c}\}$ bijectively and continuously with a continuous inverse onto the punctured plane $\mathbb{C} \backslash \{\frac{a}{c}\}$.*

Next we show

Lemma 16.24. *The composition of two Möbius transformations $w = \frac{az+b}{cz+d}$ and $\omega = \frac{\alpha z + \beta}{\gamma z + \delta}$ is again a Möbius transformation given by*

$$W(z) = \omega(w(z)) = \frac{Az + B}{Cz + D} \tag{16.41}$$

where

$$A = \alpha\, a + \beta\, c, B = \alpha\, b + \beta\, d, C = \gamma\, a + \delta\, c, D = \gamma\, b + \delta\, d, \tag{16.42}$$

i.e. we have $\begin{pmatrix} A & B \\ C & D \end{pmatrix} = \begin{pmatrix} \alpha & \beta \\ \gamma & \delta \end{pmatrix} \begin{pmatrix} a & b \\ c & d \end{pmatrix}$. Moreover the Möbius transformation W maps the punctured plane $\mathbb{C} \backslash \{-\frac{D}{C}\}$ onto the punctured place $\mathbb{C} \backslash \{\frac{A}{C}\}$.

Proof. First we note that

$$\det \begin{pmatrix} A & B \\ C & D \end{pmatrix} = AD - BC = (a\,d - b\,c)(\alpha\,\delta - \beta\,\gamma) \neq 0,$$

and since by assumption $a\,d - b\,c \neq 0$ and $\alpha\,\delta - \beta\,\gamma \neq 0$ it follows that by (16.41) a Möbius transformation with the claimed mapping properties is given. Moreover we have

$$W(z) = \omega(w(z)) = \frac{\alpha\,w\,(z) + \beta}{\gamma\,w\,(z) + \delta}$$

$$= \frac{\alpha\left(\frac{a\,z+b}{c\,z+d}\right) + \beta\left(\frac{c\,z+d}{c\,z+d}\right)}{\gamma\left(\frac{a\,z+b}{c\,z+d}\right) + \delta\left(\frac{c\,z+d}{c\,z+d}\right)}$$

$$= \frac{(\alpha\,a + \beta\,c)z + \alpha\,b + \beta\,d}{(\gamma\,a + \delta\,c)z + \gamma\,b + \delta\,d}$$

implying (16.41) and (16.42). $\qquad\square$

Corollary 16.25. *For the two Möbius transformations* $w = \frac{az+b}{cz+d}$ *and* $\omega = -\frac{dz+b}{cz-a}$ *we find* $W(z) = \omega\,(w\,(z)) = z$.

Proof. This follows immediately from (16.42). $\qquad\square$

Remark 16.26. It seems that the combined content of Proposition 16.23 and Corollary 16.25 is that the Möbius transformation form a group. However there is a problem with the domain of definition and we will resolve this shortly.

Remark 16.27. The reader might have noticed a change in some formulations and notations, for example we write $w = \frac{az+b}{cz+b}$ and $z = \frac{-dw+b}{cw-a}$, i.e. we use w as a symbol for a mapping, the image of one point under a mapping and as independent variable of a new mapping, as we do with z. This is due to the fact that many traditional notations in the theory of complex variables are still used and no adaption to more modern notation rarely took place. In order to develop the students ability to read the standard literature we shall stick to this common notation.

Every Möbius transformation associated with $a, b, c, d \in \mathbb{C}$ can be written as a composition of simple transformations. For $c \neq 0$ we have

$$w = w_1 + \frac{a}{c}, \quad w_1 = \frac{1}{w_2}, \quad w_2 = \frac{-c\,(cz+d)}{ad - bc}, \tag{16.43}$$

and for $c = 0$ we find

$$w = \frac{a}{d}\, w_1, \quad w_1 = z + \frac{b}{a}. \tag{16.44}$$

The transformations $w = \alpha\, z$, $\alpha \neq 0$, and $w = z + \beta$ map straight lines onto straight lines, which is trivial, and they map circles onto circles. Indeed, if $|z - z_0| = \varrho_0$, then we find

$$\left|\frac{1}{\alpha}\, w - z_0\right| = \varrho \quad \text{or} \quad |w - \alpha\, z_0| = |\alpha|\varrho_0,$$

and

$$|w_1 - \beta - z_0| = |w_1 - (z_0 + \beta)| = \varrho_0.$$

However the transformation $w = \frac{1}{z}$ has the same property. Consider equation (15.16) which gives circles and straight lines, i.e. looking at

$$A\, z\bar{z} + \overline{C}z + C\bar{z} + B = 0, \quad A, B \in \mathbb{R},\ C \in \mathbb{C}, \tag{16.45}$$

we find with $w = \frac{1}{z}$

$$A\, \frac{1}{w}\frac{1}{\bar{w}} + \overline{C}\,\frac{1}{w} + C\,\frac{1}{\bar{w}} + B = 0,$$

or

$$A + \overline{C}\,\bar{w} + C\, w + B\, w\bar{w} = 0, \tag{16.46}$$

i.e. the image of the set of all z satisfying (16.45) is a set of points in the w-plane satisfying a similar equation. Now, if $A = 0$, then (16.46) yields that the straight line defined by (16.45) is mapped onto a circle containing the point $w = 0$, and if $B = 0$, then a circle in the z-plane passing through $z = 0$ is mapped onto a straight line in the w-plane. In all other cases we find that circles are mapped onto circles. Combining these results we obtain

Proposition 16.28. *Every Möbius transformation $w = \frac{az+b}{cz+d}$ maps a straight line or a circle in the z-plane onto a straight line or a circle in the w-plane. In particular, if a circle in the z-plane is given by (16.45) with $A \neq 0$ and $B \neq 0$, then w maps this circle onto a circle in the w-plane.*

Thus we may look at complex-valued mappings of a complex variable also from the following point of view: they map certain nice domains onto nice domains and by this certain properties of geometric objects are left invariant, i.e. unchanged. Eventually we will see in Chapter 28 that a corner stone of classical complex analysis, the Riemann mapping theorem, is exactly concerned with such a statement.

In Appendix III we will discuss the Möbius transformation further, in particular how we can extend them to the Riemann sphere and how to interpret this extension geometrically.

Problems

1. a) Prove that the series $\sum_{n=1}^{\infty} \frac{1}{n^4+z^4}$ converges uniformly for $1 < |z| < 2$.

    b) Does the series $\sum_{n=1}^{\infty} \frac{\sin nz}{n^3}$ converge uniformly for $|z| < 1$? Compare with the result for the series $\sum_{n=1}^{\infty} \frac{\sin nx}{n^3}, x \in \mathbb{R}$.

2. Find the radius of convergence of the following power series:

    a) $\sum_{n=0}^{\infty} \frac{(z-i)^n}{(n+2)(n+3)}$;

    b) $\sum_{n=1}^{\infty} \frac{(-1)^{n-1}(n+1)(z+i)^n}{3^n(n^2+1)^{\frac{1}{2}}}$;

    c) $\sum_{n=0}^{\infty} \frac{(n!)^2}{(2n)!} z^n$.

    Hint to part c): rewrite the Sterling formula as $n! = n^n e^{-n} r_n$ with $\lim_{n\to\infty} r_n^{\frac{1}{n}} = 1$.

3. Prove that the **Mittag-Leffler function** $E_\alpha(z) := \sum_{k=0}^{\infty} \frac{z^k}{\Gamma(\alpha k+1)}$, $\alpha \in \mathbb{N}$ converges in $\mathbb{C}$, but for $\alpha = 0$ it converges in $|z| < 1$. Verify that

$$E_0(z) = \frac{1}{1-z},$$
$$E(z^2) = \cosh x,$$
$$E(-z^2) = \cos z.$$

4. Prove that the **generalised Mittag-Leffler function** $E_{\alpha,\beta}(z) := \sum_{k=0}^{\infty} \frac{z^k}{\Gamma(\alpha k+\beta)}$, $\alpha, \beta \in \mathbb{N}$, converges in $\mathbb{C}$ and verify that:

$$E_{1,1}(z) = e^z,$$
$$E_{1,2}(z) = \frac{e^z - 1}{z},$$
$$E_{2,2}(z^2) = \frac{\sinh z}{z}.$$

Remark: once we have extended the Γ-function as a meromorphic function to $\mathbb{C}$ we may increase the range of the parameters to $\operatorname{Re}\alpha > 0$, $\beta \in \mathbb{C}$, compare with [31].

5. Assume that for $|z| < 2\pi$ the equality

$$(*) \qquad \frac{z}{e^z - 1} = \sum_{k=0}^{\infty} \frac{B_k}{k!} z^k$$

holds where the **Bernoulli numbers** are defined by this equality. (This can be thought of as expanding $x \mapsto \frac{x}{e^x-1}$ into a Taylor series about 0 for $x \in (-2\pi, 2\pi)$, then the corresponding Taylor coefficients are $\frac{B_k}{k!}$.)

a) For $N \in \mathbb{N}, N > 1$, prove

$$\frac{B_0}{0!N!} + \frac{B_1}{1!(N-1)!} + \cdots + \frac{B_{N-1}}{(N-1)!1!} = 0$$

as well as

$$\binom{N}{0} B_0 + \binom{N}{1} B_1 + \cdots + \binom{N}{N-1} B_{N-1} = 0.$$

Further find B_1, B_2, B_3 and B_4.

b) Prove that $B_N = 0$ if $N > 1$ is an odd number.
Hint: show that $g(x) = \frac{x}{e^x-1} + \frac{x}{2} = \sum_{k=0}^{\infty} \frac{B_{2k}}{(2k)!} x^{2k}$.

6. Use that $\frac{z}{e^z-1} + \frac{z}{2} = \sum_{k=0}^{\infty} \frac{B_{2k}}{(2k)!} z^{2k}$ to show that

$$\pi z \cos t(\pi z) := \pi z \frac{\cos(\pi z)}{\sin(\pi z)} = \sum_{k=0}^{\infty} (-1)^k \frac{(2\pi)^{2k}}{(2k)!} z^{2k}.$$

7. Prove that for $a \in \mathbb{N}$ it follows that $\binom{a}{k} = 0$ for $k > a$. Moreover show that the binomial series extends the binomial theorem to $(1+x)^a, a \in \mathbb{R}$, $x \in \mathbb{R}, |x| < 1$, by expanding $x \mapsto (1+x)^a$ into a Taylor series about 0. From the latter result derive the formula $B_a(x)B_b(x) = B_{a+b}(x)$ for $|x| < 1, a, b \in \mathbb{R}$.

8. For $a \in \mathbb{C}$ and $k \geq 1$ prove the identity

$$\binom{a-1}{k} + \binom{a-1}{k-1} = \binom{a}{k}$$

and show that with $B_a(z) := \sum_{k=0}^{\infty} \binom{a}{k} z^k$ it follows that $(1+z)B_{a-1}(z) = B_a(z)$.

9.   a) Denote by $B(\alpha, \beta)$ the beta-function and prove that for $\alpha, \beta > 0$ we have:

$$(\alpha)_n = \frac{\Gamma(\alpha+n)}{\Gamma(\alpha)} \quad \text{and} \quad \frac{(\alpha)_n}{(\beta)_n} = \frac{B(\alpha+n, \beta-\alpha)}{B(\alpha, \beta-\alpha)}.$$

b) Verify:

$$_2F_1(1, 1; 1; z) = \sum_{k=0}^{\infty} z^k,$$

and

$$z\,_2F_1(1, 1; 2; -z) = \widehat{\ln}(1+z).$$

10. Using the series representation of $_2F_1$ prove that for $|z| < 1$ we have

a) $_2F_1(\alpha, \beta; \gamma; z) = {}_2F_1(\beta, \alpha; \gamma; z)$;

b) $(\gamma - \alpha - \beta)_2F_1(\alpha, \beta; \gamma; z) + \alpha(1-z)_2F_1(\alpha+1, \beta; \gamma; z)$
$- (\gamma - \beta)_2F_1(\alpha, \beta-1; \gamma; z) = 0$.

11.   a) Let $z_1, z_2, z_3 \in \mathbb{C}$ be three distinct points and $w_1, w_2, w_2 \in \mathbb{C}$ be three further distinct points. Find the Möbius transformation which maps the point z_k onto w_k, $1 \leq k \leq 3$. Thus we have to find $a, b, c, d, ac - bd \neq 0$ such that $w_k = \frac{az_k+b}{cz_k+d}$, $k = 1, 2, 3$.

b) Find the Möbius transformation mapping 0 onto i, $-i$ onto 1 and 1 onto $-i$.

12. Find the image of $\mathbb{R} \subset \mathbb{C}$ under the following two Möbius transformations:

a) $w(z) = \frac{z-i}{z+i}$;

b) $w(z) = \frac{z-i}{z+i}, z \neq -1$.

13. Let $w(z) = \frac{az+b}{cz+d}$ be a Möbius transformation with $c \neq 0$. Find conditions for a, b, c, d such that w admits exactly two fixed points $\zeta_1, \zeta_2 \in \mathbb{C}$, i.e. $w(\zeta_k) = \zeta_k$ for $k = 1, 2$.

17 Complex Differentiation

Let $f : D \to \mathbb{C}$ be a function defined on a non-empty open subset of $\mathbb{C}$. For $z, z_0 \in D$, $z \neq z_0$, the **difference quotient**

$$\frac{f(z) - f(z_0)}{z - z_0} \tag{17.1}$$

is well defined. This observation makes the main difference between complex analysis, i.e. the analysis of a complex-valued function depending on a complex variable and real analysis of mappings from an open set in $\mathbb{R}^2$ to $\mathbb{R}^2$. For the latter mapping the difference quotient cannot be defined. We can divide by a non-zero element in the field $\mathbb{C}$, but we cannot divide by a non-zero element in the vector space $\mathbb{R}^2$.

Once (17.1) makes sense we may try to pass in (17.1) to the limit $z \to z_0$ as we do this in the case of a real-valued function defined on an open set in $\mathbb{R}$. Before starting to investigate the limit of (17.1) as z tends to z_0 let us agree to the following conventions commonly used in complex analysis:

A non-empty open set $D \subset \mathbb{C}$ is called a **domain** in $\mathbb{C}$;
A connected domain is called a **region** in $\mathbb{C}$.

Note that $\mathbb{C}$ always carries the topology generated by the metric induced by the modulus $|.|$, hence connectivity is well-defined, compare with Definition II.3.30. If not otherwise stated we assume functions to be defined on a domain.

Definition 17.1. *Let $f : D \to \mathbb{C}$ be defined on the domain $D \subset \mathbb{C}$.*

A. *We call f (complex) **differentiable** of $z_0 \in D$, or we say that f has the (complex) **derivative** $f'(z_0)$ at z_0 if*

$$\frac{df}{dz}(z_0) := f'(z_0) := \lim_{z \to z_0} \frac{f(z) - f(z_0)}{z - z_0} \tag{17.2}$$

exists.

B. *We call f **holomorphic** in the domain D if f is for every $z \in D$ differentiable. If f is holomorphic on $\mathbb{C}$ we call f an **entire function**.*

Obviously, if $f(z) = c \in \mathbb{C}$ is a constant function then $f'(z) = 0$ for all $z \in \mathbb{C}$. The following results are proved by using results on limits in metric spaces, more precisely in $\mathbb{C}$, and do not differ in their structure from those given for the case of real-valued functions of a real variable.

Corollary 17.2. *If $f : D \to \mathbb{C}$ is differentiable of $z_0 \in \mathbb{C}$ then f is continuous at z_0. In particular, holomorphic functions are continuous.*

Proposition 17.3. *For functions $f, g : D \to \mathbb{C}$ defined on a domain $D \subset \mathbb{C}$ and differentiable at $z_0 \in D$ the functions $f \pm g$, $af (a \in \mathbb{C})$, and $f \cdot g$ are differentiable at z_0 and the following hold:*

$$(f \pm g)'(z_0) = f'(z_0) \pm g'(z_0); \tag{17.3}$$

$$(af)'(z_0) = af'(z_0); \tag{17.4}$$

$$(f \cdot g)'(z_0) = f'(z_0)\, g(z_0) + f(z_0)\, g'(z_0) \quad \textbf{\textit{(Leibniz rule)}}. \tag{17.5}$$

In addition, if $g(z_0) \neq 0$ then $\frac{f}{g}$ is defined in an open neighbourhood of z_0 and differentiable at z_0 with derivative

$$\left(\frac{f}{g}\right)'(z_0) = \frac{f'(z_0)\, g(z_0) - f(z_0)\, g'(z_0)}{g(z_0)^2} \qquad \textbf{(quotient rule)}. \tag{17.6}$$

We deduce from Proposition 17.3 that the set of all holomorphic functions on a domain is with the natural operations an algebra over $\mathbb{C}$ which we denote by $\mathcal{H}(D)$. A further consequence of Proposition 17.3 is that all polynomials $p : \mathbb{C} \to \mathbb{C}$, $p(z) = \sum_{k=0}^{N} a_k z^k$, are (complex) differentiable. For this we need only to note that for $f(z) = z$ we have

$$\lim_{z \to z_0} \frac{f(z) - f(z_0)}{z - z_0} = \lim_{z \to z_0} \frac{z - z_0}{z - z_0} = 1.$$

Further, rational functions, i.e. quotients $h = \frac{f}{g}$ of polynomials f and g are differentiable on the set $\{z \in \mathbb{C} \mid g(z) \neq 0\}$. Since g is continuous it follows that $\{z \in \mathbb{C} \mid g(z) \neq 0\} = g^{-1}(\mathbb{C}\backslash\{0\})$ is open, i.e. a domain. The proof of the **chain rule** for real-valued functions of a real variable also extends to the complex case:

Proposition 17.4. *Let $D_1, D_2 \subset \mathbb{C}$ be two domains and $f : D_1 \to \mathbb{C}$ and $g : D_2 \to \mathbb{C}$ be two functions. Suppose that $f(D_1) \subset D_2$ and f is differentiable at $z_0 \in D_1$ as well as g is differentiable at $w_0 := f(z_0) \in D_2$. Then $g \circ f : D_1 \to \mathbb{C}$ is well defined, differentiable at z_0 and we have*

$$(g \circ f)'(z_0) = g'(f(z_0)) f'(z_0). \tag{17.7}$$

We now want to discuss the difference between complex differentiability and differentiability of mappings $h : U \to \mathbb{R}^2$, $U \subset \mathbb{R}^2$ open. For this we write $z = x + iy$ and identify sometimes $\mathbb{C}$ with $\mathbb{R}^2$. For a function $f : D \to \mathbb{C}, D \subset \mathbb{C}$, we obtain two new functions $u, v : D \to \mathbb{R}$ by decomposing f into real and imaginary part.

$$f(z) = u(z) + i v(z), \quad u = \mathrm{Re} f, \quad v = \mathrm{Im} f.$$

We also may write

$$f(x, y) = u(x, y) + i v(x, y)$$

or even

$$f(z) = u(x, y) + i v(x, y).$$

for $z = x + iy \in D$. This allows us for example to take partial derivatives of u and v with respect to x and y.

Now let $f : D \to \mathbb{C}$ be complex differentiable at $z \in D$. Since D is open we can find a disc $D_r(z) \subset D$ for $r > 0$ sufficiently small. Let $h \in \mathbb{R}$ such that $z + h, z + ih \in D$. Thus we can consider the two difference quotients

$$\frac{f(x + h, y) - f(x, y)}{h} = \frac{f(z + h) - f(z)}{h} \tag{17.8}$$

and

$$\frac{f(x, y + h) - f(x, y)}{ih} = \frac{f(z + ih) - f(z)}{ih}. \tag{17.9}$$

Passing in (17.8) to the limit as h tends to 0 we obtain

$$\begin{aligned} f'(z) &= \lim_{h \to 0} \frac{f(z + h) - f(z)}{h} = \lim_{h \to 0} \frac{f(x + h, y) - f(x, y)}{h} \\ &= \lim_{h \to 0} \frac{u(x + h, y) - u(x, y) + i(v(x + h, y) - v(x, y))}{h} \\ &= u_x(x, y) + i v_x(x, y), \end{aligned}$$

while passing in (17.9) to the limit as h tends to 0 we find

$$f'(z) = \lim_{h \to 0} \frac{f(z + ih) - f(z)}{ih} = \lim_{h \to 0} \frac{f(x, y + h) - f(x, y)}{ih}$$

$$= \lim_{h \to 0} \frac{(u(x, y + h) - u(x, y)) + i(v(x, y + h) - v(x, y))}{ih}$$

$$= \frac{1}{i}(u_y(x, y) + i v_y(x, y)) = v_y(x, y) - i u_y(x, y).$$

By calculating these limits we have used the fact that a limit in $\mathbb{C}$ exists if and only if the corresponding limits for the real and imaginary part exists. Further we used our standard notation of partial derivatives, i.e. $u_x = \frac{\partial u}{\partial x}$ etc. From the considerations made above we get

Theorem 17.5. *If $f : D \to \mathbb{C}$, $f = u + iv$, is differentiable at $z = x + iy$ then the **Cauchy-Riemann differential equations***

$$u_x(x, y) = v_y(x, y) \ \text{ and } \ v_x(x, y) = -u_y(x, y) \tag{17.10}$$

must hold at $z = x + iy$.

Proof. We know that

$$u_x(x, y) + i v_x(x, y) = f'(z) = v_y(x, y) - i u_y(x, y),$$

but the real and imaginary part of the complex number $f'(z)$ is uniquely determined. $\square$

Since $f : D \to \mathbb{C}$ can be viewed as a function defined on a subset $D \subset \mathbb{R}^2$ with values in $\mathbb{R}^2$ and its differential at $(x, y) \in D$ would be the Jacobi matrix

$$\begin{pmatrix} u_x(x, y) & u_y(x, y) \\ v_x(x, y) & v_y(x, y) \end{pmatrix}.$$

In order that this matrix represents a complex number we need to have (as discussed in Chapter 13)

$$u_x(x, y) = v_y(x, y) \ \text{ and } \ v_x(x, y) = -u_y(x, y),$$

i.e. the Cauchy-Riemann differential equations must hold. Thus complex differentiability is a much more restrictive notion compared with differentiability for functions $f : D \to \mathbb{R}^2$, $D \subset \mathbb{R}^2$. Indeed we have

Theorem 17.6. *Let $f : D \to \mathbb{C}$ be a mapping. Then f is complex differen-tiable at $z = x + iy \in D$ if and only if $\begin{pmatrix} u \\ v \end{pmatrix} : D \to \mathbb{R}^2$ is real differentiable at $(x, y) \in D$ and the Cauchy-Riemann differential equations hold.*

Proof. We have seen that complex differentiability implies differentiability in $\mathbb{R}^2$. We have also seen that the Jacobi matrix of $\begin{pmatrix} u \\ v \end{pmatrix}$ at (x, y) represents a complex number if and only if the Cauchy-Riemann differential equations hold. □

We know in the real-valued case that the existence of partial derivatives does not imply the differentiability of function $f : D \to \mathbb{R}^2$, $D \subset \mathbb{R}^2$. Howe-ver if all first order partial derivatives are continuous, then the function is differentiable. This yields

Theorem 17.7. *Let $u, v : D \to \mathbb{R}$ be functions having continuous first order partial derivatives and suppose that u, v satisfy the Cauchy-Riemann diffe-rential equations. Then the function $f : D \to \mathbb{C}$, $D \subset \mathbb{C}$, $f(z) = u(z) + iv(z)$ is complex differentiable.*

We want to study some examples for the Cauchy-Riemann differential equa-tion.

Example 17.8. Consider the function $f : \mathbb{C} \to \mathbb{C}$, $f(z) = z$. Thus with $u(x, y) = x$ and $v(x, y) = y$ we have $f(z) = u(z) + iv(z)$, or $f(x, y) = u(x, y) + iv(x, y)$, and moreover

$$u_x(x, y) = 1, \; v_x(x, y) = 0, \; u_y(x, y) = 0, \; v_y(x, y) = 1$$

and we find as expected

$$u_x = 1 = v_y, \qquad\qquad v_x = 0 = -u_y.$$

Now let us consider $\overline{f}(z) = \overline{z} = x - iy$. The real part is again $u(x, y) = x$ while the imaginary part is $\tilde{v}(x, y) = -y$. We also find

$$u_x(x, y) = 1, \; \tilde{v}_x(x, y) = 0, \; u_y(x, y) = 0, \; \tilde{v}_y(x, y) = -1$$

and consequently $u_x(x, y) \neq \tilde{v}_y(x, y)$. Therefore the function $z \mapsto \overline{z}$ is **not** complex differentiable. Nonetheless the real function $\begin{pmatrix} u \\ \tilde{v} \end{pmatrix} : D \to \mathbb{R}^2$

is differentiable with differential $\begin{pmatrix} 1 & 0 \\ 0 & -1 \end{pmatrix}$. One consequence of this observation is that while $\mathbb{C}$ and the set $C(D)$ of all continuous functions $f : D \to \mathbb{C}$, $D \subset \mathbb{C}$, are algebras with involutions, namely $z \mapsto \overline{z}$ and $f \mapsto \overline{f}$, respectively, for the algebra of holomorphic functions $\mathcal{H}(D)$ the mapping $f \mapsto \overline{f}$ is not an involution since in general $\overline{f} \notin \mathcal{H}(D)$ for $f \in \mathcal{H}(D)$.

Example 17.9. For the complex exponential function we know the equality, $z = x + iy$,

$$\exp(z) = e^x(\cos y + i \sin y) = e^x \cos y + i e^x \sin y.$$

With $u(x, y) = e^x \cos y$ and $v(x, y) = e^x \sin y$ we find

$$u_x(x, y) = e^x \cos y, \ u_y(x, y) = -e^x \sin y, \ v_x(x, y) = e^x \sin y, \ v_y(x, y) = e^x \cos y,$$

and it follows that

$$u_x(x, y) = e^x \cos y = v_y(x, y)$$
$$v_x(x, y) = e^x \sin y = -u_y(x, y).$$

Thus $\exp : \mathbb{C} \to \mathbb{C}$ is complex differentiable.

Example 17.10. Since the line $\{z = x + iy \in \mathbb{C} \mid \mathrm{Re}\, z = x = 0\}$ is closed the set $\mathbb{C} \backslash \{z \in \mathbb{C} \mid \mathrm{Re}\, z = 0\}$ is a domain on which we can define the function

$$L(z) := \frac{1}{2} \ln(x^2 + y^2) + i \arctan \frac{y}{x} = u(x, y) + i\, v(x, y), \qquad (17.11)$$

where we still use $\ln$ and $\arctan$ as defined in Volume I as inverse functions to the real-valued exponential function $\xi \mapsto e^\xi$, $\xi \in \mathbb{R}$, and inverse to the tangent function restricted to $(-\frac{\pi}{2}, \frac{\pi}{2})$. Taking partial derivatives of u and v we obtain

$$u_x(x, y) = \frac{x}{x^2 + y^2}, \qquad u_y(x, y) = \frac{y}{x^2 + y^2}$$

as well as

$$v_x(x, y) = -\frac{y}{x^2 + y^2}, \qquad v_y = \frac{x}{x^2 + y^2}.$$

Hence we find $u_x = v_y$ and $v_x = -v_y$ on $\mathbb{C} \backslash \{z \in \mathbb{C} \mid \mathrm{Re}\, z = 0\}$ implying that L is on this domain complex differentiable.

Example 17.11. Let $G \subset \mathbb{C}$ be a region and $f : G \to \mathbb{C}, f = u + iv$, a function. Assume that $\operatorname{Re} f = u = c \in \mathbb{R}$ in G, i.e. u is a constant function. If f is not constant then f is not complex differentiable. Indeed, if f was complex differentiable then the Cauchy-Riemann differential equations imply $u_x = v_y$ and $u_y = -v_x$, but since $u = c$ it follows that $u_x = u_y = v_x = v_y = 0$. Since G is connected this implies that v must be constant too, see Problem 14 of Chapter II.6. Note that if G is not a region but a domain, then f must be constant on every connectivity component of G but on each component the constant may have a different value. Such a function is **locally constant** in the sense that every point of its domain has an open neighbourhood on which it is constant.

Exercise 17.12. *Let $f : G \to \mathbb{C}$ be a function which on the region $G \subset \mathbb{C}$ satisfies either $\operatorname{Im} f = c \in \mathbb{R}$ or $|f| = c \in \mathbb{R}$. Prove that if f is complex differentiable then f must be constant.*

Example 17.13. Let $f : D \to \mathbb{C}, D \subset \mathbb{C}$, be a function with $f = u + iv$. We can also interpret f as a mapping $g = \begin{pmatrix} u \\ v \end{pmatrix}$ from $D \subset \mathbb{R}^2$ to $\mathbb{R}^2$. The Jacobi determinant of g has some geometric meaning, for example when we think of the transformation theorem for integrals. Assuming that f is complex differentiable, hence u and v satisfy the Cauchy-Riemann differential equations we find

$$\det J_g(x) = \det \begin{pmatrix} u_x & u_y \\ v_x & v_y \end{pmatrix} = u_x v_y - u_y v_x.$$
$$= u_x^2 + u_y^2 = v_x^2 + v_y^2.$$

Using further that $u_x^2 = v_y^2$ and $u_y^2 = v_x^2$ and our calculation leading to the Cauchy-Riemann differential equations we find

$$|f'(z)|^2 = f'(z) \overline{f'(z)} = (u_x + i v_x)(u_x - i v_x)$$
$$= u_x^2 + v_x^2 = u_x^2 + u_y^2.$$

Thus we arrive at the remarkable fact that

$$|f'(z)|^2 = \det \begin{pmatrix} u_x(x,y) & u_y(x,y) \\ v_x(x,y) & v_y(x,y) \end{pmatrix} \geq 0. \tag{17.12}$$

Polar coordinates are quite helpful in many situations and if $f : D \to \mathbb{C}$, $D = B_\varrho(0) = \{z \in \mathbb{C} \,|\, |z| < \varrho\} = \{(x, y) \in \mathbb{R}^2 \,|\, x^2 + y^2 < \varrho^2\}$, is complex diffe-rentiable with real part u and imaginary part v we may consider u and v as functions of the polar coordinates (r, φ). The Cauchy-Riemann differential equations in polar coordinates read as

$$\frac{\partial u}{\partial r} = \frac{1}{r} \frac{\partial v}{\partial \varphi} \quad \text{and} \quad \frac{\partial v}{\partial r} = -\frac{\partial u}{\partial \varphi}, \tag{17.13}$$

see Problem 6.

We want to enlarge the class of examples of complex differentiable functions by looking at power series with postive radius of convergence. Recall that by Abel's theorem, Theorem 16.6 the convergence of $\sum_{k=0}^{\infty} a_k(z - c)^k$ in $\overline{B_\varrho(c)}$ en-tails the convergence of the series $\sum_{k=1}^{\infty} a_k k(z-c)^{k-1}$ in $\overline{B_\varrho(c)}$ and in both cases the convergence is uniform if $\varrho < \varrho_0$ where ϱ_0 is the radius of convergence of $\sum_{k=0}^{\infty} a_k (z - c)^k$. By the Cauchy-Hadamard theorem, Theorem 16.10, we further know that $\varrho_0 = \lim_{k \to \infty} \inf \frac{1}{\sqrt[k]{|a_k|}}$. It is worth noting that $\lim_{k \to \infty} \sqrt[k]{k} = 1$ implies that

$$\lim_{k \to \infty} \inf \frac{1}{\sqrt[k]{|a_k| k}} = \lim_{k \to \infty} \inf \frac{1}{\sqrt[k]{|a_k|}},$$

i.e. ϱ_0 is also the radius of convergence of the series $\sum_{k=1}^{\infty} a_k k(z - c)^{k-1}$. Since $\frac{d}{dz}(a_k (z - c)^k) = a_k k (z - c)^{k-1}$ we are led to conjecture that

$$\frac{d}{dz} \sum_{k=0}^{\infty} a_k (z - c)^k = \sum_{k=0}^{\infty} \frac{d}{dz}(a_k (z - c)^k) = \sum_{k=1}^{\infty} a_k k (z - c)^{k-1},$$

which we are now going to prove. For every complex differentiable function

$$\frac{d}{dz} f(z - c) = \left(\frac{df}{dz}\right)(z - c) = f'(z - c)$$

holds and therefore we may assume without loss of generality that $c = 0$.

Theorem 17.14. *Let* $T_{(a_k)}(z) = \sum_{k=0}^{\infty} a_k z^k$ *have radius of convergence* $\varrho_0 > 0$. *Then the function* $z \mapsto T_{(a_k)}(z)$ *is in* $B_{\varrho_0}(0) \subset \mathbb{C}$ *complex differentiable and* *we have*

$$T'_{(a_k)}(z) = \sum_{k=1}^{\infty} a_k \, k \, z^{k-1}. \tag{17.14}$$

Moreover, the series $T'_{(a_k)}(z)$ has again ϱ_0 as radius of convergence.

Remark 17.15. We already know this result for power series with real coefficients of a real variable, compare with Corollary I.29.7. However the proof of this corollary indirectly depends on the fundamental theorem of calculus which is now not at our disposal. Hence we need a different proof which we have adapted from [67].

Proof. We set $f(z) := T_{(a_k)}(z)$ and $g(z) := \sum_{k=1}^{\infty} k \, a_k \, z^{k-1}$. We have to prove that $f'(z) = g(z)$ for $z \in B_{\varrho_0}(0)$. Fix $z_0 \in B_{\varrho_0}(0)$ and define for $k \in \mathbb{N}$

$$g_k(z) := z^{k-1} + z^{k-2} \, z_0 + \dots + z^{k-j} \, z_0^{j-1} + \dots + z_0^{k-1}, \quad z \in \mathbb{C}.$$

Note that

$$g_k(z_0) = k \, z_0^{k-1}.$$

It follows that

$$f(z) - f(z_0) = \sum_{k-1}^{\infty} a_k \, (z^k - z_0^k)$$

$$= (z - z_0) \sum_{k=1}^{\infty} a_k \, g_k(z), \quad z \in B_{\varrho_0}(0).$$

With

$$f_1(z) := \sum_{k=1}^{\infty} a_k \, g_k(z)$$

we find

$$f(z) = f(z_0) + (z - z_0) f_1(z), \quad z \in B_{\varrho_0}(0),$$

and

$$f_1(z_0) = \sum_{k=1}^{\infty} k \, a_k \, z_0^{k-1} = g(z_0).$$

Suppose for a moment that f_1 is continuous at z_0. Then, since

$$\frac{f(z) - f(z_0)}{z - z_0} = f_1(z)$$

we find that

$$f'(z_0) = \lim_{z \to z_0} \frac{f(z) - f(z_0)}{z - z_0} = f_1(z_0) = g(z_0)$$

and the result was proved.

The continuity of f_1 is shown as follows. For $B_\varrho(0), |z_0| < \varrho < \varrho_0$ it follows that

$$\|a_k g_k(.)\|_{\infty, \overline{B_\varrho(0)}} \le |a_k| k \varrho^{k-1}$$

and therefore

$$\sum_{k=1}^{\infty} \|a_k g_k(.)\|_{\infty, B_\varrho}(0) \le \sum_{k=1}^{\infty} k |a_k| \varrho^{k-1} < \infty$$

by our previous considerations. Hence the series $\sum_{k=1}^{\infty} a_k g_k(z)$ converges uniformly in $B_\varrho(0)$ to a continuous function. Since $z_0 \in B_\varrho(0)$ the result follows. □

Suppose that $f : D \to \mathbb{C}$ is holomorphic in D. Then $f' : D \to \mathbb{C}$ defines a new complex-valued function in D which might be complex differentiable. Thus as in the case of real-valued functions of a real variable we can define for complex-valued functions of a complex variable higher order derivatives which we denote with $\frac{d^k f}{dz^k}(z)$ or $f^{(k)}(z)$ or for k small by $f''(z), f'''(z)$, etc. As we can derive Corollary I.29.8 from Corollary I.29.7, we can now derive from Theorem 17.14.

Corollary 17.16. *Let* $f(z) = \sum_{k=0}^{\infty} a_k (z - c)^k$ *be a power series with radius of convergence* $\varrho_0 > 0$. *The function* f *is arbitrarily complex differentiable in* $B_\varrho(c)$ *and for the coefficients* $a_k \in \mathbb{C}$ *we find*

$$a_k = \frac{1}{k!} f^{(k)}(c). \tag{17.15}$$

Proof. We may iterate the arguments of the proof of Theorem 17.14 to find for $z \in B_{\varrho_0}(0)$ and $m \in \mathbb{N}$ that

$$f^{(m)}(z) = \frac{d^m}{dz^m} f(z) = \sum_{k=m}^{\infty} k(k-1) \cdot \ldots \cdot (k - m + 1) a_k (z - c)^{k-m} \tag{17.16}$$

which implies already that f is arbitrarily often differentiable and setting in (17.16) $z = c$ formula (17.15) follows. □

Using the definition of the binomial coefficients we can rewrite (17.16) as

$$f^m(z) = \sum_{k=m}^{\infty} m! \binom{k}{m} a_k (z-c)^{k-m}. \tag{17.17}$$

Remark 17.17. In Chapter 22 we will prove that if $f : D \to \mathbb{C}$ is holomorphic then for every $z_0 \in D$ there exists a radius $\rho > 0$ such that in $B_\rho(z_0)$ the function f has a representation as a convergent power series. This implies that any holomorphic function is arbitrary often differentiable, i.e. one time differentiable implies arbitrarily often differentiable. Such a result is clearly false in the case of real-valued functions.

We now can deduce that all functions defined in (16.14)-(16.20) and (16.30) are in $B_{\varrho_0}(0)$, ϱ_0 being the radius of convergence, complex differentiable, in fact they are arbitrarily often complex differentiable in $B_{\varrho_0}(0)$. For $\exp, \cos, \sin$ and the Bessel functions $J_l, l \in \mathbb{N}_0$, this implies that they are entire functions. At the same time the question arises whether a given power series $f(z) = T^c_{(a_k)}(z)$ with radius of convergence $\varrho_0 > 0$ has a holomorphic extensions to a larger domain. For example the power series $\sum_{k=0}^{\infty} z^k$ has radius of convergence 1 and $\sum_{k=0}^{\infty} z^k = \frac{1}{1-z}$ holds for $|z| < 1$. The function $z \mapsto \frac{1}{1-z}$ is a rational function being defined and complex differentiable for all $z \in \mathbb{C}\backslash\{1\}$. Thus we may consider $z \mapsto \frac{1}{1-z}$ as a **holomorphic extension** or **continuation** of the power series $\sum_{k=0}^{\infty} z^k$. We will return to the problem of the existence of a holomorphic extension of a given holomorphic function at several occasions.

Once we know that a certain function $f : D \to \mathbb{C}$, $D \subset \mathbb{C}$, is complex differentiable we can apply our rules from Proposition 17.3 and Proposition 17.4 to calculate the complex derivatives. In particular we obtain for f extending a power series $\sum_{k=0}^{\infty} a_k(x-c)^k$, $a_k, c \in \mathbb{R}$, converging in the disc $B_\varrho(c) \subset \mathbb{C}$ that $f'(z)$ is the complex extension of $\frac{d}{dx}(\sum_{k=0}^{\infty}(x-c)^k)$. Therefore we find immediately that

$$\frac{d}{dz} \exp(z) = \exp(z), \tag{17.18}$$

$$\frac{d}{dz} \cos(z) = -\sin(z), \tag{17.19}$$

$$\frac{d}{dz} \sin(z) = \cos(z) \qquad (17.20)$$

hold for all $z \in \mathbb{C}$.

Definition 17.18. *Let G_1 and G_2 be two regions and $f : G_1 \to G_2$ a holomorphic mapping. We call f **biholomorphic** if f is bijective with holomorphic inverse $f^{-1} : G_2 \to G_1$. Two regions $\tilde{G}_1$ and $\tilde{G}_2$ are called **holomorphically equivalent** if there exists a biholomorphic mapping $\tilde{f} : \tilde{G}_1 \to \tilde{G}_2$.*

The Riemann mapping theorem, see Chapter 28, will classify all regions holomorphically equivalent to the unit disc $B_1(0)$. This theorem is one of the major results of complex analysis.

For the moment we note a further result which is essentially obtained by carrying over the proof of the corresponding result for a real-valued function of a real variable.

Proposition 17.19. *A mapping $f : G_1 \to G_2$ between two regions is biholomorphic if and only if f is holomorphic, bijective, f^{-1} is continuous and $f'(z) \neq 0$ for all $z \in G_1$. In this case we have*

$$(f^{-1})'(w) = \frac{1}{f'(z)}, \qquad w = f(z), \qquad (17.21)$$

i.e.

$$(f^{-1})'(w) = \frac{1}{f'(f^{-1}(w))}. \qquad (17.22)$$

For a proof we refer to Problem 9.

We want to return to the Cauchy-Riemann differential equations. Let $f : D \to \mathbb{C}, f = u + iv$, be a complex differentiable, i.e. a holomorphic function. If f has higher order complex derivatives, for example in the case where f is given by a convergent power series, then u and v must also have higher order partial derivatives. In light of Remark 17.17, eventually the assumption that u and v have continuous second order partial derivatives u_{xx}, $u_{xy} = u_{yx}$, u_{yy} and v_{xx}, $v_{x,y} = v_{y,x}$, v_{yy} will be no restriction. Recall that $w \in C^2(G), G \subset \mathbb{R}^2$, is called **harmonic** in G if $\Delta_2 w = \frac{d^2 w}{dx^2} + \frac{d^2 w}{dy^2} = w_{xx} + w_{yy} = 0$ in G, compare with Chapter II.9. Since in this part of the Course we deal essentially only with functions of one complex variable or two real variables we will write here Δ for Δ_2.

Theorem 17.20. *We assume that the real part and the imaginary part of the holomorphic function $f : D \to \mathbb{C}, f = u + iv$, have all second order partial derivatives in D and that they are continuous. Both, u and v are then harmonic functions in D, i.e. in D we have*

$$\Delta u = u_{xx} + u_{yy} = 0 \ \ and \ \ \Delta v = v_{xx} + v_{yy} = 0. \tag{17.23}$$

Proof. Since f is holomorphic in D the Cauchy-Riemann differential equations hold in D, i.e. $u_x = v_y$ and $u_y = -v_x$. Further, by assumption all second order partial derivatives of u and v exist and are continuous and therefore it follows that

$$u_{xx} = v_{yx}, \ u_{xy} = v_{yy} \ \text{ and } \ u_{yy} = -v_{xy}, \ u_{yx} = -v_{xx}$$

and in addition $u_{xy} = u_{yx}$ and $v_{xy} = v_{yx}$. Thus we find

$$u_{xx} + u_{yy} = v_{yx} - v_{xy} = 0$$

and

$$v_{xx} + v_{yy} = -u_{yx} + u_{xy} = 0.$$

$\square$

Example 17.21. A. Since $z \mapsto \exp(z)$ is an entire function $\operatorname{Re} \exp$ and $\operatorname{Im} \exp$ are harmonic functions in $\mathbb{R}^2$. Hence $\operatorname{Re} \exp(x, y) = e^x \cos y$ and $\operatorname{Im} \exp(x, y) = e^x \sin y$ are harmonic on $\mathbb{R}^2$.
B. From Example 17.10 we deduce that

$$(x, y) \mapsto \frac{1}{2} \log(x^2 + y^2) \ \text{ and } \ (x, y) \mapsto \arctan \frac{y}{x}, \ x \neq 0,$$

are harmonic functions.
C. If $p(z) = \sum_{k=0}^{N} a_k z^k$, $a_k \in \mathbb{C}$, is a polynomial then $\operatorname{Re} p(x, y)$ and $\operatorname{Im} p(x, y)$ are polynomials in two real variables which are on $\mathbb{R}^2$ harmonic functions. For $p(z) = z^k$, $k = 1, 2, 3, 4$ we find the following harmonic polynomials:

$$P_{1r}(x, y) = x, \ P_{1i}(x, y) = y, \tag{17.24}$$

$$P_{2r}(x, y) = x^2 - y^2, \ P_{2i}(x, y) = 2xy, \tag{17.25}$$

$$P_{3r}(x, y) = x^3 - 3xy^2, \ P_{3i}(x, y) = -y^3 + 2x^2 y, \tag{17.26}$$

$$P_{4r}(x, y) = x^4 + y^4 - 6x^2 y^2, \ P_{4i}(x, y) = 4x^3 y - 4xy^3. \tag{17.27}$$

As it turns out there is a close relation between harmonic functions of two real variables and holomorphic functions. We will discuss this relation in much more detail once we can resolve the following central question: given a harmonic function u on the open set $D \subset \mathbb{R}^2$. Can we find a second harmonic function v in D such that $f(z) = u(z) + iv(z) = u(x,y) + iv(x,y)$ is holomorphic in D? When this is true we call v a **conjugate harmonic function** to u in D.

Before we continue to develop further the general theory of holomorphic functions, in the next chapter we will discuss in more detail some concrete functions $f : D \to \mathbb{C}$. This is followed by deepening our understanding of some topological concepts in Chapter 19.

Problems

In the following $G \subset \mathbb{C}$ is always a region. Where appropriate we consider G as a subset of $\mathbb{R}^2$, $z = x + iy \in G$, $(x,y) \in G$.

1. Let $f, g : G \to \mathbb{C}$ be m-times complex differentiable functions. Prove that $f \cdot g : G \to \mathbb{C}$ is also m-times complex differentiable and we have

$$\frac{d^m}{dz^m}(f \cdot g)(z) = \sum_{l=0}^{m} \binom{m}{l} f^{(m-l)}(z) g^{(l)}(z).$$

2. Solve Exercise 17.12.

3. Let $f : [0, \infty) \to \mathbb{R}$ be a C^∞-function and consider the function $g : \mathbb{C} \to \mathbb{C}$, $g(z) := f(|z|^2)$. Is g complex differentiable?

4. Let $f : G \to \mathbb{C}$ be a complex differentiable function. Denote by G^* the set $G^* := \{x \in \mathbb{C} | z = \bar{w}, w \in G\}$. Sketch the set G^* and prove that $H : G^* \to \mathbb{C}, h(z) := \overline{f(\bar{z})}$ is complex differentiable.

5. Decompose $f : \mathbb{C} \to \mathbb{C}$, $f(z) = e^{z^2} + z^3$, into its real and imaginary parts and verify the Cauchy-Riemann differential equations for f.

6. Find the Cauchy-Riemann differential equations in polar coordinates, i.e. verify (17.13).

7. Prove that the following functions are complex differentiable:

  a) $\sum_{n=1}^{\infty} \frac{e^{inz}}{n^4}$, $\operatorname{Im} z > 0$;

  b) $\sum_{n=1}^{\infty} \frac{2^n}{z^n}$, $|z| > 2$.

8. Prove that $w : \mathbb{C} \setminus \{-1\} \to \mathbb{C} \setminus \{1\}$, $w(z) = \frac{z-1}{z+1}$, is a biholomorphic mapping.

9. Provide a proof for Proposition 17.19.

10. For $z = x + iy \in B_1(0)$ we define with $\zeta \in \partial B_1(0)$ fixed the **Poisson kernel** $P(\zeta, z)$ by

$$P(\zeta, z) := \frac{1}{2\pi} \frac{1 - |z|^2}{|\zeta - z|^2}.$$

With $z = re^{i\varphi}, \zeta = e^{i\vartheta}$, prove that

$$P(\zeta, z) = \frac{1}{2\pi} \frac{1 - r^2}{1 + r^2 - 2r \cos(\vartheta - \varphi)}.$$

Moreover, show that for $\zeta \in \partial B_1(0)$ fixed $P(\zeta, z) = P(\zeta, (x, y))$ is a harmonic function in $B_1(0) \subset \mathbb{R}^2$.
Hint: consider the real part of $z \mapsto \frac{\zeta + z}{\zeta - z}$.

11. Let $f : G_1 \to G_2$ be a holomorphic function and consider G_1 and G_2 also as subsets of $\mathbb{R}^2$. Let $h : G_2 \to \mathbb{R}$ be a harmonic function. Is the function $h \circ f : G_1 \to \mathbb{R}$ a harmonic function where with $f = u + iv$ and $z = x + iy$ the meaning of $h \circ f$ is $(h \circ f)(z) = (h \circ f)(x + iy) = h(u(x, y), v(x, y))$?

18 Some Important Functions

In this chapter we want to discuss some of the so called **elementary tran-scendental functions** defined on complex domains. We start with the exponential function exp defined by the power series

$$\exp(z) := e^z := \sum_{k=0}^{\infty} \frac{z^k}{k!} \tag{18.1}$$

which converges in $\mathbb{C}$ and hence exp is an entire function. By differentiating in (18.1) term by term we find that exp satisfies the differential equation

$$\frac{d}{dz} e^z = e^z \tag{18.2}$$

and further we have $\exp(0) = 1$. Since

$$\frac{d}{dz}(e^z e^{-z}) = e^z e^{-z} - e^z e^{-z} = 0$$

and since $\mathbb{C}$ is connected it follows that

$$e^z e^{-z} = 1$$

or

$$(e^z)^{-1} = e^{-z}, \tag{18.3}$$

which implies that $e^z \neq 0$ for all $z \in \mathbb{C}$. Moreover, with the same arguments as given in the proof of Lemma I.9.7 we obtain the well-known functional equation for the exponential function, but now for complex values of the argument:

$$e^{z+w} = e^z e^w. \tag{18.4}$$

We have already seen that on $\mathbb{C}$ the exponential function is periodic with purely imaginary period $2\pi i$:

$$e^{z+2\pi i} = e^z \text{ for all } z \in \mathbb{C}. \tag{18.5}$$

Thus, while on $\mathbb{R}$ the function $x \mapsto \exp(x)$ is injective with range $(0, \infty)$, on $\mathbb{C}$ the function $z \mapsto \exp(z)$ is not injective and hence we cannot expect to extend $\ln : (0, \infty) \to \mathbb{R}$ to become the inverse function of the complex exponential function. Indeed the logarithmic function is a much more difficult object on $\mathbb{C}$ which we start to investigate now.

Definition 18.1. *Let $G \subset \mathbb{C}\backslash\{0\}$ be a region. A continuous function $f : G \to \mathbb{C}$ is called a* **branch of the logarithmic function** *if for all $z \in G$ we have*

$$e^{f(z)} = z. \tag{18.6}$$

Remark 18.2. If f is a branch of the logarithmic function then $\exp|_{f(G)}$ is the inverse to f, in particular f is injective. Furthermore, the functions $f_k : G \to \mathbb{C}$, $f_k(z) = f(z) + 2k\pi i$, $k \in \mathbb{Z}$, are branches of the logarithmic function on G.

Corollary 18.3. *If $f : G \to \mathbb{C}$ is a branch of the logarithmic function then all other branches of the logarithmic function on G are of the type $f_k(z) = f(z) + 2k\pi i$, $k \in \mathbb{Z}$.*

Proof. Let $g : G \to \mathbb{C}$ be a further branch of the logarithmic function on G. For all $z \in G$ it follows that $e^{f(z)} = z = e^{g(z)}$, i.e. $e^{f(z)-g(z)} = 1$ implying that $f(z) - g(z) = 2\pi i\, h(z)$ with $h(z) \in \mathbb{Z}$. But h must be a continuous function on the region G, hence it must be constant. $\qquad\square$

Theorem 18.4. *Every branch $f : G \to \mathbb{C}$, $G \subset \mathbb{C}\backslash\{0\}$ a region, is on its domain a holomorphic function and we have*

$$f'(z) = \frac{1}{z}, \qquad z \in G. \tag{18.7}$$

Conversely, if $f : G \to \mathbb{C}$ is holomorphic and satisfies (18.7) and if $e^{f(z_0)} = z_0$ holds for some $z_0 \in G$, then f is a branch of the logarithmic function in G.

Proof. The mapping $f : G \to f(G)$ is bijective with inverse $\exp : f(G) \to G$. For $z, z_0 \in G$, $z \neq z_0$, we set $w = f(z)$ and $w_0 = f(z_0)$ and we observe that

$$\frac{f(z) - f(z_0)}{z - z_0} = \frac{w - w_0}{e^w - e^{w_0}} = \frac{1}{\frac{e^w - e^{w_0}}{w - w_0}}. \tag{18.8}$$

Since f is continuous, $z \to z_0$ implies $w \to w_0$ and therefore the limit $z \to z_0$ in (18.8) exists and it follows that

$$\lim_{z \to z_0} \frac{f(z) - f(z_0)}{z - z_0} = \frac{1}{e^{w_0}} = \frac{1}{z_0}.$$

We now prove the converse statement. For this let $g(z) := z\, e^{-f(z)}$ which is in G holomorphic and satisfies $g'(z) = e^{-f(z)} - z\, f'(z)\, e^{-f(z)} = 0$ since $z\, f'(z) = 1$. Since G is a region it follows that $g = c \in \mathbb{C}\backslash\{0\}$, i.e. $z = c\, e^{f(z)}$ in G. By assumption we have $z_0 = e^{f(z_0)}$ implying that $c = 1$. $\qquad\square$

Note that so far we do not know whether for some region $G \subset \mathbb{C}\backslash\{0\}$ a branch of the logarithmic function exists. This problem is solved by the following existence result:

Theorem 18.5. *On $B_1(1) = \{z \in \mathbb{C} \,|\, |z-1| < 1\}$ a branch of the logarithmic function is given by $\tilde{L} : B_1(1) \to \mathbb{C}$, $\tilde{L}(z) = \sum_{k=1}^{\infty} \frac{(-1)^{k-1}}{k}(z-1)^k$.*

Proof. We know that $\tilde{\ln}(1+z) = \sum_{k=1}^{\infty} \frac{(-1)^{k-1}}{k} z^k$ converges in $B_1(0)$, hence it is holomorphic in $B_1(0)$, and differentiating term by term yields

$$\frac{d}{dz}\tilde{\ln}(1+z) = \sum_{k=1}^{\infty}(-1)^{k-1} z^{k-1} = \sum_{k=0}^{\infty}(-z)^k = \frac{1}{1+z}.$$

For the function $\tilde{L}$ we find $\tilde{L}(z) = \tilde{\ln}(z-1)$, $z \in B_1(1)$, which is holomorphic in $B_1(1)$ and satisfies $\tilde{L}'(z) = \frac{1}{z}$. In addition we have $\tilde{L}(1) = 0$ or $e^{\tilde{L}(1)} = 1$. Now the second part of Theorem 18.4 yields the result. $\qquad\square$

Exercise 18.6. *For $a \in \mathbb{C}\backslash\{0\}$ define $L_{a,b} : B_{|a|}(a) \to \mathbb{C}$ by $L_{a,b}(z) := b + \tilde{L}(\frac{z}{a})$ where b is a logarithm of a, in particular $e^b = a$. Prove that $L_{a,b}$ is a branch of the logarithmic function.*

Our aim is to now introduce the principal branch of the logarithmic function as the mapping $\log : \mathbb{C}\backslash(-\infty, 0] \to \mathbb{C}$. For this we represent $z \in \mathbb{C}\backslash(-\infty, 0]$ as $z = |z|\,e^{i\varphi}$, $-\pi < \varphi < \pi$. Note that the domain of φ is now different to the case of polar coordinates and it is chosen such that the representation of z is unique for every $z \in \mathbb{C}\backslash(-\infty, 0]$. Moreover the mapping $z \mapsto \ln|z|$ is continuous from $\mathbb{C}\backslash(-\infty, 0]$ to $\mathbb{R}$ as the mapping $z \mapsto i\varphi$ is continuous from $\mathbb{C}\backslash(-\infty, 0]$ to $\{i\varphi \,|\, \varphi \in (-\pi, \pi)\}$. The set $\mathbb{C}\backslash(-\infty, 0]$ is often called the **cut plane**.

Theorem 18.7. *The function $\log : \mathbb{C}\backslash(-\infty, 0] \to \mathbb{C}$ defined for $z = |z|\,e^{i\varphi}$, $-\pi < \varphi < \pi$, by*

$$\log z := \ln|z| + i\varphi \tag{18.9}$$

where $\ln : (0, \infty) \to \mathbb{R}$ is the natural logarithm as defined in Chapter I.9 is a branch of the logarithmic function which extends the function $\tilde{L} : B_1(1) \to \mathbb{C}$ from Theorem 18.5.

Proof. The continuity of $z \mapsto \ln|z|$ and $z \mapsto i\varphi$, $z = |z| e^{i\varphi}$ with $-\pi < \varphi < \pi$, implies that $\log z$ is continuous from $\mathbb{C}\backslash(-\infty, 0]$ to $\mathbb{C}$. Furthermore

$$e^{\log z} = e^{(\ln|z|+i\varphi)} = e^{\ln|z|} e^{i\varphi} = |z| e^{i\varphi} = z$$

holds, hence $\log$ is in the region $\mathbb{C}\backslash(-\infty, 0]$ a branch of the logarithmic function. Since $B_1(1) \subset \mathbb{C}\backslash(-\infty, 0]$ and $\tilde{L}(z) = \sum_{k=1}^{\infty} \frac{(-1)^{k-1}}{k} (z - 1)^k$ is a branch of the logarithmic function we know that $\log z = \tilde{L}(z) + c$, $c \in \mathbb{C}$, for all $z \in B_1(1)$. However for $z = 1$ we find $\log 1 = \tilde{L}(1) = 0$ implying that $c = 0$. $\square$

Definition 18.8. *The function* $\log : \mathbb{C}\backslash(-\infty, 0] \to \mathbb{C}$ *defined by (18.9) is called the* **principal branch of the logarithmic function**.

Remark 18.9. A. We will use the following conventions:

$$\ln : (0, \infty) \to \mathbb{R}$$

denotes the natural logarithm;

$$\log : \mathbb{C}\backslash(-\infty, 0] \to \mathbb{C}$$

denotes the principal branch of the logarithmic function;

$$\tilde{\ln} : B_1(0) \to \mathbb{C}$$

is defined as $\tilde{\ln}(1 + z) = \sum_{k=1}^{\infty} \frac{(-1)^{k-1}}{k} z^k$. Further we use

$$\tilde{L} : B_1(1) \to \mathbb{C}, \tilde{L}(z) = \tilde{\ln}(1 - z).$$

B. A consequence of (18.9) is the equality

$$\log i = \frac{1}{2} \pi i, \tag{18.10}$$

since $|i| = 1$, i.e. $\ln|i| = 0$, and $i = \cos \frac{\pi}{2} + i \sin \frac{\pi}{2} = e^{\frac{\pi}{2} i}$.
C. Further branches of the logarithmic function on $\mathbb{C}\backslash(-\infty, 0]$ are given by $z \mapsto \log z + 2\pi i k$, $k \in \mathbb{Z}$.

Next we want to study the functional equation for log. Let $z_1, z_2 \in \mathbb{C}\backslash(-\infty, 0]$ such that $z_1 \cdot z_2 \in \mathbb{C}\backslash(-\infty, 0]$ with representation $z_1 = |z_1|e^{i\varphi_1}$, $z_2 = |z_2|e^{i\varphi_2}$ and $z_1 \cdot z_2 = |z_1 \cdot z_2|e^{i\varphi_3}$, $-\pi < \varphi_j < \pi$, $j = 1, 2, 3$. For φ_3 we find $\varphi_3 = \varphi_1 + \varphi_2 + \alpha$ where $\alpha \in \{-2\pi, 0, 2\pi\}$ and consequently we have

$$\log (z_1 \cdot z_2) = \ln |z_1||z_2| + i\,\varphi_3$$
$$= (\ln |z_1| + i\,\varphi_1) + (\ln |z_2| + i\,\varphi_2) + i\,\alpha,$$

i.e.

$$\log (z_1 \cdot z_2) = \log z_1 + \log z_2 + i\,\alpha. \qquad (18.11)$$

For $\operatorname{Re} z_j > 0$, $j = 1, 2$, we have $\varphi_j \in (\frac{-\pi}{2}, \frac{\pi}{2})$, hence $\varphi_1 + \varphi_2 \in (-\pi, \pi)$ and $\alpha = 0$ implying

$$\log (z_1 \cdot z_2) = \log z_1 + \log z_2, \qquad \operatorname{Re} z_j > 0. \qquad (18.12)$$

Thus the functional equation for $\ln$ is in general to be replaced by (18.11) and for $\operatorname{Re} z_1, \operatorname{Re} z_2 > 0$ it reduces to the one for $\ln$.

Next we want to study the relation $\log (\exp (z))$. First of all the definition of log requires that $\exp z \neq (-\infty, 0]$ which means $z = x + iy$ with $x \leq 0$ and $y = 0$ must be excluded. Since $e^z = e^x \cos y + i\, e^x \sin y$ these conditions imply that $y = (2k + 1)\pi$, $k \in \mathbb{Z}$, must be excluded. Otherwise, in $G := \mathbb{C}\backslash\{z \in \mathbb{C} \mid \operatorname{Im} z = (2k + 1)\pi, \, k \in \mathbb{Z}\}$ the term $\log (\exp (z))$ is defined, but note that G is **not** a region since it is not connected. Indeed we have $G_k \cap G_l = \emptyset$ for $k \neq l$ and

$$G = \bigcup_{k \in \mathbb{Z}} G_k, \quad G_k := \{z \in \mathbb{C} \mid (2k - 1)\pi < \operatorname{Im} z < (2k + 1)\pi\}.$$

However each G_k, $k \in \mathbb{Z}$, is a region, namely a strip in the complex plane parallel to the real axis and with width 2π. For $z \in G_k$ we have $e^z = e^x\, e^{iy} = e^x\, e^{i(y - 2k\pi)}$ with $y - 2k\pi \in (-\pi, \pi)$. This implies for $z \in G_k$ that

$$\log(e^z) = \ln e^x + i\,(y - 2k\,\pi)$$
$$= x + iy - 2k\,\pi\,i = z - 2\pi\,k\,i.$$

In particular, for $k = 0$ we have

$$\log (e^z) = z, \qquad z \in G_0.$$

Hence we have proved the result.

Lemma 18.10. *The exponential functions maps the region $G_0 = \{z \in \mathbb{C} \mid -\pi < \operatorname{Im} z < \pi\}$ biholomorphically onto the cut plane $\mathbb{C} \backslash (-\infty, 0]$ with the principal branch of the logarithmic function being the inverse.*

Once the logarithmic function is at our disposal we can turn to power functions, i.e. we want to define $z \mapsto z^\alpha$, $\alpha \in \mathbb{C}$, for z belonging to a region of $\mathbb{C}$. Of course for $n \in \mathbb{Z}$ we have no problem to define z^n. But already for $a > 0$ and $b \in \mathbb{R}$ we need for the definition of a^b the function ln, recall

$$a^b := e^{b \ln a}. \tag{18.13}$$

Since exp is an entire function, whenever $\log z$ is defined, for $\beta \in \mathbb{C}$ we can use (18.13) as guide and define

$$z^\beta := e^{\beta \log z}, \tag{18.14}$$

but we know that some care is needed to deal with the logarithmic function and to have the relation $e^{\log z} = z$. The general and proper definition of a power function is therefore

Definition 18.11. *Let $G \subset \mathbb{C} \backslash (-\infty, 0]$ be a region and $L : G \to \mathbb{C}$ be a branch of the logarithmic function. Further let $\beta \in \mathbb{C}$. We define a **branch of the power function** $z \mapsto z^\beta$ as $p_\beta : G \to \mathbb{C}$, $p_\beta(z) := z^\beta := e^{\beta L(z)}$.*

The functional equation of the exponential function yields

$$z^{\alpha+\beta} = e^{(\alpha+\beta) L(z)} = e^{\alpha L(z)} e^{\beta L(z)} = z^\alpha \cdot z^\beta, \tag{18.15}$$

and further we have

$$\frac{d}{dz} p_\beta(z) = \frac{d}{dz} e^{\beta L(z)} = \beta L'(z) e^{\beta L(z)} = \frac{\beta}{z} z^\beta,$$

i.e. we have

$$\frac{d}{dz} p_\beta(z) = \beta z^{\beta-1}, \quad z \in G, \ L : G \to \mathbb{C}, \ G \subset \mathbb{C} \backslash (-\infty, 0]. \tag{18.16}$$

For $\beta = n \in \mathbb{N}$ we observe that p_n has of course the holomorphic continuation to $\mathbb{C}$ given by $z \mapsto z^n$, and for $k \in \mathbb{Z}$, $k < 0$, we can extend p_k to $\mathbb{C} \backslash \{0\}$ by $z \mapsto z^k = \frac{1}{z^{-k}}$, $-k \in \mathbb{N}$, holomorphically. The key point to note is that the general power function is not defined on all of $\mathbb{C}$, but as the logarithmic function it has branches, in fact for every β every branch of the logarithmic function can be used to define a branch of $z \mapsto z^\beta$. For certain values of $\beta \in \mathbb{C}$ the function p_β has extensions onto larger sets than the region on which its defining branch of the logarithmic function is originally given, recall the cases $\beta \in \mathbb{N}$ or $\beta \in \mathbb{Z}$, $\beta < 0$.

An interesting special result, an observation made already by L. Euler, is the following: Since $i = 0 + i$, i.e. $\operatorname{Re} i = 0$ and $\operatorname{Im} i = 1 \in (-\pi, \pi)$, it follows that $i \in G_0$. Hence i is in the domain of the principal branch of the logarithmic function and we have $i^i = e^{i \log i}$.

Using (18.10) we find further

$$i^i = e^{i \log i} = e^{i \left(\frac{1}{2} \pi i \right)} = e^{\frac{-\pi}{2}}, \qquad (18.17)$$

and the surprising fact is that i^i is a real number.

Of special interest are roots or **root functions**, i.e. power functions of type $p_{\frac{1}{n}}$, $n \in \mathbb{N}$. We choose a fixed branch $L : G \to \mathbb{C}$ of the logarithmic function and observe that with $\zeta_n := e^{\frac{2\pi i}{n}}$ further branches of $p_{\frac{1}{n}}(z) = e^{\frac{1}{n} L(z)}$ are given by $p_{\frac{1}{n}, k}(z) = \zeta_n^k p_{\frac{1}{n}}(z)$, $k = 1, \ldots, n-1$. Thus given L we obtain in the region where L is defined at least n branches of $p_{\frac{1}{n}}$.

Now we can better understand our definition of the square root function on $(0, \infty)$. We have picked the principal branch of the logarithmic function to define $x \mapsto \sqrt{x}$ on $(0, \infty)$ and by this we made the square root of a positive real number uniquely defined. For $n = 2$ we have $e^{\frac{2\pi i}{2}} = e^{\pi i} = -1 = \zeta_2^1$ and $p_{\frac{1}{2}, 1}(z) = -p_{\frac{1}{n}}(z)$. Thus for $x \in (0, \infty)$ we obtain now $p_{\frac{1}{2}, 1}(x) = -\sqrt{x}$. In Problem 3 we will discuss the formula

$$(1 + z)^\beta = \sum_{k=0}^{\infty} \binom{\beta}{k} z^k, \quad z \in B_1(0), \ \beta \in \mathbb{C}. \qquad (18.18)$$

The functions sin and cos can be easily extended to $\mathbb{C}$ by their power series representations and the relation

$$e^{iz} = \cos z + i \sin z \qquad (18.19)$$

follows for all $z \in \mathbb{C}$. Moreover the two symmetries $\cos z = \cos(-z)$ and $\sin z = -\sin(-z)$ continue to hold, hence we have

$$e^{-iz} = \cos z - i \sin z \qquad (18.20)$$

which with (18.19) yields

$$\cos z = \frac{e^{iz} + e^{iz}}{2} \quad \text{and} \quad \sin z = \frac{e^{iz} - e^{-iz}}{2i}. \tag{18.21}$$

From (18.21) it is easy to extend typical addition theorem for trigonometrical functions defined on $\mathbb{R}$ to $\mathbb{C}$. Furthermore we find with $z = x + iy$ and taking into account the power series representations of sinh and cosh that

$$\text{Re}\,(\cos\,z) = \cos x \cosh y, \qquad \text{Im}\,(\cos z) = -\sin x \sinh y \tag{18.22}$$

and

$$\text{Re}\,(\sin z) = \sin x \cosh y, \qquad \text{Im}(\sin z) = \cos x \sinh y. \tag{18.23}$$

For $z \in \mathbb{C} \backslash \{(k + \frac{1}{2})\pi \,|\, k \in \mathbb{Z}\}$ the extension of tan is given by

$$\tan z = \frac{\sin z}{\cos z} = \frac{1}{i}\frac{e^{2iz} - 1}{e^{2iz} + 1} \tag{18.24}$$

and for $z \in \mathbb{C} \backslash \{k\pi \,|\, k \in \mathbb{Z}\}$ we get

$$\cot z = \frac{\cos z}{\sin z} = i\frac{e^{2iz} + 1}{e^{2iz} - 1}. \tag{18.25}$$

Again, making use of the exponential function and the power series representations of sin, cos, sinh and cosh we find

$$\sinh z = \frac{e^z - e^{-z}}{2} = \frac{1}{i}\,\sin(iz) \tag{18.26}$$

and

$$\cosh z = \frac{e^z + e^{-z}}{2} = \cos(iz) \tag{18.27}$$

as well as

$$\sin z = \frac{1}{i}\sinh(iz) \quad \text{and} \quad \cos z = \cosh(iz). \tag{18.28}$$

More complicated is the discussion of the inverse functions for trigonometrical and hyperbolic functions, therefore we only want to discuss arctan in more detail. One message we get from our considerations made above on concrete (holomorphic) functions such as Möbius transformations, the exponential function, the logarithmic function, etc., is that much care is needed to study their mapping properties. For example the discussion of the principal branch of the logarithmic function requires some knowledge of the image of $G_0 :=$ $\{z \in \mathbb{C} | -\pi < \operatorname{Im} z < \pi\}$ under the exponential mapping. In case of $w = \tan z$, using (18.24), it is formally easy to find z as a function of w. We start with

$$w = \tan z = \frac{1}{i} \frac{e^{2iz} - 1}{e^{2iz} + 1}$$

and with $v := e^{2iz}$ we find

$$w = \frac{1}{i} \frac{v - 1}{v + 1}$$

or

$$v = \frac{1 + iw}{1 - iw}$$

and therefore

$$z = \frac{1}{2i} \log \frac{1 + iw}{1 - iw}.$$

Thus the formula for arctan suggested by this calculation is

$$\arctan w = \frac{1}{2i} \log \frac{1 + iw}{1 - iw},$$

but we know that there are problems to define $\log \frac{1+iw}{1-iw}$. Note that $w = \frac{1}{i} \frac{v-1}{v+1}$ is a Möbius transformation with corresponding matrix $\begin{pmatrix} 1 & -1 \\ i & i \end{pmatrix}$ with determinant $2i \neq 0$ and inverse $v = \frac{1+iw}{1-iw}$ with corresponding matrix $\begin{pmatrix} i & 1 \\ -i & 1 \end{pmatrix}$.

From Problem 8 of Chapter I.29 we know that for $|x| < 1$

$$\arctan x = \sum_{k=0}^{\infty} (-1)^k \frac{x^{2k+1}}{2k + 1}$$

and from Problem 6 of Chapter I.29 we know that for $|x| < 1$

$$\frac{1}{2} \ln \frac{1 + x}{1 - x} = \sum_{k=0}^{\infty} \frac{x^{2k+1}}{2k + 1}.$$

Both power series converge absolutely and uniformly for $|z| < 1$ and with $z = ix$, $|x| < 1$, we find

$$\sum_{k=0}^{\infty} \frac{(ix)^{2k+1}}{2k + 1} = i \sum_{k=0}^{\infty} (-1)^k \frac{x^{2k+1}}{2k + 1},$$

or

$$\arctan(ix) = \frac{1}{2} \ln \frac{1 + ix}{1 - ix}, \quad x \in (-1, 1).$$

We want to extend the equality to $z \in B_1(0) \subset \mathbb{C}$. Consider the Möbius transformation $w(z) = \frac{1+z}{1-z}$. In Problem 5 we will show that $w(B_1(0)) = \{\zeta \in \mathbb{C} | \operatorname{Re} \zeta > 0\}$. This implies that for the principal part of the logarithmic function log we can define

$$W(z) := \log w(z), \quad z \in B_1(0),$$

and W is in $B_1(0)$ a holomorphic function with

$$W'(z) = \frac{w'(z)}{w(z)} = \frac{z}{1 - z^2}.$$

With $\arctan z$ defined for $z \in B_1(0)$ by the convergent power series we consider

$$F(z) := W(iz) - 2i \arctan z$$

which is holomorphic in $B_1(0)$ and has the derivative

$$F'(z) = iW'(iz) + 2i(\arctan z)' = \frac{2i}{1 + z^2} - \frac{2i}{1 + z^2} = 0.$$

Hence F must be constant on $B_1(0)$ and since $F(0) = 0$ it follows that

$$\arctan z = \frac{1}{2i} \log \frac{1 + iz}{1 - iz}, \quad z \in B_1(0). \tag{18.29}$$

Note that (18.29) is, at the moment, an equality for power series converging in $B_1(0)$, but in Chapter 22 we will identify the power series on the left hand side with a unique holomorphic extension of arctan as defined for a real-valued argument.

With similar arguments one can first obtain "algebraic" formulae for arcsin, arccos, arccot as well as arcsinh, arccosh and arctanh and then one can also determine the domain of the corresponding principal branch:

$$\arcsin w = \frac{1}{i} \log \left(iw + \sqrt{1 - w^2}\right) \text{ on } \mathbb{C}\backslash\{w \in \mathbb{C} \,|\, |\operatorname{Re} w| \geq 1\}, \qquad (18.30)$$

$$\arccos w = \frac{1}{i} \log \left(iw + \sqrt{w^2 - 1}\right) \text{ on } \mathbb{C}\backslash\{w \in \mathbb{C} \,|\, |\operatorname{Re} w| \leq 1\}, \qquad (18.31)$$

$$\operatorname{arccot} w = \frac{1}{2i} \log \left(\frac{w + i}{w - i}\right) \text{ on } \mathbb{C}\backslash\{w \in \mathbb{C} \,|\, |\operatorname{Im} w| \leq 1\}, \qquad (18.32)$$

$$\operatorname{arcsinh} w = \log \left(w + \sqrt{w^2 + 1}\right) \text{ on } \mathbb{C}\backslash\{w \in \mathbb{C} \,|\, |\operatorname{Im} w| \geq 1\}, \qquad (18.33)$$

$$\operatorname{arccosh} w = \log \left(w + \sqrt{w^2 - 1}\right) \text{ on } \mathbb{C}\backslash\{w \in \mathbb{C} \,|\, \operatorname{Re} w \leq 1\}, \qquad (18.34)$$

$$\operatorname{arctanh} w = \frac{1}{2} \log \left(\frac{1 + w}{1 - w}\right) \text{ on } \mathbb{C}\backslash\{w \in \mathbb{C} \,|\, |\operatorname{Re} w| \geq 1\}. \qquad (18.35)$$

At this point it is worth reminding the reader that there are many tools available to find results such as the ones listed. Of course there are fast developing (and changing) internet resources. More classical resources are books, for example the one edited by M.Abramowitz and F.A.Stegun [1]. In order to learn new mathematics proofs, and typical examples must be worked through. Working mathematicians cannot do every calculation by themselves, they need to use tools from sources they can rely on - of course they must quote the resources they have used.

Problems

1. Use the principal branch of the logarithmic function to find $(\frac{1}{2}\sqrt{2}e^{2\pi} + \frac{1}{2}\sqrt{2}e^{2\pi}i)^i$.

2. Prove that in general for the principal branch of the logarithmic function we have $\log(z_1 \cdot z_2) \neq \log z_1 + \log z_2$.

3. For $x \in \mathbb{R}$, $|x| < 1$, and $\beta \in \mathbb{C}$, $\beta = s + it$, show that

$$\sum_{k=0}^{\infty} \binom{\beta}{k} x^k = (1 + x)^s (\cos(t \ln(1 + x)) + i \sin(t \ln(1 + x))).$$

   Hint: use Problem 7 of Chapter 16.

4. Solve Exercise 18.6.

5. Prove that $w(z) := \frac{1+z}{1-z}$ maps $B_1(0)$ onto the set $\{\zeta \in \mathbb{C} | \operatorname{Re} \zeta > 0\}$.

19   Some More Topology

Line integrals of holomorphic and related functions turn out to be very helpful tools in investigating these functions. In fact they lead to completely
new insights, including, surprisingly, about the relation between analysis
and topology. For understanding these relations we first need to study in
more detail the notion of a connected set in the plane. However, since many
ideas and results hold for more general topological spaces and since we want
to make use of these results when we study differential geometry or differentiable manifolds later on, we discuss them here either in the context of
topological spaces or at least in the context of metric spaces. But there will
also be a few results especially related to the plane. Note that from the
topological or metrical point of view $\mathbb{C}$ and $\mathbb{R}^2$ are identical, but sometimes
we may prefer to use complex numbers or complex-valued functions for the
formulation of topological statements in the plane.
In the first part of this chapter we mainly recollect definitions and results
from Volume I and especially from Volume II. We start with

Definition 19.1. *Let $(X, \mathcal{O})$ be a topological space.*
A. *A pair (O_1, O_2) of non-empty open subsets of X is called a **splitting** of
X if $O_1 \cup O_2 = X$ and $O_1 \cap O_2 = \emptyset$.*
B. *A topological space is called **connected** if it does not admit a splitting.*
C. *A subset of a topological space is called connected if it is connected as a
topological space equipped with the relative topology induced by X.*
D. *A **region** $G \subset X$ is a non-empty open subset which is connected.*

We have seen these definitions before in I.19.24 for the real line and in II.3.30
for general metric spaces. Since only topological notions are involved in these
definitions they extend easily to topological spaces. In addition many of the
results proved in Chapter II.3 do hold with the same proof in topological
spaces. Here is a collection of such results.

Theorem 19.2. *Let $(X, \mathcal{O})$ be a topological space.*
A. *The space is connected if and only if X and $\emptyset$ are the only sets which are
both, open and closed. (See Remark II.3.3.1.)*
B. *The image of a connected space under a continuous mapping is connected,
or in other words, the image of a connected subset of a topological space under
a continuous mapping is connected. (See Theorem II.3.32.)*
C. *Let $(A_k)_{k \in I}$ be a family of connected subsets of X. If $\bigcap_{k \in I} A_k \neq \emptyset$ then*

347

$\bigcup_{k \in I} A_k$ is connected too. (See Theorem II.3.34.)

D. If $Y \subset X$ is connected and $Y \subset Z \subset \overline{Y}$ then Z is connected. In particular the closure $\overline{Y}$ of a connected set Y is connected. (See Proposition II.3.37.)

We have already seen examples of connected sets:

Example 19.3. A. The connected sets in $\mathbb{R}$ are the intervals (see Theorem I.19.25) and a set in $\mathbb{R}$ is open if and only if it is the denumerable union of mutually disjoint open intervals.

B. Every **star-shaped set** in $\mathbb{R}^n$ is connected. (See Problem 12.b in Chapter II.3.) Recall that $S \subset \mathbb{R}^n$ is called star-shaped if for some $x_0 \in \mathbb{R}^n$ we have $S = \bigcup_{x \in S} S_{x,x_0}$ where S_{x,x_0} denotes the closed line segment connecting x_0 with x.

Using Theorem 19.2.C we can introduce the notion of the (connectivity) component of a point.

Definition 19.4. *In a topological space $(X, \mathcal{O})$ the union $C(x)$ of all connected sets containing the point $x \in X$ is called the **connectivity component** of x. (Compare with Corollary II.3.36.)*

Clearly $C(x)$ is the largest connected subset of X which contains x. If $y \in C(x)$ then $C(x) = C(y)$, and if $y \notin C(x)$ then $C(x) \cap C(y) = \emptyset$. Thus the components $(C(x))_{x \in X}$ form a partition of X.

A further notion we have already introduced before is that of a **pathwise** or **arcwise connected** set. Since later on we will work in $\mathbb{C}$ and since it is customary to use in the plane the name arcwise connected set we will do the same. We need some preparations, again we can easily extend some definitions from metric spaces (or even $\mathbb{R}^n$) to general topological spaces.

Definition 19.5. *Let $(X, \mathcal{O})$ be a topological space.*
A. *For a closed interval $I \subset \mathbb{R}$ we call a continuous mapping $\gamma : I \to X$ an **arc** or a **path** or a **parametric curve** in X. If $I = [a, b]$, $a < b$, then $\gamma(a)$ is called the **initial point** of γ and $\gamma(b)$ is called the **terminal point** of γ. The image of I under γ is called the **trace** of γ and sometimes denoted by $\mathrm{tr}(\gamma)$. The interval I is called the **parameter interval** of γ and $t \in I$ is the **arc** or **curve parameter**.*
B. *We say that $\gamma : [a, b] \to X$ **connects** the points $x, y \in X$ if $x, y \in$*

$\{\gamma(a), \gamma(b)\}$, *i.e. x is either initial or terminal point of γ and then y is terminal or initial point of γ, respectively.*
C. *We call $\gamma : [a, b] \to X$ closed if $\gamma(a) = \gamma(b)$. If γ is closed and in addition $\gamma|_{[a,b)}$ is injective we call γ a **simply closed curve** in X.*

These definitions are completely analogous to those given in Chapter II.4 (Definition II.4.11, Definition II.4.12) and we refer also to examples given there. Further, following Definition II.4.16 we give

Definition 19.6. *Let $(x, \mathcal{O})$ be a topological space and $G \subset X$. We call G **pathwise** or **arcwise connected** if every pair of points $x, y \in G$ can be connected by an arc, i.e. there exists a continuous mapping $\gamma : [a, b] \to X$ such that $\gamma(a) = x$ and $\gamma(b) = y$.*

The proofs given in Chapter II.4 for the case $G \subset \mathbb{R}^n$ carry over to the general situation:

Proposition 19.7. A. *Every arcwise connected subset $G \subset X$ is connected. (See Proposition II.4.18.)*
B. *Let $G \subset X$ and $x_0 \in G$. The set*

$$G_{x_0} := \bigcup \{\text{tr}(\gamma) \mid \gamma : [a, b] \to G \text{ is an arc with } \gamma(a) = x_0\}$$

is arcwise connected. (See Lemma II.4.19.)

Note that Proposition 19.7.B implies that a star-shaped set is arcwise connected.
However, Theorem II.4.20 does not hold for arbitrary topological spaces, therefore we state it here once more as in Chapter II.4:

Theorem 19.8. *If $G \subset \mathbb{R}^n$ is a non-empty open set which is connected then G is arcwise connected, i.e. regions in $\mathbb{R}^n$ are arcwise connected.*

We state explicitly

Corollary 19.9. *A region $G \subset \mathbb{C}$ is arcwise connected.*

Since intervals are connected sets in $\mathbb{R}$ the trace $\text{tr}(\gamma)$ of an arc (a path, a parametric set) is by Proposition 19.7.A connected. In Problem 5 we prove

Proposition 19.10. *A continuous function between two topological spaces maps arcwise connected sets onto arcwise connected sets.*

We invite the reader to reconsider Example II.4.21 - II.4.23 to get a better understanding of all these notions and results. The careful reader will have noticed that the proofs of some of the results stated above will need the notion of a sum of two arcs (paths or curves) which we will discuss now in more detail. We start with a modification of Definition II.4.17.

Definition 19.11. *Given a topological space* $(X, \mathcal{O})$ *and two paths* $\gamma_j :$ $[a_j, b_j] \to X$ *such that* $\gamma_1(b_1) = \gamma_2(a_2)$, *i.e. the terminal point of* γ_1 *is the initial point of* γ_2. *We define the* ***sum*** $\gamma_1 \oplus \gamma_2$ *of* γ_1 *and* γ_2 *as the mapping*

$$\gamma_1 \oplus \gamma_2 : [a_1, b_1 + b_2 - a_2] \to X, \tag{19.1}$$

$$(\gamma_1 \oplus \gamma_2)(t) = \begin{cases} \gamma_1(t) & for\ t \in [a_1, b_1] \\ \gamma_2(t + a_2 - b_1) & for\ t \in [b_1, b_1 + b_2 - a_2]. \end{cases}$$

Obviously we have that $\mathrm{tr}\,(\gamma_1 \oplus \gamma_2) = \mathrm{tr}\,(\gamma_1) \cup \mathrm{tr}(\gamma_2)$, however it is not obvious that $\gamma_1 \oplus \gamma_2$ is a path, i.e. continuous. If X were a metric space this statement is easy to prove by using the characterisation of continuity with the help of limits of sequences. For the general situation we need

Proposition 19.12. *Let* $(X, \mathcal{O}_X)$ *be a topological space and* $A_1, \ldots, A_N \subset X$ *closed sets such that* $X = \bigcup\limits_{k=1}^{N} A_k$. *A mapping* $f : X \to Y$ *into a topological space* $(Y, \mathcal{O}_Y)$ *is continuous if and only if the mappings* $f|_{A_k} : A_k \to Y, k = 1, \ldots, N$, *are continuous.*

Proof. Of course, if $f : X \to Y$ is continuous, each of the mappings $f|_{A_k}$ is continuous. Now let $C \subset Y$ be any closed set. It follows that $(f|_{A_k})^{-1}(C) = A_k \cap f^{(-1)}(C)$ is closed in A_k equipped with the relative topology induced by X. Since A_k is closed it follows that $A_k \cap f^{(-1)}(C)$ is closed in X. Using that $X = \bigcup\limits_{k=1}^{N} A_k$ we deduce further that $f^{-1}(C) = \bigcup\limits_{k=1}^{N} (f|_{A_k})^{-1}(C)$ and this is a closed set in X. Thus we have proved that the pre-image of every closed set in Y is under f closed in X, i.e. f is continuous. $\square$

Corollary 19.13. *The sum of two paths* $\gamma_j : [a_j, b_j] \to X$ *is a path, i.e. a continuous mapping* $\gamma_1 \oplus \gamma_2 : [a_1, b_1 + b_2 - a_2] \to X$.

Clearly we can extend Definition 19.11 to a finite number of paths $\gamma_1, \ldots, \gamma_N$ whenever the terminal point of γ_j is the initial point of $\gamma_{j+1}, j = 1, \ldots, N-1$, and in Problem 6 we will see that the operation $\oplus$ is associative. We also need a generalisation of Definition II.4.15.

Definition 19.14. *Let* $\gamma : [\gamma, \beta] \to X$ *be a path in the topological space* (X, θ).
A. *We call a strictly increasing continuous function* $\varphi : [a, b] \to [\alpha, \beta]$ *with* $\varphi(a) = \alpha$ *and* $\varphi(b) = \beta$ *an* **orientation preserving parameter transformation** *for* γ.
B. *A strictly decreasing continuous function* $\psi : [a, b] \to [\alpha, \beta]$ *with* $\psi(a) = \beta$ *and* $\psi(b) = \alpha$ *is called an* **orientation reversing parameter transformation**.
C. *The paths* $\gamma \circ \varphi$ *and* $\gamma \circ \psi$ *are said to be obtained from* γ *by a change or transformation of the parameter.*

If φ is an orientation preserving parameter transformation for γ then $\gamma \circ \varphi :$ $[a, b] \to X$ is a path with $\operatorname{tr}(\gamma \circ \varphi) = \operatorname{tr}(\gamma)$ and $(\gamma \circ \varphi)(a) = \gamma(\alpha)$ as well as $(\gamma \circ \varphi)(b) = \gamma(\beta)$, i.e. the initial and the terminal points of γ and $\gamma \circ \varphi$ coincide. If however ψ is an orientation reversing parameter transformation then we still have $\operatorname{tr}(\gamma \circ \psi) = \operatorname{tr}(\gamma)$ but $(\gamma \circ \psi)(a) = \gamma(\beta)$ and $(\gamma \circ \psi)(b) = \gamma(\alpha)$, i.e. the initial point of $\gamma \circ \psi$ is the terminal point of γ and the terminal point of $\gamma \circ \psi$ is the initial point of γ. In other words on $\gamma \circ \psi$ we are running through $\operatorname{tr}(\gamma)$ in the reversed direction namely from $\gamma(\beta)$ to $\gamma(\alpha)$.

Definition 19.15. *We call two paths* $\gamma_j : [a_j, b_j] \to X$, $j = 1, 2$, *in a topological space* $(X, \mathcal{O})$ **equivalent paths** *if there exists an orientation preserving parameter transformations* $\varphi : [a_2, b_2] \to [a_1, b_1]$ *such that* $\gamma_2 = \gamma_1 \circ \varphi$.

Proposition 19.16. *The equivalence of paths is an equivalence relation which we denote by* "$\sim_\gamma$" *or just by* "$\sim$".

Proof. Since the identity id $: [a, b] \to [a, b]$ satisfies all conditions of an orientation preserving parameter transformation it follows always $\gamma \sim \gamma$, and since the inverse of an orientation preserving parameter transformation is such a mapping as well as the composition of two orientation preserving parameter transformations is of this type, it follows $\gamma_1 \sim \gamma_2$ implies $\gamma_2 \sim \gamma_1$ and $\gamma_1 \sim \gamma_2$ together with $\gamma_2 \sim \gamma_3$ yields $\gamma_1 \sim \gamma_3$. $\qquad \square$

We can now lift the definition of the sum $\gamma_1 \oplus \gamma_2$ of two paths γ_1 and γ_2 to the set of equivalence classes by setting

$$[\gamma_1] \oplus [\gamma_2] := [\gamma_1 \oplus \gamma_2]. \tag{19.2}$$

Clearly, we have to check that this definition is independent of the parameterization which we will do in Problem 7. Of course we can extend (19.2) to finite number of paths $\gamma_1, ..., \gamma_N$ with the terminal point of γ_j being the initial point of γ_{j+1} for $j = 1, ..., N - 1$. For the extension (19.2) of (19.1) we can again prove the associative law, see Problem 7, i.e. we have

$$([\gamma_1] \oplus [\gamma_2]) \oplus [\gamma_3] = [\gamma_1] \oplus ([\gamma_2] \oplus [\gamma_3]). \tag{19.3}$$

For $x_0 \in X$ we denote by γ_{x_0} a path $\gamma_{x_0} : [a, b] \to X$, $\gamma_{x_0}(t) = x_0$ for all $t \in [a, b]$. Moreover, for $\gamma : [a, b] \to X$ we can define $\gamma^{-1} : [a, b] \to X$, $\gamma(t) = \gamma(a+b-t)$, which starts at $\gamma(b)$ and terminates at $\gamma(a)$. Note that $\psi : [a, b] \to [a, b]$, $\psi(t) = a+b-t$ is a strictly decreasing continuous function with $\psi(a) = b$ and $\psi(b) = a$. Hence γ^{-1} is obtained from γ by applying an orientation reversing parameter transformation. Here is an illustration of these operations

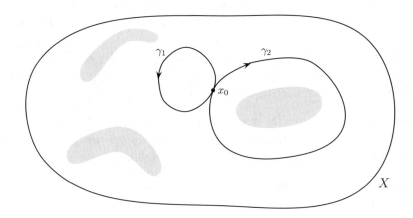

Figure 19.1

Figure 19.2

Figure 19.3

352

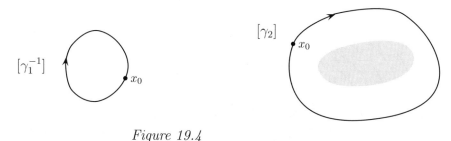

Figure 19.4

Figure 19.5

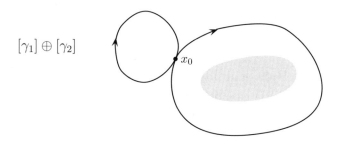

Figure 19.6

We have seen that it is possible to define some algebraic operations for (equivalence classes of) paths. This idea will now become more important when discussing homotopies and fundamental groups, concepts needed to study connected sets. The basic idea is to study the continuous deformation of continuous mappings. For example we may ask whether we can shrink a simply closed curve to a point. In $\mathbb{C}$ we may consider the circle $h : [0, 2\pi] \to \mathbb{C}$, $h(t) = e^{it}$, and the point $0 \in \mathbb{C}$. The family $H : [0, 2\pi] \times [0, 1] \to \mathbb{C}$, $H(t, r) = r\, e^{it}$ is continuous as function on $[0, 2\pi] \times [0, 1]$ and $H(t, 1) = h(t)$, $H(t, 0) = h_0 : [0, 2\pi] \to \mathbb{C}$, $h_0(t) = 0$ for all $t \in [0, 2\pi]$. Thus by $H(\cdot, r)$, $r \in [0, 1]$, a family of curves is given shrinking continuously to the point, i.e. the constant curve 0, see Figure 19.7.

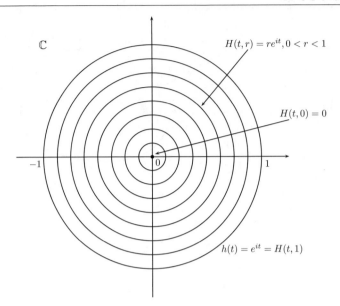

$$\mathbb{C}$$

$$H(t,r) = re^{it}, 0 < r < 1$$

$$H(t,0) = 0$$

$$h(t) = e^{it} = H(t,1)$$

Figure 19.7

Now we consider the set $\mathbb{C} \backslash \overline{B_{\frac{1}{2}}(0)}$ and start again with $h : [0, 2\pi] \to \mathbb{C}$, $h(t) = e^{it}$. In this case it seems that we cannot anymore shrink h continuously to a point, the set $\overline{B_{\frac{1}{2}}(0)}$ is in our way, see Figure 19.8.

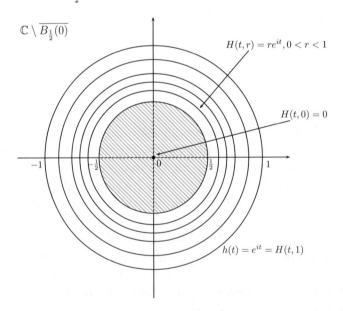

$$\mathbb{C} \backslash \overline{B_{\frac{1}{2}}(0)}$$

$$H(t,r) = re^{it}, 0 < r < 1$$

$$H(t,0) = 0$$

$$h(t) = e^{it} = H(t,1)$$

Figure 19.8

In the following we use much from the discussion in H.Schubert [77].

Definition 19.17. *Let $(X, \mathcal{O}_X)$ and $(Y, \mathcal{O}_Y)$ be two topological spaces and $f, g : Y \to X$ two continuous mappings. We call f **homotopic** to g if there exists a continuous mapping $H : Y \times [0, 1] \to X$ such that $H(\cdot, 0) = f$ and $H(\cdot, 1) = g$. The mapping H is called a **homotopy** between f and g.*

Remark 19.18. A. It is important to note that H is supposed to be continuous on $Y \times [0, 1]$, i.e. with respect to the product topology. It is not sufficient to assume that $H(\cdot, t)$ and $H(y, \cdot)$ are separately continuous. For an example of a separately continuous function which is not continuous we refer to Example II.4.1.
B. If $H : Y \times [0, 1] \to X$ is a homotopy then for every fixed $y \in Y$ the mapping $H(y, \cdot) : [0, 1] \to X$ is a path in X.
C. We may in particular consider the case where $Y = [a, b]$. In this case f and g are two paths in X and a homotopy between f and g deforms continuously the path $f = H(\cdot, 0)$ through a family of paths $H(\cdot, t), 0 < t < 1$, into the path $g = H(\cdot, 1)$, see Figure 19.9. Of special interest is the case where all paths $H(\cdot, t), t \in [0, 1]$, have the same initial point $x_0 = H(a, t)$ for $t \in [0, 1]$ and the same terminal point $x_1 = H(b, t)$ for all $t \in [0, 1]$, see Figure 19.10. The latter case includes also closed paths with $H(a, t) = H(b, t)$ for all $t \in [0, 1]$, see Figure 19.11.

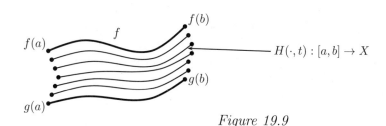

Figure 19.9

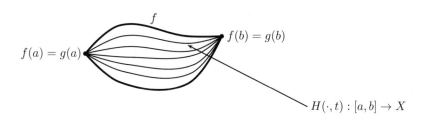

Figure 19.10

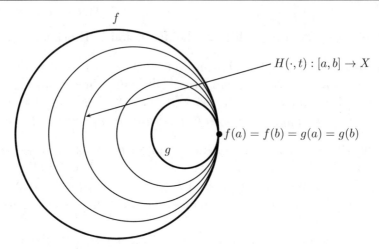

$$f$$

$$H(\cdot, t) : [a, b] \to X$$

$$f(a) = f(b) = g(a) = g(b)$$

$$g$$

Figure 19.11

Theorem 19.19. *On the set of all continuous mappings $f : Y \to X$ homotopy is an equivalence relation.*

Proof. Denote by π_Y the projection $\pi_Y : Y \times [0, 1] \to Y$, $\pi_Y (y, t) = y$. Since π_Y is continuous it follows that $H : Y \times [0, 1] \to X$, $H = f \circ \pi_Y$ is continuous and $H(y, t) = f(y)$ for all $t \in [0, 1]$, i.e. f is homotopic to itself. Let $f, g : Y \to X$ be two continuous mappings which are homotopic and $H : Y \times [0, 1] \to X$ be a homotopy between f and g. The mapping $\tau : Y \times [0, 1] \to Y \times [0, 1]$, $\tau (y, t) = (y, 1 - t)$ is continuous implying that $\tilde{H} : Y \times [0, 1] \to X$, $\tilde{H} := H \circ \tau$, $(y, t) \mapsto \tilde{H}(y, t) = H(y, 1 - t)$, is continuous and further $\tilde{H}(y, 0) = g$ holds as well as $\tilde{H}(y, 1) = f$. Thus if f is homotopic to g then g is homotopic to g. Finally suppose that f is homotopic to g and that g is homotopic to h with homotopies H_1 and H_2 respectively. We define $H_3 : Y \times [0, 1] \to X$ by

$$H_3 (y, t) := \begin{cases} H_1 (y, 2t), & 0 \le t \le \frac{1}{2} \\ H_2 (y, 2t - t), & \frac{1}{2} \le t \le 1. \end{cases}$$

First we note that $H_3 (y, 0) = f(y)$ and $H_3 (y, 1) = h(y)$ and in addition $H_3 (y, \frac{1}{2}) = g(y)$. Further, since H_3 is continuous on $Y \times [0, \frac{1}{2}]$ and on $Y \times [\frac{1}{2}, 1]$, and both sets are closed in $Y \times [0, 1]$, Proposition 19.12 implies the continuity of H_3 on $Y \times [0, 1]$. $\qquad \square$

We denote the equivalence relation "f and g are homotopic" by $f \sim_H g$ and we call the equivalence class of f the **homotopy class** of f.

Example 19.20. Let $(X, \mathcal{O})$ be a topological space and $Y = \{y_0\}$ be the topological space having just one point y_0. Every (continuous) mapping $f : Y \to X$ can be identified with the point $x_f := f(y_0)$. We note further that $\{y_0\} \times [0, 1]$ and $[0, 1]$ are homomorphic topological spaces and it follows that $f \sim_H g$ if and only if we can join x_f with x_g by a path $\gamma : [0, 1] \to X$. Indeed using a homotopy H between f and g we may choose $\gamma(t) = H(y_0, t)$. Conversely, given γ we consider $H(y_0, t) := \gamma(t)$ as homotopy.

Definition 19.21. *A continuous mapping* $f : Y \to X$ *is called* **null-homotopic** *if it is homotopic to a constant mapping* $g : Y \to X$, $g(y) = x_0$ *for all* $y \in Y$.

Definition 19.22. *Two topological spaces* X_1 *and* X_2 *are called* **homotopically equivalent** *or of the same* **homotopy type** *if and only if there exist continuous mappings* $f : X_1 \to X_2$ *and* $g : X_2 \to X_1$ *such that* $g \circ f$ *is homotopic to the identity on* X_1 *and* $f \circ g$ *is homotopic to the identity on* X_2.

Remark 19.23. In the case that X_1 and X_2 are of the same homotopy type it does not follow that f and g are uniquely determined. Moreover, if $f_1, f_2 : X_1 \to X_2$ both induce a homotopy equivalence of X_1 with X_2 it need not follow that they are homotopic.

In Problem 8 we will see that homotopy equivalence for topological spaces is an equivalence relation.

Example 19.24. Let $S = \bigcup_{x \in S} S_{x, x_0} \subset \mathbb{R}^n$ be a star-shaped set as in Example 19.3.B. We claim that S with the relative topology from $\mathbb{R}^n$ is homotopically equivalent to any topological space $Z = \{z\}$. Points in S can be described by affine coordinates, or using the standard basis $\{e_1, ..., e_n\}$ every $x \in S$ is given by $x_0 + \sum_{j=1}^{n} \xi_j e_j$ where $\xi = (\xi_1, ..., \xi_n)$ are the coordinates of $x - x_0$. We consider the two continuous mappings $f : S \to Z$, $f(x) = z$ for all $x \in S$, and $g : Z \to S$, $g(z) = x_0$. It follows that $f \circ g : Z \to Z$, $(f \circ g)(z) = z$, is the identity on Z, and for $g \circ f : S \to S$ we have $(g \circ f)(x) = x_0$ for all $x \in S$. We define the mapping $H : S \times [0, 1] \to S$ by $H(x, t) = (t\xi_1 + x_{0,1}, ..., t\xi_n + x_{0,n})$. This mapping is clearly continuous, $H(x, 0) = x_0$, i.e. $H(\cdot, 0) = g \circ f$, and $H(x, 1) = x$, i.e. $H(\cdot, 1) = \text{id}_S$. Since $f \circ g : Z \to Z$ is the identity on Z it follows that S is indeed homotopically equivalent to $Z = \{z\}$.

The mappings we are interested in are paths which we can always assume to be parametrized by the interval $[0, 1]$. So let X be a topological space and $\gamma : [0, 1] \to X$ be a path. The mapping $H : [0, 1] \times [0, 1] \to X$ defined by $H(s, t) = \gamma(s - st)$ is continuous with $H(s, 0) = \gamma(s)$ and $H(s, 1) = \gamma(0)$. Thus every path is null-homotopic in the sense of Definition 19.21 and therefore homotopy for paths is not a far reaching notion. Matters change if we assume that the initial and the terminal points must be fixed under the action of a homotopy.

Definition 19.25. A. *Let $F \subset Y$ be a subset of the topological space Y and let $f, g : Y \to X$ be two continuous mappings from Y into the topological space X. We call f and g **homotopic relative to F** if there exists a homotopy $H : Y \times [0, 1] \to X$ between f and g such that $H(y, t) = H(y, 0)$ for all $y \in F$ and $t \in [0, 1]$, i.e. $H(y, t) = f(y)$ for $y \in F$ and $t \in [0, 1]$.*
B. *We call two paths $\gamma_1, \gamma_2 : [0, 1] \to X$ **homotopic relative to their intial and terminal points** if there exists a homotopy $H : [0, 1] \times [0, 1] \to X$ between γ_1 and γ_2 such that $H(0, t) = \gamma_1(0)$ and $H(1, t) = \gamma_1(1)$ for all $t \in [0, 1]$. In particular we must have $\gamma_1(0) = \gamma_2(0)$ and $\gamma_1(1) = \gamma_2(1)$, see Figure 19.10.*
C. *We call two closed paths $\gamma_1, \gamma_2 : [0, 1] \to X$ **homotopic relative to x_0** if $\gamma_1(0) = \gamma_1(1) = \gamma_2(0) = \gamma_2(1) = x_0$ and if γ_1 and γ_2 are homotopic with respect to their initial and terminal points, see Figure 19.11.*
D. *If a closed path $\gamma : [0, 1] \to X$ is homotopic to the constant path γ_{x_0}, $x_0 = \gamma(0)$, we call γ **null-homotopic**.*

Remark 19.26. In order to have a shorter façon de parler we call two paths γ_1 and γ_2 **homotopic paths** if they are homotopic relative to their initial and terminal points.

It is easy to check that homotopy relative to F is again an equivalence relation, as is homotopy of paths or homotopy of closed paths relative to a point x_0. Furthermore we have

Proposition 19.27. *Let $\varphi : [0, 1] \to [0, 1]$ be an orientation preserving parameter transformation and $\gamma : [0, 1] \to X$ a path. The two paths γ and $\gamma \circ \varphi$ are homotopic.*

Proof. The mapping $\tau : [0, 1] \times [0, 1] \to [0, 1]$, $\tau[s, t] = s(1 - t) + t\varphi(s)$, is continuous from $[0, 1] \times [0, 1]$ to $[0, 1]$. We define $H : [0, 1] \times [0, 1] \to X$ by $H(s, t) = (\gamma \circ \tau)(s, t) = \gamma(s(1 - t) + t\varphi(s))$. This is a continuous mapping from $[0,1] \times [0,1]$ to X. For all $t \in [0,1]$ we have $H(0,t) = \gamma(t\varphi(0)) = \gamma(0)$ as well as $H(1,t) = \gamma(1 - t + t\varphi(1)) = \gamma(1)$. Moreover we have $H(s,0) = \gamma(s)$ and $H(s,1) = \gamma(\varphi(s))$ proving the proposition. □

For a path $\gamma : [0,1] \to X$ we consider as before the path $\gamma^{-1} \to X$, $\gamma^{-1}(s) = \gamma(1-s)$.

Lemma 19.28. *If γ_1 and γ_2 are two homotopic paths then γ_1^{-1} and γ_2^{-1} are homotopic too.*

Proof. Let $H : [0,1] \times [0,1] \to X$ be a homotopy between γ_1 and γ_2. Then the mapping $\tilde{H} : [0,1] \times [0,1] \to X$, $\tilde{H}(s,t) = H(1-s,t)$ is a homotopy between γ_1^{-1} and γ_2^{-1}. □

We want to study the sum of paths in light of their behaviour under homotopies. Since we now let all paths be defined on $[0,1]$ (which after a parameter transformation is always possible) we have to adapt Definition 19.11 accordingly. If $\gamma_1, \gamma_2 : [0,1] \to X$ are two paths with $\gamma_1(1) = \gamma_2(0)$ we define $\gamma_1 \oplus \gamma_2 : [0,1] \to X$ by

$$(\gamma_1 \oplus \gamma_2)(s) = \begin{cases} \gamma_1(s), & 0 \leq s \leq \frac{1}{2} \\ \gamma_2(2s-1), & \frac{1}{2} \leq s \leq 1, \end{cases} \tag{19.4}$$

and as above we set $\gamma_1^{-1} : [0,1] \to X$, $\gamma_1^{-1}(s) = \gamma_1(1-s)$. It follows that

$$(\gamma_1 \oplus \gamma_2)^{-1} = \gamma_2^{-1} \oplus \gamma_1^{-1}. \tag{19.5}$$

Now, if $\gamma_3 : [0,1] \to X$ is a third path with $\gamma_2(1) = \gamma_3(0)$ we find

$$((\gamma_1 \oplus \gamma_2) \oplus \gamma_3)(s) = \begin{cases} \gamma_1(4s), & 0 \leq s \leq \frac{1}{4} \\ \gamma_2(4s-1), & \frac{1}{4} \leq s \leq \frac{1}{2} \\ \gamma_3(2s-1), & \frac{1}{2} \leq s \leq 1 \end{cases} \tag{19.6}$$

$$(\gamma_1 \oplus (\gamma_2 \oplus \gamma_3))(s) = \begin{cases} \gamma_1(2s), & 0 \leq s \leq \frac{1}{2} \\ \gamma_2(4s-1), & \frac{1}{2} \leq s \leq \frac{3}{4} \\ \gamma_2(4s-3), & \frac{3}{4} \leq s \leq 1, \end{cases} \tag{19.7}$$

and these are different paths having however the same trace. This implies that $\oplus$ as introduced by (19.4) is not giving rise to an associative operation.

We may resolve this problem by looking at equivalence classes with respect to orientation preserving parameter transformations, however we prefer to look at equivalence classes with respect to homotopy of paths. Since (19.6) and (19.7) are obtained from each other by an orientation preserving parameter transformation they are homotopic if we impose the correct conditions on their initial and terminal points, see Proposition 19.26.

For a path $\gamma : [0, 1] \to X$ with $x_0 = \gamma(0)$ and $x_1 = \gamma(1)$ we can always form $\gamma_{x_0} \oplus \gamma$ and $\gamma \oplus \gamma_{x_1}$ and from Proposition 19.26 we deduce that $\gamma_{x_0} \oplus \gamma$, γ and $\gamma \oplus \gamma_{x_1}$ are homotopic paths. Furthermore we can always form $\gamma \oplus \gamma^{-1}$ according to

$$(\gamma \oplus \gamma^{-1})(s) = \begin{cases} \gamma(2\,s), & 0 \le s \le \frac{1}{2} \\ \gamma^{-1}(2\,s - 1) = \gamma(2 - 2s), & \frac{1}{2} \le s \le 1. \end{cases} \tag{19.8}$$

Clearly, $\gamma \oplus \gamma^{-1}$ is a closed path. Consider the continuous mapping $H : [0, 1] \times [0, 1] \to X$ defined by

$$H(s, t) := \begin{cases} \gamma(2\,s\,(1 - t)), & 0 \le s \le \frac{1}{2} \\ \gamma(2\,(1 - s)\,(1 - t)), & \frac{1}{2} \le s \le 1. \end{cases}$$

For $t = 0$ we find $H(\cdot, 0) = \gamma \circ \gamma^{-1}$ and for $t = 1$ we have $H(\cdot, 1) = \gamma(0)$. Thus $\gamma \circ \gamma^{-1}$ is null-homotopic to γ_{x_0}. Next let $\gamma_1, \gamma_2 : [0, 1] \to X$ be two paths such that $\gamma_1(1) = \gamma_2(0)$ and let $\tilde{\gamma}_j$ be homotopic to γ_j, $j = 1, 2$. Denote by H_j a homotopy between γ_j and $\tilde{\gamma}_j$ and define

$$H(s, t) := \begin{cases} H_1(2\,s, t), & 0 \le s \le \frac{1}{2} \\ H_2(2\,s - 1, t), & \frac{1}{2} \le s \le 1 \end{cases} \tag{19.9}$$

which turns out to be a homotopy between $\gamma_1 \oplus \gamma_2$ and $\tilde{\gamma}_1 \oplus \tilde{\gamma}_2$. With this preparation we can lift the operation $\oplus$ to equivalence classes of homotopic paths. The definition is of course the obvious one and we do not change the notation:

$$[\gamma_1] \oplus [\gamma_2] := [\gamma_1 \oplus \gamma_2] \tag{19.10}$$

and

$$[\gamma]^{-1} := [\gamma^{-1}]. \tag{19.11}$$

From our previous investigation we conclude:

$$[\gamma_1] \oplus ([\gamma_2] \oplus [\gamma_3]) = ([\gamma_1] \oplus [\gamma_2]) \oplus [\gamma_3], \tag{19.12}$$

$$[\gamma] + [\gamma_{x_0}] = [\gamma] = [\gamma] \oplus [\gamma_{x_1}], \ \gamma(0) = x_0 \text{ and } \gamma(1) = x_1, \tag{19.13}$$

and

$$[\gamma] \oplus [\gamma^{-1}] = [\gamma_{x_0}], \ [\gamma^{-1}] \oplus [\gamma] = [\gamma_{x_1}]. \tag{19.14}$$

We denote by Γ_{x_0,x_1} the set of all homotopy classes of paths $\eta : [0,1] \to X$ with $\eta(0) = x_0$ and $\eta(1) = x_1$, and by Γ_{x_0} we denote the set Γ_{x_0,x_0}, i.e. all homotopy classes of closed path $\eta : [0,1] \to X$ with $\eta(0) = x_0$.

Theorem 19.29. *Let $\gamma : [0,1] \to X$ be a fixed path with initial point $\gamma(0) = x_0$ and terminal point $\gamma(1) = x_1$. Then the mapping $\sigma : \Gamma_{x_0,x} \to \Gamma_{x_0}$, $\sigma([\eta]) := [\eta] \oplus [\gamma^{-1}]$ is a bijection with inverse $\tau : \Gamma_{x_0} : \Gamma_{x_0 x}$, $\tau(w) := [w] \oplus [\gamma]$.*

Proof. For $[\eta] \in \Gamma_{x_0,x_1}$ it follows that

$$\begin{aligned}
(\tau \circ \sigma)([\eta]) &= \tau(\sigma[\eta]) = \tau([\eta] \oplus [\gamma^{-1}]) \\
&= ([\eta] \oplus [\gamma^{-1}]) \oplus [\gamma] = [\eta],
\end{aligned}$$

and with $\omega \in \Gamma_{x_0}$ we have

$$\begin{aligned}
(\sigma \circ \tau)([\omega]) &= \sigma(\tau([\omega])) = \sigma([\omega] \oplus [\gamma]) \\
&= ([\omega] \oplus [\gamma]) \oplus [\gamma^{-1}] = [\omega],
\end{aligned}$$

i.e. σ and τ are inverse to each other which implies the theorem. $\qquad\square$

Corollary 19.30. *Let γ_1 and γ_2 be two paths with $\gamma_1(0) = \gamma_2(0) = x_0$ and $\gamma_1(1) = \gamma_2(1) = x_1$. These two paths are homotopic if and only if $\gamma_1 \oplus \gamma_2^{-1}$ is null-homotopic.*

In other words, the investigations on homotopy of paths can be reduced to the study of the homotopy of closed paths. From our preparations we now deduce

Theorem 19.31. *The set Γ_{x_0} with the operation $\oplus$ forms a group with neutral element $[\gamma_{x_0}]$ and inverse to $[\gamma]$ given by $[\gamma]^{-1} = [\gamma^{-1}]$.*

Definition 19.32. *The group $(\Gamma_{x_0}, \oplus)$ is called the **first fundamental group** of X with respect to x_0 and is denoted by $\pi_1(X, x_0)$.*

Theorem 19.33. *For an arcwise connected topological space X the groups $\pi_1(X, x_0)$ and $\pi_1(X, x_1)$, $x_0, x_1 \in X$, are isomorphic.*

Proof. Let $\gamma : [0,1] \to X$ be a path connecting x_0 with x_1, i.e. $\gamma(0) = x_0$ and $\gamma(1) = x_1$. We define the mapping

$$h_\gamma : \pi_1 (X, x_0) \to \pi_1 (X, x_1), \ h_\gamma ([\eta]) := [\gamma] \oplus [\eta] \oplus [\gamma]^{-1},$$

and claim that h_γ is an isomorphism. For $[\eta_1], [\eta_2] \in \pi_1 (X, x_0)$ we find

$$\begin{aligned} h_\gamma ([\eta_1] \oplus [\eta_2]) &= [\gamma] \oplus [\eta_1] \oplus [\eta_2] \oplus [\gamma]^{-1} \\ &= [\gamma] \oplus [\eta_1] \oplus [\gamma]^{-1} \oplus [\gamma] \oplus [\eta_2] \oplus [\gamma]^{-1} \\ &= h_\gamma ([\eta_1]) \oplus h_\gamma ([\eta_2]), \end{aligned}$$

i.e. h_γ is a homomorphism. Now we can use $[\gamma]^{-1}$ to define $h_{\gamma^{-1}} : \pi_1 (X, x_1) \to \pi_1(X, x_0)$ by $h_{\gamma^{-1}}([\eta]) = [\gamma]^{-1} \oplus [\eta] \oplus [\gamma]$ and it follows that $h_{\gamma^{-1}} \circ h_\gamma = \mathrm{id}_{\pi_1(x,x_0)}$ and $h_\gamma \circ h_{\gamma^{-1}} = \mathrm{id}_{\pi_1 (x,x_1)}$, i.e. h_γ (and hence $h_{\gamma^{-1}}$) is an isomorphism. $\qquad \square$

Of central importance is

Definition 19.34. *A pathwise connected topological space X is said to be **simply connected** if for one point $x_0 \in X$ the first fundamental group $\pi_1(X, x_0)$ is trivial, i.e. consist only of the neutral element.*

Remark 19.35. From Theorem 19.32 we deduce that all groups $\pi_1 (X, x)$, $x \in X$, are trivial if $\pi_1 (X, x_0)$ is trivial. Hence the notion of simple connectivity of a space does not dependent on the special point x_0 in Definition 19.33.

Example 19.36. Every convex set $K \subset \mathbb{R}^n$ is simply connected. For $\gamma_1, \gamma_2 : [0,1] \to K$ we define $H(s, t) = (1 - t)\gamma_1 (s) + t\gamma_2 (s)$ which is continuous and due to the convexity of K well defined. Moreover we have $H(\cdot, 0) = \gamma_1$ and $H(\cdot, 1) = \gamma_2$. Thus if we choose $x_0 \in K$ and γ to be any closed path in K with initial point x_0 then we find that γ is null-homotopic to γ_{x_0}, i.e. $\pi_1(K, x_0)$ is trivial.

A classical result related to (simply) connected sets in the plane is the **Jordan curve theorem**. Although we will not make use of it, we want to state it in the refined version due to Schoenflies and for a proof we refer to [8], [33], [69] or [83].

Theorem 19.37. *Let $\gamma : [0, 1] \to \mathbb{C}$ be a simply closed arc. The trace $\mathrm{tr}(\gamma)$ divides $\mathbb{C}$ into three mutually disjoint sets Γ_+, Γ_- and $\mathrm{tr}(\gamma)$, $\mathbb{C} = \Gamma_+ \cup \Gamma_- \cup \mathrm{tr}(\gamma)$, where Γ_+ is bounded, Γ_- is unbounded and $\partial \Gamma_+ = \partial \Gamma_- = \mathrm{tr}(\gamma)$. The sets Γ_+ and Γ_- are connected and Γ_+ is simply connected.*

For our final result in this chapter we need

Definition 19.38. *An arc $\gamma : [a, b] \to \mathbb{C}$ is called a **polygonal arc** (or curve or path) if its trace consists of a finite number of line segments. If all these line segments are parallel either to the real axis or to the imaginary axis we call γ an **axis parallel polygonal arc** (or curve or path).*

By definition for a polygonal arc $\gamma : [a, b] \to \mathbb{C}$ there exists a partition of $[a, b]$, $a = t_0 < t_1 < \ldots < t_n = b$ such that the mappings $\gamma|_{[t_{j-1}, t_j]}$, $j = 1, \ldots, n$, have as traces line segments. For example we may look at $t \mapsto \frac{t_j - t}{t_j - t_{j-1}} \gamma(t_{j-1}) + \frac{t - t_{j-1}}{t_j - t_{j-1}} \gamma(t_j)$, $t \in [t_{j-1}, t_j]$, as a further parametrization of $\gamma|_{[t_{j-1}, t_j]}$.

Proposition 19.39. *In a region $G \subset \mathbb{C}$ we can connect two points $z_1, z_2 \in G$ always by a polygonal arc.*

Proof. Let $\gamma : [a, b] \to G$ be a curve connecting z_1 and z_2. Since G is open and $\mathrm{tr}(\gamma) \subset G$ is compact we can cover $\mathrm{tr}(\gamma)$ by a finite number of balls $B_{\epsilon_j}(w_j)$, $j = 1, \ldots, N$, $w_j \in \mathrm{tr}(\gamma)$, $w_1 = z_1$ and $w_N = z_2$. Each pair of points in $B_{\epsilon_j}(w_j)$ can be joined by a line segment since these are convex sets. Further we have $B_{\epsilon_j}(w_j) \cap B_{\epsilon_{j+1}}(w_{j+1}) \neq \emptyset$ when we choose the enumeration of the balls appropriately. Denote by $w_{k, k+1}$ a point in $B_{\epsilon_k}(w_k) \cap B_{\epsilon_{k+1}}(w_{k+1})$. The line segments joining $z_1 = w_1$ with $w_{1,2}$, $w_{1,2}$ with $w_{2,3}$, $w_{2,3}$ with $w_{3,4}$, $\ldots$, $w_{N-1,N}$ with $w_N = z_2$ form a polygonal arc in G connecting z_1 and z_2. $\square$

Problems

1.   a) Let $(X, \mathcal{O})$ be a discrete topological space, i.e. $\mathcal{O} = \mathcal{P}(X)$. When is $(X, \mathcal{O})$ connected?

   b) Let $(X, \mathcal{O}_X)$ be a connected topological space and $(Y, \mathcal{O}_Y)$ a topological space with at least two connectivity components. Does there exist a surjective mapping $f : X \to Y$ which is continuous?

2. Let $A \subset \mathbb{R}^n$ and $B \subset \mathbb{R}^m$ be two connected non-empty open sets. Is the set $A \times B \subset \mathbb{R}^n \times \mathbb{R}^m$ connected?

3. Consider the simply closed curve γ in $\mathbb{C}$ the trace $\mathrm{tr}(\gamma)$ of which is given by the figure below:

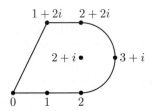

Find a piecewise continuous differentiable, anticlockwise parametrization of γ, $\gamma : [0,1] \to \mathbb{C}$, such that $\gamma(0) = 2 = \gamma(1)$.

4. Find a bijective continuous mapping $f : \mathbb{R}^2 \setminus \{0\} \to \mathbb{R}^2 \setminus \{0\}$ with continuous inverse which maps $\partial B_2(0)$ onto $\partial B_1(0) = S^1$.

5. Show that a continuous mapping between two topological spaces maps arcwise connected sets onto arcwise connected sets.

6. Prove that for three paths $\gamma_j : [a_j, b_j] \to X$, $j = 1, 2, 3$, such that $\gamma_1(b_1) = \gamma_2(a_2)$ and $\gamma_2(b_2) = \gamma_3(a_3)$ we have $(\gamma_1 \oplus \gamma_2) \oplus \gamma_3 = \gamma_1 \oplus (\gamma_2 \oplus \gamma_3)$.

7. Let γ_1 and γ_2 be two paths in the topological space X such that $\gamma_1 \oplus \gamma_2$ is defined. Prove that the definition $[\gamma_1] \oplus [\gamma_2] := [\gamma_1 \oplus \gamma_2]$ is independent of the paratmetrization. Now use Problem 6 to deduce that

$$([\gamma_1] \oplus [\gamma_2]) \oplus [\gamma_3] = [\gamma_1] \oplus ([\gamma_2] \oplus [\gamma_3])$$

holds.

8. Prove that homotopy equivalence for topological spaces is an equivalence relation.

9. Is $\pi_1(A_{\frac{1}{2},2}(0), 1)$ where $A_{\frac{1}{2},2}(0) = \{z \in \mathbb{C} | \frac{1}{2} < |z| < 2\}$, trivial?

20 Line Integrals of Complex-valued Functions

For a parametrized curve $\gamma : [a, b] \to \mathbb{C}$ with $\operatorname{tr} \gamma \subset D$, $D \subset \mathbb{C}$ a domain, we want to define for complex-valued functions $f : D \to \mathbb{C}$ the line integral $\int_\gamma f(z)\, dz$. While in the case of holomorphic functions the properties of $\int_\gamma f(z)\, dz$ will turn out to be quite different to those of general line integrals (in $\mathbb{R}^n$), the basic definition of the line integral $\int_\gamma f(z)\, dz$ is similar, in fact it is an adaptation of the theory developed in Chapter II.15 where we now identify $\mathbb{R}^2$ with $\mathbb{C}$. Therefore we start by recollecting definitions and results from Chapter II.15 but now we will use complex notation. From this chapter on we will also use the standard notation from complex variable theory and we will often write γ for the curve (as mapping) as well as for its trace.

Definition 20.1. *(II.15.1) Let $\gamma : [a, b] \to \mathbb{C}$, $a < b$, be a continuous parametrized curve and $Z = Z(t_0, ..., t_m)$, $t_0 = a$, $t_m = b$, be a partition of $[a, b]$.*
A. *We call*

$$V_Z(\gamma) := \sum_{k=0}^{m-1} |\gamma(t_{k+1}) - \gamma(t_k)| \tag{20.1}$$

*the **Z-variation** of γ and*

$$V(\gamma) := \sup_Z V_Z(\gamma) \tag{20.2}$$

*the **total variation** of γ where the supremum is taken over all partitions Z of $[a, b]$.*
B. *If $V(\gamma) < \infty$ we call γ **rectifiable** and define its length as*

$$l_\gamma := V(\gamma). \tag{20.3}$$

Proposition 20.2. *(II.15.2) A continuous parametric curve $\gamma : [a, b] \to \mathbb{C}$, $\gamma = \operatorname{Re} \gamma + i \operatorname{Im} \gamma$, is rectifiable if and only if the functions $\operatorname{Re} \gamma$ and $\operatorname{Im} \gamma$ are of bounded variations.*

Lemma 20.3. *(II.15.5) If γ_1 and γ_2 are rectifiable and $\gamma_1 \oplus \gamma_2$ is defined then $\gamma_1 \oplus \gamma_2$ is rectifiable too.*

Definition 20.4. *(II.15.7) A continuous curve $\gamma : [a, b] \to \mathbb{C}$ is **piecewise continuously differentiable** if there exists a partition $Z = Z(t_0, \ldots, t_m)$ of $[a, b]$ such that $\gamma|_{[t_j, t_{j+1}]}$, $j = 0, \ldots, m - 1$, is continuously differentiable.*

Corollary 20.5. *(II.15.9) A piecewise continuously differentiable curve $\gamma : [a, b] \to \mathbb{C}$ is rectifiable.*

We know from Example II.15.8 that every polygon is a piecewise continuously differentiable curve and if γ_1 and γ_2 are two piecewise continuously differentiable curves for which $\gamma_1 \oplus \gamma_2$ is defined, then $\gamma_1 \oplus \gamma_2$ is piecewise continuously differentiable too. Of a more technical nature is

Theorem 20.6. *(II.15.11) Let $\gamma : [a, b] \to \mathbb{C}$ be a rectifiable curve. For every $\epsilon > 0$ there exists $\delta > 0$ such that for every partition Z of $[a, b]$ with mesh size less than δ it follows that $V(\gamma) - V_Z(\gamma) < \epsilon$. If $(Z^{(\nu)})_{\nu \in \mathbb{N}}$ is a sequence of partition of $[a, b]$ with mesh size η^ν and $\lim_{\nu \to \infty} \eta^{(\nu)} = 0$ then $\lim_{\nu \to \infty} V_{Z^{(\nu)}}(\gamma) = V(\gamma) = l_\gamma$.*

Theorem 20.7. *(II.15.12) For a C^1 curve $\gamma : [a, b] \to \mathbb{C}$ we have*

$$l_\gamma = V(\gamma) = \int_a^b |\dot{\gamma}(t)| \, dt = \int_a^b ((\operatorname{Re} \dot{\gamma}(t))^2 + (\operatorname{Im} \dot{\gamma}(t))^2)^{\frac{1}{2}} \, dt. \qquad (20.4)$$

Corollary 20.8. *(II.15.13) Let $\gamma : [a, b] \to \mathbb{C}$ be a piecewise continuously differentiable curve, i.e. with $Z = Z(t_0, \ldots, t_m)$ we have that $\gamma|_{[t_j, t_{j+1}]}$, $j = 0, \ldots, m - 1$, is a C^1 curve. Then γ is a rectifiable curve and*

$$l_\gamma = \sum_{j=0}^{m-1} l_\gamma|_{[t_j, t_{j-1}]} = \sum_{j=0}^{m-1} \int_{t_j}^{t_{j+1}} |\dot{\gamma}(t)| \, dt. \qquad (20.5)$$

Now we can develop the theory of line integrals for complex-valued functions. In the remaining part of this chapter we deal with the more elementary theory using the books [26] and [67] a bit as a guide. In analogy to Definition II.15.15 we give

Definition 20.9. *Let $D \subset \mathbb{C}$ be a domian and $f : D \to \mathbb{C}$ be a continuous complex-valued function. Further let $\gamma : [a, b] \to D$ be a piecewise continuously differentiable curve, $\gamma = \gamma_1 \oplus \ldots \oplus \gamma_N$, where $\gamma_j : [t_{j-1}, t_j] \to D$, $a =$*

$t_0 < \ldots < t_N = b$, is a C^1- curve. The **line integral** of f along γ is defined by

$$\int_\gamma f(z)\, dz := \sum_{j=1}^N \int_{\gamma_j} f(z)\, dz \tag{20.6}$$

$$:= \sum_{j=1}^N \int_{t_{j-1}}^{t_j} f(\gamma(t))\, \dot{\gamma}(t)\, dt = \sum_{j=1}^N \int_{t_{j-1}}^{t_j} f(\gamma_j(t))\, \dot{\gamma}(t)\, dt.$$

Remark 20.10. Sometimes it has an advantage to write $z(t)$ for the curve and then to look at

$$\int_{t_{j-1}}^{t_j} f(z(t))\, z'(t)\, dt.$$

With this notation we can rewrite (20.4) as

$$l_\gamma = \int_a^b |z'(t)|\, dt. \tag{20.7}$$

Furthermore we note that for $\gamma : [a, b] \to \mathbb{C}, \gamma(t) = t$, we find

$$\int_\gamma f(z)\, d_z = \int_a^b f(\gamma(t))\dot{\gamma}(t)\, dt = \int_a^b f(t)\, dt,$$

i.e. our notations are consistent.

Quite a few properties of $\int_\gamma f(z)\, d_z$ are straightforward to prove:

If $f : [a, b] \to \mathbb{C}, [a, b] \subset \mathbb{R}$ is continuous then

$$\mathrm{Re} \int_a^b f(t)\, dt = \int_a^b \mathrm{Re} f(t)\, dt \text{ and } \mathrm{Im} \int_a^b f(t)\, dt = \int_a^b \mathrm{Im} f(t)\, dt, \tag{20.8}$$

as well as

$$\overline{\int_a^b f(t)\, dt} = \int_a^b \overline{f(t)}\, dt. \tag{20.9}$$

Next let $f, g : D \to \mathbb{C}$ be continuous functions and γ, γ_1 and γ_2 be curves satisfying each of the conditions of Definition 20.9 and assume that $\gamma_1 \oplus \gamma_2$ is defined. Further let $c \in \mathbb{C}$. The following hold:

$$\int_\gamma (f \pm g)(z) dz = \int_\gamma f(z) dz + \int_\gamma g(z) dz; \qquad (20.10)$$

$$\int_\gamma cf(z) dz = c \int_\gamma f(z) dz; \qquad (20.11)$$

and

$$\int_{\gamma_1 \oplus \gamma_2} f(z) dz = \int_{\gamma_1} f(z) dz + \int_{\gamma_2} f(z) dz. \qquad (20.12)$$

If $\operatorname{tr} \gamma = \{z_0\}$, i.e. $\gamma(t) = z_0$ for all t, then $\int_\gamma f(z)\, dz = 0$. If γ^{-1} is the inverse to γ, i.e. $\gamma^{-1} : [a, b] \to \mathbb{C}$, $\gamma^{-1}(t) = \gamma(a + b - t)$, then $\dot\gamma^{-1}(t) = -\dot\gamma(t)$ and it follows that

$$\int_{\gamma^{-1}} f(z)\, dz = \int_a^b f(\gamma^{-1}(t)) \,(\dot\gamma^{-1})\,(t)\, dt$$

$$= -\int_a^b f(\gamma(a + b - t))\, \dot\gamma\,(a + b - t)\, dt$$

$$= \int_a^b f(s)\, \dot\gamma(s)\, ds = -\int_a^b f(s)\, \dot\gamma(s)\, ds,$$

and we have proved

$$\int_{\gamma^{-1}} f(z)\, dz = -\int_\gamma f(z)\, dz. \qquad (20.13)$$

Example 20.11. Let $\gamma : [0, 2\pi] \to \mathbb{C}$, $\gamma(t) = z_0 + r\, e^{it}$, be the circle with centre z_0 and radius $r > 0$. Since $z_0 \notin \operatorname{tr}\gamma$ the function $z \mapsto \frac{1}{z - z_0}$ is defined and continuous in a neighbourhood of γ and we find

$$\int_{|z - z_0| = r} \frac{1}{z - z_0}\, dz = \int_0^{2\pi} \frac{1}{r\, e^{it}}\, ir\, e^{it}\, dt = 2\pi i, \qquad (20.14)$$

or

$$\frac{1}{2\pi i} \int_{|z-z_0|=r} \frac{1}{z-z_0} \, dz = 1. \tag{20.15}$$

Note that this integral is independent of z_0 and r. In (20.14) and (20.15) we have used the rather common notation

$$\int_{|z-z_0|=r} f(z) \, dz := \int_\gamma f(z) \, dz \text{ for } \gamma(t) = z_0 + r \, e^{it}. \tag{20.16}$$

Example 20.12. The upper half circle with radius 1 can be parametrized by $\gamma : [0, \pi] \to \mathbb{C}$, $\gamma(t) = e^{i(\pi - t)}$. The initial point of γ is $\gamma(0) = e^{i\pi} = -1$ and the terminal point of γ is $\gamma(\pi) = e^{i0} = 1$, see Figure 20.1. For $f(z) = |z|$ we find

$$\int_\gamma f(z) \, dz = \int_\gamma |z| \, dz = \int_0^\pi 1 \, \dot\gamma(t) \, dt = \gamma(\pi) - \gamma(0) = 2.$$

We may also integrate $f(z) = |z|$ along the interval $[-1, 1]$, i.e. the line segment σ connecting $(-1, 0)$ and $(1, 0)$, $\sigma : [-1, 1] \to \mathbb{C}$, $\sigma(t) = t$, see Figure 20.2. For this curve we get

$$\int_\sigma f(z) \, dz = \int_\sigma |z| \, dz = \int_{-1}^1 |t| \, dt = -\int_{-1}^0 t \, dt + \int_0^1 t \, dt = 1.$$

Thus in general $\int_\gamma f(z) \, dz$ will depend on γ and not only on the initial and terminal points of γ.

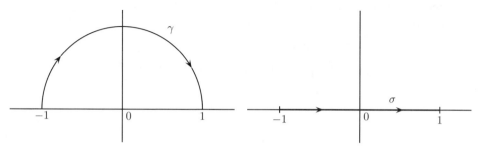

Figure 20.1   Figure 20.2

369

The following proposition gives basic estimates for line integrals.

Proposition 20.13. A. *For a continuous function $f : [a,b] \to \mathbb{C}$ the following holds*

$$\left| \int_a^b f(t)\, dt \right| \leq \int_a^b |f(t)|\, dt. \tag{20.17}$$

B. *If for a domain D the curve $\gamma : [a,b] \to D$, $D \subset \mathbb{C}$, is piecewise continuously differentiable and $f : D \to \mathbb{C}$ is continuous on $\operatorname{tr} \gamma$ then*

$$\left| \int_\gamma f(z)\, dz \right| \leq \|f\|_{\infty,\operatorname{tr}\gamma}\, l_\gamma \tag{20.18}$$

where $\|f\|_{\infty,\operatorname{tr}\gamma} = \sup\limits_{z \in \operatorname{tr}\gamma} |f(z)|$.

Proof. **A.** If f is real-valued, (20.17) is just a consequence of the triangle inequality and is well known to us. The case where $\int_a^b f(t)\, dt = 0$ is trivial. For $\int_a^b f(t)\, dt \neq 0$ we can find $s \in \mathbb{R}$ such that $e^{is} \int_a^b f(t)\, dt = \int_a^b \operatorname{Re}\left(e^{is} f(t)\right) dt$ as well as

$$\left| \int_a^b f(t)\, dt \right| = \left| e^{is} \int_a^b f(t)\, dt \right| = \left| \int_1^b \operatorname{Re}\left(e^{is} f(t)\right) dt \right|$$

$$\leq \int_a^b \left| \operatorname{Re}\left(e^{is} f(t)\right) \right| dt \leq \int_a^b |f(t)|\, dt.$$

B. Since γ is continuous and $[a,b]$ is compact, $\operatorname{tr}\gamma$ is compact too and $f|_{\operatorname{tr}\gamma}$ is bounded, i.e. $\|f\|_{\infty,\operatorname{tr}\gamma}$ is finite. Moreover, using part A we have

$$\left| \int_\gamma f(z)\, dz \right| = \left| \int_a^b f(\gamma(t))\, \dot{\gamma}(t)\, dt \right| \leq \int_a^b |f(\gamma(t))|\, |\dot{\gamma}(t)|\, dt$$

$$\leq \|f\|_{\infty,\operatorname{tr}\gamma} \int_a^b |\dot{\gamma}(t)|\, dt = \|f\|_{\infty,\operatorname{tr}\gamma}\, l_\gamma.$$

$\square$

Next we want to prove that $\int\limits_{\gamma} f(z)\,dz$ is independent of the parametrization, i.e. it is unchanged under an orientation preserving parameter transformation or change of parameter. Since we are working with piecewise continuously differentiable curves we can reformulate Definition 19.14.A as

Definition 20.14. *Let* $I := [a, b]$ *and* $J := [c, d]$ *be non-degenerate compact intervals and* $\varphi : [c, d] \to [a, b]$ *a strictly increasing piecewise continuously differentiable function, i.e.* $\varphi' > 0$ *wherever it is defined, with* $\varphi(c) = a$ *and* $\varphi(d) = b$. *Then we call* φ *a piecewise continuously differentiable orientation preserving parameter transformation or just a* **change of parameter**.

It is easy to see that if φ and ψ are changes of parameter then the same holds for $\varphi \circ \psi$ and φ^{-1}. Let $\gamma : I \to \mathbb{C}$ be a piecewise continuously differentiable curve and $\varphi : J \to I$ a change of parameter. Then $\gamma \circ \varphi : J \to \mathbb{C}$ is a piecewise continuously differentiable curve such that $\mathrm{tr}\,\gamma = \mathrm{tr}\,\gamma \circ \varphi$ and the initial and terminal points of both curves are the same. We say that $\gamma \circ \varphi$ is obtained from γ by a change of parameter.

Theorem 20.15. *Let* $D \subset \mathbb{C}$ *be a domain and* $\gamma_1 : [a, b] \to D$ *be a piecewise continuously differentiable curve and* $f : D \to \mathbb{C}$ *a function such that* $f|_{\mathrm{tr}\,\gamma_1}$ *is continuous. If* $\gamma_2 : [c, d] \to \mathbb{C}$ *is obtained from* γ_1 *by a change of parameter* $\varphi : [c, d] \to [a, b]$ *then we have*

$$\int\limits_{\gamma_1} f(z)\,dz = \int\limits_{\gamma_2} f(z)\,dz. \qquad (20.19)$$

Proof. By a straightforward calculation we get

$$\int\limits_{\gamma_1} f(z)\,dz = \int\limits_{a}^{b} f(\gamma_1(t))\,\dot{\gamma}_1(t)\,dt$$

$$= \int\limits_{c}^{d} f((\gamma_1 \circ \varphi)(s))\,\dot{\gamma}_1(\varphi(s))\,\varphi'(s)\,ds$$

$$= \int\limits_{c}^{d} f(\gamma_2(s))\,\dot{\gamma}_2(s)\,ds = \int\limits_{\gamma_2} f(z)\,dz.$$

$\square$

The central result of calculus of real-valued function of a real variable is the fundamental theorem and for this the notion of a primitive is essential.

Definition 20.16. *A holomorphic function $F : D \to \mathbb{C}$ defined as a domain $D \subset \mathbb{C}$ is called a **primitive** of the continuous function $f : D \to \mathbb{C}$ if $F' = f$.*

In Chapter 18 we have seen that holomorphic functions might have more complicated domains and therefore it is worth giving

Definition 20.17. *A continuous function $f : D \to \mathbb{C}$ has a **local primitive** if for every $z_0 \in D$ there exists an open neighbourhood $U(z_0) \subset D$ such that $f|_{U(z_0)}$ has a primitive.*

One formulation of the fundamental theorem is, see Theorem I.26.4, that for $f : [a, b] \to \mathbb{R}$ continuous with primitive F we have

$$\int_a^b f(x)\, dx = F(b) - F(a). \tag{20.20}$$

Interpreting $[a, b]$ as trace of a curve with initial point a and terminal point b we may seek to generalise (20.20) to

$$\int_\gamma f(z)\, dz = F(z_1) - F(z_0) \tag{20.21}$$

where now $f : D \to \mathbb{C}$, $\gamma : [a, b] \to \mathbb{C}$ with $\gamma(a) = z_0$ and $\gamma(b) = z_1$, and F is a primitive of f.

Theorem 20.18. *Let $f : D \to \mathbb{C}$ be a continuous function which has a primitive on the domain D. Further let $\gamma : [a, b] \to D$ be a piecewise continuously differentiable curve such that $\gamma(a) = z_0$ and $\gamma(b) = z_1$ then we indeed have (20.21), i.e.*

$$\int_\gamma f(z)\, dz = F(z_1) - F(z_0).$$

In particular the integral depends only on the initial and terminal point of γ.

Proof. Let $\gamma : [a, b] \to D$ and $a = t_0 < t_1 < \ldots < t_N = b$ be a partition of $[a, b]$ such that $\gamma_j = \gamma|_{[t_{j-1}, t_j]}$, $j = 1, \ldots, N$, is continuously differentiable. It follows that

$$\int_\gamma f(z)\, dz = \int_a^b f(\gamma(t))\, \dot\gamma(t)\, dt = \sum_{j=1}^N \int_{t_{j-1}}^{t_j} f(\gamma(t))\, \dot\gamma(t)\, dt$$

$$= \sum_{j=1}^N \int_{t_{j-1}}^{t_j} F'(\gamma(t))\, \dot\gamma(t)\, dt = \sum_{j=1}^N \int_{t_{j-1}}^{t_j} (F \circ \gamma)'(t)\, dt$$

$$= \sum_{j=1}^N ((F \circ \gamma)(t_j) - (F \circ \gamma)(t_{j-1})) = F(z_1) - F(z_0).$$

$\square$

Corollary 20.19. *Suppose that $f : D \to \mathbb{C}$ is continuous and has a primitive. If $\gamma : [a, b] \to D$ is a closed curve satisfying the assumption of Theorem 20.18 we have*

$$\int_\gamma f(z)\, dz = 0. \tag{20.22}$$

Example 20.20. A. For $n \in \mathbb{N}$ the function $f : \mathbb{C} \to \mathbb{C}$, $f(z) = z^n$, has the primitive $F(z) = \frac{1}{n+1} z^{n+1}$ and for every piecewise continuously differentiable curve γ with initial point z_0 and terminal point z_1 we have

$$\int_\gamma z^n\, dz = \frac{1}{n+1} (z_1^{n+1} - z_0^{n+1}).$$

This implies for every polynomial $p(z) = \sum_{j=0}^N a_j z^j$, $a_j \in \mathbb{C}$, that

$$\int_\gamma p(z)\, dz = \sum_{j=0}^N \frac{a_j}{j+1} z_1^{j+1} - \sum_{j=0}^N \frac{a_j}{j+1} z_0^{j+1}.$$

B. For the entire functions $\exp, \cos$ and $\sin$ we have

$$\exp'(z) = \exp(z),\ \cos'(z) = -\sin z \text{ and } \sin'(z) = \cos(z)$$

implying for any piecewise continuously differentiable curve $\gamma : [a, b] \to \mathbb{C}$ with $\gamma(a) = z_0$ and $\gamma(b) = z_1$

$$\int_\gamma e^z \, dz = e^{z_1} - e^{z_0},$$

$$\int_\gamma \cos(z) \, dz = \sin z_1 - \sin z_0,$$

$$\int_\gamma \sin(z) \, dz = -\cos z_1 + \cos z_0.$$

C. Let $F(z) = \sum_{k=0}^\infty a_k (z - c)^k$ be a power series converging in $B_r(c)$, $r > 0$.

We know by Theorem 17.14 that $F'(z) = \sum_{k=1}^\infty a_k k (z - c)^k$ converges in $B_r(c)$ too, hence it converges uniformly on every compact subset. Thus F is in $B_r(c)$ a primitive of $\sum_{k=1}^\infty a_k k (z - c)^{k-1}$. From Theorem 18.5 and its proof we now find in $B_1(1)$ that for a branch of the logarithmic function given by $\tilde{L} : B_1(1) \to \mathbb{C}$, $\tilde{L}(z) = \tilde{\ln}(z - 1)$, $\tilde{L}'(z) = \frac{1}{z}$, and therefore we have for a piecewise continuously differentiable curve γ with $\operatorname{tr} \gamma \subset B_1(1)$ and initial point z_0 as well as terminal point z_1 the formula

$$\int_\gamma \frac{1}{z} \, dz = \tilde{\ln}(z_1 - 1) - \tilde{\ln}(z_0 - 1),$$

where as before, see (16.18), $\tilde{\ln}(1 + z) = \sum_{k=1}^\infty \frac{(-1)^k}{k} z^k$, $|z| < 1$.

Example 20.21. A. In Example 20.12 we have seen that for $z \mapsto |z|$ the line integral $\int_\gamma |z| \, dz$ depends on γ and not only on the initial and terminal point of γ. Consequently $z \mapsto |z|$ has no primitive. Note that this differs completely from the situation of real-valued functions. If $g : [a, b] \to \mathbb{R}$ is continuous then $G(t) := \int_a^t g(s) \, ds$ is a primitive of g.

B. We know by Example 20.11 that $\int_{|z|=1} \frac{1}{z} \, dz = 2\pi i$ and therefore by Corollary 20.12 the function $z \mapsto \frac{1}{z}$ defined for all $\mathbb{C} \backslash \{0\}$ can not have a primitive in $\mathbb{C}\setminus\{0\}$. However we have seen in Example 20.20.C that it has a primitive (for example) in $B_1(1)$.

We want to find conditions which imply the converse to Corollary 20.19. The main constraint is of topological nature: D must be a region, i.e. connected.

Theorem 20.22. *Let* $G \subset \mathbb{C}$ *be a region and* $f : G \to \mathbb{C}$ *a continuous function. If* $\int_{\gamma} f(z)\,dz = 0$ *for every piecewise continuously differentiable and closed curve* γ *with* $\operatorname{tr}\gamma \subset D$, *then* f *has a primitive in* G.

Proof. We fix a point $a \in G$ and for $z \in G$ let γ_{az} be a piecewise continuously differentiable curve connecting a and z. Note that by Proposition 19.38 such a curve exists. We define on G the function F by

$$z \mapsto F(z) := \int_{\gamma_{az}} f(w)\,dw \tag{20.23}$$

and note first of all that F is independent of the choice of γ_{az}. Indeed if $\tilde{\gamma}_{az}$ is a further piecewise continuously differentiable curve connecting a and z with corresponding reversed curve $\tilde{\gamma}_{az}^{-1}$ we find that $\gamma_{az} \oplus \tilde{\gamma}_{az}^{-1}$ is a piecewise continuously differentiable closed curve, hence

$$\int_{\gamma_{az} \oplus \tilde{\gamma}_{az}^{-1}} f(z)\,dz = 0,$$

or

$$0 = \int_{\gamma_{az} \oplus \tilde{\gamma}_{az}^{-1}} f(z)\,dz = \int_{\gamma_{az}} f(z)\,dz + \int_{\tilde{\gamma}_{az}^{-1}} f(z)\,dz = \int_{\gamma_{az}} f(z)\,dz - \int_{\tilde{\gamma}_{az}} f(z)\,dz,$$

i.e.

$$\int_{\gamma_{az}} f(z)\,dz = \int_{\tilde{\gamma}_{az}} f(z)\,dz.$$

We claim that F is a primitive of f. Let $z_0 \in G$ such that for some $r > 0$ we have $z_0 \in B_r(z) \subset G$. We denote that line segment, connecting z_0 and z by

$[z_0, z]$ and we consider

$$F(z) - F(z_0) = \int_{\gamma_{az}} f(w)\,dw - \int_{\gamma_{az_0}} f(w)\,dw$$

$$= \int_{[z_0,z]} f(w)\,dw$$

$$= \int_0^1 f(z_0 + t(z - z_0))\,(z - z_0)\,dt,$$

or

$$\frac{F(z) - F(z_0)}{z - z_0} - f(z_0) = \int_0^1 f(z_0 + t(z - z_0))\,dt - f(z_0).$$

The function f is continuous. Thus, given $\epsilon > 0$ there exists $\delta > 0$ such that $|z - z_0| < \delta$ implies $|f(z) - f(z_0)| < \epsilon$. For $0 \le t \le 1$ we have $|z_0 + t(z - z_0) - z_0| = t|z - z_0| \le |z - z_0| < \delta$ implying $|f(z_0 + t(z - z_0)) - f(z_0)| < \epsilon$ and further

$$\sup_{0 \le t \le 1} |f(z_0 + t(z - z_0)) - f(z_0)| < \epsilon.$$

Thus we have proved that

$$F'(z_0) = \lim_{z \to z_0} \frac{F(z) - F(z_0)}{z - z_0} = f(z_0),$$

i.e. F is a primitive of f. $\qquad\square$

In order to apply Theorem 20.22 we need to verify $\int_\gamma f(z)\,dz = 0$ for all piecewise continuously differentiable curves to decide whether f has a primitive. A natural question is whether we can obtain the same conclusion if the condition $\int_\gamma f(z)\,dz = 0$ holds for a smaller set of curves. It turns out that in the case where the region G has additional geometric properties it is sufficient to verify $\int_\gamma f(z)\,dz = 0$ for boundaries of triangles only. Since triangles and their boundaries will enter more and more into our considerations we first give some preparations.

For $z_1, z_2 \in \mathbb{C}$ we denote by $[z_1, z_2]$ the line segment connecting z_1 and z_2

and also the curve $t \mapsto z_1 + t(z_2 - z_1)$, $t \in [0, 1]$. A triangle $\Delta(z_1, z_2, z_3)$ with vertices $z_1, z_2, z_3 \in \mathbb{C}$ is the convex hull of these three points in the plane, compare with Theorem II.13.24, i.e.

$$\Delta(z_1, z_2, z_3) = \{z \in \mathbb{C} \mid z = t_1 z_1 + t_2 z_2 + t_3 z_3, \, t_1, t_2, t_3 \geq 0, \, t_1 + t_2 + t_3 = 1\} \tag{20.24}$$

$$= \{z \in \mathbb{C} \mid z = z_1 + s(z_2 - z_1) + t(z_3 - z_1), \, s \geq 0, \, t \geq 0, \, s + t \leq 1\}$$

The boundary $\partial\Delta(z_1, z_2, z_3)$ is first of all the topological boundary of $\Delta(z_1, z_2, z_3) \subset \mathbb{C}$ and we can write it as

$$\partial\Delta(z_1, z_2, z_3) = [z_1, z_2] \cup [z_2, z_3] \cup [z_3, z_1]. \tag{20.25}$$

Interpreting however $[z_j, z_k]$ as a parametrized curve as mentioned above we can and will also interpret $\partial\Delta(z_1, z_2, z_3)$ as a closed, piecewise continuously differentiable curve being the sum of the three curves $[z_1, z_2], [z_2, z_3]$ and $[z_3, z_1]$, i.e. as a curve we consider $\partial\Delta(z_1, z_2, z_3)$ as

$$\partial\Delta(z_1, z_2, z_3) = [z_1, z_2] \oplus [z_2, z_3] \oplus [z_3, z_1], \tag{20.26}$$

compare with Figure 20.3:

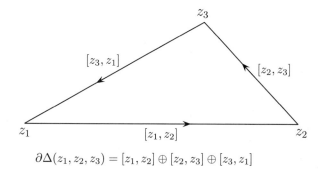

$$\partial\Delta(z_1, z_2, z_3) = [z_1, z_2] \oplus [z_2, z_3] \oplus [z_3, z_1]$$

Figure 20.3

The starting point to reduce the set of curves γ for which $\int_\gamma f(z) \, d(z) = 0$ must hold should be the definition of the primitive F, i.e.

$$F(z) = \int_{\gamma_{az}} f(w) \, dw,$$

377

see (20.23). If every $z \in G$ can be connected by a straight line $[z_0, z]$ to a fixed point $z_0 \in G$ and then define

$$F(z) = \int_{[z_0, z]} f(w)\, dw \qquad (20.27)$$

we are a step closer to our goal. Next we note that in order to form $\frac{F(z) - F(\zeta)}{z - \zeta}$ and pass to the limit $z \to \zeta$ we only need to consider values of z with $|z - \zeta|$ being small. So if z_0 is fixed as above, $\zeta \in G$ and $r > 0$ such that $B_r(\zeta) \subset G$, for all $z \in B_r(\zeta)$ the curve $[z_0, \zeta] \oplus [\zeta, z] \oplus [z, z_0]$ is a piecewise continuously differentiable curve which is closed and its trace is entirely contained in G. However $[z_0, \zeta] \oplus [\zeta, z] \oplus [z, z_0] = \partial \Delta(z_0, \zeta, z)$. Thus if for every triangle $\Delta(z_0, z_1, z_2) \subset G$ we have $\int_{\partial \Delta(z_0, z_1, z_2)} f(z)\, dz = 0$ then the proof of Theorem 20.22 works and establishes that F defined by (20.27) is a primitive of f. It remains to recall the definition of a star-shaped domain to conclude

Theorem 20.23. *Let $G \subset \mathbb{C}$ be a star-shaped domain with respect to $z_0 \in G$. Further let $f : G \to \mathbb{C}$ be a continuous function such that for every triangle $\Delta(z_0, z_1, z_2) \subset G$ we have $\int_{\partial \Delta(z_0, z_1, z_2)} f(z)\, dz = 0$. Then by*

$$F(z) := \int_{[z_0, z]} f(w)\, dw \qquad (20.28)$$

a primitive of f is given and $\int_\gamma f(z)\, dz = 0$ for all closed, piecewise continuously differentiable curves in G.

Corollary 20.24. *If $G \subset \mathbb{C}$ is a convex set, $f : G \to \mathbb{C}$ a continuous function and $\int_{\partial \Delta(z_1, z_2, z_3)} f(z)\, dz = 0$ holds for every triangle $\Delta(z_1, z_2, z_3) \subset G$ then f has a primitive F in G.*

Proof. Every convex set is star-shaped with respect to each of its points. $\square$

Remark 20.25. Note that a star-shaped set need not be convex, see Figure 20.4 or Figure 20.5.

Star-shaped but
non-convex domain

Figure 20.4

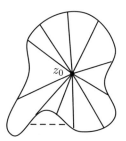

Star-shaped but
non-convex domain

Figure 20.5

Corollary 20.26. *Let $D \subset \mathbb{C}$ be a domain and suppose that $f : D \to \mathbb{C}$ is a continuous function such that for every piecewise continuously differentiable curve γ we have $\int_{\gamma} f(z)\, dz = 0$. Then f has a local primitive.*

Proof. Every point $z \in D$ has a convex open neighbourhood entirely contained in D, for example $B_{\varrho}(z)$, $\varrho > 0$ sufficiently small. □

Remark 20.27. Since the proof of Corollary 20.26 relies on Theorem 20.23, we can therefore replace $\int_{\gamma} f(z)\, dz = 0$ by $\int_{\partial \Delta(z_1, z_2, z_3)} f(z)\, dz$ for all triangles $\Delta(z_1, z_2, z_3) \subset D$.

Problems

1. Let $\gamma : [0, 2\pi k] \to \mathbb{C}$, $k \in \mathbb{N}$, be the curve $\gamma(t) = e^{it}$. When is γ a simply closed curve? Find the integral $\int_{\gamma} z^n dz$ for $n \in \mathbb{Z}$.

2. With γ and σ as in Example 20.12 find the integrals:

$$\int_{\gamma} \text{Re}\, z\, dz; \quad \int_{\sigma} \text{Re}\, z\, dz; \quad \int_{\gamma} \text{Im}\, z\, dz; \quad \int_{\sigma} \text{Im}\, z\, dz.$$

3. Consider the triangle ABC with $A = -2$, $B = 2$, $C = 2i$. Parametrize ∂ABC by $\gamma : [0, 1] \to \mathbb{C}$ such that $\gamma(0) = A$, $\gamma\left(\frac{1}{3}\right) = B$, $\gamma\left(\frac{2}{3}\right) = C$, $\gamma(1) = A$ and γ is piecewise linear, see the figure below:

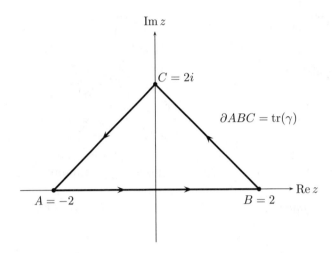

Now find by a direct calculation that $\int_\gamma |z| dz = 0$.

4.   a) In $\mathbb{R}$ we have for every interval $[a, b]$, $a < b$, the equality $\int_a^b 1 dx = b - a$, i.e. the integral of the constant function $x \mapsto 1$ over $[a, b]$, gives the interval length. Can we expect that $\left| \int_\gamma 1 dz \right| = l_\gamma$ holds?

   b) Let $R \subset \mathbb{C}$ be a region. For $k \in \mathbb{N}$ let $\gamma_k : [0, 1] \to \mathbb{R}$ be a piecewise continuously differentiable curve. We suppose that $\lim_{k \to \infty} l_{\gamma_k} = 0$. Prove that for every continuous function $f : R \to \mathbb{C}$ we have

$$\limsup_{k \to \infty} \left| \frac{1}{l_{\gamma_k}} \int_{\gamma_k} f(z) dz \right| < \infty$$

provided $\operatorname{tr} \gamma_k \subset K$ for all $k \in \mathbb{N}$ where $K \subset D$ is a compact set.

5. Consider the curve $\gamma : [0, 2\pi] \to \mathbb{C}$, $\gamma(t) = 4e^{it} + e^{-it}$. Find the trace of γ and the integral $\int_\gamma z^3 dz$.

6. Let $f : B_R(z_0) \to \mathbb{C}$ be given by a convergent power series. Show that for every $0 < r < R$ the function $F|_{B_r}(z_0)$ admits a primitive.

7. Find the following integrals:

   a) $\int_{[-1-2\pi i, 1+2\pi i]} e^{2z} dz$, where $[-1 - 2\pi i, 1 + 2\pi i]$ is the line segment connecting $-1 - 2\pi i$ and $1 + 2\pi i$;

b) $\int_\gamma \sinh(z)dz$, where the trace of γ is the arc of the circle with centre 0 and radius 2 connecting the points 2 and $-2i$;

c) $\int_\gamma p(z)e^z dz$ where p is a polynomial and γ is the curve from Problem 5.

Hint: prove that $z \mapsto p(z)e^z$ has a power series representation in $\mathbb{C}$.

8. a) Prove that the functions $z \mapsto \operatorname{Re} z$, $z \mapsto \operatorname{Im} z$ and $z \mapsto |z|$ do not have a primitive.

b) With γ and σ as in Example 20.12 find $\int_\gamma f(|z|)dz$ and $\int_\sigma f(|z|)dz$ where $f : [0, \infty) \to \mathbb{R}$ is a continuous function. Does $z \mapsto f(|z|)$ have a primitive when $f(r) = r^k$, $k \in \mathbb{N}$.

21 The Cauchy Integral Theorem and Integral Formula

The results in this Chapter can be viewed as the starting point as well as the essence of Cauchy's approach to complex variable theory. We start with a more or less heuristic (but still correct) consideration. Let $G \subset \mathbb{C} \cong \mathbb{R}^2$ be a domain for which Green's theorem in the form of (II.26.13) holds. In particular ∂G is the trace of a parametric curve and G can be considered as a normal domain with respect to both coordinate axes. Since we identify $\mathbb{R}^2$ with $\mathbb{C}$ the canonical basis in $\mathbb{R}^2$ can be identified with $\{1, i\}$, a vector z has components $z = x + iy$ and $f : G \to \mathbb{C}$ is interpreted as vector field $f(z) = u(x, y) + iv(x, y)$. Suppose that f is in a neighbourhood of G holomorphic, i.e. $f'(z)$ exists. We assume in addition that f' is continuous implying that u_x, u_y, v_x and v_y are continuous too. Now Green's theorem, i.e. (II.26.13) yields

$$\int_{\partial G} f(z)\, dz = \int_{\partial G} (u + iv)(dx + i\, dy)$$

$$= \left(\int_{\partial G} u\, dx - \int_{\partial G} v\, dy \right) + i \left(\int_{\partial G} v\, dx + \int_{\partial G} u\, dy \right)$$

$$= \int_G \left(-\frac{\partial v}{\partial x} - \frac{\partial u}{\partial y} \right) dx\, dy + i \int_G \left(\frac{\partial u}{\partial x} - \frac{\partial v}{\partial y} \right) dx\, dy.$$

Since f is holomorphic the Cauchy-Riemann differential equations hold, i.e. $u_x = v_y$ and $v_x = -u_y$, implying that

$$\int_{\partial G} f(z)\, dz = 0. \tag{21.1}$$

Thus for a function f holomorphic in a neighbourhood of G and continuous complex derivative f' we expect that (21.1) holds for at least all domains for which Green's theorem holds. We want to prove (21.1) without the assumption that f' is continuous and we start with **Goursat's Theorem** which some authors refer to as Goursat's lemma.

Theorem 21.1 (E. Goursat). *Let $\Delta \subset \mathbb{C}$ be a closed triangle and $U \subset \mathbb{C}$ an open neighbourhood of Δ. For a holomorphic function $f : U \to \mathbb{C}$ it* *follows that*

$$\int_{\partial\Delta} f(z)\,dz = 0, \tag{21.2}$$

where $\partial\Delta$ is interpreted as trace of a piecewise continuously differentiable parametric curve.

Proof. The following proof is standard and essentially identical in every book about complex analysis. Let $\Delta = \Delta(a,b,c)$ and U be as in the assumptions, see Figure 21.1.

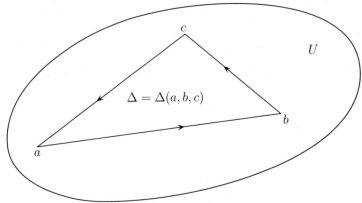

Figure 21.1

We now divide Δ into four triangles $\Delta_1^{(1)}, \ldots, \Delta_1^{(4)}$ by connecting the midpoints of the sides of Δ, see Figure 21.2.

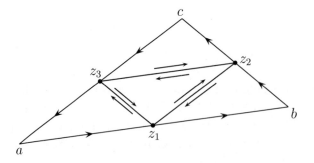

Figure 21.2

If the midpoints are z_1, z_2, z_3 we can consider $\Delta(z_1, z_2, z_3)$ and assume that $\partial\Delta(z_1, z_2, z_3)$ interpreted as parametric curve has anticlockwise orientation.

We also assume that the boundaries of the other three triangles interpreted as parametric curves have anticlockwise orientation. If the line segment $[w_1, w_2]$ (again interpreted as oriented curve) connects two midpoints $w_1, w_2 \in \{z_1, z_2, z_3\}$, then $[w_1, w_2]$ is a side in a triangle $\Delta_1^{(k)}$ and (the trace of) the reversed curve $[w_2, w_1]$ is a side of triangle $\Delta_1^{(l)}$ for suitable $k, l \in \{1, 2, 3, 4\}$. Since

$$\int_{[w_1, w_2]} f(z)\, dz = - \int_{[w_2, w_1]} f(z)\, dz$$

we find

$$\int_{\partial \Delta} f(z)\, dz = \sum_{k=1}^{4} \int_{\partial \Delta_1^{(k)}} f(z)\, dz, \tag{21.3}$$

and it follows that

$$\left| \int_{\partial \Delta} f(z)\, dz \right| \leq 4 \max_{1 \leq k \leq 4} \left| \int_{\partial \Delta_1^{(k)}} f(z)\, dz \right|. \tag{21.4}$$

Denote by $\Delta_1^{\max}$ a triangle in the set $\{\Delta_1^{(1)}, \Delta_1^{(2)}, \Delta_1^{(3)}, \Delta_1^{(4)}\}$ such that

$$\left| \int_{\partial \Delta_1^{\max}} f(z)\, dz \right| = \max_{1 \leq k \leq 4} \left| \int_{\partial \Delta_1^{(k)}} f(z)\, dz \right|. \tag{21.5}$$

For $\Delta_1^{\max}$ we repeat the procedure, see Figure 21.3.

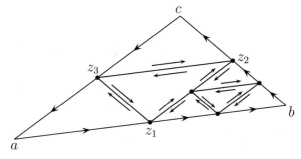

Figure 21.3

Thus we get a triangle $\Delta_2^{\max}$ for which we find

$$\left| \int_{\partial \Delta_1^{\max}} f(z)\, dz \right| \leq 4 \left| \int_{\partial \Delta_2^{\max}} f(z)\, dz \right| \tag{21.6}$$

and therefore

$$\left| \int_{\partial \Delta} f(z)\, dz \right| \leq 4^2 \left| \int_{\partial \Delta_2^{\max}} f(z)\, dz \right|. \tag{21.7}$$

Now we iterate this process to obtain triangles

$$\Delta = \Delta_0^{\max} \supset \Delta_1^{\max} \supset \Delta_2^{\max} \supset \Delta_3^{\max} \supset \ldots\ldots\ldots\ldots$$

such that

$$\left| \int_{\partial \Delta} f(z)\, dz \right| \leq 4^n \left| \int_{\partial \Delta_n^{\max}} f(z)\, dz \right| \tag{21.8}$$

holds as well as

$$l_{\partial \Delta_n^{\max}} = \frac{1}{2} l_{\partial \Delta_{n-1}^{\max}} = \ldots\ldots\ldots = 2^{-n} l_{\partial \Delta}, \tag{21.9}$$

where $l_{\partial \Delta_k^{\max}}$ denotes the length of the curve $\partial \Delta_k^{\max}$. Since $\Delta_n^{\max}$ is compact and $\operatorname{diam}(\Delta_n^{\max}) \to 0$ as $n \to \infty$ there exists a unique point z_0 such that $\{z_0\} = \bigcap_{k \in \mathbb{N}} \Delta_k^{\max}$. As f is complex differentiable at z_0 there exists a continuous function h such that $\lim_{z \to z_0} h(z) = 0$ and

$$f(z) = f(z_0) + (z - z_0)\, f'(z_0) + (z - z_0)\, h(z). \tag{21.10}$$

Furthermore we know that $z \mapsto f(z_0) + (z - z_0)\, f'(z_0)$ has a primitive which yields for all $n \in \mathbb{N}$

$$\int_{\partial \Delta_n^{\max}} (f(z_0) + (z - z_0)\, f'(z_0))\, dz = 0. \tag{21.11}$$

Now we find

$$\left| \int_{\partial \Delta_n^{\max}} f(z)\, dz \right| = \left| \int_{\partial \Delta_n^{\max}} (z - z_0)\, h(z)\, dz \right|$$

$$\leq l_{\partial \Delta_n^{\max}} \max_{z \in \partial \Delta_n^{\max}} (|z - z_0| |h(z)|)$$

$$\leq (l_{\partial \Delta_n^{\max}})^2 \max_{z \in \Delta_n^{\max}} |h(z)|.$$

386

By (21.8) and (21.9) it follows that

$$\left| \int_{\partial \Delta} f(z)\, dz \right| \le l_{\partial \Delta}^2 \max_{z \in \Delta_n^{\max}} |h(z)|. \tag{21.12}$$

From $\lim_{z \to z_0} h(z) = 0$ we deduce that for $\epsilon > 0$ we can find $N \in \mathbb{N}$ such that $n \ge N$ implies $\max_{z \in \Delta_n^{\max}} |h(z)| < \frac{\epsilon}{l_{\partial \Delta}^2}$ which yields that for every $\epsilon > 0$

$$\left| \int_{\partial \Delta} f(z)\, dz \right| < \epsilon$$

holds, i.e. we have

$$\int_{\partial \Delta} f(z)\, dz = 0, \tag{21.13}$$

and the theorem is proved. $\square$

It turns out that a small generalisation of Theorem 21.1 will have enormous implications.

Theorem 21.2. *Let $\Delta \subset \mathbb{C}$ be a closed triangle and $z_0 \in \Delta$. If $U \subset \mathbb{C}$ is an open neighbourhood of Δ and $f : U \to \mathbb{C}$ a continuous function which is holomorphic in $U \backslash \{z_0\}$ then we have*

$$\int_{\partial \Delta} f(z)\, dz = 0. \tag{21.14}$$

Proof. We will show the result in three steps. First suppose that z_0 is a vertex of Δ, see Figure 21.4

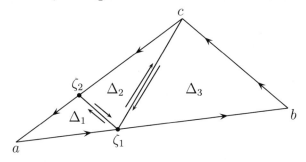

Figure 21.4

We pick a point ζ_1 on a side of $\Delta(a, b, c)$ with vertex z_0 and determine ζ_2 on the other side of $\Delta(a, b, c)$ with vertex z_0 by drawing the parallel to the side of $\Delta(a, b, c)$ opposite to z_0. Using Theorem 21.1 and its proof we get with $\Delta = \Delta(a, b, c)$

$$\int_{\partial\Delta} f(z)\, dz = \int_{\partial\Delta_1} f(z)\, dz + \int_{\partial\Delta_2} f(z)\, dz + \int_{\partial\Delta_3} f(z)\, dz = \int_{\partial\Delta_1} f(z)\, dz \quad (21.15)$$

where Δ_1 has the vertices z_0, ζ_1 and ζ_2, Δ_2 has the vertices ζ_1, ζ_2 and one vertex $z_1 \neq z_0$ of $\Delta(a, b, c)$, and Δ_3 has the vertices z_1, z_2 of $\Delta(a, b, c)$ not equal to z_0 and one of the points ζ_1 or ζ_2. This gives a decomposition of $\Delta(a, b, c)$ into triangles $\Delta_1, \Delta_2, \Delta_3$ and to Δ_2, Δ_3 we may apply Theorem 21.1, i.e. we get (21.15). Furthermore we have

$$\left| \int_{\partial\Delta_1} f(z)\, dz \right| \leq l_{\partial\Delta_1} \max_{z\in\Delta_1} |f(z)| \leq l_{\partial\Delta_1} ||f||_{\infty,\Delta} ,$$

where $||f||_{\infty,\Delta} = \max_{z\in\Delta} |f(z)|$. The compactness of Δ together with the continuity of f implies that $||f||_{\infty,\Delta} < \infty$.

Since ζ_1 was arbitrary but $[\zeta_1, \zeta_2]$ is parallel to the side of $\Delta(a, b, c)$ opposite to z_0 it follows that $\zeta_1 \to z_0$ implies $l_{\partial\Delta_1} \to 0$ and therefore

$$\lim_{\zeta_1\to z_0} \int_{\partial\Delta_1} f(z)\, dz = 0$$

which yields the result for the first case.

In the second case we assume that $z_0 \in \partial\Delta(a, b, c)$ but $z_0 \notin \{a, b, c\}$, i.e. z_0 is not a vertex of $\Delta(a, b, c)$. We now decompose $\Delta(a, b, c)$ into two triangles Δ_1 and Δ_2, where in each triangle z_0 is a vertex, see Figure 21.5.

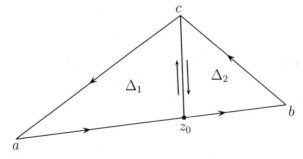

Figure 21.5

Since for Δ_1 and Δ_2 we can apply the first case it follows that

$$\int_{\partial\Delta} f(z)\,dz = \int_{\partial\Delta_1} f(z)\,dz + \int_{\partial\Delta_2} f(z)\,dz = 0.$$

Finally we assume that $z_0 \in \mathring{\Delta}$, i.e. z_0 is point in the interior of Δ. By drawing a line passing through z_0 and one vertex of $\Delta(a,b,c)$ we decompose $\Delta(a,b,c)$ into two triangles $\tilde{\Delta}_1$ and $\tilde{\Delta}_2$, see Figure 21.6.

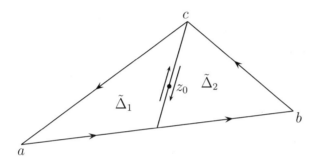

Figure 21.6

For the two triangles $\tilde{\Delta}_1$ and $\tilde{\Delta}_2$ the second case applies and we deduce

$$\int_{\partial\Delta} f(z)\,dz = \int_{\partial\tilde{\Delta}_1} f(z)\,dz + \int_{\partial\tilde{\Delta}_2} f(z)\,dz = 0$$

and the theorem is proved.   □

Combining Theorem 21.2 with Theorem 20.23 we obtain

Theorem 21.3. *Let $G \subset \mathbb{C}$ be a star-shaped domain with respect to some point $\zeta \in G$. Further let $f : G \to \mathbb{C}$ be a continuous function which is holomorphic in G or in $G\backslash\{z_0\}$, $z_0 \in G$. Then f admits a primitive in G.*

As a final consequence of these considerations we derive

Theorem 21.4 (Cauchy's Integral Theorem). *Let $G \subset \mathbb{C}$ be a star-shaped domain and $f : G \to \mathbb{C}$ a continuous function which is with the possible exception of one point $z_0 \in G$ holomorphic in G. For every closed piecewise continuously differentiable curve $\gamma : [a,b] \to G$ the following holds:*

$$\int_{\gamma} f(z)\,dz = 0. \tag{21.16}$$

Proof. Since f admits a primitive in G and γ is closed the theorem follows.

$\square$

Example 21.5. Let $z = r\,e^{i\varphi} \in \mathbb{C}\backslash(-\infty, 0]$. We can reach z starting from 1 by following the line segment $[1, z]$, but also by following $\gamma = \gamma_1 \oplus \gamma_2$ where γ_1 is the line segment from 1 to $r = |z|$ and γ_2 is the arc of the circle with centre 0 and radius r connecting r and $z = r\,e^{i\varphi}$, see Figure 21.7:

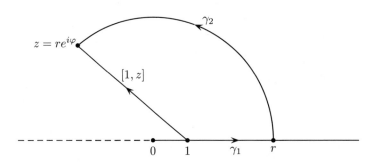

Figure 21.7

The set $\mathbb{C}\backslash(-\infty, 0]$ is star-shaped with respect to 1 and therefore $z \mapsto \int_{[1,z]} \frac{1}{\zeta}\, d\zeta$

is a primitive to $\frac{1}{z}$ in $\mathbb{C}\backslash(-\infty, 0]$. Since by Cauchy's integral theorem applied to $\frac{1}{\zeta}$ and the curve $\gamma_1 \oplus \gamma_2 \oplus [z, 1]$ it follows that

$$\int_{[1,z]} \frac{1}{\zeta}\, d\zeta = \int_{\gamma_1} \frac{1}{\zeta}\, d\zeta + \int_{\gamma_2} \frac{1}{\zeta}\, d\zeta$$

$$= \int_1^r \frac{1}{t}\, dt + \int_0^\varphi \frac{i\,r\,e^{it}}{r\,e^{it}}\, dt$$

$$= \ln r + i\varphi$$

$$= \log z,$$

i.e. the function $z \mapsto \int_{[1,z]} \frac{1}{\zeta}\, d\zeta$ gives in $\mathbb{C}\backslash(-\infty, 0]$ the principal branch of the logarithmic function, compare with Definition 18.8.

Example 21.6. In Problem 9 of Chapter 8 we calculated the **Fresnel integrals** using real variable methods, in particular a type of mollifying technique. We will now evaluate these integrals with the help of Cauchy's integral

390

theorem. This does not give only a new method for finding these two particular integrals, but will lead eventually to a complete new tool for calculating integrals, which we will discuss in great detail in Chapter 25 once we have proved the residue theorem.

We start with the observation that the function $z \mapsto e^{iz^2}$ is an entire function, hence $\int_\gamma e^{iz^2} dz = 0$ for every piecewise continuously differentiable, simply closed curve $\gamma : [a, b] \to \mathbb{C}$. For $r > 0$ we consider the curve $\gamma = \gamma_1 \oplus \gamma_2 \oplus \gamma_3$, see Figure 21.8, where γ_1 is the line segment connecting 0 with r, γ_2 is the arc of the circle with centre 0 and radius r connecting r and $re^{i\frac{\pi}{4}}$, whereas γ_3 is the line segment connecting $re^{i\frac{\pi}{4}}$ and 0.

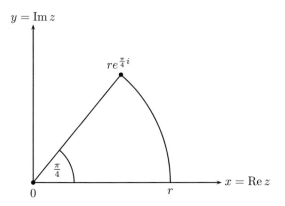

Figure 21.8

It follows that

$$0 = \int_\gamma e^{iz^2} dz = \int_{\gamma_1} e^{iz^2} dz + \int_{\gamma_2} e^{iz^2} dz + \int_{\gamma_3} e^{iz^2} dz, \quad z = x + iy.$$

On γ_1 we have $z = x$, on γ_2 we have $z = re^{i\varphi}$, $0 \le \varphi \le \frac{\pi}{4}$, and on γ_3 we have $z = \varrho e^{\frac{\pi}{4} i}$ where ϱ moves from r to 0. For the corresponding integrals this yields

$$\int_{\gamma_1} e^{iz^2}\, dz = \int_0^r e^{ix^2}\, dx = \int_0^r \cos x^2\, dx + i \int_0^r \sin x^2\, dx,$$

$$\int_{\gamma_2} e^{iz^2}\, dz = \int_0^{\frac{\pi}{4}} e^{ir^2 e^{2i\varphi}}\, ire^{i\varphi}\, d\varphi = \int_0^{\frac{\pi}{4}} e^{ir^2 \cos 2\varphi - r^2 \sin 2\varphi}\, ire^{i\varphi}\, d\varphi,$$

$$\int_{\gamma_3} e^{iz^2}\, dz = \int_r^0 e^{i\varrho^2} e^{\frac{\pi}{2}i} e^{\frac{\pi}{4}i}\, d\varrho = -e^{\frac{\pi}{4}i} \int_0^r e^{-\varrho^2}\, d\varrho,$$

or

$$\int_0^r \cos x^2\, dx + i \int_0^r \sin x^2\, dx = \frac{1}{2}\sqrt{2} \int_0^r e^{-\varrho^2}\, d\varrho + \frac{i}{2}\sqrt{2} \int_0^r e^{-\varrho^2}\, d\varrho \qquad (21.17)$$

$$- \int_0^{\frac{\pi}{4}} e^{ir^2 \cos 2\varphi - r^2 \sin 2\varphi}\, i\, re^{i\varphi}\, d\varphi.$$

Next we observe

$$\left| \int_0^{\frac{\pi}{4}} e^{ir^2 \cos 2\varphi - r^2 \sin 2\varphi}\, ir\, e^{i\varphi}\, d\varphi \right| \leq \int_0^{\frac{\pi}{4}} e^{-r^2 \sin 2\varphi} r\, d\varphi$$

$$= \frac{r}{2} \int_0^{\frac{\pi}{2}} e^{-r^2 \sin \varphi}\, d\varphi \leq \frac{r}{2} \int_0^{\frac{\pi}{2}} e^{-r^2 \frac{\varphi}{\pi}}\, d\varphi$$

$$= \frac{\pi}{4r}(1 - e^{-r^2}),$$

where we used first the change of variable $2\varphi \to \varphi$ and then the estimate $\frac{2\varphi}{\pi} \leq \sin \varphi$ for $0 \leq \varphi \leq \frac{\pi}{2}$. The latter estimate we see as follows: on $[0, \frac{\pi}{2})$ we have

$$1 \leq \frac{1}{\cos^2 y} = \frac{d}{dy}\tan y$$

or

$$\varphi = \int_0^\varphi 1 dy \leq \int_0^\varphi \frac{d}{dy}\tan y\, dy = \tan \varphi$$

implying $\varphi \cos \varphi \leq \sin \varphi$ on $[0, \frac{\pi}{2})$. Since $\frac{d}{d\varphi}\left(\frac{\sin \varphi}{\varphi}\right) = \frac{\varphi \cos \varphi - \sin \varphi}{\varphi^2}$, we derive that $\frac{\sin \varphi}{\varphi}$ is on $[0, \frac{\pi}{2}]$ a decreasing function with $\frac{\sin \varphi}{\varphi} \geq \frac{2}{\pi}$, i.e. $\frac{2\varphi}{\pi} \leq \sin \varphi$ for $\varphi \in [0, \frac{\pi}{2}]$. Now we pass in (21.17) to the limit as r tends to ∞ to obtain

$$\int_0^\infty \cos x^2 \, dx + i \int_0^\infty \sin x^2 \, dx = \frac{1}{2}\sqrt{2} \int_0^\infty e^{-\varrho^2} \, d\varrho + \frac{i}{2}\sqrt{2} \int_0^\infty e^{-\varrho^2} \, d\varrho,$$

and since $\int_0^\infty e^{-\varrho^2} \, d\varrho = \frac{\sqrt{\pi}}{2}$ we eventually have

$$\int_0^\infty \cos x^2 \, dx = \frac{1}{2}\sqrt{2} \, \frac{\sqrt{\pi}}{2} = \sqrt{\frac{\pi}{8}}$$

as well as

$$\int_0^\infty \sin x^2 \, dx = \frac{1}{2}\sqrt{2} \, \frac{\sqrt{\pi}}{2} = \sqrt{\frac{\pi}{8}}.$$

Looking more closely at our calculation we see the power of complex integration combined with the Cauchy theorem: in order to evaluate certain line integrals (including integrals of functions defined on a segment of the real axis) we may replace the original integral by an integral more convenient to handle.

In Example 20.11 a simple calculation has shown that

$$\frac{1}{2\pi i} \int_{|z-z_0|=r} \frac{1}{z - z_0} \, dz = 1.$$

This result has the following extension

Lemma 21.7. *For $z \in B_r(z_0)$ we have*

$$\frac{1}{2\pi i} \int_{|\zeta-z_0|=r} \frac{1}{\zeta - z} \, d\zeta = 1, \qquad (21.18)$$

where as usual the circle $|\zeta - z_0| = r$ is parametrized by $t \mapsto z_0 + re^{it}, t \in [0, 2\pi]$.

Proof. For $z \in B_r(z_0)$ we find $q := |z - z_0| \, r^{-1} \in [0, 1)$ implying for $z \in B_r(z_0)$ and $\zeta \in \partial B_r(z_0)$ that

$$\max_{\zeta \in \partial B_r(z_0)} \left| \left(\frac{z - z_0}{\zeta - z_0} \right)^n \right| = q^n, \; n \in \mathbb{N}.$$

With $w := \frac{z - z_0}{\zeta - z_0}$, $z \in B_r(z_0)$ and $\zeta \in \partial B_r(z_0)$ this implies

$$\frac{1}{\zeta - z} = \frac{1}{\zeta - z_0} \frac{1}{1 - w} = \frac{1}{\zeta - z_0} \sum_{k=0}^{\infty} w^k = \frac{1}{\zeta - z_0} \sum_{k=0}^{\infty} \left(\frac{z - z_0}{\zeta - z_0} \right)^k$$

and this geometric series converges absolutely and uniformly in ζ. Now it follows for $z \in B_r(z_0)$

$$\frac{1}{2\pi i} \int_{|\zeta - z_0| = r} \frac{1}{\zeta - z} d\zeta = \frac{1}{2\pi i} \int_{|\zeta - z_0| = r} \frac{1}{\zeta - z_0} \sum_{k=0}^{\infty} \left(\frac{z - z_0}{\zeta - z_0} \right)^k d\zeta$$

$$= \sum_{k=0}^{\infty} (z - z_0)^k \frac{1}{2\pi i} \int_{|\zeta - z_0| = r} \frac{1}{(\zeta - z_0)^{k+1}} d\zeta.$$

For $k + 1 = 1$, i.e. $k = 0$, we know that the value of the integral above is 1, while for $k + 1 > 1$ it follows that

$$\int_{|\zeta - z_0| = r} \frac{1}{(\zeta - z_0)^{k+1}} d\zeta = \int_0^{2\pi} \frac{1}{r^{k+1} e^{(k+1) it}} i r e^{it} \, dt$$

$$= \frac{1}{r^k} \int_0^{2\pi} e^{-ikt} \, dt = 0,$$

which yields

$$\frac{1}{2\pi i} \int_{|\zeta - z_0| = r} \frac{1}{\zeta - z} d\zeta = \sum_{k=0}^{\infty} (z - z_0)^k \frac{1}{2\pi i} \int_{|\zeta - z_0| = r} \frac{1}{(\zeta - z_0)^{k+1}} d\zeta$$

$$= \frac{1}{2\pi i} \int_{|\zeta - z_0| = r} \frac{1}{\zeta - z_0} d\zeta = 1.$$

$\square$

394

With the help of Lemma 21.7 and the Cauchy integral theorem we can obtain a formula expressing a holomorphic function in the interior of a disc by its values on the boundary of the disc.

Theorem 21.8 (Cauchy's Integral Formula). *Let $G \subset \mathbb{C}$ be a region and $f : G \to \mathbb{C}$ a holomorphic function. For every $z \in B_r(z_0) \subset G$, $r > 0$, $z_0 \in G$, we have*

$$f(z) = \frac{1}{2\pi i} \int_{|\zeta - z_0| = r} \frac{f(\zeta)}{\zeta - z} \, d\zeta \tag{21.19}$$

where the circle $|\zeta - z_0| = r$ is parametrized by $\gamma(t) = z_0 + re^{it}$, $t \in [0, 2\pi]$.

Proof. We choose $\epsilon > 0$ such that $B_{r+\epsilon}(z_0) \subset G$. Since $B_{r+\epsilon}(z_0)$ is an open, convex, hence star-shaped neighbourhood of $\overline{B_r(z_0)}$ we can apply Cauchy's integral theorem to the function $g : B_{r+\epsilon}(z_0) \to \mathbb{C}$ defined by

$$g(\zeta) := \begin{cases} \frac{f(\zeta) - f(z)}{\zeta - z}, & \zeta \neq z \\ f'(z), & \zeta = z \end{cases} \tag{21.20}$$

since g is continuous in $B_{r+\epsilon}(z_0)$ and at least holomorphic in $B_{r+\epsilon}(z_0) \backslash \{z\}$. It follows that

$$0 = \int_{|\zeta - z_0| = r} g(\zeta) \, d\zeta = \int_{|\zeta - z_0| = r} \frac{f(\zeta) - f(z)}{\zeta - z} \, d\zeta$$

$$= \int_{|\zeta - z_0| = r} \frac{f(\zeta)}{\zeta - z} \, d\zeta - f(z) \int_{|\zeta - z_0| = r} \frac{1}{\zeta - z} \, dz,$$

and by Lemma 21.7 we find

$$f(z) = \frac{1}{2\pi i} \int_{|\zeta - z_0| = r} \frac{f(\zeta)}{\zeta - z} \, d\zeta.$$

$\square$

Using the parametrization $\gamma(t) = z_0 + re^{it}$ formula (21.19) yields

Corollary 21.9 (Mean-value Theorem). *Under the assumptions of Theorem 21.8 the following hold*

$$f(z_0) = \frac{1}{2\pi} \int_0^{2\pi} f(z_0 + re^{it})\, dt;$$ (21.21)

and

$$|f(z_0)| \le \|f|_{\partial B_r(z_0)}\|_\infty = \|f\|_{\infty, \partial B_r(z_0)}.$$ (21.22)

Example 21.10. A. We want to find the values of the two integrals

$$\int_{|z-\frac{i}{2}|=1} \frac{1}{z^2+1}\, dz \quad \text{and} \quad \int_{|z+\frac{3i}{2}|=1} \frac{1}{z^2+1}\, dz,$$

where in each case the circle is parametrized as in Theorem 21.8. First we note that $\frac{1}{z^2+1} = \frac{1}{z+i}\cdot\frac{1}{z-i}$ and this function is holomorphic in $\mathbb{C}\setminus\{i, -i\}$. The function $\frac{1}{z+i}$ is holomorphic in $\mathbb{C}\setminus\{-i\}$ and therefore we can apply the Cauchy integral formula to $g(z) = \frac{1}{z+i}$ in the disc $|z - \frac{i}{2}| = 1$ to find

$$\int_{|z-\frac{i}{2}|=1} \frac{1}{z^2+1}\, dz = \int_{|z-\frac{i}{2}|=1} \frac{g(z)}{z-i}\, dz = 2\pi i\, g(i) = 2\pi i \left(\frac{1}{2i}\right) = \pi.$$

Analogously we find for $h(z) = \frac{1}{z-i}$ which is holomorphic in $\mathbb{C}\setminus\{i\}$ that

$$\int_{|z+\frac{3i}{2}|=1} \frac{1}{z^2+1}\, dz = \int_{|z+\frac{3i}{2}|=1} \frac{h(z)}{z+i}\, dz = 2\pi i\, h(-i) = 2\pi i \left(\frac{1}{-2i}\right) = -\pi.$$

B. Since the function $z \mapsto \cos z$ is an entire function we find for the circle $|z| = 3$, given with the usual parametrization that

$$\int_{|z|=3} \frac{\cos z}{z-i}\, dz = 2\pi i\, \cos i = \pi\, i(e^i - e^{-i}).$$

For developing the theory further we need to reformulate our results on the continuity and differentiability of parameter dependent integrals. The proof of the following results are the same as in the cases discussed in Chapter II.17 or in Chapter 8 of this volume.

Theorem 21.11. *Let $\gamma : [a, b] \to \mathbb{C}$ be a piecewise continuously differentiable curve, $G \subset \mathbb{R}^n$ and $f : \mathrm{tr}\,(\gamma) \times G \to \mathbb{C}$ a continuous function.*
A. *The function*

$$F : G \to \mathbb{C}, \qquad F(x) := \int_\gamma f(z, x)\, dz \qquad (21.23)$$

is on G continuous.
B. *If G is open and if on $\mathrm{tr}\,(\gamma) \times G$ the partial derivative $\frac{\partial f}{\partial x_j}$ exists and is continuous, then F has a continuous partial derivative with respect to x_j and*

$$\frac{\partial F}{\partial x_j}(x) = \int_\gamma \frac{\partial f}{\partial x_j}(z, x)\, dz. \qquad (21.24)$$

C. *Now let $G \subset \mathbb{C}$ be open and assume that for every $z \in \mathrm{tr}\,(\gamma)$ the function $\zeta \mapsto f(z, \zeta)$ is complex differentiable with continuous complex (partial) derivative $f_\zeta(z, \zeta)$. Then F is holomorphic on G and we have*

$$F'(\zeta) = \int_\gamma f_\zeta(z, \zeta)\, dz. \qquad (21.25)$$

We will now apply Theorem 21.11 to (21.19). Taking in (21.19) the derivative under the integral sign which is by Theorem 21.11.C justified we find

$$f'(z) = \frac{1}{2\pi i} \int_{|\zeta - z_0| = r} \frac{f(\zeta)}{(\zeta - z)^2}\, d\zeta. \qquad (21.26)$$

We observe that $z \mapsto \frac{f(\zeta)}{(\zeta - z)^2}$ again satisfies all assumptions of Theorem 21.11.C, so we can iterate the process. In fact we obtain the **generalised Cauchy integral formula**:

Theorem 21.12. *Let $G \subset \mathbb{C}$ be a region and $f : G \to \mathbb{C}$ a holomorphic function. Then f has all higher order complex derivates and for $n \in \mathbb{N}$, $\overline{B_r(z_0)} \subset G$, $z \in B_r(z_0)$ it follows*

$$f^{(n)}(z) = \frac{n!}{2\pi i} \int_{|\zeta - z_0| = r} \frac{f(\zeta)}{(\zeta - z)^{n+1}}\, d\zeta, \qquad (21.27)$$

where we use the parametrization $t \mapsto z_0 + re^{it}$, $t \in [0, 2\pi]$ for the circle $|\zeta - z_0| = r$.

We leave the formal, short proof by induction of Theorem 21.12 to the reader, see Problem 8.

Remark 21.13. Theorem 21.12 is quite a remarkable result as it states that if a function $f : G \to \mathbb{C}$ is one time complex differentiable, then all complex derivatives exist in G and hence they are holomorphic too. As a consequence we deduce that $u = \mathrm{Re}f$ and $v = \mathrm{Im}f$ are in $G \subset \mathbb{C} \cong \mathbb{R}^2$ arbitrary often differentiable. This raises the following two questions: 1. Suppose $u, v : G \to \mathbb{R}$ are one time partially differentiable functions in a region $G \subset \mathbb{R}^2$ and satisfy the Cauchy-Riemann differential equations. Does this imply that u and v are arbitrarily differentiable in G? 2. Suppose that $u : G \to \mathbb{R}$ have the second order partial derivative $\frac{\partial^2 u}{\partial x^2}$ and $\frac{\partial^2 u}{\partial y^2}$ and is harmonic in the region $G \subset \mathbb{R}^2$, i.e. $\frac{\partial^2 u}{dx^2} + \frac{\partial^2 u}{\partial y^2} = 0$. Does this imply that u is arbitrarily often differentiable in G? We will later see that both questions have an affirmative answer which is well understood within the theory of linear elliptic partial differential equations.

As in the case of Theorem 21.8 we can use the generalised Cauchy integral formulae to evaluate certain line integrals.

Example 21.14. A. Consider the integral

$$\int_{|\zeta|=4} \frac{e^{\zeta^2}}{(\zeta+1)^4} \, d\zeta \tag{21.28}$$

with the parametrization $t \mapsto 4\,e^{it}$, $t \in [0, 2\pi]$. Since $\zeta \mapsto e^{\zeta^2}$ is an entire function, when taking in (21.27) as f the function $f(\zeta) = e^{\zeta^2}$, $n = 3$ and $z = -1$ we get

$$f^{(3)}(-1) = \frac{3!}{2\pi i} \int_{|\zeta|=4} \frac{e^{\zeta^2}}{(\zeta+1)^4} \, d\zeta.$$

Since $f^{(3)}(\zeta) = (12\zeta + 8\zeta^3)\,e^{\zeta^2}$ we find $f^{(3)}(-1) = -20\,e$ and therefore

$$\int_{|\zeta|=4} \frac{e^{\zeta^2}}{(\zeta+1)^4} \, d\zeta = -\frac{20\,e\pi i}{3}.$$

B. For $k \geq 1$ and $r \geq 8$ the following holds with the parametrization of $|\zeta| = r$ given as $\zeta = re^{i\varphi}$, $\varphi \in [0, 2\pi]$,

$$\frac{1}{\Gamma(k+1)} = \frac{1}{2\pi i} \int\limits_{|\zeta|=r} \frac{e^{\zeta}}{(\zeta - 2\pi i)^{k+1}} \, d\zeta. \tag{21.29}$$

Since $r \geq 8$ it follows that $2\pi i \in \{z \in \mathbb{C} \,|\, |z| < r\}$, further we have $\frac{d^k}{d\zeta^k} e^{\zeta} = e^{\zeta}$ and with $\zeta = 2\pi i$ we find $e^{2\pi i} = 1$. Thus by (21.27) we have

$$1 = \frac{k!}{2\pi i} \int\limits_{|\zeta|=r} \frac{e^{\zeta}}{(\zeta - 2\pi i)^{k+1}} \, d\zeta$$

and since $k! = \Gamma(k+1)$ formula (21.29) follows.

Let $G \subset \mathbb{C}$ be a region and let $f : G \to \mathbb{C}$ be a continuous function. Let $\gamma_j : [a_j, b_j] \to G$, $j = 1, 2$, be two simply closed piecewise continuously differentiable curves with disjoint traces $\mathrm{tr}\,(\gamma_1) \cap \mathrm{tr}\,(\gamma_2) = \emptyset$. Since $\mathrm{tr}\,(\gamma_1)$ and $\mathrm{tr}\,(\gamma_2)$ are compact, $\mathrm{dist}\,(\mathrm{tr}\,(\gamma_1), \mathrm{tr}\,(\gamma_2)) > 0$. Pick $z_1 \in \mathrm{tr}\,(\gamma_1)$ and $z_2 \in \mathrm{tr}\,(\gamma_2)$ and assume that for the line segment $[z_1, z_2]$ connecting z_1 and z_2 we have $[z_1, z_2] \cap \mathrm{tr}\,(\gamma_1) = \{z_1\}$ and $[z_1, z_2] \cap \mathrm{tr}\,(\gamma_2) = \{z_2\}$, see Figure 21.9.

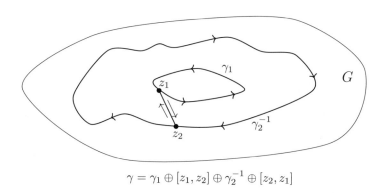

$$\gamma = \gamma_1 \oplus [z_1, z_2] \oplus \gamma_2^{-1} \oplus [z_2, z_1]$$

Figure 21.9

If necessary by changing the parametrization we may assume that $z_1 = \gamma_1(a_1)$ and $z_2 = \gamma_2(a_2)$. Denote by γ_2^{-1} the curve inverse to γ_2 and consider the simply closed, piecewise continuously differentiable curve

$$\gamma := \gamma_1 \oplus [z_1, z_2] \oplus \gamma_2^{-1} \oplus [z_2, z_1],$$

see again Figure 21.9. It follows that

$$\int_{\gamma} f(z)\, dz = \int_{\gamma_1} f(z)\, dz + \int_{[z_1, z_2]} f(z)\, dz + \int_{\gamma_2^{-1}} f(z)\, dz + \int_{[z_2, z_1]} f(z)\, dz$$

$$= \int_{\gamma_1} f(z)\, dz - \int_{\gamma_2} f(z)\, dz,$$

since $\int_{[z_1, z_2]} f(z)\, dz = - \int_{[z_2, z_1]} f(z)\, dz$ and $\int_{\gamma_2^{-1}} f(z)\, dz = - \int_{\gamma_2} f(z)\, dz$. In particular for $\int_{\gamma} f(z)\, dz = 0$ we conclude

$$\int_{\gamma_1} f(z)\, dz = \int_{\gamma_2} f(z)\, dz, \tag{21.30}$$

i.e. we can replace γ_1 by γ_2 to evaluate $\int_{\gamma_1} f(z)\, dz$.

Let us have again a look at

$$\int_{\gamma} \frac{1}{z}\, dz \tag{21.31}$$

where $\operatorname{tr}(\gamma) = S^1$. For example we may take $\gamma = \gamma_k : [0, 2\pi k] \to \mathbb{C}, \gamma_k(t) = e^{it}$, $k \in \mathbb{N}$. From the very definition of the line integral we find that

$$\int_{\gamma_k} \frac{1}{z}\, dz = \int_0^{2\pi k} \frac{1}{e^{it}}\, ie^{it}\, dt = 2\pi k\, i.$$

Thus this integral depends on γ, i.e. the parametrization, and not only on $\operatorname{tr}(\gamma)$. We also note that $\frac{1}{2\pi i} \int_{\gamma_k} \frac{1}{z}\, dz$ counts how often γ_k is winding around $z_0 = 0 \in \mathbb{C}$. This and related observation will soon become of more importance.

From Theorem 21.12 we have learnt that if $f : G \to \mathbb{C}$ is holomorphic, i.e. has a complex derivative f', then it has complex derivation $f^{(k)}$ of all order.

We even have a formula for the derivative $f^{(k)}(z)$ provided $z \in \overline{B_r(z_0)} \subset G$, namely (21.27). We want to derive some useful estimates, called **standard estimates** or **Cauchy's inequalities** for $f^{(k)}(z)$.

Theorem 21.15. *Let $G \subset \mathbb{C}$ be a region and $f : G \to \mathbb{C}$ a holomorphic function. Let $z_0 \in G$ and $\overline{B_r(z_0)} \subset G$. For every d, $0 < d \le r$, and all $z \in \overline{B_{r-d}(z_0)}$, we find for $k \in \mathbb{N}_0$*

$$|f^{(k)}(z)| \le \frac{r}{d}\frac{k!}{d^k} \max_{|\zeta - z_0| = r} |f(\zeta)|. \tag{21.32}$$

Proof. Using Cauchy's integral formula (21.27) we find for $|z - z_0| < r$ that

$$f^{(k)}(z) = \frac{k!}{2\pi i} \int\limits_{|\zeta - z_0| = r} \frac{f(\zeta)}{(\zeta - z)^{k+1}} \, d\zeta.$$

For $|z - z_0| \le r - d$ it follows that

$$d = r + d - r \le |\zeta - z_0| - |z - z_0| \le |\zeta - z|,$$

i.e. $|\zeta - z| \ge d$ or $\frac{1}{|\zeta - z|^{k+1}} \le \frac{1}{d^{k+1}}$, which implies

$$|f^{(k)}(z)| \le \frac{k!}{2\pi} \left| \int\limits_{|\zeta - z_0| = r} \frac{f(\zeta)}{(\zeta - z)^{k+1}} \, d\zeta \right|$$

$$\le \frac{k!}{2\pi} \, 2\pi r \, \frac{1}{d^{k+1}} \max_{|\zeta - z_0| = r} |f(\zeta)|$$

$$= \frac{r}{d}\frac{k!}{d^k} \max_{|\zeta - z_0| = r} |f(\zeta)|.$$

$\square$

Corollary 21.16. *Under the assumptions of Theorem 21.15 the following holds*

$$|f^{(k)}(z_0)| \le \frac{k!}{r^k} \max_{|\zeta - z_0| = r} |f(\zeta)| \tag{21.33}$$

and for $z \in \overline{B_{\frac{r}{2}}(z_0)}$ we find a constant C independent of f such that

$$|f^{(k)}(z)| \le C \frac{k!}{r^k} \max_{|\zeta - z_0| = r} |f(\zeta)|. \tag{21.34}$$

Proof. In the first case we take $d = r$, while the second estimate follows for $d = \frac{r}{2}$. $\qquad\qquad\qquad\qquad\qquad\qquad\qquad\qquad\qquad\qquad\qquad\qquad$ □

From (21.33) we deduce for every $z \in \mathbb{C}$ such that $|z - z_0| = \varrho < r$ that

$$\left| \frac{f^{(k)}(z_0)}{k!} (z - z_0)^k \right| \le ||f||_{\infty, \partial B_r(z_0)} \left(\frac{\varrho}{r} \right)^k$$

and since $0 < \frac{\varrho}{r} < 1$ it follows that for $z \in B_\varrho(z_0)$, $\varrho < r$, the series

$$\sum_{k=0}^{\infty} \frac{f^{(k)}(z_0)}{k!} (z - z_0)^k \tag{21.35}$$

converges in $\overline{B_\varrho(z_0)}$ absolutely and uniformly. We want to identify (21.35) with the Taylor series of f about z_0. Our starting point is Cauchy's integral formula for the circle $\partial B_r(z_0)$ with $|z - z_0| = r$ being parametrized by $z(t) = z_0 + re^{it}$, $t \in [0, 2\pi]$. Given a holomorphic function f on some region $G \subset \mathbb{C}$ and let $z_0 \in G$. Let $B_R(z_0) \subset G$ and $0 < r < R$. Cauchy's integral formula reads as

$$f(z) = \frac{1}{2\pi i} \int_{|\zeta - z_0| = r} \frac{f(\zeta)}{\zeta - z} d\zeta \tag{21.36}$$

and we can expand $\frac{1}{\zeta - z}$ into a geometric series according to

$$\frac{1}{\zeta - z} = \frac{1}{1 - \frac{z - z_0}{\zeta - z_0}} \cdot \frac{1}{\zeta - z_0} = \sum_{k=0}^{\infty} \frac{(z - z_0)^k}{(\zeta - z_0)^{k+1}}$$

which converges for $z \in B_r(z_0)$ uniformly in ζ, $\zeta \in \partial B_r(z_0)$. Substituting this expansion into (21.36) we obtain

$$f(z) = \frac{1}{2\pi i} \int_{|\zeta - z_0| = r} \frac{f(\zeta)}{\zeta - z} d\zeta = \frac{1}{2\pi i} \int_{|\zeta - z_0| = r} \sum_{k=0}^{\infty} \frac{f(\zeta)}{(\zeta - z_0)^{k+1}} (z - z_0)^k \, d\zeta$$

$$= \sum_{k=0}^{\infty} \left(\frac{1}{2\pi i} \int_{|\zeta - z_0| = r} \frac{f(\zeta)}{(\zeta - z_0)^{k+1}} d\zeta \right) (z - z_0)^k,$$

and with

$$a_k := \frac{1}{2\pi i} \int_{|\zeta - z_0| = r} \frac{f(\zeta)}{(\zeta - z_0)^{k+1}} d\zeta \tag{21.37}$$

it follows for $|z - z_0| < r$ that

$$f(z) = \sum_{k=0}^{\infty} a_k (z - z_0)^k \qquad (21.38)$$

and this series converges in $\overline{B_{r_0}(z_0)}$ for every $r_0 < r$. Since by Cauchy's general integral formula we have

$$a_k = \frac{f^{(k)}(z_0)}{k!}$$

it follows

Theorem 21.17. *Let $G \subset \mathbb{C}$ be a region, $z_0 \in G$, and $f : G \to \mathbb{C}$ a holomorphic function. Then there exists $R > 0$ such that in every disc $B_r(z_0) \subset B_R(z_0) \subset G$ the function f has a **Taylor expansion***

$$f(z) = \sum_{k=0}^{\infty} a_k (z - z_0)^k \qquad (21.39)$$

representing f and for the Taylor coefficients a_k we find for $0 < \varrho < R$

$$a_k = \frac{f^{(k)}(z_0)}{k!} = \frac{1}{2\pi i} \int_{|\zeta - z_0| = \varrho} \frac{f(\zeta)}{(\zeta - z_0)^{k+1}} \, d\zeta. \qquad (21.40)$$

The coefficients a_k, $k \in \mathbb{N}_0$, are uniquely determined and satisfy the estimate

$$|a_k| \leq \frac{1}{r^k} \max_{|\zeta - z_0| = r} |f(\zeta)|. \qquad (21.41)$$

Proof. It remains to prove the uniqueness of the coefficients and estimate (21.41). The uniqueness problem is solved by differentiating and evaluating $f^{(k)}$ at z_0, whereas the estimate is a consequent of Corollary 21.16. $\qquad \square$

As first consequence of Theorem 21.17 is

Corollary 21.18. *Let*

$$T_{(a_k)}^{z_0}(z) := \sum_{k=0}^{\infty} a_k (z - z_0)^k, \ z_0 \in \mathbb{C}, \ a_k \in \mathbb{C}, \qquad (21.42)$$

403

be a convergent power series with radius of convergence $R > 0$. For every $z_1 \in B_R(z_0)$ we can expand the holomorphic function $z \mapsto T^{z_0}_{(a_k)}(z)$ into a convergent power series

$$T^{z_1}_{(b_k)}(z) = \sum_{k=0}^{\infty} b_k(z - z_1)^k \qquad (21.43)$$

where the radius of convergence of $T^{z_1}_{(b_k)}(z)$ is at least $R - |z_1 - z_0|$.

Proof. Since $T^{z_0}_{(a_k)}(\,\cdot\,)$ is a holomorphic function we just need to apply Theorem 21.17. $\qquad\square$

From Corollary 21.18 we deduce further

Corollary 21.19. *If a holomorphic function is defined as a convergent power series, i.e. $f(z) = T^{z_0}_{(a_k)}(z)$ with radius of convergence $R > 0$, then $T^{z_0}_{(a_k)}(z)$ is also the Taylor series of f in $B_\varrho(z_0)$, $\varrho < R$.*

In Chapter 16 and 17 we have already encountered several well-known holomorphic functions defined by power series. The next chapter will give a further insight to holomorphic functions and power series.

Problems

1. Suppose that $f, g : G \to \mathbb{C}$ are complex differentiable functions defined on a region $G \subset \mathbb{C}$ and let $\gamma : [a, b] \to \mathbb{R}$ be a piecewise continuously differentiable curve. Prove that

   $$\int_\gamma f(z)g'(z)dz = f(\gamma(b))g(\gamma(b)) - f(\gamma(a))g(\gamma(a)) - \int_\gamma f'(z)g(z)dz$$

   and deduce for a simply closed curve that

   $$\int_\gamma f(z)g'(z)dz = -\int_\gamma f'(z)g(z)dz.$$

2.   a) For the parametric curve $\gamma(t) = ae^{it} + be^{-it}$, $t \in [0, 2\pi]$ and $0 < b < a$ and every polynomial $p(z) = \sum_{k=1}^{M} a_k z^k$ prove that $\int_\gamma p(z)dz = 0$.

   b) Sketch the curve $\gamma : [0, 2\pi] \to \mathbb{C}$ defined by $\gamma|_{[0,\pi]}(t) = t + i\sin t$ and $\gamma_{[\pi,2\pi]}(t) = 2\pi - t$. Let $Q(\zeta, \eta) = \sum_{|\alpha| \le k} a_\alpha \zeta^{\alpha_1} \eta^{\alpha_2}$, $a_\alpha \in \mathbb{C}$,

$(\alpha_1, \alpha_2) \in \mathbb{N}_0^2$ and $\zeta, \eta \in \mathbb{C}$, be a polynomial in two complex variables. Find

$$\int_\gamma Q(\cos z, \sin z)dz.$$

c) Let $g : B_1(0) \to \mathbb{C}$ be a continuous function and $h : B_1(0) \to \mathbb{C}$ a holomorphic function. For $0 < \rho < 1$ prove that

$$\left| \int_{|\zeta|=\rho} (g(\zeta) + h(\zeta))d\zeta \right| \leq 2\pi ||g||_{\infty, \partial B_\rho(0)}.$$

3.   a) Let $G \subset \mathbb{C}$ be a region and $\gamma : [a, b] \to G$ be a curve such that with $a < c < b$ the curves $\gamma_1 := \gamma|_{[a,c]}$ and $\gamma_2 := \gamma|_{[c,b]}$ are piecewise continuously differentiable curves which are closed and further assume that $\text{tr}_{\gamma_1} \cap \text{tr}_{\gamma_2} = \{\gamma(c)\}$. For a holomorphic function $h : G \to \mathbb{C}$ show that $\int_\gamma h(z)dz = 0$.

b) Consider the set $\Gamma := \partial B_1(0) \cup \partial B_1(2)$. Find a curve $\eta : [0, 4\pi] \to \mathbb{C}$ which has trace Γ and with $c = 2\pi$ satisfies all requirements made for γ in part a). Find the integral

$$\frac{1}{2\pi i} \int_\eta \frac{1}{z - \frac{3}{2}} dz.$$

4.   Find the integrals

$$\frac{1}{2\pi i} \int_{|\zeta-4|=1} \frac{\left(\zeta - \frac{\pi}{4}\right) \sin \zeta + (\zeta - \pi) \cos \zeta}{\zeta^2 - \frac{5\pi}{4}\zeta - \frac{\pi^2}{4}} d\zeta$$

and

$$\frac{1}{2\pi i} \int_{|\zeta+\frac{1}{2}|=\frac{5}{2}} \frac{\left(\zeta - \frac{\pi}{4}\right) \sin \zeta + (\zeta - \pi) \cos \zeta}{\zeta^2 - \frac{5\pi}{4}\zeta - \frac{\pi^2}{4}} d\zeta$$

where in both cases we use the standard parametrization of the corresponding circles.

5.   Prove that

$$\sum_{k=1}^{\infty} \frac{1}{2\pi i} \int_{|\zeta|=k^2} \frac{e^{-\zeta}}{\zeta - \ln(k^2 + k)} d\zeta = 1.$$

Hint: use Example I.16.3.

6.* Prove that for f as in Corollary 21.9 the estimate

$$|f(z)| \leq \|f\|_{\infty, \partial B_r(z_0)}$$

holds for all $z \in B_r(z_0)$.

Hint: apply (21.19) to f^k and use $\lim_{k \to \infty} \left(r \left\| \frac{1}{\zeta - z} \right\|_{\infty, \partial B_r(z_0)} \right)^{\frac{1}{k}} = 1$.

7. Let $G \subset \mathbb{C}$ be a region.

a) Prove that for a holomorphic function $f = u + iv$ in G the real part and the imaginary part satisfy the following mean-value equation:

$$w(z_0) = \frac{1}{2\pi} \int_0^{2\pi} w(z_0 + re^{it}) dt$$

for every $B_r(z_0) \subset G$.

b) Let $f_k : G \to \mathbb{C}$ be a sequence of holomorphic functions converging on every compact subset $K \subset G$ uniformly to a function $f : G \to \mathbb{C}$. Prove that f must satisfy the mean-value property, i.e.

$$f(z_0) = \frac{1}{2\pi} \int_0^{2\pi} f(z_0 + re^{it}) dt.$$

8. Prove the generalised Cauchy integral formula, i.e. Theorem 21.12.

9. Use the generalised Cauchy integral formula to evaluate the following integrals:

a) $\int_{|\zeta - z| = 1} \frac{\cos \zeta^2}{(\zeta - \sqrt{\pi})^2} d\zeta$;

b) For the hypergeometric function $_2F_1$ find

$$\frac{1}{2\pi i} \int_{|z| = \frac{1}{2}} \frac{_2F_1(z)}{z^{n+1}} dz, \quad n \in \mathbb{N}.$$

10. Let f be a holomorphic function in a neighbourhood of $\overline{B_1(z_0)}$. Suppose in addition that for all $z \in B_1(z_0)$ we have

$$\frac{1}{2\pi i} \int_{|\zeta - z_0| = 1} \frac{1 + \zeta - z + (\zeta - z)^2}{(\zeta - z)^4} f(\zeta) d\zeta = 0.$$

Prove that in $B_1(z_0)$ the function f satisfies the differential equation

$$f'''(z) + 3f''(z) + 6f'(z) = 0.$$

406

22 Power Series, Holomorphy and Differential Equations

Let us summarize what we know about holomorphic functions so far.

Theorem 22.1. *Let $G \subset \mathbb{C}$ be a region. The following statements are equivalent:*

i) *$f : G \to \mathbb{C}$ is holomorphic;*

ii) *f has locally a primitive;*

iii) *for every $z_0 \in G$ the function f admits an expansion into a convergent power series with strictly positive radius of convergence which represents f, this power series is the Taylor series of f about z_0;*

iv) *for every closed triangle $\triangle \subset G$ the following holds*

$$\int_{\partial\triangle} f(z)\, \mathrm{d}z = 0. \tag{22.1}$$

Proof. It remains only to prove that *iv*) implies *i*) which however follows from Remark 20.27. $\qquad\qquad\square$

Remark 22.2. A. The result that statement *iv*) implies holomorphy is often called the **Theorem of Morera**.
B. We have seen in the beginning of Chapter 21 that if two functions $u, v : G \to \mathbb{R}$, $G \subset \mathbb{C} \cong \mathbb{R}^2$, satisfy the Cauchy-Riemann differential equations and their partial derivatives are continuous then Cauchy's theorem holds for $f = u + iv$, and hence f is holomorphic. It takes some more effort to prove that the Cauchy-Riemann differential equations without the continuity condition on the partial derivatives already imply the holomorphy of f. Furthermore, given a harmonic function $\tilde{u}$ in G, i.e. $\Delta\tilde{u} = 0$, then one can find a conjugate harmonic function $\tilde{v}$ in G, i.e. $\Delta\tilde{v} = 0$, such that $\tilde{f} = \tilde{u} + i\tilde{v}$ is holomorphic. We will deal with these results later, in particular when studying partial differential equations.

Let us consider the following

Example 22.3. Given the two power series

$$T_1(z) = \sum_{k=0}^{\infty} \frac{1}{2^{k+1}} z^k \tag{22.2}$$

and

$$T_2(z) = \sum_{k=0}^{\infty} \frac{1}{(2-i)^{k+1}} (z-i)^k. \tag{22.3}$$

For T_1 we find in the disc $|z| < 2$ that

$$T_1(z) = \frac{1}{2} \sum_{k=0}^{\infty} \left(\frac{z}{2}\right)^k = \frac{1}{2} \frac{1}{1 - \frac{z}{2}} = \frac{1}{2-z}$$

and for T_2 it follows in the disc $|z - i| < \sqrt{5}$, i.e. $\left|\frac{z-i}{2-i}\right| < 1$, that

$$T_2(z) = \frac{1}{2-i} \sum_{k=0}^{\infty} \left(\frac{z-i}{2-i}\right)^k$$

$$= \frac{1}{2-i} \frac{1}{1 - \frac{z-i}{2-i}} = \frac{1}{2-z}.$$

Both, T_1 and T_2, are Taylor series and they represent the function $z \mapsto \frac{1}{2-z}$ in different domains, but these domains overlap. We can put this differently: Suppose we start with the holomorphic function $f : B_2(0) \to \mathbb{C}$, $f(z) := T_1(z)$. Then the function $g : B_2(0) \cup B_{\sqrt{5}}(i) \to \mathbb{C}$ defined by

$$g(z) := \begin{cases} T_1(z), & z \in B_2(0) \\ T_2(z), & z \in B_{\sqrt{5}}(i) \end{cases} \tag{22.4}$$

is a holomorphic extension or continuation of f to $B_2(0) \cup B_{\sqrt{5}}(i)$.

The above example shows that a holomorphic function $f : G \to \mathbb{C}$ might have a holomorphic continuation to a larger domain and we are of course interested to know whether we can determine the maximal holomorphic continuation of f. The latter problem is much more far reaching than it initially looks, eventually one is led to Riemannian surfaces. A further problem which arises is the question of the uniqueness of a holomorphic continuation. As the following example shows we cannot expect uniqueness of C^∞-extensions of C^∞-functions defined on open sets of the real line.

Example 22.4. The two functions $f_1, f_2 : \mathbb{R} \to \mathbb{R}$,

$$f_1(x) := \begin{cases} e^{-\frac{1}{x^2}}, & x > 0 \\ 0, & x \le 0 \end{cases}, \qquad f_2(x) := \begin{cases} e^{-\frac{1}{x^2}}, & x \ne 0 \\ 0, & x = 0 \end{cases}$$

are C^∞-functions, $f_1|_{[0,\infty)} = f_2|_{[0,\infty)}$, $f_1^{(k)}(0) = f_2^{(k)}(0) = 0$ for all $k \in \mathbb{N}_0$. Hence f_1 is an extension of $f_2|_{[0,\infty)}$ as f_2 is an extension of $f_1|_{[0,\infty)}$. Both functions have even the same Taylor series about 0 which however does not represent any of them. Thus C^∞-extensions of C^∞-functions need not be unique at all.

The absolute value $|.| : \mathbb{R} \to \mathbb{R}$ is a continuous function and its restriction to $\mathbb{R}\backslash\{0\}$ is even a C^∞-function. For a continuous function $f : G \to \mathbb{C}$ defined on a region $G \subset \mathbb{C}$ it is not possible for $f|_{G\backslash\{z_0\}}$, $z_0 \in G$, to be holomorphic when f is not, a result we prove next. We need

Definition 22.5. *Let (X, d) be a metric space and $M \subset X$, $M \ne \emptyset$. We call M a **discrete subset** of X if every point $x \in M$ has a neighbourhood $U(x)$ such that $U(x) \cap (M\backslash\{x\}) = \emptyset$.*

Points in a discrete set M are in some sense isolated points. If $y \in M^\complement$ then for every $x \in M$ we have $d(x, y) > 0$ and hence y has a neighbourhood completely contained in $M^\complement$, i.e. $M^\complement$ is open. Therefore a discrete set is closed. Moreover, every discrete subset M of a compact metric space is a finite set.

Theorem 22.6. *Let $G \subset \mathbb{C}$ be a region and $M \subset G$ a discrete set. Suppose that $f : G \to \mathbb{C}$ is continuous and $f|_{G\backslash M}$ is holomorphic. Then f is holomorphic in G.*

Proof. Let $z_0 \in M$ and choose $B_\epsilon(z_0) \subset G$ such that $B_\epsilon(z_0) \cap (M\backslash\{z_0\}) = \emptyset$. Theorem 21.3 implies that f has a primitive in $B_\epsilon(z_0)$ implying that f is holomorphic in z_0. $\square$

The next result allows us to remove certain singularities for holomorphic functions in the sense that originally f is defined and holomorphic on some set $D\backslash\{z_0\}$ but admits a holomorphic continuation to z_0.

Theorem 22.7 (Riemann's theorem on removable singularities). *Let $G \subset \mathbb{C}$ be a region and $z_0 \in G$. Further let $f : G\backslash\{z_0\} \to \mathbb{C}$ be a holomorphic function which is bounded on $B_r(z_0)\backslash\{z_0\} \subset G$. Then there exists a holomorphic function $h : G \to \mathbb{C}$ such that $h|_{G\backslash\{z_0\}} = f$.*

Proof. We define on G the function

$$g(z) := \begin{cases} (z - z_0)f(z), & z \in G\backslash\{z_0\} \\ 0, & z = z_0 \end{cases} \tag{22.5}$$

which is holomorphic in $G\backslash\{z_0\}$ and due to the boundedness of f in $B_r(z_0)\backslash\{z_0\}$ it is continuous at z_0, hence continuous on G. Now we can apply Theorem 22.6 to deduce that g is holomorphic in G, i.e. complex differentiable. It follows the existence of a function $h : G \to \mathbb{C}$ such that

$$g(z) = g(z_0) + h(z)(z - z_0),$$

where h is continuous at z_0. On $G\backslash\{z_0\}$ we have $h = f$, in particular h is holomorphic on $G\backslash\{z_0\}$. A further application of Theorem 22.6 yields that $h : G \to \mathbb{C}$ is holomorphic and we know already that $h|_{G\backslash\{z_0\}} = f$. $\square$

Theorem 22.6 and in particular Theorem 22.7 tell us that at certain (isolated) points we can extend a holomorphic function and the extension is holomorphic too. Our next goal is to understand how many values we need to know from f and its derivatives to determine f. As Example 22.4 shows, for C^∞-functions this question does not make much sense. For holomorphic functions the situation is quite different. We start with

Definition 22.8. A. *Let $D \subset \mathbb{C}$ be a domain and $f : D \to \mathbb{C}$ in a neighbourhood of $z_0 \in D$ a complex differentiable function. We call z_0 a **zero of order n** of f if*

$$f^{(k)}(z_0) = 0 \quad \text{for } k = 0, 1, \dots, n-1 \tag{22.6}$$

and

$$f^{(n)}(z_0) \neq 0. \tag{22.7}$$

B. *A complex number $w \in \mathbb{C}$ is a **value of order n** for f at $z_0 \in D$ if $f - w$ has at z_0 a zero of order n.*
C. *The function f has a **zero of order** ∞ at z_0 if $f^{(k)}(z_0) = 0$ for all $k \in \mathbb{N}_0$.*

We can transfer Definition 22.8 easily and in the obvious way to function $f : I \to \mathbb{R}$ where $I \subset \mathbb{R}$ is an open interval. The functions f_1 and f_2 in

Example 22.4 are now examples of C^∞-functions having at 0 a zero of order ∞. On the other hand, if the Taylor series

$$T^{z_0}(z) = \sum_{k=0}^{\infty} \frac{f^{(k)}(z_0)}{k!}(z - z_0)^k$$

has a positive radius of convergence and represents a function not identically 0, then z_0 cannot be a zero of order ∞. This observation prepares our main result. It turns out that if a holomorphic function f has a zero at z_0 then its Taylor expansion about z_0 allows us to determine the order of the zero at z_0. More precisely we have

Lemma 22.9. *A holomorphic function $f : G \to \mathbb{C}$ has a zero of order n at $z_0 \in G$ if and only if its Taylor series about z_0 is given by $\sum_{k=n}^{\infty} \frac{f^{(k)}(z_0)}{k!}(z-z_0)^k$ where $f^{(n)}(z_0) \neq 0$. This holds if and only if we can find a neighbourhood of z_0 and a holomorphic function g in this neighbourhood such that $g(z_0) \neq 0$ and $f(z) = (z - z_0)^n g(z)$.*

Proof. If f has a zero of order n at z_0 then $f^{(k)}(z_0) = 0$ for $k = 0, \ldots, n - 1$, and hence the Taylor series of f about z_0 will start with the nth term since $f^{(n)}(z_0) \neq 0$. Conversely, if the Taylor series of f about z_0 starts with the nth term (assumed to be non-zero) then by differentiating the series we obtain $f^{(k)}(z_0) = 0$ for $k = 0, 1, \ldots, n - 1$. Further, if the Taylor series of f about z_0 starts with the nth term, the function $g(z) := \sum_{k=n}^{\infty} \frac{f^{(k)}(z_0)}{k!}(z - z_0)^{k-n}$ is holomorphic in a neighbourhood of z_0 and $f(z) = (z-z_0)^n g(z)$. On the other hand, if $f(z) = (z - z_0)^n g(z)$ then Leibniz's rule yields that $f^{(k)}(z_0) = 0$ for $k = 0, 1, \ldots, n - 1$ implying that the Taylor series of f starts with the nth term. $\square$

Let us introduce a rather convenient notation. If $\varphi : X \to Y$ is any mapping and $y \in Y$ we write

$$\varphi(x) \equiv y \tag{22.8}$$

if $\varphi(x) = y$ for all $x \in X$, i.e. φ is the constant mapping $x \mapsto y$. Now we prove

Theorem 22.10. *Let $G \subset \mathbb{C}$ be a region and $f : G \to \mathbb{C}$ a holomorphic function. The function f is identically zero, i.e. $f(z) \equiv 0$ if and only if f has in G a zero of order ∞. Equivalent to this condition is the existence of a set $A \subset G$ which is not discrete such that $f(z) = 0$ for all $z \in A$.*

Proof. Clearly, if $f(z) \equiv 0$ then there exists a non-discrete set $A \subset G$ such that $f|_A = 0$. Just take $A = G$. Now let $A \subset G$ be a non-discrete set such that $f|_A = 0$. In this case there exists $z_0 \in G$ and a sequence $(z_l)_{l \in \mathbb{N}}$, $z_l \in A$, such that $z_l \neq z_0$ and $\lim_{l \to \infty} z_l = z_0$. We now prove that z_0 must be a zero of order ∞ of f. Consider the Taylor series of f about z_0, i.e. $f(z) = \sum_{k=0}^{\infty} \frac{f^{(k)}(z_0)}{k!}(z - z_0)^k$. By continuity of f we find that $f(z_0) = \lim_{l \to \infty} f(z_l) = 0$, i.e. $f^{(0)}(z_0) = 0$. Now assume that we have already proved that $f^{(0)}(z_0) = \cdots = f^{(n-1)}(z_0) = 0$. We want to show that then $f^{(n)}(z_0) = 0$ too, and by induction this would imply that z_0 is a zero of order ∞. In a neighbourhood of z_0 we have

$$f(z) = \frac{f^{(n)}(z_0)}{n!}(z - z_0)^n + \sum_{k=n+1}^{\infty} \frac{f^{(k)}(z_0)}{k!}(z - z_0)^k,$$

and for $z = z_l$ (if necessary we have to assume that l is sufficiently large) it follows that

$$0 = f(z_l) = \frac{f^{(n)}(z_0)}{n!}(z_l - z_0)^n + \sum_{k=n+1}^{\infty} \frac{f^{(k)}(z_0)}{k!}(z_l - z_0)^k,$$

or after dividing by $(z_l - z_0)^n \neq 0$

$$0 = \frac{f^{(n)}(z_0)}{n!} + \sum_{k=n+1}^{\infty} \frac{f^{(k)}(z_0)}{k!}(z_l - z_0)^{k-n}$$

$$= \frac{f^{(n)}(z_0)}{n!} + (z_l - z_0) \sum_{k=n+1}^{\infty} \frac{f^{(k)}(z_0)}{k!}(z_l - z_0)^{k-n-1}.$$

Since $z \mapsto \sum_{k=n+1}^{\infty} \frac{f^{(k)}(z_0)}{k!}(z_l - z_0)^{k-n-1}$ is continuous in a neighbourhood of z_0, in fact it is holomorphic, it follows that

$$f^{(n)}(z_0) = -(n!) \lim_{l \to \infty} (z_l - z_0) \sum_{k=n+1}^{\infty} \frac{f^{(k)}(z_0)}{k!}(z_l - z_0)^{k-n-1} = 0,$$

proving that z_0 is a zero of order ∞ of f.

Next suppose that $z_0 \in G$ is a given zero of order ∞. We want to prove that $f(z) \equiv 0$. For this we consider the set

$$M := \left\{ z \in G \,\middle|\, f^{(k)}(z) = 0 \ \text{ for all } k \in \mathbb{N}_0 \right\}.$$

We will show that M is both closed and open. Since $z_0 \in M$ it is not an empty set and therefore the connectivity of G implies $M = G$, hence $f(z) \equiv 0$. First we prove that M is closed. Let $(z_l)_{l \in \mathbb{N}}$, $z_l \in M$, be a sequence in M converging to $z' \in G$. Since f is holomorphic it follows for all $k \in \mathbb{N}_0$ that

$$0 = \lim_{l \to \infty} f^{(k)}(z_l) = f^{(k)}(z'),$$

i.e. $z_l \in M$, implying that M is closed. Now we show that M is open. Let $z_1 \in M$. In this case it follows that

$$f(z) = \sum_{k=0}^{\infty} \frac{f^{(k)}(z_0)}{k!}(z - z_1)^k = 0$$

for all z in a neighbourhood U of z_1. Thus all derivatives of f vanish in U, i.e. $U \subset M$, and therefore M is open which proves the theorem. $\qquad\square$

Corollary 22.11. *The number 0 does not belong to the range of the exponential function, i.e. $e^z \neq 0$ for all $z \in \mathbb{C}$.*

Proof. Since $(e^z)' = e^z$, if $e^{z_0} = 0$ then z_0 would be a zero of order ∞. $\qquad\square$

An application of Theorem 22.10 to $f - g$ gives the following **uniqueness theorem for holomorphic functions**.

Theorem 22.12. *Let $f, g : G \to \mathbb{C}$ be holomorphic functions on the region $G \subset \mathbb{C}$. The following statements are equivalent:*

i) $f(z) = g(z)$ for all $z \in G$, i.e. $f = g$;

ii) there exists $z_0 \in G$ such that $f^{(k)}(z_0) = g^{(k)}(z_0)$ for all $k \in \mathbb{N}_0$;

iii) there exists a non-discrete set $A \subset G$ such that $f(z) = g(z)$ for all $z \in A$.

Corollary 22.13. *Let $T^{z_1}_{(a_k)}(z) = \sum_{k=0}^{\infty} a_k(z-z_1)^k$ and $T^{z_2}_{(b_k)}(z) = \sum_{k=0}^{\infty} b_k(z-z_2)^k$ be two power series converging in $B_{r_1}(z_1)$ and $B_{r_2}(z_2)$, respectively, $r_1, r_2 > 0$. Suppose that $B_{r_1}(z_1) \cap B_{r_2}(z_2) \neq \emptyset$. Then there exists a holomorphic function $f : G \to \mathbb{C}$, $B_{r_1}(z_1) \cup B_{r_2}(z_2) \subset G$, such that $f|_{B_{r_1}(z_1)} = T^{z_1}_{(a_k)}$ and $f|_{B_{r_2}(z_2)} = T^{z_2}_{(b_k)}$.*

Of course it is possible to extend the procedure suggested in Corollary 22.13 to several power series. We may choose a sequence $(T^{z_\nu})_{\nu \in A}$, $A \in \{\mathbb{N}_n \,|\, n \in \mathbb{N}\} \cup \{\mathbb{N}\}$, $\mathbb{N}_n = \{1, \ldots, n\}$, of power series with positive radii $r_\nu > 0$ of convergence such that $B_{r_\nu}(z_\nu) \cap B_{r_{\nu+1}}(z_{\nu+1}) \neq \emptyset$, $B_{r_\nu}(z_\nu) \cap B_{r_\mu}(z_\mu) = \emptyset$ for $|\nu - \mu| \geq 2$, and $T^{z_\nu}|_{B_{r_\nu}(z_\nu) \cap B_{r_{\nu+1}}(z_{\nu+1})} = T^{z_{\nu+1}}|_{B_{r_\nu}(z_\nu) \cap B_{r_{\nu+1}}(z_{\nu+1})}$, see Figure 22.1 and Figure 22.2.

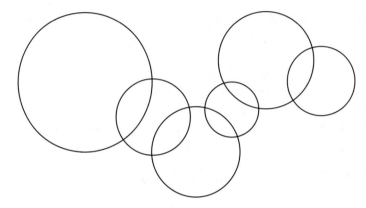

Allowed configuration   Figure 22.1

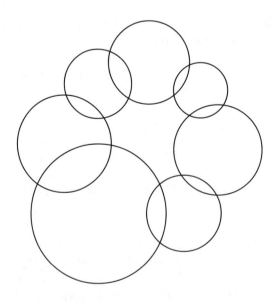

Forbidden configuration   Figure 22.2

In this case there exists a holomorphic function $g : \bigcup_{\nu \in A} B_{r_\nu}(z_\nu) \to \mathbb{C}$ such that $g|_{B_{r_\nu}(z_\nu)} = T^{z_\nu}$. Note that the condition $B_{r_\nu}(z_\nu) \cap B_{r_\mu}(z_\mu) = \emptyset$ for $|\nu - \mu| \geq 2$ assures that $\bigcup_{\nu \in A} B_{r_\nu}(z_\nu)$ is simply connected. However some care is still needed when trying to identify g with a known function, a problem we in fact encounter already with a single power series: the power series

$$T(z) := \log(-1 + i) + \sum_{k=1}^{\infty} \frac{(-1)^{k-1}}{k} \frac{1}{(-1+i)^k}(z + 1 - i)^k$$

converges for $|z + 1 - i| < \sqrt{2}$ but $(\mathbb{C} \backslash \{\operatorname{Re} z \leq 0\}) \cap B_{\sqrt{2}}(-1 + i)$ is not a connected set, see Figure 22.3.

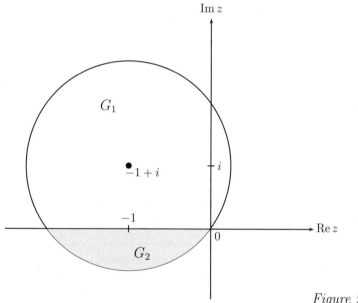

Figure 22.3

For $G_1 := \{z \in B_{\sqrt{2}}(-1 + i) | \operatorname{Im} z > 0\}$ and $G_2 = \{z \in B_{\sqrt{2}}(-1 + i) | \operatorname{Im} z < 0\}$ the functions $T|_{G_1}$ and $T|_{G_2}$ correspond to different branches of log.

In Chapter 17 we have seen that if a C^∞-function $g : (-r + a, a + r) \to \mathbb{R}$, $r > 0$, has a representation as convergent power series about a, i.e.

$$g(x) = \sum_{k=0}^{\infty} \frac{f^{(k)}(a)}{k!}(x - a)^k, \qquad |x - a| < r,$$

then we can extend g to a holomorphic function f defined by

$$f(z) := \sum_{k=0}^{\infty} \frac{f^{(k)}(a)}{k!}(z-a)^k, \quad |z-a| < r.$$

Since $(-r+a, a+r) \cap B_r(a) \subset \mathbb{C}$ is a non-discrete set in $B_r(a)$, we conclude that f is uniquely determined by g. This observation allows us to transfer certain results holding for g to f. For example the exponential function on $\mathbb{R}$ satisfies

$$\exp(s+t) = \exp(s)\exp(t), \quad s, t \in \mathbb{R}. \tag{22.9}$$

For $t \in \mathbb{R}$ fixed we consider $s \mapsto \exp(s+t)$ and $s \mapsto \exp(s)\exp(t)$, both defined on $\mathbb{R}$. From (22.9) and the uniqueness theorem we deduce that

$$\exp(z+t) = \exp(z)\exp(t) \tag{22.10}$$

for all $z \in \mathbb{C}$. Now we consider $z \in \mathbb{C}$ to be fixed and conclude by the same argument that

$$\exp(z+w) = \exp(z)\exp(w)$$

for all $z, w \in \mathbb{C}$.

In a similar way we can prove that cos or sin are 2π-periodic. But note that the holomorphic extension f of $g : (-r+a, a+r) \to \mathbb{R}$ might have additional properties. For example we know that the complex exponential function is also periodic with period $2\pi i$. It is obvious that a Taylor series expansion of a holomorphic function $f : G \to \mathbb{C}$ about a point $z_0 \in G$ of its domain is a powerful tool to study its local behaviour. However it is in general not easy to determine all coefficients $\frac{f^{(k)}(z_0)}{k!}$ of the expansion. There are some tools at our disposal which might be helpful in concrete situations. We first give

Theorem 22.14. *Let $G \subset \mathbb{C}$ be a region and $\gamma : [a,b] \to G$ be a piecewise continuously differentiable curve. Further let $h : \text{tr}(\gamma) \times G \to \mathbb{C}$ be a continuous function such that for every $t \in [a,b]$ the function $z \mapsto h(\gamma(t), z)$ is holomorphic in G. Then the function*

$$f(z) := \int_{\gamma} h(\zeta, z)\,d\xi = \int_a^b h(\gamma(t), z)\dot{\gamma}(t)\,dt \tag{22.11}$$

is holomorphic in G.

Proof. Let $\triangle \subset G$ be a closed triangle and note that

$$\int_{\partial\triangle} f(z)\,dz = \int_{\partial\triangle} \left(\int_\gamma h(\zeta,z)\,d\zeta\right) dz$$

$$= \int_\gamma \left(\int_{\partial\triangle} h(\zeta,z)\,dz\right) d\zeta = 0,$$

where we have used our standard results for interchanging the order of iterated integrals of continuous functions over the product of compact sets, and
the fact that $z \mapsto g(\zeta,z)$ is for $\zeta \in \mathrm{tr}(\gamma)$ fixed holomorphic, i.e. $\int_{\partial\triangle} g(\zeta,z)\,dz$
$= 0$ for every closed triangle $\triangle \subset G$. From Theorem 22.1 we deduce now
that f is holomorphic in G. $\qquad\square$

The reader should note the difference between this and Theorem 21.11.C
where we have assumed that $f_\zeta(\zeta,z)$ is continuous which we do not need
anymore.

Returning to our problem, we now find that if

$$f(z) = \int_\gamma h(\zeta,z)\,d\zeta \tag{22.12}$$

or

$$f(z) = \frac{d^k g(z)}{dz^k}, \tag{22.13}$$

and we know the Taylor expansion about $z_0 \in G$ of either $z \mapsto h(\zeta,z)$ for
every $\zeta \in \mathrm{tr}(\gamma)$ or the one of $z \mapsto g(z)$, then we may use (22.12) or (22.13)
to find the Taylor expansion of f about z_0.

Example 22.15. For example we know that

$$\sum_{k=0}^\infty \left(\frac{z}{a}\right)^k = \frac{a}{a-z}, \quad |z| < |a|.$$

Since $\frac{d}{dz}\left(\frac{a}{a-z}\right) = -\frac{a}{(a-z)^2}$ we find

$$\frac{1}{(a-z)^2} = -\frac{1}{a}\frac{d}{dz}\left(\frac{a}{a-z}\right) = -\sum_{k=0}^\infty \frac{d}{dz}\left(\frac{z}{a}\right)^k$$

implying

$$\frac{1}{(a-z)^2} = -\sum_{k=1}^\infty \frac{k}{a^k}z^{k-1}.$$

We will discuss similar problems in Problem 8.

We want to study how we can use the results of this and previous chapters to find solutions of certain differential equations. More precisely, for some linear differential equations we try to obtain a solution as a convergent power series.

In Problem 7 to Chapter I.29 we encountered for $l \in \mathbb{N}_0$ the **Bessel function** of order l

$$J_l(x) = \sum_{k=0}^{\infty} \frac{(-1)^k}{k!(k+1)!} \left(\frac{x}{2}\right)^{2k+1} = x^l \sum_{k=0}^{\infty} \frac{(-1)^k}{2^{2k+l}k!(k+1)!} x^{2k} \qquad (22.14)$$

as power series converging for all $x \in \mathbb{R}$ and solving the **Bessel differential equation**

$$x^2 u''(x) + x u'(x) + (x^2 - l^2)u(x) = 0, \quad x \in \mathbb{R}. \qquad (22.15)$$

From the results developed in this chapter we deduce that J_l has a holomorphic continuation to an entire function J_l with now complex power series representation

$$J_l(z) = z^l \sum_{k=0}^{\infty} \frac{(-1)^k}{2^{2k+l}k!(k+1)!} z^{2k}, \quad z \in \mathbb{C}, \qquad (22.16)$$

and this entire function solves in $\mathbb{C}$ the equation

$$z^2 u''(z) + z u'(z) + (z^2 - l)u(z) = 0. \qquad (22.17)$$

In the above mentioned problem of Chapter I.29 the series was just given and the task was to check the convergence as well as the fact that J_l solves (22.15). By using the **hypergeometric differential equation**

$$z(1-z)u''(x) + (\gamma - (\alpha + \beta + 1)z)u'(z) - \alpha\beta u(z) = 0, \qquad (22.18)$$

$\alpha, \beta, \gamma \in \mathbb{R}$, as an example we want to study now a method, namely a (power) series Ansatz, which allows us to find solutions for differential equations such as (22.17) or (22.18).

We try to obtain a solution to (22.18) as a series, more precisely we start with

$$u(z) = z^\rho \sum_{k=0}^{\infty} c_k z^k \qquad (22.19)$$

where ρ and the coefficients c_k, $k \in \mathbb{N}_0$, are to be determined. Since (22.18) is a linear homogeneous equation, if u solves this equation, every scalar multiple of u will do the same. We use this freedom to normalise in (22.19) the coefficient c_0 by setting it equal to 1, i.e. $c_0 := 1$. Now we substitute the series (22.19) into the equation (22.18) and try to determine ρ and c_k, $k \in \mathbb{N}$. Our approach is to use first formal calculations and once ρ and c_k, $k \in \mathbb{N}$, are determined we prove that the series converges and verify that it is indeed a solution of (22.18). Differentiating in (22.19) yields

$$u'(z) = \sum_{k=0}^{\infty} c_k(k + \rho)z^{k+\rho-1},$$

$$u''(z) = \sum_{k=0}^{\infty} c_k(k + \rho)(k + \rho - 1)z^{k+\rho-2},$$

leading to

$$\begin{aligned}
0 &= z(1 - z)u''(z) + (\gamma - (\alpha + \beta + 1)z)u'(z) - \alpha\beta u(z) \\
&= zu''(z) + \gamma u'(z) - z^2 u''(z) - (\alpha + \beta + 1)zu'(x) - \alpha\beta u(z) \\
&= \sum_{k=0}^{\infty} c_k(k + \rho)(k + \rho - 1)z^{k+\rho-1} + \sum_{k=0}^{\infty} c_k\gamma(k + \rho)z^{k+\rho-1} \\
&\quad - \sum_{k=0}^{\infty} c_k(k + \rho)(k + \rho - 1)z^{k+\rho} - \sum_{k=0}^{\infty} c_k(\alpha + \beta + 1)(k + \rho)z^{k+\rho} - \sum_{k=0}^{\infty} c_k\alpha\beta z^{k+\rho}.
\end{aligned}$$

We are longing for a solution which has no singularity at $z = 0$ and therefore we require

$$\rho(\gamma + \rho - 1) = 0 \tag{22.20}$$

and

$$(k + \rho)(\rho + k + \gamma - 1)c_k = (\rho + k + \alpha - 1)(\rho + k + \beta - 1), \quad k \in \mathbb{N}. \tag{22.21}$$

For $\gamma \notin -\mathbb{N}_0$ we choose $\rho = 0$ which gives

$$c_k = \frac{(k + \alpha - 1)(k + \beta - 1)}{k(k + \gamma - 1)}c_{k-1} \tag{22.22}$$

and with $c_0 = 1$ we arrive at

$$c_k = \frac{\alpha(\alpha + 1) \cdot \ldots \cdot (\alpha + k - 1)\beta(\beta + 1) \cdot \ldots \cdot (\beta + k - 1)}{k!\gamma(\gamma + 1) \cdot \ldots \cdot (\gamma + k - 1)},$$

or with the Pochhammer symbol notation, see (16.29), we get

$$c_k = \frac{(\alpha)_k(\beta)_k}{k!(\gamma)_k}, \quad k \in \mathbb{N}_0, \tag{22.23}$$

which yields

$$u(z) = \sum_{k=0}^{\infty} \frac{(\alpha)_k(\beta)_k}{k!(\gamma)_k} z^k. \tag{22.24}$$

However, this series we know already as the **hypergeometric series** associated with α, β, γ, compare with Definition 16.20.

Theorem 22.16. *The hypergeometric series*

$$_2F_1(\alpha, \beta; \gamma; z) := \sum_{k=0}^{\infty} \frac{(\alpha)_k(\beta)_k}{k!(\gamma)_k} z^k \tag{22.25}$$

has radius of convergence 1 and satisfies in $B_1(0)$ the differential equation (22.18).

Proof. The convergence of $_2F_1(\alpha, \beta; \gamma; z)$ was already investigated in Theorem 16.19. We will verify that $_2F_1(\alpha, \beta; \gamma; z)$ indeed solves (22.18) in Part 9. $\qquad\square$

In the case where $\gamma \notin \mathbb{N}\setminus\{1\}$ we can also choose $\rho = 1 - \gamma$ to find

$$\tilde{c}_k = \frac{(\alpha - \gamma + 1)_k(\beta - \gamma + 1)_k}{k!(2 - \gamma)_k} \tag{22.26}$$

and it follows that

$$\tilde{u}(z) = z^{1-\gamma}\,_2F_1(\alpha - \gamma + 1, \beta - \gamma + 1; 2 - \gamma; z) \tag{22.27}$$

is a further solution to (22.18) if $\gamma \notin \mathbb{N}\setminus\{1\}$, again see Part 9.

This method to solve the hypergeometric differential equation works for a much larger class of linear differential equations, for example for equations with polynomials as coefficients and we will return to these questions in Volume IV. Here it is important to learn that solutions of certain ordinary differential equations in the plane are holomorphic functions which we can find by a series Ansatz, $u(z) = z^\rho \sum_{k=0}^{\infty} c_k z^k$, $c_0 = 1$.

There is an obvious way to generalise the hypergeometric series, and more importantly, this generalisation is quite useful. For real numbers α_j, $1 \leq j \leq p$, and γ_l, $1 \leq l \leq q$, we define

$$_pF_q(\alpha_1, \ldots, \alpha_p; \gamma_1, \ldots, \gamma_q; z) := \sum_{k=0}^{\infty} \frac{(\alpha_1)_k \cdot \ldots \cdot (\alpha_p)_k}{k!(\gamma_1)_k \cdot \ldots \cdot (\gamma_q)_k} z^k \qquad (22.28)$$

and call $_pF_q$ for $(p, q) \neq (2, 1)$ the **generalised hypergeometric series** whereas $_2F_1$ is often called the **Gauss hypergeometric series** and sometimes just denoted by F.

Theorem 22.17. *Let $p, q \in \mathbb{N}_0$, $\alpha_1, \ldots, \alpha_p \in \mathbb{R}$ and $\gamma_1, \ldots, \gamma_q \in \mathbb{R} \backslash \{-n \mid n \in \mathbb{N}_0\}$.*

i) For $p > q + 1$ the series $_pF_q$ converges only at $z = 0$.

ii) For $p = q + 1$ the series $_pF_q$ converges in $B_1(0)$, i.e. 1 is its radius of convergence.

iii) For $p \leq q$ the radius of convergence of $_pF_q$ is ∞.

Proof. With $a_k := \frac{(\alpha_1)_k \cdot \ldots \cdot (\alpha_p)_k}{k!(\gamma_1)_k \cdot \ldots \cdot (\gamma_q)_k}$ we find

$$\left| \frac{a_{k+1} z^{k+1}}{a_k z^k} \right| = \left| \frac{(\alpha_1 + k) \cdot \ldots \cdot (\alpha_p + k)}{(\gamma_1 + k) \cdot \ldots \cdot (\gamma_q + k)(k + 1)} \right| |z| \qquad (22.29)$$
$$= \frac{k^p |(1 + \sigma(k))|}{k^{q+1} |(1 + \tau(k))|} |z|,$$

where $\sigma(x)$ is a polynomial of degree at most $p - 1$ and $\tau(x)$ is a polynomial of degree at most q. Hence, if $p > q + 1$, i.e. $p - 1 > q$, then the quotient (22.29) becomes unbounded, hence $_pF_q(\alpha_1, \ldots, \alpha_p; \gamma_1, \ldots, \gamma_q; z)$ diverges for $p > q + 1$ and $z \neq 0$. If $p = q + 1$ it follows that $\lim_{k \to \infty} \left| \frac{a_{k+1} z^{k+1}}{a_k z^k} \right| = |z|$ and therefore for $|z| < 1$ the series $_pF_{p+1}(\alpha_1, \ldots, \alpha_p, \gamma_1, \ldots, \gamma_{p+1}; z)$ converges in $B_1(0)$. Finally, if $p \leq q$ then for every $z \in \mathbb{C}$ we have $\lim_{k \to \infty} \left| \frac{a_{k+1} z^{k+1}}{a_k z_k} \right| = 0$ implying that $_pF_q(z)$ represents for $p \leq q$ an entire function. $\square$

Here are some examples

$$_0F_0(z) = \sum_{k=0}^{\infty} \frac{z^k}{k!} = e^z; \qquad (22.30)$$

$$_1F_0(\alpha, z) = \sum_{k=0}^{\infty} \frac{(\alpha)_k}{k!} z^k = \frac{1}{(1-z)^\alpha}. \qquad (22.31)$$

Using the solution to Problem 11, i.e. the identities

$$\frac{1}{(2k)!} = \frac{1}{4^k \left(\frac{1}{2}\right)_k k!} \quad \text{and} \quad \frac{1}{(2k+1)!} = \frac{1}{4^k \left(\frac{3}{2}\right)_k k!} \qquad (22.32)$$

we find further

$$\cos z = {}_0F_1\left(\frac{1}{2}; -\frac{1}{4}z^2\right); \qquad (22.33)$$

$$\cosh z = {}_0F_1\left(\frac{1}{2}; \frac{1}{4}z^2\right); \qquad (22.34)$$

$$\sin z = z\, {}_0F_1\left(\frac{3}{2}; -\frac{1}{4}z^2\right); \qquad (22.35)$$

$$\sinh z = z\, {}_0F_1\left(\frac{3}{2}; \frac{1}{4}z^2\right). \qquad (22.36)$$

Moreover, for $l \in \mathbb{N}_0$ we obtain

$$J_l(z) = \frac{\left(\frac{z}{2}\right)^l}{l!}\, {}_0F_1\left(l+1; -\frac{1}{4}z^2\right). \qquad (22.37)$$

As $_2F_1(z)$ satisfies a differential equation, namely the hypergeometric differential equation (22.18), it can be shown that $_pF_q$ is also a solution of a certain differential equation. However this will go far beyond an introduction to complex variables. For a good theoretical treatment we refer to [12], however it is worthwhile to become familiar with highly valuable sources such as [1] or [55].

Problems

1. Let $G \subset \mathbb{C}$ be a bounded domain and $M \subset G$ a discrete set. Prove that M is a finite set.

2. Let $G \subset \mathbb{C}$ be a bounded region and $M \subset G$ a discrete set. Let $f : G \setminus M \to \mathbb{C}$ be a holomorphic function which is bounded on $G \setminus M$. Prove that f has a holomorphic extension to G.
   Hint: use Theorem 22.7 and the uniqueness result for holomorphic functions.

3. Let $f(z) = \sum_{k=0}^{\infty} a_{2k+1} z^{2k+1}$ be a power series converging in $B_r(0)$, $r > 0$. Consider on $B_r(0) \setminus \{0\}$ the function $g(z) := \frac{f(z)}{z}$. Prove that g has a removable singularity at $z = 0$. Deduce that $z \mapsto \frac{\sin z}{z}$ is an entire function if appropriately extended at $z_0 = 0$.

4. Let $h : G \to \mathbb{C}$ be a holomorphic function which has at $z_0 \in G$ a zero of order $k \geq 1$. Obviously it follows that $h^2(z) = (h(z))^2$ has a zero at z_0 too. Find the order of the zero of h^2.

5. Let $p(z)$ be a polynomial of order k. Prove that the entire function $z \mapsto p(z)e^{az}, a \in \mathbb{C}$, cannot have a zero of order $k + 1$.

6. Use the fact that for all $x, y \in \mathbb{R}$ the equality $\cos(x - y) = \cos x \cos y + \sin x \sin y$ holds to derive that this equality also holds for all $z, w \in \mathbb{C}$.

7. Let $G \subset \mathbb{C}$ be a region and $\gamma : [0, 1] \to G$ a piecewise continuously differentiable curve. Further let $f : G \to \mathbb{C}$ be a continuous function. Suppose that for every $z_0 \in \text{tr}(\gamma)$ there exists $r_{z_0} > 0$ such that in $B_{r_{z_0}}(z_0) \subset G$ the function f has a representation as a convergent Taylor series, i.e. $f(z) = \sum_{k=0}^{\infty} a_k(z_0)(z - z_0)^k$, $z \in B_{r_{z_0}}(z_0)$. Prove that there exists a domain $D \subset G$ such that $f|_D$ is holomorphic.
   Hint: note that $\text{tr}(\gamma)$ is compact and use the uniqueness result for holomorphic functions.

8.   a) Use the series $\sum_{k=0}^{\infty} z^{2k}, |z| < 1$, to find the Taylor series of $h(z) = \frac{z+6z^2}{(1-z^2)^3}$.

   b) Use the equality $x_2 F_1 \left(\frac{1}{2}, 1; \frac{3}{2}, -x^2 \right) = \arctan x, |x| < 1$, to find a power series expansion of $x \mapsto \frac{1}{1+x^2}$.

9. Let $u : (a, b) \to \mathbb{R}$, $a < b$, be a twice continuously differentiable functions satisfying the differential equation $u''(x) + x^2 u'(x) + x^4 u(x) = 0$ in (a, b). Suppose that for some $x_0 \in (a, b)$ the function u has the convergent Taylor series representation $u(x) = \sum_{k=0}^{\infty} a_k(x - x_0)^k, a_k \in \mathbb{R}$,

$x \in (-\eta + x_0, x_0 + \eta) \subset (a, b), \eta > 0$. Show that there exists a holomorphic function $h : B_\eta(x_0) \to \mathbb{C}$ such that h satisfies $h''(z) + z^2 h'(z) + z^4 h(z) = 0$ in $B_\eta(x_0)$.

10.* For $l \in \mathbb{N}$ prove the integral representation of the Bessel function $J_l(z)$:

$$J_l(z) = \frac{1}{\Gamma\left(l + \frac{1}{2}\right) \Gamma\left(\frac{1}{2}\right)} \left(\frac{z}{2}\right)^l \int_{-1}^{1} (1 - t^2)^{l - \frac{1}{2}} \cos zt \, dt.$$

Hint: use the integral representation of the beta-function and the power series expansion of J_l, as well as the doubling formula for the Γ-function, i.e. (I.31.35).

11. Prove the identities (22.32) and then derive (22.33) and (22.34).

23 Further Properties of Holomorphic Functions

In this chapter we want to discuss several properties of holomorphic functions, including some of their global mapping properties, as well as properties of their range, the question of how we can approximate holomorphic functions, and infinite products of holomorphic functions. We start with some mapping properties and we need the result below as preparation.

Lemma 23.1. *Let $D \subset \mathbb{C}$ be a domain, $z_0 \in D$ and $\overline{B_r(z_0)} \subset D$, $r > 0$. Further let $f : D \to \mathbb{C}$ be a holomorphic function. If*

$$|f(z_0)| < \min_{|z-z_0|=r} |f(z)| \tag{23.1}$$

then f has a zero in $B_r(z_0)$.

Proof. Suppose that f has no zero in $B_r(z_0)$. Then the function $z \mapsto g(z) := \frac{1}{f(z)}$ is holomorphic in some neighbourhood of $\overline{B_r(z_0)}$ since (23.1) implies $f(z) \neq 0$ for $|z - z_0| = r$. The Cauchy inequalities, Theorem 21.15, for $k = 0$ now imply

$$|g(z_0)| = \max_{|z-z_0|=r} |g(z)|,$$

or

$$\frac{1}{|f(z_0)|} \leq \max_{|z-z_0|=r} \frac{1}{|f(z)|} = \frac{1}{\min_{|z-z_0|=r} |f(z)|}$$

which yields

$$|f(z_0)| \geq \min_{|z-z_0|=r} |f(z)| > 0,$$

hence a contradiction. $\qquad \square$

Our first main result is surprising when compared with the situation of real-valued C^∞-functions. We have proved, for example in Problem 5 of Chapter 10, the existence of a C^∞-function $\varphi : \mathbb{R} \to \mathbb{R}$ such that $0 \leq \varphi(t) \leq 1$, $\varphi|_{(-\infty,0]} = 0$ and $\varphi|_{[1,\infty)} = 1$. In particular $\varphi(\mathbb{R}) = [0,1]$ is a closed set. Non-constant holomorphic functions however will always map regions onto regions:

Theorem 23.2. *The image of a region $G \subset \mathbb{C}$ under a non-constant holomorphic function $f : G \to \mathbb{C}$ is a region, i.e. $f(G)$ is open and connected.*

Proof. Since f is continuous it follows that $f(G)$ is connected. The strong statement of the theorem is that $f(G)$ is open which we are going to prove now. Let $w_0 \in f(G)$, i.e. $w_0 = f(z_0)$ for some $z_0 \in G$. Since f is not constant we can find some $r > 0$ such that $\overline{B_r(z_0)} \subset G$ and $f(z) \neq w_0$ for $z \in \overline{B_r(z_0)} \backslash \{z_0\}$. Otherwise the uniqueness theorem, Theorem 22.12, would imply that $f(z) = w_0$ for all $z \in G$. For $z \in \partial B_r(z_0)$ we can now deduce that

$$|f(z) - w_0| \geq 3\epsilon > 0. \tag{23.2}$$

Our aim is to prove that $B_\epsilon(w_0) \subset f(G)$, i.e. for $\tilde{w} \in B_\epsilon(w_0)$ there exists $\tilde{z} \in G$ such that $f(\tilde{z}) = \tilde{w}$, from which the openness of $f(G)$ will follow. Now, for $|w - w_0| < \epsilon$ and $|z - z_0| = r$, r as above, it follows that

$$|f(z) - w| \geq |f(z) - w_0| - |w - w_0| \geq 3\epsilon - \epsilon = 2\epsilon,$$

i.e.

$$\min_{|z-z_0|=r} |f(z) - w_0| \geq 2\epsilon.$$

However for $z = z_0$ we find

$$|f(z_0) - w| = |w_0 - w| < \epsilon,$$

i.e.

$$|f(z_0) - w| = |w_0 - w| < \epsilon,$$

hence

$$|f(z_0) - w| < \min_{|z-z_0|=r} |f(z) - w|,$$

implying that $z \mapsto f(z) - w$ has at least one zero in $B_r(z_0)$, i.e. for every w, $|w - w_0| < \epsilon$, there exists $z \in B_r(z_0)$ such that $w = f(z)$ which means that $B_\epsilon(w_0) \subset f(G)$ implying that $f(G)$ is open. $\square$

Corollary 23.3. *Let $f : G \to \mathbb{C}$ be a holomorphic function defined on a region $G \subset \mathbb{C}$. If $|f|$, $\mathrm{Re}\, f$ or $\mathrm{Im}\, f$ is constant then f must be constant too.*

Proof. In each case $f(G)$ cannot be open. $\square$

In Chapter 17 we introduced biholomorphic functions and in Proposition 17.19 we characterised biholomorphic function $f : G_1 \to G_2$ between two regions. Now we want to prove a holomorphic version of the inverse function theorem, Theorem II.10.12.

Definition 23.4. *Let $D \subset \mathbb{C}$ be a domain and $f : D \to \mathbb{C}$ a holomorphic function. We call f **locally biholomorphic** at $z_0 \in D$ if there exists an open neighbourhood $U = U(z_0) \subset D$ such that $f|_U : U \to f(U)$ is biholomorphic.*

We aim to prove that $f'(z_0) \neq 0$ implies f to be locally biholomorphic at z_0. For this we need some auxillary results.

Lemma 23.5. *Let $g(z) = \sum_{k=0}^{\infty} c_k (z - z_0)^k$ be a power series converging in $B_r(z_0)$, $r > 0$. If $|c_1| > \sum_{k=2}^{\infty} k|c_k|r^{k-1}$, then $g : B_r(z_0) \to \mathbb{C}$ is injective.*

Proof. Let $w_1, w_2 \in B_r(z_0)$ such that $g(w_1) = g(w_2)$, and define $z_1 := w_1 - z_0$, $z_2 := w_2 - z_0$. It follows that

$$\sum_{k=1}^{\infty} c_k \left(z_1^k - z_2^k \right) = 0.$$

Since $z_1^k - z_2^k = (z_1 - z_2)\left(z_1^{k-1} + z_1^{k-2}z_2 + \cdots + z_1 z_2^{k-2} + z_2^{k-1} \right)$ we find

$$0 = \sum_{k=1}^{\infty} c_k \left(z_1^k - z_2^k \right)$$

$$= (z_1 - z_2)c_1 + \sum_{k=2}^{\infty} c_k(z_1 - z_2)\left(z_1^{k-1} + z_1^{k-2}z_2 + \cdots + z_1 z_2^{k-2} + z_2^{k-1} \right),$$

or for $z_1 \neq z_2$

$$-c_1 = \sum_{k=2}^{\infty} c_k \left(z_1^{k-1} + z_1^{k-2}z_2 + \cdots + z_1 z_2^{k-2} + z_2^{k-1} \right).$$

Recall that $w_1, w_2 \in B_r(z_0)$, i.e. $|z_1|, |z_2| < r$, and we conclude now

$$|c_1| \leq \sum_{k=2}^{\infty} |c_k| k r^{k-1},$$

which is a contradiction. $\square$

Corollary 23.6. *Let $D \subset \mathbb{C}$ be a domain and $f : D \to \mathbb{C}$ a holomorphic function. If $f'(z_0) \neq 0$ for some $z_0 \in D$ then there exists a neighbourhood $U = U(z_0) \subset D$ such that $f|_U$ is injective.*

Proof. We consider the Taylor expansion $f(z) = \sum_{k=0}^{\infty} \frac{f^{(k)}(z_0)}{k!}(z - z_0)^k$ of f about z_0 which represents f in some open ball $B_r(z_0)$, $r > 0$. Since $f'(z_0) \neq 0$ and since the function $\rho \mapsto \sum_{k=2}^{\infty} k \frac{|f^{(k)}(z_0)|}{k!} \rho^{k-1}$ is continuous at zero with value 0 for $\rho = 0$, it follows the existence of $r_1 < r$ such that

$$|f'(z_0)| > \sum_{k=2}^{\infty} k \frac{|f^{(k)}(z_0)|}{k!} r^{k-1},$$

and Lemma 23.5 implies that $f|_{B_r(z_0)}$ is injective. $\qquad\square$

Now we can prove

Theorem 23.7. *Let $D \subset \mathbb{C}$ be a domain and $f : D \to \mathbb{C}$ a holomorphic function with $f'(z_0) \neq 0$ for some $z_0 \in D$. Then there exists an open neighbourhood $U = U(z_0) \subset D$ of z_0 such that $f|_U : U \to f(U)$ is biholomorphic, i.e. f is locally biholomorphic at z_0.*

Proof. We know that we can find an open neighbourhood $U = U(z_0) \subset D$ of $z_0 \in D$ which we can choose to be a region such that $f|_U$ is injective, and we know further that $f(U)$ is a region too. An injective mapping $f|_U : U \to f(U)$ is bijective. Shrinking U if necessary we can use Theorem II.3.38 to deduce that f^{-1} is continuous. Now Proposition 17.19 implies that $f|_U$ is biholomorphic from U to $f(U)$. $\qquad\square$

A further consequence of Theorem 23.2 is the following maximum principle.

Theorem 23.8 (Boundary Maximum Principle). *Let $f : G \to \mathbb{C}$ be a holomorphic function defined on a region $G \subset \mathbb{C}$. If $|f|$ has a local maximum at $z_0 \in G$ then f is constant on G. If G is bounded and f is defined and continuous on $\overline{G}$ then $|f|$ attains its maximum on ∂G, i.e. for all $z \in \overline{G}$ we have*

$$|f(z)| \leq \max_{\zeta \in \partial G} |f(\zeta)|. \tag{23.3}$$

Proof. Let $U \subset G$ be a neighbourhood of z_0 such that $|f(z_0)| \geq |f(z)|$ for $z \in U$. The image of U under f is a subset of $\{w \,|\, |w| \leq |f(z_0)|\}$, i.e. $f(U) \subset \{w \,|\, |w| \leq f(z_0)\}$, but $\{w \,|\, |w| \leq f(z_0)\}$ is not a neighbourhood of $f(z_0)$. Thus by Theorem 23.2 f must be constant equal to $f(z_0)$ in a neighbourhood of z_0, and hence by the uniqueness theorem for holomorphic functions f must be constant in G.

The second part is seen as follows. The continuous function $|f| : \overline{G} \to \mathbb{R}$ attains its maximum on $\overline{G}$ since $\overline{G}$ is now compact. The first part implies that for a non-constant function this maximum cannot be attained in G, so it must be attained on ∂G. □

Corollary 23.9. *Let $G \subset \mathbb{C}$ be a bounded region and $f, g : \overline{G} \to \mathbb{C}$ two continuous functions such that $f, g : G \to \mathbb{C}$ are holomorphic and $f|_{\partial G} = g|_{\partial G}$. Then $f = g$ in $\overline{G}$.*

Proof. The function $f - g : \overline{G} \to \mathbb{C}$ is continuous and holomorphic on G. By the boundary maximum principle it follows that

$$|f(z) - g(z)| \leq \max_{\zeta \in \partial G} |f(\zeta) - g(\zeta)| = 0 \tag{23.4}$$

for all $z \in \overline{G}$. □

We can also derive from Theorem 23.8 a minimum principle.

Theorem 23.10 (Boundary Minimum Principle). *Let $G \subset \mathbb{C}$ be a region and $f : G \to \mathbb{C}$ be a holomorphic function. **A.** If $|f|$ has a local minimum at $z_0 \in G$ then either $f(z_0) = 0$ or f is constant. **B.** Now suppose that $\overline{G}$ is compact and f has a continuous extension to $\overline{G}$, i.e. $f : \overline{G} \to \mathbb{C}$ is continuous and $f : G \to \mathbb{C}$ is holomorphic. Then f has either a zero in G or $|f|$ attains its minimum on ∂G, i.e.*

$$|f(z)| \geq \min_{\zeta \in \partial G} |f(\zeta)| \tag{23.5}$$

for all $z \in \overline{G}$.

Proof. If $f(z) \neq 0$ in G we can consider the function $g(z) = \frac{1}{f(z)}$ which satisfies all assumptions of Theorem 23.8. □

As an application of the maximum principle we prove the **Lemma of Schwarz** which we will later on, see Chapter 28, find to be very useful.

Theorem 23.11. *For a holomorphic function $f : B_1(0) \to B_1(0)$ with $f(0) = 0$ the estimates*

$$|f(z)| \leq |z| \quad \text{for all } z \in B_1(0) \tag{23.6}$$

and

$$|f'(0)| \leq 1 \tag{23.7}$$

hold.

Proof. Since $f(0) = 0$ we have for f the power series representation $f(z) = \sum_{k=1}^{\infty} c_k z^k$ and g defined by $g(z) = \sum_{k=1}^{\infty} c_k z^{k-1}$ is a further holomorphic function defined on $B_1(0)$. Clearly $f(z) = zg(z)$ for all $z \in B_1(0)$ as well as $g(0) = c_1 = f'(0)$. Since $f(B_1(0)) \subset B_1(0)$ it follows that $|f(z)| < 1$ for all $z \in B_1(0)$, hence $r \max_{|z|=r} |g(z)| \le 1$ for every $r \in (0,1)$. Now the maximum principle gives $|g(z)| \le \frac{1}{r}$ for all $z \in B_r(0)$, $0 < r < 1$. For $r \to 1$ this implies $|g(z)| \le 1$ for all $z \in B_r(0)$ and therefore we find $|f(z)| \le |z|$ for all $z \in B_1(0)$ as well as $|f'(0)| = |g(0)| \le 1$. $\qquad\square$

For later purposes we add

Corollary 23.12. *In the situation of Theorem 23.11 assume the existence of a point $a \in B_1(0)\backslash\{0\}$ such that $|f(a)| = |a|$ or that $|f'(0)| = 1$. Then we can find $\gamma \in S^1 = \partial B_1(0)$ such that $f(z) = \gamma z$ for all $z \in B_1(0)$, i.e. f is a rotation of $B_1(0)$, in fact of $\mathbb{C}$.*

Proof. With the notations of the proof of Theorem 23.11 we find for $a \in B_1(0)\backslash\{0\}$ with $|f(a)| = |a|$ or in the case that $|f'(0)| = 1$ that either $|g(a)| = 1$ or $|g(0)| = 1$. This however means that f attains its supremum in $B_1(0)$ and hence it must be a constant. $\qquad\square$

Next we want to use our knowledge of the growth of $|f|$ for a holomorphic function f to study, in some cases, the range of f. Our first goal is to prove the fundamental theorem of algebra which states in its original version that every polynomial $p(z) = \sum_{k=0}^{n} a_k z^k$ of order n, i.e. $a_n \ne 0$, with complex coefficients $a_k \in \mathbb{C}$ has n zeroes $z_1, \ldots, z_n$ counted according to their multiplicity. In other words, given $w \in \mathbb{C}$ we can always solve the equation $p(z) = w$ and the set of solutions has at most n distinct elements. Another way of wording the result is that the range of a polynomial function is $\mathbb{C}$, and with arguments employed later on one easily derives the fundamental theorem of algebra in its classical form.

We start by estimating the growth of a polynomial.

Lemma 23.13. *Let $p(z) = \sum_{k=0}^{\infty} a_k z^k$, $a_k \in \mathbb{C}$, $a_n \ne 0$, be a polynomial of degree n. For $|z| \ge 1$ the estimate*

$$|p(z)| \le \left(\sum_{k=0}^{n} |a_k| \right) |z|^n \qquad (23.8)$$

holds. Furthermore, for every $\epsilon \in (0, 1)$ there exists $\rho(\epsilon)$ such that $|z| \geq \rho(\epsilon)$ implies

$$(1 - \epsilon)|a_n||z|^n \leq |p(z)| \leq (1 + \epsilon)|a_n||z|^n. \tag{23.9}$$

Proof. For $|z| \geq 1$ we find immediately that

$$|p(z)| \leq \sum_{k=0}^{n} |a_k||z|^k \leq \left(\sum_{k=0}^{n} |a_k|\right)|z|^n,$$

i.e. (23.8). With $q(z) = \sum_{k=0}^{n-1} a_k z^k$ we have $p(z) = a_n z^n + q(z)$ and for $|z| \geq 1$ we get

$$|q(z)| \leq \left(\sum_{k=0}^{n-1} |a_k|\right)|z|^{n-1}.$$

For $\epsilon \in (0, 1)$ we choose

$$\rho(\epsilon) := \max\left\{1, \frac{1}{\epsilon|a_n|}\sum_{k=0}^{n-1} |a_k|\right\}. \tag{23.10}$$

Now $|z| \geq \rho(\epsilon)$ implies that

$$|q(z)| \leq \left(\frac{1}{|z|}\sum_{k=0}^{n-1} |a_k||z|^n\right) \leq \epsilon|a_n||z|^n,$$

which yields

$$(1 - \epsilon)|a_n||z|^n \leq |a_n||z|^n - |q(z)| \leq |p(z)|$$
$$\leq |a_n||z|^n + |q(z)| \leq (1 + \epsilon)|a_n||z|^n$$

proving (23.9). $\qquad \square$

Corollary 23.14. *The zeroes of the polynomial $p(z) = \sum_{k=0}^{n} a_k z^k$, $a_n \neq 0$, must lie in the set $\left\{z \in \mathbb{C} \,|\, |z| \leq \max\left\{1, \frac{1}{|a_n|}\sum_{k=0}^{n-1} |a_k|\right\}\right\}$.*

Proof. From (23.9) we deduce that the zeroes of $p(z)$ must belong for every $\epsilon \in (0, 1)$ to $B_{\rho(\epsilon)}(0)$, $\rho(\epsilon)$ as in (23.10), and passing to the limit as ϵ tends to 1 yields the result. $\qquad \square$

Theorem 23.15 (Fundamental Theorem of Algebra). *Every non-constant polynomial with complex coefficients has a zero in $\mathbb{C}$.*

Proof. For the polynomial $p(z) = \sum_{k=0}^{n} a_k z^k$, $a_n \neq 0$, of degree $n \geq 1$ we choose $\epsilon = \frac{1}{2}$ in Lemma 23.13 and $R \geq \max\left\{1, \rho\left(\frac{1}{2}\right)\right\}$ such that

$$|p(0)| \leq \frac{|a_n|}{2} R^n.$$

Now (23.9) implies that $|p(0)| \leq \min_{|z|=R} |p(z)|$ and the boundary minimum principle, Theorem 23.10, yields that p must have a zero in $B_R(0)$. □

Corollary 23.16. *A polynomial $p(z) = \sum_{k=0}^{n} a_k z^k$, $a_n \neq 0$, factorises into linear factors, i.e. there exists complex numbers $z_1, \ldots, z_n \in \mathbb{C}$ (not necessarily all different) such that with some $c \in \mathbb{C}$*

$$p(z) = c \prod_{j=1}^{n} (z - z_j) \tag{23.11}$$

holds.

Proof. According to the fundamental theorem there exists $z_1 \in \mathbb{C}$ such that $p(z_1) = 0$, hence $p(z) = (z - z_1)q(z)$ where $q(z)$ is a polynomial of degree $n - 1$. Now we can iterate the argument. □

The next result is a further one which relates growth properties of an entire function to its structure.

Theorem 23.17. *Let $f : \mathbb{C} \to \mathbb{C}$ be an entire function and suppose that for some $n \in \mathbb{N}$ there exists $R > 0$ and $M > 0$ such that*

$$|f(z)| \leq M|z|^n, \quad |z| \geq R.$$

Then f is a polynomial of degree less than or equal to n.

Proof. We consider the Taylor expansion of f about $z_0 = 0$,

$$f(z) = \sum_{k=0}^{\infty} \frac{f^{(k)}(0)}{k!} z^k.$$

Using the Cauchy inequalities from Theorem 21.15 we find for $r \geq R$ that

$$\frac{|f^{(k)}(0)|}{k!} \leq \frac{1}{r^k} \max_{|z|=r} |f(z)| \leq r^{-k} M r^n = M r^{n-k}.$$

For $r \to \infty$ it follows that if $k > n$ then $f^{(k)}(0) = 0$, i.e. $f(z) = \sum_{k=0}^{n} \frac{f^{(k)}(0)}{k!} z^k$. □

Corollary 23.18 (Theorem of Liouville). *A bounded entire function is constant.*

Proof. Apply Theorem 23.17 for $n = 0$. □

We now turn to convergence and approximation properties of holomorphic functions. For continuous functions we know that uniform convergence is needed to assure that in general the limit function of a sequence of continuous functions is continuous too, however see also the remark following Theorem 23.20. For real-valued differentiable functions we need the pointwise convergence of the sequence and the uniform convergence of the sequence of derivatives, compare with Theorem I.24.6 and Theorem I.26.19. In order to assure that limit functions also have a higher order degree of differentiability appropriate conditions of the approximating sequence and the mode of convergence of the functions and their derivatives are needed. While this suggests strong conditions to assure that the limit function of a sequence of holomorphic functions is holomorphic too, we need in fact a less strong mode of convergence.

Definition 23.19. *Let $D \subset \mathbb{C}$ be a domain and $f_k : D \to \mathbb{C}$, $k \in \mathbb{N}$, a sequence of functions. We say that $(f_k)_{k \in \mathbb{N}}$ **converges locally uniform** to $f : D \to \mathbb{C}$ if every point $z_0 \in D$ has a neighbourhood $U(z_0)$ such that $\left(f_k|_{U(z_0)}\right)_{k \in \mathbb{N}}$ converges uniformly to $f|_{U(z_0)}$.*

Theorem 23.20. *Let $G \subset \mathbb{C}$ be a region and $f_k : G \to \mathbb{C}$, $k \in \mathbb{N}$, be a sequence of holomorphic functions converging locally uniform to a function $f : G \to \mathbb{C}$. Then f is holomorphic and for every $n \in \mathbb{N}$ the sequence $\left(f_k^{(n)}\right)_{k \in \mathbb{N}}$ converges locally uniformly to $f^{(n)}$.*

Proof. Clearly f is continuous since for every point $z_0 \in G$ there exists a ball $B_\epsilon(z_0) \subset G$, $\epsilon > 0$, such that $(f_k)_{k \in \mathbb{N}}$ converges on $B_\epsilon(z_0)$ uniformly to f. Let $\triangle \subset G$ be a triangle and $\partial\triangle$ anti-clockwise oriented. The uniform convergence of $(f_k|_{\partial\triangle})_{k \in \mathbb{N}}$ to $f|_{\partial\triangle}$ follows from the compactness of $\partial\triangle \subset G$ and it implies

$$\int_{\partial\triangle} f(z)\,\mathrm{d}z = \lim_{k \to \infty} \int_{\partial\triangle} f_k(z)\,\mathrm{d}z = 0.$$

Now the theorem of Morera, compare with Theorem 22.1, yields the holomorphy of f. In order to prove the second statement it is sufficient to show that $(f_k')_{k \in \mathbb{N}}$ converges locally uniform to f', the general case follows then by iterating the argument. Let $r > 0$ and $\overline{B_r(z_0)} \subset G$. Applying the Cauchy inequalities to the holomorphic function $f - f_k$ we find for all $z \in B_{\frac{r}{2}}(z_0)$ that

$$|f'(z) - f_k'(z)| \leq c\frac{1}{r} \max_{|\zeta - z_0| = r} |f(\zeta) - f_k(\zeta)|$$

$$\leq \frac{c}{r} \|f - f_k\|_{\infty, \overline{B_r(z_0)}},$$

implying the uniform convergence of $(f_k')_{k \in \mathbb{N}}$ on $\overline{B_{\frac{r}{2}}(z_0)}$ to f' and the result follows. $\qquad\square$

Remark 23.21. Note that the proof of Theorem 23.20 yields that already the locally uniform limit of continuous functions is continuous, and this holds at least for every metric space once we modify Definition 23.19 in the obvious way. We can even apply such a kind of argument to derivatives, we still need convergence of the derivatives which we do not need in the case of Theorem 23.20, i.e. holomorphic functions.

The Weierstrass approximation theorem, Theorem II.14.3, assures that the restriction of polynomials on the real line are dense in $(C([0, 1]; \mathbb{R}), \|.\|_\infty)$. Extensions such as Stone's approximation theorem, Theorem II.14.9, give further results on the approximation of continuous functions defined on a compact metric space. We are longing for some approximation results for holomorphic functions on compact sets. For this we first need

Definition 23.22. *Let $K \subset \mathbb{C}$ be a compact set. We call $f : K \to \mathbb{C}$ **holomorphic on K** if there exists an open neighbourhood U of K, $K \subset U$, and a holomorphic function $g : U \to \mathbb{C}$ such that $f = g|_K$.*

In the following, when dealing with holomorphic functions f on K, if no confusion may arise, we denote the function g in Definition 23.22 with f too, i.e. we assume that f is already given as holomorphic function $f : U \to \mathbb{C}$, $K \subset U$.

The first idea to approximate a holomorphic function on K, K compact, by polynomials turns out to be not always successful. Consider the function h, $h(z) = \frac{1}{z}$, defined and holomorphic on $\mathbb{C} \backslash \{0\}$. In particular h is holomorphic on the compact set S^1. For the unit circle S^1 we have when using our standard parametrization that $\frac{1}{2\pi i} \int_{|z|=1} \frac{1}{z} dz = 1$. Now suppose that there exists a sequence of polynomials converging locally uniformly in a neighbourhood of S^1 to h. It follows that $(p_k)_{k\in\mathbb{N}}$ converges uniformly on S^1 to h, but $\frac{1}{2\pi i}\int_{|z|=1} p_k(z)\,dz = 0$. Since by uniform convergence of $(p_k)_{k\in\mathbb{N}}$ on S^1 to h we get

$$\lim_{k\to\infty}\int_{|z|=1} p_k(z)\,dz = \int_{|z|=1} h(z)\,dz$$

we have a contradiction.

Thus, in general, to try to approximate holomorphic functions on a compact set by polynomials seems not to be a valid option. Instead we try to use rational functions and we will prove Runge's approximation theorem. As preparation we prove

Theorem 23.23. *Let $K \subset \mathbb{C}$ be a compact set and $U \subset \mathbb{C}$ an open neighbourhood of K, $K \subset U$. Then there exists finitely many axis-parallel line segments $\gamma_1, \ldots, \gamma_N$ such that $\gamma := \gamma_1 \oplus \cdots \oplus \gamma_N$ is a simply closed, positive oriented, piecewise continuously differentiable curve with $\mathrm{tr}(\gamma) \subset U\backslash K$ and*

$$f(z) = \frac{1}{2\pi i}\int_\gamma \frac{f(\zeta)}{\zeta - z}\,d\zeta \tag{23.12}$$

holds for every $z \in K$ and every holomorphic function $f : U \to \mathbb{C}$.

Remark 23.24. It is important to observe that γ, i.e. γ_j, $j = 1, \ldots, N$, depends on K and U but it is independent of f.

Proof of Theorem 23.23. Since K is compact and U is open it follows that $\delta := \mathrm{dist}(K, \partial U) > 0$. We cover U with axis-parallel closed square of side length h where we assume that $\sqrt{2}h < \delta$, see Figure 23.1.

435

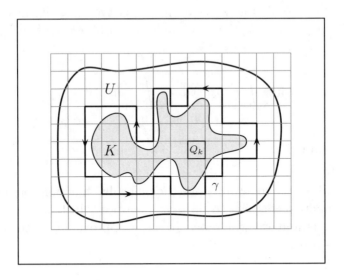

Figure 23.1

Since K is compact we can find a finite number $Q_1, \ldots, Q_M$ of these squares with the property that

$$K \subset \bigcup_{k=1}^{M} Q_k \subset U. \tag{23.13}$$

The choice of h in relation to δ assures that for $z_0 \in K \cap Q_k$ we have $B_\delta(z_0) \subset U$ and since $\operatorname{diam} Q_k = \sqrt{2}h < \delta$ it follows that $Q_k \subset U$, hence (23.13) holds. We consider further ∂Q_k as positive oriented simply closed parametric curves and we denote by $\gamma_1, \ldots, \gamma_N$ those segments which are part of the boundary of some Q_k but never of two adjacent squares Q_k and Q_l. The enumeration of $\gamma_1, \ldots, \gamma_N$ is chosen such that $\gamma := \gamma_1 \oplus \cdots \oplus \gamma_N$ is a simply closed positive oriented paramteric curve. By construction we also find that $\operatorname{tr}(\gamma) \subset K^{\complement} \cap U$. Indeed, if $\gamma_l \cap K \neq \emptyset$ then γ_l would be part of the boundary of two adjacent squares, see again Figure 23.1. Now let $f : U \to \mathbb{C}$ be a holomorphic function and $z \in K$, hence $z \in Q_{k_0}$ for some k_0. Suppose that $z \in \mathring{Q}_{k_0}$. In this case we have

$$f(z) = \frac{1}{2\pi i} \int_{\partial Q_{k_0}} \frac{f(\zeta)}{\zeta - z} \, d\zeta.$$

Since for $k \neq k_0$ we have now $\frac{1}{2\pi i} \int_{\partial Q_k} \frac{f(\zeta)}{\zeta - z} \, d\zeta = 0$ it follows that

$$f(z) = \sum_{k=1}^{M} \frac{1}{2\pi i} \int_{\partial Q_{k_0}} \frac{f(\zeta)}{\zeta - z} \, d\zeta = \frac{1}{2\pi i} \int_{\gamma} \frac{f(\zeta)}{\zeta - z} \, dz,$$

where for the last equality we used that if the trace of a line segment γ_l belongs to the boundary of two adjacent squares we integrate the same function over γ_l and γ_l^{-1} and hence the sum of these two integrals vanishes. This proves (23.12) for $z \in Q_{k_0}$. If $z \in \partial Q_{k_0}$ it is in the intersection of two squares Q_{k_0} and Q_{l_0}, hence $z \notin \operatorname{tr}(\gamma)$. Hence by approximating z by a sequence $(z_j)_{j \in \mathbb{N}}$, $z_j \in \mathring{Q}_{k_0} \cap K$, the result will follow. $\qquad \square$

With the help of Theorem 23.23 we can now prove the **approximation theorem of Runge**.

Theorem 23.25 (Runge). *Let $K \subset \mathbb{C}$ be a compact set and $f : K \to \mathbb{C}$ a holomorphic function. For every $\epsilon > 0$ we can find a rational function $r_\epsilon(z) = \frac{p_\epsilon(z)}{q_\epsilon(z)}$, $q_\epsilon(w) \neq 0$ for $w \in K$, such that $|f(z) - r_\epsilon(z)| < \epsilon$ for all $z \in K$, i.e. $\|f - r_\epsilon\|_{\infty, K} < \epsilon$.*

Proof. Let $f : K \to \mathbb{C}$ be holomorphic. Choose $U \subset \mathbb{C}$ open such that $K \subset U$ and $f = g|_K$ for a holomorphic function $g : U \to \mathbb{C}$, which we now denote by f too. Now we determine the curve γ according to Theorem 23.23 and we note that on the compact set $\operatorname{tr}(\gamma) \times K$ the function $(\zeta, z) \mapsto \frac{f(\zeta)}{\zeta - z}$ is continuous. As a continuous function on a compact set this function is uniformly continuous. Hence for $\epsilon > 0$ there exists $\delta > 0$ such that for $z \in K$ and $z_1, z_2 \in \operatorname{tr}(\gamma)$ with $|z_1 - z_2| < \delta$ it follows that

$$\left| \frac{f(z_1)}{z_1 - z} - \frac{f(z_2)}{z_2 - z} \right| < \frac{2\pi\epsilon}{l_\gamma}$$

where as usual l_γ denotes the length of γ. By construction γ is rectifiable, in fact it is an axis-parallel polygonal curve. Therefore we can divide γ into segments κ_l, $l_{\kappa_l} < \delta$, $l = 1, \ldots, M$, and then choose points $\zeta_l \in \operatorname{tr}(\kappa_l)$ such that for all $z \in K$

$$\left| \frac{1}{2\pi i} \int_{\kappa_l} \frac{f(\zeta)}{\zeta - z} \, d\zeta - \frac{1}{2\pi i} \int_{\kappa_l} \frac{f(\zeta_l)}{\zeta_l - z} \, d\zeta \right| < \frac{\epsilon}{l_\gamma} l_{\kappa_l}.$$

The function

$$r_\epsilon(z) := \frac{1}{2\pi i} \sum_{l=1}^{M} \frac{f(\zeta_l)}{\zeta_l - z} \int_{\kappa_l} 1 \, d\zeta = \frac{1}{2\pi i} \sum_{l=1}^{M} \frac{f(\zeta_l)}{\zeta_l - z} l_{\kappa_l}$$

is a rational function which is not holomorphic at the points $\zeta_l \in \kappa_l \subset K^C$, $l = 1, \ldots, M$, only. Note that of course the points ζ_l depend indirectly on ϵ. Now we observe that

$$|f(z) - r_\epsilon(z)| = \left| \sum_{l=1}^{M} \frac{1}{2\pi i} \int_{\kappa_l} \frac{f(\zeta)}{\zeta - z} \, d\zeta - \sum_{l=1}^{M} \frac{1}{2\pi i} \int_{\kappa_l} \frac{f(\zeta_l)}{\zeta_l - z} \, d\zeta \right|$$

$$< \sum_{l=1}^{M} \frac{\epsilon}{l_\gamma} l_{\kappa_l} = \epsilon,$$

and the theorem is proven. □

Remark 23.26. In proving Runge's theorem we found the presentations in [30] and [83] quite helpful.

We now turn to infinite products of holomorphic functions and infinite product representations of holomorphic functions. For a polynomial $p(z) = \sum_{k=0}^{n} a_k z^k$, $a_n \neq 0$, we have a finite product representation given by (23.11). In Chapter I.30 and I.31 we discussed infinite products of real numbers as well as of real-valued functions. For example in Theorem I.31.14 we have proved for $x \in \mathbb{R}$ the following representation of the sine function

$$\sin \pi x = \pi x \prod_{k=1}^{\infty} \left(1 - \frac{x^2}{k^2} \right), \quad x \in \mathbb{R}. \tag{23.14}$$

A natural question is whether (23.14) extends to $z \in \mathbb{C}$. We have also obtained a product representation of the Γ-function, namely the Weierstrass representation, Theorem I.31.2,

$$\Gamma(x) = \frac{e^{-\gamma x}}{x} \prod_{k=1}^{\infty} \frac{e^{\frac{x}{k}}}{1 + \frac{x}{k}}, \quad x > 0, \tag{23.15}$$

where $\gamma = \lim_{N \to \infty} \left(\sum_{k=1}^{N} \frac{1}{k} - \ln N \right)$ is the Euler constant, also see Theorem I.18.24. Now we may ask whether we can extend the product on the right hand side to some domain in $\mathbb{C}$ which could allow us to extend the Γ-function to subsets of $\mathbb{C}$ larger than $\mathbb{R}_+\backslash\{0\}$.

We start by looking at the definition of infinite products of sequences of complex numbers. However it is more convenient for our purposes to start with $(a_k)_{k\in\mathbb{N}}$, $a_k \in \mathbb{C}$, and look at

$$\prod_{k=1}^{\infty}(1 + a_k). \tag{23.16}$$

In Problem 11 we will handle $\prod_{k=1}^{\infty} c_k$, $c_k \in \mathbb{C}$, in light of some of the results of Chapter I.30.

Definition 23.27. *For a sequence $(a_k)_{k\in\mathbb{N}}$ of complex numbers $a_k \in \mathbb{C}$ the infinite product (23.16) converges if*

$$\lim_{N\to\infty} \prod_{k=1}^{N}(1 + a_k) \tag{23.17}$$

exists. In this case the limit is denoted by (23.16).

Analogously to Proposition I.30.10 we have

Proposition 23.28. *Let $(a_k)_{k\in\mathbb{N}}$ be a sequence of complex numbers for which $\sum_{k=1}^{\infty} a_k$ converges absolutely. Then the infinite product $\prod_{k=1}^{\infty}(1 + a_k)$ converges. It diverges to 0 if and only if one of its factors is 0.*

Remark 23.29. In Definition I.30.2 it was defined that if $\lim_{N\to\infty} \prod_{k=0}^{\infty} c_k = 0$ then we call $\prod_{k=0}^{\infty} c_k$ divergent to 0. In order to be consistent we use in Proposition 23.28 now the wording "diverges to 0" instead of "converges to 0" as for example in [83].

Proof of Proposition 23.28. Since $\sum_{k=1}^{\infty} |a_k|$ converges there exists $M \in \mathbb{N}$ such that $|a_k| < \frac{1}{2}$ for $k \geq M$. Since finitely many terms do not affect the convergence we assume $|a_k| < \frac{1}{2}$ for all $k \in \mathbb{N}$. Now we can use the branch of the logarithmic function given as in Chapter 18 by $\widetilde{\ln}(1 + z) = \sum_{k=1}^{\infty} \frac{(-1)^{k-1}}{k}$, $|z| < 1$, and we obtain in particular

$$1 + z = e^{\widetilde{\ln}(1+z)}, \quad |z| < 1.$$

This implies further that

$$\prod_{k=1}^{N}(1 + a_k) = \prod_{k=1}^{N} e^{\tilde{\ln}(1+a_k)} = \exp\left(\sum_{k=1}^{N} \tilde{\ln}(1 + a_k)\right).$$

Since $|z| < \frac{1}{2}$ implies

$$|\tilde{\ln}(1 + z)| \le \sum_{k=1}^{\infty} \frac{|z|^k}{k} \le \sum_{k=1}^{\infty} |z|^k = \frac{1}{1 - |z|} - 1 = \frac{|z|}{1 - |z|} < 2|z|,$$

we deduce $|\tilde{\ln}(1 + a_k)| \le 2|a_k|$ implying that $\left(\sum_{k=1}^{N} \tilde{\ln}(1 + a_k)\right)_{N\in\mathbb{N}}$ converges. The continuity of the exponential function yields the convergence of $\left(\exp\left(\sum_{k=1}^{N} \tilde{\ln}(1 + a_k)\right)\right)_{N\in\mathbb{N}}$ and it follows that the convergence of $\sum_{k=1}^{\infty} |a_k|$ entails the convergence of $\prod_{k=1}^{\infty}(1 + a_k)$. If $1 + a_{k_0} = 0$ for some k_0, then obviously this product is zero. However, if for all $k \in \mathbb{N}$ we have $1 + a_k \ne 0$, then the product can not be zero since it is of type e^A, $A \in \mathbb{C}$, compare with Corollary 22.11. $\qquad\square$

As in the case of a sequence of real-valued functions we can introduce absolutely and/or uniformly convergent infinite products of complex-valued functions defined in the plane.

Definition 23.30. *Let $A \subset \mathbb{C}$ be a set and $(u_k)_{k\in\mathbb{N}}$, $u_k : A \to \mathbb{C}$, a sequence of functions. We call $\prod_{k=1}^{\infty}(1 + u_k)$ **absolutely convergent** if the product $\prod_{k=1}^{\infty}(1 + |u_k|)$ converges. If the sequence $\left(\prod_{k=1}^{N}(1 + u_k)\right)_{N\in\mathbb{N}}$ of functions converges uniformly on A to a limit function denoted by $\prod_{k=1}^{\infty}(1 + u_k)$, then we call $\prod_{k=1}^{\infty}(1 + u_k)$ **uniformly convergent** on A.*

Theorem 23.31. *Let $D \subset \mathbb{C}$ be a domain and $f_k : \Omega \to \mathbb{C}$, $k \in \mathbb{N}$, a sequence of holomorphic functions. Suppose that there exists a sequence $(M_k)_{k\in\mathbb{N}}$, $M_k \ge 0$, such that $\|f_k - 1\|_{\infty,0} \le M_k$ and $\sum_{k=1}^{\infty} M_k < \infty$. Then the product $\prod_{k=1}^{\infty} f_k(z)$ converges uniformly on D to a holomorphic function $f(z) = \prod_{k=1}^{\infty} f_k(z)$.*

Proof. With $h_k(z) = f_k(z) - 1$ we have to investigate the product $\prod_{k=1}^{\infty}(1 + h_k(z))$ and it follows that $|h_k(z)| \le M_k$ for all $z \in D$. Now we can adapt the proof of Theorem I.30.18 to the new situation to deduce the uniform convergence. Since the uniform limit of a sequence of holomorphic functions

440

is holomorphic - note that we now speak about the sequence $\left(\prod_{k=1}^{N} f_k\right)_{N \in \mathbb{N}}$ - it follows that $\prod_{k=1}^{\infty} f_k(z)$ is holomorphic. $\qquad\square$

Remark 23.32. If we can prove for a sequence $f_k : D \to \mathbb{C}$, $k \in \mathbb{N}$, of holomorphic functions that $\left(\prod_{k=1}^{N} f_k\right)_{N \in \mathbb{N}}$ converges locally uniform, it still follows that the limit is holomorphic, see Theorem 23.20.

Example 23.33 (Product representation of $\sin z$). For all $z \in \mathbb{C}$ the sine function admits the product representation

$$\sin \pi z = \pi z \prod_{k=1}^{\infty} \left(1 - \frac{z^2}{k^2}\right), \quad z \in \mathbb{C}. \tag{23.18}$$

For $R > 0$ we note the estimate $\left|\frac{z^2}{k^2}\right| \le \frac{R^2}{k^2}$, $z \in B_R(0)$, and since the series $\sum_{k=1}^{\infty} \frac{R^2}{k^2}$ converges, we conclude that $\left(\prod_{k=1}^{N} \left(1 - \frac{z^2}{k^2}\right)\right)_{N \in \mathbb{N}}$ converges locally uniform on $\mathbb{C}$. This implies that $\pi z \prod_{k=1}^{\infty} \left(1 - \frac{z^2}{k^2}\right)$ is an entire function which coincides by Theorem I.31.14 on $\mathbb{R}$ with $\sin x$. Hence by the uniqueness theorem, Theorem 22.12, we deduce (23.18).

Theorem 23.34. *Let $D \subset \mathbb{C}$ be a domain and the sequence $f_k : D \to \mathbb{C}$, $k \in \mathbb{N}$, of holomorphic functions is assumed to satisfy the assumptions of Theorem 23.31. Suppose further that for all $k \in \mathbb{N}$ the function f_k has no zero, i.e. $f_k(z) \neq 0$ for all $z \in D$. If we denote by $f(z)$ the infinite product of $h_k(z) = f_k(z) - 1$, then we have*

$$\frac{f'(z)}{f(z)} = \sum_{k=1}^{\infty} \frac{f_k'(z)}{f_k(z)} = \sum_{k=1}^{\infty} \frac{h_k'(z)}{1 + h_k(z)}. \tag{23.19}$$

Proof. Once we can assure the uniform convergence of the products and series entering in (23.19), the formula itself follows as in the proof of Theorem I.30.19. Now, from Theorem 23.31 we deduce that $\left(\prod_{k=1}^{N} f_k(z)\right)_{N \in \mathbb{N}}$ converges locally uniformly to $f(z)$ and by Theorem 23.20 it follows also the locally uniform convergence of $\left(\left(\prod_{k=1}^{N} f_k(z)\right)'\right)_{N \in \mathbb{N}}$ to $f'(z)$. Let $K \subset D$ be a compact set, hence $\left(\prod_{k=1}^{N} f_k(z)\right)_{N \in \mathbb{N}}$ and $\left(\left(\prod_{k=1}^{N} f_k(z)\right)'\right)_{N \in \mathbb{N}}$ converge

441

uniformly on K and on K the sequence $\left(\prod_{k=1}^{N} f_k(z)\right)_{N \in \mathbb{N}}$ is uniformly boun-
ded away from 0, i.e. for $z \in K$ we have $\left|\prod_{k=1}^{N} f_k(z)\right| \geq \eta(K) > 0$ for all
$N \in \mathbb{N}$. From this we can deduce that

$$\frac{\left(\prod_{k=1}^{N} f_k(z)\right)'}{\prod_{k=1}^{N} f_k(z)} \to \frac{f'(z)}{f(z)} \quad \text{uniformly on } K,$$

hence for all $z \in D$. Now it remains to argue as in Theorem I.30.19 that

$$\frac{\left(\prod_{k=1}^{N} f_k(z)\right)'}{\prod_{k=1}^{N} f_k(z)} = \sum_{k=1}^{N} \frac{f'(z)}{f_k(z)}$$

and the theorem follows. $\qquad\square$

We will return to product representatives, especially to the Weierstrass representation of the Γ-function, once we have a better understanding of so
called meromorphic functions.

Problems

1. a) Give a counter example to the statement: if $f'(z) \neq 0$ for all
    $z \in \mathbb{C}$, then the holomorphic function f maps $\mathbb{C}$ biholomorphically
    onto its range.

    b) Let $g : B_R(0) \to \mathbb{C}$ be a holomorphic function such that $g(0) \neq 0$.
    Prove that the function $f : B_R(0) \to \mathbb{C}$, $f(z) = g(z) \sin z$, is locally
    biholomorphic at 0.

2. Revisit the proof of Theorem 23.2 and prove the following refined sta-
    tement: let $B_R(z_0) \subset G$ and $\delta := \frac{1}{2} \min_{z \in \partial B_R(z_0)} |f(z) - f(z_0)| > 0$.
    Then it follows that $B_R(f(z)) \subset f(B_R(z_0))$.

3. Give a counter example to the boundary maximum principle in the case
    where G is unbounded.
    Hint: you may consider on $G = \{z \in \mathbb{C} | \operatorname{Re} z > 0, \operatorname{Im} z > 0\}$ the
    function $z \mapsto f(z) := e^{iz^4}$.

4. Let $(f_k)_{k\in\mathbb{N}}$, $f_k : G \to \mathbb{C}$, be a sequence of holomorphic functions on the bounded region G with continuous extension to $\overline{G}$. Suppose that there exists a continuous function $f : \overline{G} \to \mathbb{C}$ such that $f_k(z) \to f(z)$ for all $z \in G$ as k tends to infinity and $\lim_{k\to\infty} ||f_k - f||_{\infty,\partial G} = 0$. Prove that $\lim_{k\to\infty} ||f_k - f||_{\infty,G} = 0$ Is $f|_G$ a holomorphic function?

5. Let $D \subset \mathbb{C}$ be a domain and $\overline{B_R(0)} \subset D$. On the set $\mathcal{F} := \{f|_{\overline{B_R(0)}} | f : D \to \mathbb{C}$ is holomorphic$\}$ define $||f||_{\mathcal{F}} := \max_{z\in\partial B_R(0)} |f(z)|$. Prove that $(\mathcal{F}, ||\cdot||_{\mathcal{F}})$ is a Banach space.

6. Let $f : B_R(0) \to \mathbb{C}$ be a holomorphic function and $M : [0, R) \to \mathbb{R}$ be defined by $M(r) := ||f||_{\infty,\partial B_r(0)}$. Prove that M is monotone increasing.

7. (This problem is taken from [67]). Let $G \subset \mathbb{C}$ be a relatively compact region and $f_k : G \to \mathbb{C}$, $k \in \mathbb{N}$, be a sequence of holomorphic functions converging locally uniform to $f : G \to \mathbb{C}$. Then for every $z_0 \in G$ we can find $N(z_0) \in \mathbb{N}$ and a sequence $(z_n)_n \geq N(z_0)$ such that $\lim_{n\to\infty} z_n = z_0$ and $f_n(z_n) = f(z_0)$ for all $n \geq N(z_0)$.
Hint: we may assume that $f(z_0) = 0$. Now deduce that f has no further zero in a neighbourhood of $z_0 = 0$ and use the boundary minimum principle.

8. Prove that the range of a non-constant polynomial is closed and now use Theorem 23.2 to deduce the fundamental theorem of algebra.

9. Does there exist a biholomorphic function $f : \mathbb{C} \to B_1(0)$?

10. Let $f : \mathbb{C} \to \mathbb{C}$ be a continuous function and $a, b \in \mathbb{C}$ be linearly independent over $\mathbb{R}$, i.e. a and b span a parallelogram in the plane. Suppose that for all $z \in \mathbb{C}$ we have $f(a+z) = f(z)$ and $f(b+z) = f(z)$. Prove that if f is holomorphic then f is constant.

11.   a) Prove that if $\prod_{k=1}^{\infty} c_k = c \neq 0$ then $\lim_{k\to\infty} c_k = 1$, $c_k \in \mathbb{C}$.

b) Formulate and prove the Cauchy criterion for infinte products $\prod_{k=1}^{\infty} c_k, c_k \in \mathbb{C}$.
Hint: follow the ideas of Proposition I.30.7.

12. For $\pi \cot \pi z := \pi \frac{\cos \pi z}{\sin \pi z}$ defined on $\{z \in \mathbb{C}| \sin \pi z \neq 0\}$ prove the product representation

$$\pi \cot \pi z = \frac{1}{z} + \sum_{k=1}^{\infty} \frac{2z}{z^2 - k^2}.$$

24 Meromorphic Functions

Rational functions $R = \frac{P}{Q}$, P and Q are polynomials, are defined on $\mathbb{C}\backslash\{z \in \mathbb{C} \,|\, Q(z) = 0\}$ and in this set they are holomorphic. Since Q has only a finite number of zeroes, in the case where Q is of degree m the set $\{z \in \mathbb{C} \,|\, Q(z) = 0\}$ has by the fundamental theorem of algebra at most m elements, the function R is holomorphic in $\mathbb{C}$ except at a finite number of points. For reasons which will become clear later we write

$$\mathrm{Sing}(R) := \{z \in \mathbb{C} \,|\, Q(z) = 0\}. \tag{24.1}$$

The fact that R is holomorphic in $\mathbb{C}\backslash \mathrm{Sing}(R)$ has immediate consequences for the Taylor expansion of R. Let $\zeta_0 \in \mathbb{C}\backslash \mathrm{Sing}(R)$, hence $\delta := \mathrm{dist}(\zeta_0, \mathrm{Sing}(R)) > 0$. In every disc $B_r(\zeta_0)$, $r < \delta$, the function R is holomorphic and admits a Taylor expansion about every point of $B_r(\zeta_0)$. However we cannot expect that a Taylor expansion of R about a point $\zeta \in B_r(\zeta_0)$ has radius of convergence greater than $\eta := \mathrm{dist}(\zeta, \mathrm{Sing}(R))$ otherwise R would be holomorphic in a neighbourhood of a point of $\mathrm{Sing}(R)$.

The question arises whether we can obtain a local and convergent series expansion of R in some set $A \subset \mathbb{C}\backslash\{z_0\}$ where $R|_A$ is holomorphic and $z_0 \in \mathrm{Sing}(R)$. Of course, this expansion in general cannot be a Taylor or power series. We may try to obtain a series expansion which also contains terms of negative powers in $(z - z_0)$, and certainly we shall expect such a series not to converge at z_0 nor to represent R at z_0 since R is not defined at z_0.

Definition 24.1. *For $0 \le r < R \le \infty$ the open **annulus** with centre $z_0 \in \mathbb{C}$ is defined by*

$$A_{r,R}(z_0) := \{z \in \mathbb{C} \,|\, r < |z - z_0| < R\}, \tag{24.2}$$

*see Figure 24.1. For $r = 0$ we obtain the **punctured disc***

$$A_{0,R}(z_0) = B_R(z_0)\backslash\{z_0\}, \tag{24.3}$$

see Figure 24.2

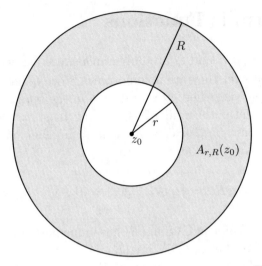

$$A_{r,R}(z_0)$$

Figure 24.1

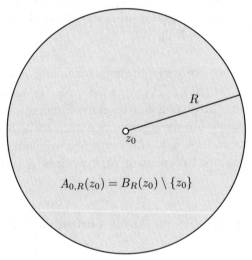

$$A_{0,R}(z_0) = B_R(z_0) \setminus \{z_0\}$$

Figure 24.2

Obviously we have

$$A_{r,R}(z_0) = B_R(z_0) \setminus \overline{B_r(z_0)} \tag{24.4}$$

and

$$A_{r,\infty}(z_0) = B_r(z_0)^{\complement}. \tag{24.5}$$

446

There are many holomorphic functions on $A_{r,R}(z_0)$ which are not holomorphic on $B_r(z_0)$. Just look at $z \mapsto \frac{1}{(z-z_0)^k}$, $k \in \mathbb{N}$, or even at $z \mapsto \frac{1}{(z-z_1)^k}$, $z_1 \in B_r(z_0)$, and now take linear combinations of such functions, e.g. $z \mapsto \sum_{k=1}^{N} \frac{c_k}{(z-z_k)^k}$, $z_k \in B_r(z_0)$. A further example, not being a rational function is $z \mapsto \frac{1}{\sin z}$, $z_0 = 0$, $r = \frac{1}{2}$ and $R = \pi$.

In general we shall not expect that a function holomorphic in $A_{r,R}(z_0)$ has a holomorphic extension to $B_R(z_0)$. However in the case $A_{0,R}(z_0)$ we can rely on Theorem 22.7 and we know that if $h : A_{0,R}(z_0) \to \mathbb{C}$ is a holomorphic function bounded on some set $B_\rho(z_0)\backslash\{z_0\}$, $0 < \rho < R$, then h admits a holomorphic extension to $B_R(z_0)$. The next Theorem 24.2 shows that every holomorphic function $f : A_{r,R}(z_0) \to \mathbb{C}$ has a decomposition $f = f_1 + f_2$ where f_1 is holomorphic in $A_{r,\infty}(z_0)$ and f_2 is holomorphic in $B_R(z_0)$, see Figure 24.3.

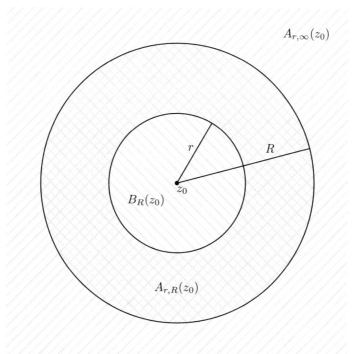

Figure 24.3

For its proof we need the following considerations: let $G \subset \mathbb{C}$ be a region and $h : G \to \mathbb{C}$ a holomorphic function. For $z_0 \in \mathbb{C}$, note that z_0 need not belong to G, and $0 < r_2 < r_1$ assume that $\overline{A_{r_2,r_1}(z_0)} \subset G$, see Figure 24.4.

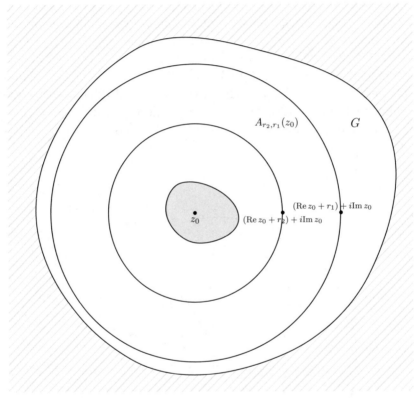

Figure 24.4

Let $\gamma_1(\varphi) = z_0 + r_1 e^{i\varphi}$ and $\gamma_2(\varphi) = z_0 + r_2 e^{i\varphi}$, $\varphi \in [0, 2\pi]$, the standard parametrization of $\partial B_{r_1}(z_0)$ and $\partial B_{r_2}(z_0)$, respectively. For $\epsilon \in \left(-\frac{\pi}{8}, \frac{\pi}{8} \right)$ let $z_{1,\pm\epsilon} = r_1 e^{\pm i\epsilon}$ and $z_{2,\pm\epsilon} = r_2 e^{\pm i\epsilon}$. By $\gamma_{j,\epsilon}$, $j = 1, 2$, we denote the curves $\gamma_{j,\epsilon} : [\epsilon, 2\pi - \epsilon] \to \mathbb{C}$, $\gamma_{j,\epsilon}(\varphi) = \gamma_j(\varphi)$, $\varphi \in [\epsilon, 2\pi - \epsilon]$. Further we denote by $\sigma_{\pm\epsilon}$ the line segments connecting $z_{1,\epsilon}$ with $z_{2,\epsilon}$ and $z_{1,-\epsilon}$ with $z_{2,-\epsilon}$, respectively, see Figure 24.5.

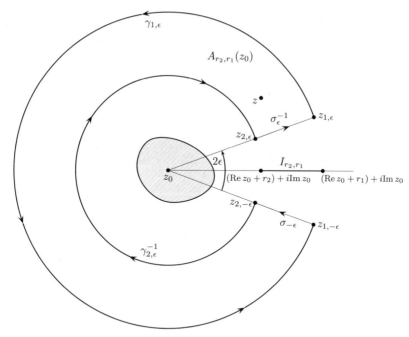

Figure 24.5

Now we consider the curve

$$\gamma_\epsilon := \gamma_{1,\epsilon} \oplus \sigma_{-\epsilon} \oplus \gamma_{2,\epsilon}^{-1} \oplus \sigma_\epsilon^{-1}, \tag{24.6}$$

compare again with Figure 24.5. The curve γ_ϵ is a simply closed, piecewise continuously differentiable curve the trace of which belongs entirely to a domain on which h is holomorphic and it is the boundary of a set $A_{r_2,r_1,\epsilon}(z_0) \subset A_{r_2,r_1}(z_0)$. Therefore the Cauchy integral formula implies for $z \in A_{r_2,r_1,\epsilon}(z_0)$

$$h(z) = \frac{1}{2\pi i} \int_\gamma \frac{h(\zeta)}{\zeta - z}\, d\zeta \tag{24.7}$$

$$= \frac{1}{2\pi i} \left(\int_{\gamma_{1,\epsilon}} \frac{h(\zeta)}{\zeta - z}\, d\zeta + \int_{\sigma_{-\epsilon}} \frac{h(\zeta)}{\zeta - z}\, d\zeta + \int_{\gamma_{2,\epsilon}^{-1}} \frac{h(\zeta)}{\zeta - z}\, d\zeta + \int_{\sigma_\epsilon^{-1}} \frac{h(\zeta)}{\zeta - z}\, d\zeta \right).$$

We introduce, see Figure 24.5,

$$I_{r_1,r_2} := \left\{ w \in \mathbb{C} \,\middle|\, \mathrm{Re}\, z_0 + r_2 \le \mathrm{Re}\, w \le \mathrm{Re}\, z_0 + r_1,\ \mathrm{Im}\, w = \mathrm{Im}\, z_0 \right\}$$

449

and note that for every $z \in A_{r_2,r_1}(z_0)\backslash I_{r_1,r_2}$ there exists $\epsilon_0 = \epsilon_0(z) > 0$ such that $z \in A_{r_2,r_1,\epsilon}(z_0)$ for all $\epsilon < \epsilon_0$. Therefore it follows for all $z \in A_{r_2,r_1,\epsilon}(z_0)\backslash I_{r_1,r_2}$, $\epsilon < \epsilon_0$, that

$$\lim_{\epsilon \to 0} \int_{\gamma_{1,\epsilon}} \frac{h(\zeta)}{\zeta - z}\, d\zeta = \int_{\gamma_1} \frac{h(\zeta)}{\zeta - z}\, d\zeta \qquad (24.8)$$

and

$$\lim_{\epsilon \to 0} \int_{\gamma_{2,\epsilon}^{-1}} \frac{h(\zeta)}{\zeta - z}\, d\zeta = \int_{\gamma_2^{-1}} \frac{h(\zeta)}{\zeta - z}\, d\zeta = -\int_{\gamma_2} \frac{h(\zeta)}{\zeta - z}\, d\zeta. \qquad (24.9)$$

We also claim for these values of z that

$$\lim_{\epsilon \to 0} \left(\int_{\sigma_{-\epsilon}} \frac{h(\zeta)}{\zeta - z}\, d\zeta + \int_{\sigma_\epsilon^{-1}} \frac{h(\zeta)}{\zeta - z}\, d\zeta \right) = 0. \qquad (24.10)$$

In order to prove (24.10) we introduce the curves $\eta_{j,\epsilon} : [-\epsilon, \epsilon] \to \mathbb{C}$, $\eta_{j,\epsilon}(\varphi) = r_j e^{i\varphi}$, $\varphi \in [-\epsilon, \epsilon]$, see Figure 24.6, and consider the simply closed, piecewise continuously differentiable curve

$$\eta_\epsilon := \sigma_\epsilon^{-1} \oplus \eta_{1,\epsilon}^{-1} \oplus \sigma_{-\epsilon} \oplus \eta_{2,\epsilon}.$$

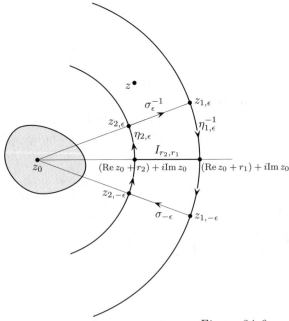

Figure 24.6

For $z \in A_{r_2,r_1}(z_0) \backslash I_{r_1,r_2}$ and $\epsilon < \epsilon_0(z)$ we now deduce

$$0 = \int_{\eta_\epsilon} \frac{h(\zeta)}{\zeta - z} \, d\zeta$$

and consequently

$$\int_{\sigma_{-\epsilon}} \frac{h(\zeta)}{\zeta - z} \, d\zeta + \int_{\sigma_\epsilon^{-1}} \frac{h(\zeta)}{\zeta - z} \, d\zeta = \int_{\eta_{1,\epsilon}} \frac{h(\zeta)}{\zeta - z} \, d\zeta - \int_{\eta_{2,\epsilon}} \frac{h(\zeta)}{\zeta - z} \, d\zeta,$$

or

$$\left| \int_{\sigma_{-\epsilon}} \frac{h(\zeta)}{\zeta - z} \, d\zeta + \int_{\sigma_\epsilon^{-1}} \frac{h(\zeta)}{\zeta - z} \, d\zeta \right| \leq M(\epsilon_0) \left(l_{\eta_{1,\epsilon}} + l_{\eta_{2,\epsilon}} \right),$$

where we used that for $z \in A_{r_2,r_1,\epsilon_0}(z_0)$ and $\zeta \in \mathrm{tr}(\eta_\epsilon)$ the term $\left| \frac{h(\zeta)}{\zeta - z} \right|$ is bounded and the bound is independent of ϵ, but may depend on ϵ_0. Since for $\epsilon \to 0$ both, $l_{\eta_{1,\epsilon}}$ and $l_{\eta_{2,\epsilon}}$ tend to zero, we arrive at

$$h(z) = \frac{1}{2\pi i} \int_{\gamma_1} \frac{h(\zeta)}{\zeta - z} \, d\zeta - \frac{1}{2\pi i} \int_{\gamma_2} \frac{h(\zeta)}{\zeta - z} \, d\zeta \qquad (24.11)$$

for all $z \in A_{r_2,r_1}(z_0) \backslash I_{r_1,r_2}$. By an analogous argument we obtain (24.11) for all $z \in A_{r_2,r_1}(z_0) \backslash \tilde{I}_{r_1,r_2}$ where

$$\tilde{I}_{r_1,r_2} := \left\{ w \in \mathbb{C} \mid -\mathrm{Re}\, z_0 - r_1 \leq \mathrm{Re}\, w \leq -\mathrm{Re}\, z_0 - r_2, \, \mathrm{Im}\, w = \mathrm{Im}\, z_0 \right\},$$

and hence we have proved (24.11) to hold for all $z \in A_{r_2,r_1}(z_0)$.

In Chapter 25 we will have a closer look at this type of result and argument after we have developed the concept of null-homologous cycles and related integrals.

Now we prove

Theorem 24.2. *For a holomorphic function $f : A_{r,R}(z_0) \to \mathbb{C}$ there exists a holomorphic function $f_1 : A_{r,\infty}(z_0) \to \mathbb{C}$ and a holomorphic function $f_2 : B_R(z_0) \to \mathbb{C}$ such that on $A_{r,R}(z_0)$ we have*

$$f = f_1 + f_2. \qquad (24.12)$$

If in addition we require $\lim_{|z| \to \infty} |f_1(z)| = 0$ then the decomposition (24.12) is unique.

Proof. We start by defining f_2. For $z \in B_\rho(z_0)$, $r < \rho < R$, we set

$$f_{2,\rho}(z) := \frac{1}{2\pi i} \int_{|\zeta - z_0| = \rho} \frac{f(\zeta)}{\zeta - z} \, d\zeta \qquad (24.13)$$

which is a holomorphic function on $B_\rho(z_0)$. The Cauchy integral yields that for $r < \rho_1 < \rho_2 < R$ we have on $B_{\rho_1}(z_0)$ the equality $f_{2,\rho_1} = f_{2,\rho_2}$. It follows that a holomorphic function f_2 is defined on $B_R(z_0)$ by

$$f_2(z) := \frac{1}{2\pi i} \int_{|\zeta - z_0| = \rho} \frac{f(\zeta)}{\zeta - z} \, d\zeta \qquad (24.14)$$

provided $\max\{r, |z - z_0|\} < \rho < R$. Furthermore, for $r < \sigma < \min\{R, |z - z_0|\}$ a holomorphic function f_1 is defined on $A_{r,\infty}(z_0)$ by

$$f_1(z) := -\frac{1}{2\pi i} \int_{|\zeta - z_0| = \sigma} \frac{f(\zeta)}{\zeta - z} \, d\zeta. \qquad (24.15)$$

From the Cauchy inequalities, Theorem 21.15, we derive that

$$\lim_{|z| \to \infty} |f_1(z)| = 0, \qquad (24.16)$$

note that for $|z| \to \infty$ we can choose a sequence $(\sigma_l)_{l \in \mathbb{N}}$ with $\lim_{l \to \infty} \sigma_l = \sigma$. We are now in a situation where we can apply (24.11) to f, γ_1 is the circle $|z - z_0| = \rho$ and γ_2 is the circle $|z - z_0| = \sigma$, $r < \sigma < |z - z_0| < \rho < R$. Thus we obtain

$$f(z) = \frac{1}{2\pi i} \int_{|\zeta - z_0| = \rho} \frac{f(\zeta)}{\zeta - z} \, d\zeta - \frac{1}{2\pi i} \int_{|\zeta - z_0| = \sigma} \frac{f(\zeta)}{\zeta - z} \, d\zeta = f_2(z) + f_1(z),$$

i.e. the decomposition of f is established. It remains to prove the uniqueness result. Suppose that $f = g_1 + g_2$ is a second decomposition of f with $\lim_{|z| \to \infty} |g_1(z)| = 0$. Then we find that $f_1 - g_1 = g_2 - f_2$ on $A_{r,R}(z_0)$. Thus

$$h(z) := \begin{cases} f_1 - g_1 & \text{on } A_{r,\infty}(z_0), \\ f_2 - g_2 & \text{on } B_R(z_0) \end{cases}$$

is an entire function vanishing at infinity, i.e. $\lim_{|z| \to \infty} |h(z)| = 0$. Hence h is bounded and by Louiville's theorem h must be constant and therefore identically zero. $\qquad \Box$

Let $f : A_{r,R}(z_0) \to \mathbb{C}$ be as in Theorem 24.2 admitting the decomposition $f = f_1 + f_2$. The function f_2 has in $B_R(z_0)$ an expansion into a Taylor series representing f_2, i.e.

$$f_2(z) = \sum_{k=0}^{\infty} c_k(z - z_0)^k \tag{24.17}$$

where

$$c_k = \frac{f^{(k)}(z_0)}{k!} = \frac{1}{2\pi i} \int_{|\zeta - z_0| = \rho} \frac{f(\zeta)}{(\zeta - z_0)^{k+1}} \, d\zeta, \quad r < \rho < R, \ k \in \mathbb{N}_0. \tag{24.18}$$

In order to handle f_1 we consider the biholomorphic function $h : B_{\frac{1}{r}}(0) \backslash \{0\} \to \mathbb{C} \backslash \overline{B_r(z_0)}$, $h(w) = z_0 + \frac{1}{w}$. We leave it as an exercise to prove that h is indeed a biholomorphic function, see Problem 1. Note that $B_{\frac{1}{r}}(0) \backslash \{0\} = A_{0, \frac{1}{r}}(0)$ and $\mathbb{C} \backslash \overline{B_r(z_0)} = A_{r,\infty}(z_0)$, so h maps biholomorphically $A_{0, \frac{1}{r}}(0)$ onto $A_{r,\infty}(z_0)$. Since f_1 has the property that $\lim_{|z| \to \infty} f_1(z) = 0$ we conclude that for $f_1 \circ h$ it follows that $\lim_{w \to 0}(f_1 \circ h)(w) = 0$. Therefore, by Theorem 22.7 we may extend $f_1 \circ h$ to a holomorphic function on $B_{\frac{1}{r}}(0)$ which admits in $B_{\frac{1}{r}}(0)$ a Taylor representation

$$(f_1 \circ h)(w) = \sum_{l=1}^{\infty} \gamma_l w^l, \tag{24.19}$$

where we used that $\lim_{w \to 0}(f_1 \circ h)(w) = 0$ implies $\gamma_0 = 0$. The convergence of the series in (24.19) is uniform and absolute on every disc $\overline{B_{\frac{1}{\rho}}(0)}$, $\rho > r$. Now we recall that for $z = h(w)$ it follows that $w = \frac{1}{z - z_0}$ and we can rewrite (24.19) as

$$f_1(z) = \sum_{l=1}^{\infty} \gamma_l(z - z_0)^{-l} = \sum_{l=1}^{\infty} \gamma_l \frac{1}{(z - z_0)^l}, \tag{24.20}$$

and due to the mapping properties of h we know that this series converges in $\mathbb{C} \backslash \overline{B_\rho(z_0)}$ uniformly for every $\rho > r$. We define

$$c_{-k} := \gamma_k, \quad k \in \mathbb{N} \tag{24.21}$$

and we arrive at

$$f_1(z) = \sum_{k=-1}^{-\infty} c_k(z - z_0)^k. \tag{24.22}$$

453

Combining (24.17) with (24.21) we find for $z \in A_{r,R}(z_0)$

$$f(z) = f_1(z) + f_2(z) = \sum_{k=-1}^{-\infty} c_k(z - z_0)^k + \sum_{k=0}^{\infty} c_k(z - z_0)^k, \qquad (24.23)$$

where the first series converges locally uniformly in $A_{r,\infty}(z_0)$ and the second in $B_R(z_0)$, i.e. both converge locally uniformly in $A_{r,R}(z_0)$ and the sum of both series in $A_{r,R}(z_0)$ is f. We claim further that (24.18) also holds for $k \in -\mathbb{N}$. Indeed, for $r < \rho < R$ and $k \in \mathbb{Z}$ we find

$$(z - z_0)^{-k-1} f(z) = \sum_{l=-1}^{\infty} c_{l+k+1}(z - z_0)^l + \sum_{l=0}^{\infty} c_{l+k+1}(z - z_0)^l$$

and these series converge uniformly on $|z - z_0| = \rho$. Thus we may integrate term by term which yields

$$\int_{|z-z_0|=\rho} \frac{f(\zeta)}{(z - z_0)^{k+1}} \, d\zeta = c_k \int_{|z-z_0|=\rho} \frac{dz}{z - z_0} = 2\pi i c_k, \qquad (24.24)$$

as usual in (24.24) we used the standard parametrization of the circle $|z - z_0| = \rho$. Thus we have proved

Theorem 24.3. *A holomorphic function $f : A_{r,R}(z_0) \to \mathbb{C}$, $0 \leq r < R \leq \infty$, admits the representation (24.23) with coefficients given by*

$$c_k = \frac{1}{2\pi i} \int_{|\zeta-z_0|=\rho} \frac{f(\zeta)}{(\zeta - z_0)^{k+1}} \, d\zeta, \quad r < \rho < R. \qquad (24.25)$$

Furthermore we can differentiate the series in (24.23) term by term.

Definition 24.4. *A*. *A series*

$$\sum_{k=-\infty}^{\infty} c_k(z - z_0)^k = \sum_{k=-1}^{-\infty} c_k(z - z_0)^k + \sum_{k=0}^{\infty} c_k(z - z_0)^k, \qquad (24.26)$$

*is called a **Laurent series** about $z_0 \in \mathbb{C}$. It converges at $z \in \mathbb{C}$ if both series in (24.26) converge for z.*
***B**. The first series in (24.26) is called the **principal part** of the series (24.26).*

Remark 24.5. A. Now we can read Theorem 24.3 as follows: a holomorphic function $f : A_{r,R}(z_0) \to \mathbb{C}$ admits a locally uniform convergent Laurent series expansion in $A_{r,R}(z_0)$ and the principal part of this expansion is given by the expansion of f_1 according to (24.22). Hence we can also call f_1 the principal part of f.

Consider the series (24.26) and assume that it converges locally uniformly in some annulus $A_{r,R}(z_0)$. For every $k \in \mathbb{Z}\backslash\{-1\}$ the function $z \mapsto (z - z_0)^k$ has a primitive in $A_{r,R}(z_0)$, a fact which we indirectly used to derive (24.25). This gives the coefficient c_{-1} a special role. Let $\gamma : [a, b] \to A_{r,R}(z_0)$ be a simply closed, piecewise continuously differentiable curve which is anticlockwise oriented. We assume in addition that the set bounded by $\mathrm{tr}(\gamma) \subset A_{r,R}(z_0)$ contains $B_r(z_0)$, see Figure 24.7. Note that due to the Jordan curve theorem, Theorem 19.37, such an assumption is justified.

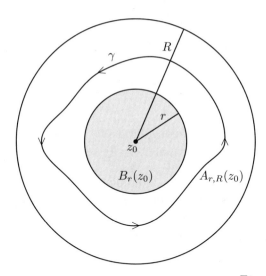

Figure 24.7

Since $\mathrm{tr}(\gamma)$ is compact the locally uniform convergence of (24.25) in $A_{r,R}(z_0)$ allows us to interchange integration and summation and we find for such a curve that

$$\int_\gamma \sum_{k=-\infty}^\infty c_k(z - z_0)^k \, dz = \int_\gamma c_{-1}(z - z_0)^{-1} \, dz = 2\pi i c_{-1}. \qquad (24.27)$$

In particular if $f : A_{r,R}(z_0) \to \mathbb{C}$ is holomorphic with Laurent series (24.23) then we have for every curve as described above

$$c_{-1} = \frac{1}{2\pi i} \int_\gamma f(z)\, dz = \frac{1}{2\pi i} \int_\gamma f_1(z)\, dz. \qquad (24.28)$$

Of course, for (24.27) or (24.28) it is important that z_0 lies in the set bounded by γ.

Definition 24.6. *Given the punctured disc $A_{0,R}(z_0) \subset \mathbb{C}$ and a holomorphic function $f : A_{0,R}(z_0) \to \mathbb{C}$. The number*

$$\mathrm{res}(f, z_0) := \frac{1}{2\pi i} \int_{|\zeta - z_0| = \rho} f(\zeta)\, d\zeta, \quad 0 < \rho < R, \qquad (24.29)$$

is called the **residue** *of f at z_0. Hence $\mathrm{res}(f, z_0)$ is the first coefficient of the principal part of the Laurent expansion of f in $A_{0,R}(z_0)$.*

Remark 24.7. From the considerations preceding Definition 24.6 it is clear that $\mathrm{res}(f, z_0)$ is independent of the choice of ρ. In fact we may choose instead $|\zeta - z_0| = \rho$ a simply closed piecewise continuously differentiable curve γ with $\mathrm{tr}(\gamma) \subset A_{0,R}(z_0)$ and z_0 in the interior of γ.

Before we continue with our theoretical considerations we want to look at some examples.

Example 24.8. A. The function $z \mapsto \frac{1}{z}$ is holomorphic in $\mathbb{C} \backslash \{0\}$ and its Laurent series about $z_0 = 0$ is just the function itself, the same applies to $z \mapsto \frac{1}{z^l}, l \in \mathbb{N}$.
B. Let $h : \mathbb{C} \to \mathbb{C}$ be an entire function with Taylor expansion

$$h(z) = \sum_{k=0}^\infty c_k (z - z_0)^k$$

about $z_0 \in \mathbb{C}$. The function $f(z) := h\left(\frac{1}{z - z_0}\right)$ is holomorphic in $A_{0,\infty}(z_0)$ and its Laurent expansion about z_0 is given by

$$f(z) = \sum_{k=1}^\infty c_k (z - z_0)^{-k} + c_0.$$

In particular we find for the Laurent expansion of $e^{\frac{1}{z}}$, $z \neq 0$,

$$e^{\frac{1}{z}} = \sum_{k=1}^{\infty} \frac{1}{k!} z^{-k} + 1.$$

C. Let us consider once more the function $z \mapsto \frac{1}{z}$ but this time in the annulus $A_{1,2}\left(\frac{1}{2}\right)$. Clearly the function is holomorphic in $A_{1,2}\left(\frac{1}{2}\right)$ and we want to find its Laurent series about $\frac{1}{2}$. For this we write

$$\frac{1}{z} = \frac{1}{z - \frac{1}{2} + \frac{1}{2}} = \frac{2}{1 + 2\left(z - \frac{1}{2}\right)}$$

$$= 2\sum_{k=0}^{\infty} (-2)^k \left(z - \frac{1}{2}\right)^k$$

which converges for $\left|z - \frac{1}{2}\right| < \frac{1}{2}$. Hence the Laurent expansion of $\frac{1}{z}$ about $\frac{1}{2}$ is a Taylor expansion and the principal part vanishes. However the residue $\mathrm{res}\left(\frac{1}{z}\right)(0) = 1$ and not equal to 0. For determining the residue of a function f at z_0 we must take its Laurent expansion about z_0, not about a different point.

D. In this example, essentially taken from [26] we want to calculate the Laurent expansion for $h : \mathbb{C}\backslash\{0, i\} \to \mathbb{C}$, $h(z) = \frac{1}{z(z-i)^2}$, for three different values of z_0 and corresponding annuli: $A_{0,1}(0)$, $A_{1,\infty}(1)$ and $A_{0,1}(i)$. In each case a reduction to a geometric or related series is essential, and in fact this is a major tool to find Taylor as well as Laurent expansions.

i) In $A_{0,1}(0)$ we find

$$\frac{1}{z(z-i)^2} = -\frac{1}{z} \frac{1}{\left(1 - \frac{z}{i}\right)^2} = -\frac{1}{z}\sum_{k=0}^{\infty}(k+1)\left(\frac{z}{i}\right)^k$$

$$= -\frac{1}{z} + i\sum_{k=0}^{\infty}(k+2)\left(\frac{z}{i}\right)^k = -\frac{1}{z} + \sum_{k=0}^{\infty}\frac{k+2}{i^{k-1}}z^k.$$

ii) In $A_{1,\infty}(1)$ we have

$$\frac{1}{z(z-i)^2} = \frac{1}{z^3}\frac{1}{\left(1 - \frac{i}{z}\right)^2} = \sum_{k=-3}^{-\infty} -i^{-k-1}(k+2)z^k = \sum_{k=-3}^{-\infty}\frac{k+2}{i^{k+2}}z^k.$$

iii) In $A_{0,1}(i)$ we obtain

$$\frac{1}{z(z-i)^2} = \frac{-i}{(z-i)^2} + \frac{1}{z-i} - \frac{1}{z} = \frac{-i}{(z-i)^2} + \frac{1}{z-i} + \frac{i}{1-i(z-1)}$$

$$= \frac{-i}{(z-i)^2} + \frac{1}{z-i} + \sum_{k=0}^{\infty} i(i(z-i))^k$$

$$= \frac{-i}{(z-i)^2} + \frac{1}{z-i} + \sum_{k=0}^{\infty} i^{k+1}(z-i)^k.$$

Proposition 24.9. A. Let $\sum_{k=-\infty}^{\infty} a_k(z-z_0)^k$ and $\sum_{k=-\infty}^{\infty} c_k(z-z_0)^k$ be two Laurent series representing on some annulus $A_{r,R}(z_0)$ a holomorphic function h, then $a_k = c_k$ for all $k \in \mathbb{Z}$.
B. Suppose that $f(z) = \sum_{k=-\infty}^{\infty} c_k(z-z_0)^k$, $z \in A_{r,R}(z_0)$ and let $r < \rho < R$. It follows for all $k \in \mathbb{Z}$ that

$$|c_k| \leq \rho^{-k} \sup_{|z-z_0|=\rho} |f(z)|. \tag{24.30}$$

Proof. **A.** Fix $k_0 \in \mathbb{Z}$ and note that

$$(z-z_0)^{-k_0-1} \sum_{k=-\infty}^{\infty} a_k(z-z_0)^k = (z-z_0)^{-k_0-1} \sum_{k=-\infty}^{\infty} c_k(z-z_0)^k.$$

Integrating this equality over the circle $|z-z_0| = \rho$, $r < \rho < R$, yields $a_{k_0} = c_{k_0}$.
B. The coefficients c_k are given by (24.25) and it follows

$$|c_k| \leq \frac{1}{2\pi} \int_{|z-z_0|=\rho} \frac{|f(z)|}{\rho^{k+1}} \, d\zeta \leq \frac{1}{2\pi\rho^{k+1}} \sup_{|z-z_0|=\rho} |f(z)| \int_{|z-z_0|=\rho} 1 \, d\zeta,$$

i.e. $|c_k| \leq \rho^{-k} \sup_{|z-z_0|=\rho} |f(z)|$. $\qquad\square$

Let $f : A_{0,R}(z_0) \to \mathbb{C}$ be a holomorphic function with a non-trivial principal part. In this case f cannot be bounded in a neighbourhood of z_0 and z_0 cannot be a removable singularity of f, compare with Theorem 22.7. We now want to study certain singularities of holomorphic functions.

Definition 24.10. Let $z_0 \in \mathbb{C}$ and $U \subset \mathbb{C}$ a neighbourhood of z_0. Further let $f : U \backslash \{z_0\} \to \mathbb{C}$ be a holomorphic function. We call z_0 an **isolated singularity** of f.

Note that Definition 24.10 also covers the case where f is bounded in a neighbourhood of z_0 and hence admits a holomorphic extension $\tilde{f}$ to U by Theorem 22.7. Clearly $\tilde{f}$ will be holomorphic in U, i.e. will have no isolated singularity. In such a case having an isolated singularity is mainly a problem of the domain and not a property of the function.

We can extend Definition 24.10 in the following sense: if $M \subset D$ is a discrete set in a domain $D \subset \mathbb{C}$ and $f : D\backslash M \to \mathbb{C}$ is a holomorphic function, then we call the points of M the isolated singularities of f.

Here are some examples of holomorphic functions with isolated singularities.

Example 24.11. A. For $z_0 \in \mathbb{C}$ and $k \in \mathbb{N}$ the function $f(z) = \frac{1}{(z-z_0)^k}$ has an isolated singularity at z_0 and is holomorphic in $\mathbb{C}\backslash\{z_0\}$.
B. The rational function $R = \frac{P}{Q}$ has isolated singularities for all points belonging to $\mathrm{Sing}(R)$ as defined by (24.1).
C. The function $z \mapsto e^{\frac{a}{z^k}}$, $a \neq 0$, $k \in \mathbb{N}$, has an isolated singularity at $z_0 = 0$, otherwise it is holomorphic on $\mathbb{C}\backslash\{0\}$.
D. The function $z \mapsto \frac{\sin z}{z}$ has an isolated singularity at $z_0 = 0$ as does the function $\frac{1-\cos z}{z^2}$.
E. The set $\mathbb{Z}$ is the set of isolated singularities of the function $\cot : \mathbb{C}\backslash\mathbb{Z} \to \mathbb{C}$, $\cot \pi z = \frac{\cos \pi z}{\sin \pi z}$.

The natural programme to follow is to classify isolated singularities and to study the behaviour of a holomorphic function in a neighbourhood of an isolated singularity. Of course, Theorem 22.7 on removable singularities is such a result. In our discussion of singularities we follow partly [26].

Definition 24.12. *Let z_0 be an isolated singularity of the holomorphic function $f : U\backslash\{z_0\} \to \mathbb{C}$ where $U \subset \mathbb{C}$ is a neighbourhood of z_0.*
A. *If there exists a neighbourhood $V \subset U$ of z_0 such that $f|_{V\backslash\{z_0\}}$ is bounded then we call z_0 a **removable singularity** of f.*
B. *Suppose that $\lim_{z \to z_0} |f(z)| = \infty$, then we call z_0 a **pole** of f.*
C. *In the case that $z_0 \in \mathbb{C}$ is an isolated singularity of the holomorphic function $f : U\backslash\{z_0\} \to \mathbb{C}$ which is neither removable nor a pole, then we call z_0 an **essential singularity** of f.*

We do already understand removable singularities. Now we turn to poles. Let $f : U\backslash\{z_0\} \to \mathbb{C}$ be a holomorphic function where U is a neighbourhood of z_0 and assume that z_0 is a pole of f. Since in this case $\lim_{z \to z_0} |f(z)| = \infty$, we can find a neighbourhood V of z_0 such that $|f(z)| \geq 1$ for $z \in$

$V\backslash\{z_0\}$. In particular $f|_{V\backslash\{z_0\}}$ does not vanish implying that $z \mapsto \frac{1}{f(z)}$ is a holomorphic function on $V\backslash\{z_0\}$ and $\lim_{z\to z_0} \frac{1}{f(z)} = 0$. Thus $\frac{1}{f}$ is bounded in a neighbourhood of z_0 and hence z_0 is a removable singularity of $\frac{1}{f} : V\backslash\{z_0\} \to$ $\mathbb{C}$ and we can extend $\frac{1}{f}$ to V by defining $\left(\frac{1}{f}\right)(z_0) = 0$. By Theorem 22.7 this extension is holomorphic. Thus there exists a holomorphic function $g : V \to \mathbb{C}$ and $n \in \mathbb{N}$ such that

$$\frac{1}{f(z)} = (z - z_0)^n g(z) \tag{24.31}$$

holds and g does not vanish on V. Therefore we can find a neighbourhood $W \subset U$ of z_0 and a holomorphic function $h(z) = \frac{1}{g(z)}$ not vanishing at z_0 such that in $W\backslash\{z_0\}$ we have

$$f(z) = (z - z_0)^{-n} h(z). \tag{24.32}$$

The number $n \in \mathbb{N}$ in (24.32) is called the **order** or the **multiplicity** of the pole z_0: since $|f(z)| = |z - z_0|^{-n}|h(z)|$ and h does not vanish in $W\backslash\{z_0\}$ it follows that $\lim_{z\to z_0} |f(z)| = \infty$, i.e. if (24.32) holds, then f has a pole (of order n) at z_0.

Corollary 24.13. A. *Let $g : U \to \mathbb{C}$ be a holomorphic function and suppose that its set of zeroes $N(f) := \{z \in U \,|\, g(z) = 0\}$ is a discrete set. In this case the function $\frac{1}{g} : U\backslash N \to \mathbb{C}$ is holomorphic and has poles at N. The order of a pole $z_0 \in N$ of g equals the order of the zero z_0 of f.*
B. *Let $R = \frac{P}{Q}$ be a rational function and $\mathrm{Sing}(R)$ defined as before, i.e.* $\mathrm{Sing}(R) = \{w \in \mathbb{C} \,|\, Q(w) = 0\}$. *Then $\mathrm{Sing}(R)$ is a subset of the poles of R.*

Part B of Corollary 24.13 stimulates a closer study of the isolated singularities of rational functions. Suppose that $R = \frac{P}{Q}$ and P has order n, Q has order m. According to the fundamental theorem of algebra we have the factorizations

$$p(z) = c\prod_{j=1}^{N}(z - z_j)^{\alpha_j} \quad \text{and} \quad Q(z) = d\prod_{l=1}^{M}(z - w_l)^{\beta_l}$$

where z_j, $1 \leq j \leq N$, are the zeroes of P with multiplicity α_j, and w_l, $1 \leq l \leq M$, are the zeroes of Q with multiplicity β_l. Hence $\sum_{j=1}^{N} \alpha_j = n$ and $\sum_{l=1}^{M} \beta_l = m$. Thus we can write R on $\mathbb{C}\backslash\{w_1, \ldots, w_n\}$ as

$$R(z) = cd^{-1}(z - z_1)^{\alpha_1} \cdot \ldots \cdot (z - z_N)^{\alpha_N}(z - w_1)^{-\beta_1} \cdot \ldots \cdot (z - w_M)^{-\beta_M}.$$

In the case where $\{z_1, \ldots, z_N\} \cap \{w_1, \ldots, w_M\} = \emptyset$, the function R has poles of order β_l at w_l. However, whenever $z_{j_0} = w_{l_0}$ for some j_0 and l_0 we have to have a closer look at

$$(z - z_{j_0})^{\alpha_{j_0}} (z - w_{l_0})^{-\beta_{l_0}} = (z - w_{l_0})^{\alpha_{j_0} - \beta_{l_0}}.$$

If $\alpha_{j_0} - \beta_{l_0} \geq 0$, then R has a removable singularity at w_{l_0}, whereas if $\alpha_{j_0} - \beta_{l_0} < 0$ then R has a pole of order $\beta_{l_0} - \alpha_{j_0}$ at w_{l_0}. Hence for a rational function we know that all singularities are isolated singularities or removable singularities. They are located at the zeroes of $Q(z)$ and their type is determined by the relations of the zeroes of $Q(z)$ to the zeroes of $P(z)$:

If $Q(z)$ and $P(z)$ have no common zeroes, then R has at every zero of $Q(z)$ a pole the order of which is the multiplicity of the corresponding zero of $Q(z)$. For a common zero z_0 of $P(z)$ and $Q(z)$ the function R has a removable singularity of the order of z_0 as zero of $P(z)$ is larger or equal to the order of z_0 as zero of $Q(z)$. In the other case R has a pole of order equal to the absolute value of the difference of the multiplicities.

We can characterise a pole in terms of estimates.

Theorem 24.14. *Let $f : U \backslash \{z_0\} \to \mathbb{C}$ be a holomorphic function, where U is a neighbourhood of z_0. The isolated singularity z_0 of f is a pole of order n if and only if in some neighbourhood $V \subset U$ of z_0 the estimates*

$$\kappa_0 |z - z_0|^{-n} \leq |f(z)| \leq \kappa_1 |z - z_0|^{-n} \tag{24.33}$$

hold for constants $\kappa_0, \kappa_1 > 0$.

Proof. Suppose that z_0 is a pole of order n. By (24.32) we can find $\kappa_0, \kappa_1 > 0$ such that $0 < \kappa_0 \leq |h(z)| \leq \kappa_1$ which immediately implies (24.33). Now suppose that (24.33) holds. In this case $h(z) := (z - z_0)^n f(z)$ is bounded in a neighbourhood of z_0, hence z_0 is for h a removable singularity. Since by (24.33) the function h is bounded from below by κ_0 it follows that $h(z_0) \neq 0$ and we obtain for f the representation (24.32) which yields that f has a pole of order n at z_0. $\square$

Definition 24.15. *A complex-valued function $f : D \to \mathbb{C}$ defined on a domain $D \subset \mathbb{C}$ is called **meromorphic** in D if there exists a discrete subset $\mathrm{Pol}(f) \subset D$ such that $f|_{D \backslash \mathrm{Pol}(f)}$ is a holomorphic function and every $z \in \mathrm{Pol}(f)$ is a pole of f.*

Example 24.16. A. If $\operatorname{Pol}(f) = \emptyset$, which is not excluded by our definition, then f is holomorphic in D, i.e. holomorphic functions are meromorphic.
B. The rational functions are meromorphic on $\mathbb{C}$ but in general they are not holomorphic.
C. We claim that the function $z \mapsto \cot \pi z$ is meromorphic on $\mathbb{C}$. Since $\cot \pi z = \frac{\cos \pi z}{\sin \pi z}$ the integers $\mathbb{Z}$ forms the set of isolated singularities of $\cot \pi z$. For $z_0 = 0$ we find

$$\pi(z - z_0) \cot \pi z = \pi z \cot \pi z = \frac{\pi z}{\sin \pi z} \cos \pi z$$

and for $z \to z_0$, i.e. $z \to 0$, it follows

$$\lim_{z \to 0} \pi z \cot \pi z = \lim_{z \to 0} \left(\frac{\pi z}{\sin \pi z} \cdot \cos \pi z \right)$$
$$= \frac{\lim_{z \to 0} \cos \pi z}{\lim_{z \to 0} \frac{\sin \pi z}{\pi z}} = 1,$$

thus $\cot \pi z$ has a pole of order 1 at $z_0 = 0$. Using either the periodicity of $\cot \pi z$ or the Taylor expansion of $\sin \pi z$ about $k\pi$, $k \in \mathbb{Z}$, we deduce that every $k \in \mathbb{Z}$ is a pole of order 1 of $\cot \pi z$.

Example 24.16.C is of particular interest since it shows firstly that there exists meromorphic functions which are not rational functions. Secondly it suggests that meromorphic functions are quotients of holomorphic functions. At least locally we can already prove

Theorem 24.17. *Let $f : D \to \mathbb{C}$, $D \subset \mathbb{C}$ a domain, be a meromorphic function. For every $z_0 \in D$ we can find a neighbourhood $U \subset D$ of z_0 and two holomorphic functions $g, h : U \to \mathbb{C}$ such that on V we have $f(z) = \frac{g(z)}{h(z)}$.*

Proof. Let $z_0 \in D$. If z_0 is not a singularity of f we choose $U = D$, $g = f$ and $h = 1$. In the case where z_0 is an isolated singularity it must be a pole, say of order n. Now we may use (24.32), i.e. the representation of f in some neighbourhood U of z_0 as $f(z) = (z - z_0)^{-n} g(z)$, i.e. we take as $h(z)$ the function $z \mapsto (z - z_0)^n$. $\qquad\square$

Remark 24.18. The general global results, i.e. the statement that every meromorphic function $f : G \to \mathbb{C}$ defined on a region $G \subset \mathbb{C}$ admits a representation on G as quotient of two holomorphic functions $g, h : G \to \mathbb{C}$, i.e. $f(z) = \frac{g(z)}{h(z)}$ for all $z \in D$ we do not prove in our treatise and we refer to [26] or [67].

We turn now our attention to essential singularities.

Theorem 24.19 (F. Casorati and K. Weierstrass). *Let $z_0 \in D$ and $D \subset \mathbb{C}$ be a domain. Further let $f : D \backslash \{z_0\} \to \mathbb{C}$ be a holomorphic function with isolated singularity at z_0. This isolated singularity is an essential singularity if and only if for every $w_0 \in \mathbb{C}$ there exists a sequence $(z_k)_{k \in \mathbb{N}}$, $z_k \in D \backslash \{z_0\}$, converging to z_0 and $\lim_{k \to \infty} f(z_k) = w_0$. In other words, the image $f(U \backslash \{z_0\})$ is for every neighbourhood $U \subset D$ of z_0 dense in $\mathbb{C}$, i.e. $\overline{f(U \backslash \{z_0\})} = \mathbb{C}$.*

Proof. The "only if" part immediately implies that z_0 can neither be a removable singularity nor a pole, hence it must be an essential singularity. Now let z_0 be an essential singularity of f. Suppose that there exists a neighbourhood $U \subset D$ of z_0 such that $f(U \backslash \{z_0\})$ is not dense in $\mathbb{C}$. In this case we can find some $w_0 \in \mathbb{C}$ and $\epsilon > 0$ such that $B_\epsilon(w_0) \cap f(U \backslash \{z_0\}) = \emptyset$, i.e. $|f(z) - w_0| > \epsilon$ for all $z \in U \backslash \{z_0\}$. It follows that $z \mapsto g(z) := \frac{1}{f(z) - w_0}$ is on $U \backslash \{z_0\}$ holomorphic and bounded by $\frac{1}{\epsilon}$, hence g has a removable singularity at z_0. If now $\lim_{z \to z_0} g(z) \neq 0$, then $z \mapsto f(z) = w_0 + \frac{1}{g(z)}$ has a removable singularity at z_0, while if $\lim_{z \to z_0} g(z) = 0$ then $z \mapsto f(z) = w_0 + \frac{1}{g(z)}$ has a pole at z_0. Both cases contradict our assumption that z_0 is an essential singularity of f. $\square$

Now we combine the classifications of isolated singularities with the Laurent series expansion of a holomorphic function.

Theorem 24.20. *Let $f : D \backslash \{z_0\} \to \mathbb{C}$ be a holomorphic function which has an isolated singularity at $z_0 \in D$, $D \subset \mathbb{C}$ is a domain. Let $\epsilon > 0$ such that the punctured disc $A_{0,\epsilon}(z_0)$ is a subset of $D \backslash \{z_0\}$, i.e. $A_{0,\epsilon}(z_0) \subset D \backslash \{z_0\}$ and let $f(z) = \sum_{k=-\infty}^{\infty} c_k (z - z_0)^k$ be the Laurent expansion of f about z_0 converging in $A_{0,\epsilon}(z_0)$.*

  i) The singularity z_0 is removable if and only if $c_k = 0$ for $k < 0$.

  ii) The singularity z_0 is a pole of order n if and only if $c_{-n} \neq 0$ and $c_k = 0$ for $k < n$.

  iii) The singularity z_0 is an essential singularity if $c_k \neq 0$ for infinitely many $k \in \mathbb{Z}$, $k < 0$.

Proof. If z_0 is a removable singularity the Taylor series of its holomorphic extension to $B_\epsilon(z_0)$ must coincide in $A_{0,\epsilon}(z_0)$ with the Laurent expansion of f, hence $c_k = 0$ for $k < 0$. Conversely, if all c_k, $k < 0$, are zero, then the principal part of the Laurent expansion of f about z_0, vanishes identically, implying that f is bounded in a neighbourhood of z_0, hence z_0 is removable. Suppose next that z_0 is a pole of order n. In this case we can write $f(z) = (z-z_0)^{-n}h(z)$ where h is a holomorphic function in $B_\epsilon(z_0)$ not vanishing at z_0, i.e. $h(z_0) \neq 0$. For the Laurent expansion of f about z_0 this implies $c_k = 0$ for $k < -n$ and $c_{-n} \neq 0$. The converse is trivial, i.e. if $f(z) = \sum_{k=-n}^\infty c_k(z-z_0)^k$, $c_{-n} \neq 0$, then f has a pole of order n at z_0.

Finally, if neither $i)$ or $ii)$ hold f must have an essential singularity at z_0 and at the same time infinitely many coefficients c_k, $k < 0$, cannot vanish. $\qquad\square$

Example 24.21. A. Consider the function $f(z) = \frac{\cos z - 1}{z^2}$ defined on $\mathbb{C}\backslash\{0\}$. Using the Taylor expansion of cos we find

$$\frac{\cos z - 1}{z^2} = \frac{1}{z^2}\left(\sum_{k=0}^\infty \frac{(-1)^k}{(2k)!}z^{2k} - 1\right)$$

$$= \frac{1}{z^2}\sum_{k=1}^\infty \frac{(-1)^k}{(2k)!}z^{2k} = \sum_{k=1}^\infty \frac{(-1)^k}{(2k)!}z^{2(k-1)}$$

$$= \sum_{k=0}^\infty \frac{(-1)^{k+1}}{2(k+1)!}z^{2k},$$

and it follows that f has a removable singularity at $z_0 = 0$.

B. Using the results from Example 24.8.D we find that $z \mapsto \frac{1}{z(z-i)^2}$ has a pole of order 2 at i.

C. Since for $z \neq 0$ and $n \in \mathbb{N}$

$$e^{\frac{1}{z^n}} = \sum_{l=0}^\infty \frac{1}{l!}\frac{1}{z^{nl}}$$

it follows that $z \mapsto e^{\frac{1}{z^n}}$ has an essential singularity at $z_0 = 0$.

We want to explore the structure of the set of all meromorphic functions f on a region (or a domain with finitely many connectivity components) $G \subset \mathbb{C}$. For f there exists a discrete set $\mathrm{Pol}(f) \subset G$ and $w_0 \in \mathrm{Pol}(f)$ is a pole of f. In $G\backslash\mathrm{Pol}(f)$ the function f is holomorphic and since $G\backslash\mathrm{Pol}(f)$ is a region, if f is not identically equal to 0, then by Theorem 22.10 f has no zero of order ∞ and the set $N(f)$ of all zeroes is discrete. It follows that $\mathrm{Pol}(f) \cup N(f)$ is a discrete set in G and the functions $\frac{1}{f} : G \setminus (\mathrm{Pol}(f) \cup N(f)) \to \mathbb{C}$ is holomorphic. In a neighbourhood of a pole $w_0 \in \mathrm{Pol}(f)$ we have the representation

$$f(z) = (z - w_0)^{-n} g(z)$$

where $n \in \mathbb{N}$ is the order of the pole w_0 and g is holomorphic with $g(w_0) \neq 0$. In a neighbourhood of a zero $z_0 \in N(f)$ we have

$$f(z) = (z - z_0)^m h(z)$$

where $m \in \mathbb{N}$ is the multiplicity of order of the zero z_0 and h is a holomorphic function with $h(z_0) \neq 0$. It follows that $\frac{1}{f}$ is meromorphic in G and $\mathrm{Pol}\left(\frac{1}{f}\right) = N(f)$. Moreover we have that $N\left(\frac{1}{f}\right) = \mathrm{Pol}(f)$, i.e. $\frac{1}{f}$ has a holomorphic extension to $G \setminus N(f)$.

Now we may look at algebraic operations for meromorphic functions defined on a region $G \subset \mathbb{C}$. It is important to recall that a meromorphic function on G is a holomorphic function defined on $G \setminus X$ where $X \subset G$ is a discrete set.

Let f_j, $j = 1, 2$, be meromorphic functions with sets of poles $\mathrm{Pol}(f_j)$. On $G \setminus (\mathrm{Pol}(f_1) \cup \mathrm{Pol}(f_2))$ we can define $f_1 + f_2$, $f_1 - f_2$ and $f_1 \cdot f_2$ and these are holomorphic functions on the given domain, hence meromorphic functions on G since $\mathrm{Pol}(f_1) \cup \mathrm{Pol}(f_2)$ is a discrete set in G. In addition we can define for a meromorphic function f not identically zero in G with set of poles $\mathrm{Pol}(f)$ and set of zeroes $N(f)$ the holomorphic function $\frac{1}{f} : G \setminus (\mathrm{Pol}(f) \cup N(f)) \to \mathbb{C}$, then $\frac{1}{f}$ is a meromorphic function on G. Note that $\mathrm{Pol}(f_1 \pm f_2)$ and $\mathrm{Pol}(f_1 \cdot f_2)$ are in general subsets of $\mathrm{Pol}(f_1) \cup \mathrm{Pol}(f_2)$ since singularities may cancel or become removable. We already know that $\mathrm{Pol}\left(\frac{1}{f}\right) = N(f)$. We agree that when operating with meromorphic functions we always extend these functions into their removable singularities as holomorphic functions. The following result is now easy to check although it takes some time and we leave this to the reader.

Theorem 24.22. *The meromorphic functions on $G \subset \mathbb{C}$, where G is a region or a domain with finitely many connectivity components, form with the operations introduced above a field containing the holomorphic functions.*

The neutral element with respect to addition is the function 0, *i.e.* $z \mapsto 0$
for all $z \in G$, *and the neutral element with respect to multiplication is the*
function 1, *i.e.* $z \mapsto 1$ *for all* $z \in G$.

This theorem determines the algebraic structure of the meromorphic functions on G. There are other operations we may apply to obtain a new meromorphic function from a given one. Differentiation is such an operation.

Lemma 24.23. *For a meromorphic function* f *on a domain* $D \subset \mathbb{C}$ *the*
derivative f' *is a meromorphic function on* G *too and* $\mathrm{Pol}(f') = \mathrm{Pol}(f)$.

Proof. On $D \backslash \mathrm{Pol}(f)$ the function f, hence f', is holomorphic and it remains
to prove that $z_0 \in \mathrm{Pol}(f)$ is also a pole of f'. For some $\epsilon > 0$ we have in
$A_{0,\epsilon}(z_0)$ the Laurent series representation

$$f(z) = \sum_{k=-n}^{\infty} c_k(z - z_0)^k \tag{24.34}$$

where $n \in \mathbb{N}$ is the order of the pole z_0, $c_{-n} \neq 0$. Since we can differentiate
this series term by term we find in $A_{0,\epsilon}(z_0)$ that

$$f'(z) = \sum_{k=-n}^{\infty} kc_k(z - z_0)^{k-1} = \sum_{k=-n-1}^{\infty} (k+1)c_{k+1}(z - z_0)^k \tag{24.35}$$

and it follows that f' has a pole of order $-n - 1$ at z_0. Thus we conclude
that $\mathrm{Pol}(f') = \mathrm{Pol}(f)$ and the lemma is proved. $\qquad\square$

Since with f also f' is meromorphic, under the assumption that f is not
identically zero we may consider on G the term $\frac{f'}{f}$. Suppose that f has a
zero $z_0 \in G$ of order n, i.e. with some holomorphic function g, $g(z_0) \neq 0$,
we have in a neighbourhood of z_0 the equality $f(z) = (z - z_0)^n g(z)$, hence
$f'(z) = n(z - z_0)^{n-1}g(z) + (z - z_0)^n g'(z)$, implying

$$\frac{f'(z)}{f(z)} = \frac{n(z - z_0)^{n-1}g(z) + (z - z_0)^n g'(z)}{(z - z_0)^n g(z)}$$

$$= \frac{n}{z - z_0} + \frac{g'(z)}{g(z)}. \tag{24.36}$$

Thus $\frac{f'}{f}$ has a pole of order 1 at z_0 and $\mathrm{res}\left(\frac{f'}{f}, z_0\right) = n$. Now, if w_0 is
a pole of order m of f then we find in a neighbourhood of w_0 the equality

466

$f(z) = (z - w_0)^{-m} h(z)$ where h is a holomorphic function such that $h(w_0) \neq 0$. Now we find

$$\frac{f'(z)}{f(z)} = \frac{-m}{z - w_0} + \frac{h'(z)}{h(z)}, \qquad (24.37)$$

i.e. $\frac{f'}{f}$ has a pole of order 1 at w_0 and $\mathrm{res}\left(\frac{f'}{f}, w_0\right) = -m$. We have proved that if f is meromorphic in $G \subset \mathbb{C}$ with set of poles $\mathrm{Pol}(f)$ and set of zeroes $N(f)$, then $\frac{f'}{f}$ is meromorphic in G having poles of order 1 for all points in $\mathrm{Pol}(f) \cup N(f)$. In order to exploit this observation further we first need to extend (24.29) which leads to a first version of the **residue theorem**.

Theorem 24.24. *Let $D \subset \mathbb{C}$ be a domain and $\overline{B_r(a)} \subset D$. Assume that f is meromorphic in D and $\mathrm{Pol}(f) = \{w_1, \ldots, w_n\} \subset B_r(a)$. Then we have*

$$\frac{1}{2\pi i} \int_{\partial B_r(a)} f(z) \, dz = \sum_{k=1}^{N} \mathrm{res}(f, w_k), \qquad (24.38)$$

where we used for the circle $\partial B_r(a)$ our standard parametrization.

Proof. Consider Figure 24.8. With an argument similar to that leading to (24.11) we obtain that

$$\frac{1}{2\pi i} \int_{\partial B_r(a)} f(z) \, dz = \sum_{k=1}^{N} \frac{1}{2\pi i} \int_{\partial B_{\epsilon_k}(w_k)} f(z) \, dz, \qquad (24.39)$$

where the ϵ_k's are chosen such that $\overline{B_{\epsilon_k}(w_k)}$ does not contain any further pole of f, and all circles are parametrized in the standard way. Now we can apply (24.29) to each integral $\int_{\partial B_{\epsilon_k}(w_k)} f(z) \, dz$ and (24.38) follows.

467

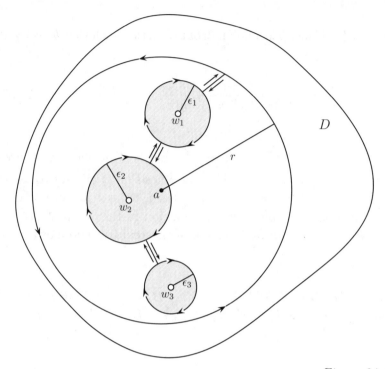

Figure 24.8

In the next chapter we will provide many applications of Theorem 24.24 and its generalisation. We will also see that $\partial B_r(a)$ in (24.38) can be replaced by a more general curve γ. This applies also for the following result, the **argument principle**:

Theorem 24.25. *For a domain $D \subset \mathbb{C}$ and $\overline{B_r(a)} \subset D$ let f be a meromorphic function in D. We denote the zeroes of f in $B_r(a)$ by $z_1, \ldots, z_M$ and z_k has multiplicity α_k. The poles of f in $B_r(a)$ are $w_1, \ldots, w_N$ and w_l has order β_l. If f has no zero and no pole on $\partial B_r(a)$ then we have*

$$\frac{1}{2\pi i} \int_{\partial B_r(a)} \frac{f'(z)}{f(z)} \, dz = \sum_{k=1}^{M} \alpha_k - \sum_{l=1}^{N} \beta_l. \tag{24.40}$$

Proof. We need only combine (24.36) and (24.37) with Theorem 24.24. □

We end this chapter with an interesting application of the argument principle.

468

Theorem 24.26 (Theorem of E. Rouché). *Let $D \subset \mathbb{C}$ be a domain and $\overline{B_r(a)} \subset D$. Suppose that $f, g : D \to \mathbb{C}$ are holomorphic functions and that $|f(z)| > |g(z)|$ on $\partial B_r(a)$. In this situation the functions f and $f \pm g$ have the same number of zeroes in $B_r(a)$.*

Proof. For $s \in [0, 1]$ we consider the family $f_s := f + sg$ of holomorphic functions in D, $f_0 = f$ and $f_1 = f + g$. Denote by $n(s)$ the number of zeroes of f_s in $B_r(a)$ counted according to their multiplicity. On $\partial B_r(a)$ we have

$$|f_s(z)| = |f(z) + sg(z)| \geq |f(z)| - s|g(z)| > 0$$

since $|f(z)| > |g(z)|$ on $\partial B_r(a)$. By the argument principle we obtain

$$n(s) = \frac{1}{2\pi i} \int_{\partial B_r(a)} \frac{f'(z)}{f(z)} \, dz$$

where we use the standard parametrization for $\partial B_r(a)$. Since $n(s) \in \mathbb{Z}$ for every $s \in [0, 1]$, once we have proved that $n(\cdot)$ is continuous it follows that $n(\cdot)$ is constant, i.e. $n(s) = n(0)$ for all s implying that $n(1) = n(0)$, i.e. the theorem will follow. The function $(z, s) \mapsto \frac{f'_s(z)}{f_s(z)}$ is continuous on $\partial B_r(a) \times [0, 1]$ since $f_s(z) \neq 0$ for all $z \in \partial B_r(a)$. Now the compactness of $\partial B_r(a)$ implies, see Theorem 21.11, the continuity of $n(\cdot)$. $\square$

Remark 24.27. In Problem 11 we will see how we can use the Theorem of Rouché to localise zeroes of holomorphic functions.

We need a better "calculus for parametric curves" and this will be developed in the following chapter.

Problems

1. Prove that $h : B_{\frac{1}{r}}(0) \setminus \{0\} \to \mathbb{C} \setminus \overline{B_r(z_0)}$, $h(w) = z_0 + \frac{1}{w}$, is a biholomorphic function.

2. Find Laurent expansions of the following functions and identify the type of singularity:

   a) $\frac{z - \frac{z^3}{6} - \sin z}{z^5}$ at $z_0 = 0$;

   b) $(z - 4) \sin \frac{1}{z+3}$ at $z_0 = -3$;

   c) $\frac{\cos 2z}{(z-1)^3}$ at $z_0 = 1$.

3. Expand $f(z) = \frac{1}{(z+2)(z+4)}$ as a Laurent series for:

   a) $z \in A_{2,4}(0)$;

   b) $z \in A_{4,\infty}(0)$;

   c) $z \in A_{1,2}(-2)$;

   d) $z \in B_2(0)$.

4. Determine the singularities of each of the following functions:

   a) $f(z) = \frac{1}{(4\sin z - 2)^2}$;

   b) $h(z) = \frac{z}{e^{\frac{1}{2z}} - 1}$.

5. Give an example of a meromorphic function which is not a rational function and has a pole of order 2 at $z_0 = 1$.

6. Let $f : U \setminus \{z_0\} \to \mathbb{C}$ be a holomorphic function where U is a neighbourhood of z_0. Suppose that in some neighbourhood $V \subset U$ of z_0 we have the estimate

$$\kappa_0 e^{\frac{1}{|z-z_0|}} \le |f(z)|, \quad \kappa_0 > 0.$$

What type of singularity does f have at z_0?

7. Consider on $B_{\frac{1}{2}}(0) \setminus \{0\}$ the holomorphic function $h(z) = \exp\left(\frac{1}{z}\right)$. Let $w_0 \in \mathbb{C} \setminus \{0\}$. Prove that there exists a countable set $\{z_k \mid k \in \mathbb{N}\} \subset B_{\frac{1}{2}}(0) \setminus \{0\}$ such that $h(z_k) = w_0$.

8. Let $f, g : U \setminus \{z_0\} \to \mathbb{C}$ be holomorphic functions and assume that f as well as g has a pole of order N at z_0. Prove that for some neighbourhood $V \subset U$ of z_0 the function $\frac{f}{g}$ is defined in $V \setminus \{z_0\}$ and has a removable singularity at z_0. Further prove that

$$h(z) := \begin{cases} \frac{f}{g}(z), & z \in V \setminus \{z_0\} \\ \frac{a_{-N}}{b_{-N}}, & z = z_0 \end{cases}$$

is a holomorphic extension of $\frac{f}{g} : V \setminus \{z_0\} \to \mathbb{C}$, where a_{-N} and b_{-N} are the coefficients of the leading terms in the Laurent expansion of f and g about z_0.

470

9. Give a further proof of the fundamental theorem of algebra by using the argument principle.

   Hint: for $p(z) = z^n + a_1 z^{n-1} + \cdots + a_n$, $n \geq 1$, choose $R > 0$ such that $|p(z)| \geq 1$ for $z \in B_R^{\complement}(0)$ and consider

$$\frac{1}{2\pi i} \int_{\partial B_R(0)} \frac{p'(z)}{p(z)} dz.$$

10. Use the Theorem of Rouché to prove that all zeroes of the polynomial $z^5 - 2z^3 + 10$ belong to $\overline{A_{1,2}(0)}$.

    Hint: consider $f(z) = 10$ and $g(z) = z^5 - 2z^3$ to prove that there are no zeroes in $B_1(0)$. Then consider $f(z) = z^5$ and $g(z) = -2z^3 + 10$ to prove that all zeroes are in $B_2(0)$.

25 The Residue Theorem

So far when working with closed curves we emphasised that they were simply closed, in particular for the circle $|z_0 - z| = r$ we used the standard (anti-clockwise) parametrization $t \mapsto z(t) = z_0 + re^{it}$, $t \in [0, 2\pi]$. A very useful observation was that with this parametrization we have

$$\frac{1}{2\pi i} \int_{|\zeta_0 - z_0| = r} \frac{1}{\zeta - z} \, d\zeta = 1, \quad z \in B_r(z_0), \tag{25.1}$$

compare with Lemma 21.7. In preparing the proof of the Runge theorem, Theorem 23.25, we gave already some extension of (25.1), see Theorem 23.23. We now want to investigate the integral $\int_\gamma \frac{d\zeta}{\zeta - z}$ for general (i.e. piecewise continuously differentiable) closed curves $\gamma : [a, b] \to \mathbb{C}$.

Proposition 25.1. *For a piecewise continuously differentiable closed curve $\gamma : [a, b] \to \mathbb{C}$ and $z \notin \mathrm{tr}(\gamma)$ the integral*

$$\mathrm{ind}_\gamma(z) := \frac{1}{2\pi i} \int_\gamma \frac{1}{\zeta - z} \, d\xi \tag{25.2}$$

has the following properties:

 i) $\mathrm{ind}_\gamma(z) \in \mathbb{Z}$;

 ii) $\mathrm{ind}_\gamma(z_1) = \mathrm{ind}_\gamma(z_2)$ for z_1, z_2 belonging to the same open connectivity component of $\mathrm{tr}(\gamma)^\complement$;

 iii) for z in the unbounded component of $\mathrm{tr}(\gamma)^\complement$ it follows that $\mathrm{ind}_\gamma(z) = 0$.

Proof. For $t \in [a, b]$ and $z \notin \mathrm{tr}(\gamma)$ fixed consider the integral

$$h_z(t) := \int_a^t \frac{\dot\gamma(t)}{\gamma(t) - z} \, dt.$$

Since γ is piecewise continuously differentiable we find with the possible exception of finitely many values of $t \in [a, b]$ that

$$\dot h_z(t) = \frac{\dot\gamma(t)}{\gamma(t) - z}.$$

With the possible exception of finitely many values the function $t \mapsto ((\gamma(t) - z)e^{-h_z(t)})$ is continuously differentiable and we find

$$\frac{d}{dt}\left((\gamma(t) - z)e^{-h_z(t)}\right) = \dot{\gamma}(t)e^{-h_z(t)} + (\gamma(t) - z)\left(-\dot{h}_z(t)e^{-h_z(t)}\right)$$

$$= \left(\dot{\gamma}(t) - (\gamma(t) - z)\frac{\dot{\gamma}(t)}{\gamma(t) - z}\right)e^{-h_z(t)} = 0.$$

It then follows that $(\gamma(\cdot) - z)e^{-h_z(\cdot)}$ must be equal to a constant $A \neq 0$, note that $\gamma(t) \neq z$ for all $t \in [a, b]$ by our assumptions. Using the fact that $\gamma(a) = \gamma(b)$ we get

$$1 = e^{h_z(a)} = \frac{1}{A}(\gamma(a) - z) = \frac{1}{A}(\gamma(b) - z)e^{h_z(b)},$$

or $h_z(b) = 2\pi i k$, $k \in \mathbb{Z}$. Thus we arrive at

$$\text{ind}_\gamma(z) = \frac{1}{2\pi i}\int_\gamma \frac{1}{\zeta - z}\, d\zeta$$

$$= \frac{1}{2\pi i}\int_a^b \frac{\dot{\gamma}(t)}{\gamma(t) - z}\, dt = \frac{1}{2\pi i}h_z(b) \in \mathbb{Z},$$

and i) is proved.

The function $z \mapsto \text{ind}_\gamma(z)$ is continuous in $\mathbb{C}\backslash\text{tr}(\gamma)$ and an integer-valued continuous function on an (open) connected set must be constant which implies ii). Finally we note that

$$\lim_{z \to \infty} \text{ind}_\gamma(z) = \lim_{z \to \infty} \frac{1}{2\pi i}\int_\gamma \frac{1}{\zeta - z}\, d\zeta = 0,$$

which now yields iii). $\square$

Definition 25.2. *The integer* $\text{ind}_\gamma(z)$ *is called the **winding number** or the **index** of γ with respect to $z \in \mathbb{C}\backslash\text{tr}(\gamma)$.*

Example 25.3. A. Let γ be the circle $|\zeta - z_0| = r$ with our standard parametrization. Then we have $\text{ind}_\gamma(z) = 1$ for $z \in B_r(z_0)$ and $\text{ind}_\gamma(z) = 0$ for $z \in \overline{B_r(z_0)}^c$. **B.** Let $\gamma_k : [0, 2\pi k] \to \mathbb{C}$, $\gamma_k(t) = z_0 + re^{it}$, $k \in \mathbb{N}$. The trace of γ_k is again the circle with centre z_0 and radius r, i.e. $\partial B_r(z_0)$, and for $z \in B_r(z_0)$ we find

$$\text{ind}_{\gamma_k}(z) = \text{ind}_{\gamma_k}(z_0) = \frac{1}{2\pi i}\int_0^{2\pi k}\frac{ire^{it}}{z_0 + re^{it} - z_0}\, dt = \frac{1}{2\pi i}\int_0^{2\pi k} i\, dt = k.$$

In the case where we use the parametrization $\eta_k(t) = z_0 + re^{-it}$, $t \in [0, 2\pi k]$, i.e. $\eta_k = \gamma_k^{-1}$ we find

$$\text{ind}_{\eta_k}(z) = \text{ind}_{\eta_k}(z_0) = \frac{1}{2\pi i} \int_0^{2\pi k} (-i)\, dt = -k, \quad z \in B_r(z_0).$$

Note that we have used in both calculations the fact that $\text{ind}_\gamma(\cdot)$ is constant on $B_r(z_0)$.

Further note that $\text{ind}_{\gamma_k^{-1}}(z) = -\text{ind}_{\gamma_k}(z)$, for $z \in \mathbb{C}\backslash \partial B_r(z_0)$, and this is indeed a general result.

Lemma 25.4. *Let* $\gamma_j : [a_j, b_j] \to \mathbb{C}$ *be two closed piecewise continuously differentiable curves such that* $\gamma_1(b_1) = \gamma_2(a_2)$. *For* $z \notin \text{tr}(\gamma_1) \cup \text{tr}(\gamma_2)$ *we have*

$$\text{ind}_{\gamma_1 \oplus \gamma_2}(z) = \text{ind}_{\gamma_1}(z) + \text{ind}_{\gamma_2}(z), \tag{25.3}$$

and

$$\text{ind}_{\gamma_1}(z) = -\text{ind}_{\gamma_1^{-1}}(z). \tag{25.4}$$

Proof. From our assumptions we deduce that $\gamma_1 \oplus \gamma_2$ is a well defined closed piecewise continuously differentiable curve. Since

$$\frac{1}{2\pi i} \int_{\gamma_1 \oplus \gamma_2} \frac{1}{\zeta - z}\, d\zeta = \frac{1}{2\pi i} \int_{\gamma_1} \frac{1}{\zeta - z}\, d\zeta + \frac{1}{2\pi i} \int_{\gamma_2} \frac{1}{\zeta - z}\, d\zeta,$$

we have proved (25.3) whereas (25.4) follows from the fact that

$$\frac{1}{2\pi i} \int_{\gamma_1} \frac{1}{\zeta - z}\, d\zeta = -\frac{1}{2\pi i} \int_{\gamma_1^{-1}} \frac{1}{\zeta - z}\, d\zeta.$$

$\square$

For a simply closed curve we can rely on the Jordan curve theorem, Theorem 19.37, to define the interior and the exterior of its trace, where the interior is bounded and simply connected. Such a result cannot be expected for an arbitrary closed curve, see Figure 25.1 below,

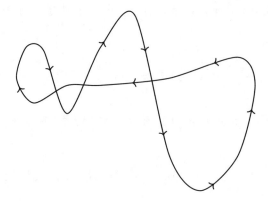

Figure 25.1

Therefore we give

Definition 25.5. *Let* $\gamma : I \to \mathbb{C}$ *be a closed, piecewise continuously differentiable curve. The **interior** of* γ *is defined as*

$$\text{int } \gamma := \left\{ z \in \mathbb{C} \backslash \text{tr}(\gamma) \,\middle|\, \text{ind}_\gamma(z) \neq 0 \right\}, \tag{25.5}$$

*and the **exterior** of* γ *is defined as*

$$\text{ext } \gamma := \left\{ z \in \mathbb{C} \backslash \text{tr}(\gamma) \,\middle|\, \text{ind}_\gamma(z) = 0 \right\}. \tag{25.6}$$

With these definitions we find the decomposition

$$\mathbb{C} = \text{int } \gamma \cup \text{tr}(\gamma) \cup \text{ext } \gamma, \tag{25.7}$$

and we have

Lemma 25.6. *For a closed, piecewise continuously differentiable curve* $\gamma :$ $I \to \mathbb{C}$ *the sets* $\text{int } \gamma$ *and* $\text{ext } \gamma$ *are open and* $\partial(\text{int } \gamma) \subset \text{tr}(\gamma)$ *as well as* $\partial(\text{ext } \gamma) \subset \text{tr}(\gamma)$. *Moreover,* $\text{int } \gamma \subset B_R(z_0)$ *and* $\mathbb{C} \backslash B_R(z_0) \subset \text{ext } \gamma$ *for some open disc such that* $\text{tr}(\gamma) \subset B_R(z_0)$.

Proof. The first part is trivial since ind_γ is locally constant and $\text{tr}(\gamma)$ is compact. Furthermore, since $\emptyset \neq B_R(z_0)^{\complement}$ is a connected set the function ind_γ must be constant on $B_R(z_0)^{\complement}$, but for $z \to \infty$ we have $\text{ind}_\gamma(z) \to 0$, so $B_R(z_0)^{\complement} \subset \text{ext } \gamma$ which implies by (25.7) that $\text{int } \gamma \subset B_R(z_0)$. $\square$

476

Remark 25.7. If $\mathrm{tr}(\gamma)$ is also the trace of a simply connected curve, then we have of course the Jordan curve theorem at our disposal. The following figure shows that we cannot expect $\partial(\mathrm{ext}\,\gamma) = \mathrm{tr}(\gamma)$.

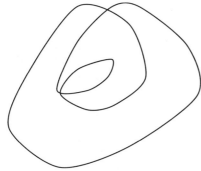

Figure 25.2

In previous chapters we often have replaced a given path of integration by another one without changing the value of the integral, see for example (24.11) when preparing the proof of Theorem 24.2. Of course we can interpret the Cauchy integral theorem, Theorem 21.4, in this sense: we replace $\partial\triangle$ by γ. We want to study integrals over finite systems of curves γ_k, $k = 1, \ldots, M$, in such a way that the "arithmetic" for integrals is reflected in operations in systems of curves. We start with

Definition 25.8. *Let $G \subset \mathbb{C}$ be a domain and let $\gamma_k : I_k \to G$, $k = 1, \ldots, M$, be a finite family of piecewise continuously differentiable curves. The formal sum*

$$\gamma := a_1\gamma_1 \tilde{+} \cdots \tilde{+} a_M\gamma_M, \qquad a_k \in \mathbb{Z}, \tag{25.8}$$

*is called a **chain** of curves in G with integer coefficients.*

Remark 25.9. A. As long as no integrals are involved, we can allow in all considerations below to work just with continuous curves. **B.** It is important to note that so far "$\tilde{+}$" in (25.8) has no meaning, we just use it to define a new class of objects. In particular one shall not mix up (25.8) with the operation $\eta_1 \oplus \eta_2$ for two curves with the terminal point of η_1 being the initial point of η_2.

We now want to introduce an operation on the set of all chains in G. Let $\gamma = a_1\gamma_1 \tilde{+} \cdots \tilde{+} a_M\gamma_M$. We agree that

$$a_1\gamma_1 \tilde{+} \cdots \tilde{+} a_M\gamma_M = a_{\sigma(1)}\gamma_{\sigma(1)} \tilde{+} \cdots \tilde{+} a_{\sigma(M)}\gamma_{\sigma(M)}$$

for any permutation $\sigma \in S_M$. For $\gamma = a_1\gamma_1 \tilde{+} \cdots \tilde{+} a_M\gamma_M$ and $\eta = b_1\gamma_1 \tilde{+} \cdots \tilde{+} b_M\gamma_M$ we define

$$\gamma \tilde{+} \eta := (a_1 + b_1)\gamma_1 \tilde{+} \cdots \tilde{+} (a_M + b_M)\gamma_M,$$

which immediately yields $\gamma \tilde{+} \eta = \eta \tilde{+} \gamma$, and with $-\gamma := -a_1\gamma_1 \tilde{+} \cdots \tilde{+}(-a_M)\gamma_M$ we have $\gamma \tilde{+}(-\gamma) = 0\gamma_1 \tilde{+} \cdots \tilde{+} 0\gamma_M$, and furthermore we have $(\gamma \tilde{+} \eta) \tilde{+} \tau = \gamma \tilde{+}(\eta \tilde{+} \tau)$ where $\tau = c_1\gamma_1 \tilde{+} \cdots \tilde{+} c_M\gamma_M$. We will write $0 := 0\gamma_1 \tilde{+} \cdots \tilde{+} 0\gamma_M$ and $\gamma \tilde{-} \eta$ for $\gamma \tilde{+}(-\eta)$, in particular we write $\tilde{-}\gamma = 0 \tilde{-} \gamma$.

Next, for $\gamma = a_1\gamma_1 \tilde{+} \cdots \tilde{+} a_M\gamma_M$ and $\eta = b_1\eta_1 \tilde{+} \cdots \tilde{+} b_N\eta_N$ we define

$$\gamma \tilde{+} \eta = a_1\gamma_1 \tilde{+} \cdots \tilde{+} a_M\gamma_M \tilde{+} b_1\eta_1 \tilde{+} \cdots \tilde{+} b_N\eta_N, \tag{25.9}$$

and we note that if $\gamma_j = \eta_l$ for some j and l, then we can rewrite (25.9) as

$$\gamma \tilde{+} \eta = a_1\gamma_1 \tilde{+} \cdots \tilde{+} a_{j-1}\gamma_{j-1} \tilde{+}(a_j - b_l)\gamma_j \tilde{+} a_{j+1}\gamma_{j+1} \tilde{+} \cdots \tilde{+} \gamma_M$$
$$\tilde{+} b_1\eta_1 \tilde{+} \cdots \tilde{+} b_{l-1}\eta_{l-1} \tilde{+} b_{l+1}\eta_{l+1} \tilde{+} \cdots \tilde{+} b_N\eta_N.$$

With these definitions we have introduced on the set of all chains in G the structure of an Abelian group.

For a chain $\gamma = a_1\gamma_1 \tilde{+} \cdots \tilde{+} a_M\gamma_M$ we define its trace as union of the traces of the curves γ_k, $1 \le k \le M$, i.e.

$$\operatorname{tr}(\gamma) = \bigcup_{k \in Z} \operatorname{tr}(\gamma_k), \quad Z = \{l \in \{1, \ldots, M\} \,|\, a_l \ne 0\}.$$

Now let $f : \operatorname{tr}(\gamma) \to \mathbb{C}$ be a continuous function. We set

$$\int_\gamma f(z)\,\mathrm{d}z = \sum_{k=1}^M a_k \int_{\gamma_k} f(z)\,\mathrm{d}z, \tag{25.10}$$

where $\gamma = a_1\gamma_1 \tilde{+} \cdots \tilde{+} a_M\gamma_M$.

If for example all curves γ_k are simply closed and $G \subset \mathbb{C}$ starshaped as well as f holomorphic on G then we have by the Cauchy integral theorem

$$\int_\gamma f(z)\,\mathrm{d}z = \sum_{k=1}^M a_k \int_{\gamma_k} f(z)\,\mathrm{d}z = 0.$$

478

In the case where we can form $\gamma_1 \oplus \cdots \oplus \gamma_M$ then it follows for $\gamma = \gamma_1 \tilde{+} \cdots \tilde{+} \gamma_M$ that

$$\int_\gamma f(z)\,\mathrm{d}z = \int_{\gamma_1 \oplus \cdots \oplus \gamma_M} f(z)\,\mathrm{d}z.$$

For a curve η_1 we denote its initial point by $\eta_1^{(0)}$ and its terminal point by $\eta_1^{(1)}$. For a chain $\gamma = a_1\gamma_1 \tilde{+} \cdots \tilde{+} a_M\gamma_M$ we can form $\sum_{k,z=\gamma_k^{(0)}} a_k$ which counts how often z, taking multiplicity into account, is an initial point of some curve γ_k of the chain γ. Analogously $\sum_{k,z=\gamma_k^{(1)}} a_k$ counts, again multiplicity taken into account, how often z is a terminal point of γ. For $z \in \mathbb{C}$ not an initial point (terminal point) of any of the curves γ_k of the chain we have of course $\sum_{k,z=\gamma_k^{(0)}} a_k = 0 \left(\sum_{k,z=\gamma_k^{(1)}} a_k = 0 \right)$.

Definition 25.10. *We call a chain* $\gamma = a_1\gamma_1 \tilde{+} \cdots \tilde{+} a_M\gamma_M$ *in* G ***closed*** *or a* ***cycle*** *in* G *if for all* $z \in \mathbb{C}$ *we have*

$$\sum_{k,z=\gamma_k^{(0)}} a_k = \sum_{k,z=\gamma_k^{(1)}} a_k, \tag{25.11}$$

i.e. every $z \in \mathbb{C}$ *is as often initial as terminal point (multiplicity taken into account) of some* γ_k.

Example 25.11. In the following, by curve we always mean a piecewise continuously differentiable curve.
A. Every curve is of course a chain and every simply closed curve is a cycle. Moreover, if γ is a curve for which every point is an m-fold point in the sense of Definition II.7.12.B, then γ is a cycle. The last example has some further interesting features. Consider the simply closed curve $\gamma : [0,1] \to G \subset \mathbb{C}$, and now extend γ periodically to $\gamma^{(m)} : [0,m] \to G$, $\gamma^{(m)}(t+1) = \gamma(t)$, $t \in [0, m-1]$. The curve $\gamma^{(m)}$ is a cycle in G. The chain $m\gamma$ we may write as $m\gamma = \gamma \tilde{+} \cdots \tilde{+} \gamma$ and we may interpret this as running m-times through γ, which is the same as running once through $\gamma^{(m)}$ or through $\gamma \oplus \cdots \oplus \gamma$ (m terms).
B. If γ is a curve, hence a chain, then $\gamma \tilde{+} \gamma^{-1}$ is a cycle.
C. Let γ_k, $1 \le k \le M$, be curves such that $\gamma_{k+1}^{(0)} = \gamma_k^{(1)}$, $k = 1, \ldots, M-1$, and $\gamma_M^{(1)} = \gamma_1^{(0)}$. Then $\gamma_1 \tilde{+} \cdots \tilde{+} \gamma_M$ is a cycle.

As a first application we can now extend Theorem 20.22.

Theorem 25.12. *A continuous function $f : G \to \mathbb{C}$ defined on the region $G \subset \mathbb{C}$ admits a primitive in G if and only if for all cycles γ in G we have*

$$\int_\gamma f(z)\,dz = 0. \qquad (25.12)$$

Proof. Let F be a primitive of f and $\gamma = a_1\gamma_1 \tilde{+} \cdots \tilde{+} a_M\gamma_M$ a cycle. It follows that

$$\int_\gamma f(z)\,dz = \sum_{k=1}^M a_k \int_{\gamma_k} f(z)\,dz = \sum_{k=1}^M a_k \left(F\left(\gamma_k^{(1)}\right) - F\left(\gamma_k^{(0)}\right) \right)$$

$$= \sum_z \left(\sum_{k,z=\gamma_k^{(1)}} a_k - \sum_{k,z=\gamma_k^{(0)}} a_k \right) F(z),$$

where the summation is over all $z \in \mathbb{C}$ which are initial or terminal points of some γ_k. Since γ is a cycle the inner sum is however equal to zero. To see the converse direction, we just need to note that simply closed piecewise continuously differentiable curves are cycles and therefore Theorem 20.22 yields the existence of a primitive. $\qquad \square$

To proceed further we need

Definition 25.13. A. *Let $\gamma = a_1\gamma_1 \tilde{+} \cdots \tilde{+} a_M\gamma_M$ be a cycle. For $z \notin \operatorname{tr}(\gamma)$ we define its* **index** *as*

$$\operatorname{ind}_\gamma(z) := \frac{1}{2\pi i} \int_\gamma \frac{d\zeta}{\zeta - z}. \qquad (25.13)$$

B. *We call a cycle γ in a domain $G \subset \mathbb{C}$* **null-homologous** *if for every $z \in G^{\complement}$ the index of γ at z is zero, i.e.*

$$\operatorname{ind}_\gamma(z) = 0 \quad \text{for } z \in G^{\complement}. \qquad (25.14)$$

C. *If the difference of two cycles γ and η in G is null-homologous we call γ and η* **homologous.**

Since for every cycle $\gamma = a_1\gamma_1 \tilde{+} \cdots \tilde{+} a_M\gamma_M$ we have

$$\frac{1}{2\pi i} \int_\gamma \frac{d\zeta}{\zeta - z} = \sum_{k=1}^M a_k \int_{\gamma_k} \frac{d\zeta}{\zeta - z}$$

we deduce that

$$\operatorname{ind}_{\tilde{-}\gamma}(z) = -\operatorname{ind}_\gamma(z), \quad z \notin \operatorname{tr}(\gamma) \tag{25.15}$$

and

$$\operatorname{ind}_\gamma(z) = \sum_{k=1}^{M} a_k \operatorname{ind}_{\gamma_k}(z), \quad z \notin \operatorname{tr}(\gamma). \tag{25.16}$$

From (25.16) we deduce further for two cycles γ and η and $z \notin \operatorname{tr}(\gamma \tilde{+} \eta)$ that

$$\operatorname{ind}_{\gamma \tilde{+} \eta}(z) = \operatorname{ind}_\gamma(z) + \operatorname{ind}_\eta(z). \tag{25.17}$$

The formulae (25.15) - (25.17) allow us to immediately transfer results holding for the index of a curve to the index of a cycle.

Proposition 25.14. *The index of a cycle is an integer-valued function and on every connectivity component $\mathbb{C} \setminus \operatorname{tr}(\gamma)$ it is constant. Moreover, if $R > 0$ is such that $\operatorname{tr}(\gamma) \subset B_R(0)$, then $\operatorname{ind}_\gamma|_{\overline{B_R(0)}^{\complement}}$ is equal to the zero function, i.e.* $\operatorname{ind}_\gamma(z) = 0$ *for* $z \in \overline{B_R(0)}^{\complement}$.

Let γ be a cycle, $\gamma = a_1\gamma_1 \tilde{+} \cdots \tilde{+} a_M\gamma_M$, and consider the cycle $\tilde{-}\gamma = (-a_1)\gamma_1 \tilde{+} \cdots \tilde{+} (-a_M)\gamma_M$. Both cycles have the same trace namely $\operatorname{tr}(\gamma) = \operatorname{tr}(\tilde{-}\gamma) = \bigcup_{k=1}^{M} \operatorname{tr}(\gamma)$, and for $z \in \mathbb{C} \setminus \operatorname{tr}(\gamma)$ we find

$$\operatorname{ind}_{\tilde{-}\gamma}(z) = \sum_{k=1}^{M} (-a_k)\operatorname{ind}_{\gamma_k}(z) = \sum_{k=1}^{M} a_k \operatorname{ind}_{\tilde{-}\gamma_k}(z)$$

$$= \sum_{k=1}^{M} a_k \operatorname{ind}_{\gamma_1^{-1}}(z) = \operatorname{ind}_{\gamma^{-1}}(z),$$

with $\gamma^{-1} = a_1\gamma_1^{-1} \tilde{+} \cdots \tilde{+} a_M\gamma_M^{-1}$. In this sense we can identify for calculating indices of cycles $\tilde{-}\gamma$ with γ^{-1}.

Example 25.15. A. For γ given as $\gamma(t) = z_0 + re^{ikt}$, $t \in [0, 2\pi]$, we find for $z \in B_r(z_0)$, $z \notin \operatorname{tr}(\gamma) = \partial B_r(z_0)$, that $\operatorname{ind}_\gamma(z) = k$ since $\operatorname{ind}_\gamma(z_0) = k$ as is shown by a simple calculation.
B. Let γ_j, $j = 1, 2$, be the standard parametrization of the circle $|\zeta - z_j| = r_j$ and suppose $B_{r_1}(z_1) \subset B_{r_2}(z_2)$. We consider the cycle $\gamma := \gamma_2 \tilde{+} (-\gamma_1)$ and we know that $\operatorname{ind}_\gamma = \operatorname{ind}_{\gamma_2 \tilde{+} \gamma_1^{-1}}$. Therefore we find for $z \in \operatorname{tr}(\gamma)$

i) $\operatorname{ind}_\gamma(z) = 0$ for $z \in B_{r_2}^{\complement}(z_2)$;

ii) $\operatorname{ind}_\gamma(z) = \operatorname{ind}_{\gamma_2}(z) + \operatorname{ind}_{\gamma_1^{-1}}(z) = 1$ for $z \in B_{r_2}(z_2)\backslash\overline{B_{r_1}(z_1)}$, since in this case $z \in \overline{B_{r_1}(z_1)}^C$ and hence $\operatorname{ind}_{\gamma_1^{-1}}(z) = -\operatorname{ind}_{\gamma_1}(z) = 0$, but $\operatorname{ind}_{\gamma_2}(z) = 1$.

iii) $\operatorname{ind}_\gamma(z) = \operatorname{ind}_{\gamma_2}(z) - \operatorname{ind}_{\gamma_1}(z) = 0$ for $z \in B_r(z_1)$, since now $\operatorname{ind}_{\gamma_2}(z) = \operatorname{ind}_{\gamma_1}(z) = 1$.

C. From part B we deduce that in $\mathbb{C}\backslash\{z_0\}$ the cycles γ_1, $\gamma_1(t) = z_0 + r_1 e^{it}$, $t \in [0, 2\pi]$, and γ_2, $\gamma_2(t) = z_0 + r_2 e^{it}$, $t \in [0, 2\pi]$, $r_1 > r_2 > 0$, are homologous since $\gamma_1 \dot{-} \gamma_2 = \gamma_1 \dot{+} \gamma_2^{-1}$ is null-homologous in $\mathbb{C}\backslash\{z_0\}$.

Lemma 25.16. *On the set of all cycles in G homology is an equivalence relation.*

Proof. Clearly γ is homologous to itself since $\operatorname{ind}_{\gamma \dot{-} \gamma}(z) = \operatorname{ind}_{\gamma \dot{+} \gamma^{-1}}(z) = 0$, $z \in G^C$. Further, since

$$\operatorname{ind}_{\gamma \dot{-} \eta}(z) = \operatorname{ind}_{\gamma \dot{+} \eta^{-1}}(z) = \operatorname{ind}_\gamma(z) - \operatorname{ind}_\eta(z)$$
$$= -(\operatorname{ind}_\eta(z) - \operatorname{ind}_\gamma(z)) = -\operatorname{ind}_{\eta \dot{-} \gamma}(z),$$

we deduce $\operatorname{ind}_{\gamma \dot{-} \eta}(z) = 0$ if and only if $\operatorname{ind}_{\eta \dot{-} \gamma}(z) = 0$, i.e. homology is a symmetric relation. Now let $z \in G^C$ and $\gamma \dot{-} \eta$ as well as $\eta \dot{-} \tau$ be null-homologous cycles, i.e. $\operatorname{ind}_{\gamma \dot{+} \eta^{-1}}(z) = 0$ and $\operatorname{ind}_{\eta \dot{+} \tau^{-1}}(z) = 0$. For $\operatorname{ind}_{\gamma \dot{-} \tau^{-1}}(z) = \operatorname{ind}_{\gamma \dot{+} \tau^{-1}}(z)$ we find

$$\operatorname{ind}_{\gamma \dot{+} \tau^{-1}} = \operatorname{ind}_\gamma(z) - \operatorname{ind}_\tau(z)$$
$$= \operatorname{ind}_\gamma(z) - \operatorname{ind}_\eta(z) + \operatorname{ind}_\eta(z) - \operatorname{ind}_\tau(z)$$
$$= \operatorname{ind}_{\gamma \dot{+} \eta^{-1}} + \operatorname{ind}_{\eta \dot{+} \tau^{-1}}(z) = 0,$$

i.e. homology is transitive and the lemma is proved. □

We are now in a position to prove one of the main results in complex analysis, the general **Cauchy integral formula** for null-homologous cycles.

Theorem 25.17. *Let γ be a null-homologous cycle in the domain $D \subset \mathbb{C}$ and $f : D \to \mathbb{C}$ a holomorphic function. For all $k \in \mathbb{N}_0$ and all $z \in D\backslash\operatorname{tr}(\gamma)$ we have*

$$\operatorname{ind}_\gamma(z) f^{(k)}(z) = \frac{k!}{2\pi i} \int_\gamma \frac{f(\zeta)}{(\zeta - z)^{k+1}} \, d\zeta. \tag{25.18}$$

First we want to draw some consequences of Theorem 25.17 and then we will provide the proof.

Theorem 25.18. *Under the assumptions of Theorem 25.17 we find*

$$\int_\gamma f(\zeta)\, d\zeta = 0. \tag{25.19}$$

Proof. Let $z_1 \in D \setminus \operatorname{tr}(\gamma)$ and consider $h(z) = f(z)(z - z_1)$. This function is holomorphic on D and $h(z_1) = 0$. Now, using (25.18) we get

$$0 = \operatorname{ind}_\gamma(z_1) h(z_1) = \frac{1}{2\pi i} \int_\gamma \frac{h(\zeta)}{\zeta - z_1}\, d\zeta = \frac{1}{2\pi i} \int_\gamma f(\zeta)\, d\zeta.$$

$\square$

Theorem 25.19. *For two homologous cycles γ and η in D and every holomorphic function $f : D \to \mathbb{C}$ we have*

$$\int_\gamma f(\zeta)\, d\zeta = \int_\eta f(\zeta)\, d\zeta. \tag{25.20}$$

Proof. By assumption $\gamma \dot{-} \eta$ is null-homologous and therefore we have by Theorem 25.17

$$0 = \int_{\gamma \dot{-} \eta} f(z)\, dz = \int_\gamma f(z)\, dz - \int_\eta f(z)\, dz.$$

$\square$

Note that (25.20) is exactly the type of statement we are searching for: we replace the closed curve (the cycle) and the integral remains unchanged for all holomorphic functions.

Proof of Theorem 25.17. (Following [26]). First we note that if (25.18) holds for $k = 0$ it holds for all $k \in \mathbb{N}_0$. Indeed we can differentiate in

$$\operatorname{ind}_\gamma(z) f(z) = \frac{1}{2\pi i} \int_\gamma \frac{f(\zeta)}{(\zeta - z)}\, d\zeta, \tag{25.21}$$

on the right hand side k-times to obtain the right hand side in (25.18) and on the left hand side we only need to note that $\operatorname{ind}_\gamma(z)$ is constant on connectivity components. Thus we aim to prove (25.21). Since

$$\operatorname{ind}_\gamma(z) f(z) = \frac{1}{2\pi i} \int_\gamma \frac{f(z)}{\zeta - z}\, d\zeta,$$

483

we can rewrite (25.21) as

$$\int_\gamma \frac{f(\zeta) - f(z)}{\zeta - z} \, \mathrm{d}\zeta = 0, \quad z \in D \backslash \mathrm{tr}(\gamma).$$

We first study the function

$$g(\zeta, z) := \begin{cases} \frac{f(\zeta) - f(z)}{\zeta - z}, & \zeta \neq z, \\ f'(z), & \zeta = z \end{cases}$$

as function defined on $D \times D$. For $(\zeta_0, z_0) \in D \times D$ and $\zeta_0 \neq z_0$ the function g is continuous as function of two variables at (ζ_0, z_0). Now let $\zeta_0 = z_0$ and $\delta > 0$ to the determined later. We study the function $(\zeta, z) \mapsto g(\zeta, z) - g(\zeta_0, z_0)$ on $B_\delta(z_0) \times B_\delta(z_0)$ where we assume that $\overline{B_\delta(z_0)} \subset D$. In the case that $\zeta = z$ $(\in B_\delta(z_0))$ we note that

$$g(\zeta, z) - g(\zeta_0, z_0) = g(z, z) - g(z_0, z_0) = f'(z) - f'(z_0),$$

whereas for $\zeta \neq z$ we have

$$g(\zeta, z) - g(\zeta_0, z_0) = g(\zeta, z) - g(\zeta_0, z_0) = \frac{f(\zeta) - f(z)}{\zeta - z} - f'(z_0)$$

$$= \frac{1}{\zeta - z} \int_{[z,\zeta]} (f'(w) - f'(z_0)) \, \mathrm{d}w.$$

Since f is holomorphic, f' is continuous, in particular at z_0. Given $\epsilon > 0$ we can now choose $\delta > 0$ in such a way that for $w \in B_\delta(z_0)$ it follows that $|f'(w) - f'(z_0)| < \epsilon$. This implies in the case where $\zeta = z$ that

$$|g(z, z) - g(z_0, z_0)| = |f'(z) - f'(z_0)| < \epsilon,$$

and for $\zeta \neq z$ we find

$$|g(\zeta, z) - g(\zeta_0, z_0)| \leq \frac{1}{|\zeta - z|} |\zeta - z| \sup_{w \in [z,\zeta]} |f'(w) - f'(z_0)| < \epsilon.$$

Thus we have proved that g is on $D \times D$ continuous. Since $\mathrm{tr}(\gamma)$ is compact we immediately find that the function

$$h_0(z) := \int_\gamma g(\zeta, z) \, \mathrm{d}\zeta$$

484

is on D continuous. For a triangle $\triangle \subset D$ with anticlockwise orientated boundary $\partial \triangle \subset D$ we find

$$\int_{\partial \triangle} h_0(z)\, dz = \int_{\partial \triangle} \left(\int_\gamma g(\zeta, z)\, d\zeta \right) dz = \int_\gamma \left(\int_{\partial \triangle} g(\zeta, z)\, dz \right) d\zeta = 0,$$

since for all ζ fixed $z \mapsto g(\zeta, z)$ is holomorphic in $D\backslash\{\zeta\}$, but since g is continuous on D, this is a removable singularity, hence we have indeed by Goursat's theorem

$$\int_{\partial \triangle} h_0(z)\, dz = 0$$

for every such a triangle implying that h_0 is holomorphic in D. Let $\tilde{D} := \{z \in \mathbb{C} \,|\, \text{ind}_\gamma(z) = 0\}$. Since γ is null-homologous it follows that $D^{\complement} \subset \tilde{D}$ and therefore we have $\tilde{D} \cup D = \mathbb{C}$. On $\tilde{D} \cap D$ we have

$$h_0(z) = \int_\gamma \frac{f'(\zeta)}{\zeta - z}\, d\zeta$$

and since $\text{tr}(\gamma) \subset \tilde{D}^{\complement}$ we conclude that

$$h_1(z) := \int_\gamma \frac{f(\zeta)}{\zeta - z}\, d\zeta$$

is on $\tilde{D}$ holomorphic. Thus by

$$h(z) := \begin{cases} h_0(z), & z \in D, \\ h_1(z), & z \in \tilde{D} \end{cases}$$

an entire function is given and if $\text{tr}(\gamma) \subset B_R(0) \subset \overline{B_R(0)} \subset D$, then we have for $z \in \overline{B_R(0)}^{\complement}$ the representation

$$h(z) = \int_\gamma \frac{f(\zeta)}{\zeta - z}\, d\zeta. \tag{25.22}$$

We can estimate $h(z)$ in (25.22) according to

$$|h(z)| \leq \frac{1}{\text{dist}(z, \gamma)} l_\gamma \|f\|_{\infty, \text{tr}(\gamma)} \tag{25.23}$$

where for $\gamma = a_1\gamma_1 \tilde{+} \cdots \tilde{+} a_M\gamma_M$ we have $l_\gamma = \sum_{k=1}^{M} |a_k| l_{\gamma_k}$. On the compact set $\overline{B_R(0)}$ the holomorphic function h is bounded and by (25.23) it is also bounded on $\overline{B_R(0)}^{\complement}$. Hence h is a bounded entire function and by Liouville's theorem h is a constant. Our aim is to prove that this constant is zero, which will imply h_0 is identically zero, hence $\int_\gamma g(\zeta, z) \, d\zeta = 0$ for $z \notin \mathrm{tr}(\gamma)$ and therefore (25.18) will follow. Again the compactness of $\mathrm{tr}(\gamma)$ helps. For a sequence $(z_k)_{k\in\mathbb{N}}$ of complex numbers such that $z_k \in \overline{B_R(0)}^{\complement}$ and $\lim_{k\to\infty} |z_k| = \infty$ the estimate (25.23) implies $\lim_{k\to\infty} |h(z_k)| = 0$. $\qquad\square$

We can prove the **residue theorem** for null-homologous cycles.

Theorem 25.20. *Let $G \subset \mathbb{C}$ be a region and f a function which is on G holomorphic with the exception of isolated singularities. For every null-homologous cycle γ in G on the trace of which are no singularities of f we have*

$$\frac{1}{2\pi i} \int_\gamma f(\zeta) \, d\zeta = \sum_{z\in G} \mathrm{ind}_\gamma(z) \, \mathrm{res}(f, z). \tag{25.24}$$

Remark 25.21. Since $\mathrm{tr}(\gamma)$ is compact, hence $\mathrm{tr}(\gamma) \subset \overline{B_R(0)}$ for some $R > 0$, outside of the compact set $\overline{B_R(0)}$ we have $\mathrm{ind}_\gamma(z) = 0$ which implies that in the sum of the right hand side of (25.24) only finitely many terms do not vanish, i.e. the sum is a finite one.

Proof of Theorem 25.20. Let γ be a null-homologous cycle in G. We decompose the set $\mathrm{sing}(f)$ of singularities of f according to

$$\mathrm{sing}(f) = \mathrm{sing}(f, \gamma) \cup \mathrm{sing}\left(f, \gamma^{\complement}\right)$$

where

$$\mathrm{sing}(f, \gamma) := \left\{ z \in \mathrm{sing}(f) \,\middle|\, \mathrm{ind}_\gamma(z) \neq 0 \right\}$$

and

$$\mathrm{sing}(f, \gamma^{\complement}) := \left\{ z \in \mathrm{sing}(f) \,\middle|\, \mathrm{ind}_\gamma(z) = 0 \right\}.$$

By our assumptions it follows that $\mathrm{sing}(f, \gamma)$ is a finite set, say $\mathrm{sing}(f, \gamma) = \{z_1, \ldots, z_M\}$. For $z_k \in \mathrm{sing}(f, \gamma)$ we denote the principal part of its Laurent expansion about z_k by f_{z_k}. On $\mathbb{C}\backslash\{z_k\}$ the function f_{z_k} is holomorphic and

486

therefore $f - \sum_{k=1}^{M} f_{z_k}$ is holomorphic in $G \backslash \operatorname{sing}(f, \gamma^C)$. Since γ is null-homologous in G it is also null-homologous in $G \backslash (f, \gamma^C)$ and by Theorem 25.17 (or Theorem 25.18) we find

$$0 = \int_{\gamma} \left(f - \sum_{k=1}^{M} f_{z_k} \right)(\zeta) \, d\zeta$$

or

$$\int_{\gamma} f(\zeta) \, d\zeta = \sum_{k=1}^{M} \int_{\gamma} f_{z_k}(\zeta) \, d\zeta.$$

By compactness of $\operatorname{tr}(\gamma)$ the Laurent series $f_{z_k}(z) = \sum_{l=1}^{\infty} a_{-l}^{(k)}(z - z_l)^{-l}$ converges uniformly on $\operatorname{tr}(\gamma)$ and consequently

$$\frac{1}{2\pi i} \int_{\gamma} f_{z_k}(\zeta) \, d\zeta = \sum_{l=1}^{\infty} \frac{a_{-l}^{(k)}}{2\pi i} \int_{\gamma} (\zeta - z_k)^{-l} \, d\zeta$$

$$= a_{-1}^{(k)} \frac{1}{2\pi i} \int_{\gamma} \frac{1}{\zeta - z_k} \, d\zeta = \operatorname{ind}_{\gamma}(z_k) \operatorname{res}(f, z_k)$$

implying the theorem. $\qquad\square$

With Theorem 25.17 and Theorem 25.20 we have proved some of the main results of classical complex analysis in their most general form. Of course there is a need to explore the notion of homology, in particular of null-homology in more detail. In particular the relation of simply connected sets, i.e. sets with a trivial first fundamental group, and null-homology is of interest. This leads to more topological discussions and we will postpone this to our final volume.

We will now apply the residue theorem to evaluate certain integrals. For this we need first of all an easier way to find residues. By its very definition the residue of f at z_0 is the coefficient c_{-1} of the principal part of the Laurent expansion of f about z_0, provided z_0 is an isolated singularity of f and we have

$$\operatorname{res}(f, z_0) = c_{-1} = \frac{1}{2\pi i} \int_{|\zeta - z_0| = \rho} f(\zeta) \, d\zeta, \tag{25.25}$$

for every circle $|\zeta - z| = \rho$, such that z_0 is the only singularity of f in $\overline{B_{\rho}(z_0)}$. Since we want to use residues to calculate integrals we are looking for a different way to find $\operatorname{res}(f, z_0)$. Note that finding the Laurent expansion might also be troublesome. Here comes an easier way for poles. First suppose that the pole is of first order. Then it follows that in some annulus $A_{0,r}(z_0)$ we have

$$f(z) = \sum_{k=-1}^{\infty} c_k(z - z_0)^k \tag{25.26}$$

and

$$(z - z_0)f(z) = \sum_{k=-1}^{\infty} c_k(z - z_0)^{k+1}$$

implying

$$\lim_{z \to z_0} (z - z_0)f(z) = \sum_{k=-1}^{\infty} c_k \lim_{z \to z_0} (z - z_0)^{k+1} = c_{-1}. \tag{25.27}$$

In the case that z_0 is a pole of order n we find

$$f(z) = \sum_{k=-n}^{\infty} c_k(z - z_0)^k \tag{25.28}$$

and therefore

$$(z - z_0)^n f(z) = \sum_{k=-n}^{\infty} c_k(z - z_0)^{k+n} = \sum_{k=0}^{\infty} a_{k-n}(z - z_0)^k.$$

This is a holomorphic function in $B_r(z_0)$ and therefore we get

$$\frac{d^{n-1}}{dz^{n-1}}((z - z_0)^n f(z)) = \sum_{k=n-1}^{\infty} a_{k-n}(n - 1)!(z - z_0)^{k-n+1},$$

implying

Proposition 25.22. *Let the holomorphic function f have a pole of order n at z_0. For the residue $\mathrm{res}(f, z_0)$ we find*

$$\mathrm{res}(f, z_0) = \lim_{z \to z_0} \frac{d^{n-1}}{dz^{n-1}}\left(\frac{1}{(n - 1)!}(z - z_0)^n f(z)\right). \tag{25.29}$$

Example 25.23. A. The rational function $R(z) = \frac{z}{(z-1)(z+1)^2}$ has the pole $z_{0,1} = 1$ of order 1 and the pole $z_{0,2} = -1$ of order 2. The corresponding residues are

$$\mathrm{res}(R, 1) = \lim_{z \to 1} \left((z-1) \frac{z}{(z-1)(z+1)^2} \right) = \frac{1}{4}$$

and

$$\mathrm{res}(R, -1) = \lim_{z \to -1} \left(\frac{1}{1!} \frac{\mathrm{d}}{\mathrm{d}z} \left((z+1)^2 \frac{z}{(z-1)(z+1)^2} \right) \right) = -\frac{1}{4}.$$

B. The functions $z \mapsto \exp\left(-\frac{1}{z}\right)$ has an essential singularity at $z_0 = 0$, so we cannot apply Proposition 25.22. However we know its Laurent series about $z_0 = 0$ which is $\sum_{k=0}^{\infty} (-1)^k \frac{1}{k!} z^{-k}$ with principal part $\sum_{k=1}^{\infty} (-1)^k \frac{1}{k!} z^{-k}$. This implies that

$$\mathrm{res}\left(\exp\left(-\frac{1}{z}\right), 0 \right) = -1.$$

A further useful rule with trivial proof is

Proposition 25.24. *Let the holomorphic f have a pole of first order at z_0 and let g be holomorphic in a neighbourhood of z_0 with $g(z_0) \neq 0$. In this case we have*

$$\mathrm{res}((f \cdot g), z_0) = g(z_0) \, \mathrm{res}(f, z_0). \tag{25.30}$$

Proof. First we note that in this case $f \cdot g$ has at z_0 a pole of first order and it follows that

$$\lim_{z \to z_0} (z - z_0)(f(z)g(z)) = \lim_{z \to z_0} g(z) \lim_{z \to z_0} (z - z_0)f(z) = g(z_0) \, \mathrm{res}(f, z_0).$$

$\square$

We now will evaluate quite a few integrals of (mainly) real-valued functions defined on an interval using the residue theorem. We do not long for a systematical treatment which can be found in [26] or [67], to mention just some sources, but we want to provide some of the basic ideas.

Convention: in all examples, if not otherwise stated, circles or parts of a circle are parametrized in the standard way, i.e. $|z - z_0| = \rho$ is parametrized by $\gamma(t) = z_0 + \rho e^{it}$, $t \in [0, 2\pi]$ (or a subinterval). Furthermore, whenever we consider a simply closed curve its parametrization is anti-clockwise oriented. Moreover, as before $[z_1, z_2]$ stands for the line segment connecting z_1 with z_2 usually parametrized as $z_1 + t(z_2 - z_1)$, $t \in [0, 1]$.

Example 25.25. A. We want to show that

$$\int_0^{2\pi} \frac{\cos 2\vartheta}{2 - \cos \vartheta} \, d\vartheta = \pi \left(\frac{1 + (2 - \sqrt{3})^4}{2\sqrt{3}(2 - \sqrt{3})^2} - 8 \right). \tag{25.31}$$

With $z = e^{i\vartheta}$ we find $\cos \vartheta = \frac{z + z^{-1}}{2}$ and $\cos 2\vartheta = \frac{z^2 - z^{-2}}{2}$ and it follows that

$$\int_0^{2\pi} \frac{\cos 2\vartheta}{2 - \cos \vartheta} \, d\vartheta = \int_{|z|=1} \frac{z^2 + z^{-2}}{2} \left(2 - \frac{z + z^{-1}}{2} \right)^{-1} \frac{1}{iz} \, dz$$

$$= \int_{|z|=1} \frac{i(z^4 + 1)}{z^2(z^2 - 4z + 1)} \, dz = \int_{|z|=1} \frac{i(z^4 + 1)}{z^2(z - 2 - \sqrt{3})(z - 2 + \sqrt{3})} \, dz.$$

On the circle $|z| = 1$ the integral has no singularity and in the interior of $|z| = 1$, i.e. in $B_1(0)$, we have a pole of order 2 at $z_1 = 0$ and a pole of order 1 at $z_2 = 2 - \sqrt{3}$. Note that $2 + \sqrt{3} > 1$, so $z_3 = 2 + \sqrt{3}$ is a pole of order 1 outside of $B_1(0)$. The residues can be determined by using Proposition 25.22 and we find with $f(z) = \frac{i(z^4 + 1)}{z^2(z^2 - 4z + 1)}$

$$\text{res}(f, 0) = \lim_{z \to 0} \frac{d}{dz} \left(\frac{1}{1!} z^2 f(z) \right) = \lim_{z \to 0} \frac{d}{dz} \left(\frac{i(z^4 + 1)}{z^2 - 4z + 1} \right) = 4i$$

and

$$\text{res}(f, 2 - \sqrt{3}) = \lim_{z \to 2 - \sqrt{3}} (z - 2 + \sqrt{3}) f(z) = \lim_{z \to 2 - \sqrt{3}} \left(\frac{i(z^4 + 1)}{z^2(z - 2 - \sqrt{3})} \right)$$

$$= \left(-\frac{1 + (2 - \sqrt{3})^4}{2\sqrt{3}(2 - \sqrt{3})^2} \right) i$$

which yields

$$\int_0^{2\pi} \frac{\cos 2\vartheta}{2 - \cos \vartheta} \, d\vartheta = 2\pi i \left(\text{res}(f, 0) + \text{res}(f, 2 - \sqrt{3}) \right)$$

$$= \pi \frac{1 + (2 - \sqrt{3})^4}{2\sqrt{3}(2 - \sqrt{3})^2} - 8\pi.$$

B. The following holds for $a > |b|$

$$\int_0^{2\pi} \frac{d\varphi}{a + b \sin \varphi} = \frac{2\pi}{\sqrt{a^2 - b^2}}.$$

Again we substitute $z = e^{i\varphi}$ and now with $\sin\varphi = \frac{z-z^{-1}}{2i}$ we arrive at

$$\int_0^{2\pi} \frac{d\varphi}{a+b\sin\varphi} = \int_{|z|=1} \frac{1}{iz\left(a+b\left(\frac{z-z^{-1}}{2i}\right)\right)}\,dz = \int_{|z|=1} \frac{2}{b^2z^2 + 2ai - b}\,dz.$$

The integrand has poles of first order at

$$z_{1,2} = -\frac{a}{b}i \pm \frac{i}{b}\sqrt{a^2 - b^2}$$

and $|z_1| < 1$ whereas $|z_2| > 1$ which follows from $|b| < a$. With $h(z) = \frac{2}{b^2z^2+2ai-b}$ we find

$$\int_0^{2\pi} \frac{d\varphi}{a+b\sin\varphi} = \int_{|z|=1} \frac{2}{b^2z^2 + 2ai - b}\,dz$$

$$= 2\pi i\,\mathrm{res}(h, z_1)$$

$$= 2\pi i \lim_{z\to z_1}\left((z - z_1)h(z)\right) = 2\pi i\left(\frac{1}{i\sqrt{a^2 - b^2}}\right)$$

$$= \frac{2\pi}{\sqrt{a^2 - b^2}}.$$

The integrals in Example 25.25 are of the type

$$\int_0^{2\pi} R(\cos\varphi, \sin\varphi)\,d\varphi$$

where $R(x, y)$ is a rational function in two real variables. With the substitution $z = e^{i\varphi}$, $\varphi \in [0, 2\pi]$, they transform to

$$\int_0^{2\pi} R\left(\frac{1}{2}\left(e^{i\varphi} + e^{-i\varphi}\right), \frac{1}{2i}\left(e^{i\varphi} - e^{-i\varphi}\right)\right)d\varphi$$

$$= \frac{1}{i}\int_{|z|=1} R\left(\frac{1}{2}\left(z + \frac{1}{z}\right), \frac{1}{2i}\left(z - \frac{1}{z}\right)\right)\frac{dz}{z} = \int_{|z|=1} h(z)\,dz.$$

Applying the residue theorem we find

$$\int_0^{2\pi} R(\cos\varphi, \sin\varphi)\,d\varphi = 2\pi \sum_{|z|<1} \mathrm{res}(h, z) \qquad (25.32)$$

where the sum is taken over all poles of $h(z)$ in $B_1(0)$. Note since we have assumed that the (improper) integrals exists, there cannot be a non-integrable singularity of $R(\cos\varphi, \sin\varphi)$ in $[0, 2\pi]$.

491

Example 25.26. We want to give a further proof for

$$\int_{-\infty}^{\infty} \frac{dx}{1+x^2} = \pi.$$

We know the result already since a primitive of $\frac{1}{1+x^2}$ is arctan. We extend $x \mapsto \frac{1}{1+x^2}$ to the rational function $z \mapsto \frac{1}{1+z^2} = \frac{1}{(z-i)(z+i)}$ on $\mathbb{C}$ which has in the upper half plane $\operatorname{Im} z \geq 0$ the pole of first order $z_1 = i$. For the curve $[-R, R] + \kappa_R^+$, $\operatorname{tr}(\kappa_R^+)$ is the upper half circle with centre 0 and radius R, see Figure 25.3, we find provided that $R > 1$

$$\int_{[-R,R]\oplus\kappa_R^+} \frac{1}{1+z^2} \, dz = 2\pi i \operatorname{res}\left(\frac{1}{1+z^2}, i\right)$$

$$= 2\pi i \lim_{z \to i}\left((z-i)\frac{1}{(z-i)(z+i)}\right) = \pi.$$

On the other hand we find

$$\int_{[-R,R]\oplus\kappa_R^+} \frac{1}{1+z^2} \, dz = \int_{[-R,R]} \frac{1}{1+z^2} \, dz + \int_{\kappa_R^+} \frac{1}{1+z^2} \, dz$$

$$= \int_{-R}^{R} \frac{1}{1+x^2} \, dx + \int_{0}^{\pi} \frac{Rie^{it}}{1+(Re^{it})^2} \, dt.$$

For $R \to \infty$ the first integral tends to $\int_{-\infty}^{\infty} \frac{1}{1+x^2} \, dx$, while for the second integral we use the estimate

$$\left|\frac{Rie^{it}}{1+(Re^{it})^2}\right| = \frac{R}{(1+R^4+2R^2\cos 2t)^{\frac{1}{2}}} \leq \frac{R}{(1+R^4-2R^2)^{\frac{1}{2}}} = \frac{R}{R^2-1}$$

which holds for $R > 1$. This estimate implies however

$$\lim_{R\to\infty} \int_{0}^{R} \frac{Rie^{it}}{1+(Re^{it})^2} \, dt = 0.$$

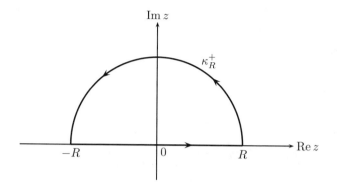

Figure 25.3

Example 25.27. We claim

$$\int_0^\infty \frac{\cos x}{\alpha^2 + x^2}\, dx = \frac{\pi}{2\alpha} e^{-\alpha}, \quad \alpha > 0. \tag{25.33}$$

Due to the symmetry of the integrand we need to prove that

$$\int_0^\infty \frac{\cos x}{\alpha^2 + x^2}\, dx = \frac{1}{2}\int_{-\infty}^\infty \frac{\cos x}{\alpha^2 + x^2}\, dx = \frac{1}{2}\operatorname{Re}\int_{-\infty}^\infty \frac{e^{ix}}{\alpha^2 + x^2}\, dx = \frac{\pi}{2\alpha}e^{-\alpha}.$$

For $R > \alpha$ we consider the simply closed, piecewise continuously differentiable curve $\gamma_R := [-R, R] \oplus [R, R + iR] \oplus [R + iR, -R + iR] \oplus [-R + iR, -R]$ see Figure 25.4

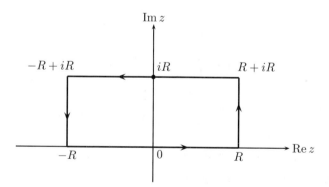

Figure 25.4

With $h(z) = \frac{e^{iz}}{(z-i\alpha)(z+i\alpha)}$ it follows that

$$\int_{\gamma_R} \frac{e^{iz}}{\alpha^2 + z^2}\, dz = \int_{\gamma_R} \frac{e^{iz}}{(z - i\alpha)(z + i\alpha)}\, dz = 2\pi i \operatorname{res}(h, i\alpha) = \frac{\pi}{\alpha}e^{-\alpha},$$

493

where we used that $i\alpha$ is the only pole (and of first order) of h in the interior of γ_R and therefore

$$\operatorname{res}(h, i\alpha) = \lim_{z \to i\alpha} ((z - i\alpha)h(z)) = \lim_{z \to i\alpha} \frac{e^{iz}}{z + i\alpha} = \frac{e^{-\alpha}}{2i\alpha}.$$

Now we note that

$$\int_{-R}^{R} \frac{e^{iz}}{\alpha^2 + z^2}\, dz = \int_{-R}^{R} \frac{\cos x + i \sin x}{\alpha^2 + x^2}\, dx,$$

or

$$\lim_{R \to \infty} \operatorname{Re} \int_{-R}^{R} \frac{e^{iz}}{\alpha^2 + z^2}\, dz = \int_{-\infty}^{\infty} \frac{\cos x}{\alpha^2 + x^2}\, dx.$$

We want to show that for $R \to \infty$ the remaining three integrals, i.e. $\int_{\gamma_j} \frac{e^{iz}}{\alpha^2+z^2}\, dz,\ j = 2, 3, 4$, vanish. For γ_2, $\gamma_2(t) = R + itR,\ t \in [0, 1]$, we have

$$\int_{\gamma_2} \frac{e^{iz}}{\alpha^2 + z^2}\, dz = \int_0^1 \frac{e^{i(R+itR)}}{\alpha^2 + (R + itR)^2}(iR)\, dt.$$

Since for $R > \alpha$ it follows that

$$|\alpha^2 + (R(1 + it)^2)| \geq |R(1 + it)|^2 - \alpha^2 = R^2(1 + t^2) - \alpha^2 \geq R^2 - \alpha^2,$$

we find

$$\left| \int_{\gamma_2} \frac{e^{it}}{\alpha^2 + z^2}\, dz \right| \leq \int_0^1 \frac{Re^{-tR}}{|\alpha^2 + R^2(1 + it)^2|}\, dt$$

$$\leq \int_0^1 \frac{Re^{-tR}}{R^2 - \alpha^2}\, dt = \frac{1 - e^{-R}}{R^2 - \alpha^2},$$

which tends to 0 for $R \to \infty$. Analogously we find that $\left| \int_{\gamma_4} \frac{e^{iz}}{\alpha^2+z^2}\, dz \right|$ tends to 0 as $R \to \infty$. For the final integral we have

$$\int_{\gamma_3} \frac{e^{iz}}{\alpha^2 + z^2}\, dz = \int_0^1 \frac{e^{i(R+iR-2tR)}}{\alpha^2 + (R + iR - 2tR)^2}(-2R)\, dt$$

$$= -2 \int_0^1 \frac{e^{i(R-2tR)}e^{-R}R}{\alpha^2 + R^2(1 - 2t + i)^2}\, dt.$$

494

Now we use that

$$|R^2(1 - 2t + i)^2 + \alpha^2| \geq R^2|1 - 2t + i|^2 - \alpha^2 \geq R^2 - \alpha^2$$

to obtain

$$\left| \int_{\gamma_3} \frac{e^{iz}}{\alpha^2 + z^2} \, dz \right| \leq 2 \int_0^1 \frac{Re^{-R}}{R^2 - \alpha^2} \, dt = 2 \frac{Re^{-R}}{R^2 - \alpha^2} \to 0 \ \text{ as } R \to \infty.$$

Hence we have proved

$$\lim_{R \to \infty} \text{Re} \int_{\gamma_R} \frac{e^{iz}}{\alpha^2 + z^2} \, dz = \int_{-\infty}^{\infty} \frac{\cos x}{\alpha^2 + x^2} \, dx = \frac{\pi}{\alpha} e^{-\alpha}$$

implying (25.33).

The insight we get from Example 25.26 and 25.27 is the following: let $g : I \to \mathbb{R}$ be a function defined on an unbounded interval which is improperly Riemann integrable. Suppose that there exists a region $G \subset \mathbb{C}$ such that with the exception of finitely many isolated singularities $z_1, \ldots, z_N$ the function g has a holomorphic extension to G. For simplicity we denote this extension again by g. Suppose further that for every $R \geq R_0$ we can find a simply closed piecewise continuously differentiable curve γ_R such that the interior of γ_R contains all the singularities of g and $\text{tr}(\gamma_R) \subset G$. Moreover assume that

$$\lim_{R \to \infty} \int_{\gamma_R} g(z) \, dz = \int_I g(x) \, dx.$$

In this case we have

$$\int_I g(x) \, dx = 2\pi i \sum_{k=1}^{\infty} \text{res}(g, z_j). \tag{25.34}$$

Of course, the problem is to find a good family of curves γ_R, but classical books in complex analysis, see [36] or [93] just to mention two of them, provide a lot of results in this direction.

It is also possible to use the residue theorem in some cases for integrals having a singularity on the interval or the endpoints of the interval of integration.

Example 25.28. Using the residue theorem we will prove

$$\int_0^\infty \frac{\sin x}{x}\, dx = \frac{\pi}{2}. \tag{25.35}$$

The integral suggests to look at an integral over some curve for the function $z \mapsto \frac{e^{iz}}{z}$. While $\lim_{x \to 0} \frac{\sin x}{x} = 1$ assures that $x \mapsto \frac{\sin x}{x}$ is at $x = 0$ a continuous function, for $z \mapsto \frac{e^z}{z}$ the value $z_0 = 0$ is a pole and it is located at the boundary of $[0, \infty)$ (or $(0, \infty)$) and hence it causes at first glance a problem. For $\epsilon > 0$ and $R > 0$ we consider the curve $\gamma_{\epsilon, R} = \kappa_R^+ \oplus [-R, -\epsilon] \oplus (\kappa_\epsilon^+)^{-1} + [\epsilon, R]$, see Figure 25.5, where as before κ_R^+ is the upper half-circle with centre 0 and radius R parametrized as $\kappa_R^+(\varphi) = Re^{i\varphi}$, $\varphi \in [0, \pi]$.

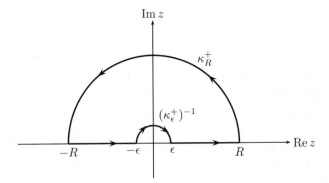

Figure 25.5

Since $z \mapsto \frac{e^{iz}}{z}$ is in $\mathbb{C}\setminus\{0\}$ holomorphic and 0 is not a point on $\gamma_{\epsilon, R}$ we find $\int_{\gamma_{\epsilon, R}} \frac{e^{iz}}{z}\, dz = 0$, or

$$\int_\epsilon^R \frac{e^{ix}}{x}\, dx + \int_{-R}^{-\epsilon} \frac{e^{ix}}{x}\, dx = -\int_{(\kappa_\epsilon^+)^{-1}} \frac{e^{iz}}{z}\, dz - \int_{\kappa_R^+} \frac{e^{iz}}{z}\, dz$$

leading to

$$\int_\epsilon^R \frac{e^{ix} - e^{-ix}}{x}\, dx = \int_{\kappa_\epsilon^+} \frac{e^{iz}}{z}\, dz - \int_{\kappa_R^+} \frac{e^{iz}}{z}\, dz$$

i.e.

$$\int_\epsilon^\infty \frac{\sin x}{x}\, dx = \frac{1}{2i}\left(\int_{\kappa_\epsilon^+} \frac{e^{iz}}{z}\, dz - \int_{\kappa_R^+} \frac{e^{iz}}{z}\, dz \right).$$

Now we observe that

$$\frac{1}{2i}\int_{\kappa_\epsilon^+}\frac{e^{iz}}{z}\,\mathrm{d}z = \frac{1}{2\pi i}\int_0^\pi \frac{e^{i\epsilon e^{i\varphi}}}{\epsilon e^{i\varphi}}\,i\epsilon e^{i\varphi}\,\mathrm{d}\varphi$$

$$= \frac{1}{2}\int_0^\pi e^{i\epsilon e^{i\varphi}}\,\mathrm{d}\varphi$$

and in the limit $\epsilon \to 0$ we get

$$\lim_{\epsilon\to 0}\frac{1}{2\pi}\int_{\kappa_\epsilon^+}\frac{e^z}{z}\,\mathrm{d}z = \pi.$$

It remains to prove that

$$\lim_{R\to\infty}\frac{1}{2i}\int_{\kappa_R^+}\frac{e^{iz}}{z}\,\mathrm{d}z = 0.$$

For this we note

$$\left|\frac{1}{2i}\int_{\kappa_R^+}\frac{e^{iz}}{z}\,\mathrm{d}z\right| = \left|\frac{1}{2i}\int_0^\pi \frac{e^{iRe^{i\varphi}}}{Re^{i\varphi}}iRe^{i\varphi}\,\mathrm{d}\varphi\right|$$

$$\leq \frac{1}{2}\int_0^\pi \left|\frac{e^{iR\cos\varphi - R\sin\varphi}}{Re^{i\varphi}}Re^{i\varphi}\right|\,\mathrm{d}\varphi$$

$$= \frac{1}{2}\int_0^\pi e^{-R\sin\varphi}\,\mathrm{d}\varphi = \int_{\frac{\pi}{2}}e^{-R\sin\vartheta}\,\mathrm{d}\vartheta,$$

where we used the symmetry of the sine function with respect to the axis passing through $\frac{\pi}{2}$ and being orthogonal to the ϑ-axis. Using again the estimate $\sin\vartheta \geq \frac{2\vartheta}{\pi}$ for $0 \leq \vartheta \leq \frac{\pi}{2}$, see Example 21.6, we find

$$\int_0^{\frac{\pi}{2}}e^{-R\sin\vartheta}\,\mathrm{d}\vartheta \leq \int_0^{\frac{\pi}{2}}e^{-2R\frac{\vartheta}{\pi}}\,\mathrm{d}\vartheta = \frac{\pi}{2R}(1 - e^{-R})$$

and we have indeed proved that

$$\lim_{R\to\infty}\frac{1}{2i}\int_{\kappa_R^+}\frac{e^z}{z}\,\mathrm{d}z = 0$$

implying eventually (25.35).

Finally we want to discuss an integral of the type

$$\frac{1}{\sqrt{2\pi}} \int_{\mathbb{R}} e^{-ix\xi} g(x)\, dx. \tag{25.36}$$

We will learn in Part 8 that (25.36) is the Fourier transform of g. In Example II.22.19 we have already seen that

$$\frac{1}{\sqrt{2\pi}} \int_{\mathbb{R}} e^{-ix\xi} e^{-\frac{x^2}{2}}\, dx = \frac{1}{\sqrt{2\pi}} \int_{\mathbb{R}} \cos x\xi\, e^{-\frac{x^2}{2}}\, dx = e^{-\frac{\xi^2}{2}}.$$

Now we want to find the Fourier transform of $f(x) = \frac{1}{x^2+a^2}$.

Example 25.29. The Fourier transform of $f(x) = \frac{1}{x^2+a^2}$, $a > 0$, is the function

$$\frac{1}{\sqrt{2\pi}} \int_{\mathbb{R}} e^{-ix\xi} \frac{1}{x^2+a^2}\, dx = \sqrt{\frac{\pi}{2}} \frac{e^{-|\xi|a}}{a}. \tag{25.37}$$

We can extend $f_\xi(x) := \frac{e^{-ix\xi}}{x^2+a^2}$, $x \in \mathbb{R}$, to the function $f_\xi(z) = \frac{e^{-iz\xi}}{z^2+a^2} = \frac{e^{-iz\xi}}{(z-ia)(z+ia)}$ which is holomorphic in $\mathbb{C}\setminus\{-ia, ia\}$ and at $z_1 = ia$, $z_2 = -ia$ we have poles of first order. However we have a problem with the growth of $z \mapsto e^{-iz\xi}$ depending on ξ. With $z = x + iy$ we find $e^{-iz\xi} = e^{y\xi} e^{-ix\xi}$. In order to proceed further we must note that $e^{y\xi}$ need not be bounded, but it will be so long as $y\xi < 0$. Therefore we discuss two cases, first let $\xi < 0$ and $y = \operatorname{Im} z > 0$. In the upper half plane $\operatorname{Im} z > 0$ we have the pole ia and we choose the curve $\gamma_R := [-R, R] \oplus \kappa_R^+$, $R > ia$. It follows that

$$\frac{1}{\sqrt{2\pi}} \int_{\gamma_R} \frac{e^{-iz\xi}}{z^2+a^2}\, dz = \frac{1}{\sqrt{2\pi}} \int_{\gamma_R} \frac{e^{-iz\xi}}{(z-ia)(z+ia)}\, dz = \sqrt{2\pi} i \operatorname{res}(f_\xi, ia)$$

and

$$\operatorname{res}(f_\xi, ia) = \lim_{z \to ia} \left((z-ia)\frac{e^{-iz\xi}}{(z-ia)(z+ia)} \right) = \frac{e^{\xi a}}{2ia} = \frac{e^{-|\xi|a}}{2ia}$$

or, for $\xi < a$

$$\frac{1}{\sqrt{2\pi}} \int_{\gamma_R} \frac{e^{-iz\xi}}{z^2+a^2}\, dz = \frac{\sqrt{\pi}}{\sqrt{2a}} e^{-|\xi|a}.$$

On the other hand we find

$$\frac{1}{\sqrt{2\pi}} \int_{\gamma_R} \frac{e^{-iz\xi}}{z^2+a^2}\, dz = \frac{1}{\sqrt{2\pi}} \int_{-R}^{R} \frac{e^{-ix\xi}}{x^2+a^2}\, dx + \frac{1}{\sqrt{2\pi}} \int_{\kappa_R^+} \frac{e^{-iz\xi}}{z^2+a^2}\, dz$$

498

and further with arguments as in Example 25.28

$$\left| \int_{\kappa_R^+} \frac{e^{-iz\xi}}{z^2+a^2} \, dz \right| = \left| \int_0^\pi \frac{e^{-iR\xi\cos\varphi+\xi R\sin\varphi}}{(Re^{i\varphi})^2+a^2} iRe^{i\varphi} \, d\varphi \right|$$

$$\leq \int_0^\pi \frac{e^{-|\xi|R\sin\varphi}}{R^2-a^2} R \, d\varphi \leq \frac{R}{R^2-a^2} \int_0^{\frac{\pi}{2}} e^{-2|\xi|R\frac{\vartheta}{\pi}} \, d\vartheta$$

$$= \frac{R}{R^2-a^2} \cdot \frac{1}{2|\xi|R} \left(1 - e^{-|\xi|R} \right),$$

i.e.

$$\lim_{R\to\infty} \int_{\kappa_R^+} \frac{e^{-iz\xi}}{z^2+a^2} \, dz = 0,$$

implying for $\xi < 0$ that

$$\frac{1}{\sqrt{2\pi}} \int_{\mathbb{R}} \frac{e^{-ix\xi}}{x^2+a^2} \, dx = \sqrt{\frac{\pi}{2}} \frac{e^{-|\xi|a}}{a}.$$

For $\xi > 0$ we choose the pole $-ia$ and a similar calculation now yields

$$\frac{1}{\sqrt{2\pi}} \int_{\mathbb{R}} \frac{e^{-ix\xi}}{x^2+a^2} \, dx = \sqrt{\frac{\pi}{2}} \frac{e^{-\xi a}}{a} = \sqrt{\frac{\pi}{2}} \frac{e^{-|\xi|a}}{a}$$

and (25.37) is proved. This results should be compared with Example 25.27.

In Part 8 we will find further Fourier transforms by applying the residue theorem.

Problems

1. Let $K \subset \mathbb{C}$ be a convex set such that ∂K is a polygon with vertices $A_1, \ldots, A_N$. Prove that we can interpret ∂K as a cycle.

2. In the following all line segments are parametrized with the help of the unit interval $[0, 1]$. Consider the following figure

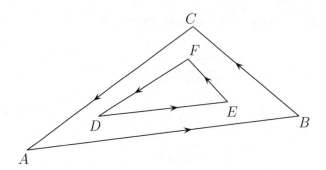

Why can we interpret this figure as the trace of a chain? Is this chain a cycle?

3. Consider the following figure in $G = B_2(z_0)$:

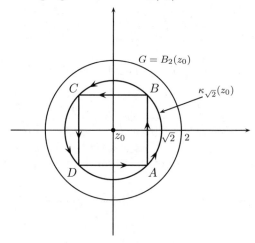

where $ABCD = \mathrm{tr}(\eta)$, $\eta = [A, B] \oplus [B, C] \oplus [C, D] \oplus [D, A]$. Are $\kappa_{\sqrt{2}}(z_0)$ and η cycles? Are they homologous? (For all curves appearing in the figure we use our standard parametrizations.)

4. Let $\Delta = \Delta(A, B, C) \subset \mathbb{C}$ be the open triangle with boundary $ABC = \mathrm{tr}(\gamma) = \mathrm{tr}([A, B] \oplus [B, C] \oplus [C, A])$, see the figure below.

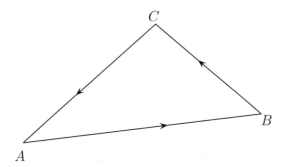

For $z \in \mathbb{C} \setminus \partial \Delta$ find $\text{ind}_\gamma(z)$.

5. Denote by $\tau_a : \mathbb{C} \to \mathbb{C}$, $\tau_a(z) = z + a$, $a \in \mathbb{C}$, the translation by a. Let $g : U \setminus \{z_0\} \to \mathbb{C}$ be a holomorphic function. Define the function $h := g \circ \tau_a$ and prove that $\text{res}(g, z_0) = \text{res}(g \circ \tau_a, z_0 - a)$.

6. Find the residue of

   a) $f(z) = \frac{z^3 - z^2 + 2z}{(z+1)^2(z^2+4)}$ at $z_1 = -1$ and $z_{2,3} = \pm 2i$;

   b) $g(z) = \frac{e^{2z}}{\sin^2 z}$ at $z = k\pi i$, $k \in \mathbb{Z}$.

7. Prove the following:

   Proposition. Let $p(z) = \sum_{k=0}^{N} a_k z^k$ and $q(z) = \sum_{l=0}^{M} b_l z^l$ be two polynomials with $M \geq N + 2$. Suppose that q has no real zeroes and $c_1, \ldots, c_k$ are the poles of $f := \frac{p}{q}$ in the upper plane $\text{Im } z > 0$. Then we have

   $$\int_{-\infty}^{\infty} f(x)dx = 2\pi i \sum_{j=1}^{K} \text{Res}(f, c_j).$$

8. Find the integral

   $$\int_0^{\infty} \frac{1}{x^6 + 1} dx.$$

9. Show that for $k \in \mathbb{N}$ we have

   $$\int_{-\infty}^{\infty} \frac{1}{(s^2 + 1)^k} ds = \frac{\pi}{2^{2k-2}} \frac{(2k-2)!}{((k-1)!)^2}.$$

10. For $\mu, \nu \in \mathbb{Z}$, $0 \leq \mu < \nu$, verify

    $$\int_{-\infty}^{\infty} \frac{t^{2\mu}}{1 + t^{2\nu}} dt = \frac{\pi}{\nu \sin((2\mu + 1)\frac{\pi}{2\nu})}.$$

11. Use formula (25.33) to find for all $r \in \mathbb{R} \setminus \{1, -1\}$ the integral

$$\int_0^{2\pi} \frac{d\varphi}{1 - 2r \cos \varphi + r^2}.$$

12. For $0 < \alpha < 1$ prove that

$$\int_0^\infty \frac{x^\alpha}{1 + x} dx = \frac{\pi}{\sin \alpha \pi}.$$

Hint: consider the following curve γ and use the fact that for $x \in [a, b] \subset \mathbb{R}$, $\epsilon > 0$ and $\beta \in \mathbb{R}$ we have

$$\lim_{\epsilon \to 0} (x + i\epsilon)^\beta = x^\beta \quad \text{and} \quad \lim_{\epsilon \to 0} (x - i\epsilon)^\beta = x^\beta e^{2\pi i \beta}.$$

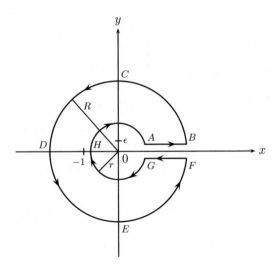

13. Use the residue theorem to prove for $\eta \in \mathbb{R}$ that

$$\frac{1}{\sqrt{2\pi}} \int_{-\infty}^\infty \frac{e^{-iy\eta}}{\cosh \left(\sqrt{\frac{\pi}{2}} y \right)} dx = \frac{1}{\cosh \left(\sqrt{\frac{\pi}{2}} \eta \right)}.$$

Hint: first transform the identity to

$$\int_{-\infty}^\infty \frac{e^{-2\pi i x \xi}}{\cosh \pi x} dx = \frac{1}{\cosh \pi \xi}.$$

502

Now integrate along the following curve:

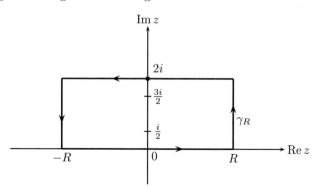

26 The Γ-function, the ζ-function and Dirichlet Series

This chapter is devoted to some special functions which are of central importance in many branches of mathematics. In Volume I we encountered the Γ-function and related functions such as the beta-function, but on that occasion we considered these functions as real-valued depending on real variables. Now we want to study them as complex-valued functions of a complex variable.

We start with defining for $a > 0$ the function a^z by

$$z \mapsto a^z := e^{z \ln a}. \tag{26.1}$$

The usual rules for dealing with exponentials still apply and further we have

$$\frac{\mathrm{d}}{\mathrm{d}z} a^z = \ln a e^{z \ln a} = (\ln a) a^z. \tag{26.2}$$

We also note that for $z = x + iy$

$$a^z = e^{(x+iy)\ln a} = e^{x \ln a} e^{iy \ln a} = a^x a^{iy}. \tag{26.3}$$

For $x > 0$ the Γ-function was defined as the improper integral

$$\Gamma(x) = \int_0^\infty t^{x-1} e^{-t} \,\mathrm{d}t, \tag{26.4}$$

and in Lemma I.28.21 the convergence of the integral in (26.4) was proved. We extend this result first in

Lemma 26.1. *For all $z \in \mathbb{C}$, $\mathrm{Re}\, z > 0$, the integral*

$$\Gamma(z) = \int_0^\infty t^{z-1} e^{-t} \,\mathrm{d}t \tag{26.5}$$

converges and we have $\Gamma(\bar{z}) = \overline{\Gamma(z)}$.

Proof. With $z = x + iy$, $x > 0$, we find for $t > 0$

$$t^{z-1} e^{-t} = t^{x-1} e^{-t} e^{-iy \ln t}$$

and therefore
$$\left| t^{z-1}e^{-t} \right| = t^{x-1}e^{-t}.$$

This implies already the convergence of $\Gamma(z)$ in the half plane $\operatorname{Re} z > 0$, and using the proof of Lemma I.28.21 we can also deduce that in every plane $\operatorname{Re} z \geq x_0 > 0$ the convergence is uniform, indeed this follows from
$$\left| t^{z-1}e^{-t} \right| \leq t^{x_0-1}e^{-t} \quad \text{for } t < 1,$$

whereas for $t > 1$ the exponential term takes care for the convergence. $\qquad\square$

Theorem 26.2. *The Γ-function is on $\operatorname{Re} z > 0$ a holomorphic function.*

Proof. Since
$$\frac{d}{dz}\left(t^{z-1}e^{-t} \right) = (\ln t)t^{z-1}e^{-t},$$
for $\epsilon > 0$ and $R > \epsilon$ we find
$$\frac{d}{dz}\int_{\epsilon}^{R} t^{z-1}e^{-t}\,dt = \int_{\epsilon}^{R} (\ln t)t^{z-1}e^{-t}\,dt$$

and the integral on the right hand side converges uniformly in z for $\operatorname{Re} z \geq x_0 > 0$ and by Theorem 22.14 the function $z \mapsto \int_{\epsilon}^{R}(\ln t)t^{z-1}e^{-t}\,dt$ is holomorphic in $\operatorname{Re} z \geq x_0 > 0$ for every $x_0 > 0$, hence in $\operatorname{Re} z > 0$. Now we use the uniform convergence in $\operatorname{Re} z \geq x_0 > 0$ of $\int_{\epsilon}^{R}(\ln t)t^{z-1}e^{-t}\,dt$ to $\int_{0}^{\infty}(\ln t)t^{z-1}e^{-t}\,dt$ which follows from the fact that at 0 the term t^{z-1} controls $\ln t$ as does e^{-t} at ∞. This however implies that $\Gamma(z)$ is differentiable for $\operatorname{Re} z > 0$ and
$$\Gamma'(z) = \int_{0}^{\infty} (\ln t)t^{z-1}e^{-t}\,dt. \tag{26.6}$$

$\qquad\square$

Using the integration by parts argument as in the proof of Theorem I.28.23 we arrive at

Lemma 26.3. *For $\operatorname{Re} z > 0$*
$$\Gamma(z+1) = z\Gamma(z) \tag{26.7}$$

holds and for $n \in \mathbb{N}$, $\operatorname{Re} z > 0$ we have
$$\Gamma(z+n) = z(z+1)\cdot\ldots\cdot(z+n-1)\Gamma(z). \tag{26.8}$$

From Lemma 26.3 we deduce as in Chapter I.31.

Theorem 26.4. *The gamma function is a meromorphic function on $\mathbb{C}\backslash(-\mathbb{N}_0)$ with simple poles at $-n$, $n \in \mathbb{N}_0$.*

Proof. For $-n$, $n \in \mathbb{N}_0$, we define for $\operatorname{Re} z > -n$ but $z \neq 0, -1, \ldots, -n$, using (26.8)

$$\Gamma(z) = \frac{\Gamma(z+n)}{z(z+1)\cdot\ldots\cdot(z+n-1)}. \tag{26.9}$$

We note that for $z = -k$, $k \in \mathbb{N}_0$, $0 \leq k \leq n$, we have $\Gamma(z+n) = \Gamma(n-k) = (n-k-1)! \neq 0$, hence $\Gamma(z)$ has poles of first order at $-k$, $k \in \mathbb{N}_0$. □

Corollary 26.5. *The residue of $\Gamma(z)$ at $-k$ is*

$$\operatorname{res}(\Gamma, -k) = \frac{(-1)^k}{k!}, \qquad k \in \mathbb{N}_0. \tag{26.10}$$

Proof. Since $-k$ is a pole of order 1 we have

$$\operatorname{res}(\Gamma, -k) = \lim_{z \to -k} (z+k)\Gamma(z)$$

$$= \lim_{z \to -k} \frac{\Gamma(z+k+1)}{(-k)(-k+1)\cdot\ldots\cdot(-1)}$$

$$= \frac{\Gamma(1)}{(-k)(-k+1)\cdot\ldots\cdot(-1)} = \frac{(-1)^k}{k!},$$

where we used (26.8). □

Using our uniqueness result for meromorphic functions, Theorem I.31.13 implies immediately

Theorem 26.6. *For all $z \in \mathbb{C}\backslash\mathbb{Z}$ we have*

$$\Gamma(z)\Gamma(1-z) = \frac{\pi}{\sin \pi z}. \tag{26.11}$$

Formula (26.11) has some interesting consequences. We can write for $z \in \mathbb{C}\backslash\mathbb{Z}$

$$\frac{1}{\Gamma(z)} = \Gamma(1-z)\frac{\sin \pi z}{\pi}. \tag{26.12}$$

The function $z \mapsto \Gamma(1-z)$ has poles of first order for $z \in \mathbb{N}_0$ and is holomorphic on $\mathbb{C}\backslash\mathbb{N}_0$. However $z \mapsto \frac{\sin \pi z}{\pi}$ has simple zeroes, i.e. zeroes of multiplicity 1, at $z \in \mathbb{N}_0$. Hence the singularities of $z \mapsto \Gamma(1-z)\frac{\sin \pi z}{\pi}$ are all removable and this entails

Theorem 26.7. *The function* $z \mapsto \frac{1}{\Gamma(z)}$ *is an entire function and the Γ-function has no zero, i.e. $\Gamma(z) \neq 0$ for all $z \in \mathbb{C}\backslash(-\mathbb{N}_0)$.*

Next we turn to the product representation (I.31.12). For all $x > 0$ we have

$$\frac{1}{\Gamma(x)} = xe^{\gamma x} \prod_{k=1}^{\infty} \left(1 + \frac{x}{k}\right) e^{-\frac{x}{k}} \tag{26.13}$$

where γ is the Euler constant $\gamma = \lim_{N\to\infty} \left(\sum_{k=1}^{N} \frac{1}{k} - \ln N\right)$. First we prove

Lemma 26.8. *For every $n \in \mathbb{N}$ we have*

$$\sum_{k=1}^{\infty} \left\|1 - \left(1 + \frac{\cdot}{k}\right) e^{-\frac{\cdot}{k}}\right\|_{\infty, B_n(0)} < \infty. \tag{26.14}$$

Proof. For $a \in \mathbb{C}$ we have

$$1 - (1-a)e^a = a^2 \left(\left(1 - \frac{1}{2!}\right) + \left(\frac{1}{2!} - \frac{1}{3!}\right)a + \cdots + \left(\frac{1}{l!} - \frac{1}{(l+1)!}\right)a^{l-1} + \cdots\right)$$

$$= a^2 \sum_{l=1}^{\infty} \left(\frac{1}{l!} - \frac{1}{(l+1)!}\right) a^{l-1}.$$

Since $\frac{1}{l!} - \frac{1}{(l+1)!} > 0$ it follows for $|a| \leq 1$ that

$$|1 - (1-a)e^a| \leq |a|^2 \sum_{l=1}^{\infty} \left(\frac{1}{l!} - \frac{1}{(l+1)!}\right) = |a|^2.$$

For $a = -\frac{z}{k}$ and $|z| \leq k$ we find

$$\left|\left(1 - \left(1 + \frac{z}{k}\right)\right) e^{-\frac{z}{k}}\right| \leq \frac{|z|^2}{k^2}$$

implying now

$$\sum_{k\geq n} \sup_{z \in B_n(0)} \left|1 - \left(1 + \frac{z}{k}\right) e^{-\frac{z}{k}}\right| \leq n^2 \sum_{k\geq n} \frac{1}{k^2} < \infty$$

and (26.14) is proved. $\qquad\square$

By Theorem 23.31 a consequence of Lemma 26.8 is that the right hand side of (26.13) extends to an entire function and since (26.13) holds for $x > 0$ the uniqueness result for holomorphic functions yields

Theorem 26.9. *For* $\frac{1}{\Gamma(z)}$, $z \in \mathbb{C}$, *the following product representation*

$$\frac{1}{\Gamma(z)} = ze^{\gamma z} \prod_{k=1}^{\infty} \left(1 + \frac{z}{k}\right) e^{-\frac{z}{k}} \tag{26.15}$$

holds uniformly on compact subsets on $\mathbb{C}$.

Finally we deduce from (26.15) the product representations for the Γ function.

Theorem 26.10. *For* $z \in \mathbb{C}\backslash(-\mathbb{N}_0)$ *the* Γ-*function has the presentation*

$$\Gamma(z) = \frac{e^{-\gamma z}}{z} \prod_{k=1}^{\infty} \frac{e^{\frac{z}{k}}}{1 + \frac{z}{k}}, \tag{26.16}$$

where the convergence is uniform on compact subsets of $\mathbb{C}\backslash(-\mathbb{N}_0)$.

Proof. With

$$\frac{1}{\Gamma_k(z)} := ze^{z\left(1 + \frac{1}{2} + \cdots + \frac{1}{k} - \ln k\right)} \left(1 + \frac{z}{1}\right) e^{-\frac{z}{1}} \cdot \ldots \cdot \left(1 + \frac{z}{k}\right) e^{-\frac{z}{k}}$$

we have as $k \to \infty$ formula (26.15). Since

$$\Gamma_k(z) = \frac{1}{z} e^{-z\left(1 + \frac{1}{2} + \cdots + \frac{1}{k} - \ln k\right)} \frac{e^{\frac{z}{1}}}{1 + \frac{z}{1}} \cdot \ldots \cdot \frac{e^{\frac{z}{k}}}{1 + \frac{z}{k}}$$

we find that as $k \to \infty$ formula (26.16) follows. □

In Definition I.31.9 we introduce the beta-function for $x, y > 0$ as

$$B(x, y) = \int_0^1 t^{x-1}(1 - t)^{y-1} \, dt, \tag{26.17}$$

and we proved in Theorem I.31.11 that

$$B(x, y) = \frac{\Gamma(x)\Gamma(y)}{\Gamma(x + y)}. \tag{26.18}$$

Since this is a function of two real variables we can only try to find an extension of $B(x, y)$ as a separately holomorphic or meromorphic function $B(z, w)$, i.e. as a function which is holomorphic in z (in w) for $w(z)$ being fixed. Since $\Gamma(z)$ is holomorphic in the half plane $\operatorname{Re} z > 0$, formula (26.18) allows us to identify

$$(z, w) \mapsto B(z, w) := \frac{\Gamma(z)\Gamma(w)}{\Gamma(z + w)} \tag{26.19}$$

as a separately holomorphic function on $\{\operatorname{Re} z > 0\} \times \{\operatorname{Re} w > 0\}$. In Problem 3 we will see that in $\{\operatorname{Re} z > 0\} \times \{\operatorname{Re} w > 0\}$ the integral

$$\int_0^1 t^{z-1}(1 - t)^{w-1}\, dt \tag{26.20}$$

converges on compact sets uniformly and absolutely to a separate holomorphic function representing $B(z, w)$.

Now we turn to the Riemann ζ-function and in close relation with it to Dirichlet series. The importance of these functions lies in their relation to number theory. In particular the Riemann ζ-function has deep connections with prime numbers. Once we have studied the ζ-function we can also formulate the best known among all challenging open problems in Mathematics, the Riemann conjecture which links the (non-trivial) zeroes of the ζ-function with the distribution of prime numbers. While the results we are going to discuss are interesting in their own right, for us they serve mainly to see complex analysis in action. But we need some basic facts from elementary number theory to proceed and we collect them first. A good reference is for example [5] in which the proofs are also given.

Number theory is concerned with the structure of $\mathbb{N}$ and we begin with the following two classical results:

Theorem 26.11. A. (*Euclid*) *There exists infinitely many prime numbers.* **B. (*Fundamental Theorem of Arithmetic*)** *Every $n \in \mathbb{N}$, $n > 1$, admits a unique representation as a product of prime numbers when the order of the factors is not taken into account.*

Note that our formulation of the fundamental theorem of arithmetic allows and must allow that some prime numbers occur as a factor more than once in the product representations of n, i.e. we have

$$n = p_{(1)}^{a_1} \cdot \ldots \cdot p_{(r)}^{a_r} = \prod_{j=1}^{r} p_{(j)}^{a_j},$$

with $a_j \in \mathbb{N}$ and the numbers $p_{(1)}, \ldots, p_{(r)}$ are the distinct prime factors.

As number theory is concerned with the set $\mathbb{N}$, we expect functions $f : \mathbb{N} \to \mathbb{C}$ will play a role in number theory and they are called **arithmetical functions**. But the reader should note that this is just a new name for a sequence $(f(n))_{n \in \mathbb{N}}$ of complex numbers.

Definition 26.12. *Let f be an arithmetical function not identically equal to zero.* **A.** *We call f **multiplicative** if*

$$f(mn) = f(m)f(n) \quad \text{for } m \text{ and } n \text{ relative prime.} \tag{26.21}$$

B. *If*

$$f(mn) = f(m)f(n) \quad \text{for all } m, n \in \mathbb{N} \tag{26.22}$$

*we call f **completely multiplicative**.*

Recall that $m, n \in \mathbb{N}$ are called **relatively prime** if their greatest common divisor is equal to 1.

Example 26.13. For $n \in \mathbb{N}$ and $s = \sigma + it$, $\sigma, t \in \mathbb{R}$, we have $n^s = n^\sigma e^{i \ln t}$ and for the arithmetical function $f_s(n) = n^s$ we find

$$\begin{aligned} f_s(mn) &= (mn)^s = (mn)^\sigma e^{it \ln(mn)} \\ &= m^\sigma n^\sigma e^{it(\ln m + \ln n)} \\ &= \left(m^\sigma e^{it \ln m}\right)\left(n^\sigma e^{it \ln n}\right) = m^s n^s = f_s(m) f_s(n), \end{aligned}$$

i.e. n^s, hence n^{-s}, is completely multiplicative.

In Example 26.13 we have already picked up the standard notation from analytical number theory and in particular the theory of Dirichlet series: complex numbers are in some typical situations, for example in terms as n^{-s}, denoted by $s = \sigma + it$, $s, t \in \mathbb{R}$.

Let f be an arithmetical function and $s = \sigma + it \in \mathbb{C}$. If $\operatorname{Re} s = \sigma \geq \sigma_0$ we find

$$\left| \frac{f(n)}{n^s} \right| \leq \frac{|f(n)|}{n^{\sigma_0}}$$

implying that if $\sum_{n=1}^{\infty} \frac{|f(n)|}{n^{\sigma_0}}$ converges then $\sum_{n=1}^{\infty} \frac{f(n)}{n^s}$ converges absolutely in the half plane $\operatorname{Re} s \geq \sigma_0$ and in fact this convergence is also uniform.

Proposition 26.14. *Let f be an arithmetical function such that the series $\sum_{n=1}^{\infty} \left| \frac{f(n)}{n^s} \right|$ neither converges for all $s \in \mathbb{C}$ nor diverges for all $s \in \mathbb{C}$. Then we can find $\sigma_a \in \mathbb{R}$ called the **abscissa of convergence** of $\sum_{n=1}^{\infty} \frac{f(n)}{n^s}$ such that for $\operatorname{Re} s > \sigma_a$ the series $\sum_{n=1}^{\infty} \frac{f(n)}{n^s}$ converges absolutely whereas for $\operatorname{Re} s < \sigma_a$ the series does not converge absolutely.*

Proof. Consider the set $A_f := \left\{ \sigma \in \mathbb{R} \mid \sum_{n=1}^{\infty} \left| \frac{f(n)}{n^s} \right| \text{ diverges, } s = \sigma + it \right\}$. Since by assumption our series does not converge absolutely for all $s \in \mathbb{C}$ the set A_f is not empty. On the other hand, since the series converges absolutely for at least one σ_0, by our preceding consideration, A_f is bounded from above. For $\sigma_a := \sup A_f$ the result follows: if $\sigma < \sigma_a$ then $\sigma \in A_f$, hence the series does not converge absolutely, if $\sigma > \sigma_a$, then $\sigma \in A_f$ and the absolute convergence follows. $\qquad \square$

Definition 26.15. A. *Let $f : \mathbb{N} \to \mathbb{C}$ be an arithmetical function. We call*

$$s \mapsto \sum_{n=1}^{\infty} \frac{f(n)}{n^s}$$

*the **Dirichlet series** associated with or generated by f.*
B. *The Dirichlet series associated with the constant function $f(n) = 1$ is by definition the **Riemann ζ-function**, i.e.*

$$\zeta(s) = \sum_{n=1}^{\infty} \frac{1}{n^s}. \tag{26.23}$$

Corollary 26.16. A. *The abscissa of convergence of the Riemann ζ-function is 1, i.e. for $\sigma > 1$, $s = \sigma + it$, the series (26.23) converges absolutely while it does not converge absolutely for $\sigma \leq 1$.*
B. *An arbitrary Dirichlet series converges absolutely in the half plane $\operatorname{Re} s = \sigma > \sigma_a$ with the understanding that $\sigma_a = -\infty$ if the series converges for all $\sigma \in \mathbb{R}$ absolutely and $\sigma_a = +\infty$ if the series converges nowhere absolutely.*

Proof. Part B follows from Proposition 26.14. In order to see Part A we note that $\sum_{n=1}^{\infty} \frac{1}{n^{\sigma}}$ converges for $\sigma > 1$ and diverges for $\sigma \leq 1$. □

Let us agree to write

$$D_f(s) := \sum_{n=1}^{\infty} \frac{f(n)}{n^s}, \qquad s = \sigma + it \in \mathbb{C}, \tag{26.24}$$

with the understanding that $D_f(s)$ denotes the formal series as well as the corresponding function in the half plane $\operatorname{Re} s > \sigma_a$, $\sigma_a < \infty$.

Corollary 26.17. *If the abscissa of absolute convergence σ_a of the Dirichlet series $D_f(s)$ satisfies $\sigma_a < \infty$ then $D_f(s)$ is in the half-plane $\operatorname{Re} s > \sigma_a$ a holomorphic function and for its derivative we find*

$$D_f'(s) = -\sum_{n=1}^{\infty} \frac{f(n) \ln n}{n^s}. \tag{26.25}$$

Proof. Let $\sigma_0 > \sigma_a$. From $\left| \frac{f(n)}{n^s} \right| \leq \frac{|f(n)|}{n^{\sigma_0}}$ we deduce that $D_f(s)$ converges in each half-plane $\operatorname{Re} s \geq \sigma_0 > \sigma_a$ uniformly, hence locally uniformly and from Theorem 23.20 we deduce that $D_f(s)$ is in the open half-plane $\operatorname{Re} s > \sigma_a$ holomorphic. The formula (26.25) follows now since we are allowed to differentiate the series term by term. □

Corollary 26.18. *The Riemann ζ-function is in the half-plane $\operatorname{Re} s > 1$ holomorphic with derivative*

$$\zeta'(s) = -\sum_{n=2}^{\infty} \frac{\ln n}{n^s}. \tag{26.26}$$

Proposition 26.19. *For $D_f(s)$ as in (26.24) assume $\sigma_a < \infty$. Then we have uniformly for $t \in \mathbb{R}$*

$$\lim_{\sigma \to \infty} D_f(\sigma + it) = f(1). \tag{26.27}$$

Proof. We have to prove that

$$\lim_{\sigma \to \infty} \sum_{n=2}^{\infty} \frac{f(n)}{n^s} = 0. \tag{26.28}$$

For $\sigma \geq \sigma_0 > \sigma_a$ we find with $s = \sigma + it$

$$\left| \sum_{n=2}^{\infty} \frac{f(n)}{n^s} \right| \leq \sum_{n=2}^{\infty} \frac{|f(n)|}{n^\sigma}$$

$$\leq 2^{-(\sigma - \sigma_0)} \sum_{n=2}^{\infty} \frac{|f(n)|}{n^{\sigma_0}},$$

and since $\sum_{n=2}^{\infty} \frac{|f(n)|}{n^{\sigma_0}} < \infty$ we arrive at (26.28). $\qquad \square$

As in the case of power series the "coefficients" $f(n)$ determine a Dirichlet series.

Theorem 26.20. *Let* $D_{f_1}(s)$ *and* $D_{f_2}(s)$ *be two Dirichlet series both converging absolutely for* $\operatorname{Re} s = \sigma > \sigma_a$. *If* $D_{f_1}(s) = D_{f_2}(s)$ *in the half-plane* $\operatorname{Re} s > \sigma_a$ *then* $f_1(n) = f_2(n)$ *for all* $n \in \mathbb{N}$.

Proof. By linearity we need to prove that if the Dirichlet series $D_f(s)$ converges absolutely for $\operatorname{Re} s = \sigma > \sigma_a$ to the zero function, then $f(n) = 0$ for all $n \in \mathbb{N}$. Suppose not all values $f(n)$, $n \in \mathbb{N}$, are zero. Let n_0 be the smallest number such that $f(n_0) \neq 0$. It follows that

$$0 = D_f(s) n_0^s = \sum_{n=n_0}^{\infty} f(n) \left(\frac{n}{n_0} \right)^{-s} = f(n_0) \sum_{n=n_0+1}^{\infty} f(n) \left(\frac{n}{n_0} \right)^{-s}.$$

The convergence of the series and the fact that $\frac{n}{n_0} > 1$ for $n \geq n_0 + 1$ implies that $\lim_{\sigma \to \infty} \sum_{n=n_0+1}^{\infty} f(n) \left(\frac{n}{n_0} \right)^{-\sigma} = 0$, i.e. $f(n_0) = 0$ which is a contradiction. $\qquad \square$

Exercise 26.21. *Prove the following refinement of the uniqueness result, i.e. Theorem 26.20. Two Dirichlet series* $D_{f_1}(s)$ *and* $D_{f_2}(s)$ *converging absolutely in the half-plane* $\operatorname{Re} \sigma > \sigma_a$ *are equal if and only if there exists a sequence* $(s_k)_{k \in \mathbb{N}}$, $\operatorname{Re} s_k = \sigma_k > \sigma_a$ *and* $\lim_{k \to \infty} \sigma_k = \infty$, *such that* $D_{f_1}(s_k) = D_{f_2}(s_k)$. ***Hint:*** *Prove that these conditions imply* $f_1(n) = f_2(n)$ *for all* $n \in \mathbb{N}$. *For this the following result may be used: for* $k \geq 1$ *and* $\sigma \geq c \geq \sigma_a$ *the following holds:*

$$\left| \sum_{n=k}^{\infty} g(n) n^{-s} \right| \leq k^{-(\sigma - c)} \sum_{n=k}^{\infty} |g(n)| n^{-c}$$

where the Dirichlet series $\sum_{k=1}^{\infty} \frac{g(n)}{n^s}$ *has the abscissa of convergence* σ_a.

Using the result of Exercise 26.21 we find

Corollary 26.22. *Let the Dirichlet series $D_f(s)$ converge absolutely in* $\mathrm{Re}\, s = \sigma > \sigma_a$ *and assume that for some* s_0, $\mathrm{Re}\, s_0 > \sigma_a$, *we have* $D_f(s_0) \neq 0$. *Then there exists* $\sigma_0 \geq \sigma_a$ *such that* $D_f(s) \neq 0$ *for all* s *such that* $\mathrm{Re}\, s > \sigma_0$.

Proof. Assume that no such a half-plane exists. In this case, for $k \in \mathbb{N}$ we can pick $s_k \in \mathbb{C}$, $\mathrm{Re}\, s_k = \sigma_k > k$ such that $D_f(s_k) = 0$. Since $\lim_{k \to \infty} \sigma_k = \infty$ the uniqueness result in form of Exercise 26.21 yields for $f(n) = 0$ for all $n \in \mathbb{N}$, i.e. $D_f(s) = 0$ for all s, which of course contradicts the assumption.   □

Our next goal is to prove a product representation of certain Dirichlet series, in particular we want to show that

$$\zeta(s) = \prod_p \frac{1}{1 - p^{-s}}, \qquad \mathrm{Re}\, s = \sigma > 1,$$

where the product is taken over all prime numbers. We need some preparations.

Our presentation of the following theorem and proof follows closely [5].

Theorem 26.23. *Let f be a multiplicative arithmetical function and assume that $\sum_{n=1}^{\infty} |f(n)|$ converges. It follows that*

$$\sum_{n=1}^{\infty} f(n) = \prod_p \left(\sum_{k=0}^{\infty} f(p^k) \right) \qquad (26.29)$$

where the product is taken over all prime numbers p. If in addition f is completely multiplicative then we have

$$\sum_{n=1}^{\infty} f(n) = \prod_p \frac{1}{1 - f(p)}. \qquad (26.30)$$

Proof. For $x \in \mathbb{R}$ we define

$$P(x) := \prod_{p \leq x} \left(\sum_{k=0}^{\infty} f(p^k) \right). \qquad (26.31)$$

This is a finite product and all the series $\sum_{k=0}^{\infty} f\left(p^k\right)$ converge absolutely which follows from the absolute convergence of $\sum_{k=1}^{\infty} f(n)$. Thus the finite product of the series $\sum_{k=0}^{\infty} f\left(p^k\right)$ can be taken in any order and we may arrange the sum as we need. A typical term in $P(x)$ can be written as

$$f\left(p_1^{a_1}\right) f\left(p_2^{a_2}\right) \cdot \ldots \cdot f\left(p_r^{a_r}\right) = f\left(p_1^{a_1} \cdot \ldots \cdot p_r^{a_r}\right),$$

where we used the fact that f is multiplicative and we assume $p_1 < p_2 < \cdots < p_r$. Denote by $A(x)$ the set of all $n \in \mathbb{N}$ with all prime factors being less or equal to x and let $B(x)$ be the set of all $n \in \mathbb{N}$ with at least one prime factor strictly larger than x. By the fundamental theorem of arithmetic it follows that

$$P(x) = \sum_{n \in A(x)} f(n)$$

and therefore

$$\left| \sum_{n=1}^{\infty} f(n) - P(x) \right| \leq \sum_{n \in B(x)} |f(n)| \leq \sum_{n>x} |f(n)|.$$

Since $\sum_{n=1}^{\infty} |f(n)|$ converges we find that $\lim_{x \to \infty} \sum_{n>x} |f(n)| = 0$ implying that

$$\prod_p \left(\sum_{k=0}^{\infty} f\left(p^k\right) \right) = \lim_{x \to \infty} \prod_{p \leq x} \left(\sum_{k=0}^{\infty} f\left(p^k\right) \right) = \sum_{n=1}^{\infty} f(n).$$

In order to see the absolute convergence of the product in (26.29) we note that the product $\prod_{n=1}^{\infty} (1 + c_n)$ converges absolutely if and only if $\sum_{n=1}^{\infty} c_n$ converges absolutely. Since

$$\sum_{p \leq x} \left| \sum_{k=1}^{\infty} f\left(p^k\right) \right| \leq \sum_{p \leq x} \sum_{k=1}^{\infty} |f\left(p^k\right)| \leq \sum_{n=2}^{\infty} |f(n)|$$

the absolute convergence of the product follows. Finally we note that if f is completely multiplicative we have $f\left(p^k\right) = f(p)^k$ and $\sum_{k=0}^{\infty} f\left(p^k\right) = \sum_{k=1}^{\infty} f(p)^k$. Since this geometric series converges its limit is $\frac{1}{1-f(p)}$ and (26.30) is proved. $\square$

Corollary 26.24. *Suppose that* $D_f(s) = \sum_{n=1}^{\infty} \frac{f(n)}{n^s}$ *converges absolutely for* $\operatorname{Re} s = \sigma > \sigma_a$. *For f being multiplicative we have*

$$\sum_{n=1}^{\infty} \frac{f(n)}{n^s} = \prod_p \left(\sum_{k=0}^{\infty} \frac{f\left(p^k\right)}{p^{ks}} \right), \qquad \operatorname{Re} s > \sigma_a, \tag{26.32}$$

and if f is completely multiplicative

$$\sum_{n=1}^{\infty} \frac{f(n)}{n^s} = \prod_{p} \frac{1}{1 - f(p)p^{-s}} \qquad (26.33)$$

holds.

Corollary 26.25. *For the Riemann ζ-function we have the Euler product representation*

$$\zeta(s) = \sum_{n=1}^{\infty} \frac{1}{n^s} = \prod_{p} \frac{1}{1 - p^s}, \qquad \mathrm{Re}\, s = \sigma > 1, \qquad (26.34)$$

which implies in particular $\zeta(s) \neq 0$ *for* $\mathrm{Re}\, s > 0$.

A remark to further examples seems to be appropriate. Of course we can start to "play" with arithmetic functions f, i.e. complex sequences $(f(n))_{n \in \mathbb{N}}$ to construct Dirichlet series. However it is much more important to note that there are classical arithmetic functions such that

$$
\begin{array}{ll}
\varphi(n) & \text{the Euler } \varphi\text{-function} \\
\mu(n) & \text{the Möbius function} \\
\lambda(n) & \text{the Liouville function} \\
\chi(n) & \text{the Dirichlet character}
\end{array}
$$

$$\vdots$$

and each give rise to a Dirichlet series which in turn is an important tool in number theory. For example we have

$$\frac{1}{\zeta(s)} = \sum_{n=1}^{\infty} \frac{\mu(n)}{n^s} = \prod_{p} \left(1 - p^{-s}\right), \qquad \mathrm{Re}\, s = \sigma > 1,$$

or

$$L(s, \chi) := \sum_{n=1}^{\infty} \frac{\chi(n)}{n^s} = \prod_{p} \frac{1}{1 - \chi(p)p^{-s}}, \qquad \mathrm{Re}\, s = \sigma > 1.$$

The latter function is the Dirichlet L-function which is of great importance when studying for example arithmetic progression. For the definitions and the properties of these classical arithmetical functions as well as for their

517

applications to number theory we refer to standard textbooks in number theory, [5] or the classics by G. Hardy and E. Wright [32]. In the following we only deal further with the Riemann ζ-function. Our first result is an integral representation for the ζ-function which gives also a first link to the Γ-function. We remind the reader on Lemma 26.1 stating that

$$\Gamma(s) = \int_0^\infty y^{s-1} e^{-y} \, dy \tag{26.35}$$

holds for all $s \in \mathbb{C}$ with $\operatorname{Re} s > 0$.

Theorem 26.26. *For $s \in \mathbb{C}$, $\operatorname{Re} s > 1$, we have*

$$\zeta(s) = \frac{1}{\Gamma(s)} \int_0^\infty \frac{x^{s-1}}{e^x - 1} \, dx = \frac{1}{\Gamma(s)} \int_0^\infty \frac{x^{s-1} e^{-x}}{1 - e^{-x}} \, dx. \tag{26.36}$$

Proof. First we show that for $\sigma = \operatorname{Re} s > 1$ the integral exists as an improper, absolutely convergent integral. For $0 < \eta \le 1 < R$ we have for $s = \sigma + it$

$$\left| \int_\eta^R \frac{x^{s-1}}{e^x - 1} \right| = \left| \int_\eta^R \frac{x^{\sigma-1} e^{it \ln x}}{e^x - 1} \, dx \right|$$

$$\le \int_\eta^R \frac{x^{\sigma-1}}{e^x - 1} \, dx = \int_\eta^1 \frac{x^{\sigma-1}}{e^x - 1} \, dx + \int_1^R \frac{x^{\sigma-1}}{e^x - 1} \, dx.$$

Since $\sigma - 2 > -1$ and $\lim_{x \to 0} \frac{x}{e^x - 1} = 1$ we find that

$$\int_0^1 \frac{x^{\sigma-1}}{e^x - 1} \, dx = \lim_{\eta \to 0} \int_\eta^1 \frac{x^{\sigma-1}}{e^x - 1} \, dx < \infty.$$

Furthermore, from $\frac{1}{1 - e^{-x}} \le \frac{1}{1 - e^{-1}}$ for $x \ge 1$ we deduce

$$\int_1^R \frac{x^{\sigma-1}}{e^x - 1} \, dx = \lim_{R \to \infty} \int_1^R \frac{x^{\sigma-1}}{e^x - 1} \, dx = \lim_{R \to \infty} \int_1^\infty \frac{e^{-x} x^{\sigma-1}}{1 - e^{-x}} \, dx < \infty.$$

Hence for $\sigma > 1$ the integral $\int_0^\infty \frac{x^{s-1}}{e^x - 1} \, dx$ exists both as an improper complex-valued Riemann integral and as a Lebesgue integral. Now we use in (26.35) the substitution $y = nx$ to find

$$\Gamma(s) = n^s \int_0^\infty x^{s-1} e^{-nx} \, dx$$

518

or

$$n^{-s} = \frac{1}{\Gamma(s)} \int_0^\infty x^{s-1} e^{-nx} \, dx,$$

which yields

$$\sum_{k=1}^N n^{-s} = \frac{1}{\Gamma(s)} \sum_{k=1}^N \int_0^\infty x^{s-1} e^{-nx} \, dx$$

$$= \frac{1}{\Gamma(s)} \int_0^\infty x^{s-1} \frac{e^{-x} \left(1 - e^{-Nx}\right)}{1 - e^{-x}} \, dx$$

$$= \frac{1}{\Gamma(s)} \int_0^\infty x^{s-1} \frac{1 - e^{-Nx}}{e^x - 1} \, dx,$$

where we have used that $\Gamma(s)$ has no zeroes and

$$\sum_{n=1}^N e^{-nx} = \sum_{n=1}^N \left(e^{-x}\right)^n = \sum_{n=0}^N \left(e^{-x}\right)^n - 1$$

$$= \frac{1 - e^{-(N+1)x}}{1 - e^{-x}} - 1 = \frac{e^{-x} \left(1 - e^{-Nx}\right)}{1 - e^{-x}}.$$

We can split the integral into two absolutely convergent integrals to obtain

$$\sum_{n=1}^N n^{-s} = \frac{1}{\Gamma(s)} \left(\int_0^\infty \frac{x^{s-1}}{e^x - 1} \, dx - \int_0^\infty \frac{e^{-Nx} x^{s-1}}{e^x - 1} \, dx \right).$$

For the second integral we find for $c > 0$ that

$$\int_0^\infty \frac{e^{-Nx} x^{s-1}}{e^x - 1} \, dx = \int_0^c \frac{e^{-Nx} x^{s-1}}{e^x - 1} \, dx + \int_c^\infty \frac{e^{-Nx} x^{s-1}}{e^x - 1} \, dx.$$

We note that for $\operatorname{Re} s = \sigma > 1$ we have

$$\left| \int_0^c \frac{e^{-Nx} x^{s-1}}{e^x - 1} \, dx \right| \leq \int_0^c \frac{e^{-N} x^{\sigma-1}}{e^x - 1} \, dx < \int_0^c \frac{x^{\sigma-1}}{e^x - 1} \, dx$$

and therefore, given $\epsilon > 0$ we can find $c_0 \leq 1$ such that $c \leq c_0$ implies

$$\left| \int_0^c \frac{e^{-Nx} x^{s-1}}{e^x - 1} \, dx \right| < \int_0^c \frac{x^{\sigma-1}}{e^x - 1} \, dx < \epsilon.$$

For the second integral we observe that for $\operatorname{Re} s = \sigma > 1$

$$\left| \int_c^\infty \frac{e^{-Nx} x^{s-1}}{e^x - 1} \, dx \right| \leq \int_c^\infty \frac{e^{-Nx} x^{\sigma-1}}{e^x - 1} \, dx$$

$$\leq e^{-Nx} \int_c^\infty \frac{x^{\sigma-1}}{e^x - 1} \, dx$$

and since $\int_c^\infty \frac{x^{\sigma-1}}{e^x-1} \, dx$ is finite for $\sigma > 1$ and $c \geq 0$ it follows as $N \to \infty$ that for $s = \sigma + it$, $\operatorname{Re} s = \sigma > 1$, we have

$$\zeta(s) = \lim_{N \to \infty} \sum_{n=1}^N \frac{1}{n^s} = \frac{1}{\Gamma(s)} \int_0^\infty \frac{x^{s-1}}{e^x - 1} \, dx.$$

$\square$

Remark 26.27. In [61] the proof of (26.36) is first given for $s \in \mathbb{R}$, $s > 1$, and then it is shown that the integral $s \mapsto \int_0^\infty \frac{x^{s-1}}{e^x-1} \, dx$ admits a holomorphic extension to the half-plane $\operatorname{Re} s > 1$. Now the uniqueness result for holomorphic functions entails that if (26.36) holds for $s \in \mathbb{R}$, $s > 1$, then it holds also for $\sigma \in \mathbb{C}$, $\operatorname{Re} \sigma > 1$.

By Theorem 26.7 we know that $s \mapsto \frac{1}{\Gamma(s)}$ is an entire function. Hence if we can extend $s \mapsto \int_0^\infty \frac{x^{s-1}}{e^x-1} \, dx = \int_0^\infty \frac{x^{s-1} e^{-x}}{1-e^{-x}} \, dx$ to a larger subset of $\mathbb{C}$ than the half-plane $\operatorname{Re} s = \sigma > 1$ as a holomorphic function we can extend the ζ-function. We start with formula (26.36)

$$\Gamma(s)\zeta(s) = \int_0^\infty \frac{x^{s-1}}{e^x - 1} \, dx = \int_0^1 \frac{x^{s-1}}{e^x - 1} \, dx + \int_1^\infty \frac{x^{s-1}}{e^x - 1} \, dx$$

which holds for $\operatorname{Re} s = \sigma > 1$. For $R > 1$ we find that

$$s \mapsto \int_1^R \frac{x^{s-1}}{e^x - 1} \, dx = \int_1^R \frac{e^{(s-1)\ln x}}{e^x - 1} \, dx$$

is a holomorphic function and for s in a compact set we have uniformly

$$\lim_{R \to \infty} \int_1^R \frac{x^{s-1}}{e^x - 1} \, dx = \int_1^\infty \frac{x^{s-1}}{e^x - 1} \, dx$$

implying that $s \mapsto \int_1^\infty \frac{x^{s-1}}{e^x-1} \, dx$ is an entire function. Now we turn to the first integral in the decomposition and observe that $x \mapsto \frac{x}{e^x-1}$ admits a Taylor expansion about $x = 0$ given by

$$\frac{x}{e^x-1} = 1 - \frac{1}{2}x + \sum_{k=1}^{\infty} \frac{(-1)^{k-1}\beta_k}{(2k)!} x^{2k} \qquad (26.37)$$

where the β_k's are related to the Bernoulli numbers by $B_{2k} = (-1)^{k-1}\beta_k$, also see Problem 5 in Chapter 16. The radius of convergence of this series is 2π. Indeed, for $z \mapsto \frac{z}{e^z-1}$ we have poles at $2\pi i m$, $m \in \mathbb{Z}\backslash\{0\}$, so 2π must be a bound for the radius of convergence. For the following the actual values of the Bernoulli numbers are not important. However, since there is some nice part of mathematics to learn we discuss the expansion (26.37) and the Bernoulli numbers in some detail in Appendix IV.

We multiply (26.37) with x^{s-2}, $\operatorname{Re} s > 1$, to get

$$\frac{x^{s-1}}{e^x-1} = x^{s-2} - \frac{1}{2}x^{s-1} + \sum_{k=1}^{\infty} \frac{(-1)^{k-1}\beta_k}{(2k)!} x^{2k+s-1}$$

and integration over $[0, 1]$ yields

$$\int_0^1 \frac{x^{s-1}}{e^x-1} \, dx = \frac{1}{s-1} - \frac{1}{2s} + \sum_{k=1}^{\infty} \frac{(-1)^{k-1}\beta_k}{(2k)!} \frac{1}{2k+s}. \qquad (26.38)$$

For $s_0 \neq 1, 0, -1, -3, -5, \ldots$ we can always find a ball $B_r(s_0)$, $r > 0$, such that the right hand side in (26.38) is in $B_r(s_0)$ a holomorphic function in s, hence the right hand side is a meromorphic function in $\mathbb{C}$ with isolated singularities at $s_0 = 1, 0, -1, -3, \ldots$, which are all poles of first order.

We know that $\frac{1}{\Gamma(s)}$ is an entire function with simple zeroes at $0, -1, -2, -3 \ldots$. This implies now

Theorem 26.28. *The ζ-function originally defined for $\operatorname{Re} s > 0$ can be extended to a meromorphic function on $\mathbb{C}$ by*

$$\zeta(s) = \frac{1}{\Gamma(s)} \int_0^1 \frac{x^{s-1}}{e^x-1} \, dx + \frac{1}{\Gamma(s)} \int_1^\infty \frac{x^{s-1}}{e^x-1} \, dx = \frac{1}{\Gamma(s)} \int_0^\infty \frac{x^{s-1}}{e^x-1} \, dx \qquad (26.39)$$

which has only one singularity at $s_0 = 1$. This is a pole of first order with residue $\operatorname{res}(\zeta, 1) = 1$.

In order to study $\zeta(s)$ in more detail we need the following functional equation for the ζ-function,

$$\zeta(s) = 2(2\pi)^{s-1} \sin \frac{\pi s}{2} \Gamma(1-s)\zeta(1-s) \tag{26.40}$$

or with s replaced by $1 - s$

$$\zeta(1-s) = 2^{1-s}\pi^{-s} \cos \frac{\pi s}{2} \Gamma(s)\zeta(s). \tag{26.41}$$

There are different ways to derive this functional equation, one uses Fourier analysis and we will provide this proof in Part 8 of our Course. Other proofs are given in [20], [33], [61] or [73]. Once (26.40) or (26.41) are established we can draw further consequences about the Riemann ζ-function.

So far we know that the ζ-function is a meromorphic function with a pole of first order at 1 with residue 1. Further we know that for $\operatorname{Re} s > 1$ the ζ-function has no zeroes, a fact which follows from the product representation given in Corollary 26.25. The functional equations are identities for meromorphic functions hence they must hold everywhere (except at the pole of ζ at 1). For $\operatorname{Re} s = \sigma < 0$ the function $s \mapsto \Gamma(1 - s)$ is holomorphic as is the function $s \mapsto \zeta(1 - s)$ and both functions are non-zero in the half-plane $\operatorname{Re} s < 0$. The same holds for $s \mapsto (2\pi)^{-s}$ which now implies that in the half-plane $\operatorname{Re} s < 0$ the ζ-function has the same zeroes as $s \mapsto \sin \frac{\pi s}{2}$, i.e. we have

Corollary 26.29. *In the half-plane* $\operatorname{Re} s < 0$ *the Riemann* ζ*-function has simple zeroes for* $s = -2, -4, \ldots$, *i.e.* $s \in -2\mathbb{N}$. *These zeroes of the* ζ*-functions are called the* ***trivial zeroes*** *of* ζ.

Thus we know that outside the strip $0 \leq \sigma \leq 1$, $\operatorname{Re} s = \sigma$, the ζ-function has only the trivial zeroes. The strip $0 \leq \sigma \leq 1$ is called the **critical strip** of the Riemann ζ-function. It needs further efforts to prove that on $\operatorname{Re} s = 1$ and $\operatorname{Re} s = 0$ the ζ-function has no zeroes, however in the critical strip there are infinitely many zeroes of the ζ-function. We can now formulate the **Riemann Conjecture** or as it should be called historically correct the **Riemann Hypothesis**:

All non-trivial zeroes of $\zeta(s)$ lie on the critical line $\operatorname{Re} s = \frac{1}{2}$.

We refer to [20], [39], [61], [63], [73], [82] or [86] where partly details of the above mentioned results and partly the relation of the non-trivial zeroes of the Riemann ζ-function to the distribution of prime numbers are discussed. A fascinating topic which however goes beyond our Course.

Problems

1.   a) Find the value $\Gamma\left(-k+\frac{1}{2}\right)$ for $k \in \mathbb{N}$.

   b) Prove $|\Gamma(iy)|^2 = \frac{\pi}{y \sinh \pi y}$ and $\left|\Gamma\left(\frac{1}{2} + iy\right)\right|^2 = \frac{\pi}{\cosh \pi y}$.
   Hint: use (26.11), $\Gamma(\bar{z}) = \overline{\Gamma(z)}$, $-i \sin iy = \sinh y$ and $\cos iy = \cosh y$.

2.   Prove the Legendre duplication formula (I.31.35) for $z \in \mathbb{C}$, Re $z > 0$:

$$\Gamma(2z) = \frac{2^{2z-1}}{\sqrt{\pi}} \Gamma(z) \Gamma\left(z + \frac{1}{2}\right).$$

3.   For Re $z > 0$ and Re $w > 0$ justify

$$B(z, w) = \int_0^1 t^{z-1}(1 - t)^{w-1} dt.$$

4.   For Re $z > 0$ prove

$$\Gamma(z) = 2 \int_0^\infty t^{2z-1} e^{-t^2} dt.$$

5.   Consider the logarithmic derivative $\psi(z) := \frac{\Gamma'(z)}{\Gamma(z)}$ of the Γ-function. Find its singularities and in some natural domain prove the functional equation

$$\psi(z + 1) = \frac{1}{z} + \psi(z)$$

and

$$\psi(1 - z) - \psi(z) = \pi \cot \pi z.$$

Hint: use the functional equation $\Gamma(z + 1) = z\Gamma(z)$ and $\Gamma(z)\Gamma(1 - z) = \frac{\pi}{\sin \pi z}$, respectively.

6. With ψ as in Problem 5 prove in a natural domain

$$\psi(z) = -\gamma - \frac{1}{z} - \sum_{k=1}^{\infty} \left(\frac{1}{z+k} - \frac{1}{k} \right).$$

Hint: use the product representation of the Γ-function.

7. The **Möbius function** $\mu : \mathbb{N} \to \mathbb{R}$ is defined as follows. For $n = 1$ we set $\mu(1) = 1$. Now let $n \in \mathbb{N}, n \geq 2$, have the prime number decomposition $u = p_1^{\nu_1} \cdot \ldots \cdot p_k^{\nu_k}$. We define $\mu(n) = (-1)^k$ if $\nu_1 = \cdots = \nu_k = 1$, $\mu(n) = 0$ otherwise.

    a) Prove that $\mu(n) = 0$ if and only if n has a square factor larger than 1.

    b) Prove that μ is a multiplicative arithmetic function which is not completely multiplicative.

8. Solve Exercise 26.21.

9.* Let $(a_k)_{k \in \mathbb{N}}$ be a bounded sequence of complex numbers. Prove for $\sigma > 1$ that

$$\lim_{R \to \infty} \frac{1}{2R} \int_{-R}^{R} \left| \sum_{n=1}^{\infty} \frac{a_n}{n^{\sigma + it}} \right|^2 dt = \sum_{n=1}^{\infty} \frac{|a_n|^2}{n^{2\sigma}}.$$

(This problem is taken from [83].)

10.* Finding special values of $\zeta(z)$ is non-trivial, we will discuss some ideas in Appendix IV and in Part 8. The strategy as indicated in the following hint to prove

$$\zeta(2) = \frac{\pi^2}{6}$$

is taken from [43].
Hint. Prove:

(i) For $0 < x < \frac{\pi}{2}$ it follows that $\cot^2 x < \frac{1}{x^2} < 1 + \cot^2 x$.

(ii) Derive for $x = \frac{k\pi}{2m+1}, k = 1, 2, \ldots, n$, that

$$\sum_{k=1}^{m} \cot^2 \frac{k\pi}{2m+1} < \frac{(2m+1)^2}{\pi^2} \sum_{k=1}^{m} \frac{1}{k^2} < m + \sum_{k=1}^{m} \cot^2 \frac{k\pi}{2m+1}.$$

(iii) Show by using de Moivres' law that

$$\sum_{k=1}^{m} \cot^2 \frac{k\pi}{2m+1} = \frac{m(2m-1)}{3}.$$

(iv) Combine the inequalities of step (ii) with the result of step (iii) and pass to the limit as m tends to ∞.

27 Elliptic Integrals and Elliptic Functions

The purpose of this chapter is to indicate the existence of an entirely new area of Mathematics, i.e. the theory of elliptic integrals and elliptic functions. We cannot even start to explore this in our Course, however the reader should learn about its existence, its historical importance and that today many branches of Mathematics originate from it. In our treatment of the history of analysis we will address some of the major developments of the theory of elliptic functions and the influence they had and still have.

Among the most surprising results in the classical calculus of real-valued functions of a real variable are the relations between trigonometrical functions and their inverses of integrals of certain algebraic functions, e.g. roots of rational functions. For example we have

$$\arcsin x = \int_0^x \frac{1}{\sqrt{1 - t^2}} \, dt, \quad x \in [-1, 1], \tag{27.1}$$

or

$$\arctan x = \int_0^x \frac{1}{1 + t^2} \, dt, \quad x \in \mathbb{R}. \tag{27.2}$$

Ignoring historical developments we can start to introduce trigonometrical functions by (27.1) or (27.2) and after some time we might spot that a better approach is to start with the inverse function of $x \mapsto \int_0^x \frac{1}{\sqrt{1-t^2}} dt$, etc. The inverse function can be extended to $\mathbb{R}$ as a C^∞-function, it is a periodic function and it is much easier to handle. Knowing that (27.1) was derived from $\frac{d}{dx} \arcsin x = \frac{1}{\sqrt{1-x^2}}$, the substitution $t = \sin \varphi$, i.e. $\varphi = \arcsin t$ gives of course

$$\arcsin x = \int_0^{\arcsin x} 1 \, d\varphi,$$

and hence this substitution gives in this case no further insight as does the formula

$$\sin x = \arcsin^{-1} \left(\int_0^x \frac{1}{\sqrt{1 - t^2}} \, dt \right).$$

But now we change the game and we want to study integrals of the type

$$\int_0^x R(t, q(t)) \, dt \tag{27.3}$$

where $R(t, s)$ is a rational function in two variables and $q(t) = \sqrt{p(t)}$ with a real polynomial of order 3 and 4. Included are integrals such as

$$F(r) = \int_0^r \frac{dt}{\sqrt{1 - t^4}} \tag{27.4}$$

which gives the arc length of the lemniscate, see Problem 2, or

$$l(\mathcal{E})(x) = \int_0^x \frac{\sqrt{1 - k^2 t^2}}{\sqrt{1 - t^2}} \, dt, \quad 0 < k < 1, \tag{27.5}$$

which gives the arc length of an ellipse $\mathcal{E}$, see Problem 1, and for this reason the integrals considered in this chapter are called elliptic integrals.

The substitution $t = \sin \varphi$ is now less trivial and transforms these integrals into integrals involving trigonometrical functions. One of the first results in the theory of elliptic integrals goes back to A. M. Legendre who reduced the investigation of integrals of type (27.3) to the study of the following three integrals where we always assume $|k| < 1$:

$$\int_0^x \frac{dt}{\sqrt{(1 - t^2)(1 - k^2 t^2)}} = \int_0^\varphi \frac{d\vartheta}{\sqrt{1 - k^2 \sin^2 \vartheta}}, \tag{27.6}$$

$$\int_0^x \sqrt{\frac{1 - k^2 t^2}{1 - t^2}} \, dt = \int_0^\varphi \sqrt{1 - k^2 \sin^2 \vartheta} \, d\vartheta \tag{27.7}$$

and with $n \in \mathbb{N}$

$$\int_0^x \frac{dt}{(1 + nt^2)\sqrt{(1 - t^2)(1 - k^2 t^2)}} = \int_0^\varphi \frac{d\varphi}{(1 + n \sin^2 \vartheta)\sqrt{1 - k^2 \sin^2 \vartheta}}. \tag{27.8}$$

These integrals are called **elliptic integrals** of **1**st (27.6), **2**nd (27.7) and **3**rd **kind**, (27.8). While Legendre made remarkable progress in investigating these integrals, in some sense he can be viewed almost as the founder of the subject, the insight of N. H. Abel and C. G. J. Jacobi to take the inverse functions of elliptic integrals as starting point and to extend the considerations to complex arguments transformed the whole field. It is worth noting that while not publishing any results, C. F. Gauss had essentially all of Legendre's, Abel's and Jacobi's insights and results before them.

To get a flavour of the subject let us introduce the function sn by its inverse, i.e.

$$\operatorname{sn}^{-1}(w) := \int_0^w \frac{dt}{\sqrt{(1-t^2)(1-k^2 t^2)}}, \tag{27.9}$$

where we allow w to be complex valued and do not specify for the moment the domain of w. We define further

$$\operatorname{cn} z = \sqrt{1 - \operatorname{sn}^2 z} \tag{27.10}$$

and

$$\operatorname{dn} z = \sqrt{1 - k^2 \operatorname{sn}^2 z}. \tag{27.11}$$

The three functions sn, cn, dn are **elliptic functions in the sense of Jacobi**.

Please note the analogy

$$\sin^{-1} w = \int_0^w \frac{dt}{\sqrt{1-t^2}}$$

and

$$\cos z = \sqrt{1 - \sin^2 z}.$$

We can even define in analogy to $\tan z = \frac{\sin z}{\cos z}$ the function $\operatorname{tn} z = \frac{\operatorname{sn} z}{\operatorname{cn} z}$. Again, for the moment we leave domain issues aside.

Lemma 27.1. *The function* sn *is an odd and* cn *as well as* dn *are even functions.*

Proof. In light of (27.10) and (27.11) we only need to prove that sn is an odd function. For

$$z = \int_0^w \frac{dt}{\sqrt{(1-t^2)(1-k^2 t^2)}}$$

we have $\operatorname{sn} z = w$. Substituting $t = -s$ we find

$$z = -\int_0^{-w} \frac{ds}{\sqrt{(1-s^2)(1-k^2 s^2)}},$$

i.e.

$$-z = \int_0^{-w} \frac{ds}{\sqrt{(1-s^2)(1-k^2 s^2)}}$$

which means $\operatorname{sn}(-z) = -w = -\operatorname{sn}(z)$. $\quad\square$

Corollary 27.2. *At* $z = 0$ *we have the values*

$$\operatorname{sn}(0) = 0, \quad \operatorname{cn}(0) = 1, \quad \operatorname{dn}(0) = 1. \tag{27.12}$$

Lemma 27.3. *For the derivative of the functions* sn, cn *and* dn *we find:*

$$\frac{d}{dz} \operatorname{sn}(z) = \operatorname{cn}(z) \operatorname{dn}(z); \tag{27.13}$$

$$\frac{d}{dz} \operatorname{cn}(z) = -\operatorname{sn}(z) \operatorname{dn}(z); \tag{27.14}$$

$$\frac{d}{dz} \operatorname{dn}(z) = -k^2 \operatorname{sn}(z) \operatorname{cn}(z). \tag{27.15}$$

Proof. In order to find $\frac{d}{dz} \operatorname{sn}(z)$ we apply our rule to find the derivative of an inverse function. Since

$$w = \operatorname{sn}(z) \quad \text{if} \quad z = \int_0^w \frac{dt}{\sqrt{(1 - t^2)(1 - k^2 t^2)}}$$

we have

$$\frac{d}{dz} \operatorname{sn}(z) = \frac{dw}{dz} = \frac{1}{\frac{dz}{dw}} = \sqrt{(1 - w^2)(1 - k^2 w^2)}$$

$$= \sqrt{1 - \operatorname{sn}^2 z} \sqrt{1 - k^2 \operatorname{sn}^2 z} = \operatorname{cn}(z) \operatorname{dn}(z).$$

Now we use (27.10) to find

$$\frac{d}{dz} \operatorname{cn}(z) = \frac{d}{dz} \sqrt{1 - \operatorname{sn}^2 z} = \frac{1}{2\sqrt{1 - \operatorname{sn}^2(z)}} \frac{d}{dz} \left(-\operatorname{sn}^2(z) \right)$$

$$= \frac{-2 \operatorname{sn}(z) \operatorname{cn}(z) \operatorname{dn}(z)}{2 \operatorname{cn}(z)} = -\operatorname{sn}(z) \operatorname{dn}(z).$$

Moreover, for $\operatorname{dn}(z)$ we have

$$\frac{d}{dz} \operatorname{dn}(z) = \frac{d}{dz} \sqrt{1 - k^2 \operatorname{sn}^2 z} = \frac{1}{2\sqrt{1 - k^2 \operatorname{sn}^2(z)}} \frac{d}{dz} \left(-k^2 \operatorname{sn}^2 z \right)$$

$$= \frac{-2k^2 \operatorname{sn}(z) \operatorname{cn}(z) \operatorname{dn}(z)}{2 \operatorname{dn}(z)} = -k^2 \operatorname{sn}(z) \operatorname{cn}(z).$$

Of course, we need to be more careful with domains, but at the moment we postpone this. $\square$

The following **addition formula** can be proved by using the general theory of elliptic functions, in particular the Weierstrass $\wp$-function. In Problem 7 we will give a proof of (27.16) using some results of the theory of ordinary differential equations, see [80].

Lemma 27.4. *The functions* sn, cn *and* dn *satisfy the following equalities*

$$\operatorname{sn}(z_1 + z_2) = \frac{\operatorname{sn} z_1 \operatorname{cn} z_2 \operatorname{dn} z_2 + \operatorname{cn} z_1 \operatorname{dn} z_1 \operatorname{sn} z_2}{1 - k^2 \operatorname{sn}^2 z_1 \operatorname{sn}^2 z_2}; \tag{27.16}$$

$$\operatorname{cn}(z_1 + z_2) = \frac{\operatorname{cn} z_1 \operatorname{cn} z_2 - \operatorname{sn} z_1 \operatorname{sn} z_2 \operatorname{dn} z_1 \operatorname{dn} z_2}{1 - k^2 \operatorname{sn}^2 z_1 \operatorname{sn}^2 z_2}; \tag{27.17}$$

$$\operatorname{dn}(z_1 + z_2) = \frac{\operatorname{dn} z_1 \operatorname{cn} z_2 - k^2 \operatorname{sn} z_1 \operatorname{sn} z_2 \operatorname{cn} z_1 \operatorname{cn} z_2}{1 - k^2 \operatorname{sn}^2 z_1 \operatorname{sn}^2 z_2}. \tag{27.18}$$

A classical notation is that of the **amplitude**

$$\varphi = \operatorname{am} z \quad \text{if} \quad z = \int_0^\varphi \frac{d\vartheta}{\sqrt{1 - k^2 \sin^2 \vartheta}}. \tag{27.19}$$

Since the integrals

$$z = \int_0^w \frac{dt}{\sqrt{(1 - t^2)(1 - k^2 t^2)}} = \int_0^\varphi \frac{d\vartheta}{\sqrt{1 - k^2 \sin^2 \vartheta}}$$

are connected by the change of variables $t = \sin \varphi$, we find

$$\operatorname{sn} z = \sin(\operatorname{am} \varphi) \tag{27.20}$$

which yields by (27.10) that

$$\operatorname{cn} z = \cos(\operatorname{am} \varphi). \tag{27.21}$$

These two formulae justifying the classical names **sines amplitudines** and **cosines amplitudines** for sn and cn, respectively.

The next deep discovery was that the Jacobi elliptic functions have two (independent) periods, i.e. they are **double periodic** functions.

Definition 27.5. *The number k is called the **modulus** and the number $\tilde{k} = \sqrt{1 - k^2}$ is called the **complementary modulus** of the integral (27.9).*

We set further

$$K = K(k) := \int_0^1 \frac{dt}{\sqrt{(1-t^2)(1-k^2t^2)}} = \int_0^{\frac{\pi}{2}} \frac{d\vartheta}{\sqrt{1-k^2\sin^2\vartheta}} \qquad (27.22)$$

and

$$\tilde{K} := \int_0^1 \frac{dt}{\sqrt{(1-t^2)(1-\tilde{k}^2t^2)}} = \int_0^{\frac{\pi}{2}} \frac{d\vartheta}{\sqrt{1-\tilde{k}^2\sin^2\vartheta}}. \qquad (27.23)$$

Note that we can consider K as a function of k. The integral or function $K(k)$ is called the **complete elliptic integral of the first kind**, the integral

$$E(k) = \int_0^1 \frac{\sqrt{1-k^2t^2}}{\sqrt{1-t^2}} \, dt \qquad (27.24)$$

is called the **complete elliptic integral of the second kind**.

We leave it as an exercise to prove that these integrals as improper Riemann integrals are well defined on $(-1, 1)$. Without proof, see [12], we state in the interest of the reader a connection between the complete elliptic integrals and the hypergeometric function:

$$K(k) = \frac{\pi}{2} \, {}_2F_1\left(\frac{1}{2}, \frac{1}{2}; 1; k^2\right); \qquad (27.25)$$

and

$$E(k) = \frac{\pi}{2} \, {}_2F_1\left(-\frac{1}{2}, \frac{1}{2}; 1; k^2\right). \qquad (27.26)$$

Lemma 27.6. *For* sn *and* cn *the following formulae hold:*

$$\mathrm{sn}(z + 2k) = -\mathrm{sn}\, z; \qquad (27.27)$$

and

$$\mathrm{cn}(z + 2k) = -\mathrm{cn}\, z. \qquad (27.28)$$

Proof. First we note that

$$\int_0^{\varphi+\pi} \frac{d\vartheta}{\sqrt{1-k^2\sin^2\vartheta}} = \int_0^{\pi} \frac{d\vartheta}{\sqrt{1-k^2\sin^2\vartheta}} + \int_{\pi}^{\varphi+\pi} \frac{d\vartheta}{\sqrt{1-k^2\sin^2\vartheta}}$$

$$= 2\int_0^{\frac{\pi}{2}} \frac{d\vartheta}{\sqrt{1-k^2\sin^2\vartheta}} + \int_0^{\varphi} \frac{d\vartheta}{\sqrt{1-k^2\sin^2\vartheta}};$$

where we used the symmetry of the integrand about the axis $x = \frac{\pi}{2}$ and its π-periodicity. Using now the definition of the amplitude (27.19) and further (27.20) we arrive at $\mathrm{am}(z + 2k) = \varphi + \pi$ and therefore

$$\mathrm{sn}(z + 2k) = \sin(\mathrm{am}(z + 2k)) = \sin(\varphi + \pi)$$
$$= -\sin\varphi = -\sin(\mathrm{am}\, z) = -\mathrm{sn}\, z,$$

and analogously

$$\mathrm{cn}(z + 2k) = \cos(\mathrm{am}(z + 2k)) = \cos(\varphi + \pi)$$
$$= -\cos(\varphi) = -\cos(\mathrm{am}\, z) = -\mathrm{cn}\, z.$$

$\square$

From Lemma 27.6 we deduce immediately

Theorem 27.7. *The functions* sn *and* cn *are periodic with period* $4k$ *and* dn *is periodic with period* $2k$.

Proof. By Lemma 27.6 we have

$$\mathrm{sn}(z + 4k) = -\mathrm{sn}(z + 2k) = \mathrm{sn}(z),$$
$$\mathrm{cn}(z + 4k) = -\mathrm{cn}(z + 2k) = \mathrm{cn}(z)$$

and

$$\mathrm{dn}(z + 2k) = \sqrt{1 - k^2 \mathrm{sn}^2(z + 2k)} = \sqrt{1 - k^2 \mathrm{sn}^2(z)} = \mathrm{dn}(z).$$

$\square$

Consider the integral (27.23), i.e.

$$\tilde{K} = \int_0^1 \frac{dt}{\sqrt{(1 - t^2)(1 - \tilde{k}^2 t^2)}}, \quad k = \sqrt{1 - k^2}, \quad |k| < 1.$$

With the substitution $u := \frac{1}{\sqrt{1 - \tilde{k}^2 t^2}}$, $0 < t < 1$, we find as range of u the interval $\left(1, \frac{1}{k}\right)$ and it follows that

$$t = \frac{\sqrt{u^2 - 1}}{\tilde{k} u}$$

and

$$\frac{du}{dt} = \tilde{k}^2 u^3 t = \tilde{k} u^2 \sqrt{u^2 - 1}.$$

Moreover we find using the relation $\tilde{k}^2 = 1 - k^2$ that

$$\frac{1}{\sqrt{1 - t^2}} = \frac{\tilde{k} u}{\sqrt{1 - \tilde{k}^2 u^2}}$$

which now yields

$$\tilde{K} = \int_0^1 \frac{dt}{\sqrt{1 - t^2}\sqrt{1 - \tilde{k}^2 t^2}} = \int_1^{\frac{1}{\tilde{k}}} \frac{\tilde{k} u^2}{\sqrt{1 - k^2 u^2}\, \tilde{k} u^2 \sqrt{u^2 - 1}} \frac{du}{}$$

$$= \int_1^{\frac{1}{\tilde{k}}} \frac{du}{\sqrt{u^2 - 1}\sqrt{1 - k^2 u^2}}.$$

This is a well justified calculation for the improper Riemann integral of a real-valued function. Now we note that

$$\frac{1}{\sqrt{u^2 - 1}} = \frac{1}{i\sqrt{1 - u^2}} = -i \frac{1}{\sqrt{1 - u^2}}$$

and arrive at

$$\tilde{K} = -i \int_1^{\frac{1}{\tilde{k}}} \frac{du}{\sqrt{1 - u^2}\sqrt{1 - k^2 u^2}}. \tag{27.29}$$

Here however a problem is hidden, namely how to define and handle the square root in the complex plane. In fact this is also a more general problem when extending sn, cn and dn to the complex plane $\mathbb{C}$. To fully understand this process we need to investigate so called multi-valued functions and their uniformization, i.e. their Riemannian surface. Since we just want to give some basic first ideas of the subject following its historical development and since we will soon give a different approach to elliptic functions, we leave some gaps in the argument, but still follow the historical development a bit further.

Now we note that (27.29) implies

$$K + i\tilde{K} = \int_0^1 \frac{du}{\sqrt{(1 - u^2)(1 - k^2 u^2)}} + \int_1^{\frac{1}{\tilde{k}}} \frac{du}{\sqrt{(1 - u^2)(1 - k^2 u^2)}}$$

$$= \int_0^{\frac{1}{\tilde{k}}} \frac{du}{\sqrt{(1 - u^2)(1 - k^2 u^2)}}, \tag{27.30}$$

or sn $(K + iK') = \frac{1}{k}$, and we have proved (leaving aside the problem of the correct definition of $\sqrt{1 - u^2}$ for $u > 1$).

Lemma 27.8. *For* sn, cn *and* dn *we have*

$$\operatorname{sn}\left(K + i\tilde{K}\right) = \frac{1}{k}, \tag{27.31}$$

$$\operatorname{cn}\left(K + i\tilde{K}\right) = -i\frac{\tilde{k}}{k}, \tag{27.32}$$

and

$$\operatorname{dn}\left(K + i\tilde{K}\right) = 0. \tag{27.33}$$

Proof. As already mentioned (27.31) follows from (27.30) when noting

$$\operatorname{sn}(z) = w \quad \text{if} \quad z = \int_0^w \frac{dt}{\sqrt{(1 - t^2)(1 - k^2 t^2)}}.$$

For (27.32) we note that

$$\operatorname{cn}\left(K + i\tilde{K}\right) = \sqrt{1 - \operatorname{sn}^2(K + i\tilde{K})} = \sqrt{1 - \frac{1}{k^2}} = -i\frac{\tilde{k}}{k},$$

and (27.33) follows from (27.31) and the definition of dn:

$$\operatorname{dn}(K + i\tilde{K}) = \sqrt{1 - k^2 \operatorname{sn}^2(K + i\tilde{K})} = 0. \qquad \square$$

Combining the addition formulae with Lemma 27.8 we find

$$\operatorname{sn}\left(2K + 2i\tilde{K}\right) = \frac{2\operatorname{sn}\left(K + i\tilde{K}\right)\operatorname{cn}\left(K + i\tilde{K}\right)\operatorname{dn}\left(K + i\tilde{K}\right)}{1 - k^2 \operatorname{sn}^4\left(K + i\tilde{K}\right)} = 0, \tag{27.34}$$

$$\operatorname{cn}\left(2K + 2i\tilde{K}\right) = \frac{\operatorname{cn}^2\left(K + i\tilde{K}\right) - \operatorname{sn}^2\left(K + i\tilde{K}\right)\operatorname{dn}^2\left(K + i\tilde{K}\right)}{1 - k^2 \operatorname{sn}^4\left(K + i\tilde{K}\right)} = 1, \tag{27.35}$$

and

$$\operatorname{dn}\left(2K + 2i\tilde{K}\right) = \frac{\operatorname{dn}^2\left(K + i\tilde{K}\right) - k^2\operatorname{sn}^2\left(K + i\tilde{K}\right)\operatorname{cn}^2\left(K + i\tilde{K}\right)}{1 - k^2\operatorname{sn}^4\left(K + i\tilde{K}\right)} = -1. \tag{27.36}$$

We may iterate the arguments which led to (27.34) - (27.36) and then we get

$$\mathrm{sn}\left(4K + 4i\tilde{K}\right) = 0, \tag{27.37}$$

$$\mathrm{cn}\left(4K + 4i\tilde{K}\right) = 1, \tag{27.38}$$

and

$$\mathrm{dn}\left(4K + 4i\tilde{K}\right) = 1. \tag{27.39}$$

Now we can prove

Theorem 27.9. *The functions* sn, cn *and* dn *are double periodic functions and we have*

$$4K \text{ and } 2i\tilde{K} \text{ are the periods of sn} \tag{27.40}$$

$$4K \text{ and } 2K + 2i\tilde{K} \text{ are the periods of cn} \tag{27.41}$$

$$2K \text{ and } 4i\tilde{K} \text{ are the periods of dn}. \tag{27.42}$$

Here K and $\tilde{K}$ are given by (27.22) and (27.23) respectively.

Proof. We have already seen that $4K$ is a period of sn and cn as well as $2K$ is a period of dn. Using the addition formulae and our previous results on special values of sn, cn and dn we find for sn

$$\mathrm{sn}(z + 2i\tilde{K}) = \mathrm{sn}(z - 2K + 2K + 2i\tilde{K})$$
$$= \frac{\mathrm{sn}(z - 2K)\,\mathrm{cn}(2K + 2i\tilde{K})\,\mathrm{dn}(2K + 2i\tilde{K}) + \mathrm{sn}(2K + 2i\tilde{K})\,\mathrm{cn}(z - 2K)\,\mathrm{dn}(z - 2K)}{1 - k^2\,\mathrm{sn}^2(z - 2K)\,\mathrm{sn}^2(2K + 2i\tilde{K})}$$
$$= -\mathrm{sn}(z - 2K) = -\mathrm{sn}(z + 2K) = \mathrm{sn}(z).$$

For cn it now follows

$$\mathrm{cn}(z + 2K + 2i\tilde{K}) = \frac{\mathrm{cn}(z)\,\mathrm{cn}(2K + 2i\tilde{K}) - \mathrm{sn}(z)\,\mathrm{sn}(2K + 2i\tilde{K})\,\mathrm{dn}(z)\,\mathrm{dn}(2K + 2i\tilde{K})}{1 - k^2\,\mathrm{sn}^2(z)\,\mathrm{sn}(2K + 2i\tilde{K})}$$
$$= \mathrm{cn}(z),$$

and the calculations for dn yields

$$\mathrm{dn}(z + 4i\tilde{K}) = \mathrm{dn}(z - 4K + 4K + 4i\tilde{K})$$
$$= \frac{\mathrm{dn}(z - 4K)\,\mathrm{dn}\left(4K + 4i\tilde{K}\right) - k^2\,\mathrm{sn}(4K + 4i\tilde{K})\,\mathrm{cn}(z - 4K)\,\mathrm{cn}(4K + 4i\tilde{K})}{1 - k^2\,\mathrm{sn}^2(z - 4K)\,\mathrm{sn}^2(4K + 4i\tilde{K})}$$
$$= \mathrm{dn}(z - 4K) = \mathrm{dn}(z - 2K) = \mathrm{dn}(z).$$

$\square$

Remark 27.10. We appreciate very much the useful and very explicit calculations in [80] which we employed to derive Theorem 27.9, and which are much easier to follow than those in [41].

The theory of elliptic integrals and elliptic functions had changed completely when first F. G. M. Eisenstein realised that starting with double periodic meromorphic functions is an alternative approach. Without noting or at least without mentioning Eisenstein's work Weierstrass developed the whole theory largely depending on his $\wp$-function. In the following we give some introduction to this theory.

Before we turn to double periodic functions we want to make a few remarks about simply periodic functions. Let f be a meromorphic function on $\mathbb{C}$ satisfying for some $w \neq 0$

$$f(z + w) = f(z) \quad \text{for all} \quad z \in \mathbb{C}. \tag{27.43}$$

In this case we call w a period of f. Clearly with w every number kw, $k \in \mathbb{Z}\backslash\{0\}$, is a further period of f. In [61] it is proved that the moduli of all periods have an infimum bounded away from zero and that periods have no accumulation point in $\mathbb{C}$. Hence for a periodic meromorphic function there exists a period w_0 of minimal modulus, $|w_0| \neq 0$. Of course, with w_0 the number $-w_0$ is a further period of minimal modulus, otherwise w_0 is uniquely determined. Clearly, all periods of the type kw_0, $k \in \mathbb{Z}\backslash\{0\}$, are on the straight line g satisfying $z \in g$ if and only if $\arg z = \arg w_0$. On g we do not have any further period w_1 of f. Suppose $w_1 \in g$ is such a period lying between kw_0 and $(k+1)w_0$, $k \in \mathbb{Z}$. It follows that $w_1 - kw_0$ is a further period of f on the line g with $|w_1 - kw_0| < |w_0|$ which is a contradiction. We call w_0 a **primitive period** of f.

Definition 27.11. *A meromorphic function f on $\mathbb{C}$ which admits only one primitive period is called **simply periodic**.*

We refer again to [61] where the following theorem is proved:

Theorem 27.12. *Suppose that f is a simply periodic function with primitive period $w_0 = 2\pi i$. If f satisfies the growth condition*

$$|f(z)| \leq M e^{\kappa|x|}, \quad |x| \geq x_0, \quad z = x + iy, \tag{27.44}$$

then there exists a rational function R such that $f(z) = R(e^z)$.

We now turn to meromorphic functions on $\mathbb{C}$ with exactly two periods $w_1 \neq 0$ and $w_2 \neq 0$, i.e. for all $z \in \mathbb{C}$ we have

$$f(z + w_1) = f(z) \text{ and } f(z + w_2) = f(z). \tag{27.45}$$

The case of interest is where w_1 and w_2 are linearly independent over $\mathbb{R}$. If $w_1 = \lambda w_2$, $\lambda \in \mathbb{R}$, then both w_1 and w_2 lie on the same line and depending on whether λ is rational or irrational the function f is either simply periodic or constant. (An exercise with hints leading to this result can be found in [83]). In the following we assume that w_1 and w_2 are linearly independent over $\mathbb{R}$, hence they span a parallelogram $P = P(w_1, w_2)$ in $\mathbb{C}$, which we call the **fundamental parallelogram**, i.e.

$$P = P(w_1, w_2) := \left\{ z \in \mathbb{C} \,\middle|\, z = \lambda_1 w_1 + \lambda_2 w_2 \,\middle|\, 0 \leq \lambda_1, \lambda_2 < 1 \right\}. \tag{27.46}$$

In order to establish some type of uniqueness for w_1 and w_2 we assume that they have minimal modulus among all periods lying on the line they generate. Further we assume that w_1 belongs to the first quadrant, i.e. $\operatorname{Re} w_1 \geq 0$ and $\operatorname{Im} w_1 \geq 0$, and that $\{w_1, w_2\}$ is positive (anticlockwise) orientated and that w_2 belongs to the upper half plane, see Figure 27.1 below.

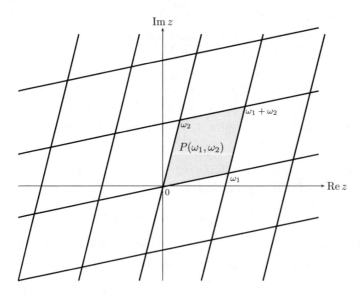

Figure 27.1

With $w_1, w_2 \in \mathbb{C}$ as above we also introduce the (**periodicity**) **lattice** $L = L(w_1, w_2)$ generated by w_1 and w_2 as

$$L = L(w_1, w_2) = \{mw_1 + nw_2 \,|\, m, n \in \mathbb{Z}\}, \qquad (27.47)$$

see Figure 27.1.

Some authors define the fundamental parallelogram to be $\overline{P(w_1, w_2)}$ which has sometimes an advantage, however our definition has the advantage that we obtain a cover of $\mathbb{C}$ by mutually disjoint sets with the help of the family $(P_{m,n}(w_1, w_2))_{m,n\in\mathbb{Z}}$, where

$$P_{m,n}(w_1, w_2) = \{z = \lambda_1 w_1 + \lambda_2 w_2 \,|\, m \leq \lambda_1 < m + 1, \, n \leq \lambda_2 < n + 1\}.$$

It is clear that for $w \in L$ and z in the domain of f we have

$$f(z + w) = f(z). \qquad (27.48)$$

Since f is meromorphic it might have poles on a discrete set $\mathrm{Pol}(f)$. For $w \in L$ and $a \in \mathrm{Pol}(f)$ we find that $a + w \in \mathrm{Pol}(f)$, thus (27.48) holds for all $z \in \mathbb{C}$ and $w \in L$.

We can now give the definition of an elliptic function as it is used nowadays.

Definition 27.13. *An **elliptic function** associated with a periodicity lattice $L = L(w_1, w_2)$ is a meromorphic function on $\mathbb{C}$ with the property (27.48), i.e. for all $w \in L$ and all $z \in \mathbb{C}$ we have $f(z + w) = f(z)$.*

Remark 27.14. A. Instead of $f(z + w) = f(z)$ for all $w \in L$ and $z \in \mathbb{C}$ it is sufficient to require $f(z + w_1) = f(z)$ and $f(z + w_2) = f(z)$ for all $z \in \mathbb{C}$ and the generators w_1 and w_2 of L.
B. Elliptic functions are sometimes called **double periodic functions**.
C. If f is double periodic then f' is double periodic too.

The following three results are often referred to as the three Liouville theorems.

Theorem 27.15 (1st **Liouville theorem**). *An elliptic function f without pole is constant.*

Proof. Suppose that f has no pole. Then f is continuous and on the compact set $\overline{P(w_1, w_2)}$ the continuous function $|f|$ is bounded. If $z \in \mathbb{C}$ then we can find integers $m, n \in \mathbb{Z}$ such that $z - mw_1 - nw_2 \in P(w_1, w_2)$ implying

$$|f(z)| = |f(z - mw_1 - nw_2)| \le \max_{w \in P(w_1, w_2)} |f(w)| < \infty.$$

This yields however that $\sup_{z \in \mathbb{C}} |f(z)| \le \max_{w \in \overline{P(w_1, w_2)}} |f(w)|$, i.e. f is a bounded entire function and by Liouville's theorem, Corollary 23.18, f must be a constant. $\qquad\square$

Theorem 27.16 (2nd **Liouville theorem**)**.** *Let w_1 and w_2 generate the lattice L and let f be a non-constant elliptic function with periodicity lattice L. Taking multiplicity into account, the total numbers of poles f has in $P(w_1, w_2)$ is strictly larger than 1 and their residues add up to zero.*

Proof. First we note that in the compact set $\overline{P(w_1, w_2)}$ the meromorphic function f can have only a finite number of poles, hence there are only finitely many poles in $P(w_1, w_2)$. We want to integrate f over $\partial P(w_1, w_2)$ and to apply the residue theorem. However some of the poles of f may lie on $\partial P(w_1, w_2)$. In this case, since we are dealing only with finitely many poles we can translate $P(w_1, w_2)$ by some $z_0 \in \mathbb{C}$ such that on $\partial(z_0 + P(w_1, w_2))$ there are no poles of f. Hence we may assume that on $\partial P(w_1, w_2)$ are no poles and now by the residue theorem we find

$$\int_{\partial P(w_1, w_2)} f(z)\, \mathrm{d}z = 2\pi i \sum_{\nu} \operatorname{res}(f, z_\nu),$$

where the sum goes over all residues of f in $P(w_1, w_2)$. We claim that the integral on the left hand side is zero. Taking the periodicity of f into account this is almost trivial. We may write

$$\partial P(w_1, w_2) = [0, w_1] \oplus [w_1, w_1 + w_2] \oplus [w_1 + w_2, w_2] \oplus [w_2, 0]$$

and we note that $[0, w_1]$ and $[w_1 + w_2, w_2]$ as well as $[w_1, w_1 + w_2]$ and $[w_2, 0]$ have opposite orientation whereas $f|_{[0, w_1]} = f|_{[w_1 + w_2, w_2]}$ and $f|_{[w_1, w_1 + w_2]} = f|_{[w_2, 0]}$, see Figure 27.2 below.

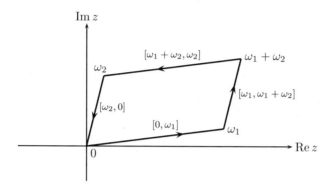

Figure 27.2

Thus we have proved that $\sum_\nu \operatorname{res}(f, z_\nu) = 0$ which excludes that f has only a single pole of first order. □

For the third Liouville theorem we need

Definition 27.17. *The **order** $\operatorname{ord}(f)$ of an elliptic function f with periodicity lattice $L = L(w_1, w_2)$ is the number of its poles in $P(w_1, w_2)$ counted according to their multiplicity.*

Theorem 27.18 (3rd **Liouville theorem**). *Suppose that the elliptic function f has order $\operatorname{ord}(f) = M$. When we take multiplicity into account then f also has M zeroes in $P(w_1, w_2)$, i.e. the number of poles and the number of zeroes (each counted with its multiplicity) of an elliptic function in its fundamental parallelogram is the same.*

Proof. First we note that the argument principle, Theorem 24.25, can be proved for boundaries of parallelograms. Secondly we can argue as in the proof of Theorem 27.16 and assume that on $\partial P(w_1, w_2)$ there are no poles and no zeroes of f. Denoting again by $\operatorname{N}(f)$ the number of zeroes of f in $P(w_1, w_2)$, of course counted according to their multiplicity, we find by the argument principle that

$$\int_{\partial P(w_1, w_2)} \frac{f'(z)}{f(z)}\, dz = 2\pi i (\operatorname{N}(f) - \operatorname{ord}(f)). \tag{27.49}$$

Since $\frac{f'}{f}$ has also $L(w_1, w_2)$ as periodicity lattice, with the same argument as in the proof of Theorem 27.16 we find that the integral in (27.49) must be zero and the theorem follows. $\qquad \square$

For a given periodicity lattice $L = L(w_1, w_2)$ we want to determine, if possible, all associated elliptic functions. In Problem 9b of Chapter 1 and in Example 5.36 we discussed the counting measure $\mu_\mathbb{Z} = \sum_{k \in \mathbb{Z}} \epsilon_k$ and we have seen its invariance under the action $\mathbb{Z}$. This observation is helpful to construct periodic functions. If the series

$$G(z) := \sum_{k \in \mathbb{Z}} g(z + kw) \qquad (27.50)$$

converges on $\mathbb{C}$ and hence represents a function then we find

$$G(z + w) = \sum_{k \in \mathbb{Z}} g(z + w + kw) = \sum_{k \in \mathbb{Z}} g(z + (k+1)w)$$
$$= \sum_{l \in \mathbb{Z}} g(z + lw) = G(z),$$

i.e. G is periodic with period w. This construction works of course for two periods w_1 and w_2: the function

$$G(z) = \sum_{(k_1, k_2) \in \mathbb{Z}^2} g(z + k_1 w_1 + k_2 w_2) \qquad (27.51)$$

has two periods w_1 and w_2. The problem is of course the convergence of a series such as (27.51), in particular if we need to construct a function with poles.

Consider the series

$$\sum_{\substack{(k_1, k_2) \in \mathbb{Z}^2 \\ (k_1, k_2) \neq (0,0)}} \frac{1}{(k_1^2 + k_2^2)^\alpha}, \qquad (27.52)$$

$\alpha \in \mathbb{R}$. We define the partial sums

$$S_{\alpha, N} := \sum_{(k_1, k_2) \in K_N} \frac{1}{(k_1^2 + k_2^2)^\alpha}$$

where $K_N := \{(k_1, k_2) \in \mathbb{Z}^2 \backslash \{(0, 0)\} \mid \max \{|k_1|, |k_2|\} < N\}$. The limit of $(S_{\alpha, N})_{N \in \mathbb{N}}$, if it exists, is also denoted by (27.52). Since all terms in (27.52) are non-negative, convergence of (27.52) will be absolute convergence and hence any other way of forming partial sums will converge to the same limit. Note that this is essentially depending on a "one-dimensional" proof working with different enumerations of $\mathbb{Z}^2$, see also Theorem I.18.27. The following lemma, extending the one-dimensional result, i.e. Theorem I.18.21, gives a helpful criterion to decide on the convergence of (27.52).

Lemma 27.19. *The sequence* $(S_{\alpha,N})_{N \in \mathbb{N}}$ *converges if and only if the integral* $\int_K \frac{dx\,dy}{(x^2+y^2)^\alpha}$, $K = \{(x,y) \in \mathbb{R}^2 \mid \|(x,y)\|_\infty \geq 1\}$, *converges as improper Riemann integral.*

This proof works as that of Theorem I.18.21 by approximating the integral with special Riemann sums and is left to the reader. Of course the convergence of $\int_K \frac{dx\,dy}{(x^2+y^2)^\alpha}$ is equivalent to the convergence of the integral $\int_{B_1^\complement(0)} \frac{dx\,dy}{(x^2+y^2)^\alpha}$ and now with the help of polar coordinates we get

$$\int_{B_1^\complement(0)} \frac{dx\,dy}{(x^2+y^2)^\alpha} = 2\pi \int_1^\infty \frac{1}{r^{2\alpha}} r\,dr = \int_1^\infty r^{1-2\alpha}\,dr$$

which is finite if and only if $2\alpha > 2$, i.e. $\alpha > 1$.

The function

$$z \mapsto \sum_{w \in L} \frac{1}{(z-w)^2}$$

is a good candidate (see below) for an elliptic function with a pole of order 2 at $z = 0$ and periodicity lattice $L = L(w_1, w_2)$. The summation is to be understood as

$$\sum_{m,n \in \mathbb{Z}} \frac{1}{(z - mw_1 - nw_2)^2}. \tag{27.53}$$

However, for $w_1 = 1$ and $w_2 = i$ we have with $z = 0$ that

$$\left| \frac{1}{(z - mw_1 - nw_2)^2} \right| = \frac{1}{m^2 + n^2}$$

and from Lemma 27.19 we deduce that (27.53) will not converge absolutely on $\mathbb{C}$. In general we have

Lemma 27.20. *Let* $L = L(w_1, w_2)$ *be a periodicity lattice and* $\alpha > 2$. *Then the series* $\sum_{w \in L \setminus \{0\}} |w|^{-\alpha}$ *converges where* 0 *stands for the lattice point* $(0,0)$.

Proof. The function $h(\xi, \eta) = \frac{|\xi w_1 + \eta w_2|^2}{\xi^2 + \eta^2}$ is homogeneous of degree 0 since

$$h(t\xi, t\eta) = \frac{|t\xi w_1 + t\eta w_2|^2}{t^2 \xi^2 + t^2 \eta^2} = \frac{|\xi w_1 + \eta w_2|^2}{\xi^2 + \eta^2} = h(\xi, \eta).$$

As a continuous function it obtains its minimum κ on S^1. Since w_1 and w_2 are linearly independent over $\mathbb{R}$, for $(\xi, \eta) \in S^1$ we have $|\xi w_1 + \eta w_2|^2 > 0$, implying that $h(\xi, \eta) \geq \kappa > 0$. The homogeneity of h now implies

$$|m w_1 + n w_2|^2 \geq \kappa(m^2 + n^2)$$

for all $m, n \in \mathbb{Z}^2$ which yields $\frac{1}{|m w_1 + n w_2|^2} \leq \frac{1}{\kappa} \frac{1}{m^2 + n^2}$ and from Lemma 27.19 we deduce for $\alpha > 2$ that

$$\sum_{w \in L \backslash \{0\}} |w|^{-\alpha} \leq \frac{1}{\kappa} \sum_{m,n \in \mathbb{Z}^2 \backslash \{0\}} \frac{1}{(m^2 + n^2)^{\frac{\alpha}{2}}} < \infty.$$

$\square$

Of greatest interest is now

Lemma 27.21. *Let $L = L(w_1, w_2)$ be a periodicity lattice. The series*

$$\sum_{w \in L \backslash \{0\}} \left(\frac{1}{(z - w)^2} - \frac{1}{w^2} \right) \tag{27.54}$$

converges absolutely and locally uniformly in every subset of $\mathbb{C} \backslash (L \backslash \{0\})$.

Proof. Since

$$\frac{1}{(z-w)^2} - \frac{1}{w^2} = \frac{w^2 - (z-w)^2}{(z-w)^2 w^2} = \frac{-z(z - 2w)}{(z-w)^2 w^2}$$

we have

$$\left| \frac{1}{(z-w)^2} - \frac{1}{w^2} \right| = \frac{|z| |z - 2w|}{|z - w|^2 |w|^2}.$$

Now let $z \in \overline{B_r(0)}$ and $|w| \geq 2r$. It follows that $|z| \leq \frac{|w|}{2}$ and therefore $|z| |z - 2w| \leq 3r|w|$ and $|z - w| \geq |w| - |z| \geq \frac{|w|}{2}$, i.e.

$$\frac{|z| |z - 2w|}{|z - w|^2 |w|^2} \leq 12r|w|^{-3},$$

which implies the locally uniform and absolute convergence of (27.54) for all $z \in \mathbb{C} \backslash (L \backslash \{0\})$. $\square$

Definition 27.22. *The **Weierstrass $\wp$-function** associated with the periodicity lattice $L = L(w_1, w_2)$ is the function defined by*

$$\wp(z, L) = \wp(z) = \frac{1}{z^2} + \sum_{w \in L \setminus \{0\}} \left(\frac{1}{(z-w)^2} - \frac{1}{w^2} \right), \quad z \notin L. \qquad (27.55)$$

For $z \in L$ we may define $\wp(z) = \infty$ when we understand by ∞ the point at infinity obtained by the one point compactification of $\mathbb{C}$. Our considerations so far are summarised in

Theorem 27.23. *Given a periodicity lattice $L = L(w_1, w_2)$. The associated Weierstrass $\wp$-function is a function meromorphic on $\mathbb{C}$ with poles of second order at the lattice points. In the complement of L it is a holomorphic function which is even, i.e. $\wp(z) = \wp(-z)$ and its Laurent expansion about 0 is of the type*

$$\wp(z) = \frac{1}{z^2} + \sum_{k=1}^{\infty} a_{2k} z^{2k}. \qquad (27.56)$$

Proof. We only need a remark as to why $\wp$ is an even function, which however follows when we replace z by $-z$ in (27.55) and note that $-(L \setminus \{0\}) = L \setminus \{0\}$. $\qquad \square$

Since we can differentiate (27.55) in $\mathbb{C} \setminus (L \setminus \{0\})$ term by term we obtain

Corollary 27.24. *The derivative of $\wp$ as in Theorem 27.23 is given by*

$$\wp'(z) = -2 \sum_{w \in L} \frac{1}{(z-w)^3}. \qquad (27.57)$$

This is meromorphic function on $\mathbb{C}$ with poles of order 3 at the lattice points and holomorphic in $\mathbb{C} \setminus L$. In addition $\wp'$ is an odd function, i.e. $\wp'(-z) = -\wp'(z)$.

Our goal is to prove that $\wp$ and $\wp'$ are elliptic functions and every elliptic function f with periodicity lattice $L = L(w_1, w_2)$ is of the type

$$f = R_1(\wp) + \wp' R_2(\wp) \qquad (27.58)$$

where R_1 and R_2 are rational functions. Hence the whole theory of elliptic functions is encoded in the functions $\wp$. It is of some interest to note that

Eisenstein was the first who noted that such an approach might work, his starting point was the series (27.53). He was well aware of the convergence problems and could manage related problems as A. Weil [91] has pointed out. Nonetheless, his work was not picked up and it is one of the merits of Weierstrass to have developed the theory of elliptic functions as we handle them today.

Theorem 27.25. *The function $\wp$ is an elliptic function of order 2 and $\wp'$ is an elliptic function of order 3.*

Proof. It remains to prove the periodicity of $\wp$ and $\wp'$ and this is easier for $\wp'$. Indeed, for $w_0 \in L$ we have

$$\wp'(z + w_0) = -2 \sum_{w \in L} \frac{1}{(z + w_0 - w)^3} = \wp'(z)$$

since $-w \mapsto w_0 - w$ is a bijection of L onto itself. Now we prove that for every $w_0 \in L$ the function $z \mapsto \wp(z + w_0) - \wp(z)$ is constant and equal to zero. Since $\wp'(z + w_0) = \wp'(z)$ this function must be constant, note that $\mathbb{C} \backslash L$ is arcwise connected. Take $w_0 = w_1$, recall $L = L(w_1, w_2)$, i.e. w_1 is one of the generators of the periodicity lattice. It follows that $-\frac{w_1}{2} \notin L$ and

$$\wp\left(-\frac{w_1}{2} + w_1\right) - \wp\left(-\frac{w_1}{2}\right) = \wp\left(\frac{w_1}{2}\right) - \wp\left(-\frac{w_1}{2}\right) = 0$$

where we used that $\wp$ is an even function. Hence $\wp(z + w_0) = \wp(z)$. $\qquad\square$

Let $w \in \{w_1, w_2\}$. It follows that

$$\wp'\left(\frac{w}{2}\right) = \wp'\left(\frac{w}{2} - w\right) = \wp'\left(-\frac{w}{2}\right) = -\wp'\left(\frac{w}{2}\right)$$

implying that $\wp'\left(\frac{w_1}{2}\right) = \wp'\left(\frac{w_2}{2}\right) = 0$. Moreover, for $w = w_1 + w_2$ we have

$$\wp'\left(\frac{w_1 + w_2}{2}\right) = \wp'\left(\frac{w_1 + w_2}{2} - (w_1 + w_2)\right) = -\wp'\left(\frac{w_1 + w_2}{2}\right),$$

i.e. $\wp\left(\frac{w_1 + w_2}{2}\right) = 0$ and we have found in $P(w_1, w_2)$ three mutually distinct zeroes of $\wp'$: $\frac{w_1}{2}, \frac{w_2}{2}$ and $\frac{w_1 + w_2}{2}$. Since by the 3rd Liouville theorem, Theorem 27.18, $\wp'$ has exactly three zeroes in $P(w_1, w_2)$ we know all zeroes of $\wp'$. The values

$$e_1 := \wp\left(\frac{w_1}{2}\right), \quad e_2 := \wp\left(\frac{w_2}{2}\right), \quad e_3 = \wp\left(\frac{w_1 + w_2}{2}\right) \qquad (27.59)$$

are playing some important role in the theory of elliptic functions.
We want to determine the coefficients of the Laurent expansion of $\wp$ about
$z_0 = 0$. For this we note first that by Lemma 27.20 for $n \geq 3$ the series

$$G_n = G_n(L) = G_n(w_1, w_2) := \sum_{w \in L \backslash \{0\}} w^{-n} \qquad (27.60)$$

converges absolutely. Moreover for n odd it follows that

$$G_n = \sum_{w \in L \backslash \{0\}} w^{-n} = \sum_{w \in L \backslash \{0\}} (-w)^n = - \sum_{w \in L \backslash \{0\}} w^{-n}$$

implying

$$G_{2n+1} = 0 \quad \text{for all } n \in \mathbb{N}. \qquad (27.61)$$

Theorem 27.26. *The Laurent expansion of the Weierstrass $\wp$-function associated with the periodicity lattice $L = L(w_1, w_2)$ is given by*

$$\wp(z) = \frac{1}{z^2} + \sum_{k=1}^{\infty} (2k+1) G_{2(k+1)} z^{2k}. \qquad (27.62)$$

Proof. We know already from Theorem 27.23 that

$$\wp(z) = \frac{1}{z^2} + \sum_{k=1}^{\infty} a_{2k} z^{2k}.$$

Consider the function

$$g(z) = \wp(z) - \frac{1}{z^2}$$

which must have Taylor coefficients $a_0 = 0$ and

$$a_{2k} = \frac{g^{(2k)}(0)}{(2k)!}.$$

For $l > 1$ we have

$$g^{(l)}(z) = (-1)^l (l+1)! \sum_{w \in L \backslash \{0\}} \frac{1}{(z - w)^{l+2}}, \qquad (27.63)$$

i.e. $g^{(l)}(z) = (-1)^l (l+1)! G_{l+2}$. For l odd we know by (27.61) that $G_{l+2} = 0$,
i.e. $g^{(l)}(0) = 0$. For $l = 2k$ we find

$$g^{(2k)}(0) = (2k+1)! G_{2(k+1)}$$

implying that

$$a_{2k} = \frac{g^{(2k)}(0)}{(2k)!} = \frac{(2k+1)!\, G_{2(k+1)}}{(2k)!} = (2k+1)G_{2(k+1)}.$$

$\square$

Remark 27.27. The series G_n are called **Eisenstein series** and they are of great importance in several branches of Mathematics. Recall that G_n depends on L, i.e. $G_n = G_n(L)$.

We can now prove the first main result in Weierstrass' theory of elliptic functions.

Theorem 27.28. *Let $L = L(w_1, w_2)$ be a periodicity lattice and f an elliptic function with periodicity lattice L. Then there exists two rational functions R_1 and R_2 such that with $\wp(z) = \wp(z, L)$ we have*

$$f(z) = R_1(\wp(z)) + \wp'(z)R_2(\wp(z)). \tag{27.64}$$

Proof. We prove this theorem in several steps similar to the structure of the proof given in [28].

1. Assume that f is not constant, even and that all poles of f are contained in L. This implies that f must have a pole at 0 and the Laurent expansion of f about 0 is of the form

$$f(z) = a_{-2n}^{(n)} z^{-2n} + a_{-2(n-1)}^{(n)} z^{-2(n-1)} + \cdots.$$

The Laurent expansion of $\wp(z)^n$ about 0 is of the form

$$\wp(z)^n = z^{-2n} + \cdots.$$

This implies that the function $f - a_{-2n}^{(n)}\wp^n$ is an even elliptic function with poles only on L which have order strictly less than $\mathrm{ord}(f)$. Hence for some $m < n$ we have that

$$f - a_{-2n}^{(n)}\wp^n - a_{-2m}^{(m)}\wp^m$$

is an elliptic function with poles only on L where $a_{-2m}^{(m)}$ is determined from

$$f - a_{-2n}^{(n)}\wp^n = a_{-2m}^{(m)} z^{-2m} + a_{-2(m-1)}^{(m)} z^{-2(m-1)} + \cdots.$$

Thus, after at most n steps we find that

$$f - \sum_{k=1}^{n} a_{-2k}^{(k)} \wp^k$$

is an even elliptic function with no pole, i.e. it must be constant. Hence we have proved:

If f is an even elliptic function with poles only on L then f is a polynomial in $\wp(\cdot, L)$ of order $\frac{\operatorname{ord}(f)}{2}$.

2. Suppose that f is still even and not constant but f may have poles not only on L. Let $a \notin L$ be such a pole of f and consider

$$h(z) = (\wp(z) - \wp(a))^N f(z),$$

where N is determined such that the singularity of h at a is removable. Since f can have only a finite number of poles in $P(w_1, w_2)\backslash L$, say $a_1, \ldots, a_M$, we deduce that with suitable numbers $N_j \in \mathbb{N}$,

$$g(z) := f(z) \prod_{j=1}^{M} (\wp(z) - \wp(a_j))^{N_j}$$

has no pole outside of L. This implies that g is a polynomial in $\wp$ and f is a rational function of $\wp$, i.e. $f(z) = R(\wp(z))$ for some rational function R.

3. Now let f be an arbitrary non-constant elliptic function with periodicity lattice L and $\wp = \wp(\cdot, L)$ the Weierstrass $\wp$-function associated with L. We decompose $f = f_{\text{even}} + f_{\text{odd}}$ into its even and odd part where $f_{\text{even}}(z) = \frac{1}{2}(f(z) + f(-z))$. Both are elliptic functions with periodicity lattice L. For f even we have already a representation

$$f_{\text{even}}(z) = R_1(\wp(z))$$

with a rational function R_1. Now, $\wp'$ is an odd function and therefore $\frac{f_{\text{odd}}}{\wp'}$ is an even function which is an elliptic function with periodicity lattice L. Hence we can find a rational function R_2 such that $\frac{f_{\text{odd}}}{\wp'} = R_2(\wp)$ which eventually yields

$$f(z) = f_{\text{even}}(z) + f_{\text{odd}}(z) = R_1(\wp(z)) + \wp'(z)R_2(\wp(z)).$$

$\square$

As the Jacobi elliptic functions sn, cn and dn are double periodic functions with periodicity lattice $L_{\text{sn}} = L_{\text{sn}}(4K, K + 2i\tilde{K})$, $L_{\text{cn}} = L_{\text{cn}}(4K, 2K + 2i\tilde{K})$ and $L_{\text{dn}} = L_{\text{dn}}(2K, 4i\tilde{K})$, respectively, to each of these functions corresponds a Weierstrass $\wp$-function with the respective periodicity lattice. This allows us to represent the Jacobi elliptic function with the help of the $\wp$-function, its derivative and two rational functions in $\wp$. We do not want to enter into a more detailed discussion and refer for example to [88].

Finally we want to derive a differential equation for $\wp$. For this consider the function

$$H(z) = (\wp(z) - e_1)(\wp(z) - e_2)(\wp(z) - e_3) \tag{27.65}$$

where e_j is as in (27.59). Since $\wp'(\tilde{w}) = 0$ for $\tilde{w} = \{\frac{w_1}{2}, \frac{w_2}{2}, \frac{w_1+w_2}{2}\}$ and these are simple roots of $\wp'$, it follows that $\frac{w_1}{2}$, $\frac{w_2}{2}$ and $\frac{w_1+w_2}{2}$ are zeroes of multiplicity two of H. Moreover $(\wp')^2$ has roots of order two at $\frac{w_1}{2}$, $\frac{w_2}{2}$ and $\frac{w_1+w_2}{2}$. In addition H has poles of order 6 at w_1, w_2 and w_1+w_2 as does $(\wp')^2$. This implies that the elliptic function $\frac{(\wp')^2}{H}$ which has periodicity lattice L is in fact holomorphic, hence constant. By using the Laurent expansion about zero, i.e.

$$\wp(z) = \frac{1}{z^2} + \cdots, \quad \wp'(z) = -\frac{2}{z^3} + \cdots,$$

we deduce that $\frac{(\wp'(z))^2}{H(z)} = 4$, and we have proved

Theorem 27.29. *The Weierstrass $\wp$-function satisfies the differential equation*

$$(\wp')^2 = 4(\wp - e_1)(\wp - e_2)(\wp - e_3). \tag{27.66}$$

Note that solving the differential equation

$$(g')^2 = 4(g - e_1)(g - e_2)(g - e_3)$$

for a function $g = g(t)$, $t \in \mathbb{R}$, leads by the method of separating variables, formally to

$$\int \frac{dg}{2\sqrt{(g - e_1)(g - e_2)(g - e_3)}} = t - t_0$$

and we encounter an elliptic integral!

There is another way to express (27.66). For this we use a very formal but justifiable way of calculation. Given the Laurent expansion about 0 of two meromorphic functions

$$f(z) = a_{-k}z^{-k} + a_{-(k-1)}z^{-(k-1)} + \cdots$$

and

$$g(z) = b_{-m}z^{-m} + b_{-(m-1)}z^{-(m-1)} + \cdots .$$

We obtain the Laurent expansion of the product by multiplying these two series and then arranging the product series according to increasing powers:

$$(f \cdot g)(z) = a_{-k}b_{-m}z^{-k-m} + (a_{-k}b_{-(m-1)} + a_{-(k-1)}b_{-m})z^{-k-m-1} + \cdots .$$

Thus we find for $\wp$ and $\wp'$

$$\wp(z) = z^{-2} + 3G_4 z^2 + 5G_6 z^4 + \cdots$$
$$\wp'(z) = -2z^{-3} + 6G_4 z + 20G_6 z^3 + \cdots$$
$$\wp(z)^2 = z^{-4} + 6G_4 + 10G_6 z^2 + \cdots$$
$$\wp(z)^3 = -z^{-6} + 9G_4 z^{-2} + 15G_6 + \cdots$$

and

$$\wp'(z)^2 = 4z^{-6} - 24G_4 z^{-2} - 80G_6 + \cdots$$

which implies that

$$\wp'(z) - 4\wp(z)^3 + 60G_4\wp(z) + 140G_6$$

is an elliptic function with periodicity lattice L which is holomorphic, hence it must be constant. Since for $z = 0$ this holomorphic function is 0, we arrive at

Theorem 27.30. *The Weierstrass $\wp$-function associated with the periodicity lattice $L = L(w_1, w_2)$ satisfies the differential equation*

$$(\wp')^2 = 4\wp^3 - 60G_4\wp - 140G_6. \qquad (27.67)$$

Comparing (27.66) with (27.67) allows us to establish relations between e_1, e_2 and e_3 and the Eisenstein series G_4 and G_6. Finally we state a fundamental result linking certain elliptic integrals with elliptic functions.

Theorem 27.31. *Let $p(t)$ be a polynomial of degree 3 or 4 without multiple zeroes. Then there exists a non-constant elliptic function f and an open set $D \subset \mathbb{C}$ such that $f|_D$ admits an inverse $g : f(D) \to \mathbb{C}$ and (with an appropriate choice of the branch of the square root function) we have $g'(z) = \frac{1}{\sqrt{p(z)}}$, i.e. g is a primitive of $\frac{1}{\sqrt{p(\cdot)}}$, or $g(z) = \int_a^z \frac{1}{\sqrt{p(t)}} dt$.*

Here we end our discussion of elliptic integrals and elliptic functions, but we emphasise once again that the theory and its application just starts with these results. In addition to the sources already mentioned, e.g. [12], [28], [83] and [88] we recommend [53] as a further reading.

Problems

1. For $0 < b < a$ let $\gamma : [0, 2\pi] \to \mathbb{R}^2$, $\gamma(t) = (a \sin t, b \sin t)$, be a parametrization of the ellipse $\frac{x^2}{a^2} + \frac{y^2}{b^2} = 1$. Find the length l_γ of the ellipse in terms of an elliptic integral.

2. The **lemniscate** is the algebraic curve in the plane determined by the equation $(x^2 + y^2)^2 = x^2 - y^2$, see the corresponding figure in the solution. In polar coordinates the lemniscate is given by the equation $r^2 = 2 \cos 2\varphi$. Find the length of an arc of the lemniscate starting at 0 and terminating at a point z with $\arg(z) < \frac{\pi}{4}$. Use symmetry and a consideration for the limit $\varphi \to \frac{\pi}{4}$ to find the length l of the lemniscate.

3. Prove formula (27.6).

4. With $K(k)$ as in (27.22) prove

$$\int_0^{\frac{\pi}{2}} \frac{dx}{\sqrt{\sin x}} = \sqrt{2} \int_0^{2\pi} \frac{dt}{\sqrt{1 - \frac{1}{2} \sin^2 t}} = \sqrt{2} K \left(\frac{1}{2} \right).$$

5. Let $a, b \in \mathbb{R}$ and $a + b \neq 0$ and $K(k)$ as in (27.22). Using the substitution $\vartheta = 2\varphi$ prove

$$J := \int_0^\pi \frac{d\vartheta}{\sqrt{a^2 + 2ab \cos \vartheta + b^2}} = \frac{2}{a + b} \int_0^{\frac{\pi}{2}} \frac{d\varphi}{\sqrt{1 - \frac{4ab}{(a+b)^2} \sin^2 \varphi}}.$$

Now apply the substitution $a\sin(\vartheta - \psi) = b\sin\psi$ to derive

$$\int_0^\pi \frac{d\vartheta}{\sqrt{a^2 + 2ab\cos\vartheta + b^2}} = \frac{2}{a}\int_0^{\frac{\pi}{2}} \frac{1}{\sqrt{1 - \frac{b^2}{a^2}\sin^2\psi}} d\psi.$$

Deduce

$$K\left(\frac{b}{a}\right) = \frac{2}{a+b}K\left(\frac{2\sqrt{ab}}{a+b}\right),$$

or with $k = \frac{2\sqrt{ab}}{a+b}$ and $k' = \sqrt{1-k^2}$

$$K(k) = \frac{2}{1+k'}K\left(\frac{1-k'}{1+k}\right).$$

6. Prove the equation

$$\operatorname{dn}^2 z - k^2 \cos^2 z = k'^2$$

and

$$\operatorname{sn}^2 z = \frac{1-\cos 2z}{1+\operatorname{dn}2z}.$$

7. Verify (27.16) - (27.18).
   Hint: work through the solution we have provided as a proof of (27.16).

8. For the Jacobi elliptic functions sn, cn and dn derive the following differential equations:

$$\left(\frac{d\operatorname{sn}z}{dz}\right)^2 = (1 - \operatorname{sn}^2 z)(1 - k^2\operatorname{sn}^2 z);$$

$$\left(\frac{d\operatorname{cn}z}{dz}\right)^2 = (1 - \operatorname{cn}^2 z)(k'^2 + k^2\operatorname{cn}^2 z);$$

$$\left(\frac{d\operatorname{dn}z}{dz}\right)^2 = -(1 - \operatorname{dn}^2 z)(k'^2 - \operatorname{dn}^2 z).$$

9. Give a more detailed proof of Lemma 27.19. Formulate an extension of this lemma to $\mathbb{R}^n$ and indicate the proof.

10. Consider the Weierstrass $\wp$-function $\wp(z, L)$ as a function of z and $w_1, w_2, L = L(w_1, w_2)$, i.e. write $\wp(z, L) = \wp(z; w_1, w_2)$. Prove that $(z, w_1, w_2) \mapsto \wp(z; w_1, w_2)$ is homogeneous of degree -2.

11. Let $L = L(1,\tau)$, $\operatorname{Im}\tau > 0$. For the corresponding Eisenstein series G_n, $n \geq 3$, prove with $G_n(\tau) := G_n(1;\tau)$ that

$$G_n(\tau+1) = G_n(\tau) \quad \text{and} \quad G_n(\tau) = \tau^{-n} G_n\left(-\frac{1}{\tau}\right).$$

28 The Riemann Mapping Theorem

Our approach to complex-valued functions of a complex variable is in some sense an analytic approach putting properties of holomorphic functions on the centre stage. However on several occasions we encountered the need to think more (complex-) geometrical. In fact, once basic notions and tools have been introduced, we can look more at topological and most of all geometric or metric mapping properties of these functions. Here the aspect of mapping sets onto sets becomes more prominent. One such result is Theorem 23.2 stating that holomorphic functions are open mappings, i.e. they map open sets onto open sets, provided they are not constant. In this chapter we want to look at simply connected open sets in the plane and study their biholomorphic images.

Geometry has entered somewhat hidden into our considerations in some other context: it is a problem to find "good" domains of definition for so called multi-valued functions such as $z \mapsto z^\alpha$, $\alpha \notin \mathbb{Z}$, or $z \to \log z$, etc. Our way of dealing with these functions was to shrink their domains in such a way that they become well defined holomorphic functions admitting a primitive in this domain. This led to the introduction of branches of functions, for example in the case of the logarithmic function. Of similar nature was our problem when we dealt with elliptic integrals and Jacobi elliptic functions and tried to determine their periods.

B. Riemann suggested (in our modern interpretation) that instead of shrinking the domains of a multi-valued function to enlarge the domains and go beyond subsets of the complex plane (or its one-point-compactification, the Riemann sphere). This resulted in the theory of Riemannian surfaces, thanks to H. Weyl [92] in the theory of differentiable manifolds, but also in the theory of uniformisation of (algebraic) functions. Eventually it turns out that both problems are closely connected and Koebe's uniformisation theorem, according to L. Ahlfors [4] "perhaps the single most important theorem in the whole theory of analytic [=holomorphic] functions of one variable", does for Riemann surfaces what the Riemann mapping theorem does for regions in $\mathbb{C}$. The Koebe theorem and the theory of uniformisation will not be discussed here, see for example [60], and Riemann surfaces we will briefly handle in our final volume when dealing with differentiable manifolds. However the Riemann mapping theorem and some of its consequences we can treat and this is the aim of this Chapter.

By Definition 17.18 two regions G_1 and G_2 are **holomorphically equivalent** if there exists a biholomorphic function $f : G_1 \to G_2$. By Proposition 17.19 a mapping $f : G_1 \to G_2$ is biholomorphic if and only if f is holomorphic, bijective, f^{-1} is continuous and $f'(z) \neq 0$ for all $z \in G_1$. Since the identity is biholomorphic and since with $f : G_1 \to G_2$ the function $f^{-1} : G_2 \to G_1$ is biholomorphic it follows that biholomorphy is an equivalence relation in the set of all regions. From Liouville's theorem we deduce that there cannot exist a biholomorphic mapping from $\mathbb{C}$ onto the unit disc $\mathbb{D} := B_1(0)$. Indeed, if $f : \mathbb{C} \to \mathbb{D}$ were biholomorphic it would be a bounded entire function, hence constant. Thus the problem to classify all biholomorphically equivalent regions is a non-trivial one. Since biholomorphic mappings are also homeomorphisms we restrict ourselves to the topologoical simplest situation namely to simply connected regions. Before we turn to the Riemann mapping theorem and its proof we first want to look at some concrete examples and then we want to look at biholomorphic mappings from a simply connected region into itself and discuss the underlying group structure.

It is helpful to use in this chapter as notation for special sets

$$\mathbb{D} := B_1(0) \quad \text{and} \quad \mathbb{H} := \{z \in \mathbb{C} \,|\, \operatorname{Im} z > 0\} \tag{28.1}$$

which are both simply connected regions. We next use our results on Möbius transformations from Chapter 16. In Proposition 16.23 we have proved the following:

The Möbius transform $w(z) = \frac{az+b}{cz+d}$ maps the punctured plane $\mathbb{C} \backslash \{-\frac{d}{c}\}$ bijectively and continuously with continuous inverse onto the punctured plane $\mathbb{C} \backslash \{\frac{a}{c}\}$.

For $z \neq -\frac{d}{c}$ the function w is holomorphic. Further we find

$$w'(z) = \frac{\det \begin{pmatrix} a & b \\ c & d \end{pmatrix}}{(cz+d)^2} \neq 0 \tag{28.2}$$

since for a Möbius transform $\frac{az+b}{cz+d}$ by assumption $\det \begin{pmatrix} a & b \\ c & d \end{pmatrix} \neq 0$. Thus, by Proposition 17.19 the Möbius transform w is a biholomorphic mapping between the two simply connected regions $\mathbb{C} \setminus \{-\frac{d}{c}\}$ and $\mathbb{C} \setminus \{\frac{a}{c}\}$. As a corollary we obtain

Corollary 28.1. *Two punctured planes are biholomorphic equivalent.*

We can exploit Corollary 28.1 further. Let $G_1 \subset \mathbb{C} \setminus \{-\frac{d}{c}\}$ be a simply connected region and $w(z) = \frac{az+b}{cz+d}$ a Möbius transformation. Then $w(G_1)$ and G_1 are biholomorphic equivalent provided $\frac{a}{c} \notin w(G_1)$.

Consider the Möbius transformation $W_1(z) = \frac{i-z}{i+z}$ corresponding to the matrix $\begin{pmatrix} -1 & i \\ 1 & i \end{pmatrix}$ with determinant $-2i$ and inverse $W_1^{-1}(w) = V_1(w) = \frac{i-iw}{1+w}$.

Recall that if a Möbius transformation is associated with $\begin{pmatrix} a & b \\ c & d \end{pmatrix}$ then its inverse is associated with $\begin{pmatrix} -d & b \\ c & -a \end{pmatrix}$. We know already that W_1 maps $\mathbb{C} \setminus \{-i\}$ biholomorphically onto $\mathbb{C} \setminus \{-1\}$. For $z \in \mathbb{H}$ it follows that $|i - z| < |z + i|$, hence $W_1(\mathbb{H}) \subset \mathbb{D}$. On the other hand for $w = u + iv \in \mathbb{D}$ it follows that

$$
\begin{aligned}
\operatorname{Im} V_1(w) &= \operatorname{Im} \left(\frac{i - i(u + iv)}{i + u + iv} \right) = \operatorname{Re} \left(\frac{1 - u - iv}{1 + u + iv} \right) \\
&= \operatorname{Re} \left(\frac{(1 - u - iv)(1 + u - iv)}{(1 + u)^2 + v^2} \right) \\
&= \frac{1 - u^2 - v^2}{(1 + u^2) + v^2} > 0,
\end{aligned}
$$

thus $V_1(\mathbb{D}) \subset \mathbb{H}$ and we have proved

Lemma 28.2. *The mapping $W_1(z) = \frac{i-z}{i+z}$ maps the upper half-plane $\mathbb{H}$ biholomorphically onto the unit disc $\mathbb{D}$ with inverse mapping $V_1(w) = \frac{i-iw}{i+w}$.*

In Problem 1 we will see that W_1 maps $\partial\mathbb{H} \cong \mathbb{R}$ continuously onto $\partial D \setminus \{-1\}$, recall that $\partial\mathbb{D} \cong S^1$.

Exercise 28.3. *Prove that the Möbius transformation $w(z) = \frac{1+z}{1-z}$ maps $\mathbb{D} \cap \mathbb{H}$ onto the first quadrant $\{u + iv \in \mathbb{C} \,|\, u > 0, v > 0\}$.*

For $a \in \mathbb{D}$ we define the Möbius transform

$$
W_a(z) := \frac{z - a}{1 - \bar{a}z}. \tag{28.3}
$$

557

The matrix associated with W_a is $\begin{pmatrix} 1 & -a \\ -\bar{a} & 1 \end{pmatrix}$ with determinant $1-|a|^2 > 0$ for $a \in \mathbb{D}$.

Lemma 28.4. *For $a \in \mathbb{D}$ the Möbius transformation W_a is a bijective mapping from $\mathbb{D}$ onto itself as well as from S^1 onto itself. The inverse transformation to W_a is W_{-a} and we have*

$$W_a'(0) = 1 - |a|^2, \quad W_a'(a) = \frac{1}{1 - |a|^2}. \tag{28.4}$$

Proof. The inverse to W_a is the Möbius transformation corresponding to the matrix $\begin{pmatrix} -1 & -a \\ -\bar{a} & -1 \end{pmatrix}$ and hence is given by

$$z \mapsto \frac{-z - a}{-\bar{a}z - 1} = \frac{z + a}{1 + \bar{a}z} = W_{-a}(z).$$

Since

$$W_a'(z) = \frac{\det \begin{pmatrix} 1 & -a \\ -\bar{a} & 1 \end{pmatrix}}{(1 - \bar{a}z)^2} = \frac{1 - |a|^2}{(1 - \bar{a}z)^2},$$

we find for $z = 0$ that $W_a'(0) = 1 - |a|^2$ and $W_a'(a) = \frac{1}{1-|a|^2}$. Now let $z = e^{it}$, $t \in [0, 2\pi]$. It follows that

$$|W_a(z)| = \left| \frac{e^{it} - a}{1 - \bar{a}e^{-t}} \right| = \left| \frac{e^{it} - a}{e^{-it} - \bar{a}} \right|$$

$$= \frac{|e^{it} - a|}{|e^{it} - a|} = 1,$$

i.e. $W_a(S^1) \subset S^1$. The same argument applies to W_{-a} implying that $W_a(S^1) = S^1$. Moreover, the boundary maximum principle, Theorem 23.8, yields now $W_a(\mathbb{D}) \subset \mathbb{D}$ as well as $W_{-a}(\mathbb{D}) \subset \mathbb{D}$ which again gives that $W_a(\mathbb{D}) = \mathbb{D}$. $\square$

Lemma 28.4 provides us with an example of a biholomorphic mapping from $\mathbb{D}$ onto itself $\mathbb{D}$. We want to have a closer look at such mappings. First we note a simple general result.

Lemma 28.5. *Let $G \subset \mathbb{C}$ be a simply connected region. The set of all biholomorphic mappings from G onto itself forms a group under the usual compositions of mappings.*

Proof. Since with f and g also $f \circ g$ is a biholomorphic mapping from G onto G, and since f^{-1} is biholomorphic if f is, we only need add the remark that id_G is a biholomorphic mapping. $\square$

Definition 28.6. *The **automorphism group** of a simply connected region $G \subset \mathbb{C}$ is the group of all biholomorphic mappings from G onto itself and is denoted by $\mathrm{Aut}(G)$.*

We want to determine $\mathrm{Aut}(\mathbb{D})$. Clearly, if $\varphi \in \mathbb{R}$ then $z \mapsto e^{i\varphi}z$ is an automorphism of $\mathbb{D}$. Combined with Lemma 28.4 we find that for every $\varphi \in \mathbb{R}$ and $a \in \mathbb{D}$ an automorphism of $\mathbb{D}$ is given by

$$z \mapsto e^{i\varphi}W_a(z). \tag{28.5}$$

For the following it is more convenient to consider $V_a(z) := -W_a(z)$. The matrix corresponding to V_a is $\begin{pmatrix} -1 & a \\ -\bar{a} & 1 \end{pmatrix}$ and now the matrix corresponding to V_a^{-1} becomes $\begin{pmatrix} -1 & a \\ -\bar{a} & 1 \end{pmatrix}$, i.e. $V_a = V_a^{-1}$. In light of (28.5) changing the sign of $W_a(z)$ can be compensated substituting φ by $\varphi + \pi$. The surprising result is that all automorphisms of $\mathbb{D}$ are of type (28.5) or equivalently of type $e^{i(\varphi+\pi)}V_a(\cdot)$.

Theorem 28.7. *For every automorphism W of $\mathbb{D}$ there exists $\vartheta \in \mathbb{R}$ and $a \in \mathbb{D}$ such that*

$$W(z) = e^{i\vartheta}\frac{a-z}{1-\bar{a}z} \tag{28.6}$$

holds.

Proof. The proof will make much use of the **Lemma of Schwarz**, i.e. Theorem 23.11. Let W be an automorphism of $\mathbb{D}$. Denote by $a \in \mathbb{D}$ the unique point with $W(a) = 0$ and consider the automorphism $U := W \circ V_a$, $V_a(z) = \frac{a-z}{1-\bar{a}z}$. We find that $V_a(0) = a$, hence $U(0) = 0$. According to Theorem 23.11 it follows that

$$|U(z)| \leq |z| \quad \text{for } z \in \mathbb{D} \tag{28.7}$$

and
$$|U'(0)| \le 1.$$

Since U^{-1} is also an automorphism of $\mathbb{D}$ and $U^{-1}(0) = 0$, we find by a further application of Schwarz' lemma that

$$|U^{-1}(w)| \le |w| \quad \text{for all } w \in \mathbb{D},$$

and with $w = U(z)$ we arrive at

$$|z| \le |U(z)| \quad \text{for all } z \in \mathbb{D}, \tag{28.8}$$

i.e. we have
$$|U(z)| = |z| \quad \text{for all } z \in \mathbb{D}.$$

Now we use Corollary 23.12, the refinement to the Lemma of Schwarz, to conclude that for some $\vartheta \in \mathbb{R}$ (or equivalently some $\gamma \in S^1$) we must have $U(z) = e^{i\vartheta z}$ (or $U(z) = \gamma z$). With z replaced by $V_a(z)$ and using the fact that $V_a \circ V_a = \text{id}$, we eventually have

$$e^{i\vartheta} V_a(z) = U(V_a(z)) = W(V_a \circ V_a(z)) = W(z).$$

$\square$

Corollary 28.8. *If an element of* $\text{Aut}(\mathbb{D})$ *has the origin as a fixed point it is a rotation.*

Lemma 28.9. *The groups* $\text{Aut}(\mathbb{D})$ *and* $\text{Aut}(\mathbb{H})$ *are isomorphic.*

Proof. We define the mapping $\Phi : \text{Aut}(\mathbb{D}) \to \text{Aut}(\mathbb{H})$ by $\Phi(W) := W_1^{-1} \circ W \circ W_1$ where $W_1(z) = \frac{i-z}{i+z}$ is the mapping from Lemma 28.2, i.e. W_1 maps $\mathbb{H}$ biholomorphically onto $\mathbb{D}$. This implies that $\Phi(W)$ is holomorphic and bijective with inverse $\Phi(V)^{-1} = W_1 \circ V \circ W_1^{-1}$ which is again a holomorphic mapping, hence $\Phi(W)$ is a biholomorphic mapping. Moreover we have

$$\begin{aligned}
\Phi(W \circ V) &= W_1^{-1} \circ (W \circ V) \circ W_1 \\
&= W_1^{-1} \circ W \circ W_1 \circ W_1^{-1} \circ V \circ W_1 \\
&= \Phi(W) \circ \Phi(V)
\end{aligned}$$

and $\Phi(\text{id}_{\mathbb{D}}) = \text{id}_{\mathbb{H}}$. Thus we have proved that Φ is an isomorphism. $\square$

Remark 28.10. Note that the proof of Lemma 28.9 yields that whenever G_1 and G_2 are biholomorphic equivalent then their automorphism groups $\mathrm{Aut}(G_1)$ and $\mathrm{Aut}(G_2)$ are isomorphic.

It is now possible to determine the structure of elements of $\mathrm{Aut}(\mathbb{H})$.

Theorem 28.11. *For $V \in \mathrm{Aut}(\mathbb{H})$ there exists $M = \begin{pmatrix} a & b \\ c & d \end{pmatrix} \in SL(2;\mathbb{R}) =$* $\{A \in GL(2;\mathbb{R}) \,|\, \det A = 1\}$ *such that*

$$V(z) = \frac{az+b}{cz+d}. \qquad (28.9)$$

Conversely, every mapping V as in (28.9) belongs to $\mathrm{Aut}(\mathbb{H})$.

We refer to [83] for a detailed proof, also compare with Problem 4.

The next preparation for the proof of the Riemann mapping theorem is a type of compactness result for subsets of holomorphic functions. The reader should compare this result, which is sometimes called **Montel's theorem**, with the theorem of Arzela-Ascoli, Theorem II.14.25, which we will use in its proof.

Definition 28.12. *Let $G \subset \mathbb{C}$ be a region and $\mathcal{H} = \mathcal{H}(G)$ be a family of holomorphic functions $f : G \to \mathbb{C}$. We call $\mathcal{H}$ a **normal family** or a **normal set** if every sequence in $\mathcal{H}$ contains a subsequence which converges uniformly on compact sets of G. The limit of this subsequence is of course holomorphic but does not necessarily belong to $\mathcal{H}$.*

Theorem 28.13. *Let $\mathcal{H}$ be a family of holomorphic functions $f : G \to \mathbb{C}$, where $G \subset \mathbb{C}$ is a region. If $\mathcal{H}$ is uniformly bounded on every compact subset K of G, i.e.*

$$\sup_{f \in \mathcal{H}} \|f\|_{\infty,K} \leq M_K < \infty \qquad (28.10)$$

where M_K is independent of $f \in \mathcal{H}$, then $\mathcal{H}$ is a normal family.

Proof. The first observation is that if $(K_n)_{n \in \mathbb{N}}$, $K_n \subset G$ compact, is a compact exhaustion of G, i.e. $K_n \subset \mathring{K}_{n+1}$ and $\bigcup_{n \in G} K_n = G$, and if we can prove that every sequence in $\mathcal{H}$ has a subsequence converging uniformly on each set K_n, then the result follows by a standard diagonal argument. Next we note that the Arzela-Ascoli theorem, Theorem II.14.25 also holds for complex-valued functions where in the proof we only need to replace the absolute value in $\mathbb{R}$ by the modulus in $\mathbb{C}$, see also [38] which we used in our proof of Theorem II.14.25. Moreover, condition (28.10) implies that $H(z) := \{f(z) \mid f \in \mathcal{H}\}$ is relative compact in $\mathbb{C}$ as long as z belongs to a compact set K or K_n, respectively. Thus, if we can prove that $\mathcal{H}|_{K_n} = \{f|_{K_n} \mid f \in \mathcal{H}\}$ is equi-continuous, then by the Arzela-Ascoli theorem every sequence in $\mathcal{H}|_{K_n}$ would have a subsequence converging uniformly on K_n implying the theorem. Now we prove the equi-continuity condition. For K_n there exists $\delta_n > 0$ such that for $z \in K_n$, we have $B_{2\delta_n}(z) \subset K_{n+1}$. Let $z_1, z_2 \in K_n$ such that $|z_1 - z_n| < \delta_n$. Let $\gamma(t) = z_1 + 2\delta_n e^{it}$, $t \in [0, 2\pi]$. Since

$$\frac{1}{\zeta - z_1} - \frac{1}{\zeta - z_2} = \frac{z_1 - z_2}{(\zeta - z_1)(\zeta - z_2)}$$

the Cauchy integral formula yields for $f \in \mathcal{H}$ that

$$f(z_1) - f(z_2) = \frac{z_1 - z_2}{2\pi i} \int_\gamma \frac{f(\zeta)}{(\zeta - z_1)(\zeta - z_2)} \, d\zeta. \tag{28.11}$$

Since $|\zeta - z_1| = 2\delta_n$ and $|\zeta - z_2| > \delta_n$ for $\zeta \in \operatorname{tr}(\gamma)$ by our assumptions, we deduce from (28.11)

$$|f(z_1) - f(z_2)| \le \frac{M_{K_{n+1}}}{\delta_n} |z_1 - z_2|$$

for all $f \in \mathcal{H}$, i.e. $\mathcal{H}|_{K_n}$ is equi-continuous and $\mathcal{H}$ is a normal family. $\square$

Remark 28.14. Our proof of the equi-continuity of $\mathcal{H}|_{K_n}$ follows [71] where a proof independent of the Arzela-Ascoli theorem is given. More precisely, once the equi-continuity of $\mathcal{H}|_{K_n}$ is proved, a proof of the Aszela-Ascoli theorem is in this special case is provided in [71].

Our final preparations for the proof of the Riemann mapping theorem states the existence of the (complex) square root of certain holomorphic functions. For this we need the existence of a branch of the logarithmic functions first.

Theorem 28.15. *Let $G \subset \mathbb{C}$ be a simply connected region and $f : G \to \mathbb{C}$ be a holomorphic function not vanishing in G. Then there exists a branch g of the logarithmic function in G such that $f(z) = e^{g(z)}$.*

Proof. Fix $z_0 \in G$ and choose $c_0 \in \mathbb{C}$ such that $e^{c_0} = f(z_0)$. For $z \in G$ let γ be a continuously differentiable path in G connecting z_0 with z. Consider the integral

$$g(z) = \int_\gamma \frac{f'(w)}{f(w)} \, dw + c_0.$$

Since $\frac{f'}{f}$ is holomorphic in G, $g(z)$ is independent of the choice of γ, recall that G is simply connected. We can now argue similarly as in the proofs leading to Theorem 20.23 and Theorem 21.4 to deduce that g is a holomorphic function. Using

$$\frac{d}{dz}\left(f(z)e^{-g(z)}\right) = \left(f'(z) - f(z)g'(z)\right)e^{-g(z)} = 0$$

we derive that $z \mapsto f(z)e^{-g(z)}$ is constant in the connected set G. Since $f(z_0)e^{-g(z_0)} = 1$ it follows that $f(z) = e^{g(z)}$. $\qquad\square$

Corollary 28.16. *Let $G \subset \mathbb{C}$ be a simply connected region and $f : G \to \mathbb{C}$ be a holomorphic function not vanishing in G. Then by $h(z) = \sqrt{f(z)} = (f(z))^{\frac{1}{2}} = e^{\frac{1}{2}g(z)}$ a holomorphic branch of the square root of f in G is defined, i.e. $h^2(z) = f(z)$ in G.*

Now we can prove a version of the **Riemann mapping theorem**.

Theorem 28.17. *A simply connected region $G \subset \mathbb{C}$, $G \neq \mathbb{C}$, is biholomorphically equivalent to $\mathbb{D}$.*

Proof. First we prove the existence of an injective holomorphic mapping $W : G \to \mathbb{D}$. For this let $w_0 \in G^{\complement}$. By Corollary 28.16 there exists a holomorphic function $V : G \to \mathbb{C}$ such that $V^2(z) = z - w_0$. This mapping is injective since $V(z_1) = V(z_2)$ implies that $z_1 - w_0 = z_2 - w_0$, i.e. $z_1 = z_2$. Furthermore suppose that for $z_1, z_2 \in G$ we have $V(z_1) = -V(z_2)$, i.e. $V^2(z_1) = V^2(z_2)$. As before we deduce now that $z_1 = z_2$. As a holomorphic mapping V is open, compare with Theorem 23.2, i.e. $V(G)$ is a region. By our previous considerations we can find $a \in V(G)$ and r, $0 < r < |a|$, such that $B_r(a) \subset V(G)$ and $B_r(-a) \cap V(G) = \emptyset$. We define now the mapping $W : G \to \mathbb{C}$ by $W(z) = \frac{r}{V(z)+a}$. Since $-a \notin V(G)$ it follows that W is a holomorphic mapping and since the distance of $V(z)$ to $-a$ is by construction larger than r we have

$$|W(z)| = \left|\frac{r}{V(z) + a}\right| < 1.$$

The injectivity of V implies that $W : G \to \mathbb{D}$ is an injective holomorphic mapping.

Next let $W : G \to \mathbb{D}$ be an injective holomorphic mapping such that $W(G) \neq \mathbb{D}$, i.e. there exists some $w \in \mathbb{D}$ not belonging to $W(G)$. We claim that in this case for $z_0 \in G$ there exists $W_1 : G \to \mathbb{D}$ holomorphic and injective such that

$$|W'(z_0)| < |W_1'(z_0)|. \tag{28.12}$$

Let $W_a(z) = \frac{z-a}{1-\bar{a}z}$ be the Möbius transformation (28.3) considered in Lemma 28.4. We know that W_a is a biholomorphic mapping from $\mathbb{D}$ into itself with inverse W_{-a}. If $W : G \to \mathbb{D}$ is an injective holomorphic mapping and $a \in \mathbb{D} \cap W(\mathbb{D})^{\complement}$ then $W_a \circ W : G \to \mathbb{D}$ is again an injective mapping and in addition $(W_0 \circ W)(z) \neq 0$ for all $z \in G$. Again by Corollary 28.16 we can find a holomorphic mapping $H : G \to \mathbb{C}$ such that for $z \in G$ we have $H^2(z) = (W_a \circ W)(z)$. With arguments along the lines used above we deduce that H is also injective and maps G into $\mathbb{D}$. We set $b := H(z_0)$ and define $W_1 := W_b \circ H : G \to \mathbb{D}$ which is an injective holomorphic mapping. With $S : \mathbb{C} \to \mathbb{C}$, $S(w) := w^2$, it follows now that

$$W = W_{-a} \circ S \circ H = W_{-a} \circ S \circ W_{-b} \circ W_1.$$

With $F := W_{-a} \circ S \circ W_{-b}$ the chain rule yields $W' = F'(W_1)W_1'$ and since $W_1(z_0) = 0$ we find

$$W'(z_0) = F'(0)W_1(z_0).$$

The mapping F maps $\mathbb{D}$ into itself but since S is not injective F is not injective either. We can apply the lemma of Schwarz in form of Corollary 23.12 to F and we deduce that $|F'(0)| < 1$, or

$$|W'(z_0)| < |W_1'(z_0)|.$$

Note that since W is injective we have $W'(z_0) \neq 0$. For $z_0 \in G$ fixed we set

$$m := \sup \left\{ |W'(z_0)| \,\middle|\, W : G \to \mathbb{D} \text{ is injective and holomorphic} \right\}. \tag{28.13}$$

In view of (28.12) if we can find an injective holomorphic mapping $W_0 : G \to \mathbb{D}$ for which the supremum in (28.13) is attained then W_0 must be biholomorphic mapping. The family of all injective mappings $W : G \to \mathbb{D}$ is by the Montel theorem, Theorem 28.13, a normal family. Note that $\sup |W(z)| \leq 1$ for all $W : G \to \mathbb{D}$ and $z \in G$. By the definition of the supremum there must exist a sequence $W_k : G \to \mathbb{D}$ of injective holomorphic mappings with $\lim_{k\to\infty} W'_k(z_0) = m$, and this sequence must have a subsequence, again denoted by $(W_k)_{k\in\mathbb{N}}$, which converges uniformly on compact subsets of G to some mapping J, hence $|J'(z_0)| = m$. We know that the set of all injective holomorphic mappings $W : G \to \mathbb{D}$ is not empty and that $m > 0$. Thus J is not a constant mapping and since $W_k(G) \subset \mathbb{D}$ it follows that $J(G) \subset \mathbb{D}$ where we used that by Theorem 23.2 the holomorphic mapping J is open. Our aim is now to prove that J is injective. Let $z_1, z_2 \in G$, $z_1 \neq z_2$, and put $a := J(z_1)$, $a_k := W_k(z_1)$, and choose $r > 0$ such that $z_1 \notin B_r(z_2)$ and such that $J - a$ has no zeroes on $\partial B_r(z_2)$. Note that in $\overline{B_r(z_2)}$ the function $J - a$ can have only finitely many zeroes. Then if we reduce r a little if necessary we can always achieve that $J - a$ has no zeros on $\partial B_r(z_2)$. The sequence $(W_k - a)_{k\in\mathbb{N}}$ converges uniformly on $\overline{B_r(z_2)}$ to $J - a$, and the functions $W_k - a$ do not have a zero in $\overline{B_r(z_2)}$ since they are injective and $W_k(z_1) - a = 0$. Now, as seen in Problem 6, we can apply the theorem of Rouché, Theorem 24.26, to conclude that $J - a$ has no zero in $\overline{B_r(z_2)}$, which yields in particular that $J(z_1) \neq J(z_2)$. $\qquad\qquad\square$

Remark 28.18. Our proof of Theorem 28.17 is the standard proof used nowadays and which goes back to C. Carathéodory and P. Koebe. We followed essentially the presentation in [71], but we have to make some adaptations to our different approach to some auxillary results.

Suppose that $G \neq \mathbb{C}$ is a simply connected region and let $W : G \to \mathbb{D}$ be biholomorphic mapping. This mapping induces a bijective mapping $H_W : \mathcal{H}(\mathbb{D}) \to \mathcal{H}(G)$ by $f \mapsto f \circ W$ with inverse $H_W^{-1} : \mathcal{H}(G) \to \mathcal{H}(\mathbb{D})$, $H_W^{-1} = H_{W^{-1}}$. Thus, once we understand $\mathcal{H}(\mathbb{D})$ we understand up to biholomorphic mappings all spaces $\mathcal{H}(G)$, $G \subset \mathbb{C}$, $G \neq \mathbb{C}$ and a simply connected region. However there might be severe problems at the boundaries. So far we have not even established that we can map $\overline{G}$ onto $\overline{\mathbb{D}}$ with the help of a continuous extension of a biholomorphic mapping $W : G \to \mathbb{D}$. In fact this is not necessarily true. The discussion of the boundary regularity of biholomorphic mappings $W : G_1 \to G_2$ is a very interesting, challenging and important subject, but it goes beyond the scope of our Course which gives a first encounter with complex variable theory.

Finally we want to give an application of biholomorphic mappings to the theory of harmonic functions.

Let $f : G_2 \to G_1$ be a holomorphic function $f = u + iv$ and with $z = x + iy$ we write $f(z) = u(x, y) + iv(x, y)$. For u and v the Cauchy-Riemann differential equations hold, i.e. $u_x = v_y$ and $u_y = -v_x$. This implies in particular that $|f'(z)|^2 = |u_x + iv_x|^2 = u_x^2 + u_y^2 = v_x^2 + v_y^2$, compare with Example 17.13. Moreover we know that u and v must be harmonic, i.e. $\Delta u = \Delta v = 0$, see Theorem 17.20 and add that we know meanwhile that u and v must be C^∞-functions. Now let $h : G_1 \to \mathbb{R}$ be a harmonic function and with $w = f(z) = u + iv$ consider the function $g : G_2 \to \mathbb{R}$, $g = h \circ f$. A straightforward calculation yields for $g(x, y) = h(u(x, y, v(x, y)))$ that

$$\frac{\partial^2 g}{\partial x^2} = \frac{\partial^2 h}{\partial u^2} u_x^2 + 2 \frac{\partial^2 h}{\partial u \partial v} v_x u_x + \frac{\partial^2 h}{\partial v^2} v_x^2 + \frac{\partial h}{\partial u} u_{xx} + \frac{\partial h}{\partial v} v_{xx}$$

and

$$\frac{\partial^2 g}{\partial y^2} = \frac{\partial^2 h}{\partial u^2} u_y^2 + 2 \frac{\partial^2 h}{\partial u \partial v} v_y u_y + \frac{\partial^2 h}{\partial v^2} v_y^2 + \frac{\partial h}{\partial u} u_{yy} + \frac{\partial h}{\partial v} v_{yy}$$

which yields

$$\frac{\partial^2 g}{\partial x^2} + \frac{\partial^2 g}{\partial y^2} = \frac{\partial^2 h}{\partial u^2} \left(u_x^2 + u_y^2 \right) + \frac{\partial^2 h}{\partial v^2} \left(v_x^2 + v_y^2 \right)$$

$$+ \frac{\partial h}{\partial u} \Delta u + \frac{\partial h}{\partial v} \Delta v + 2 \frac{\partial^2 h}{\partial u \partial v} \left(v_x u_x + v_y u_y \right)$$

$$= |f'(z)|^2 \left(\frac{\partial^2 h}{\partial u^2} + \frac{\partial^2 h}{\partial v^2} \right).$$

Since h is by assumption harmonic we have proved

Proposition 28.19. *If $f : G_2 \to G_1$ is holomorphic and $h : G_1 \to \mathbb{R}$ is harmonic, then $g = h \circ f : G_2 \to \mathbb{R}$ is harmonic too.*

Many two-dimensional problems in physics or mechanics, but also differential geometry lead to the task to find a harmonic function with special properties, for example prescribed behaviour at the boundary. Harmonic functions on $\mathbb{D}$ or $\mathbb{H}$ are easy to construct as we will see in our Course when dealing with partial differential equations. Thus if we can map a complicated (simply connected) region onto a simpler one, say $\mathbb{D}$ or $\mathbb{H}$, by a biholomorphic mapping which also behaves "nice" at the boundary we can reduce the original problem from physics or mechanics or geometry to a much simpler one. For such a strategy the Riemann mapping theorem is a good starting point.

In this chapter we dealt with biholomorphic mappings $f : G_1 \to G_2$ and by their very definition they are bijective, hence injective. Moreover their derivative f' never vanishes. A more classical way and also still very important approach is to start with **conformal mappings**. These are by definition holomorphic functions $h : G_1 \to G_2$ with $h'(z) \neq 0$ for all $z \in G_1$. This condition yields only local injectivity. The importance of conformal mappings lies in the fact that in the small they preserve angles. Conformal mappings have a large domain of real-world applications as they have many interesting mathematical properties. Again, we will not handle the theory of conformal mappings in our Course and have to refer to the literature, e.g. [3], [4], [17], [20], [37], [61], [68], [80] or [83].

Problems

1. Consider $W_1 : \mathbb{H} \to \mathbb{D}$, $W_1(z) = \frac{i-z}{i+z}$, as in Lemma 28.2 and prove that W_1 maps $\partial\mathbb{H}$ continuously onto a subset of $\partial\mathbb{D}$.

2. Solve Exercise 28.3.

3. Prove that if G_1 and G_2 are two biholomorphically equivalent regions then $\mathrm{Aut}(G_1)$ is isomorphic to $\mathrm{Aut}(G_2)$.

4. Prove that $A = \begin{pmatrix} a & b \\ c & d \end{pmatrix} \in SL(2; \mathbb{R})$ induces by

$$f_A(z) := \frac{az + b}{cz + d}$$

an automorphism of $\mathbb{H}$.

5. Let $G \subset \mathbb{C}$ be a region and $\mathcal{H}$ a normal family of holomorphic functions $f : G \to \mathbb{C}$. Prove that for every $k \in \mathbb{N}$ the family $\mathcal{H} := \{f^{(k)} | f \in H\}$ is normal too.

   Hint: use the standard estimates for derivatives.

6. Fill the gaps at the end of the proof of the Riemann mapping theorem, i.e. use the theorem of Rouché to prove that $J - a$ has no zero in $\mathbb{D}$.

7. Verify that $h(x, y) = x^2 - y^2$ is in $\mathbb{R}^2 \cong \mathbb{C}$ harmonic. Find the harmonic function $g : \{(x, y) \in \mathbb{R}^2 | y > 0\} \to \mathbb{R}$ where $g = h \circ W_1$ and W_1 is from Problem 1.

29 Power Series in Several Variables

In this chapter we discuss some properties of power series of several variables, complex or real. We do not intend to give an introduction to the theory of complex-valued functions of several variables, this is beyond our Course. However already when looking at real-valued functions of several real variables, we stopped short of treating Taylor series which should be of course power series. We have seen the very close relationship between power series of one real variable and of one complex variable. Now we will see some similarity in several variables, but convergence domains (and results) will be quite different. This is best understood in several complex variables although in [51] a readable treatment of the real case is available, and it also indicates why the theory of differentiable functions from some set $G \subset \mathbb{C}^n$ to $\mathbb{C}$ should be expected to have quite different features than the one-dimensional theory. In our presentation we make use of [29] and [66].

We will use multi-index notation as we are used to from Volume II, in particular for $z = (z_1, \ldots, z_n) \in \mathbb{C}^n$ and $\alpha \in \mathbb{N}_0^n$ we will write

$$z^\alpha := z^{\alpha_1} \cdot \ldots \cdot z_n^{\alpha_n}. \tag{29.1}$$

While in one dimension there is a natural way to form partial sums of a series $\sum_{k=0}^\infty c_k z^k$, $c_k \in \mathbb{C}$, i.e. we consider $S_N := \sum_{k=0}^N c_k z^k$, $N \in \mathbb{N}_0$, this is not the case in several variables. Every enumeration of $\mathbb{N}_0^n$ will give rise to a sequence of partial sums and there is no reason to consider one as superior to the other. Since we know that for absolutely convergent series every rearrangement will lead to the same limit, see Theorem I.18.27 for the real case which easily extends to the complex case, we discuss only absolutely convergence series in case of several variables. In order to proceed we need some definitions.

Definition 29.1. A. *Any bijective mapping $\varphi : \mathbb{N}_0 \to \mathbb{N}_0^n$ we call an **enumeration** of $\mathbb{N}_0^n$.*
B. *For a sequence $(c_\alpha)_{\alpha \in \mathbb{N}_0^n}$ of complex numbers we call the series $\sum_{\alpha \in \mathbb{N}_0^n} c_\alpha$ convergent if for every enumeration φ of $\mathbb{N}_0^n$ the series $\sum_{k=0}^\infty \left| c_{\varphi(k)} \right| = \lim_{N \to \infty} \sum_{k=0}^N \left| c_{\varphi(k)} \right|$ converges.*

Remark 29.2. Convergence of $\sum_{\alpha \in \mathbb{N}_0^n} c_\alpha$ is always absolute convergence, i.e. $\sum_{k=0}^\infty \left| c_{\varphi(k)} \right|$ converges. Since by Theorem I.18.27 the limit of $\sum_{k=0}^\infty \left| c_{\varphi(k)} \right|$ is independent of the order of summation it follows that the convergence of $\sum_{k=0}^\infty \left| c_{\varphi(k)} \right|$ for one enumeration will already imply the convergence for all other enumerations of $\mathbb{N}_0^n$. A different question is whether for all enumerations the limit $\lim_{N\to\infty} \sum_{k=1}^{N} c_{\varphi(k)}$ is the same. The result holds and can be proved along the lines of the proof of Theorem I.18.27, see Problem 1.

Definition 29.3. A. *The* ***open ball in*** $\mathbb{C}^n$ *with centre* $a \in \mathbb{C}^n$ *and radius* $r > 0$ *is denoted by* $B_r(a) := \{z \in \mathbb{C}^n \,|\, \|z - a\| < r\}$, *where* $\|z\| = (|z_1|^2 + \ldots + |z_n|^2)^{\frac{1}{2}}$.
B. *The* ***polydisc*** $P(a, \vec{r}) \subset \mathbb{C}^n$ *with centre* $a \in \mathbb{C}^n$ *and multi-radius* $\vec{r} = (r_1, \ldots, r_n)$, $r_j \geq 0$, *is the set*

$$P(a, \vec{r}) = \{z \in \mathbb{C}^n \,|\, |z_j - a_j| < r_j, \quad j = 1, \ldots, n\}. \tag{29.2}$$

C. *The* ***distinguished boundary*** $\partial_{\mathrm{d}} P(a, \vec{r})$ *is the set*

$$\partial_{\mathrm{d}} P(a, \vec{r}) = \{z \in \mathbb{C}^n \,|\, |z_j - a_j| = r_j, \quad j = 1, \ldots, n\}. \tag{29.3}$$

Remark 29.4. The reader can easily check that

$$P(a, \vec{r}) = B_{r_1}(a_1) \times \cdots \times B_{r_n}(a_n) \tag{29.4}$$

and

$$\partial_{\mathrm{d}} P(a, \vec{r}) = \partial B_{r_1}(a_1) \times \cdots \times \partial B_{r_n}(a_n). \tag{29.5}$$

In particular for $a = 0 \in \mathbb{C}^n$ and $\vec{1} = (1, \ldots, 1)$ we can identify $\partial_{\mathrm{d}} P(0, \vec{1})$ with the n-dimensional torus $\mathbb{T}^n = S^1 \times \cdots \times S^1 \subset \mathbb{C}^n$.

The topology on $\mathbb{C}^n$ induced by $\|z\|$ is of course the Euclidean topology on $\mathbb{R}^{2n}$ and the family of all open balls $B_r(a)$ as well as the family of all polydiscs $P(a, \vec{r})$ form a base of this topology.

Definition 29.5. *The set* $\mathbb{R}_+^n = \{x \in \mathbb{R}^n \,|\, x_j \geq 0\}$ *is called the* ***absolute space*** *(of* $\mathbb{C}^n$) *and* $\tau : \mathbb{C}^n \to \mathbb{R}_+^n$, $\tau(z) = \tau(z_1, \ldots, z_n) := (|z_1|, \ldots, |z_n|)$ *is called the* ***natural projections*** *of* $\mathbb{C}^n$ *onto* $\mathbb{R}_+^n$.

Note that for $\vec{r} \in \mathbb{R}_+^n$ the pre-image under τ is the set $\partial_{\mathrm{d}} P(0, \vec{r})$.

Definition 29.6. *An open connected set* $R \subset \mathbb{C}^n$ *is called a* ***Reinhardt domain*** *if* $z \in R$ *implies that* $\partial_{\mathrm{d}} P(0, \tau(z)) \subset G$.

Note that $\partial_{\mathrm{d}} P(0, \tau(z)) = \tau^{-1}(\tau(z))$. If G is Reinhardt domain then $G = \tau^{-1}(\tau(G))$.

Definition 29.7. A. *A Reinhardt domain $G \subset \mathbb{C}^n$ is called proper if $0 \in G$.*
B. *A **complete** Reinhardt domain $G \subset \mathbb{C}^n$ is a Reinhardt domain with the property that for all $z \in G \cap (\mathbb{C} \backslash \{0\})^n$ it follows that $P(0, \tau(z)) \subset G$.*

Note that $G \subset \mathbb{C}^n$ is a complete Reinhardt domain if and only if

$$G = \bigcup_{z \in G} \overline{P}(0, \tau(z)).$$

We want to discuss some examples. Of course even for $n = 2$ there is no chance to sketch an open set $G \subset \mathbb{C}^2$. However, for $n = 2$ the absolute space is a subset of $\mathbb{R}^2$ and we can at least draft $\tau(G)$ for $n = 2$.

For $B_r(0) \subset \mathbb{C}^2$ we find that $\tau(B_r(0))$ is the quarter disc in $\mathbb{R}^2$, see Figure 29.1, and for $P(0, (r_1, r_2)) \subset \mathbb{C}^2$ the natural projection $\tau(P(0, (r_1, r_2)))$ is a square, see Figure 29.2.

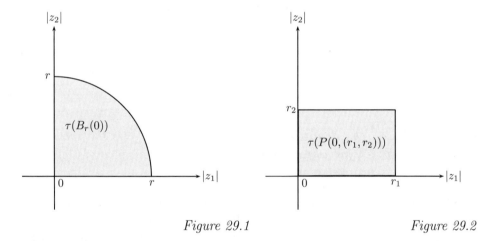

Figure 29.1   *Figure 29.2*

Note that in both cases the dotted lines do not belong to the set. It is not hard to see that polydiscs and open balls with centre $0 \in \mathbb{C}^n$ are complete and proper Reinhardt domains.

For a complete Reinhardt domain $G \subset \mathbb{C}^2$ the natural projection $\tau(G)$ looks as in Figure 29.3 while Figure 29.4 features the natural projection of a possible Reinhardt domain which however is not complete.

571

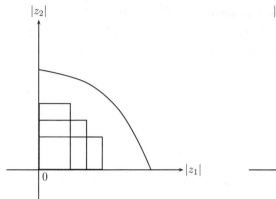

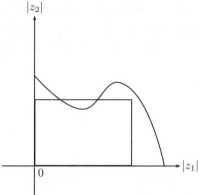

Figure 29.3   Figure 29.4

Example 29.8. Let $a = (a_1, a_2) \in \mathbb{C}^2$, $|a_j| > 1$. For every $t \in \mathbb{R}$ it follows that $\tau(e^{it}a) = (|e^{it}a_1|, |e^{it}a_2|) = (|a_1|, |a_2|) = \tau(a)$. We claim that we can find some $\epsilon > 0$ such that $P(a, (\epsilon, \epsilon))$ is not a Reinhardt domain. For this we note that $|e^{it}a - a| = |e^{it} - 1| |a| > |e^{it} - 1|$ since $|a_j| > 1$ which implies for t suitably chosen that $|e^{it}a - a| \notin P(a, (\epsilon, \epsilon))$.

We now turn to power series about the origin which is sufficient for our purposes:

$$T_{ca}(z) = \sum_{\alpha \in \mathbb{N}_0^n} c_\alpha z^\alpha. \tag{29.6}$$

Definition 29.9. *The **domain of convergence** of T_{ca} is the interior of the set of all $z \in \mathbb{C}^n$ for which (29.6) converges (in the sense of Definition 29.1).*

The key result for power series $\sum_{k=0}^{\infty} a_k(z - c)^k$ in one dimension was Abel's theorem, Theorem 16.6, with Corollary 16.9. If the series converges for some $z_0 \neq c$ then it converges in the open set $\overline{B_\rho(c)}$, $0 < \rho < |z_0|$. Combined with the Hadamard-Cauchy theorem, Theorem 16.10, we know that indeed there exists a maximal $R \geq 0$ called the radius of convergence such that locally on compact sets of $B_r(c)$ the series converges uniformly and in $\overline{B_R(c)}^C$ the series diverges.

Consider the series $\sum_{\alpha \in \mathbb{N}_0^n} \alpha_1^{\alpha_1} z_1^{\alpha_1} z_2^{\alpha_2}$. It is trivial that this series converges on the complex line $z_2 = 0$. However it diverges for every $z_1 \neq 0$ if $z_2 \neq 0$

572

implying that its domain of convergence is empty. Thus convergence of a series of the type $T_{c_\alpha}(z)$ for some point $z_0 \in C^n$ will in general not help us to find a subset of its domain of convergence. The analogous result to Abel's theorem in several variables, often called **Abel's lemma**, is

Theorem 29.10. *Let $P(0, \vec{r}_0) \subset \mathbb{C}^n$ be a polydisc with $\vec{r}_0 = (r_{0,1}, \ldots, r_{0,n})$, $r_{0,j} > 0$, and let $P(0, \vec{r})$ a further polydisc with $\vec{r} = (r_1, \ldots, r_n)$, $0 < r_j < r_{0,j}$ for $j = 1, \ldots, n$. If the power series*

$$T_{c_\alpha}(z) = \sum_{\alpha \in \mathbb{N}_0^n} c_\alpha z^\alpha \tag{29.7}$$

converges for one point z_0 in the distinguished boundary of $P(0, \vec{r}_0)$, i.e. $z_0 \in \partial_d P(0, \vec{r}_0)$ then $T_{c_\alpha}(z)$ converges for all $z \in P(0, \vec{r})$.

Proof. Let $z_0 \in \partial_d P(0, \vec{r}_0)$ and suppose that $\sum_{\alpha \in \mathbb{N}_0^n} c_\alpha z_0^\alpha$ converges. It follows that the sequence $(|c_\alpha z_0^\alpha|)_{\alpha \in \mathbb{N}_0^n}$ must be bounded, i.e. $|c_\alpha z_0^\alpha| < K$. Since $0 < r_j < r_{0,j}$ for $1 \leq j \leq n$, we can find $\eta_j \in (0,1)$ such that for all $z \in P(0, \vec{r})$ we have $|z_j| < \eta_j |z_{0,j}|$, which implies that $|c_\alpha z^\alpha| \leq \eta^\alpha$, $\eta = (\eta_1, \ldots, \eta_n)$, or $\sup_{z \in P(0, \vec{r})} |c_\alpha z^\alpha| \leq \eta^\alpha$. Now we observe for η that

$$\sum_{|\alpha| \leq N} \eta^\alpha \leq \prod_{j=1}^n \sum_{\alpha_j = 1}^N \eta_j^{\alpha_j} \leq \prod_{j=1}^n \frac{1}{1 - \eta_j}$$

and we find that $\sum_{\alpha \in \mathbb{N}_0^n} c_\alpha z^\alpha$ converges for $z \in P(0, \vec{r})$. $\quad\square$

The proof of Theorem 29.10 shows more, namely that the convergence in $P(0, \vec{r})$ is uniform. A preferred notion of convergence in this context is that of **normal convergence**: a series of functions $\sum_{k=0}^\infty f_k$, $f_k : G \to \mathbb{C}$, is said to be **normally convergent** in G if $\sum_{k=0}^\infty \|f_k\|_{\infty, G}$ converges. Furthermore, since the open polydiscs form a base of the topology in $\mathbb{C}^n$, we have

Corollary 29.11. *Let $P(0, \vec{r}_0) \subset \mathbb{C}^n$ and suppose that the power series (29.7) converges for some point $z_0 \in \partial_d P(0, \vec{r}_0)$. Then $T_{c_\alpha}(z)$ converges uniformly on every compact subset of $P(0, \vec{r}_0)$.*

The importance of Reinhardt domains is seen from the following

Theorem 29.12. *The domain D of convergence of a power series of type (29.7) is a complete Reinhardt domain.*

Proof. Let $z_0 \in D$. Then there exists a polydisc $P(z_0, \vec{\epsilon}) \subset D$, $\vec{\epsilon} = (\epsilon, \dots, \epsilon)$, i.e. $P(0, \vec{\epsilon}) = B_\epsilon(z_{0,1}) \times \cdots \times B_\epsilon(z_{0,n})$. We pick $z_1 = (z_{1,1}, \dots, z_{1,n})$, $z_{1,j} \in B_\epsilon(z_{0,j})$, such that $|z_{1,j}| > |z_{0,j}|$. It follows that $z_1 \in D$ and $z_0 \in P(0, \tau(z_1))$. For every z_0 we fix such a point z_1. We know that $T_{c_\alpha}(z_1)$ converges, but $z_1 \in \partial_{\mathrm{d}} P(0, \tau(z_1))$, i.e. $P(0, \tau(z_1)) \subset D$. Since $P(0, \tau(z_0)) \subset P(0, \tau(z_1))$ and $\partial_{\mathrm{d}} P(0, \tau(z_0)) \subset P(0, \tau(z_1))$ we deduce that D is a complete Reinhardt domain. $\square$

Note that this result already indicates implications for the domain of convergence of a real power series $\sum_{\alpha \in \mathbb{N}_0^n} a_\alpha x^\alpha$, $a_\alpha \in \mathbb{R}$, $x \in \mathbb{R}^n$. When passing the complexification $x_j \mapsto z_j = x_j + iy_j$, we must end up in $\mathbb{C}^n$ with a Reinhardt domain. As it turns out, being a Reinhardt domain is not sufficient to be a domain of convergence. We need

Definition 29.13. *A Reinhardt domain G is called* **logarithmically convex** *if* $\ln\left(\tau\left(G \cap (\mathbb{C}\backslash\{0\})^n\right)\right)$ *is a convex set in* $\mathbb{R}^n$. *Here* $\ln \vec{r} = (\ln r_1, \dots, \ln r_n)$ *and* $\ln A$ *stands for the image of the set A under* $\ln$.

Theorem 29.14. *The domain D of convergence of a power series of the type* (29.7) *is logarithmically convex.*

Proof. Let $z_1, z_2 \in D \cap (\mathbb{C}\backslash\{0\})^n$ and set $x_1 := \ln(\tau(z_1))$ as well as $x_2 = \ln(\tau(z_2))$. Since D is open, for $\eta > 0$ sufficiently small $\eta z_j \in D \cap (\mathbb{C}\backslash\{0\})^n$, hence $T_{c_\alpha}(\eta z_1)$ and $T_{c_\alpha}(\eta z_2)$ converge implying for some M that for all $\alpha \in \mathbb{N}_0^n$ we have $|c_\alpha| \eta^{|\alpha|} |z_1^\alpha| \leq M$ and $|c_\alpha| \eta^{|\alpha|} |z_2^\alpha| \leq M$. For $0 \leq \lambda \leq 1$ this implies for every $\alpha \in \mathbb{N}_0^n$ that $|c_\alpha| \eta^{|\alpha|} |z_1^\alpha|^\lambda |z_2^\alpha|^{1-\lambda} \leq M$. From Theorem 29.10 we deduce that $T_{c_\alpha}(\cdot)$ converges in a neighbourhood of $z_\lambda := \left(|z_{1,1}|^\lambda |z_{2,1}|^{1-\lambda}, \dots, |z_{1,n}|^\lambda |z_{2,n}|^{1-\lambda}\right)$, i.e. $z_\lambda \in D$ and $\lambda x_1 + (1 - \lambda)x_2 = \ln(\tau(z_\lambda))$, or $\ln \tau(z_\lambda) \in \ln\left(\tau\left(G \cap (\mathbb{C}\backslash\{0\})^n\right)\right)$ proving the theorem. $\square$

Thus only logarithmically convex Reinhardt domains are convergence domains of power series of type (29.7). It turns out that the analogous result holds for real power series, details are given in [51].

One way of defining holomorphy in $\mathbb{C}^n$, $n \geq 1$, is

Definition 29.15. *Let $D \subset \mathbb{C}^n$ be an open set. We call $f : D \to \mathbb{C}$* **holomorphic** *if for every point $z_0 \in D$ there exists an open neighbourhood $U(z_0) \subset D$ and a power series $T_{c_\alpha}^{z_0}(z) = \sum_{\alpha \in \mathbb{N}_0} c_\alpha (z - z_0)^\alpha$ converging in $U(z_0)$ to $f(z)$.*

In some sense Reinhardt domains then turn out to be maximal holomorphy domains. But here we have to stop.

Problems

1. Let $\varphi : \mathbb{N}_0 \to \mathbb{N}_0^n$ be an enumeration of $\mathbb{N}_0^n$ and let $(c_\alpha)_{\alpha \in \mathbb{N}_0^n}$ be a sequence of complex numbers for which $\left(\sum_{|\alpha| \leq N} |c_\alpha| \right)_{N \in \mathbb{N}_0}$ converges. Prove that $\sum_{k=0}^{\infty} c_{\varphi(k)} = \sum_{\alpha \in \mathbb{N}_0^n} c_\alpha$.

2.   a) Prove that the power series $\sum_{\alpha \in \mathbb{N}_0^2} z^\alpha$ converges in $B_1(0) \times B_1(0)$.

    b) For $z_1 \in \mathbb{C}$ and $z_2 \in B_1(0)$ prove that $\sum_{(1,\alpha_2) \in \mathbb{N}_0^2} z^\alpha = \frac{z_1}{1-z_2}$ and deduce that the power series converges in $\mathbb{C} \times B_1(0)$.

    c) Show that for $|z_1 z_2| < 1$ the power series $\sum_{\nu \in \mathbb{N}_0} z_1^\nu z_2^\nu$ converges.

Appendices

Appendix I: More on Point Set Topology

In this appendix we want to collect some further results from topology, in particular the construction of a product topology as well as the construction of compactifications.

Let $(X, \mathcal{O}_X)$ be a topological space. A family $(U_j)_{j \in I}$ of non-empty open sets is called a **base** of the topology (or the open sets) of X if every non-empty open set $U \subset X$ is the union of a sub-family of $(U_j)_{j \in I}$, see Definition II.3.7. Now let $(X_k, \mathcal{O}_{X_k})$, $k = 1, \ldots, N$, be a finite family of topological spaces and $X := \prod_{k=1}^{N} X_k$ the Cartesian product. We want to construct on X a topology such that all projections $\mathrm{pr}_k : X \to X_k$ are continuous and which is minimal. It turns out, see [21], or [65] that the family

$$\left\{ U \in X \,\middle|\, U = \bigcap_{j=1}^{N} \mathrm{pr}_j^{-1}(O_j), O_j \in \mathcal{O}_j \right\}$$

is a base of this (uniquely determined) topology on X and this topology is called the **product topology** on X.

In particular if (X_j, d_j), $j = 1, \ldots, N$, are metric spaces then the product topology on $X = \prod_{j=1}^{N} X_j$ induced by the metric topologies $\mathcal{O}_{d_j}$ (on X_j) and the topology on X induced by the metric $d = \sum_{j=1}^{N} d_j$ are equal.

Within this frame where we can now define a topological group $(G, \mathcal{O}_G, \circ)$ as a topological space $(G, \mathcal{O}_G)$ which carries a group operation comp : $G \times G \to G$, $\mathrm{comp}(g_1, g_2) = g_1 \circ g_2$, which together with the inverse is continuous, i.e. comp : $G \times G \to G$, $\mathrm{comp}(g_1, g_2) = g_1 \circ g_2$ is continuous from $(G \times G, \mathcal{O}_{G \times G})$ to $(G, \mathcal{O}_G)$, as the inverse, inv : $G \to G$, $g \mapsto g^{-1}$, is continuous from $(G, \mathcal{O}_G)$ into itself. Examples of topological groups are $GL(n; \mathbb{R})$, $GL(n; \mathbb{C})$ as well as the subgroups $SL(n; \mathbb{R})$, $SL(n; \mathbb{C})$, $O(n; \mathbb{R})$, $U(n, \mathbb{C})$, $SO(n; \mathbb{R})$ or $SU(n; \mathbb{C})$. Similarly we can introduce topological vector spaces over $\mathbb{R}$ or $\mathbb{C}$. The additional requirement to the vector space axioms is that vector addition and its inverse as well as scalar multiplication are continuous.

Next we turn to the problem of finding a compactification of a given (non-compact) toplogical space. We first discuss the **one-pont compactification** or **Alexandrov-compactification** of a locally compact space $(X, \mathcal{O}_X)$. Recall that $(X, \mathcal{O}_X)$ is a locally compact space if it is a Hausdorff space and if every point $x \in X$ has a compact neighbourhood. Here we call $(X, \mathcal{O}_X)$

a **Hausdorff space** if for $x \neq y$, $x, y \in X$, there exist (open) neighbour-hoods $U(x)$ and $U(y)$ of x and y, respectively, such that $U(x) \cap U(y) = \emptyset$. Every metric space (X, d) is a Hausdorff space since $B_r(x) \cap B_r(y) = \emptyset$ for $r = \frac{1}{2}d(x, y) > 0$. The aim is to find for $(X, \mathcal{O}_X)$ a compact space $(Y, \mathcal{O}_Y)$ such that Y contains a topological subspace Y_1, which is homeomorphic to X. If $(X, \mathcal{O}_X)$ is compact itself we can of course choose $(Y, \mathcal{O}_Y) = (X, \mathcal{O}_X)$.

Theorem A.I.1 (Alexandrov compactification). *Let $(X, \mathcal{O}_X)$ be a locally compact, non-compact Hausdorff space and let ∞ be a point not belonging to X, i.e. $\infty \notin X$. Then the set $Y := X \cup \{\infty\}$ admits a topology $\mathcal{O}_Y$ such that $(Y, \mathcal{O}_Y)$ is a compact space. Moreover there exists a topological subspace $Y_1 \subset Y$ such that X is homeomorphic to Y_1, $Y \backslash Y_1 = \{\infty\}$, and Y_1 is dense in Y.*

The point ∞ in Theorem A.I.1 is called the **point at infinity**. A detailed proof of Theorem A.I.1 is given in [47] or [65]. We only want to indicate the construction of $\mathcal{O}_Y : U \in \mathcal{O}_Y$ if either $U \in \mathcal{O}_X$ or $U = Y \backslash K$ where K is a compact set in X.

It is not difficult to see that the one-point compactification $\hat{\mathbb{R}}^n$ of $\mathbb{R}^n$ can be identified with the sphere S^{n-1}. In particular for $\mathbb{R}$ we obtain the circle $S^1 \subset \mathbb{R}^2$, and for $\mathbb{R}^2$ we obtain the sphere $S^2 \subset \mathbb{R}^3$. When identifying topologically $\mathbb{C}$ with $\mathbb{R}^2$ we now understand that topologically the Riemann sphere is the one-point compactification of $\mathbb{C}$ and that the north pole $(0, 0, 1) \in S^2 \subset \mathbb{R}^3$ can be looked at as the point at infinity.

A further remark is helpful. If $u \in C_\infty(\mathbb{R}^n)$, i.e. $u : \mathbb{R}^n \to \mathbb{R}$ is a continuous function which vanishes at infinity, then u has a continuous extension to $C(\hat{\mathbb{R}}^n)$ when $\hat{\mathbb{R}}^n$ is considered as one-point compactification of $\mathbb{R}^n$, we just define $u(\infty) := 0$ where ∞ denotes the point at infinity. It follows that $C_\infty(\mathbb{R}^n)$ can be considered as a subspace of $C(\hat{\mathbb{R}}^n)$.

The one-point compactification of $\mathbb{R}$ is not helpful when we want to maintain the natural order structure on $\mathbb{R}$, i.e. when dealing with $[-\infty, \infty] = \bar{\mathbb{R}}$. The solution is to pass to a two-point compactification by adding to $\mathbb{R}$ the two points $-\infty, \infty \notin \mathbb{R}$, $-\infty \neq \infty$. A natural way to define on $\bar{\mathbb{R}}$ the topology $\mathcal{O}_{\bar{\mathbb{R}}}$ is the following: choose as a base for $\mathcal{O}_{\bar{\mathbb{R}}}$ all open sets belonging to $\mathbb{R}$ (or just all open intervals) and in addition all sets of the type $\{x \in \bar{\mathbb{R}} \,|\, x > a\}$,

$a \geq 1$, and $\{x \in \mathbb{R} \mid x < b\}$, $b \leq -1$. This compactification of $\mathbb{R}$ is the one we need to handle $\overline{\mathcal{B}}^{(1)}$ as Borel σ-field in Chapter 4.

We close this appendix by extending our discussion of the Lemma of Urysohn, see [47]. A topological space $(X, \mathcal{O}_X)$ is called **normal topological space** if for any pair of disjoint closed subsets $C_1, C_2 \subset X$ there exist disjoint open sets $U_1, U_2 \subset X$ such that $C_1 \subset U_1$ and $C_2 \subset U_2$. The space $\mathbb{R}^n$ with the Euclidean topology is a normal space as is every metric space.

Theorem A.I.2. *For two disjoint closed sets C_1, C_2 in a normal space $(X, \mathcal{O}_X)$ we can find a continuous function $f : X \to [0, 1]$ such that $f|_{C_1} = 0$ and $f|_{C_2} = 1$.*

Corollary A.I.3. *Theorem A.I.2 holds for compact Hausdorff spaces.*

Corollary A.I.4. *If $(X, \mathcal{O}_X)$ is a locally compact Hausdorff space and $C_1, C_2 \subset X$ are two disjoint closed sets then there exists a continuous function $f : X \to [0, 1]$ with compact support such that $f|_{C_1} = 0$ and $f|_{C_2} = 1$.*

Recall that the **support** of a continuous function is defined as

$$\operatorname{supp} f = \overline{\{x \in X \mid f(x) \neq 0\}}.$$

Appendix II: Measure Theory, Topology and Set Theory

When starting to study analysis in most investigations we depend on the notation of set theory but we really do not use much set theory. At the beginning we do not need to understand its foundations, its axioms, more precisely the system of axioms of a model of set theory. The standard one would be ZF, i.e. the Zermelo-Fraenkel system of axioms. Nonetheless when coming to more subtle topics we need more set theory and usually a second course on measure theory, functional analysis or point set topology on a graduate level starts with some deeper discussion of set theory. There are good reasons for this, and we shall not forget that set theory emerged from the efforts to understand the convergence of trigonometrical series, i.e. problems from analysis were the starting point of Cantor's investigations leading to set theory.

Measure theory starts by introducing σ-fields, a proper theory of continuous functions starts by introducing topologies. Both are systems of subsets of a given set closed under certain operations. In the case of σ-fields only certain countable operations are permitted, for topologies we allow arbitrary unions of open sets. We can use topologies to generate σ-fields (Borel σ-fields) and now it becomes apparent that a more detailed study of Borel sets (for example) will depend on properties of sets obtained from such operations, i.e. unions or intersections of certain families of sets, hence on the underlying model of set theory.

A highly non-trivial ingredient from set theory needed to understand the relation of measure theory and topology is the axiom of choice. Let $(Y_j)_{j\in I}$, $I \neq \emptyset$, be an arbitrary family of sets and denote by $\bigtimes_{j\in I} Y_j$ the set of all mapping f with domain I such that $f(j) \in Y_j$. We call f a **choice function** for the family $(Y_j)_{j\in I}$ and $f(j)$ the jth coordinate of f.

Axiom of Choice (1st version) If $I \neq \emptyset$ and $Y_j \neq \emptyset$ for all $j \in I$ then there exists at least one choice function for the family $(Y_j)_{j\in I}$.

We can rephrase the axiom of choice as follows

Axiom of Choice (2nd version) Let $I \neq \emptyset$ and $Y_j \neq \emptyset$ for all $j \in I$. Suppose that the sets Y_j are mutually disjoint. Then there exists a set $M \subset \bigcup_{j\in I} Y_j$

which contains exactly one element from each set Y_j, $j \in I$.

It is remarkable that this axiom is independent of the other axioms of set theory, a result due to P. Cohen. The axiom of choice is equivalent to other surprising statements such as the,

- **Hausdorff maximality principle:** every non-empty partially ordered set contains a maximal chain,

- **Lemma of Zorn:** every partially ordered set in which each chain has an upper bound has a maximal element,

- **Well-ordering theorem of Zermelo:** every set can be well-ordered.

We do not need these results or notions here and therefore we refer to [35] or the nice general introduction to set theory [79]. The reader might be interested to note that the axiom of choice is in its equivalent form of Zorn's lemma is also key in linear algebra: it is needed to prove that every vector space admits an algebraic basis, i.e. a maximal linearly independent set. Following closely R. Schilling [75] we prove

Theorem A.II.1. *Under the assumption that the axiom of choice holds there exists a non-Lebesgue measurable subset of $[0, 1)$.*

Recall that the Lebesgue σ-field is the completion of the Borel σ-field with respect to the Lebesgue-Borel measure.

Proof of Theorem A.II.1. We call $x, y \in [0, 1)$ equivalent and write $x \sim y$ if $x - y \in \mathbb{Q}$. This yields for the equivalence class $[x]$ that $[x] = (x + \mathbb{Q}) \cap [0, 1)$. These equivalence classes form a partition $(M_j)_{j \in J} = ([x_j])_{j \in J}$ of $[0, 1)$, i.e. $[0, 1) = \bigcup_{j \in J} M_j$ and $M_j \cap M_l = \emptyset$ for $j \neq l$. By the axiom of choice (in the second version) there exists a set $M \subset [0, 1)$ containing exactly one element $m_j \in M_j$, $j \in J$. We claim that M is not Lebesgue measurable. Suppose that M is Lebesgue measurable. Then we can find for $x \in [0, 1)$ some $j_0 \in J$ such that $[x] \cap M = [m_{j_0}]$, hence the existence of $q \in \mathbb{Q}$ follows such that $x = m_{j_0} + q$, $-1 < q < 1$. We deduce that

$$[0, 1) \cap M + (\mathbb{Q} \cap (-1, 1)) \subset [0, 1) + (-1, 1) = [-1, 2),$$

or

$$[0, 1) \subset \bigcup_{q \in \mathbb{Q} \cap (-1,1)} (q + M) \subset [-1, 2).$$

584

By the definition of the equivalence relation "$\sim$" we conclude further that $(q + M) \cap (r + M) = \emptyset$ for $q \neq r$. Since $\lambda^{(1)}$ is σ-additive we now find

$$\lambda^{(1)} ([0, 1)) \leq \sum_{q \in \mathbb{Q} \cap (-1, 1)} \lambda^{(1)}(q + M) \leq \lambda^{(1)} ([-1, 2)) = 3.$$

The translation invariance of the Lebesgue measure yields

$$1 \leq \sum_{q \in \mathbb{Q} \cap (-1, 1)} \lambda^{(1)}(M) \leq 3$$

which cannot hold and therefore we have a contradiction. Hence M is not Lebesgue measurable. $\qquad\square$

With further efforts, compare with [75], Appendix D, we can now prove

- the existence of Lebesgue measurable sets which are not Borel sets;

- there exists a subset of $\mathbb{R}$ ($\mathbb{R}^n$) which is not a Borel set.

The proofs of these statements use comparisons of the cardinality of sets such as $\mathbb{R}$, $\mathcal{B}^{(1)}$ and the Lebesgue sets of $\mathbb{R}$.
Although the following discussion extends to more general topological spaces we restrict ours to $\mathbb{R}^n$ or even to $\mathbb{R}$. Let $\mathcal{O}^{(n)}$ denote the Euclidean topology, $\mathcal{C}^{(n)}$ the closed sets and $\mathcal{B}^{(n)}$ the Borel σ-field in $\mathbb{R}^n$. All elements of $\mathcal{O}^{(n)}$ and $\mathcal{C}^{(n)}$ belong to $\mathcal{B}^{(n)}$, the converse is of course not true. Moreover, for a sequence $(U_k)_{k \in \mathbb{N}}$, $U_k \in \mathcal{O}^{(n)}$, we can in general not expect that $\bigcap_{k \in \mathbb{N}} U_k$ is an element of $\mathcal{O}^{(n)}$, but $\bigcap_{k \in \mathbb{N}} U_k$ is always a Borel set. Analogously, for a sequence $(C_k)_{k \in \mathbb{N}}$, $C_k \in \mathcal{C}^{(n)}$, we cannot expect that $\bigcup_{k \in \mathbb{N}} C_k$ is a closed set but it is always a Borel set. This leads to

Definition A.II.2. *A set $E \subset \mathbb{R}^n$ is called a* **G_δ-set** *if it is a countable intersection of open sets. We call $F \subset \mathbb{R}^n$ an* **F_σ-set** *if it is a countable union of closed set.*

As already mentioned, G_δ- and F_σ-sets are Borel sets. In the study of real-valued functions on $\mathbb{R}^n$, especially on $\mathbb{R}$, F_σ- and G_δ-sets play an important role. Here are without proof some results highlighting the importance of F_σ- and G_δ-sets and indirectly the interplay of measurability and topology.

Theorem A.II.3. A. *Let $f : \mathbb{R} \to \mathbb{R}$ be a function. Its set of discontinuities $\{x \in \mathbb{R} \mid f$ is discontinuous at $x\}$ is an F_σ-set. Conversely, to every F_σ-set $A \subset \mathbb{R}$ there exists a function f_A such that the set of discontinuities of f_A is A.*

B. *Let $f : \mathbb{R} \to [0,1]$ be a continuous function with compact support. For every $r > 0$ the set $\{f \geq r\}$ is a compact G_δ-set. If $r \geq 0$ then the set $\{f > r\}$ is an open F_σ-set.*

C. *Let $F \subset \mathbb{R}^n$ be an F_σ-set and $g : F \to \mathbb{R}^m$ be an Hölder continuous mapping in the sense that $\|g(x) - g(y)\|_\infty \leq c\|x - y\|_\infty^\alpha$ holds for all $x, y \in F$ and some $0 < \alpha \leq 1$. Then for every F_σ-set $\tilde{F} \subset \mathbb{R}^n$ the set $g(F \cap \tilde{F})$ is an F_σ-set in $\mathbb{R}^m$.*

A proof of part A can be found in [6] or [14], part B is proved in [47], and a proof of part C is given in [75].

We now turn to Theorem 10.10. A full proof we will provide in Part 12. Any proof of this theorem is either lengthy or needs more preparation of a different type, i.e. the construction of the Lebesgue measure with the help of the Riesz representation theorem. The statement of Theorem 10.10 is best understood in the context of regular measures on locally compact metrizable Hausdorff spaces, however we stay in the context of $\mathbb{R}^n$. We call a Borel measure μ on $\mathcal{B}^{(n)}$ a **regular measure** if it is finite on the compact sets and if for every Borel set $A \in \mathcal{B}^{(n)}$ we have

$$\mu(A) = \sup \{\mu(K) \mid K \subset A, \, K \text{ compact} \} \qquad \text{(A.II.1)}$$
$$= \inf \{\mu(U) \mid A \subset U, \, U \text{ open} \}.$$

Note that (A.II.1) implies for every $A \in \mathcal{B}^{(n)}$ such that $\mu(A) < \infty$ that for $\epsilon > 0$ there exists a compact set K_ϵ and an open set U_ϵ such that $K_\epsilon \subset A \subset U_\epsilon$ and $\mu(U_\epsilon \backslash K_\epsilon) < \epsilon$. However this statement is needed to prove the regularity of μ.

One way to prove (A.II.1) is to prove that the sets of all A satisfying (A.II.1) form a σ-field containing the compact set. This can be done by using Dynkin systems and explicit proofs are given in [11] or [47]. In [75] a proof is given by using Carathéodory construction of the Lebesgue measure, in [27] or [71] a proof is given using the construction of Borel measures with the help of the Riesz representation theorem.

Finally we mention a result a proof of which can be found in [75].

Lemma A.II.4. *Let $A \in \mathcal{B}^{(n)}$ be a Borel set in $\mathbb{R}^n$. Then we can find an F_σ-set F and a G_δ-set G such that $F \subset A \subset G$ and $\lambda^{(n)}(F) = \lambda^{(n)}(A) = \lambda^{(n)}(G)$.*

Of course, this lemma implies Theorem 10.10, but again Theorem 10.10 is needed to prove this lemma.

Appendix III: More on Möbius Transformations

In Chapter 16 we introduced Möbius transformations

$$w(z) := \frac{az + b}{cz + d} \tag{A.III.1}$$

with $a, b, c, d \in \mathbb{C}$, det $\begin{pmatrix} a & b \\ c & d \end{pmatrix} \neq 0$. They have served us first of all as example of complex-valued functions of a complex variable, however we could use them later to construct certain biholomorphic mappings, for example from $\mathbb{D}$ to $\mathbb{H}$.

Furthermore we have seen in Proposition 16.28 that Möbius transformations map straight lines or circles onto straight lines or circles. In addition we could determine the inverse of a Möbius transformation and prove that the composition of two Möbius transformations is again a Möbius transformation.

In this appendix we want to briefly collect some further geometry related properties of Möbius transformations. For this topic classical texts such as [16] or [61] are still some of the best sources. A further good reference is [2]. First we extend every Möbius transformation to the one-point compactification $\hat{\mathbb{C}}$ of $\mathbb{C}$ by

$$w(z) := \begin{cases} \frac{az+b}{cz+d}, & z \neq -\frac{c}{d} \\ \infty, & z = -\frac{c}{d} \end{cases}. \tag{A.III.2}$$

We can now invert w on $\hat{\mathbb{C}}$ and we obtain

$$w^{-1}(\zeta) = \begin{cases} \frac{-d\zeta+b}{c\zeta-a}, & \zeta \neq \frac{a}{c} \\ \infty, & \zeta = \frac{a}{c} \end{cases}, \tag{A.III.3}$$

and it follows that the family of all extended Möbius transformations on $\hat{\mathbb{C}} \, (\cong S^2)$ from a non-abelian group. A first result is

Theorem A.III.1. *Let $(z_1, z_2, z_3), (w_1, w_2, w_3) \in \hat{\mathbb{C}}$ be given triples of distinct points. Then exists a unique Möbius transformation $w : \hat{\mathbb{C}} \to \hat{\mathbb{C}}$ such that $w(z_j) = w_j$, $j = 1, 2, 3$.*

It is convenient to call the set of all circles and straight lines on $\mathbb{C}$ the set of all **generalised circles**. Let $\partial B_r(z_0)$, $z_0 \in \mathbb{C}$, be a circle in $\mathbb{C}$. The mapping

$R_{z_0,r} : \mathbb{C} \setminus \{z_0\} \to \mathbb{C} \setminus \{z_0\}$ defined by

$$R_{z_0,r}(z) = z_0 + \frac{r^2}{\bar{z} - \bar{z_0}} \tag{A.III.4}$$

is called the **reflection at the circle** $\partial B_r(z_0)$ and we have

Theorem A.III.2. *Every reflection at a circle maps a generalised circle into a generalised circle.*

With the help of Theorem A.III.2 one can deduce from Theorem A.III.1.

Corollary A.III.3. *Three distinct points $z_1, z_2, z_3 \in \mathbb{C}$ always lie on a unique generalised circle.*

Proof. (Sketch) Transform the points $\{z_1, z_2, z_3\}$ by a Möbius transformation onto the points $\{0, \infty, 1\}$ which must lie by Theorem A.III.2 on a generalised circle. $\qquad\square$

Corollary A.III.4. *Two generalised circles can be transformed onto each other by a Möbius transformation.*

To proceed further we need

Definition A.III.5. *Let $z_1, z_2, z_3, z_4 \in \hat{\mathbb{C}}$. Their **cross ratio** is defined by*

$$(z_1 : z_2; z_3 : z_4) := \frac{\frac{z_1 - z_3}{z_2 - z_3}}{\frac{z_1 - z_4}{z_2 - z_4}}. \tag{A.III.5}$$

Here we use the conventions

$$(\infty : z_2; z_3 : z_4) := \frac{z_2 - z_4}{z_2 - z_3}$$

or

$$(z_1 : z_2; \infty : z_4) := \frac{z_4 - z_2}{z_4 - z_1},$$

etc, i.e. we pass in (A.III.5) to the corresponding limit as $z_j \to \infty$. Clearly, the cross ratio is translation invariant on $\mathbb{C}$. Furthermore we have

Theorem A.III.6. *If $z_1, z_2, z_3, z_4 \in \hat{\mathbb{C}}$ are distinct points and $w : \hat{\mathbb{C}} \to \hat{\mathbb{C}}$ is a Möbius transformation then we have*

$$(w(z_1) : w(z_2); w(z_3) : w(z_4)) = (z_1 : z_2; z_3 : z_4). \tag{A.III.6}$$

As a corollary we obtain

Corollary A.III.7. *Four points $z_1, z_2, z_3, z_4 \in \hat{\mathbb{C}}$ lie on a generalised circle if and only if their cross ratio is a real number.*

Proof. (Sketch) We map z_1, z_2, z_3 by a Möbius transformation w onto 0, ∞, 1 and note that now

$$(z_1 : z_2; z_3 : z_4) = (0 : \infty; 1; w(z_4)) = \frac{1}{w(z_4)}.$$

$\square$

It is now possible to use Möbius transformations more extensively as a tool in studying geometric objects in the plane or in $\hat{\mathbb{C}}$ ($\cong S^2$), a beautiful and classical topic in geometry but we refer to the sources mentioned above and end our considerations here.

Appendix IV: Bernoulli Numbers

In Chapter 26 we introduced the Bernoulli numbers $B_{2k} = (-1)^{k-1}\beta_{2k}$, $k \geq 1$, with the help of the Taylor expansion of $x \mapsto \frac{x}{e^x-1}$, i.e.

$$\frac{x}{e^x-1} = 1 - \frac{1}{2}x + \sum_{k=1}^{\infty} \frac{(-1)^{k-1}\beta_{2k}}{(2k)!} x^{2k}. \tag{A.IV.1}$$

Starting with the Ansatz

$$\frac{x}{e^x-1} = \sum_{k=0}^{\infty} \frac{B_k}{k!} x^k \tag{A.IV.2}$$

and using the Taylor expansion of the exponential function

$$e^x - 1 = x + \frac{x^2}{2!} + \frac{x^3}{3!} + \cdots = x\left(1 + \frac{x}{2!} + \frac{x^2}{3!} + \cdots\right)$$

we find

$$\frac{x}{e^x-1} = \frac{1}{1 + \frac{x}{2!} + \frac{x^3}{3!} + \cdots}.$$

For $x \in \mathbb{R}$ such that $\sum_{l=1}^{\infty} \frac{|x|^l}{(l+1)!} < 1$ we deduce further

$$\frac{x}{e^x-1} = \frac{1}{1 - \left(-\sum_{l=1}^{\infty} \frac{x^l}{(l+1)!}\right)} = \sum_{k=0}^{\infty}(-1)^k \left(\sum_{l=1}^{\infty} \frac{x^l}{(l+1)!}\right)^k$$

$$= 1 - \frac{1}{2}x + \sum_{n=2}^{\infty} \frac{B_n}{n!} x^n,$$

where the last equality now defines the **Bernoulli numbers** B_n, $n \geq 2$, and further we set $B_0 := 1$ and $B_1 := -\frac{1}{2}$. Since

$$\frac{x}{e^x-1} + \frac{1}{2}x = \frac{1}{2}x\left(\frac{e^x+1}{e^x-1}\right)$$

and

$$\frac{e^{-x}+1}{e^{-x}-1} = \frac{1+e^x}{1-e^x} = -\frac{e^x+1}{e^x-1}$$

it follows that

$$\frac{x}{e^x - 1} + \frac{1}{2}x = \sum_{n=2}^{\infty} \frac{B_n}{n!} x^n$$

is an even function and therefore $B_n = 0$ for $n \geq 2$ and odd. Moreover, from (A.IV.2) we deduce by using the Cauchy product of series

$$x = (e^x - 1) \sum_{n=0}^{\infty} \frac{B_n}{n!} x^n = \left(\sum_{k=1}^{\infty} \frac{x^k}{k!} \right) \left(\sum_{n=0}^{\infty} \frac{B_n}{n!} x^n \right)$$

$$= \sum_{m=1}^{\infty} c_m x^m$$

with

$$c_m = \sum_{l=0}^{m-1} \frac{B_l}{l!(m-l)!}$$

implying by comparing coefficients that

$$\sum_{l=0}^{m-1} \binom{m}{l} B_l = 0, \qquad \text{(A.IV.3)}$$

thus we have a recursion formula for the Bernoulli numbers B_k. Here is a list of the non-trivial Bernoulli numbers up to $k = 12$. Note that by (A.IV.3) all Bernoulli numbers are rational numbers:

$$B_0 = 1, \quad B_1 = -\frac{1}{2}, \quad B_2 = \frac{1}{6}, \quad B_4 = -\frac{1}{30},$$

$$B_6 = \frac{1}{42}, \quad B_8 = -\frac{1}{30}, \quad B_{10} = \frac{5}{66}, \quad B_{12} = -\frac{691}{2730}.$$

Since the series $\sum_{n=2}^{\infty} \frac{B_n}{n!} x^n$ has positive radius of convergence, namely 2π, it follows that there exists infinitely many non-zero Bernoulli numbers. We can also prove that the series $(B_{2k})_{k \in \mathbb{N}}$ is alternating, i.e. $\operatorname{sgn} B_{2k} = -\operatorname{sgn} B_{2(k+1)}$. Note that since $B_n = 0$ for $n \geq 2$ and odd some authors call the numbers $\beta_k := (-1)^{k-1} B_{2k}$, $k \in \mathbb{N}$, together with $\beta_0 = B_0 = 1$ and $\beta_1 = \frac{1}{2}$ the Bernoulli numbers.

Bernoulli numbers were first introduced by Jacob I Bernoulli when summing $\sum_{k=0}^{N} k^n$. Since then they have plenty appearances in Mathematics, a classical book dealing entirely with Bernoulli numbers in N. Nielsen's "Traité élémentaire des nombres de Bernoulli", [62], but still new results on Bernoulli numbers are being discovered. We want to prove next the formula

$$\sum_{k=1}^{N} k^n = \sum_{j=1}^{n+1} \frac{n!}{(n-j+1)!j!} B_{n-j+1}(N+1)^j \qquad \text{(A.IV.4)}$$

$$= \frac{1}{n+1} \sum_{j=1}^{n+1} \binom{n+1}{j} B_{n-j+1}(N+1)^j.$$

For $N = 1, 2, 3$ we find

$$\sum_{k=1}^{N} k = \frac{N(N+1)}{2},$$

$$\sum_{k=1}^{N} k^2 = \frac{N(N+1)(2N+1)}{6},$$

and

$$\sum_{k=1}^{N} k^3 = \frac{(N(N+1))^2}{4}.$$

In order to derive (A.IV.4) we first note that

$$\sum_{k=0}^{N} e^{kx} = \sum_{k=0}^{N} \sum_{n=0}^{\infty} \frac{1}{n!} (kx)^n = \sum_{n=0}^{\infty} \left(\sum_{k=0}^{N} k^n \right) \frac{x^n}{n!}, \qquad \text{(A.IV.5)}$$

but in addition we have for $x \neq 0$

$$\sum_{k=0}^{N} e^{kx} = \sum_{k=0}^{N} (e^x)^k = \frac{e^{(N+1)x} - 1}{e^x - 1}. \qquad \text{(A.IV.6)}$$

Combining (A.IV.5) and (A.IV.6) we find

$$\sum_{n=0}^{\infty} \left(\sum_{k=0}^{N} k^n \right) \frac{x^{n+1}}{n!} = \frac{x}{e^x - 1} \left(e^{(N+1)x} - 1 \right)$$

and with (A.IV.2) we arrive at

$$\sum_{n=0}^{\infty}\left(\sum_{k=0}^{N}k^n\right)\frac{x^{n+1}}{n!} = \left(\sum_{l=0}^{\infty}\frac{B_l}{l!}x^l\right)\left(\sum_{j=1}^{\infty}\frac{(N+j)^j}{j!}x^j\right)$$

$$= \sum_{n=0}^{\infty}c_n x^{n+1}$$

with

$$c_n = \sum_{j=1}^{n+1}\frac{1}{(n-j+1)!}B_{n-j+1}(N+1)^j.$$

Finally, by comparing the coefficients in

$$\sum_{n=0}^{\infty}\left(\sum_{k=0}^{N}k^n\right)\frac{x^{n+1}}{n!} = \sum_{n=0}^{\infty}c_n x^{n+1}$$

we obtain (A.IV.4). In our argument we followed [22] closely.

It was Euler who could prove that

$$\zeta(2k) = \sum_{n=1}^{\infty}\frac{1}{n^{2k}} = (-1)^{k+1}\frac{(2\pi)^{2k}}{2(2k)!}B_{2k} \qquad \text{(A.IV.7)}$$

holds for $k \in \mathbb{N}$. Hence the Bernoulli numbers are also linked to the Riemann ζ-function which in Euler's times was not yet studied as a function in the complex plane. Euler had also obtained Taylor expansions for trigonometrical series with the help of the Bernoulli numbers, see [67],

$$\cot z = \frac{1}{2} + \sum_{j=1}^{\infty}(-1)^j\frac{4^j}{(2j)!}B_{2j}z^{2j-1}, \qquad \text{(A.IV.8)}$$

$$\tan z = \sum_{j=1}^{\infty}(-1)^{j-1}\frac{4^j(4^j-1)}{(2j)!}B_{2j}z^{2j} \qquad \text{(A.IV.9)}$$

or

$$\frac{z}{\sin z} = \sum_{j=0}^{\infty}(-1)^{j-1}\frac{2^{2j}-2}{(2j)!}B_{2j}z^{2j}. \qquad \text{(A.IV.10)}$$

Finally we want to mention that there exists integral representations of the Bernoulli numbers such as

$$B_{2n} = 8n(-1)^{n+1} \int_0^\infty \frac{t^{4n} - 1}{e^{2\pi t} - 1} \, dt, \qquad \text{(A.IV.11)}$$

which is taken from [93]. More formulae of this type as well as a table of Bernoulli numbers can be found in [1].

Solutions to Problems of Part 6

Chapter 1

1.   a) By definition $\emptyset$ is denumerable, hence $\Omega \in \mathcal{A}$, and if $A \in \mathcal{A}$ then A or $A^{\complement}$ is denumerable, hence $A^{\complement}$ belongs to $\mathcal{A}$ since $\left(A^{\complement}\right)^{\complement} = A$. Finally we note for a sequence $(A_j)_{j\in\mathbb{N}}$, $A_j \in \mathcal{A}$, that if all sets A_j are denumerable, then $\bigcup_{j\in\mathbb{N}} A_j$ is denumerable. However if one of the sets, say A_{j_0}, is not denumerable, then we find

$$\left(\bigcup_{j\in\mathbb{N}} A_j\right)^{\complement} = \bigcap_{j\in\mathbb{N}} A_j^{\complement} \subset A_{j_0}^{\complement}$$

and $A_{j_0}^{\complement}$ is denumerable, hence $\left(\bigcup_{j\in\mathbb{N}} A_j\right)^{\complement}$ is denumerable implying $\bigcup_{j\in\mathbb{N}} A_j \in \mathcal{A}$.

    b) Since $\Omega' = \Omega' \cap \Omega$ and $\Omega \in \mathcal{A}$ it follows that $\Omega' \in \mathcal{A}_{\Omega'}$. Moreover, for $A' \in \mathcal{A}_{\Omega'}$ we can find $A \in \mathcal{A}$ such that $A' = \Omega' \cap A$. Now we have to find the complement of A' in Ω' which is of course $\Omega' \setminus A' = \Omega' \cap (\Omega' \cap A)^{\complement} = \Omega' \cap (\Omega'^{\complement} \cup A^{\complement}) = \Omega' \cap A^{\complement} \in \mathcal{A}_{\Omega'}$ since $A^{\complement} \in \mathcal{A}$.

For a sequence $(A'_j)_{j\in\mathbb{N}}$, $A'_j \in \mathcal{A}_{\Omega}$, it holds that $A'_j = \Omega' \cap A_j$ for some $A_j \in \mathcal{A}$.

Therefore $\bigcup_{j\in\mathbb{N}} A'_j = \bigcup_{j\in\mathbb{N}} (\Omega' \cap A_j) = \Omega' \cap \left(\bigcup_{j\in\mathbb{N}} A_j\right)$. Since $\bigcup_{j\in\mathbb{N}} A_j \in \mathcal{A}$ it follows eventually that $\mathcal{A}_{\Omega'}$ is a σ-field.

2. Proof of Lemma 1.9.A:
We note that if $\mathcal{A}$ is a σ-field in Ω such that $\mathcal{E} \subset \mathcal{A}$ then $E^{\complement} \in \mathcal{A}$ and hence $\mathcal{E}^{\complement} \subset \mathcal{A}$. Conversely, if $\tilde{\mathcal{A}}$ is a σ-field containing $\mathcal{E}^{\complement}$ then $\tilde{\mathcal{A}}$ also contains $\mathcal{E}$. Thus $\{A \mid \mathcal{E} \subset \mathcal{A} \text{ and } \mathcal{A} \text{ is a } \sigma\text{-field in } \Omega\} = \{\tilde{A} \mid \mathcal{E}^{\complement} \subset \tilde{\mathcal{A}} \text{ and } \tilde{\mathcal{A}} \text{ is a } \sigma \text{ field in } \Omega\}$ implying $\sigma(\mathcal{E}) = \sigma(\mathcal{E}^{\complement})$.
Proof of Lemma 1.9.B: Since $\mathcal{E} \subset \mathcal{I}$ it follows that $\sigma(\mathcal{E}) \subset \sigma(\mathcal{I})$. On the other side, if $E_j \in \mathcal{E}, j \in J \subset \mathbb{N}$, then $\bigcup_{j\in J} E_j \in \sigma(\mathcal{E})$, i.e. $\mathcal{I} \subset \sigma(\mathcal{E})$ implying $\sigma(\mathcal{I}) \subset \sigma(\mathcal{E})$.
Proof of Lemma 1.9.C: Obviously we have $\sigma(\mathcal{E}_1) \subset \sigma(\mathcal{E}_1 \cup \mathcal{E}_2)$. Since $\mathcal{E}_1 \cup \mathcal{E}_2 \subset \sigma(\mathcal{E}_1)$ it also follows that $\sigma(\mathcal{E}_1 \cup \mathcal{E}_2) \subset \sigma(\mathcal{E}_1)$.

3. Since $\mathcal{O}_X = \mathcal{C}_X^{\complement}$ where $\mathcal{C}_X$ denotes the closed sets in X the first part follows from Lemma 1.9.A.
Now let (X, d) be a separable **metric space**. Denote by $x_k, k \in \mathbb{N}$, the points of a dense subset of X, i.e. $X = \overline{\{x_j \mid j \in \mathbb{N}\}}$. As countable union of countable sets the set

$$\mathcal{D}_d := \{B_{\frac{1}{n}}(x_j) \mid n \in \mathbb{N}, x_j \in X\} \cup \{B_n(x_j) \mid n \in \mathbb{N}, x_j \in X\}$$

is countable and every open set in X is a union of sets from $\mathcal{D}_d$. This implies that $\sigma(\mathcal{D}_d) = \sigma(\mathcal{D}) = \sigma(\mathcal{O}_X) = \mathcal{B}(X)$.

4. Consider the Cartesian product $\mathcal{B}^1 \times \mathcal{B}^1 = \{A = B_1 \times B_2 \mid B_j \in B^{(1)}\}$.
For $[0,1] \times [0,1] \in \mathcal{B}^{(1)} \times \mathcal{B}^{(1)}$ it follows that

$([0,1] \times [0,1])^{\complement} = (-\infty, 0) \times \mathbb{R} \cup (1, \infty) \times \mathbb{R} \cup (0,1) \times (-\infty, 0) \cup (0,1) \times (1, \infty)$
which is a set not belonging to $\mathcal{B}^1 \times \mathcal{B}^1$, so $\mathcal{B}^1 \times \mathcal{B}^1$ can not be a σ-field.

5. The key observation is that the estimates $\|x\|_\infty \leq \|x\|_2 \leq \sqrt{n}\,\|x\|_\infty$ imply for all $r > 0$ and $z \in \mathbb{R}^n$

$$B_{\frac{r}{n}}^{(\infty)}(z) \subset B_r^{(2)}(z) \subset B_r^{(\infty)}(z)$$

where $B_\rho^{(2)}(y) = \{x \in \mathbb{R}^n \mid \|x - y\|_2 < \rho\}$ and $B_\rho^{(\infty)}(y) = \{x \in \mathbb{R}^n \mid \|x - y\|_\infty < \rho\}$.
Now, if $U \in \mathcal{O}_n$ then we have $U = \bigcup_{x \in U} B_{\rho(x)}^{(2)}(x)$ and therefore $\bigcup_{x \in U} B_{\frac{\rho(x)}{\sqrt{n}}}^{(\infty)}(x)$, however
it is trivial that $U \subset \bigcup_{x \in U} B_{\frac{\rho(x)}{\sqrt{n}}}^{(\infty)}(x) \subset U$. Thus we have $U = \bigcup_{x \in U} B_{\frac{\rho(x)}{\sqrt{n}}}^{(\infty)}(x)$ and
$B_{\frac{\rho(x)}{\sqrt{n}}}^{(\infty)}(x) \in J$ which now gives $U = \bigcup_{I \in J, I \subset U} I$.

6. From Problem 3 it follows that $\sigma(\mathcal{C}_n) = \mathcal{B}^{(n)}$. Further we know that $\sigma(\mathcal{K}_n) \subset \sigma(\mathcal{C}_n)$
since $\mathcal{K}_n \subset \mathcal{C}_n$. For $C \in \mathcal{C}_n$ the set $C_j := C \cap B_j(0)$ is compact and $C = \bigcup_{j \in \mathbb{N}} C_j$, hence
$C_n \subset \sigma(\mathcal{K}_n)$, i.e. $\sigma(\mathcal{C}_n) = \sigma(\mathcal{K}_n)$. Thus we have already proved $\mathcal{B}^{(n)} = \sigma(\mathcal{O}_n) = \sigma(\mathcal{C}_n) = \sigma(\mathcal{K}_n)$.
In order to prove $\sigma(\mathcal{I}_{r,n}) = \sigma(\mathcal{I}_{l,n}) = \mathcal{B}^{(n)}$ we can use the considerations of Corollary
1.11. It is helpful to introduce first some notation. For $a = (a_1, ..., a_n)$ and $b = (b_1, ..., b_n)$ we set

$$(a, b) := \bigtimes_{k=1}^n (a_k, b_k), \qquad [a, b] := \bigtimes_{k=1}^n [a_k, b_k],$$

as well as

$$[a, b) := \bigtimes_{k=1}^n [a_k, b_k), \qquad (a, b] := \bigtimes_{k=1}^n (a_k, b_k],$$

and we call these sets the n-dimensional open, closed and half-open intervals, respectively. With $c = (1, ..., 1) \in \mathbb{R}^n$ it follows that

$$[a, b) = \bigcap_{j \in \mathbb{N}} (a - \frac{1}{j}c, b) \text{ and } (a, b] = \bigcap_{j \in \mathbb{N}} (a, b + \frac{1}{j}c),$$

hence

$$\sigma(\mathcal{I}_{r,n}) \subset \mathcal{B}^{(n)} \text{ and } \sigma(\mathcal{I}_{l,n}) \subset \mathcal{B}^{(n)}.$$

On the other hand we have

$$(a, b) = \bigcup \{[c, d] \mid [c, d) \subset (a, b), c, d \in \mathbb{Q}^n\},$$

and

$$(a, b) = \bigcup \{(c, d] \mid (c, d] \subset (a, b), c, d \in \mathbb{Q}^n\}.$$

The countability of $\mathbb{Q}$ now yields that

$$\mathcal{B}^n \subset \sigma(\mathcal{I}_{r,n}) \text{ and } \mathcal{B}^n \subset \sigma(\mathcal{I}_{l,n}).$$

The remaining part consists of two problems: firstly to pass from half-open n-dimensional intervals to open or closed n-dimensional intervals, and secondly to pass from arbitrary real endpoints, to rational endpoints. Since we can approximate real numbers by rational numbers the step from n-dimensional intervals with real endpoints to those with rational endpoints is clear: replace the real endpoints $a, b \in \mathbb{R}^n$ by sequences of approximating rational endpoints say $(a_k)_{k \in \mathbb{N}}, (b_k)_{k \in \mathbb{N}}$ (and $(\tilde{a}_k)_{k \in \mathbb{N}}, (\tilde{b}_k)_{k \in \mathbb{N}})$ and now consider the intersection and union respectively, i.e.

$$(a, b) = \bigcup_{k \in \mathbb{N}} (a_k, b_k) \text{ and } [a, b] = \bigcap_{k \in \mathbb{N}} [\tilde{a}_k, \tilde{b}_k].$$

The proof of $\mathcal{B}^{(n)} = \sigma((\mathcal{I}_{r,n})$ implies already that we can use the n-dimensional open intervals to generate $\mathcal{B}^{(n)}$. Finally we remark that $[a, b] = \left(\bigcup_{k \in \mathbb{N}} [\tilde{a}_k, \tilde{b}_k]^{\complement} \right)^{\complement}$ to conclude that $\mathcal{B}^{(n)}$ is also generated by the n-dimensional closed intervals.

7.   a) If we know that f maps open intervals onto open intervals the results is proved: in this case we have $\{(a, b) \mid a < b\} \subset \sigma(f) = \{f^{-1}(V) \mid V \in \mathcal{B}^{(1)}\}$ but we have $\mathcal{B}^{(1)} = \sigma(\{(a, b) \mid a < b\}) \subset \sigma(\mathcal{B}) \subset \mathcal{B}^{(1)}$. Since f is bijective f and f^{-1} are strictly monotone (either increasing or decreasing) and since f is continuous $f((a, b))$ is an interval with midpoints A and B. By monotonicity and continuity we must have $A, B \in \{f(a), f(b)\}$. Suppose $A \in J$. Then $f^{-1}(A)$ must belong to (a, b), i.e. a or b belongs to (a, b) which is a contradiction. The same argument applies to B. Hence $J = (A \vee B, A \wedge B)$ is open and the result follows.

b) The mappings g_{11}, g_{12}, g_{13} are $\mathcal{A}_1/\mathcal{A}_3$-measurable since for every $A \subset \mathbb{R}$ it follows that $g_{1j}^{-1}(A) \in \mathcal{P}(\mathbb{R}) = \mathcal{A}_1$.
The mapping g_{21} is in general not $\mathcal{B}^1/\mathcal{P}(\mathbb{R})$-measurable, just take the identity and consider the pre-image of a non-Borel measurable set. Since continuous mappings are Borel-measurable g_{22} is measurable. If $A \in \mathcal{B}^{(1)}$ then g_{23} is $\mathcal{B}^{(1)}/\mathcal{A}_3$-measurable, otherwise in general it is not. If $B \notin \mathcal{A}_3$ then $id : \mathbb{R} \to \mathbb{R}$ is not $\mathcal{A}_3/\mathcal{P}(\mathbb{R})$-measurable, hence in general g_{31} is not $\mathcal{A}_3/\mathcal{P}(\mathbb{R})$-measurable and with the same type of argument we see that g_{32} is in general not $\mathcal{A}_3/\mathcal{B}^{(1)}$-measurable. However, since for a constant function $g : \mathbb{R} \to \mathbb{R}, g(x) = c$ for all x, we have for every set $B \subset \mathbb{R}$ that $g^{-1}(B) \in \{\emptyset, \mathbb{R}\}$, these functions are both $\mathcal{A}_3/\mathcal{P}(\mathbb{R})$-measurable and $\mathcal{A}_3/\mathcal{B}^{(1)}$-measurable. Finally, g_{33} is in general not $\mathcal{A}_3/\mathcal{A}_3$-measurable since $g_{33} := id + a, a \in \mathbb{R}$, yields for any set $g_{33}^{-1}(A) = A \neq a$.

8. Since $(g \circ f)^{-1}(A) = f^{-1}(g^{-1}(A))$ it follows that if $A \in \mathcal{A}_3$ then $g^{-1}(A) \in \mathcal{A}_2$ and consequently $f^{-1}(g^{-1}(A)) \in \mathcal{A}_1$, i.e. for every $A \in \mathcal{A}_3$ the set $(g \circ f)^{-1}(A)$ belongs to $\mathcal{A}_1$, in other words $g \circ f$ is $\mathcal{A}_1/\mathcal{A}_3$.
Now since $pr_j : \mathbb{R}^n \to \mathbb{R}, 1 \leq j \leq n$, is $\mathcal{B}^{(1)}/\mathcal{B}^{(1)}$-measurable it follows for a $\mathcal{B}^{(1)}/\mathcal{B}^{(n)}$-measurable mapping $f : \mathbb{R} \to \mathbb{R}^n$ that $f_i = pr_j \circ f$ is $\mathcal{B}^{(1)}/\mathcal{B}^{(1)}$-measurable. On the other hand if all mappings $f_j, 1 \leq j \leq n$, are $\mathcal{B}^{(1)}/\mathcal{B}^{(1)}$-measurable and $A \in \mathcal{B}^{(n)}$ then $f^{(-1)}(A) = \bigcap_{j=1}^{n} f_j^{(-1)}(A)$ belongs to $\mathcal{B}^{(1)}$ implying that f is $\mathcal{B}^{(1)} - \mathcal{B}^{(n)}$-measurable.

9. a) Since $\emptyset$ is by definition finite, hence denumerable, it follows that $\nu(\mathcal{O}) = 0$. Let $(A_j)_{j\in\mathbb{N}}$ be a sequence of mutually disjoint sets belonging to it. If all sets A_j are denumerable we find that $\bigcup_{j\in\mathbb{N}} A_j$ is denumerable and therefore

$$0 = \nu\left(\bigcup_{j\in\mathbb{N}} A_j\right) = \sum_{j=1}^{\infty} \nu(A_j) = 0.$$

Now suppose that some A_{j_0} of the family $(A_j)_{j\in\mathbb{N}}$ is not denumerable, i.e. $A_{j_0}^{\complement}$ is denumerable. We claim that no other set $A_j, j \neq j_0$, can be denumerable. Indeed if $A_{j_n}, j_n \neq j_0$ was denumerable then $A_{j_0}^{\complement}$ and $A_{j_n}^{\complement}$ were denumerable and hence $A_{j_0}^{\complement} \cup A_{j_n}^{\complement}$.
Consequently $(A_{j_0}^{\complement} \cup A_{j_n}^{\complement})^{\complement}$ was non-denumerable, but $(A_{j_0}^{\complement} \cup A_{j_n}^{\complement})^{\complement} = A_{j_0} \cap A_{j_n} = \emptyset$ by assumption. Thus at most one set of $(A_j)_{j\in\mathbb{N}}, A_j \cap A_l = \emptyset$, can be non-denumerable. If A_{j_0} is such a set we find now since $\bigcup_{j\in\mathbb{N}}(A_j)$ is non-denumerable

$$\nu\left(\bigcup_{j\in\mathbb{N}}(A_j)\right) = 1 = \nu(A_{j_0}) = \nu(A_{j_0}) + \sum_{j\neq j_0} \nu(A_j) = \sum_{j\in\mathbb{N}} \nu(A_j),$$

implying that ν is a measure. Finally we observe that $\nu(\Omega) = 1$ since Ω is denumerable, so ν is a probability measure.

b) Clearly

$$\sum_{k\in\mathbb{Z}} \epsilon_k(\emptyset) = 0$$

and for $A_j \subset \mathbb{Z}, j \in \mathbb{N}$, with $A_j \cap A_l = \emptyset$ if $j \neq l$ it follows that

$$\mu\left(\bigcup_{j\in\mathbb{N}}(A_j)\right) = \sum_{k\in\mathbb{Z}} \epsilon_k\left(\bigcup_{j\in\mathbb{N}}(A_j)\right) = \sum_{j\in\mathbb{N}}\sum_{k\in\mathbb{Z}} \epsilon_k\left(A_j\right) = \sum_{j\in\mathbb{N}} \mu(A_j).$$

This proves already that μ is a measure on $(\mathbb{Z}, \mathcal{P}(\mathbb{Z}))$. If $A \subset \mathbb{Z}$ is finite and has N elements and if $m \in \mathbb{Z}$ then $A + m$ has also N elements and therefore

$$\sum_{k\in\mathbb{Z}} \epsilon_k(A) = N = \sum_{k\in\mathbb{Z}} \epsilon_k(A + m).$$

If A is not finite then $A + m$ is not finite either and

$$\sum_{k\in\mathbb{Z}} \epsilon_k(A) = \infty = \sum_{k\in\mathbb{Z}} \epsilon_k(A + m).$$

Thus for all $A \in \mathcal{P}(\mathbb{Z})$ and all $m \in \mathbb{Z}$ we have $\mu(A) = \mu(A + m)$.

c) That μ is a measure follows from

$$\mu(\emptyset) = \sum_{j=1}^{\infty} \frac{1}{2^j}\mu_j(\emptyset) = 0$$

and

$$\mu\left(\bigcup_{k\in\mathbb{N}}A_k\right)=\sum_{j=1}^{\infty}\frac{1}{2^j}\mu_j\left(\bigcup_{k\in\mathbb{N}}A_k\right)=\sum_{j=1}^{\infty}\frac{1}{2^j}\sum_{k=1}^{\infty}\mu_j(A_k)=\sum_{k=1}^{\infty}\sum_{j=1}^{\infty}\frac{1}{2^j}\mu_j(A_k)=\sum_{k=1}^{\infty}\mu(A_k)$$

for every sequence $(A_k)_k \in \mathbb{N}, A_k \in \mathcal{A}$, of mutually disjoint sets. Since

$$\sum_{j=1}^{\infty}\frac{1}{2^j} = 1$$

it follows further that $\mu(\Omega) = 1$, i.e. μ is a probability measure.

10.   a) Obviously we have $\mu_{\Omega'}(\emptyset) = \mu(\Omega' \cap \emptyset) = \mu(\emptyset) = 0$. If $(A_k)_{k\in\mathbb{N}}, A_k \in \mathcal{A}$, is a sequence of mutually disjoint sets then $\Omega' \cap \bigcup_{k\in\mathbb{N}} A_k = \bigcup_{k\in\mathbb{N}} (\Omega' \cap A_k)$ and the sets $(\Omega' \cap A_k), k \in \mathbb{N}$, are mutually disjoint too.
Hence we have

$$\mu_{\Omega'}\left(\bigcup_{k\in\mathbb{N}}A_k\right)=\mu\left(\Omega'\cap\bigcup_{k\in\mathbb{N}}A_k\right)=\sum_{k=1}^{\infty}\mu\left(\bigcup_{k\in\mathbb{N}}(\Omega'\cap A_k)\right)=\sum_{k=1}^{\infty}\mu(\Omega'\cap A_k)=\sum_{k=1}^{\infty}\mu_{\Omega'}(A_k),$$

i.e. μ' is a measure.

b) Since $\widetilde{\mathcal{A}}$ is a σ-field, $\emptyset \in \widetilde{\mathcal{A}}$ and therefore we find $\mu|_{\widetilde{\mathcal{A}}}(\emptyset) = \mu(\emptyset) = 0$. Now let $\widetilde{A}_k \in \widetilde{\mathcal{A}}, k \in \mathbb{N}$, such that $\widetilde{A}_k \cap \widetilde{A}_l = \emptyset$ for $k \neq l$. Since $\widetilde{\mathcal{A}} \subset \mathcal{A}$ it follows that

$$\mu_{\widetilde{\mathcal{A}}}\left(\bigcup_{k\in\mathbb{N}}A_k\right) = \mu\left(\bigcup_{k\in\mathbb{N}}\widetilde{A}_k\right) = \sum_{k=1}^{\infty}\mu(\widetilde{A}_k) = \sum_{k=1}^{\infty}\mu_{\widetilde{A}_k}(\widetilde{A}),$$

hence $\mu_{\widetilde{\mathcal{A}}}$ is a measure on $(\Omega, \widetilde{\mathcal{A}})$.

Chapter 2

1. We must first recollect the way we worked with the partitions in Volume II Chapter 18. If $Q \subset \mathcal{R}^n$ is any non-degenerate hyper-rectangle, we learnt that partitions of the generating intervals led to a partition of Q. If $A_\alpha, \alpha \in I$, denotes the hyper-rectangle obtained from such a partition, then we have

$$\lambda^{(1)}(Q) = \sum_{\alpha\in I}\lambda^{(1)}(A_\alpha).$$

We also note that for this equality it does not matter whether A_α is open, closed or half-open, since $\lambda^{(n)}(\overset{\circ}{A}_\alpha) = \lambda^{(n)}(\overline{A_\alpha})$. Next we observe that

$$F = \bigcup_{k=1}^{M}\bigcup_{l=1}^{N}(A_k \cap B_l) = \bigcup_{k=1}^{M} A_k = \bigcup_{l=1}^{N}B_l$$

and $A_k \cap B_l$ is an element in $\mathcal{I}_{r,n}$ or $A_k \cap B_l$ is a set contained in a hyperplane of $\mathbb{R}^n$ and then $\lambda^{(1)}(A_k \cap B_l) = 0$. Thus we have

$$\lambda^{(n)}(F) = \sum_{k=1}^{M} \sum_{l=1}^{N} \lambda^{(n)}(A_k \cap B_l).$$

On the one hand we have

$$\sum_{k=1}^{M} \sum_{l=1}^{N} \lambda^{(n)}(A_k \cap B_l) = \sum_{k=1}^{M} \lambda^{(n)}\left(\bigcup_{l=1}^{N}(A_k \cap B_l)\right) = \sum_{k=1}^{M} \lambda^{(n)}\left(A_k \cup \left(\bigcup_{l=1}^{N} B_l\right)\right)$$

$$= \sum_{k=1}^{M} \lambda^{(n)}(A_k \cap F) = \sum_{k=1}^{M} \lambda^{(n)}(A_k)$$

while on the other hand we have

$$\sum_{k=1}^{M} \sum_{l=1}^{N} \lambda^{(n)}(A_k \cap B_l) = \sum_{l=1}^{N} \lambda^{(n)}\left(\left(\bigcup_{k=1}^{M} A_k\right) \cap B_l\right)$$

$$= \sum_{l=1}^{N} \lambda^{(n)}(F \cap B_l) = \sum_{l=1}^{N} \lambda^{(n)}(B_l),$$

and hence we have shown that

$$\sum_{k=1}^{M} \lambda^{(n)}(A_k) = \sum_{l=1}^{N} \lambda^{(n)}(B_l).$$

2. We start by proving that $\mathcal{R}$ is a ring. Since $\emptyset$ is by definition finite $\emptyset \in \mathcal{R}$. Now let $A, B \in \mathcal{R}$. If A is finite then $A \setminus B$ is finite. In general we have $A \setminus B = A \cap B^{\complement}$. Thus if A is infinite and B is infinite then $B^{\complement}$ is finite and $A \cap B^{\complement}$ too, hence $A \setminus B$ is finite. Finally, if A is infinite and B is finite, then $A^{\complement} \cup B$ is finite, but $A^{\complement} \cup B = (A \setminus B)^{\complement}$, thus $(A \setminus B)^{\complement}$ is finite and hence $A \setminus B \in \mathcal{R}$ in each case.

Moreover, for A and B finite it follows that $A \cup B$ is finite and if A is finite and B is infinite then $A \cup B$ must be infinite but $(A \cup B)^{\complement} = A^{\complement} \cap B^{\complement}$ is finite since $B^{\complement}$ is finite. The same argument holds for A infinite and B finite. If however both, A and B are infinite then $(A \cup B)^{\complement} = A^{\complement} \cap B^{\complement}$ is finite. Thus $\mathcal{R}$ is indeed a ring. Clearly we have $\mu(\emptyset) = 0$ since $\emptyset$ is finite. Now let $(A_k)_{k \in \mathbb{N}}, A_k \in \mathcal{R}$, be a sequence of mutually disjoint elements of $\mathcal{R}$ such that $\bigcup_{k \in \mathbb{N}} A_k \in \mathcal{R}$. Suppose that all A_k are finite and assume that $\bigcup_{k \in \mathbb{N}} A_k$ is not finite. Then $\bigcap_{k \in \mathbb{N}} A_k^{\complement}$ must be finite since by assumption $\bigcup_{k \in \mathbb{N}} A_k \in \mathcal{R}$. But $\bigcup_{k \in \mathbb{N}} A_k$ is denumerable, hence $\Omega = \left(\bigcup_{k \in \mathbb{N}} A_k\right) \cup \left(\bigcap_{k \in \mathbb{N}} A_k^{\complement}\right)$ must be denumerable which is a contradiction to our assumption. Thus if all A_k are finite then $\bigcup_{k \in \mathbb{N}} A_k$ must be finite too.

On the other hand, if A_{k_0} is infinite then $A_{k_0}^{\complement}$ is finite and $\left(\bigcup_{k \in \mathbb{N}} A_k\right)^{\complement} = \bigcap_{k \in \mathbb{N}} A_k^{\complement} \subset$

$A_{k_0}^C$ is finte. Suppose $A_{k_1}, k_1 \neq k_0$, is infinite too. Then $A_{k_0}^C$ and $A_{k_1}^C$ are finite and so is $A_{k_0}^C \cup A_{k_n}^C$. Consequently $\left(A_{k_0}^C \cup A_{k_1}^C\right)^C$ is not finite, but $\left(A_{k_0}^C \cup A_{k_1}^C\right)^C = A_{k_0} \cap A_{k_1} = \emptyset$.

Thus only one of the sets $A_k, k \in \mathbb{N}$, can be not finite. For μ this means that:

$$\mu(\emptyset) = 0$$

since $\emptyset$ is finite. If all $A_k, k \in \mathbb{N}$, are finite then

$$0 = \mu\left(\bigcup_{k \in \mathbb{N}} A_k\right) = \sum_{k=1}^{\infty} \mu(A_k) = 0$$

and if A_{k_0} is infinite then

$$1 = \mu\left(\bigcup_{k \in \mathbb{N}} A_k\right) = \mu(A_{k_0}) = \mu(A_{k_0}) + \sum_{k \neq k_0} \mu(A_k) = \sum_{k=1}^{\infty} \mu(A_k).$$

3. Obviously we find for $\emptyset$ that

$$\mu(\emptyset) = \sum_{k=1}^{\infty} a_k \mu_k(\emptyset) = 0.$$

Now let $(A_l)_{l \in \mathbb{N}}$ be a sequence of mutually disjoint sets $A_l \in \mathcal{R}$ such that $\bigcup_{l \in \mathbb{N}} A_l \in \mathcal{R}$. Now we find

$$\mu\left(\bigcup_{l \in \mathbb{N}} A_l\right) = \sum_{k=1}^{\infty} a_k \mu_k\left(\bigcup_{l \in \mathbb{N}} A_l\right)$$

$$= \sum_{k=1}^{\infty} a_k \sum_{l=1}^{\infty} \mu_k(A_l) = \sum_{l=1}^{\infty} \sum_{k=1}^{\infty} a_k \mu_k(A_l)$$

$$= \sum_{l=1}^{\infty} \mu_k(A_l).$$

4. For $\emptyset$ we have obviously $\mu(\emptyset) = \sup_{k \in \mathbb{N}} \mu_k(\emptyset) = 0$. Now let $(A_l)_{l \in \mathbb{N}}, A_l \in \mathcal{R}$, be a sequence of mutually disjoint sets such that $\bigcup_{l \in \mathbb{N}} A_l \in \mathcal{R}$. Then we have

$$\mu\left(\bigcup_{l \in \mathbb{N}} A_l\right) = \sup_{k \in \mathbb{N}} \mu_k\left(\bigcup_{l \in \mathbb{N}} A_l\right)$$

$$= \sup_{k \in \mathbb{N}} \sum_{l=1}^{\infty} \mu_k(A_l) = \sup_{k \in \mathbb{N}} \sup_{N \in \mathbb{N}} \sum_{l=1}^{N} \mu_k(A_l)$$

$$= \sup_{N \in \mathbb{N}} \lim_{k \to \infty} \sum_{l=1}^{N} \mu_k(A_l) = \sup_{N \in \mathbb{N}} \sum_{l=1}^{N} \lim_{k \to \infty} \mu_k(A_l)$$

$$= \sup_{N \in \mathbb{N}} \sum_{l=1}^{N} \sup_{k \in \mathbb{N}} \mu_k(A_l) = \sup_{N \in \mathbb{N}} \sum_{l=1}^{N} \mu(A_l) = \sum_{l=1}^{\infty} \mu(A_l).$$

5. We start with $B_1 := A_1$ and by induction we define $B_k := A_k \setminus (A_1 \cup ... \cup A_{k-1})$. First we note that all sets B_k belong to $\mathcal{R}$. Further we have by construction that $B_k \cap B_l = \emptyset$ for $k \neq l$. Now we set $\tilde{A}_N := \bigcup_{j=n}^N B_j$. Clearly we have $\tilde{A}_N \subset \tilde{A}_{N+1}$ and $\bigcup_{n \in \mathbb{N}} \tilde{A}_N = \bigcup_{j \in \mathbb{N}} B_j = \bigcup_k A_k = \Omega$. Finally, since $B_k \cap B_l = \emptyset$ for $k \neq l$ we find

$$\mu(\tilde{A}_N) = \mu\left(\bigcup_{j=n}^N B_j\right) = \sum_{j=1}^N \mu(B_j) < \infty$$

since $\mu(B_j) \leq \mu(A_j)$.

6. Since $\Omega \in \mathcal{D}$ for all Dynkin systems $\mathcal{D}$ it follows that $\Omega \in \delta(\mathcal{E})$. Suppose that $A \in \delta(\mathcal{E})$. Then A belongs to all Dynkin system $\mathcal{D}$ containing $\mathcal{E}$, hence A^C belongs to all Dynkin systems containing $\mathcal{E}$, i.e. $A^C \in \delta(\mathcal{E})$. Finally let $(A_k)_{k \in \mathbb{N}}, A_k \in \delta(\mathcal{E})$, be a sequence of mutually disjointed sets. These sets also belong Dynkin systems $\mathcal{D}$ such that $\mathcal{E} \subset \mathcal{D}$ and hence their union $\bigcup_{k \in \mathbb{N}} A_k$ is an element of every Dynkin systems $\mathcal{D}$ with $\mathcal{E} \subset \mathcal{D}$. Thus $\bigcup_{k \in \mathbb{N}} A_k$ is in the intersection of all Dynkin systems, i.e. $\bigcup_{k \in \mathbb{N}} A_k \in \delta(\mathcal{E})$, proving that $\delta(\mathcal{E})$ is a Dynkin system.

7. Of course we have $g \subset H \subset \delta(\mathcal{H})$. In addition $\delta(g)$ is the minimal Dynkin system containing g and $\delta(\mathcal{H})$ is a Dynkin system containing g, thus $\delta(g) \subset \delta(\mathcal{H})$. We know that every σ-field is a Dynkin system and since $\delta(g)$ is the smallest Dynkin system containing g we must have $\delta(g) \subset \sigma(g)$.

8. We can use essentially the proof of Problem 5 in Chapter 1. An open hyper-rectangle is of the type $(a, b) = X_{j=1}^n (a_j, b_j)$ and (a, b) has rational vertices if $a_i, b_j \in \mathbb{Q}$ for all $j = 1, ..., n$. If $U \subset \mathbb{R}^n$ is open then for every $x \in U$ exists an open hyper-rectangle $R(x) \subset U$ since there exists an open ball $B_\rho^{(2)}(x) \subset U$ and $B_\rho^{(2)}(x)$ contains such an open hyper-rectangle. The density of $\mathbb{Q}$ in $\mathbb{R}$ allows us to now replace $R(x)$ by an open hyper-rectangle $Q(x) \subset U$ with rational vertices.
Clearly we have $\bigcup_{x \in U} Q(x) \subset U$ and $U \subset \bigcup_{x \in U} Q(x)$ is trivial.

Chapter 3

1.   a) The measure $\mu = \sum_{k=1}^n \epsilon_k$ counts the elements of a subset $A \subset \{1, ..., n\}$, i.e. $\mu(A) = \#(A)$. Since for all $\sigma \in S_n$ the sets A and $\sigma(A)$ have the same number of elements we have for all $\sigma \in S_n$ and all $A \subset \{1, ..., n\}$ that $\mu(\sigma(A)) = \#(\sigma(A)) = \#(A) = \mu(A)$.

   b) Let $A \subset \{1, ..., n\}$ and $A_1 \subset A$ be the subset of A which contains all even numbers belonging to A and $A_2 = A_1^C$ is the set containing all odd numbers belonging to A. We set $\#(A_1) = n_1$ and $\#(A_2) = n_2$ and we have $0 \leq n_j \leq n, n_1 + n_2 = n$. We now find

$$\nu(A) = \frac{n_1}{2} + \frac{n_2}{3}.$$

For $n = 7$ and $\sigma \in S_7$ with $\sigma(1) = 1, \sigma(2) = 3, \sigma(3) = 5, \sigma(4) = 7, \sigma(5) = 2, \sigma(6) = 4, \sigma(7) = 6$, and the set $A = \{1, 2, 3, 4\}$ it follows that $\sigma(A) = \{1, 3, 5, 7\}$ and further

$$\nu(A) = \frac{2}{2} + \frac{2}{3} = \frac{5}{3}$$

but

$$\nu(\sigma(A)) = \frac{0}{2} + \frac{4}{3} = \frac{4}{3},$$

i.e. $\mu(A) \neq \nu(\sigma(A))$ which means that in general ν is not invariant under S_n. However if σ maps even numbers onto even numbers and odd numbers onto odd numbers, then for $A = A_1 \cup A_2$ and $\sigma(A) = \sigma(A)_1 \cup \sigma(A)_2$, where $\sigma(A)_1$ contains the even numbers of $\sigma(A)$ whereas $\sigma(A)_2$ contains the odd numbers of $\sigma(A)$. Moreover we have $\#(A_1) = \#(\sigma(A)_1)$ and $\#(A_2) = \#(\sigma(A)_2)$ implying that $\nu(A) = \nu(\sigma(A))$.

2. By definition we have

$$T(\mu)(A) = \mu\{T^{-1}(A)\} = \mu\{k \in \mathbb{Z} \mid T(k) \in A\}.$$

Let $A \subset \mathbb{Z}, A = A_1 \cup A_2$ where $A_1 = \{l \in A \mid l \geq 0\}$ and $A_2 = \{l \in A \mid l < 0\}$. First we note that for $l < 0$ it follows that $T^{-1}(\{l\}) = \emptyset$ since $T(k) = k^2 \geq 0$. Thus we have $T^{-1}(A) = T^{-1}(A_1) = T^{-1}\left(\bigcup_{l \in A_1}\{l\}\right) = \bigcup_{l \in A_1} T^{-1}\{l\}$.

Moreover, for $l_1, l_2 \in A_1, l_1 \neq l_2$ it follows that $T\{l_1\} \cap T\{l_2\} = \emptyset$ and therefore

$$T(\mu)(A) = \mu(T^{-1}(A)) = \mu(T^{-1}(A_1)) = \sum_{l \in A_1} \mu(T^{-1}(\{l\})).$$

Furthermore we have

$$T^{-1}(l) = \begin{cases} \emptyset, & \text{if } l \neq k^2, k \in \mathbb{Z} \\ \{-k, k\}, & \text{if } l = k^2, k \in \mathbb{Z} \end{cases}$$

This now implies that

$$T(\mu) = \sum_{\substack{l \in A_n \\ l = k^2 \in \mathbb{Z}}} (\mathcal{E}_{-\sqrt{l}} + \mathcal{E}_{\sqrt{l}}).$$

Thus $T(\mu)(A)$ is twice the number of squares of integers belonging to A.

3. Since T is continuous it is Borel measurable. For $(x, y, z) \in B_1(0)$ let $(\xi, \eta, \zeta) = T(x, y, z) = (ax, by, cz)$, i.e. $\xi = ax, \eta = by, \zeta = cz$. Now it follows that $\left(\frac{\xi}{a}\right)^2 + \left(\frac{\eta}{b}\right)^2 + \left(\frac{\zeta}{c}\right)^2 = x^2 + y^2 + z^2 < 1$, and therefore

$$T(B_1(0)) = \left\{(\xi, \eta, \zeta) \in \mathbb{R}^3 \mid \left(\frac{\xi}{a}\right)^2 + \left(\frac{\eta}{b}\right)^2 + \left(\frac{\zeta}{c}\right)^2 < 1\right\}$$

is the ellipsoid with centre $0 \in \mathbb{R}^3$, symmetry axes being the coordinate axes and the semi-axes of length $a > b > c$. Since T is a linear transformation with determinant $\det T = abc > 0$, by Theorem 3.11 we have according to (3.6)

$$T(\lambda^{(3)}) = \frac{1}{abc}\lambda^{(3)}.$$

For $\mathcal{E} = \{(x, y, z) \in \mathbb{R}^3 \mid \frac{x^2}{9} + \frac{y^2}{16} + \frac{z^2}{25} < 1\}$ we find further with $T_{(3,4,5)}(x, y, z) = (3x, 4y, 5z)$ that $\mathcal{E} = T_{(3,4,5)}(B_1(0))$ and (3.5) in Theorem 3.11 now yields

$$\lambda^{(3)}(\mathcal{E}) = \| \det T_{(3,4,5)} \| \lambda^{(3)}(B_1(0)) = 60\lambda^{(3)}(B_1(0)) = 60.\frac{4}{3}\pi = 80\pi.$$

4.   a) The Bernoulli distribution β_p^N is a measure on $\mathcal{B}^{(1)}$, so we are seeking all Borel sets $A \subset \mathbb{R}$ with $\beta_p^N(A) = 0$ where

$$\beta_p^N(A) = \sum_{k=0}^{N} \binom{N}{k} p^k (1-p)^{N-k} \epsilon_k(A).$$

It follows that $\beta_P^N(A) = 0$ for all Borel sets A not containing a subset of $\{0, 1, ..., N\}$. Whereas if A contains a subset of $\{0, 1, ..., N\}$ then $\beta_P^N(A) \neq 0$.

Similarly, we find for the Poisson distribution $\pi_\alpha = \sum_{k=0}^{\infty} e^{-\alpha} \frac{\alpha^k}{k!} \epsilon_k$ that $\pi_\alpha(A) = 0$ for all Borel sets $A \subset (\mathbb{N}_0)^{\complement}$ where $\mathbb{N}_0 = \mathbb{N} \cup \{0\}$, but for a Borel set $A \subset \mathbb{R}$ such that $A \cap \mathbb{N}_0 \neq \emptyset$ it follows $\pi_\alpha(A) \neq 0$.

b) For example every constant mapping will fit the bill. Take $T_c : \mathbb{R}^n \to \mathbb{R}^n, T_c x = c \in \mathbb{R}^n$. Then for every non-empty set $A \subset \mathbb{R}^n$ we have $T_c(A) = \{c\}$ and $\lambda^{(1)}(\{c\}) = 0$.

Thus if we choose $A \in \mathcal{B}^{(n)}$ such that $\lambda^{(n)}(A) > 0$, every open set $A \subset \mathbb{R}^n$ will have this property, then we have constructed a measurable mapping T_c and a measurable set $A \subset \mathbb{R}^n$ such that $\lambda^{(n)}(T_c(A)) = 0$ but $\lambda^{(n)}(A) > 0$.

c) For every mapping $T : \mathbb{R} \to \mathbb{R}$ the set $T(\mathcal{Q})$ is denumerable, hence measurable and $\lambda^{(1)}(T(\mathcal{Q})) = 0$.

5. Consider $(\Omega_1, \mathcal{A}_1, \mu_1)$ as any measure space admitting a non-empty set of measure zero. Now take $(\Omega_2, \mathcal{A}_2, \mu_2)$ as the space with $\Omega_2 = \{a_1, a_2\}, a_1 \neq a_2, \mathcal{A}_2 = \{\emptyset, \{a_1\}, \{a_2\}.\{a_1, a_2\}\} = \mathcal{P}(\{a_1, a_2\})$ and define μ_2 on $\mathcal{A}_2$ by

$$\mu_2(\emptyset) = 0, \mu_2(a_1) = 0, \mu(a_2) = 1, \mu_2(a_1, a_2) = 1.$$

Clearly, μ_2 is a measure on $(\Omega_2, \mathcal{A}_2)$. Now we have as a mapping h the constant mapping $h : \Omega_1 \to \Omega_2, h(w) = a_2$ for all $\omega \in \Omega_n$. It follows for every $A \subset \Omega_1, A \neq \emptyset$, that $h(A) = \{a_2\}$ and $\mu_2(h(A)) = 1$. This holds in particular for every non-empty set of measure zero in $\mathcal{A}_1$.

6. We note that $\varphi(\mathbb{R}) = S^1$ and we recall that

$$\varphi(\lambda^{(1)})(A) = \lambda^{(1)}\{t \in \mathbb{R} \mid \varphi(t) \in A\}$$

and therefore we may assume $A \subset S^1, A$ measurable.

Let $U = U(r) = \begin{pmatrix} \cos r & -\sin r \\ \sin r & \cos r \end{pmatrix} \in SO(2)$. It follows that

$$\varphi(\lambda^{(1)})|_{S^1}(U(r)A) = \lambda^{(1)}\{t \in \mathbb{R} \mid \varphi(t) \in U(r)A\}$$

608

$$= \lambda^{(1)}\{t \in \mathbb{R} \mid \begin{pmatrix} \cos t \\ \sin t \end{pmatrix} \in U(r)A\}$$

$$= \lambda^{(1)}\{t \in \mathbb{R} \mid U^{-1}(r)\begin{pmatrix} \cos t \\ \sin t \end{pmatrix} \in A\}$$

$$= \lambda^{(1)}\{t \in \mathbb{R} \mid \begin{pmatrix} \cos r & -\sin r \\ \sin r & \cos r \end{pmatrix}\begin{pmatrix} \cos t \\ \sin t \end{pmatrix} \in A\}$$

$$= \lambda^{(1)}\{t \in \mathbb{R} \mid \begin{pmatrix} \cos(r+t) \\ \sin(r+t) \end{pmatrix} \in A\}$$

$$= \lambda^{(1)}\{s - r \in \mathbb{R} \mid \begin{pmatrix} \cos s \\ \sin s \end{pmatrix} \in A\}$$

$$= \lambda^{(1)}\{s \in r + \mathbb{R} \mid \begin{pmatrix} \cos s \\ \sin s \end{pmatrix} \in A\}$$

$$= \lambda^{(1)}\{s \in \mathbb{R} \mid \begin{pmatrix} \cos s \\ \sin s \end{pmatrix} \in A\}$$

$$= \varphi(\lambda^{(1)})|_{S^1}(A),$$

hence

$$\varphi(\lambda^{(1)})|_{S^1}(U(r)A) = \varphi(\lambda^{(1)})|_{S^1}(A),$$

i.e. we have proved the invariance of $\varphi(\lambda^{(1)})|_{S^1}$ under rotations.

7. First note that a Lebesgue null set in the sense of Definition II.19.6 need not belong to $\mathcal{B}^{(n)}$, i.e. need not be on Borel measurable set.

However, if $A \subset \mathbb{R}^n$ is a Borel set and a Lebesgue null set in the sense of Definition II.19.6, then for every $\epsilon > 0$ we can find open cells $K_j \subset \mathbb{R}^n, j \in \mathbb{N}$, such that $A \subset \bigcup_{j \in \mathbb{N}} K_j$ and $\sum_{j=1}^{\infty} \lambda^{(n)}(K_j) < \epsilon$. Since now $\lambda^{(n)}(A)$ is defined it follows that

$$\lambda^{(n)}(A) \le \sum_{j=1}^{\infty} \lambda^{(n)}(K_j) < \epsilon.$$

for every $\epsilon > 0$, implying that $\lambda^{(n)}(A) = 0$, i.e. A is indeed a set of measure zero.

8. Consider the set $A = \{x \in Y_1 \cap Y_2 \mid f(x) = g(x)\}$. It follows from our assumptions that $\mu(A^{\complement}) = 0$ and all subsets of $A^{\complement}$ belong to $\mathcal{A}$. Furthermore we find for any $a \in \mathbb{R}$ that

$$g^{-1}((a, \infty)) = (g^{-1}((a, \infty)) \cap A) \cup (g^{-1}((a, \infty)) \cap A^{\complement})$$

$$= (f^{-1}((a, \infty)) \cap A) \cup (g^{-1}((a, \infty)) \cap A^{\complement}).$$

Since f is measurable, $f^{-1}((a, \infty)) \cap A \in \mathcal{A}$, and since $g^{-1}((a, \infty)) \cap A^{\complement} \subset A^{\complement}$ it follows that $g^{-1}((a, \infty)) \cap A^{\complement} \subset \mathcal{A}$ too. Hence $g^{-1}((a, \infty)) \subset \mathcal{A}$ for all $a \in \mathbb{R}$ and Remark 1.12 yields the result.

9. Let $A \in l^{(n)}$, i.e. $A = B \cup N$ where $B \in \mathcal{B}^{(n)}$ and $N \subset K \in \mathcal{B}^{(n)}$ with $\lambda^{(n)}(K) = 0$. For $x \in \mathbb{R}^n$ it follows that $x + A = (x+B) \cup (x+N)$, $x+B \in \mathcal{B}^{(n)}$ and $x+N \subset x+K$, but $\lambda^{(n)}(x+N) = \lambda^{(n)}(K) = 0$. Thus we find

$$\lambda^{(n)}(x+A) = \lambda^{(n)}((x+B) \cup (x+N)) = \lambda^{(n)}(x+B = \lambda^{(n)}(B) = \lambda^{(n)}(A).$$

10. Let $N \in \mathcal{A}$ be locally of measure zero. Since $N = N \cap \Omega = N \cap \left(\bigcup_{k \in \mathbb{N}} B_k\right) = \bigcup_{k \in \mathbb{N}}(N \cap B_k)$ we have $\mu(A) \leq \sum_{k=1}^{\infty} \mu(N \cap B_k)$. By assumption $\mu(B_k) < \infty$ and hence $\mu(N \cap B_k) = 0$ for all $k \in \mathbb{N}$ implying that $\mu(A) = 0$.
    Note that in general a set which is locally of measure zero need not be a set of measure zero.

11. Since g is bijective it is strictly monotone and for two sets $A, B \subset \mathbb{R}$ we have $g(A \cap B) = g(A) \cap g(B)$ as well as $g(A \setminus B) = g(A) \setminus g(B)$. Now consider $\mathcal{A} := \{A \subset \mathbb{R} \mid g(A) \in \mathcal{B}^{(1)}\}$ and we claim that it is a σ-field. For $A \in \mathcal{A}$ it follows that $g(A^{\complement}) = g(\mathbb{R} \setminus A) = \mathbb{R} \setminus g(A)$, hence $A^{\complement} \in \mathcal{A}$, and $\mathbb{R} \in \mathcal{A}$ is trivial since $g(\mathbb{R}) = \mathbb{R}$. If $(A_k)_{k \in \mathbb{R}}$ is a sequence in $\mathcal{A}$ then $g\left(\bigcup_{k \in \mathbb{N}} A_k\right) = \bigcup_{k \in \mathbb{N}} g(A_k)$ and therefore $\bigcup_{k \in \mathbb{N}} A_k \in \mathcal{A}$. The continuity of g implies that compact sets belong to $\mathcal{A}$ which in turn yields that $\mathcal{B}^{(1)} \subset \mathcal{A}$, i.e. the image of every Borel set in $\mathbb{R}$ under g is a Borel set.

12.   a) We can apply Corollary 3.37 to obtain first

$$\mathcal{H}_\alpha(\gamma(\mathcal{I})) \leq (\sup_{t \in \mathcal{I}} ||\dot\gamma||)^\alpha H_\alpha(\mathcal{I}),$$

where we used that $\dot\gamma$ is bounded on $\mathcal{I}$ and that

$$||\gamma(t) - \gamma(s)|| = ||\int_s^t \dot\gamma(v)dv|| \leq \sup_{v \in [s,t]} ||\dot\gamma(v)||(t - s)$$

holds for all $t, s \in \mathcal{I}, s < t$. Now $\lim_H(\mathcal{I}) = 1$ implying by Lemma 3.30 that $\mathcal{H}_\alpha(\gamma(\mathcal{I})) = 0$ for $\alpha > 1$.

   b) Our argument goes along the same lines as in part a): $f|_{\overline{B_1(0)}}$ is Lipschitz continuous and we have the estimate

$$\mathcal{H}_\beta(f(B_1(0))) \leq \left(\sup_{x \in \overline{B_1(0)}} ||\operatorname{grad} f(x)||\right)^\beta \mathcal{H}_\beta(\overline{B_1(0)}),$$

but $\mathcal{H}_2(B_1(0)) = 2$, hence $\mathcal{H}_\beta(f(\overline{B_1(0)})) = 0$ for $\beta > 2$.
    By Lemma 3.36 the following must hold

$$\mathcal{H}_{\frac{\alpha}{s}}([0, 1] \times [0, 1]) \leq L^{\frac{\alpha}{s}} \mathcal{H}_\alpha([0, 1]).$$

For $\alpha = 1$ we obtain $\mathcal{H}_{\frac{\alpha}{s}}([0, 1] \times [0, 1]) < \infty$ implying that $\frac{\alpha}{s} = \frac{1}{s} \geq 2$, i.e. $s \leq \frac{1}{s}$.

13.   a) Since T is symmetric and positive definite we find that

$$\det(T * T) = (\det T)^2 = \Pi_{j=1}^n \kappa_j^2$$

where $\kappa_j, 1 \le j \le n$, denotes the jth eigenvalue of T (multiplication taken into account).

By assumption we have $\kappa_j > 0$ and $\det T = \Pi_{j=1}^n \kappa_j$. This now implies

$$g_T = \sqrt{\det(T * T)} = \left(\prod_{j=1}^n \kappa_j^2 \right)^{\frac{1}{2}} = \Pi_{j=1}^n \kappa_j = \det T.$$

b) First note that $d_x f$ is given by

$$d_x f = \begin{pmatrix} f_{1,u}(x) & f_{1,v}(x) \\ f_{2,u}(x) & f_{2,v}(x) \\ f_{3,u}(x) & f_{3,v}(x) \end{pmatrix},$$

where $f(x) = f(u,v) = \begin{pmatrix} f_1(u,v) \\ f_2(u,v) \\ f_3(u,v) \end{pmatrix} = \begin{pmatrix} f_1(x) \\ f_2(x) \\ f_3(x) \end{pmatrix}$ and $f_{j,u} = \frac{\partial f_j}{\partial u}$ as well as $f_{j,v} = \frac{\partial f_j}{\partial v}$.

This yields for $g_f^2(x)$:

$$g_f^2(x) = g_{d_x f} = (\det(d_x f)(d_x f))$$

$$= \det \begin{pmatrix} f_{1,u}(x) & f_{2,u}(x) & f_{3,u}(x) \\ f_{1,v}(x) & f_{2,v}(x) & f_{3,v}(x) \end{pmatrix} \begin{pmatrix} f_{1,u}(x) & f_{1,v}(x) \\ f_{2,u}(x) & f_{2,v}(x) \\ f_{3,u}(x) & f_{3,v}(x) \end{pmatrix}$$

$$= \det \begin{pmatrix} \langle f_u, f_u \rangle(x) & \langle f_u, f_v \rangle(x) \\ \langle f_v, f_u \rangle(x) & \langle f_v, f_v \rangle(x) \end{pmatrix} = E(x)G(x) - F^2(x),$$

i.e. we have

$$g_f(x) = \sqrt{E(x)G(x) - F^2(x)}.$$

Chapter 4

1. Recall

$$\chi_A(\omega) = \begin{cases} 1, \omega \in A \\ 0, \omega \notin A \end{cases}$$

i) Since $A \subset B$ means $\omega \in A$ implies $\omega \in B$ it follows for $\omega \in A$ that $\chi_A(\omega) = \chi_B(\omega) = 1$. For $\omega \in B \setminus A$ we have $\chi_A(\omega) = 0$ but $\chi_B(\omega) = 1$, and for $\omega \in B^{\complement}$ it follows that $\chi_A(\omega) = \chi_B(\omega) = 0$. Hence we have indeed $\chi_A \le \chi_B$.

ii) Note that

$$\chi_{A^{\complement}}(\omega) = \begin{cases} 1, \omega \in A^{\complement} \\ 0, \omega \notin A \end{cases}$$

611

$$= 1 - \begin{cases} 0, \omega \in A^{\complement} \\ 1, \omega \notin A \end{cases}$$

$$1 - \chi_A(\omega).$$

iii) For $\omega \in \bigcup_{k \in \mathbb{N}} A_k$ there exists $k_0 \in \mathbb{N}$ such that $\omega \in A_{k_0}$ and therefore $\sup_{k \in \mathbb{N}} \chi_{A_k}(x) = 1$. If $\omega \notin \bigcup_{k \in \mathbb{N}} A_k$ then $\omega \in \bigcap_{k \in \mathbb{N}} A_k^{\complement}$ and $\chi_{A_k}(\omega) = 0$ for all $k \in \mathbb{N}$, i.e. $\sup_{k \in \mathbb{N}} \chi_{A_k}(\omega) = 0$, and we have proved that

$$\chi_{\bigcup_{k \in \mathbb{N}} A_k} = \sup_{k \in \mathbb{N}} \chi_{A_k}.$$

iv) If $\omega \in \bigcap_{k \in \mathbb{N}} A_k$ then $\omega \in A_k$ for all $k \in \mathbb{N}$ and $\chi_{A_k}(\omega) = 1$ for all $k \in \mathbb{N}$, hence $\inf_{k \in \mathbb{N}} \chi_{A_k}(\omega) = 1$. In the case that $\omega \in \left(\bigcap_{k \in \mathbb{N}} A_k \right)^{\complement}$ we have $\omega \in A_{k_1}^{\complement}$ for some $k_1 \in \mathbb{N}$ and $\chi_{A_{k_1}} = 0$, hence $\inf_{k \in \mathbb{N}} \chi_{A_k}(\omega) = 0$.

b) From (4.15) it follows that χ_A is measurable if and only if A is measurable. Thus if $\sup \chi_{A_k}$ is measurable, by iii) in part a) it follows that $\bigcup_{k \in \mathbb{N}} A_k$ is measurable. Of course this does not imply that all sets A_k are measurable. Just take a decomposition of $\Omega = B_1 \cup B_2$ with B_1 not measurable, hence B_2 is not measurable. Define $A_1 := B_1$ and $A_k := B_2$ for $k > 2$. None of the sets $A_k, k \in \mathbb{N}$, is measurable but $\bigcup_{k \in \mathbb{N}} A_k = \Omega$ is.

2.   a) The product space (X,d) is given by

$$X = \prod_{j=1}^{M} X_j \text{ and } d_X = \sum_{j=1}^{M} d_j,$$

and we know that d_X is equivalent to any of the metrics

$$d_p := \left(\sum_{j=1}^{M} d_{X_j}^p \right)^{\frac{1}{p}}, 1 \le p \le \infty,$$

which follows from the inequalities $c_{p,q} ||x||_q \le ||x||_p \le C_{p,q} ||x||_q$, where $||x||_p = \left(\sum_{j=1}^{M} |x_j|^p \right)^{\frac{1}{p}}$. In addition we know that the projection $pr_j : X \to X_j$ is continuous, hence measurable. Now Theorem 1.20 yields the solution when noting that the open sets generate the σ-fields.

b) The key observation is the validity of the estimate

$$(*) \qquad c_p \sum_{j=1}^{N} |x_j| \le \left(\sum_{j=1}^{N} |x_j|^p \right)^{\frac{1}{p}} \le C_p \sum_{j=1}^{N} |x_j|$$

which holds for all $x = (x_1, \dots, x_N) \in \mathbb{R}^N$, compare with Problem 6 of Chapter II.1. If we substitute in $(*)$ x_j by $d_{X_j}(\xi_j, \eta_j)$, $\xi_j, \eta_j \in X_j$, the equivalence of the metrics $d_1 := \sum_{j=1}^{N} d_{X_j}$ and $d_p := \left(\sum_{j=1}^{N} d_{X_j}^p \right)^{\frac{1}{p}}$ on $X = \prod_{j=1}^{N} X_j$ follows immediately. The fact that d_p is a metric on X can be seen as follows:

612

(i) $d_p(\xi, \eta) = 0$ is equivalent to $d_{X_j}(\xi_j, \eta_j) = 0$ for all $j = 1, \ldots, N$;

(ii) $d_p(\xi, \eta) = d_p(\eta, \xi)$ since $d_{X_j}(\xi_j, \eta_j) = d_{X_j}(\eta_j, \xi_j)$ for all $j = 1, \ldots, N$;

(iii) The triangle inequality for the Euclidean metric yields

$$\left(\sum_{j=1}^{N} d_{X_j}^p(\xi_j, \eta_j) \right)^{\frac{1}{p}} \le \left(\sum_{j=1}^{N} \left(d_{X_j}(\xi_j, \zeta_j) + d_{X_j}(\zeta_j, \eta_j) \right)^p \right)^{\frac{1}{p}}$$

$$\le \left(\sum_{j=1}^{N} d_{X_j}^p(\xi_j, \zeta_j) \right)^{\frac{1}{p}} + \left(\sum_{j=1}^{N} d_{X_j}^p(\zeta_j, \eta_j)^p \right)^{\frac{1}{p}}.$$

The continuity of the projection $\mathrm{pr}_j = \mathrm{pr}_{X_j} : X \to X_j$, $\mathrm{pr}_j(\xi) = \xi_j$ follows from the fact that for every sequence $\left(\xi^{(k)} \right)_{k \in \mathbb{N}} = \left(\xi_1^{(k)}, \ldots, \xi_N^{(k)} \right)_{k \in \mathbb{N}}$ the convergence $\lim_{k \to \infty} \xi^{(k)} = \xi$ implies $\lim_{k \to \infty} \xi_j^{(k)} = \xi_j$.

3. The mapping $\mathrm{mult}_M : \mathbb{R}^M \to \mathbb{R}$ defined for $x = (x_1, \ldots, x_M) \in \mathbb{R}^M$ by

$$\mathrm{mult}_M(x) = x_1 \cdot \ldots \cdot x_M = pr_1(x) \cdot \ldots \cdot pr_M(x)$$

is continuous, hence measurable. Further, with $g_j, 1 \le j \le M$, also the mapping $G : \mathbb{R}^n \to \mathbb{R}^n, G(x) = \left(g_1(x), \ldots, g_m(x) \right)$ is measurable. Finally we note $g(x) = \mathrm{mult}_M G(x)$.

4.   a) If h is concave then $-h$ is convex and by Theorem II.13.27 it follows that $-h$, hence h, is continuous. This implies in turn the measurability of h.

b) By definition we have for every $x \in \mathbb{R}$ the existence of the limit

$$g'(x) = \lim_{h \to 0} \frac{g(x+h) - g(x)}{h}.$$

For $h = \frac{1}{n}$ we find

$$g'(x) = \lim_{h \to 0} n \left(g\left(x + \frac{1}{n}\right) - g(x) \right).$$

We define $g_n : \mathbb{R} \to \mathbb{R}$ by $g_n(x) = h(g(x + \frac{1}{n}) - g(x))$ which is continuous, hence measurable. Now we see that g' is the pointwise limit of the measurable functions g_n, thus g' is measurable.

5.   a) This result is an easy consequence of the fact that $|| \cdot || : \mathbb{R}^n \to \mathbb{R}$ is continuous, note $\bigl| ||x|| - ||y|| \bigr| \le ||x - y||$, and that the composition of measurable functions is measurable. Note that if we replace $(\mathbb{R}^n, || \cdot ||)$ by any normed space $(X, || \cdot ||)$ and take on X the Borel σ-field generated by the norm topology, then the result still holds.

b) Our problem is solved if we can prove that $\sigma(\mathrm{id}_n) : \mathbb{R}^n \to \mathbb{R}^n, (x_1, \ldots, x_n) \mapsto (x_{\sigma(1)}, \ldots, x_{\sigma(n)})$ is measurable. Let $A = A_1 \times \ldots \times A_n \in \mathcal{B}^{(n)}, A_j \in \mathcal{B}^{(1)}$. It follows that

$$\sigma(\mathrm{id}_n)^{-1}(A) = \bigcap_{j=1}^{n} pr_{\sigma(j)}^{-1}(A_j)$$

613

and since each set $pr_{\sigma(j)}^{-1}(A_j)$ is in $\mathcal{B}^{(n)}$ the result follows from Theorem 1.20.

6. This result is essentially a refinement of Corollary 4.19 by looking closer to its proof. With $f = f^+ - f^-$ and $(u_k)_{k\in\mathbb{N}}, (v_k)_{k\in\mathbb{N}} \in S(\Omega)$ converging to f^+ and f^-, respectively we define $\varphi_k := u_k - v_k$. By Theorem 4.18 we may assume that $u_k \leq u_{k+1}$ and $v_k \leq v_{k+1}$. It follows that $\lim_{k\to\infty} \varphi_k(w) = f(w)$ and further, using the fact that both sequence $(u_k)_{k\in\mathbb{N}}$ and $(v_k)_{k\in\mathbb{N}}$ are increasing we find with $\varphi_k^+ = u_k$ and $\varphi_k^- = v_k$ that

$$|\varphi_k(w)| = u_k(w) + v_k(w) \leq u_{k+1}(w) + v_{k+1}(w) = |\varphi_k(w)|.$$

Note further that $|\varphi_k(w)| \leq |f(w)|$ holds for $w \in \Omega$ and $k \in \mathbb{N}$ since $u_k \leq f^+$ and $v_k \leq f^-$.

7. The first equality is just notational, $\overline{\mathcal{B}}_{\mathbb{R}}^{(n)}$ and $\mathbb{R} \cap \overline{\mathcal{B}}^{(n)}$ both denote the trace of the σ-field $\overline{\mathcal{B}}^{(1)}$ in $\mathbb{R}$. By definition of the trace σ-field we have

$$\mathbb{R} \cap \overline{\mathcal{B}}^{(1)} = \{\mathbb{R} \cap A \mid A \in \overline{\mathcal{B}}^{(1)}\}.$$

But for $A \in \overline{\mathcal{B}}^{(1)}$ it follows that $\mathbb{R} \cap A \in \overline{\mathcal{B}}^{(1)}$, i.e. $\mathbb{R} \cap \overline{\mathcal{B}}^{(1)} \subset \overline{\mathcal{B}}^{(1)}$, whereas the inclusion $\overline{\mathcal{B}}^{(1)} \subset \mathbb{R} \cap \overline{\mathcal{B}}^{(1)}$ is trivial since $\mathbb{R} \in \overline{\mathcal{B}}^{(1)}$.

8. The functions u_k are continuous, hence measurable and of Baire class zero. The function u is obviously not continuous, hence it is not of Baire class zero. However we have $\lim_{k\to\infty} u_k(x) = u(x)$ for all $x \in \mathbb{R}$. Indeed, if $|x| < 1$ then $\lim_{k\to\infty} x^{2k} = 0$ and consequently $\lim_{k\to\infty} \frac{1}{1+x^{2k}} = 1$. For $|x| = 1$ we have $x^{2k} = 1$ and $\lim_{k\to\infty} \frac{1}{1+1} = \frac{1}{2}$, and finally, for $|x| > 0$ it follows that $\lim_{k\to\infty} x^{2k} = \infty$ implying that $\lim_{k\to\infty} \frac{1}{1+x^{2k}} = 0$.

Chapter 5

1. In $\Omega = \{1, ..., N\}$ we choose of course the power set $\mathcal{P}(\Omega)$ and σ-field and hence every subset of Ω is measurable. We can represent $A \subset \Omega$ as $A = \{a_1\} \cup ... \cup \{a_k\}, a_j \in \{1, ..., N\}, 1 \leq j \leq k \leq N, a_j \neq a_l$ for $j \neq l$. It follows that for every A we have $\chi_A = \sum_{j=1}^{k} \chi_{\{a_j\}}$ and therefore every simple function function $u \in S(\Omega)$ is of type $u = \sum_{k=1}^{N} \gamma_k \chi_{\{h\}}, \gamma_k \geq 0$. Consequently we have a bijective mapping $k : (\mathbb{R}_+ \cup \{0\})^N \to S(\Omega)$ defined by $\gamma_1, ..., \gamma_N) \mapsto \sum_{k=1}^{N} \gamma_k \chi_{\{k\}}$. Every measure μ on $\mathcal{P}(\Omega)$ is determined by its values $\mu(\{k\})$ and therefore we find $\int u d\mu = \sum_{k=1}^{N} \mu(\{k\})$.

2. Let $\{\alpha_{j_1}, ... \alpha_{j_M}\}$ be the maximal subset of $\{\alpha_1, ..., \alpha_N\}$ consisting of elements which are mutually distinct, i.e. $\alpha_{j_l} \neq \alpha_{j_m}$ for $l \neq m$. The value α_{j_k} is attained by u on the set $B_k = A_1^{(k)} \cup ... \cup A_{L(j_k)}^{(k)}$ for certain sets $A_l^{(k)}, 1 \leq l \leq L(j_k)$, and by assumption $A_l^{(k)} \cap A_1 A_m^{(k)} = \emptyset$ for $l \neq m$. Thus it follows that

$$u = \sum_{j=1}^{N} \alpha_j \chi_{A_j} = \sum_{k=1}^{M} \alpha_{j_k} \chi_{B_k}$$

614

and therefore

$$\sum_{j=1}^{N} \alpha_j \mu(A_j) = \sum_{j=1}^{M} \alpha_{j_k} \mu(B_k) = \int u d\mu.$$

3. The sets $A_k, k = 1, ..., N$, are Borel sets and we have $\lambda^{(2)}(B_{r_1}(0)) = \pi r_1^2, \lambda^{(2)}(A_k) = \pi r_k^2 - \pi r_{k-1}^2 = \pi(r_k^2 - r_{k-1}^2), k = 2, ..., N$. It follows that

$$\int u d\lambda^{(2)} = \sum_{k=1}^{N} \gamma_k \lambda^{(2)}(A_k) = \frac{1}{r_1} \pi r_1^2 + \sum_{k=2}^{N} \frac{1}{r_k - r_{k-1}} \lambda^{(2)}(A_k)$$

$$= \pi r_1 + \sum \pi \frac{1}{r_k - r_{k-1}}(r_k^2 - r_{k-1}^2) = 2\pi \sum_{k=1}^{N} r_k - r_N.$$

4. The following is an illustration of the situation:
Since $\left[\frac{l-1}{k}, \frac{l}{k}\right)$ is Borel measurable, $\left[\frac{l-1}{k}\right) \cap \left[\frac{m-1}{k}, \frac{m}{k}\right) = \emptyset$ for $l \neq m$ and $\bigcup_{l=1}^{k} \left[\frac{l-1}{k}, \frac{l}{k}\right) = [0,1)$, the function u_k belongs to $S([0,1))$ and is given in normal representation. The integral $\int u d\lambda^{(1)}$ is given by

$$\int u_k d\lambda^{(1)} = \sum_{l=1}^{k} \frac{l-1}{k} \lambda^{(1)}\left(\left[\frac{l-1}{k}, \frac{l}{k}\right)\right) = \sum_{l=1}^{k} \frac{l-1}{k^2}$$

$$\frac{1}{k^2} \sum_{l=1}^{k}(l-1) = \frac{k+1}{2k} - \frac{1}{k} = \frac{k-1}{2k}.$$

Since $k^2 \geq k^2 - 1 = (k+1)(k-1)$ we have $\frac{k}{k+1} \geq \frac{k-1}{k}$ and therefore we find that $\frac{(k+1)-1}{2(k+1)} \geq \frac{k-1}{2k}$, i.e. $\left(\frac{k-1}{2k}\right)_{k \in \mathbb{N}}$ is a monotone increasing sequence and we find

$$\sup_{k \in \mathbb{N}} \int u_k d\lambda^{(1)} = \sup_{k \in \mathbb{N}} \frac{k-1}{2k} = \lim_{k \to \infty} \frac{k-1}{2k} = \frac{1}{2}.$$

The sequence $(u_k)_{k \in \mathbb{N}}$ is not monotone increasing. On $\left[0, \frac{1}{k+1}\right)$ the functions u_k and u_{k+1} are equal (with value 0), on $\left[\frac{1}{k+1}, \frac{1}{k}\right)$ the function u_k still has the value 0, but u_{k+1} equals on $\left[\frac{1}{k+1}, \frac{1}{k}\right)$ to $\frac{1}{k+1}$, so $u_k \leq u_{k+1}$ on $\left[\frac{1}{k+1}, \frac{1}{k}\right]$. However on $\left[\frac{1}{k}, \frac{2}{k+1}\right]$ we find $u_k = \frac{1}{k}$ and $u_{k+1} = \frac{1}{k+1}$, i.e. $u_{k+1} \leq u_k$. Thus the standard approximation of the integral $\int_0^1 x dx$ by the Riemann sums of the function u_k will not lead to a monotone approximation needed for our new theory. Of course the functions $v_k := u_1 \vee ... \vee u_k$ are elements of $S(\Omega)$, the sequence $(v_k)_{k \in \mathbb{N}}$ is monotone increasing and $\sup_{k \in \mathbb{N}} \int v_k d\lambda^{(1)} = \int x \lambda^{(1)}(dx)$.

5. The key observation is that by Proposition 5.2 the statement (5.17) holds for simple functions. Now let $(u_k)_{k\in\mathbb{N}}$ and $(v_k)_{k\in\mathbb{N}}$ be sequences of simple functions increasing to f and g, respectively, i.e. $u_k \leq u_{k+1}$ and $f = \sup_{k\in\mathbb{N}} u_k$ as well as $v_k \leq v_{k+1}$ and $g = \sup v_k$. The sequence $(u_k + v_k)_{k\in\mathbb{N}}$ is increasing to $f + g$ and we have $\sup_{k\in\mathbb{N}}(u_k + v_k) = \sup_{k\in\mathbb{N}} u_k + \sup_{k\in\mathbb{N}} v_k = f + g$. It follows that

$$\sup_{k\in\mathbb{N}} \int (u_k + v_k)d\mu = \sup_{k\in\mathbb{N}} \left(\int u_k d\mu + \int v_k d\mu \right) = \sup_{k\in\mathbb{N}} \int u_k d\mu + \sup_{k\in\mathbb{N}} \int v_k d\mu$$

$$= \int \sup_{k\in\mathbb{N}} u_k d\mu + \int \sup_{k\in\mathbb{N}} v_k d\mu = \int f d\mu + \int g d\mu$$

as well as

$$\sup_{k\in\mathbb{N}} \int (u_k + v_k)d\mu = \int \sup_{k\in\mathbb{N}}(u_k + v_k)d\mu = \int (f + g)d\mu.$$

Thus $f + g$ is μ-integrable and

$$\int (f + g)d\mu = \int f d\mu + \int g d\mu.$$

6. Clearly we have $\nu(A) = \int_A \rho(x)\lambda^{(1)}(dx) \geq 0$ for all $A \in \mathcal{B}^{(1)}$. Furthermore we have $\nu(\emptyset) = 0$ since $\int_\emptyset \rho(x)\lambda^{(1)}(dx) = 0$, and for a mutually disjoint collection of Borel sets $(A_k)_{k\in\mathbb{N}}$ we find

$$\nu\left(\bigcup_{k\in\mathbb{N}} A_k \right) = \int_{\bigcup_{k\in\mathbb{N}} A_k} \rho(x)\lambda^{(1)}(dx) = \sum_{k=1}^{\infty} \int_{A_k} \rho(x)\lambda^{(1)}(dx) = \sum_{k=1}^{\infty} \nu(A_k).$$

Thus ν is a measure on $\mathcal{B}^{(1)}$. It is sufficient to prove that if $g \geq 0$ is measurable and bounded then g is ν-integrable. Let u_k be a simple function, then $u_k\rho \geq 0$ is $\mathcal{B}^{(1)}$-integrable. Moreover, if $(u_k)_{k\in\mathbb{N}}$ is a sequence of simple function increasing to g, then $(u_k)_{k\in\mathbb{N}}$ is a sequence of non-negative measurable function increasing to $g\rho$. By the monotone convergence theorem, Theorems 5.13, we find that

$$\sup_{k\in\mathbb{N}} \int u_k \rho d\lambda^{(1)} = \int \sup_{k\in\mathbb{N}} \left(u_k \rho \right) d\lambda^{(1)} = \int g\rho d\lambda^{(1)},$$

which we can re-interpret as

$$\sup_{k\in\mathbb{N}} \int u_k d\nu = \int \sup_{k\in\mathbb{N}} u_k d\nu = \int g d\nu,$$

i.e. g is ν-integrable.

7. We proceed as in Example 5.19.

i)

$$\int_{\mathbb{R}} g \, d\beta_N^P = \sum_{k=0}^{N} \binom{N}{k} p^k g^{N-k} \int g \, d\epsilon_k$$

$$= \sum_{k=0}^{N} \binom{N}{k} p^k q^{N-k} g(k) = \sum_{k=0}^{N} \binom{N}{k} p^k q^{N-k} k^2$$

$$= Np \sum_{k=1}^{N} k \binom{N-1}{k-1} p^{k-1} q^{N-k} = Np(Np+q).$$

ii)

$$\int g \, d\pi_\gamma = \sum_{k=0}^{\infty} e^{-\gamma} \frac{\gamma^t}{k!} \int g \, d\epsilon_k$$

$$= \sum_{k=0}^{\infty} e^{-\gamma} \frac{\gamma^t}{k!} k^2 = \sum_{k=1}^{\infty} e^{-\gamma} \frac{\gamma}{(k-1)!} (k-1+1)$$

$$= \gamma^2 + \gamma.$$

Note that if we consider β_N^p and π_γ as distributions of random variables X and Y respectively, then the **variance** of X is given by

$$V(X) := E((X - E(X))^2) = E(X^2) - E(X)^2 = N_{pq}$$

and the variance of Y is

$$V(Y) = E(Y^2) - E(Y)^2 = \gamma.$$

8. Since $\lim_{x \to 0} \frac{\sin x}{x} = 1$ the function h is continuous on $\mathbb{R}$ and from $\left| \frac{\sin x}{x} \right| \leq \frac{1}{|x|}$ we deduce that h vanishes at infinity. The function h is even, hence it is integrable over $\mathbb{R}$ if and only if it is integrable over $[0, \infty)$. Now we can apply some ideas of the solution to Problem 8 of Chapter 1.28. The function h is integrable over $[0, \infty)$ if and only if $|h|$ is integrable over $[0, \infty)$, see Theorem 5.22. Next we note that

$$\int_0^\infty \left| \frac{\sin x}{x} \right| \lambda^{(1)}(dx) = \sum_{k=0}^{\infty} \int_{k\pi}^{(k+1)\pi} \left| \frac{\sin x}{x} \right| \lambda^{(1)}(dx).$$

With $T_k : [0, \pi] \to [k\pi, (k+1)\pi], T_k x = x + k\pi$, we find using Theorem 5.31 and the translation invariance of $\lambda^{(1)}$ that

$$\int_{k\pi}^{(k+1)\pi} \left| \frac{\sin x}{x} \right| T_k(\lambda^{(1)})(dx)$$

$$= \int_0^\pi \left| \frac{\sin(x+k\pi)}{x+k\pi} \right| \lambda^{(1)}(dx) = \int_0^\pi \left| \frac{\sin x}{x+k\pi} \right| \lambda^{(1)}(dx).$$

Since

$$\int_0^\pi \left| \frac{\sin x}{x+k\pi} \right| dx \geq \frac{1}{(k+1)\pi} \int_0^\pi \sin x \, dx$$

and

$$\int_0^\pi \sin x \, dx \geq \int_{\frac{\pi}{6}}^{\frac{5\pi}{6}} \sin x \, dx \geq \frac{1}{2} \cdot \frac{4\pi}{6} = \frac{\pi}{3}$$

617

we arrive at

$$\sum_{k=0}^{\infty} \int_{k\pi}^{(k+1)\pi} \left| \frac{\sin x}{x} \right| \lambda^{(1)}(dx) \geq \frac{\pi}{3} \sum_{k=0}^{\infty} \frac{1}{(k+1)\pi} \geq \frac{1}{3} \sum_{k=1}^{\infty} \frac{1}{k} = \infty,$$

i.e. the function h is not integrable over $\mathbb{R}$.

9. We use Corollary 5.32 and we note that $\lambda^{(n)}$ in the formula we want to prove stands for the restriction of the Lebesgue-Borel measure $\lambda^{(n)}$ onto the appropriate sets, i.e. $W_1^{(n)}$ and $T(W_1^{(n)})$, respectively. With $T \in GL(n, \mathbb{R})$ we consider $T^{-1} : T(W_1^{(n)}) \to W_1^{(n)}$ where we also consider on $T(W_1^{(n)})$ the measure $T(\lambda^{(n)})$. Applying (5.47) to $g : W_1^{(n)} \to \mathbb{R}$ we get

$$\int_{W_1^{(n)}} g \, d\lambda^{(n)} = \int_{W_1^{(n)}} g \, d(T^{-1} \circ T(\lambda^{(n)})) = \int_{T(W_1^{(n)})} (g \circ T^{(-1)}) d(T(\lambda^{(n)})).$$

By Theorem 3.11 we have that $T(\lambda^{(n)}) = \frac{1}{|\det T|} \lambda^{(n)}$, and hence we arrive at

$$\int_{W_1^{(n)}} g d\lambda^{(n)} = \frac{1}{|\det T|} \int_{T(W_1^{(n)})} (g \circ T^{-1}) d\lambda^{(n)}.$$

10. The mapping $U(\varphi)$ is a rotation around the origin by the angle φ. For $\varphi \in \{0, \frac{\pi}{2}, \pi, \frac{3\pi}{2}\}$ such a rotation leaves every square with centre $0 \in \mathbb{R}^2$ and sides parallel to the coordinate axes invariant, i.e. $U(\varphi)W_1^{(2)} = W_1^{(2)}$. Moreover, since $U(\varphi) \in SO(2)$ we have

$$U(\varphi)(\lambda^{(2)} = \lambda^{(2)}.$$

With (5.49) we find now for a continuous, hence integrable function $f : W_1^{(2)} \to \mathbb{R}$ that

$$\int_{W_1^{(2)}} f d\lambda^{(2)} = \int_{W_1^{(2)}} f dU(\varphi)(\lambda^{(2)}) = \int_{W_1^{(2)}} (f \circ U(\varphi)) d\lambda^{(2)}.$$

Chapter 6

1.   a) We define $\Phi : C(Y) \to C(X)$ by $\Phi(u) = u \circ \varphi$. For $u \in C(Y)$ it follows that $\Phi(u) \in C(X)$. Since φ is a homomorphism it is injective implying that $u \neq v$ yields $u \circ \varphi \neq v \circ \varphi$, and for $g \in C(X)$ the function $g \circ \varphi^{-1}$ belongs to $C(Y)$ and it follows that $\Phi(g \circ \varphi^{-1}) = g \circ \varphi \circ \varphi^{-1} = g$, i.e. Φ is also surjective, hence Φ is bijective. For $u, v \in C(Y)$ and $\alpha, \beta \in \mathbb{R}$ we find
$(\star)$

$$\Phi(\alpha u + \beta v) = (\alpha u + \beta v) \circ \varphi = (\alpha u) \circ \varphi + (\beta v) \circ \varphi$$

$$= \alpha u \circ \varphi + \beta v \circ \varphi = \alpha \Phi(u) + \beta \Phi(v),$$

i.e. Φ is linear. Furthermore

$$\varphi(u.v) = (u.v) \circ \varphi = (u \circ \varphi) \cdot (v \circ \varphi) = \Phi(u) \cdot \Phi(v),$$

The inverse to Φ is of course $\Phi^{-1} : C(X) \to C(Y), \Phi^{-1}(g) = g \circ \varphi^{-1}$, which has the same properties as Φ.

b) First we note that with $u \in C^k(H)$ the function $\Psi(u) = u \circ \psi$ is an element in $C^k(G)$ and with the same arguments as in part a) we see that $\Psi^{-1} : C^k(G) \to C^k(H), \Psi^{-1}(g) = g \circ \psi^{-1}$ is inverse to Ψ. The calculation $(\star)$ imply now that Ψ is indeed a vector space isomorphism.

2. We have seen already at several occasions that there exists constants $c_{a,b} > 0$ and $C_{a,b}$ such that for all $x \in \mathbb{R}^n$ we have $c_{a,b}||x||_b \leq ||x||_a \leq C_{a,b}||x||_b, 1 \leq a, b < \infty$, for example compare with Problem 6 of Chapter 1 in Volume II. Thus given p, q, r we find constants $C_{p,r}$ and $C_{p,q}$ such that

$$||g(x) - g(y)||_r \leq C_{p,r}||g(x) - g(y)||_p$$

$$\leq C_{p,r}\gamma_{p,p}||x - y||_p \leq C_{p,r}\gamma_{p,p}C_{p,q}||x - y||_q.$$

3. a) The function $f : \mathbb{R}^n \to \mathbb{R}$ is integrable with respect to ν if $\int_{\mathbb{R}^n} |f|d\nu$ exists, i.e. is finite. Since $\nu = \chi_G \lambda^{(n)}$ it follows that

$$\int_{\mathbb{R}^n} |f|d\nu = \int_{\mathbb{R}} |f|\chi_G d\lambda^{(n)} = \int_G |f|d\lambda^{(n)}.$$

Thus f is integrable over $\mathbb{R}^n$ with respect to $\nu = \chi_G \lambda^{(n)}$ if and only if $f|_G$ is integrable over G with respect to $\lambda^{(n)}_{|G}$.

b) We need to assure that $\int_{\mathbb{R}} |u(x)|g_s(x)\lambda^{(1)}(dx)$ is finite. From our assumption we deduce

$$|u(x)|g_s(x) \leq c_u(1 + |x|^2)^{\frac{r}{2}}(1 + |x|^2)^{\frac{-s}{2}}.$$

Thus we need that

$$\int_{\mathbb{R}} (1 + |x|^2)^{\frac{r-s}{2}} \lambda^{(1)}(dx) < \infty.$$

We look at the decomposition.

$$\int_{\mathbb{R}} (1 + |x|^2)^{\frac{r-s}{2}} \lambda^{(1)}(dx) = \int_{|x| \leq 1} (1 + |x|^2)^{\frac{r-s}{2}} \lambda^{(1)}(dx) + \int_{|x| > 1} (1 + |x|^2)^{\frac{r-s}{2}} \lambda^{(1)}(dx).$$

The first integrand is always finite since the integral is a bounded continuous function on a set of finite Borel-Lebesgue measure. Now we handle the second integral. We note that $x \mapsto (1 + |x|^2)^{\frac{r-s}{2}}$ is an even function, hence we only need to investigate $\int_1^\infty (1 + |x|^2)^{\frac{r-s}{2}} \lambda^{(1)}(dx)$. If $r - s \geq 0$ then $\int_1^\infty (1 + |x|^2)^{\frac{r-s}{2}} \lambda^{(1)}(dx)$ is not finite since in this case $1 \leq (1 + |x|^2)^{\frac{r-s}{2}}$ and the integral $\int_1^\infty 1 \lambda^{(1)}(dx)$ is not finite. So let $r - s < 0$. For $|x| \geq 1$ we now have

$$\frac{1}{2^{s-r}|x|^{s-r}} \leq \frac{1}{(1 + |x|^2)^{\frac{s-r}{2}}} \leq \frac{1}{|x|^{s-r}}$$

619

and therefore

$$\frac{1}{2^{s-r}}\int_1^\infty \frac{1}{x^{s-r}}\lambda^{(1)}(dx) \le \int_1^\infty (1+x^2)^{\frac{r-s}{2}}\lambda^{(1)}(dx) \le \int_1^\infty \frac{1}{x^{s-r}}\lambda^{(1)}(dx).$$

Further we have

$$\int_1^\infty \frac{1}{x^{s-r}}\lambda^{(1)}(dx) = \sum_{k=1}^\infty \int_k^{k+1}\frac{1}{x^{s-r}}\lambda^{(1)}(dx)$$

and this yields

$$\sum_{k=2}^\infty \frac{1}{k^{s-r}} = \sum_{k=1}^\infty \frac{1}{(k+1)^{s-r}} \le \sum_{k=1}^\infty \int_k^{k+1}\frac{1}{x^{s-r}}\lambda^{(1)}(dx) \le \sum_{k=1}^\infty \frac{1}{k^{s-r}},$$

which implies that for $s - r < 1$ the integral $\int_1^\infty \frac{1}{x^{s-r}}\lambda^{(1)}(dx)$ diverges whereas for $s - r > 1$ this integral converges. Thus we conclude that u is ν-integrable over $\mathbb{R}$ if $r < s - 1$.

4. a) With $\mu_j, j = 1, ..., N$, a further measure on $(\Omega, \mathcal{A})$ is given by $\mu := \sum_{j=1}^N \mu_j$ and for every $A \in \mathcal{A}$ we have

$$\mu_j(A) \le \mu(A),$$

i.e. $\mu(A) = 0$ implies $\mu_j(A) = 0$ for $j = 1, ..., N$, and hence μ_j is absolutely continuous with respect to μ.

b) Let $A \in \mathcal{A}$ and $\mu(A) = 0$. It follows that $\nu(A) = 0$ since ν is absolutely continuous with respect to μ. But now the absolute continuity of π with respect to ν implies that $\pi(A) = 0$, i.e. $\mu(A) = 0$ implies $\pi(A) = 0$, i.e. π is absolutely continuous with respect to μ.

5. From $\nu_1 = g\lambda^{(n)}$ and $0 < c_0 \le \frac{g(x)}{h(x)} \le c_1$ we conclude for $A \in \mathcal{B}^{(n)}$ that

$$\nu_1(A) = \int_A g(x)\lambda^{(n)}(dx) \le c_1 \int_A h(x)\lambda^{(n)}(dx) = c_1\nu_2(A)$$

and if $\nu_2(A) = 0$ then $\nu_1(A) = 0$, i.e. ν_2 is absolutely continuous with respect to ν_1. Furthermore we have

$$\nu_2(A) = \int_A h(x)\lambda^{(n)}(dx) \le \int_A \frac{1}{c_0}g(x)\lambda^{(n)}(dx) = \frac{1}{c_0}\int_A g(x)\lambda^{(n)}(dx) = \frac{1}{c_0}\gamma_1(A).$$

i.e. ν_2 is absolutely continuous with respect to ν_1.

6. For $A \in \mathcal{A}$ the following holds

$$|\nu_k(A) - \nu(A)| = |\int_A g_k d\mu - \int_A g d\mu|$$

$$= |\int_A (g_k - g)d\mu| \le \int_A |g_k - g|d\mu$$

$$\leq \int_\Omega \|g_k - g\|_\infty d\mu \leq \|g_k - g\|_\infty \int_\Omega 1 d\mu$$

$$= \mu(\Omega)\|g_k - g\|_\infty.$$

Thus $\lim_{k\to\infty} \|g_k - g\|_\infty = 0$ implies $\lim_{k\to\infty} |\nu_k(A) - \nu(A)| = 0$ or $\lim_{k\to\infty} \nu_k(A) = \nu(A)$ for all $A \in \mathcal{A}$.

7.   a) The set $\{x_0\}$ has Borel-Lebesgue measure zero, i.e. $\lambda^{(1)}(\{x_0\}) = 0$ but $\epsilon_{x_0}(\{x_0\}) = 1$, so ϵ_{x_0} is not absolutely continuous with respect to $\lambda^{(1)}$. Conversely, every open interval $(a, b) \subset \mathbb{R}$ has Borel-Lebesgue measure $\lambda((a, b)) = b - a \neq 0$, but for $x_0 \notin (a, b)$ it follows that $\epsilon_{x_0}((a, b)) = 0$, and therefore $\lambda^{(1)}$ is not absolutely continuous with respect to ϵ_0.

b) In order that ν is absolutely continuous with respect to μ it must follow for $A \in \mathcal{B}^{(1)}$ and $\mu(A) = 0$ that $\nu(A) = 0$.
Let $A \in \mathcal{B}^{(1)}$. Only points in $A \cap \mathbb{N}$ will contribute to $\nu(A)$ and $\mu(A)$, i.e.

$$\mu(A) = \sum_{k\in\mathbb{N}} a_k \mu(A) = \sum_{k\in A\cap\mathbb{N}} a_k$$

and

$$\mu(A) = \sum_{k\in\mathbb{N}} b_k \nu(A) = \sum_{k\in A\cap\mathbb{N}} b_k.$$

If $\mu(A) = 0$ then $a_k = 0$ for all $k \in A\cap\mathbb{N}$ and in order that $\nu(A) = 0$ holds we must have $b_k = 0$ for all $k \in A\cap\mathbb{N}$. This implies that ν is absolutely continuous with respect to μ if and only if $b_k = 0$ for all $a_k = 0$.

8. This problem is essentially solved with the calculation made with the solution to Problem 3.b). The expectation of X is given by

$$E(X) = \frac{\alpha}{\pi} \int_\mathbb{R} \frac{x}{\alpha^2 + x^2} \lambda^{(1)}(dx), \alpha > 0.$$

Now, this integral exists if and only if $\int_\mathbb{R} \frac{|x|}{\alpha^2+x^2}\lambda^{(1)}(dx)$ exists, i.e. is finite. This requires $\int_0^\alpha \frac{x}{\alpha^2+x^2}\lambda^{(1)}(dx)$ and $\int_\alpha^\infty \frac{x}{\alpha^2+x^2}\lambda^{(1)}(dx)$ to be finite. However we have

$$\frac{1}{2}\int_\alpha^\infty \frac{1}{x}\lambda^{(1)}(dx) \leq \int_\alpha^\infty \frac{x}{\alpha^2+x^2}\lambda^{(1)}(dx)$$

and we know that the integral on the left hand side does not exist (as a finite integral), compare the solution to Problem 3.b).

Chapter 7

1.   a) We determine the sets of μ-measure zero. Let $A \in \mathcal{P}(\Omega)$, i.e. $A \subset \Omega$, and suppose that $\mu(A) = 0$. This means that $\frac{1}{N}\sum_{k=1}^N \epsilon_k(A) = 0$, or for all $k \in \{1, ..., N\}$ it must hold $k \notin A$, i.e. $A = \emptyset$. Thus only the empty set has μ-measure zero implying that every μ-a.e. statement is in fact an everywhere statement.

b) First we note that for every consideration of convergence the first finitely many terms of a sequence do not count. Thus we only have to look at $(v_j)_{j\geq 5}$. Since $\bigcap_{j=5}^{\infty}[\frac{1}{4} - \frac{1}{j} + \frac{1}{2}] = [\frac{1}{4}, \frac{1}{2}]$ it follows that for all $x \in \mathbb{R}$ we have $\lim_{j\to\infty} v_j(x) = \chi_{[\frac{1}{4},\frac{1}{2}]}(x)$, i.e the pointwise convergence of $(v_j)_{j\in\mathbb{N}}$ to $\chi_{[\frac{1}{4},\frac{1}{2}]}$. From Remark 7.3.B we further deduce that $(v_j)_{j\in\mathbb{N}}$ converges ν-almost everywhere to $\chi_{[\frac{1}{4},\frac{1}{2}]}(x)$. For $\chi \in [\frac{1}{4}, \frac{1}{2}]^{\complement} = \{X < \frac{1}{4}\} \cup \{X > \frac{1}{2}\}$ the function $\chi_{[\frac{1}{4},\frac{1}{2}]}(x)$ is identically zero. However with respect $\nu = \frac{1}{N} \sum_{k=1}^{N} \epsilon_k$ the set $[\frac{1}{4}, \frac{1}{2}]$ is a set of ν-measure zero since $[\frac{1}{4}, \frac{1}{2}] \cap \mathbb{N} = \emptyset$. Thus the functions $\chi_{[\frac{1}{4},\frac{1}{2}]}$ and $\widetilde{v} = 0$ only differ on a set of ν-measure zero and therefore $(v_j)_{j\in\mathbb{N}}$ converges also ν-almost everywhere to $\widetilde{v} = 0$. Again, from the pointwise convergence of $(v_j)_{j\in\mathbb{N}}$ to $\chi_{[\frac{1}{4},\frac{1}{2}]}$ it follows that $(v_j)_{j\in\mathbb{N}}$ converges $\lambda^{(1)}$-almost everywhere to $\chi_{[\frac{1}{4},\frac{1}{2}]}$. However the set $[\frac{1}{4}, \frac{1}{2}]$ has measure $\lambda^{(1)}([\frac{1}{4}, \frac{1}{2}]) = \frac{1}{4} > 0$.

Therefore with respect to $\lambda^{(1)}$ the two functions $\chi_{[\frac{1}{4},\frac{1}{2}]}$ and $\widetilde{v}$ are not almost everywhere equal and convergence $\widetilde{v}$ cannot be an $\lambda^{(1)}$-almost everywhere limit of $(v_j)_{j\in\mathbb{N}}$.

2. From the assumption

$$\lim_{l\to\infty} N_p(u_l - u) = \lim_{l\to\infty} \left(\int_{\mathbb{R}} |u_l(x) - u(x)|^p \sum_{k=1}^{\infty} \epsilon_k(dx) \right)^{\frac{1}{p}} = 0$$

we deduce

$$\int_{\mathbb{R}} |u_l(x)-u(x)|^p \sum_{k=1}^{\infty} \epsilon_k(dx) = \sum_{k=1}^{\infty} \int_{\mathbb{R}} |u_l(x)-u(x)|^p \epsilon_k(dx) = \sum_{k=1}^{\infty} |u_l(k)-u(k)|^p \to 0.$$

Since

$$|u_l(x) - u(x)| \leq \left(\sum_{k=1}^{\infty} |u_l(k) - u(k)|^p \right)^{\frac{1}{p}} \to 0$$

the result follows.

3. Since $(u_j)_{j\in\mathbb{N}}$ converges $\lambda^{(1)}$-almost everywhere to u there exists a set $\mathcal{N} \in \mathcal{B}^{(1)}$ of $\lambda^{(1)}$-measure zero, i.e. $\lambda^{(1)}(\mathcal{N}) = 0$, such that for $x \in \mathcal{N}^{\complement}$ we have $\lim_{k\to\infty} u_k(x) = u(x)$. For $x \in \mathcal{N}^{\complement}$ it follows

$$|f(u_k(x)) - f(u(x))| \leq L|u_k(x) - u(x)| \to 0,$$

i.e. $\left(f(u_k)\right)_{k\in\mathbb{N}}$ converges in the complement of a $\lambda^{(1)}$-measure set of zero to $f(u)$, i.e. it converges $\lambda^{(1)}$-almost everywhere to $f(u)$.

4. a) We first use the Chebychev-Markov inequaltiy for f and the measure ν :

$$(\star) \quad \nu(\{|f| \geq \alpha\}) \leq \frac{1}{\alpha} \int |f| d\nu = \frac{1}{\alpha} \int |f| g\, d\lambda^{(1)}.$$

Since $\delta \in L^p(\mathbb{R}^n)$ and $g \in L^q(\mathbb{R}^n)$, $\frac{1}{p} + \frac{1}{q} = 1$, we can apply the Cauchy-Schwarz inequality to obtain

$$\nu(\{|f| \geq \alpha\}) \leq \frac{1}{\alpha} \|g\|_{L^q} \|f\|_{L^q}.$$

b) Replacing in $(\star)$ f by $f_k - h$ we find for every $\alpha > 0$

$$\nu(\{|f_k - h| \geq \alpha\}) \leq \frac{1}{\alpha} \int |f_k - h| g \, d\lambda^{(n)}$$

implying

$$\lim_{k \to \infty} \nu(\{|f_k - h| \geq \alpha\}) = 0$$

for every $\alpha > 0$, i.e. $(f_k)_{k \in \mathbb{N}}$ converges in ν-measure to h.

5. Let $g_n = f_j^{(k)}$ be given and $\alpha > 0$. It follows that

$$\{x \in [0,1) \mid |g_n| \geq \alpha\} = \left[\frac{j-1}{k}, \frac{j}{k}\right)$$

and therefore we have $\lambda^{(1)}(\{x \in [0,1) \mid |g_n| \geq \alpha\}) = \frac{1}{k}$ which implies $\lambda^{(1)}(\{x \in [0,1) \mid |g_n| \geq \alpha\}) \to 0$ as $n \to \infty$, i.e. $(g_n)_{n \in \mathbb{N}}$ converges in $\lambda_{[0,1)}^{(1)}$-measure to zero. However for any $x \in [0,1)$ we have $\lim_{k \to \infty} g_k(x) = 0$. Given $x_0 \in [0,1)$ we can find for every k some j such that $x_0 \in \left[\frac{j-1}{k}, \frac{j}{k}\right)$, i.e. $g_n(x_0) = f_j^{(k)}(x_0) = 1$. Thus for every $x_0 \in [0,1)$ the sequence $(g_n(x_0))_{n \in \mathbb{N}}$ contains a subsequence $(g_{n_l}(x_0))_{l \in \mathbb{N}}$ with $g_{k_l}(x_0) = 1$.

6. Since $\left(\frac{1}{1+|x|}\right)^2 = \frac{1}{1+2|x|+|x|^2} \leq \frac{1}{1+|x|^2}$ we find

$$N_2\left(\frac{1}{1+|\cdot|}\right) = \int_{\mathbb{R}} \left(\frac{1}{1+|x|}\right)^2 \lambda^{(1)}(dx) \leq \int_{|x| \leq 1} \frac{1}{1+|x|^2} \lambda^{(1)}(dx) + \int_{|x| \geq 1} \frac{1}{|x|^2} \lambda^{(1)}(dx)$$

$$= 2 \int_0^1 \frac{1}{1+|x|^2} \lambda^{(1)}(dx) + 2 \sum_{k=1}^{\infty} \int_k^{k+1} \frac{1}{|x|^2} \lambda^{(1)}(dx)$$

$$\leq 2 + 2 \sum_{k=1}^{\infty} \frac{1}{k^2} < \infty$$

implying that $g \in \mathbb{B}^2(\mathbb{R})$. On the other hand we have

$$\int_{\mathbb{R}} \frac{1}{1+|x|} \lambda^{(1)}(dx) = 2 \sum_{k=0}^{\infty} \int_k^{k+1} \frac{1}{1+|x|} \lambda^{(1)}(dx) \geq 2 \sum_{k=0}^{\infty} \frac{1}{k+1} = 2 \sum_{k=1}^{\infty} \frac{1}{k} = \infty,$$

i.e. $g \notin \mathbb{B}^1(\mathbb{R})$. Consequently the estimate

$$N_1(g) \leq C N_2(g)$$

cannot hold and indeed more formally $\lambda^{(1)}(\mathbb{R}) = \infty$ and with $q = 1, p = 2$ we have $\frac{p-q}{pq} = \frac{1}{2}$ and the constant in 7.8 would be $(\lambda^{(1)}(\mathbb{R}))^{\frac{1}{2}} = \infty$.

7.   a) From the Cauchy-Schwarz inequality we first deduce that

$$|\langle u, v \rangle| = |\int uvd\mu| \le \left(\int |u|^2 d\mu \right)^{\frac{1}{2}} \left(\int |v|^2 d\mu \right)^{\frac{1}{2}}$$

implying that $\langle \cdot, \cdot \rangle$ is defined on $L^2(\Omega)$. The bijectivity of $(u, v) \mapsto \langle u, v \rangle$ is now trivial by the properties of the integral. Moreover we have $\langle u, u \rangle = \int |u|^2 d\mu \ge 0$. Now suppose that $\langle u, u \rangle = \int |u|^2 d\mu = 0$. This implies that $|u|^2 = 0$ μ-a.e. from which we deduce that $u = 0$ μ-a.e.

b) Clearly, $p_1(u) \ge 0$ and for $\alpha \in \mathbb{R}$ we find

$$p_1(\alpha u) = \sup_{x \in [0,1]} |\alpha u'(x)| = |\alpha| \sup_{x \in [0,1]} |u'(x)| = |\alpha| p_1(u).$$

Moreover, for $u, v \in C_b^1([0, 1])$ we have

$$p_1(u + v) = \sup_{x \in [0,1]} |u'(x) + v'(x)| \le \sup_{x \in [0,1]} (|u'(x)| + |v'(x)|)$$

$$\le \sup_{x \in [0,1]} |u'(x)| + \sup_{x \in [0,1]} |v'(x)| = p_1(u) + p_1(v).$$

Thus p_1 is a semi-norm. For $c \in \mathbb{R}$ and $u_c(x) = c$ we get $u_c' = 0$ and hence $p_1(u_c) = 0$ but $u_c \ne 0$, implying that p_1 is not a norm. Since $(u_k - (u + c))' = (u_k - u)'$ we also deduce that $\lim_{k \to \infty} p_1(u_k - u) = \lim_{k \to \infty} p_1(u_k - (u + c))$.

c) Again, it is obvious that $q_1(u) \ge 0$ and that for $\alpha \in \mathbb{R}$

$$q_1(\alpha u) = \left(\int_{[0,1]} |\alpha u'(x)|^2 \lambda^{(1)}(dx) \right)^{\frac{1}{2}} = |\alpha| \left(\int_{[0,1]} |u'(x)|^2 \lambda^{(1)}(dx) \right)^{\frac{1}{2}} = |\alpha| q_1(u).$$

Using the Minkowski inequality we also have for $u, v \in C_b^1([a, b])$ that

$$q_1(u + v) \le q_1(u) + q_2(v),$$

implying that q_1 is a semi-norm on $C_b^1([0, 1])$. Suppose that $q_1(u) = 0$, i.e.

$$\int_{[0,1]} |u'(x)|^2 \lambda^{(1)}(dx) = 0.$$

Since by assumption u' is continuous on $[0, 1]$ this implies with the meanwhile standard argument that $u' = 0$ on $[0, 1]$ which in turn yields that $u = c, c \in \mathbb{R}$. Since for $u_c \in C_b^1([0, 1])$, $u_c(x) = c \in \mathbb{R}$, it follows that $q_1(u_c)$ we find that $q_1(u) = 0$ if and only if u is a constant function. From $\lim_{k \to \infty} p_1(u_k - u) = 0$ we deduce that

$$\left(\int_{[0,1]} |u_k'(x) - u'(x)|^2 \lambda^{(1)}(dx) \right)^{\frac{1}{2}} \le p_1(u_k - u) \left(\int_{[0,1]} 1 \lambda^{(1)}(dx) \right)^{\frac{1}{2}},$$

i.e.

$$\lim_{k \to \infty} \left(\int_{[0,1]} |u_k'(x) - u'(x)|^2 \lambda^{(1)}(dx) \right)^{\frac{1}{2}} = 0.$$

With the help Minkowski's inequality we now find

$$\left(\int_{[0,1]} |u_k'(x) - h(x)|^2 \lambda^{(1)}(dx)\right)^{\frac{1}{2}}$$

$$\leq \left(\int_{[0,1]} |u_k'(x) - u(x)|^2 \lambda^{(1)}(dx)\right)^{\frac{1}{2}} + \left(\int_{[0,1]} |u_k'(x) - h(x)|^2 \lambda^{(1)}(dx)\right)^{\frac{1}{2}},$$

and therefore we have

$$N_2(u' - h) = \left(\int_{[0,1]} |u_k'(x) - h(x)|^2 \lambda^{(1)}(dx)\right)^{\frac{1}{2}} = 0$$

which gives $u' - h = 0$ $\lambda^{(1)}_{|[0,1]}$-almost everywhere, i.e. $h = u'\lambda^{(1)}_{|[0,1]}$ - almost everyw-here.

8. Clearly we have for every $f \in \mathcal{L}^\infty(\Omega)$ that $N_\infty(f) \geq 0$ and for $\alpha \in \mathbb{R}$ it follows that

$$N_\infty(\alpha f) = \inf\{M \geq 0 | \mu(\{|\alpha f| > M\}) = 0\} = |\alpha| N_\infty(f).$$

The triangle inequality follows for $f, g \in \mathcal{L}^\infty(\Omega)$ as follows: $N_\infty(f) < \infty$ means that $|f(\omega)| \leq C_f < \infty$ for all $\omega \in \Omega \setminus N_f$, $\mu(N_f) = 0$. We call C_f an **essential bound** of f. Clearly we have

$$N_\infty(f) = \inf\{C_f | C_f \text{ is an essential bound of } f\}.$$

Thus $|f(\omega)| \leq C_f$ μ-a.e. and $|g(\omega)| \leq C_g$ μ-a.e. yields

$$|f(\omega) + g(\omega)| \leq C_f + C_g \leq N_\infty(f) + N_\infty(g)\mu - \text{a.e.}$$

which gives

$$N_\infty(f + g) \leq N_\infty(f) + N_\infty(g).$$

9.   a) For $k \in \mathbb{N}$ we have $\int u_k \lambda^{(1)}(dx) = 1$, hence $\lim_{k\to\infty} \int u_k \lambda^{(1)}(dx) = 1$. On the other hand we have for all $x \in \mathbb{R}$ that $\lim_{k\to\infty} u_k(x) = 0$. This is trivial for $x \in (0,1)^\complement$. For $x \in (0,1)$ we can find N such that $x > \frac{1}{N^2}$ implying $u_k(x) = 0$ for $k > N$. Thus

$$0 = \int_\mathbb{R} \lim_{k\to\infty} u_k(x) \lambda^{(1)}(dx) = \int_\mathbb{R} \liminf_{k\to\infty} u_k(x) \lambda^{(1)}(dx) \leq \liminf_{k\to\infty} \int_\mathbb{R} u_k(x) \lambda^{(1)}(dx) = 1.$$

b) From $v_k(x) \leq v(x)$ we deduce that $\int v_k d\mu \leq \int v d\mu$ and hence $\lim_{k\to\infty} \int v_k d\mu$ exists as limit of the monotone and bounded sequence $\left(\int v_k d\mu\right)_{k\in\mathbb{N}}$. The Lemma of Fato yields further

$$\int v d\mu = \int \lim_{k\to\infty} v_k d\mu \leq \liminf_{k\to\infty} \int v_k d\mu \leq \int v d\mu,$$

hence we have $\lim_{k\to\infty} \int v_k d\mu = \int v d\mu$.

c) This follows when taking in the lemma of Fatou the sequence $f_k = \chi_{A_k}$ and when recalling the definition of lim inf which yields

$$\chi_{\bigcup_{k=1}^{\infty} \bigcap_{l=k}^{\infty} A_l} = \liminf_{k \to \infty} \chi_{A_k}$$

and now we just have to use the fact that $\mu(A) = \int \chi_A d\mu$.

10. Since $\left(\int_\Omega |u|^p d\mu \right)^{\frac{1}{p}} \leq ||u||_\infty \mu(\Omega)^{\frac{1}{p}}$ it follows that

$$\limsup_{p \to \infty} \left(\int_\Omega |u|^p d\mu \right)^{\frac{1}{p}} \leq ||u||_\infty \lim_{p \to \infty} \mu(\Omega)^{\frac{1}{p}} = ||u||_\infty,$$

recall that for $a > 0$ we have $\lim_{p \to \infty} a^{\frac{1}{p}} = 1$. Suppose that $||u||_\infty > 0$, otherwise the statement is trivial. For $0 < \epsilon \leq ||f||_\infty$ we find $\delta := \mu(\{\omega \in \Omega \mid |f(\omega)| > ||f||_\infty - \epsilon\}) > 0$ which implies that

$$\int_\Omega |f|^p d\mu \geq \left(||f||_\infty - \epsilon \right)^p \delta$$

and therefore

$$\liminf_{p \to \infty} \left(\int_\Omega |f|^p d\mu \right)^{\frac{1}{p}} \geq (||f||_\infty - \epsilon) \liminf_{p \to \infty} \delta^{\frac{1}{p}} = ||f||_\Omega - \epsilon.$$

For $\epsilon \to 0$ if follows that

$$\liminf_{p \to \infty} \left(\int_\Omega |f|^p d\mu \right)^{\frac{1}{p}} \geq ||f||_\infty \geq \limsup_{p \to \infty} \left(\int_\Omega |f|^p d\mu \right)^{\frac{1}{p}},$$

i.e.

$$\lim_{p \to \infty} \left(\int_\Omega |f|^p d\mu \right)^{\frac{1}{p}} = ||f||_\infty.$$

11. Since $(f_k)_{k \in \mathbb{N}}$ converges on Ω in measure to f, for $m \in \mathbb{N}$ there exists k_m such that

$$\mu(\{\omega \in \Omega \mid |f_{k_m}(\omega) - f(\omega)| > \frac{1}{m}\} \leq \frac{1}{2^m}.$$

Without loss of generality we assume that the sequence $(k_m)_{m \in \mathbb{N}}$ is strictly increasing. Define

$$\Omega_l := \Omega \setminus \bigcup_{m=l}^{\infty} \{\omega \in \Omega \mid |f_{k_m}(\omega) - f(\omega)| > \frac{1}{m}\}.$$

It follows that

$$\mu(\Omega \setminus \Omega_l) \leq \sum_{m=l}^{\infty} \frac{1}{2^m} = 2^{1-l}$$

and therefore $\mu(\Omega \setminus \bigcup \Omega_l) = 0$. For $m \geq l$ and $\omega \in \Omega_l$ we have $|f_{k_m}(\omega) - f(\omega)| < \frac{1}{m}$. This implies the convergence of $(f_{k_m}(\omega))_{m \in \mathbb{N}}$ to $f(\omega)$ for $\omega \in \bigcup_{l \in \mathbb{N}} \Omega_l$, but as already stated, $\mu(\Omega \setminus \bigcup_{l \in \mathbb{N}} \Omega_l) = 0$.

626

12. The estimate $|f_k(\omega)| \le g(\omega)$ implies that all functions are f_k are integrable. From Problem 10 we deduce that a subsequence $(f_{k_m})_{k\in\mathbb{N}}$ converges μ-almost everywhere to f. To this subsequence we may apply the dominated convergence theorem to find first that f is integrable and secondly that

$$\int f d\mu = \lim_{m\to\infty} \int f_{k_m} d\mu.$$

Now we consider the sequence $(\gamma_k)_{k\in\mathbb{N}}, \gamma_k := \int f_k d\mu$. Since

$$|\gamma_k| = |\int f_u d\mu| \le \int |f_k| d\mu \le \int g d\mu$$

the sequence $(\gamma_k)_{k\in\mathbb{N}}$ is bounded. Suppose that $(\gamma_k)_{k\in\mathbb{N}}$ does not converge to $\gamma :=$ $\int f d\mu$. Then it must have a subsequence $(\gamma_{k_l})_{l\in\mathbb{N}}$ which does not converge to γ, but by the Bolzano-Weierstrass theorem a subsequence of this subsequence must have some limit $\tilde{\gamma} \ne \gamma$. We denote this subsequence again with $(\gamma_{k_l})_{l\in\mathbb{N}}$. However the corresponding sequence $(f_{k_l})_{l\in\mathbb{N}}$ has a subsequence converging μ-almost everywhere to f and the dominated convergence theorem applied to this subsequence of $(f_{k_l})_{l\in\mathbb{N}}$ yields the convergence of the corresponding integrals to γ which is a contradiction.

13. Since $|h_k| \le c$ it follows that $|h_k g|^p \le c^p |g|^p$ and the dominated convergence theorem implies that

$$\lim_{k\to\infty} \int_\Omega |h_k g|^p d\mu = \int_\Omega |hg|^p d\mu.$$

From Theorem 7.31 we deduce that for the sequence $(h_k g)_{k\in\mathbb{N}}$ that

$$\lim_{k\to\infty} N_p(h_k g - hg) = \lim_{k\to\infty} \left(\int |h_k g - hg|^p \, d\mu \right)^{\frac{1}{p}} = 0.$$

In addition we find

$$\int |h_k g_k - hg|^p d\mu \le 2^p \left(\int |h_k|^p |g_k - g|^p d\mu + \int |h_k g - hg|^p d\mu \right)$$

$$\le 2^p \left(c^p N_p^p(g_k - g) + N_p^p(g(h_k - h)) \right),$$

since $N_p(g_k - g)$ tends for k to infinity to 0 by our assumption and since we have proved before that $N_p(h_k g - hg)$ tends to 0 too, the result is proved.

Chapter 8

1.   a) We use Theorem 8.1. The function $g(x, y) = \frac{e^{-x(1+y^2)}}{1+y^2}$ defined on $[0, \infty) \times \mathbb{R}$ is continuous, in particular for $y \in \mathbb{R}$ fixed the function $x \mapsto g(x, y)$ is continuous on $[0, \infty)$. Furthermore, for $x \in [0, 1]$ and $y \in \mathbb{R}$ we have

$$\frac{e^{-x(1+y^2)}}{1+y^2} \le \frac{e^{-x|y|}}{1+y^2} \le \frac{1}{1+y^2}$$

and for $x \geq 1$ and $y \in \mathbb{R}$ we have

$$\frac{e^{-x(1+y^2)}}{1+y^2} \leq \frac{e^{-x|y|}}{1+y^2} \leq e^{-|y|}.$$

Thus for all $(x, y) \in [0, \infty] \times \mathbb{R}$ it follows that

$$|g(x,y)| = g(x,y) \leq \frac{1}{1+y^2} + e^{-|y|}$$

and the function $g \mapsto \frac{1}{1+y^2} + e^{-|y|}$ is integrable over $\mathbb{R}$ which we can deduce for example from Theorem 8.14. Thus by Theorem 8.1 we find that φ is a continuous function. In addition it follows that

$$\varphi(0) = \int_{-\infty}^{\infty} \frac{1}{1+y^2}\,dy = \lim_{R \to \infty} \int_{-R}^{R} \frac{1}{1+y^2}\,dy = \lim_{R \to \infty} \arctan y \Big|_{-R}^{R} = \pi.$$

b) Consider the function $h_k : [0, \infty] \times [0, \infty) \to \mathbb{R}$ defined for $k \geq 2$ by

$$h_k(\xi, x) := \begin{cases} e^{-\xi x} \left(\frac{\sin x}{x}\right)^k, \xi \in [0, \infty) \times (0, \infty) \\ 1, \xi \in [0, \infty) \times \{0\} \end{cases}$$

Since $\lim_{x \to 0} \frac{\sin x}{x} = 1$ it follows that h_k is on $[0, \infty) \times [0, \infty)$ continuous and

$$\int_{(0,\infty)} e^{-\xi x} \left(\frac{\sin x}{x}\right)^k \lambda^{(1)}(dx) = \int_{[0,\infty)} e^{\xi x} \left(\frac{\sin x}{x}\right)^k \lambda^{(1)}(dx).$$

For $x \geq 1$ we find $\left|\left(\frac{\sin x}{x}\right)^k\right| \leq \frac{1}{x^k} \leq 1$ and further we know that the continuous function $x \mapsto \left(\frac{\sin x}{x}\right)^k$ is bounded on $[0, 1]$. Thus it follows that

$$\left|e^{-\xi x} \left(\frac{\sin x}{x}\right)^k\right| \leq \frac{M}{1+x^k},$$

and Theorem 8.1 implies for $k \geq 2$ the continuity of ψ since $\int_0^{\infty} \frac{1}{1+x^k}\,dx < \infty$ for $k \geq 2$.

2. The proof of the extension of Theorem 8.4 follows by induction.
   So we fix $m \in \mathbb{N}$ and we assume that

   i) for all $x \in I$ the functions $\frac{\partial^k u(x_j, \cdot)}{\partial x^k} : \Omega \to \mathbb{R}, 0 \leq k \leq m-1$, are μ-integrable;

   ii) for all $\omega \in \Omega$ the function $\frac{\partial^{m-1} u(\cdot, \omega)}{\partial x^{m-n}} : I \to \mathbb{R}$ is partial differentiable with respect to x;

   iii) for μ-integrable functions $h_k : \Omega \to [0, \infty)$ the following holds on $I \times \Omega$

   $$\left|\frac{\partial^k}{\partial x^k} u(x, \omega)\right| \leq h_k(\omega), \quad 1 \leq k \leq m.$$

Then $g(x) := \int u(x,w)\mu(dw)$ is on I m-times differentiable, $w \mapsto \frac{\partial^k u}{\partial x^k}(x,w)$, $1 \le k \le m$, is for all $x \in I$ integrable and

$$\frac{\partial^k}{\partial x^k}g(x) = \int \frac{\partial^k u}{\partial x^k}(x,w)\mu(dw), \quad k \le m.$$

These conditions are such that we can take Theorem 8.4 with $m = 1$ as starting point of the induction and then we replace in the induction steps u by $\frac{\partial^k}{\partial x^k}u(x,w)$.

3. Note that we claim the differential equation holds in the open interval $(0,\infty)$. For proving that φ is in $(0,\infty)$, differentiable we fix $x_0 > 0$ and note that

$$\frac{\partial}{\partial x}\left(\frac{e^{-x(1+y^2)}}{1+y^2}\right)\Big|_{x_0}^{x} = e^{-x_0(1+y^2)} \le e^{-x_0|y|}$$

which implies by Theorem 8.4 for $x > x_0$ that

$$\varphi'(x) = \frac{d}{dx}\int_{\mathbb{R}} \frac{e^{-x(1+y^2)}}{1+y^2}\lambda^{(1)}(dy) = \int_{\mathbb{R}} \frac{d}{dx}\left(\frac{e^{-x(1+y^2)}}{1+y^2}\right)\lambda^{(1)}(dy) = \int_{\mathbb{R}} e^{-x(1+y^2)}\lambda^{(1)}(dy).$$

Since $x_0 > 0$ was arbitrary we obtain

$$\varphi'(x) = \int_{\mathbb{R}} e^{-x(1+y^2)}\lambda^{(1)}(dy) \text{ for all } x > 0,$$

and we can identify this integral as the improper Riemann integral $\int_{-\infty}^{\infty} e^{-x(1+y^2)}dy$. Using the substitution $\xi := y\sqrt{x}$ we find

$$\int_{-\infty}^{\infty} e^{-x(1+y^2)}dy = \lim_{R\to\infty} \int_{-R}^{R} e^{-x(1+y^2)}dy$$

$$= \lim_{R\to\infty} \int_{-R\sqrt{x}}^{R\sqrt{x}} e^{-x}e^{-\xi^2}\frac{1}{\sqrt{x}}d\xi = \frac{1}{\sqrt{x}}e^{-x}\int_{-\infty}^{\infty} e^{-\xi^2}d\xi = \sqrt{\frac{\pi}{x}}e^{-x}, x > 0.$$

4.   a) The following holds

$$\frac{d}{dx}g(x) = \frac{d}{dx}e^{-\frac{x^2}{2}} = -xe^{-\frac{x^2}{2}} = -xg(x)$$

or $g'(x) + xg(x) = 0$.

b) Since $|(\cos\xi x)e^{-\frac{x^2}{2}}| \le e^{-\frac{x^2}{2}}$ we can differentiate $\tilde{g}$ under the integral and we can identify Lebesgue integrals with improper Riemann integrals. Now we find

$$\frac{d}{d\xi}\tilde{g}(\xi) = \frac{1}{\sqrt{2\pi}}\int_{\mathbb{R}}(\cos\xi x)e^{-\frac{x^2}{2}}\lambda^{(1)}(dx)$$

$$= \frac{1}{\sqrt{2\pi}}\int_{\mathbb{R}}(\frac{\partial}{\partial\xi}\cos\xi x)e^{-\frac{x^2}{2}}\lambda^{(1)}(dx) = \frac{1}{\sqrt{2\pi}}\int_{-\infty}^{\infty}(\frac{\partial}{\partial\xi}\cos\xi x)e^{-\frac{x^2}{2}}\lambda^{(1)}(dx)$$

$$= \frac{1}{\sqrt{2\pi}} \lim_{R \to \infty} \int_{-R}^{R} (-x \sin \xi x) e^{-\frac{x^2}{2}} dx = \frac{1}{\sqrt{2\pi}} \lim_{R \to \infty} \int_{-R}^{R} (\sin \xi x) \left(\frac{d}{dx} e^{-\frac{x^2}{2}} \right) dx$$

$$= \frac{1}{\sqrt{2\pi}} \lim_{R \to \infty} \left((\sin \xi x) e^{-\frac{x^2}{2}} \Big|_{-R}^{R} - \int_{-R}^{R} \left(\frac{d}{dx} \sin \xi x \right) e^{-\frac{x^2}{2}} dx \right)$$

$$= -\frac{1}{\sqrt{2\pi}} \int_{-\infty}^{\infty} \xi (\cos \xi x) e^{-\frac{x^2}{2}} dx = -\xi \widetilde{g}(\xi)$$

implying that $\widetilde{g}(\xi) + \xi \widetilde{g}(\xi) = 0$, i.e. $(\star)$ holds for $\widetilde{g}$.

c) We can now conclude that $g(x) = c\widetilde{g}(x)$ which yields that $g(0) = c\widetilde{g}(0)$. However $g(0) = 1$ and $\widetilde{g}(0) = \frac{1}{\sqrt{2\pi}} \int_{\mathbb{R}} e^{-\frac{x}{2}} \lambda^{(1)}(dx) = 1$ which follows from $\int_{\mathbb{R}} e^{-x^2} \lambda^{(1)}(dx) = \sqrt{\pi}$ and the substitution $y = \frac{x}{\sqrt{2}}$. Thus $c = 1$ and eventually we have proved $g = \widetilde{g}$.

5.   a) Since $|\cos yx| \le 1$ and $x \mapsto 1$ is integrable against the bounded measure μ it follows that $\widetilde{\mu}$ is continuous.

b) We want to use Theorem 8.4. A formal application yields that

$$\frac{d^l}{dy^l} \widetilde{\mu}(y) = \int_{\mathbb{R}} \frac{d^l}{dy^l} (\cos yx) \mu(dx).$$

Since $\left| \frac{d^l}{dy^l} \cos yx \right| \le |x|^l$ we can prove the following result after introducing a further common definition. For a measure ν the number $M_l := \int |x|^l \nu(dx)$ is called the lth **absolute moment** of ν.
If a bounded measure μ on $\mathcal{B}^{(1)}$ has all absolute moments up to order N then the function $\widetilde{\mu}$ is N-times continuously differentiable and for $1 \le k \le N$ we have

$$\frac{d^k}{dy^k} \widetilde{\mu}(y) = \int_{\mathbb{R}} \left(\frac{d^k}{dy^k} \cos yx \right) \mu(dy).$$

6. For every probability measure P, every integrable random variable X and convex functions f Jensen's inequality states, see (8.13),

$$(E(X)) \le E(f \circ X)$$

or

$$f \left(\int X dP \right) \le \int f \circ X dP.$$

For $\left(\mathbb{R}, \mathcal{B}^{(1)}, \frac{1}{N} \sum_{k=1}^{N} \epsilon_k \right)$ and $X = g : \mathbb{R} \to \mathbb{R}$ measurable we find

$$\int_{\mathbb{R}} X dP = \int_{\mathbb{R}} g \frac{1}{N} \sum_{k=1}^{N} d\epsilon_k = \frac{1}{N} \sum_{k=1}^{N} g(k)$$

and

$$\int (f \circ g) \frac{1}{N} \sum_{k=1}^{N} d\epsilon_k = \frac{1}{N} \sum_{k=1}^{N} f(g(k))$$

which eventually yields

$$f\left(\frac{1}{N}\sum_{k=1}^{N}g(k)\right) \le \frac{1}{N}\sum_{k=1}^{N}f(g(k)).$$

7. The statement can be read as

$$-\left(1-\left(\int_0^1 u(x)dx\right)^2\right)^{\frac{1}{2}} \le \int_0^1 (-(1-u^2)^{\frac{1}{2}})dx$$

and will follow once $\varphi(t) = -(1-t^2)^{\frac{1}{2}}$ is verified to be a convex function. For $t \in (0,1)$ we have

$$(-\varphi')(t) = t(1-t^2)^{-\frac{1}{2}}$$

and

$$(-\varphi'')(t) = (1-t^2)^{-\frac{1}{2}} + t^2(1-t^2)^{-\frac{3}{2}} = \frac{1}{(1-t^2)^{-\frac{3}{2}}} > 0$$

implying the result.

8.   a) On $[a,b]$ equipped with the σ-field $[a,b]\cap\mathbb{B}^{(1)}$ a measure is given by $\mu(A) := \int_A w(x)\lambda^{(1)}(dx)$. Indeed, for $A = \emptyset$ it follows that $\mu(\emptyset) = \int_\emptyset w(x)\lambda^{(1)}(dx) = 0$, and for a collection $(A_k)_{k\in\mathbb{N}}$ of mutually disjoint sets $A_k \in [a,b]\cap\mathbb{B}^{(1)}$ we have

$$\mu\left(\bigcup_{k\in\mathbb{N}}A_k\right) = \int_{\bigcup_{k\in\mathbb{N}}A_k} w(x)\lambda^{(1)}(dx) = \sum_{k=1}^{\infty}\int_{A_k} w(x)\lambda^{(1)}(dx) = \sum_{k=1}^{\infty}\mu(A_k).$$

By assumption we have $\mu[a,b] = \int w(x)\lambda^{(1)}(dx) < \infty$, i.e. μ is a finite measure. Now Theorem 8.8 implies

$$\varphi\left(\frac{1}{\mu([a,b])}\int_{[a,b]} u(x)\mu(dx)\right) \le \frac{1}{\mu([a,b])}\int (\varphi\circ u)(x)\mu(dx)$$

or

$$\varphi\left(\frac{1}{\int_{([a,b]]} w(x)\lambda^{(1)}(dx)}\int_{[a,b]} u(x)w(x)\lambda^{(1)}(dx)\right)$$

$$\le \frac{1}{\int_{[a,b]} w(x)\lambda^{(1)}(dx)}\int_{[a,b]} (\varphi\circ u)(x)w(x)\lambda^{(1)}(dx).$$

b) The function $t \mapsto e^t$ is convex on $[0,1]$ and with $w(x) = 1$ we find for the measure μ in part a) that $\mu = \lambda^{(1)}_{|[0,1]}$. Further, in our situation the Riemann and the Lebesgue integral coincide and it follows that

$$e^{\int_0^1 u(x)dx} \le \int_0^1 e^{u(x)}dx$$

or

$$\int_0^1 u(x)dx \le \ln\left(\int_0^1 e^{u(x)}dx\right).$$

631

9. The suggested substitution leads for the improper Riemann integrals to

$$\int_0^1 \cos(x^2)\,dx = \frac{1}{2}\int_0^\infty \frac{\cos y}{\sqrt{y}}\,dy$$

and

$$\int_0^\infty \sin(x^2)\,dx = \frac{1}{2}\int_0^\infty \frac{\sin y}{\sqrt{y}}\,dy.$$

For the improper Riemann integral to be a Lebesgue integral we need the integrability of $\frac{|\cos y|}{\sqrt{y}}$ and $\frac{|\sin y|}{\sqrt{y}}$ over $[0,\infty)$. Suppose that $y \to \frac{|\sin y|}{\sqrt{y}}$ was integrable over $[0,\infty)$. It follows that

$$\int_0^\infty \frac{|\sin y|}{\sqrt{y}}\,dy = \sum_{k=0}^\infty \int_{k\pi}^{(k+1)\pi} \frac{|\sin y|}{\sqrt{y}}\,dy = \sum_{k=0}^\infty \int_0^\pi \frac{|\sin(y+k\pi)|}{\sqrt{y+k\pi}}\,dy$$

$$\geq \sum_{k=0}^\infty \frac{1}{\sqrt{(k+1)\pi}} \int_0^\pi |\sin(y+k\pi)|\,dy = \frac{2}{\sqrt{\pi}}\sum_{k=1}^\infty \frac{1}{\sqrt{k}},$$

but the series $\sum_{k=1}^\infty \frac{1}{\sqrt{k}}$ diverges. A similar argument holds for the second integral. Now we want to find $\int_0^\infty \sin(x^2)\,dx$ (as improper integral) and we will make use of the calculations in [44]. Note that the following calculation is in the frame of Riemann's theory, not of Lebesgue's theory. This refers in particular for applications of the change of variable and the interchanging of integrals. From $\int_0^\infty e^{-x^2}\,dx = \frac{\sqrt{\pi}}{2}$ we deduce that $\frac{1}{\sqrt{u}} = \frac{2}{\sqrt{\pi}}\int_0^\infty e^{-uy^2}\,dy$. Now consider for $\epsilon > 0$ the integral.

$$\int_0^\infty \frac{\sin u}{\sqrt{u}}e^{-\epsilon u}\,du = \frac{2}{\sqrt{\pi}}\int_0^\infty (\sin u)e^{-\epsilon u}\left(\int_0^\infty e^{-uy^2}\,dy\right)du$$

$$= \frac{2}{\sqrt{\pi}}\int_0^\infty \left(\int_0^\infty (\sin u)e^{-(\epsilon+y^2)u}\,dy\right)du$$

$$= \frac{2}{\sqrt{\pi}}\int_0^\infty \left(\int_0^\infty (\sin u)e^{-(\epsilon+y^2)u}\,du\right)dy.$$

The inner integral we can calculate explicitly using integration by parts twice:

$$\int_0^\infty (\sin u)e^{-(\epsilon+y^2)u}\,du = 1 - (\epsilon+y^2)^2 \int_0^\infty (\sin u)e^{-(\epsilon+y^2)u}\,du$$

or

$$\int_0^\infty (\sin u)e^{-(\epsilon+y^2)u}\,du = \frac{1}{1+(\epsilon+y^2)^2}.$$

Therefore we find

$$\int_0^\infty \frac{(\sin u)}{\sqrt{u}}e^{-\epsilon u}\,du = \frac{2}{\sqrt{\pi}}\int_0^\infty \frac{1}{1+(\epsilon+y^2)^2}\,dy.$$

We are allowed to pass to the limit as ϵ tends to 0 to get

$$\int_0^\infty \frac{(\sin u)}{\sqrt{u}} du = \frac{2}{\sqrt{\pi}} \int_0^\infty \frac{1}{1+y^4} dy.$$

At this stage we already know that the integral $\int_0^\infty \frac{(\sin u)}{\sqrt{u}} du = \int_0^\infty \sin(x^2) dx$ exists as an improper Riemann integral. Moreover, a primitive of $g(y) = \frac{1}{1+y^4}$ is the function

$$G(y) = \frac{1}{4\sqrt{2}} \ln \frac{y^2 + \sqrt{2}y + 1}{y^2 - \sqrt{2}y + 1} + \frac{1}{2\sqrt{2}} \Big(\arctan(\sqrt{2}y + 1) + \arctan(\sqrt{2}y - 1) \Big).$$

This yields

$$\lim_{y \to \infty} G(y) - \lim_{y \to 0} G(y) = \frac{\pi}{2\sqrt{2}}$$

and eventually we have

$$\int_0^\infty \sin(x^2) dx = \sqrt{\frac{\pi}{8}}.$$

10. For all $x \geq 0$ and $\alpha > 0$ we have $e^{-x^\alpha} \geq 0$ thus $x \mapsto e^{-x^\alpha}$ is improper Riemann integrable if and only if it is Lebesgue integrable. Since $\lim_{x \to \infty} (x^2 e^{-x^\alpha}) = 0$ we can find $R \geq 1$ such that $x \geq R$ implies $e^{-x^\alpha} \leq \frac{1}{x^2}$ and therefore

$$\int_0^\infty e^{-x^\alpha} dx \leq \int_0^R e^{-x^\alpha} dx + \int_R^\infty e^{-x^\alpha} dx \leq \int_0^R e^{-x^\alpha} dx + \int_R^\infty \frac{1}{x^2} dx.$$

Both integrals on the right hand side are finite, the first as it is an integral of a continuous function over a compact set, and the second since $\int_R^\infty x^{-2} dx = \frac{1}{R}$.

11. We define $L^\infty(\Omega)$ to consist of all real-valued functions $f : \Omega \to \mathbb{R}$ which are essentially bounded, i.e. $|f| \leq M_f < \infty$ μ-almost everywhere and further we define

$$N_\infty(f) := \inf\{M \geq 0 \mid \mu(\{|f|\} > \mu\}) = 0\}.$$

Clearly $N_\infty(f) \geq 0$ and if $\lambda \in \mathbb{R}$ and $f \in L^\infty(\Omega)$ then $|\lambda f| = |\lambda||f| \leq |\lambda| M_f$ μ-almost everywhere, i.e. λf is also essentially bounded and $N_\infty(\lambda f) \leq |\lambda| N_\infty(f)$. If $f, g \in L^\infty(\Omega)$ and $|f| \leq M_f$, $|g| \leq M_g$ μ-almost everywhere then we have $|f + g| \leq |f| + |g| \leq M_f + M_g$ μ-almost everywhere and it follows that

$$N_\infty(f + g) \leq N_\infty(f) + N_\infty(g).$$

Thus N_∞ is a semi-noun on the vector space L^∞. Now let $(f_j)_{j \in \mathbb{N}}$ be a Cauchy sequence with respect to N_∞, i.e. for every $\epsilon > 0$ there exists $N(\epsilon) \in \mathbb{N}$ such that $k, l \geq N(\epsilon)$ implies $N_\infty(f_k - f_l) < \epsilon$. Consider the sets

$$A_{k,l} = \{|f_k| \geq N_\infty(f_k)\} \cup \{|f_k - f_l| > N_\infty(f_- f_l)\}$$

and $A = \bigcup_{k,l \in \mathbb{N}} A_{k,l}$. From the definition it follows that $\mu(A_{k,l}) = 0$ and $\mu(A) = 0$ implying that $N_\infty(\chi_A f_j) = 0$ for all $j \in \mathbb{N}$. On A^C the sequence $(f_j)_{j \in \mathbb{N}}$ converges with respect to the supremums norm to a bounded function: for $\omega \in A^{\mathsf{C}}$ we have uniformly $|f_k(\omega) - f_l(\omega)| < \epsilon$, $k, l \geq N(\epsilon)$, hence for every $\omega \in A^{\mathsf{C}}$ the limit $\lim_{k \to \infty} f_k(\omega) =: f(\omega)$ exists and the convergence is again uniform with respect to $\omega \in A^{\mathsf{C}}$. In particular f is measurable and bounded on A^{C}. Hence f extends to a μ-almost everywhere bounded and measurable function on Ω and $\lim_{j \to \infty} N_\infty(f_j - f\chi_{A^{\mathsf{C}}}) = 0$. Thus $L^\infty(\Omega)$ is complete in the same that every Cauchy sequence with respect to N_∞ has a limit. When now passing to the quotient space $L^\infty(\Omega) = L^\infty(\Omega)\backslash_{\sim_\mu}$ we obtain a Banach space, the space $L^\infty(\Omega)$. In $L^\infty(\Omega)$ we often write $||u||_\infty$ for the norm instead of $N_\infty(u)$ or $\mathrm{ess\,sup}_{\omega \in \Omega} |u(\omega)|$.

12.   a) For $f \in L^q(\mathbb{R}^n)$ we consider the set $\{|f| > 1\}$ and the decomposition of f according to

$$f = g + h = f\chi_{\{|f|>1\}} + f\chi_{\{|f|>1\}^{\mathsf{C}}}.$$

Since for $1 < q$ we have $|g| \leq |f|\chi_{\{|f|>1\}}$ it follows that $g \in L^1(\mathbb{R}^n)$ and for $q \leq r$ we have $|h|^r = |f|^r\chi_{\{|f|\geq 1\}} \leq |f|^q\chi_{\{|f|>1\}^{\mathsf{C}}}$ hence $h \in L^r(\mathbb{R}^N)$. In the case of $r = \infty$ we find $||h||_\infty \leq 1$.

b) For $r < \infty$ the numbers $\frac{1}{\lambda q}$ and $\frac{r}{(1-\lambda)q}$ have the property that $\frac{\lambda q}{1} + \frac{(1-\lambda)q}{r} = 1$, i.e. they are conjugate indices in Hölder's inequality and therefore it follows that

$$||u||_{L^q}^q = \int_\mathbb{R} |u|^{\lambda q}||u||^{(1-\lambda)q}d\lambda^{(n)} \leq |||u|^{\lambda q}||_{L^{\frac{1}{\lambda q}}} |||u|^{(1-\lambda)q}||_{L^{\frac{r}{(1-\lambda)q}}}$$

$$= \left(\int |u|d\lambda^{(n)}\right)^{\lambda q}\left(\int |u|^r d\lambda^{(n)}\right)^{\frac{(1-\lambda)q}{r}} \leq ||u||_{L^1}^{\lambda q}||u||_{L^r}^{(1-\lambda)q}$$

which yields

$$||u||_{L^q} \leq ||u||_{L^1}^{\lambda}||u||_{L^r}^{(1-\lambda)}.$$

13. The sequence $(f_k - g)_{k \geq 0}$ is a sequence of non-negative integrable functions converging $\lambda^{(1)}$-almost everywhere to $f - g$. By Fato's lemma we get

$$\int \liminf_{k \to \infty}(f_k - g)d\lambda^{(n)} \leq \liminf_{k \to \infty} \int (f_k - g)d\lambda^{(n)}$$

but

$$\liminf_{k \to \infty}(f_k - g) = f - g$$

and it follows that

$$\int f d\lambda^{(n)} - \int g d\lambda^{(n)} \leq \liminf_{k \to \infty} \int (f_k - g)d\lambda^{(n)}$$

$$\leq \liminf_{k \to \infty} \int f_k d\lambda^{(n)} - \int g d\lambda^{(n)}$$

or

$$\int f d\lambda^{(n)} \leq \liminf_{k \to \infty} \int f_k d\lambda^{(n)}.$$

Chapter 9

1. The system $\mathcal{E} := \{C \times D \mid C \in \mathcal{A}_1 \cap \mathcal{A}_1 \text{ and } D \in \mathcal{A}_2 \cap \mathcal{A}_2\} = \{C \times D \mid C \in \mathcal{A}_1, C \subset \mathcal{A}_1 \text{ and } D \in \mathcal{A}_2, D \subset \mathcal{A}_2\}$ is the natural choice for a generator of $(\mathcal{A}_1 \times \mathcal{A}_2) \cap (\mathcal{A}_1 \otimes \mathcal{A}_2) = (\mathcal{A}_1 \cap \mathcal{A}_2) \otimes \mathcal{A}_2 \cap \mathcal{A}_2$. Note that we have $(\mathcal{A}_1 \times \mathcal{A}_2) \cap (\mathcal{A}_1 \otimes \mathcal{A}_2) = (\mathcal{A}_1 \cap \mathcal{A}_2) \otimes (\mathcal{A}_2 \cap \mathcal{A}_2)$.

2. We know that $\mathbb{B}^{(n)} = \otimes_{j=1}^{n} \mathcal{B}^{(1)}$ (n copies of $\mathcal{B}^{(1)}$) and for $x \in \mathbb{R}^n, A \in \mathcal{B}^{(n)}$ we have

$$\epsilon_x(A) = \begin{cases} 1, & x \in A \\ 0, & x \notin A \end{cases}$$

Furthermore we know that $\mathcal{B}^{(1)} \times ... \times \mathcal{B}^{(1)}$ (n copies of $\mathcal{B}^{(1)}$) is a generator of $\mathcal{B}^{(n)}$ which determines ϵ_x. Let $A = A_1 \times ... \times A_k, A_j \in \mathcal{B}^{(1)}$, and $x = (x_1, ..., x_n) \in \mathbb{R}^n$. Since $x \in A$ if and only if $x_j \in A_j$ it follows that $\epsilon_x(A) = \epsilon_{x_1}(A_1) \cdot ... \cdot \epsilon_{x_n}(A_n)$, in other words $\epsilon_x = \epsilon_{x_1} \otimes ... \otimes \epsilon_{x_n}$.

3.   a) Again we note that $\mathcal{A}_1 \times ... \times \mathcal{A}_N = \{A_1 \times ... \times A_N \mid A_j \in \mathcal{A}_j\}$ is a generator of $\mathcal{A}_1 \otimes ... \otimes \mathcal{A}_N$ and $\nu_1 \otimes ... \otimes \nu_N$ as well as $\mu_1 \otimes ... \otimes \mu_N$ are determined on $\mathcal{A}_1 \otimes ... \otimes \mathcal{A}_N$. It follows with $\omega = (\omega_1, ..., \omega_N), \omega_j \in \mathbb{R}$, that

$$(\nu_1 \otimes ... \otimes \nu_n)(A_1 \otimes ... \otimes A_N) = \nu_1(A_1) \cdot ... \cdot \nu_N(A_N)$$

$$= \int_{A_1} g_1(\omega_1)\mu_1(d\omega_1) \cdot ... \cdot \int_{A_N} g_N(\omega_N)\mu_N(d\omega_N)$$

$$= \int_{A_1 \otimes ... \otimes A_N} g_1(\omega_1) \cdot ... \cdot g_N(\omega_N)(\mu_1 \otimes ... \otimes \mu_N)(d\omega),$$

and therefore

$$\nu_1 \otimes ... \otimes \nu_N = g_1 \cdot ... \cdot g_N \mu_1 \otimes ... \otimes \mu_N.$$

Since by assumption all functions g_j are integrable we have

$$(\nu_1 \otimes ... \otimes \nu_N)(\Omega_1 \otimes ... \otimes \Omega_N) = \prod_{j=1}^{N} \int_{\Omega_j} g_j(\omega_j)\mu_j(d\omega_j) < \infty,$$

i.e. $\nu_1 \otimes ... \otimes \nu_N$ is finite measure.

     b) Since $\mathcal{A}_1 \otimes \mathcal{A}_2$ is generated by $\mathcal{A}_1 \times \mathcal{A}_2$ it follows for $A \in \mathcal{A}_1$ and $N \in \mathcal{N}_2$ that $A \times N \in \mathcal{A}_1 \times \mathcal{A}_2$. Moreover we have $(\mu_1 \otimes \mu_2)(A \times N) = \mu_1(A)\mu_2(N) = 0$, or $A \times N \in \mathcal{N}$.

4.   a) Consider the following figure

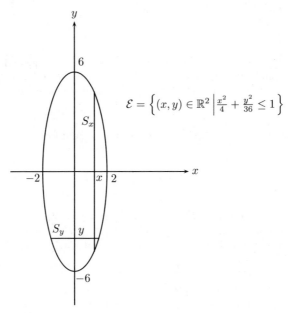

$$\mathcal{E} = \left\{ (x,y) \in \mathbb{R}^2 \,\middle|\, \tfrac{x^2}{4} + \tfrac{y^2}{36} \leq 1 \right\}$$

Since $\mathcal{E}_x := \{y \in \mathbb{R} \mid (x,y) \in \mathcal{E}\}$ for $x \in (-\infty, -2)\cup(2, \infty)$ we have obviously $\mathcal{E}_x = \emptyset$. If $x \in [-2,2]$ then $(x,y) \in \mathcal{E}$ means $\tfrac{x^2}{4} + \tfrac{y^2}{36} \leq 1$ or $y \in [-3\sqrt{4-x^2}, 3\sqrt{4-x^2}]$, i.e.

$$\mathcal{E}_x = \begin{cases} [-3\sqrt{4-x^2}, 3\sqrt{4-x^2}], & x \in [-2,2] \\ \emptyset, & \text{otherwise.} \end{cases}$$

Analogously we find

$$\mathcal{E}_y = \begin{cases} [-\tfrac{1}{3}\sqrt{36-y^2}, \tfrac{1}{3}\sqrt{36-y^2}], & y \in [-6,6] \\ \emptyset, & \text{otherwise} \end{cases}$$

b) We use the figure below

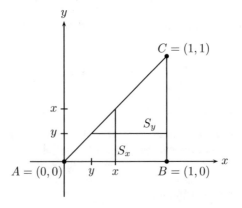

By definition we have $S_x = \{y \in \mathbb{R} \mid (x,y) \in S = ABC\}$ which yields for $x \in [0,1]^{\complement}$ that $S_x = \emptyset$. For $x \in [0,1]$ we find that $S_x = \{y \in \mathbb{R} \mid 0 \le y \le 1 - x\}$, hence

$$S_x = \begin{cases} [0, 1-x], & x \in [0,1] \\ \emptyset, & \text{otherwise,} \end{cases}$$

and analogously

$$S_y = \begin{cases} [y, 1], & y \in [0,1] \\ \emptyset, & \text{otherwise.} \end{cases}$$

It follows that $\lambda^{(1)}(S_x) = 1 - x$ and $\lambda^{(1)}(S_y) = 1 - y$ implying

$$\int_{\mathbb{R}} \lambda^{(1)}(S_x)\lambda^{(1)}(dx) = \int_{[0,1]}(1-x)dx = \frac{1}{2}$$

and

$$\int_{\mathbb{R}} \lambda^{(1)}(S_y)\lambda^{(1)}(dy) = \int_{[0,1]}(1-y)dy = \frac{1}{2}.$$

Since $\lambda^{(2)}(S) = (\lambda^{(1)} \otimes \lambda^{(1)})(S) = \text{area}(ABC) = \frac{1}{2}$ we have indeed verified (g.16)

$$(\lambda^{(1)} \otimes \lambda^{(1)})(S) = \frac{1}{2} = \int_{\mathbb{R}} \lambda^{(1)}(S_x)\lambda^{(1)}(dx) = \int_{\mathbb{R}} \lambda^{(1)}(S_y)\lambda^{(1)}(dy).$$

5. Let $\mathcal{C}_j \subset \mathcal{P}(\Omega), j \in I$, be a collection of monotone classes in Ω and $\mathcal{C} := \bigcap_{j \in I} \mathcal{C}_j$. Further let $(A_k)_{k \in \mathbb{N}}$ and $(B_k)_{k \in \mathbb{N}}$ be sequences of elements $A_k, B_k \in \mathcal{C}$ such that $A_k \subset A_{k+1}$ and $B_k \supset B_{k+1}$. We have to show that $\bigcup_{k \in \mathbb{N}} A_k \in \mathcal{C}$ and $\bigcap_{k \in \mathbb{N}} B_k \in \mathcal{C}$. Since $A_k \in \bigcap_{j \in \mathbb{N}} \mathcal{C}_j$ it follows that $A_k \in \mathcal{C}_j$ for all $j \in I$ and therefore $\bigcup_{k \in \mathbb{N}} A_k \in \bigcap_{j \in I} \mathcal{C}_j$, i.e. $\bigcup_{k \in \mathbb{N}} A_k \in \mathcal{C}$. Analogously, since $B_k \in \mathcal{C} = \bigcap_{j \in I} \mathcal{C}_j$ we deduce that $B_k \in \mathcal{C}_j$ for all $j \in I$ and therefore $\bigcap_{k \in \mathbb{N}} B_k \in \mathcal{C}_j$ for all $j \in I$ which yields $\bigcap_{k \in \mathbb{N}} B_k \in \mathcal{C}$ and we have proved that $\mathcal{C}$ is a monotone class.

6. Suppose that f is measurable and consider the function $F : \mathbb{R}^{n+1} \to \mathbb{R}$ defined by $F(x,y) = y - f(x)$. It follows that $\{F(x,y) \le 0\}$ is measurable as is $\{y \ge 0\}$ hence $\Gamma_s(f) = \{F(x,y) \le 0\} \cap \{y \ge 0\}$ is measurable. Now suppose that $\Gamma_s(f)$ is measurable. This implies that $\Gamma_s(f)_x = \{y \in \mathbb{R} \mid (x,y) \in \Gamma_s(f)\} = [0, f(x)]$ is measurable and $x \mapsto \lambda^{(1)}([0, f(x)]) = f(x)$ is measurable too by Theorem 9.15. Finally, by (9.16) we have

$$\lambda^{(n+1)}(\Gamma_s(f)) = \int \chi_{\Gamma_s(f)}(x,y)\lambda^{(n+1)}(dxdy)$$

$$= \int_{\mathbb{R}^n} \lambda^{(1)}(\Gamma_s(f)_x)\lambda^1(dx) = \int_{\mathbb{R}^n} f(x)\lambda^{(n)}(dx).$$

The interpretation of this result is that the volume of the body bounded from below by the plane $\mathbb{R}^n \times \{0\}$ and from above by the graph of f, i.e. $\Gamma(f)$, is given by $\int_{\mathbb{R}^n} f(x)\lambda^{(n)}(dx)$ as expected.

7. First we note that

$$\int_K M_{K,Q,\rho}(x)\lambda^{(3)}(dx) = \int_K \left(\int_Q \frac{(\mathrm{dist}(K,y))^\rho}{||x-y||^{3+\rho}} \lambda^{(3)}(dy) \right)\lambda^{(3)}(dx)$$

$$= \int_Q \left(\int_K \frac{(\mathrm{dist}(K,y))^\rho}{||x-y||^{3+\rho}} \lambda^{(3)}(dx) \right)\lambda^{(3)}(dy)$$

$$= \int_Q (\mathrm{dist}(K,y))^\rho \left(\int_K \frac{1}{||x-y||^{3+\rho}} \lambda^{(3)}(dx) \right)\lambda^{(3)}(dy)$$

$$= \int_{Q\setminus K} (\mathrm{dist}(K,y))^\rho \left(\int_K \frac{1}{||x-y||^{3+\rho}} \lambda^{(3)}(dx) \right)\lambda^{(3)}(dy)$$

Since $\mathrm{dist}(K,y) = 0$ for $y \in K$, for $y \in Q \setminus K$ and $x \in K$ we have $||x - y|| \geq \mathrm{dist}(K,y) > 0$ since K is closed, see also Example II. 3.28. This implies now for $y \in Q \setminus K$ fixed and $x \in K$

$$\int_K \frac{1}{||x-y||^{3+\rho}} \lambda^{(3)}(dy) \leq \int_{||x-y||\geq \mathrm{dist}(K,y)} \frac{1}{||x-y||^{3+\rho}} \lambda^{(3)}(dx)$$

$$= \int_{||z||\geq \mathrm{dist}(K,y)} \frac{1}{||z||^{3+\rho}} \lambda^{(3)}(dz)$$

$$\leq 4\pi \int_{\mathrm{dist}(K,y)}^{\infty} \frac{1}{r^{3+\rho}} r^2 dr$$

$$= 4\pi \int_{\mathrm{dist}(K,y)}^{\infty} \frac{1}{r^{1+\rho}} dr = \frac{4\pi}{\rho(\mathrm{dist}(K,y))^\rho}.$$

Now we find

$$\int_K M_{K,Q,\rho}(x)\lambda^{(3)}(dx) \leq \int_{Q\setminus K} (\mathrm{dist}(K,y))^\rho \frac{4\pi}{\rho(\mathrm{dist}(K,y))^\rho} \lambda^{(3)}(dy)$$

$$= \frac{4\pi}{\rho}\lambda^{(3)}(Q \setminus K).$$

In our calculation we needed that $x \mapsto \mathrm{dist}(K,x)$ is measurable which follows from its continuity, compare with Example II.3.28. However it is helpful to know that $\mathrm{dist}(K,\cdot)$ is Lipschitz continuous and this we see as follows: for $x, y \in \mathbb{R}^n$ fixed assume that $\mathrm{dist}(K,x) \geq \mathrm{dist}(K,y)$. Hence we can find for $\epsilon > 0$ some $z_1 \in K$ such that $\mathrm{dist}(K,y) \geq ||y - z_1|| - \epsilon$, and therefore

$$0 \leq \mathrm{dist}(K,x) - \mathrm{dist}(K,y) \leq \left(\inf_{z\in K} ||x-z|| \right) - ||y-z_1|| + \epsilon$$

$$\leq ||x-z_1|| - ||y-z_1|| + \epsilon \leq ||x-y|| + \epsilon,$$

and since $\epsilon > 0$ was arbitrary we arrive at

$$|\mathrm{dist}(K,x) - \mathrm{dist}(K,y)| \leq ||x-y||,$$

i.e. $x \mapsto \mathrm{dist}(K,x)$ is Lipschitz continuous with constant 1.

638

8. The functions $x \mapsto \frac{xy}{(x^2+y^2)^2}$ and $y \mapsto \frac{xy}{(x^2+y^2)^2}$ are on $(-1,1)$ odd functions or identically zero and therefore we have

$$\int_{-1}^{1} \frac{xy}{(x^2+y^2)^2} \lambda^{(1)}(dx) = \int_{-1}^{1} \frac{xy}{(x^2+y^2)^2} \lambda^{(1)}(dy) = 0.$$

Suppose that f is Lebesgue integrable. Then $|f|$ is Lebesgue integrable too as it was Riemann integrable and using polar coordinates we find

$$\int_{(-1,1)\times(-1,1)} |f(x,y)|\lambda^{2)}(dxdy) \geq \int_0^1 \int_0^{2\pi} \frac{(r|\cos\varphi|r|\sin\varphi|}{r^4} r d\varphi dr$$

$$= \int_0^{2\pi} |\cos\varphi||\sin\varphi|d\varphi \int_0^1 \frac{1}{r}dr.$$

However the integral $\int_0^1 \frac{1}{r}dr$ is not finite, hence $|f|$, i.e. f, cannot be Lebesgue integrable.

9. With the arguments given in Example 9.22.B we get

$$\int_{\mathbb{R}^n} e^{-t\psi(x)}\lambda^{(n)}(dx) = \int_0^\infty \lambda^{(n)}\left(\left\{e^{-t\psi} \geq y\right\}\right)dy$$

$$= \int_0^\infty \lambda^{(n)}(\{\psi \leq -\frac{1}{t}\ln y\})dy = t\int_0^\infty \lambda^{(n)}(\{\psi \leq \rho\})e^{-t\rho}ds$$

$$= t\int_0^\infty \lambda^{(n)}(B^\psi_{\sqrt{\rho}}(0))e^{-t\rho}d\rho,$$

which yields further

$$\int_{\mathbb{R}^n} e^{-t\psi(x)}\lambda^{(n)}(dx) = \int_0^\infty \lambda^{(n)}(B^\psi_{\sqrt{\frac{r}{t}}}(0))e^{-r}dr.$$

Now we observe that

$$\int_0^\infty \lambda^{(n)}(B^\psi_{\sqrt{\frac{r}{t}}}(0))e^{-r}dr \geq \int_1^\infty \lambda^{(n)}(B^\psi_{\sqrt{\frac{r}{t}}}(0))e^{-r}dr$$

$$\geq \lambda^{(n)}(B^\psi_{\sqrt{\frac{1}{t}}}(0))\int_1^\infty e^{-r}dr = \frac{1}{e}\lambda^{(n)}(B^\psi_{\sqrt{\frac{1}{t}}}(0)),$$

which gives the lower bound with $\kappa_0 = \frac{1}{e}$. For the upper bound we note that

$$\int_0^\infty \lambda^{(n)}(B^\psi_{\sqrt{\frac{r}{t}}}(0))e^{-r}dr = \int_0^1 \lambda^{(n)}(B^\psi_{\sqrt{\frac{r}{t}}}(0))e^{-r}dr + \int_1^\infty \lambda^{(n)}(B^\psi_{\sqrt{\frac{r}{t}}}(0))e^{-r}dr$$

$$\leq \left(1-\frac{1}{e}\right)\lambda^{(n)}(B_{\sqrt{\frac{1}{t}}}(0)) + \lambda^{(n)}(B_{\sqrt{\frac{1}{t}}}(0))\int_1^\infty \gamma_0(1)r^{\frac{\alpha}{2}}e^{-r}dr$$

$$= \kappa_1\lambda^{(n)}(B_{\sqrt{\frac{1}{t}}}(0)),$$

which is the upper bound. Here we have used that for $\alpha \geq 0$ the integral $\int_1^\infty r^{\frac{\alpha}{2}}e^{-r}dr$ is finite and we set $\kappa_1 := \left(1-\frac{1}{e}\right) + \gamma_0(1)\int_1^\infty r^{\frac{\alpha}{2}}e^{-r}dr$.

10. The proof is quite similar to that of Corollary 9.23. We apply again Theorem 9.20 and use the substitution $y = \varphi(s)$ to find

$$\int_\Omega (\varphi \circ u)\, d\mu = \int_0^\infty \mu(\{\varphi \circ u \geq y\})dy$$

$$= \int_0^\infty \mu(\{\varphi \circ u \geq \varphi(s)\})\varphi'(s)ds = \int_0^\infty \mu(\{u \geq y\})\varphi'(y)dy.$$

11. The following holds

$$\|K_{op}u\|_{L^2} = \left(\int_{[a,b]} | \int_{[a,b]} k(x,y)u(y)dy|^2 dx \right)^{\frac{1}{2}}$$

$$\leq \left(\int_{[a,b]} \left(\int_{[a,b]} |k(x,y)u(y)|dy \right)^2 dx \right)^{\frac{1}{2}}$$

$$(\star) \quad \leq \int_{[a,b]} \left(\int_{[a,b]} |k(x,y)|^2 |u(y)|^2 dx \right)^{\frac{1}{2}} dy$$

$$= \int_{[a,b]} \left(\int_{[a,b]} |k(x,y)|^2 dx \right)^{\frac{1}{2}} |u(y)|dy$$

$$(\star\star) \leq \left(\int_{[a,b]} \left(\int_{[a,b]} |k(x,y)|^2 \right) dy \right)^{\frac{1}{2}} \left(\int_{[a,b]} |u(y)|^2 dy \right)^{\frac{1}{2}}$$

$$= \|k\|_{L^2} \|u\|_{L^2},$$

where we used in $(\star)$ Minkowski's integral inequality and in $(\star\star)$ the Cauchy-Schwarz inequality.

Chapter 10

1. Since for $A \in O(n)$ and $x \in \mathbb{R}^n$ the mapping $y \mapsto Ty := Ay + x$ is arbitrary often differentiable it follows that $u \circ T$ is continuous for u continuous and $u \circ T \in C^k(\mathbb{R}^n)$ for $u \in C^k(\mathbb{R}^n)$. Furthermore, if $u \in C_b(\mathbb{R}^n)$, i.e. $\|u\|_\infty = M < \infty$ then $\|u \circ T\|_\infty \leq M$, i.e. $u \circ T \in C_b(\mathbb{R}^N)$. We note that T is bijective with inverse $T^{-1}z = A^{-1}(z - x) = A^{-1}z - A^{-1}x$ and that $T(B_r(0)) = A(B_r(0)) + x = B_r(x)$ since $A \in O(n)$. This implies that if $|u(y)| < \epsilon$ in the complement of some compact set K, then $|(u \circ T)(y)| < \epsilon$ in the complement of some compact set K_T, i.e. $u \in C_\infty(\mathbb{R}^n)$ yields $u \circ T \in C_\infty(\mathbb{R}^4)$. Finally, if $\operatorname{supp} u \subset \overline{B_R(0)}$ then we have $\operatorname{supp}(u \circ T) \subset \overline{B_R(x)}$, i.e. $u \in C_0(\mathbb{R}^n)$ implies $u \circ T \in C_0(\mathbb{R}^n)$.

2. We know that $(L^1(\mathbb{R}), +)$ is a vector space of $\mathbb{R}$ and by Young's inequality we have $u * v \in L^1(\mathbb{R}^n)$ for $u, v \in L^1(\mathbb{R}^n)$. Moreover, by (10.6) and (10.7) it follows that convolution is an associative and commutative operation on $L^1(\mathbb{R}^n)$. Furthermore, from the definition of convolution we get for $\alpha, \beta \in \mathbb{R}$ and $u, v, w \in L^1(\mathbb{R}^n)$

$$(\alpha u * \beta v)(x) = \int_{\mathbb{R}^n} \alpha u(x - y)\beta v(y)\lambda^{(n)}(dy) = \alpha\beta \int_{\mathbb{R}^n} u(x - y)v(y)\lambda^{(n)}(dy)$$

$$= \alpha\beta(u * v)(x),$$

and

$$((u + v) * w)(x) = \int_{\mathbb{R}^n} (u + v)(x - y)w(y)\lambda^{(n)}(dy)$$

$$= \int_{\mathbb{R}^n} u(x - y)w(y)\lambda^{(n)}(dy) + \int_{\mathbb{R}^n} v(x - y)w(y)\lambda^{(n)}(dy) = (u * w)(x) + (v * w)(x).$$

Hence $(L^1(\mathbb{R}^n), +, *)$ is a commutative $\mathbb{R}$-algebra.

3.   a) For $a \in C_b(\mathbb{R}^n)$ and $u \in C_0(\mathbb{R}^n)$ it follows that $au \in C_0(\mathbb{R}^n) \subset L^p(\mathbb{R}^n)$. Moreover we have

$$\|au\|_{L^p} = \left(\int_{\mathbb{R}^n} |a(x)u(x)|^p \lambda^{(n)}(dx)\right)^{\frac{1}{p}} = \left(\sup_{x \in \mathbb{R}^n} |a(x)|\right)\left(\int_{\mathbb{R}^n} |u(x)|^p \lambda^{(n)}(dx)\right)^{\frac{1}{p}}$$

$$= \|a\|_\infty \|u\|_{L^p}.$$

Now Proposition 10.6 combined with the density of $C_0(\mathbb{R}^n)$ in $L^p(\mathbb{R}^n)$ yields the result. Note that $C_0^\infty(\mathbb{R}^n) \subset C_0(\mathbb{R}^n) \subset L^p(\mathbb{R}^n)$ and since by Theorem 10.17 $C_0^\infty(\mathbb{R}^n)$ is dense in $L^p(\mathbb{R}^n)$ it follows that $C_0(\mathbb{R}^n)$ is dense in $L^p(\mathbb{R}^n)$ too.

b) A function $u \in C_0^\infty((0, 1))$ we can extend by zero outside of $(0, 1)$ and we obtain a C^∞-function on $\mathbb{R}$ with support in $[0, 1]$. Therefore we have $C_0^\infty((0, 1)) \subset C^1([0, 1]) \subset L^2([0, 1])$ and $C^1([0, 1])$ is dense in $L^2([0, 1])$. Consider now the sequence $(u_k)_{k \in \mathbb{N}}, u_k(x) = \sin(2\pi kx)$, of functions in $L^2([0, 1])$. It follows that $\frac{d}{dx} u_k(x) = (2\pi k)\cos(2\pi kx)$, hence $\frac{du_k}{dx} \in L^2([0, 1])$. We find further

$$\left\|\frac{du_k}{dx}\right\|_{L^2}^2 = \int_0^1 |2\pi k \cos(2\pi kx)|^2 dx = 2\pi^2 k^2,$$

whereas

$$\|u_k\|_{L^2}^2 = \int_0^1 |\sin(2\pi kx)|^2 dx = \frac{1}{2}.$$

If $\frac{d}{dx}$ was continuous in $L^2([0, 1])$ satisfying with some $c > 0$ the estimate $\left\|\frac{du}{dx}\right\|_{L^2} \leq c\|u\|_{L^2}$, then we must have

$$\left\|\frac{du_k}{dx}\right\|_{L^2} = \sqrt{2}\pi k \leq c\frac{1}{2}\sqrt{2} = c\|u_k\|_{L^2},$$

which is of course not possible. Thus $\frac{d}{dx}$ has no continuous extension from $L^2(\mathbb{R}^n)$ to $L^2(\mathbb{R}^n)$.

Remark. Note that we used for linear operators defined in a Banach space the equivalence of continuity and boundedness in the sense of $\|Tu\|_X \leq c\|u\|_X$. That a linear continuous operator is bounded can be seen as follows: for $\epsilon = 1$ there exists $\delta > 0$ such that $\|u\|_X < \delta$ implies $\|Tu\|_X = \|Tu - T0\|_X \leq 1$. If $u \neq 0$ it follows $\frac{\delta u}{2\|u\|} \in B_\delta(u)$ and therefore $\left\|T\frac{\delta u}{2\|u\|_X}\right\|_X = \frac{\delta}{2\|u\|_X}\|Tu\|_X \leq 1$, or $\|Tu\|_X \leq \frac{2}{\delta}\|u\|_X$ which is trivial for $u = 0$.

4. For $m_1, m_2 \in \mathbb{N}_0$ we find by rather rough estimates

$$(1 + ||x||^2)^{\frac{m_2}{2}} \sum_{|\alpha| \leq m_1} |\partial^\alpha u(x)| \leq \sum_{|\alpha| \leq m_1} (1 + ||x||^2)^{m_2} |\partial^\alpha u(x)|$$

$$\leq \sum_{|\alpha| \leq m_1} \sum_{l=0}^{m_2} \binom{m_2}{l} ||x||^{m_2 - l} |\partial^\alpha u(x)| = \sum_{|\alpha| \leq m_1} \sum_{l=0}^{m_2} \binom{m_2}{l} (x_1^2 + \cdots + x_n^2)^{m_2 - 1} |\partial^\alpha u(x)|,$$

and it follows

$$(1 + ||x||^2)^{\frac{m_2}{2}} \sum_{|\alpha| \leq m_1} |\partial^\alpha u(x)| \leq C \sum_{|\alpha| \leq m_1} \sum_{|\beta| \leq 2m_2} |x^{2\beta} \partial^\alpha u(x)|$$

which is finite since $p_{\alpha\beta}(u)$ is for all $\alpha, \beta \in \mathbb{N}_0^n$ finite. Thus, for $u \in \mathcal{J}(\mathbb{R}^n)$ we have for every $\alpha \in \mathbb{N}_0^n$ and every $m \in \mathbb{N}_0$ the estimate

$$(\star) \quad |\partial^\alpha u(x)| \leq C_{\alpha,m}(u)(1 + ||x||^2)^{-\frac{m}{2}}.$$

For $u, v \in \mathcal{J}(\mathbb{R}^n)$ we find now using our results on differentiating parameter dependent integrals, note that $(\star)$ implies $\partial^\alpha u \in L^1(\mathbb{R}^n) \cap L^\infty(\mathbb{R}^n)$ for all $\alpha \in \mathbb{N}_0^n$ and $u \in \mathcal{J}(\mathbb{R}^n)$, the estimate

$$x^\beta \partial_x^\alpha (u \star v)(x) = |x^\beta \partial_x^\alpha \int_{\mathbb{R}^n} u(x - y)v(y)dy|$$

$$\leq \int_{\mathbb{R}^n} |x^\beta \partial_x^\alpha u(x - y)v(y)|dy$$

$$\leq \int_{\mathbb{R}^n} (1 + ||x||^2)^{\frac{|\beta|}{2}} |(\partial^\alpha u)(x - y)| |v(y)| dy$$

$$\leq C \int_{\mathbb{R}^n} \frac{(1 + ||x||^2)^{\frac{|\beta|}{2}}}{(1 + ||x - y||^2)^{\frac{|\beta|}{2}}} (1 + ||y||^2)^{-\frac{k}{2}} dy,$$

where we used $(\star)$ for $(\partial^\alpha u)$ and v and k will determined later. From Peetre's inequality we deduce

$$\frac{(1 + ||x||^2)^{\frac{|\beta|}{2}}}{(1 + ||x - y||^2)^{\frac{|\beta|}{2}}} \leq 2^{\frac{|\beta|}{2}} (1 + ||y||^2)^{\frac{|\beta|}{2}}$$

which yields

$$|x^\beta \partial_x^\alpha (u * v)(x)| \leq C' \int (1 + ||y||^2)^{\frac{|\beta| - k}{2}} dy$$

and for $k > n + |\beta|$ we find $P_{\alpha\beta}(u * v) < \infty$, hence $u * v \in \mathcal{J}(\mathbb{R}^n)$.

5. Since $K \cap \partial G = \emptyset$ it follows that $\mathrm{dist}(K, \partial G) > 0$. We choose $\epsilon > 0$ such that $\epsilon < \frac{1}{2} \mathrm{dist}(K, \partial G)$. For

$$K_\epsilon := \{x \in \mathbb{R}^n \mid \text{ there exists } y \in K \text{ such that } ||x - y|| < \epsilon\} = K + B_\epsilon(0)$$

we consider $J_\epsilon(\chi_{K_\epsilon})$, i.e.

$$J_\epsilon(\chi_{K_\epsilon})(x) = \int_{\mathbb{R}^n} j_\epsilon(x-y)\chi_{K_\epsilon}(x-y)\lambda^{(n)}(dy) = \int_{B_\epsilon(0)} \chi_{K_\epsilon}(x-y)j_\epsilon(y)\lambda^{(n)}(dy),$$

which is nothing but the Friedrichs Mollifier applied to the characteristic functions of K_ϵ. Thus $J_\epsilon(\chi_{K_\epsilon}) \in C^\infty(\mathbb{R}^4), 0 \le J_\epsilon\left(\chi_{K_\epsilon}\right) \le 1$ and since $\epsilon < \frac{1}{2}\operatorname{dist}(K, \partial G)$ we have

$$\operatorname{supp} J_\epsilon\left(\chi_{K_\epsilon}\right) \subset \overline{B_\epsilon(0)} + \overline{K_\epsilon} \subset \overline{B_{2\epsilon}(0)} + K \subset G.$$

Finally, for $x \in K$ and $y \in B_\epsilon(0)$ we find that $\chi_{K_\epsilon}(x-y) = 1$ implying that $J_\epsilon\left(\chi_{K_\epsilon}\right)(x) = 1$.

6. Consider the sets $G_{j,\epsilon} := \{x \in G_j \mid \inf_{y\in\partial G_j} ||x-y|| > \epsilon\}$, see the Figure below.

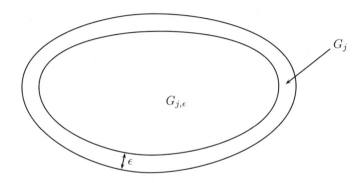

There exists $\epsilon_0 > 0$ such that for $0 < \epsilon < \epsilon_0$, the sets $G_{j,\epsilon}$ are non-empty and open. Furthermore $\overline{G_{j,\epsilon}}$ is compact and $\overline{G_{j,\epsilon}} \subset G_j$. We claim now the existence of $\epsilon_1 > 0$ such that $0 < \epsilon < \epsilon_1 \le \epsilon_0$ implies $K \subset \bigcup_{j=1}^N G_{j,\epsilon}$. If this was not the case then there would be a sequence $(x_l)_{l\in\mathbb{N}}, x_l \in K$, such that $x_l \in \bigcup_{j=1}^N G_{j,\frac{1}{l}}$. Since K is compact $(x_l)_{l\in\mathbb{N}}$ has a convergent subsequence which we denote again by $(x_l)_{l\in\mathbb{N}}$ and which has a limit $x \in K$. Since $K \subset \bigcup_{j=1}^N G_j$ there exists j_0 such that $x \in G_{j_0}$ and hence $x \in G_{j_0,\frac{1}{k}}$ for $k \ge k_0$. This however implies for $l \ge k_0$ that $x_l \in G_{j_0,\frac{1}{k}}$ and hence $x_l \in \bigcup_{j=1}^N G_{j,\frac{1}{k}}$ which is a contradiction. Hence for $0 < \epsilon < \epsilon_1$ we have $K \subset \bigcup_{j=1}^N G_{j,\epsilon}$. We now fix $\epsilon > 0$ with these properties. Since K is compact and $\bigcup_{j=1}^N G_{j,\epsilon}$ is open, there exists $\eta > 0$ such that

$$\{x \in \mathbb{R}^n \mid \text{ there exists } y \in K \text{ such that } ||x-y|| < \eta\} := K_\eta \subset \bigcup_{j=1}^N G_{j,\epsilon}.$$

Let $\psi_j \in C_0^\infty(G_j) \subset C_0^\infty(\mathbb{R}^n)$ such that $\operatorname{supp}\psi_j \subset G_j, 0 \le \psi_j \le 1$ and $\psi|_{\overline{G_{j,\epsilon}}} = 1$. By Problem 5 such a function ψ_j does exist. It follows that $\left(\sum_{j=1}^N \psi_j\right)|_K \ge 1$.

For $\psi \in C_0^\infty(K_\eta) \subset L_0^\infty(\mathbb{R}^n)$ such that $\psi|_K = 1$ we define the functions $\varphi_j(x) := \frac{\psi_j(x)}{\sum_{l=1}^\infty \psi_l(x)}\psi(x)$, which belong to $C_0^\infty(G_j) \subset C_0^\infty(\mathbb{R}^n)$ and for $x \in K$ it follows that $\sum_{j=1}^N \psi_j(x) = \psi(x) = 1$. (Our solution follows closely [89].)

7. For $A \in \mathcal{B}^{(n)}$ we have

$$(\mu_t * \mu_s)(A) = \int_\mathbb{R} \left(\int_\mathbb{R} \chi_A(x+y)\mu_s(dx) \right) \mu_t(dy)$$

$$= \int_\mathbb{R} \left(\int_\mathbb{R} \chi_A(x+y)e^{-as}\epsilon_0(dx) \right) e^{-at}\epsilon_0(dy)$$

$$= \int_\mathbb{R} e^{-as}\chi_A(y)e^{-at}\epsilon_0(dy)$$

$$= e^{-a(s+t)}\chi_A(0) = \int_\mathbb{R} \chi_A(z)e^{-a(s+t)}\epsilon_0(dz)$$

$$= \mu_{t+s}(A).$$

8. Under our assumptions the following calculation is justified

$$\left(T_s \circ T_s u\right)(x) = \int_{\mathbb{R}^n} \left(\int_{\mathbb{R}^n} u(x - z - y)\mu_t(dy) \right) \mu_s(dz)$$

$$= \int_{\mathbb{R}^n} u(x-z)(\mu_t * \mu_s)(dz) = \int_{\mathbb{R}^n} u(x-z)\mu_{t+s}(dz)$$

$$= T_{s+t}u(x).$$

9. Since k_α is even and non-negative we find

$$\int_\mathbb{R} k_\alpha(x)dx = 2\int_0^\infty \chi_{[0,1]}(x)|x|^{-\alpha}dx$$

$$= 2\int_0^1 |x|^{-\alpha}dx = \frac{2}{1-\alpha}|x|^{1-\alpha}\Big|_0^1 = \frac{2}{1-\alpha} < \infty,$$

hence $k_\alpha \in L^1(\mathbb{R})$ and by Young's inequality we find for $u \in C_0(\mathbb{R}) \subset L^p(\mathbb{R})$

$$||K_{op}u||_{L^p} \leq \frac{2}{1-\alpha}||u||_{L^p}$$

which allows us to obtain a continuous extension of K_{op} to $L^p(\mathbb{R})$ which satisfies the same estimate. Now we turn to $k_{\gamma,n}(||x||) = \chi_{\overline{B_1}(0)}(x)||x||^{-\gamma}$ in $\mathbb{R}^n$. It follows with the help of spherical coordinates that

$$\int_{\mathbb{R}^n} k_{\gamma,n}(||x||)dx = \int_{\mathbb{R}^n} \chi_{\overline{B_1}(0)\backslash\{0\}}(x)||x||^{-\gamma}dx$$

$$= C_n \int_0^1 r^{-\gamma}r^{n-1}dr = C_n \int_0^1 r^{n-1-\gamma}dr = C_n \frac{1}{n-\gamma}r^{n-\gamma}\Big|_0^1$$

$$= \frac{C_n}{n - \gamma},$$

provided $n - \gamma > 1$ i.e. $\gamma < n$. Note that by (II.12.26) we have

$$C_n = 2\pi \int_0^\pi \left(... \int_0^\pi \sin^{n-2} \vartheta_1 \cdot ... \cdot \sin \vartheta_{n-2} d\vartheta_{n-2} \right) ... d\vartheta_1.$$

Now we may argue as before to find that $K_{op}^{(n)}$ has a continuous extension to $L^p(\mathbb{R}^n)$ satisfying

$$\|K_{op}^{(n)} u\|_{L^p} \leq \frac{C_n}{n - \gamma} \|u\|_{L^p}.$$

Chapter 11

1. Since g is continuous and $g(x) \neq 0$ on $[a, b]$ it follows for some c_0 that $|g(x)| \geq c_0 > 0$ for all $x \in [a, b]$ implying that

$$\left| \frac{1}{g(x)} - \frac{1}{g(y)} \right| \leq \frac{|g(x) - g(y)|}{c_0^2}.$$

This estimate however implies the absolute continuity of $\frac{1}{g}$ and therefore it follows that $\frac{f}{g}$ is absolutely continuous too.

2.   a) For φ there exists a constant $L > 0$ such that $|\varphi(x) - \varphi(y)| \leq L|x - y|$ holds for all $x, y \in [a, b]$. This implies

$$|(\varphi \circ f)(x) - (\varphi \circ f)(y)| \leq L|f(x) - f(y)|$$

for all $x, y \in [a, b]$ and again the absolute continuity of f implies that of $\varphi \circ f$.

b) Let $\epsilon > 0$ and $\delta > 0$ such that for every finite number of intervals $(x_k, y_k) \subset [a, b]$, $k = 1, ..., N$, with $\sum_{k=1}^N (y_k - x_k) < \delta$ it follows that $\sum_{k=1}^N |f(x_k) - f(y_k)| < \epsilon$. Suppose that g is an increasing function. Since it is by assumption absolutely continuous we can find $\nu < 0$ such that for every finite collection of intervals $(s_j, t_j) \subset [a, b], j = 1, ..., M$, with $\sum_{j=1}^N (t_j - s_j) < \nu$ it follows $\sum_{j=1}^M (g(t_j) - g(s_j)) < \delta$. This implies of course

$$\sum_{j=1}^M |f(g(t_j)) - f(g(s_j))| < \epsilon.$$

The case of a decreasing function g goes analogously.

3. Let $N \subset (a, b)$ be a set of measure zero and $\epsilon > 0$. We can find $\delta > 0$ such that for every pairwise disjoint, finite family of intervals $(x_k, y_k) \subset [a, b]$, $1 \leq k \leq N$, with $\sum_{k=1}^N (y_k - x_k) < \delta$ it follows that $\sum_{k=1}^N |f(x_k) - f(y_k)| < \epsilon$. For the set $N \subset (a, b)$ of measure zero we can find an open set $U \subset (a, b)$ with $N \subset U$ and $\lambda^{(1)}(U) < \delta$. Since every open set in $\mathbb{R}$ is the the denumerable union of open intervals, compare with Theorem I.19.27, there exist pairwise disjoint open intervals (s_j, t_j), $j \in \mathbb{N}$, such and $U \subset \bigcup_{j \in \mathbb{N}} (s_j, t_j)$ and $\sum_{j \in \mathbb{N}}^\infty (t_j - s_j) < \delta$. On $[s_j, t_j]$ the continuous function f

645

attains its minimum, say at α_j, and its maximum, say at β_j, with $a_j := \alpha_j \wedge \beta_j$ and $b_j = \alpha_j \vee \beta_j$ it follows that $f([s_j, t_j]) = [f(a_j), f(b_j)]$. For every $N \in \mathbb{N}$ we have $\sum_{j=1}^{N}(b_j - a_j) < \delta$ and consequently $\sum_{j=1}^{N}|f(a_j) - f(b_j)| < \epsilon$, which yields in the limit $N \to \infty$ that $\sum_{j=1}^{\infty}|f(a_j) - f(b_j)| < \epsilon$.

However

$$f(N) \subset f(U) = \bigcup_{j \in \mathbb{N}} f((s_j, t_j)) \subset \bigcup_{j \in \mathbb{N}} f([s_j, t_j]),$$

implying

$$\lambda^{(1)}(f(N)) \leq \sum_{j=1}^{\infty} \lambda^{(1)}(f([s_j, t_j])) \leq \sum_{j=1}^{\infty} |f(a_j) - f(b_j)| \leq \epsilon,$$

i.e. $\lambda^{(1)}(f(N)) = 0$. Note that since $\lambda^{(1)}(\{a\}) = \lambda^{(1)}(\{b\}) = 0$ the assumption $N \subset (a, b)$ is no restriction.

4. Since $[a, b]$ is of finite measure it follows that $L^p([a, b])$ is a subset of $L^1([a, b])$. With $G(x) := \int_{a-h}^{x} g(t)dt$ we find that

$$g_h(x) := \frac{1}{2h}\left(G(x+h) - G(x-h)\right)$$

and by Corollary 11.10 the function G, hence g_k, is a continuous function. For $p = 1$ we have

$$|g_h(x)| \leq \frac{1}{2h}\int_{x-h}^{x+h}|g(t)|dt,$$

and for $1 \leq p \leq \infty$, $\frac{1}{p} + \frac{1}{q} = 1$, Hölder's inequality gives

$$|g_h(x)|^p \leq \frac{1}{(2h)^p}\left(\int_{x-h}^{x+h} 1dt\right)^{\frac{p}{q}}\left(\int_{x-h}^{x+h}|g(t)|^p dt\right)$$

$$= \frac{1}{2h}\int_{x-h}^{x+h}|g(t)|^p dt.$$

Now we find for $1 \leq p < \infty$ that

$$\int_{a}^{b}|g_h(x)|^p dx \leq \frac{1}{2h}\int_{a}^{b}\left(\int_{x-h}^{x+h} 1|g(t)|^p dt\right)dx$$

$$= \frac{1}{2h}\int_{a}^{b}\left(\int_{-h}^{h}|g(t+x)|^p dt\right)dx = \frac{1}{2h}\int_{-h}^{h}\left(\int_{a}^{b}|g(t+x)|^p dx\right)dt$$

$$= \frac{1}{2h}\int_{-h}^{h}\left(\int_{a+t}^{b+t}|g(s)|^p ds\right)dt.$$

Since $g|_{[a,b]^c} = 0$ we get further

$$\frac{1}{2h}\int_{-h}^{h}\left(\int_{a+t}^{b+t}|g(s)|^p ds\right)dt \leq \frac{1}{2h}\int_{-h}^{h}\left(\int_{a}^{b}|g(s)|^p ds\right)dt,$$

or

$$\int_a^b |g_k(x)|^p dx \le \int_a^b |g(s)|^p ds,$$

implying that $\|g_h\|_{L^p} \le \|g\|_{L^p}$.

5. If we set $\varphi(x) = f(a) + \int_a^x f'(t)dt$ and $s(x) = f(x) - \varphi(x)$, it follows that φ is absolutely continuity with $\varphi(a) = f(a)$, and for s it follows that s is either identically zero or s is a singular function. For two decomposition of f,

$$f(x) = \varphi(x) + s(x) = \tilde{\varphi}(x) + \tilde{s}(x)$$

we find $\varphi(x) - \tilde{\varphi}(x) = \tilde{s}(x) - s(x)$, which implies by the absolute continuously of φ and $\tilde{\varphi}$ that $(\varphi - \tilde{\varphi})' = 0$ almost everywhere, hence $(\varphi - \tilde{\varphi}) = $ constant, and from $\varphi(a) = \tilde{\varphi}(a) - f(a)$ we deduce $\varphi = \tilde{\varphi}$ which in turn implies $s = \tilde{s}$, i.e. the decomposition of f is unique.

6. Before turning to the calculation let us understand the meaning of this result: there are functions having a derivative everywhere, but still the fundamental theorem in the form of $u(x) - u(y) = \int_y^x u'(t)dt$ does not necessarily hold.

The function u is bounded on $[0, 1]$ and we have $|u(x)| \le x^2, u(0) = 0$. We extend u as an even function to $[-1, 1]$ or directly modify the solution to Problem 7 of Chapter I.21 to see that u is differentiable at $x = 0$. The differentiability at all other points follows from the chain rule. For $0 < a < b < b \le 1$ we can apply the fundamental theorem to find

$$\int_a^b u'(t)dt = b^2 \cos \frac{\pi}{b^2} - a^2 \cos \frac{\pi}{a^2}.$$

Taking $a_n := \sqrt{\frac{2}{4n+1}}$ and $b_n := \sqrt{\frac{1}{2n}}$ it follows that

$$\int_{a_n}^{b_n} u'(t)dt = \frac{1}{2u}.$$

Now, $[a_n, b_n] \cap [a_l, b_l) = \emptyset$ for $k \ne l$ and therefore we find with $A := \bigcup_{n \in \mathbb{N}} [a_n, b_n] \subset [0, 1]$ that

$$\int_A |u'(t)|dt \ge \sum_{n=1}^\infty \frac{1}{2n} = +\infty,$$

i.e. $u' \notin L^1([0, 1])$.

7. As a convex function g is continuous. By the proof of Theorem I.23.4 we know that $h \mapsto \frac{g(x+h)-g(x)}{h} = -\left(\frac{g(x)-g(x+h)}{h}\right)$ is decreasing in h. We also know that the derivatives $g'_+(x)$ from the right and $g'_-(x)$ from the left exist and

$$g'_+(x) = \lim_{h \to 0, h > 0} \frac{g(x+h) - g(x)}{h}$$

647

as well as

$$g'_-(x) = \lim_{h \to 0, h > 0} \frac{g(x) - g(x - h)}{h}$$

hold. Moreover, we have

$$\frac{g(x) - g(x - h)}{h} \leq \frac{g(x + h) - g(x)}{h}, h > 0,$$

implying that $g'_-(x) \leq g'_+(x)$ and both are finite numbers. Now we claim that $a < y < x < b$ implies that $g'_+(y) \leq g'_-(x)$. This follows from (I.23.4) which yields

$$g'_+(y) \leq \frac{g(x) - g(y)}{x - y} \leq g'_-(x).$$

Thus we have

$$(\star) \quad g'_+(y) \leq g'_-(x) \leq g'_+(x).$$

implying that g'_+ is monotone increasing. The proof that g'_- is monotone increasing is similar. By Problem 6 to Chapter I.20 we know that both g'_+ and g'_- have at most countable many points of discontinuity. For a point of continuity x of g'_+ the estimates $(\star)$ implies $g'_+(x) = g'_-(x)$. Thus g' exists at every point of continuity of g'_+, in particular $\lambda^{(1)}$-almost everywhere, and is monotone increasing.

8. For simplicity we set $\mu(s) := \lambda^{(1)}(\{\mathcal{M}(f) > s\})$ and

$$g(x) := \begin{cases} f(x), & |f(x)| \geq \frac{s}{2} \\ 0, & \text{otherwise.} \end{cases}$$

It follows that $|f(x)| \leq |g(x)| + \frac{s}{2}$ and consequently

$$\mathcal{M}(f)(x) \leq \sup_{h > 0} \frac{1}{2h} \int_{x-h}^{x+h} |g(y)| dy + \frac{s}{2} = \mathcal{M}(g)(x) + \frac{s}{2}.$$

Moreover, since $\{x \in \mathbb{R} \mid \mathcal{M}(f)(x) > s\} \subset \{x \in \mathbb{R} \mid \mathcal{M}(g) > \frac{s}{2}\}$ we find by Theorem 11.28 that

$$\mu(s) \leq \lambda^{(1)}(\{\mathcal{M}(g) > \frac{s}{2}\}) \leq \frac{6}{s} \|g\|_{L^1}$$

$$= \frac{6}{s} \int_{|f| \geq \frac{s}{2}} |f(y)| dy.$$

Now we find

$$\int_{\mathbb{R}} |\mathcal{M}(f)|^p \lambda^{(1)} = p \int_0^\infty \mu(s) s^{p-1} ds$$

$$\leq p \int_0^\infty s^{p-1} \left(\frac{6}{s} \int_{|f| \geq \frac{s}{2}} |f(y)| dy \right) ds = 6p \int_{\mathbb{R}} |f(y)| \left(\int_0^{2|f(y)|} s^{p-2} ds \right) dy,$$

where in the last step we used Tonelli's theorem. Since

$$(\star) \quad \int_0^{2|f(y)|} s^{p-2} ds = \frac{(2|f(y)|)^{p-1}}{p - 1}, p > 1,$$

we arrive at

$$\int_{\mathbb{R}} |\mathcal{M}(f)|^p d\lambda^{(1)} \leq 2^p \frac{3p}{p-1} \int_{\mathbb{R}} |f|^p d\lambda^{(1)},$$

i.e.

$$\leq ||\mathcal{M}||_{L^p} \leq 2 \left(\frac{3p}{p-1} \right)^{\frac{1}{p}} ||f||_{L^p}.$$

A remark is needed to $(\star)$. Since $f \in L^p(\mathbb{R})$, f is $\lambda^{(1)}$-almost everywhere finite, hence $(\star)$ holds almost everywhere only, which however is sufficient for our purpose. It is now easy to extend the result to $\mathbb{R}^n$, a detailed proof which we need for our case is given in [90].

9. The function g is bounded, i.e. $|g(x)| \leq 1$ for all $x \in \mathbb{R}$, and since $\sum_{k=0}^{\infty} \left(\frac{3}{4} \right)^k = 4$, the Weierstrass test implies the continuity of f.

Now let $x \in \mathbb{R}$ and $m \in \mathbb{N}$. The intervals $(4^m x, 4^m x + \frac{1}{2})$ and $(4^m x - \frac{1}{2}, 4^m x)$ cannot both contain an integer. Thus with $\nu_m := \pm \frac{1}{2} 4^{-m}$ chosen approximately, in the interval with endpoints $4^m x$ and $4^m (x + \nu_m)$ there is no integer. Note that for $m \in \mathbb{N}$ fixed the term

$$\frac{g(4^m(x + \nu_m)) - g(4^m x)}{\nu_m}, \; n \in \mathbb{N}_0,$$

has always the same sign and moreover we have

$$\left| \frac{g(4^m(x + \nu_m)) - g(4^m x)}{\nu_m} \right| = \begin{cases} 0, & n > m \\ 4^n, & 0 \leq n \leq m. \end{cases}$$

Here we need that for $n > m$ we have $4^n \nu_m = \pm \frac{1}{2} 4^{n-m} = \pm 2l$ and g has period 2. It follows that for f that

$$\left| \frac{f(x + \nu_m) - f(x)}{\nu_m} \right| = \left| \sum_{k=0}^{\infty} \left(\frac{3}{4} \right)^k \frac{g(4^k(x + \nu_m)) - g(4^k x)}{\nu_m} \right|$$

$$= \left| \sum_{k=0}^{m} \left(\frac{3}{4} \right)^k \frac{g(4^k(x + \nu_m)) - g(4^k x)}{\nu_m} \right|$$

$$= \sum_{k=0}^{m} \left(\frac{3}{4} \right)^k \cdot 4^k = \frac{3^{m+1} - 1}{2}.$$

This implies however that for $\nu_m \to 0$, i.e. $m \to \infty$, the limit $\lim_{m \to \infty} \frac{f(x+\nu_m)-f(x)}{\nu_m}$ cannot exist, i.e. $f'(x)$ does not exist.

Chapter 12

1. Let $x_1, x_2 \in G$ be two solutions of $f(x) = y_0$. It follows that $0 = f(x_2) - f(x_1) = J_f(x_1)(x_2 - x_1) + \varphi(x_2 - x_1)$ and $\lim_{x_2 \to 1} \frac{||\varphi(x_2 - x_1)||}{||x_2 - x_1||} = 0$. Since x_1 is not a critical

point, the symmetric matrix $J_f(x_1)$ has full rank and we can find some $\nu > 0$ such that $||J_f(x_1)z|| \geq \nu||z||$ for all $z \in \mathbb{R}^n$. This implies

$$\nu||x_2 - x_1|| \leq ||J_f(x_1)(x_2 - x_1)|| + ||\varphi(x_2 - x_1)||$$

or

$$0 < \nu \leq \frac{||\varphi(x_2 - x_1)||}{||x_2 - x_1||}.$$

Since $\lim_{h \to 0} \frac{||\varphi(h)||}{||h||} = 0$ it follows that any second solution of $f(x) = y_0$ must have a positive distance to x_1, i.e the solution of $f(x) = y_0$ are isolated, implying that in G we can have at most finitely many solutions to $f(x) = y_0$.

2. We know that $C \subset [0, 1]$ is a compact set which is non-denumerable and has Lebesgue measure zero, see Theorem I.32.4. Therefore g is a Lipschitz continuous function on $\mathbb{R}$ and $g \mid_C = 0$, but $g(x) > 0$ for $x \in \mathbb{R} \setminus C$. This implies that $h : [0, 1] \to \mathbb{R}$ defined by $h(x) := \int_0^x g(t)dt$ is a strictly monotone increasing function which is continuously differentiable and $h'(x) = 0$ for $x \in C$. Since f is injective the set $h(\mathrm{crit}(h|_{(0,1)})) \cup \{h(0), h(1)\} = h(C)$ is non-denumerable, whereas h has only the (local) extreme values $h(0) = 0$ and $h(1) = \int_0^1 g(t)dt$. Thus $Ex(h)$ has two elements but $h(\mathrm{crit}(h|_{(0,1)}))$ is non-denumerable.

3. We only need to apply Corollary 12.5.

4. Let $\{\alpha_1, ..., \alpha_N\} \subset \mathbb{R}$ be the range of s and set $A_j = \{x \in A | s(x) = \alpha_j\}$. Clearly, A_j is measurable. By Lemma 12.6, given $\epsilon > 0$ we can find closed sets $C_{j,\epsilon} \subset A$ such that $\lambda^{(n)}(A_j \setminus C_{j,\epsilon}) \leq \frac{\epsilon}{N}, j = 1, ..., N$. with $C_\epsilon := \bigcup_{j=1}^N C_{j,\epsilon}$ we find $\lambda^{(n)}(A \setminus C_\epsilon) \leq \epsilon$. Since $\mathrm{dist}(C_{j,\epsilon}, C_{l,\epsilon}) > 0$ and $s|_{C_{j,\epsilon}} = \alpha_j$ it follows that $s|_{C_\epsilon}$ is continuous.

5. For $k \in \mathbb{N}$ we can find closed sets $C_k \subset A$ such that $\lambda^{(n)}(A \setminus C) < \frac{1}{k}$ and $f_k := f|_{C_k}$ is continuous. The set $C := \bigcup_{k \in \mathbb{N}} C_k$ belongs to A and $\lambda^{(n)}(A \setminus C) = 0$. Furthermore, for $a \in \mathbb{R}$ we have

$$\{x \in A \mid f(x) > a\}$$

$$= \{x \in C \mid f(x) > a\} \cup \{x \in A \setminus C \mid f(x) > a\}$$

$$= \bigcup_{k \in \mathbb{N}} \{x \in C_k \mid f_k(x) > a\} \cup \{x \in A \setminus C \mid f(x) > a\}.$$

On C_k the functions f_k is continuous and therefore measurable, which yields that $\{x \in C_k \mid f_k(x) > a\}$ is measurable and since $\{x \in A \setminus C \mid f(x) > a\}$ is a set of measure zero it follows that $\{x \in A \mid f(x) > a\}$ is measurable, hence f is measurable.

6. For $\epsilon > 0$ choose $\nu > 0$ such that $\delta := \epsilon - \nu > 0$. Then we can find points $x_1, ..., x_N$ such that $Y \subset \bigcup_{j=1}^N \overline{B_\delta(x_j)}$, but $\overline{B_\delta(x_j)} \subset B_\epsilon(x_j)$ and therefore we have $Y \subset \bigcup_{j=1}^N B_\epsilon(x_j)$, i.e. Y admits a finite ϵ-net. Clearly, if Y admits for every $\epsilon > 0$ a finite ϵ-net, i.e. $Y \subset \bigcup_{j=1}^M B_\epsilon(y_j)$, then we find $Y \subset \bigcup_{j=1}^M \overline{B_\epsilon(y_j)}$, i.e. the assumptions of Problem 6 are fulfilled.

7.   a) Given $\epsilon > 0$, choose $N > \frac{b-a}{\epsilon}$. Then $\{a + \epsilon, a + 2\epsilon, ..., a + N\epsilon\}$ is a finite ϵ-net for (a, b).

b) Since $v \in Y$ implies $0 \leq v(t) \leq 1$ for all $t \in [0, 1]$ it follows that $||v||_\infty \leq 1$ for all $v \in Y$, i.e. Y is a bounded set in $C([0, 1])$, in fact it is a subset of the unit ball $\overline{B_1(0)}$, where 0 denotes the zero function on $[0, 1]$. Since for all $v \in Y$ we have $\sup_{t \in [0,1]} |v(t) - \frac{1}{2}| \leq \frac{1}{2}$ it follows from the solution of Problem 6 that Y admits a finite $\frac{1}{2}$-net. The following figure indicates how u may look like:

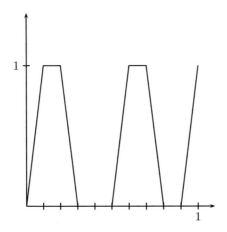

By construction we have $u \in Y$ and $||u - u_j||_\infty \geq \frac{1}{2} > \epsilon$ for all $j \in \{1, ..., m\}$ implying that $\{u_1, ...u_m\}$ cannot be an ϵ-net for Y.

651

Solutions to Problems of Part 7

Chapter 13

1. a)
$$\frac{3+5i}{2i-7} + \frac{4i}{3+i} = \frac{(3+5i)(-7-2i)}{(-7+2i)(-7-2i)} + \frac{4i(3-i)}{(3+i)(3-i)}$$
$$= \frac{-11-41i}{53} + \frac{4+12i}{10}$$
$$= \frac{51+113i}{265};$$

b)
$$\frac{(2i)^8 - 128}{(2+2i)(2-2i)} = \frac{2^8 i^8 - 128}{4+4} = \frac{256-128}{8} = 16$$
note $i^8 = i^2 \cdot i^2 \cdot i^2 \cdot i^2 = (-1)(-1)(-1)(-1) = 1$;

c)
$$\frac{\frac{1-i}{2+3i}}{\frac{2+4i}{6-2i}} = \frac{1-i}{2+3i} \cdot \frac{6-2i}{2+4i} = \frac{4-8i}{-8+10i} = \frac{2-4i}{-4+5i}$$
$$= \frac{(2-4i)(-4-5i)}{16+25} = \frac{-28+6i}{41};$$

d)
$$\left(-\frac{1}{2} + \frac{\sqrt{3}}{2}i\right)\left(-\frac{1}{2} + \frac{\sqrt{3}}{2}i\right) = \frac{1}{4} - \frac{\sqrt{3}}{4}i - \frac{\sqrt{3}}{4}i - \frac{3}{4}$$
$$= -\frac{1}{2} - \frac{\sqrt{3}}{2}i.$$

2. a)
$$\frac{(a+ib)^2 - 2iab}{i(a-b)} = \frac{a^2 + 2iab - b^2 - 2iab}{i(a-b)}$$
$$= \frac{a^2 - b^2}{i(a-b)} = -i(a+b);$$

b) With $z_+ = z_1$, $z_- = z_2$ we find
$$z_\pm^2 = \left(-\frac{p}{2} \pm i\sqrt{q - \frac{p^2}{2}}\right)^2$$
$$= \frac{p^2}{2} - q \mp ip\sqrt{q - \frac{p^2}{2}}$$
$$pz_\pm = -\frac{p^2}{2} \pm ip\sqrt{q - \frac{p^2}{4}}$$

and therefore

$$z_\pm^2 + pz_\pm + q = \frac{p^2}{2} - q \mp ip\sqrt{q - \frac{p^2}{4}} - \frac{p^2}{2} \pm ip\sqrt{q - \frac{p^2}{4}} + q = 0.$$

3. a) With $z_j = x_j + iy_j$ we have $\bar{z}_j = x_j - iy_j$ and it follows

$$\overline{z_1 \cdot z_2} = \overline{(x_1 + iy_1)(x_2 + iy_2)}$$
$$= \overline{(x_1 x_2 - y_1 y_2) + i(x_1 y_2 + x_2 y_1)}$$
$$= x_1 x_2 - y_1 y_2 - i(x_1 y_2 + x_2 y_1)$$

and

$$\bar{z}_1 \cdot \bar{z}_2 = (x_1 - iy_1)(x_2 - iy_2)$$
$$= x_1 x_2 - y_1 y_2 + i(-x_1 y_2 - x_2 y_1)$$
$$= x_1 x_2 - y_1 y_2 - i(x_1 y_2 + x_2 y_1),$$

implying the assertion.

b) $z \cdot \bar{z} = (x + iy)(x - iy) = x^2 - ixy + ixy + y^2 = x^2 + y^2.$

c)

$$\frac{z + \bar{z}}{2} = \frac{x + iy + x - iy}{2} = \frac{2x}{2} = \operatorname{Re} z$$

and

$$\frac{z - \bar{z}}{2i} = \frac{x + iy - x + iy}{2i} = \frac{2iy}{2i} = \operatorname{Im} z.$$

4. The binomial theorem, see Theorem I.3.9, has for $\mathbb{C}$ the same form as for $\mathbb{R}$: for $z, w \in \mathbb{C}$ and $n \in \mathbb{N}$ we have

$$(z + w)^n = \sum_{k=0}^{n} \binom{n}{k} z^{n-k} w^k.$$

There is also no need to modify the proof: the manipulations of terms involving z and w use only the axioms of a field and the manipulations of the binomial coefficients are unchanged.

5. a) The triangle inequality yields

$$|z| \le |z - w| + |w| \quad \text{or} \quad |z| - |w| \le |z - w|$$

as well as

$$|z| \le |z + w| + |-w| \quad \text{or} \quad |z| - |w| \le |z + w|,$$

and therefore

$$|z| - |w| \le |z - w| \wedge |z + w|.$$

Changing the role of z and w we find

$$-(|z| - |w|) = |w| - |z| \le |z - w| \wedge |z + w|,$$

654

which gives eventually

$$||z| - |w|| \leq |z - w| \wedge |z + w|.$$

b) For $z = x + iy$ and $|z| = \left(x^2 + y^2\right)^{\frac{1}{2}}$ we derive immediately $|z| \leq |x| + |y|$. Since

$$\frac{|x|^2 + 2|x||y| + |y|^2}{2} = \frac{|x|^2 + |y|^2}{2} + |x||y| \leq \frac{|x|^2 + |y|^2}{2} + \frac{|x|^2 + |y|^2}{2}$$

or

$$\frac{(|x| + |y|)^2}{2} \leq |x|^2 + |y|^2$$

implying

$$\frac{|x| + |y|}{\sqrt{2}} \leq |z|.$$

6.   a) For $f, g \in V$ we have of course $d(f, g) = \|f - g\| \geq 0$ and $d(f, g) = 0$ implies $\|f - g\| = 0$, i.e. $f = g$. Moreover it follows that

$$d(f, g) = \|f - g\| = \| - 1(g - f)\| = |- 1|\|g - f\| = \|g - f\| = d(f, g).$$

Eventually we observe that

$$d(f, g) = \|f - g\| = \|f - h + h - g\| \leq \|f - h\| + \|h - g\| = d(f, h) + d(h, g)$$

and we have proved that d is a metric.

b) Clearly $\|z\| \geq 0$ and $\|z\| = 0$ implies $|z_j| = 0$ for $j = 1, \ldots, n$, i.e. $z = 0$. For $\lambda \in \mathbb{C}$ we find that

$$\|\lambda z\| + \left(\sum_{j=1}^{n} |\lambda z_j|\right)^{\frac{1}{2}} = \left(\sum_{j=1}^{n} |\lambda|^2 |z_j|^2\right)^{\frac{1}{2}} = |\lambda| \left(\sum_{j=1}^{n} |z_j|^2\right)^{\frac{1}{2}} = |\lambda| \|z\|.$$

In order to prove the triangle inequality we first rewrite $\|z\|$. Let $z_j = x_j + i x_{n+j}$, $1 \leq j \leq n$, $x_l \in \mathbb{R}$, $1 \leq l \leq 2n$. It follows that $|z_j|^2 = x_j^2 + x_{n+j}^2$ and

$$\left(\sum_{j=1}^{n} |z_j|^2\right)^{\frac{1}{2}} = \left(\sum_{l=1}^{2n} x_l^2\right)^{\frac{1}{2}},$$

in other words $\|z\|$ is the Euclidean norm of the vector $(x_1, \ldots, x_{2n})$. With $w_j = y_j + i y_{n+j}$, $1 \leq j \leq n$, $y_l \in \mathbb{R}$, $1 \leq l \leq 2n$, we find now

$$\|z + w\| = \|(x_1 + y_1, \ldots, x_{2n} + y_{2n})\|$$

where the norm on the right hand side is the Euclidean norm in $\mathbb{R}^{2n}$ and therefore we get

$$\|z + w\| \leq \|(x_1, \ldots, x_{2n})\| + \|(y_1, \ldots, y_{2n})\| \leq \|z\| + \|w\|.$$

7. a) Note that

$$\frac{3n^2 - 5in}{n^2 i} = \frac{(3n^2 - 5in)(-i)}{n^2} = \frac{-3n^2 i - 5n}{n^2} = -\frac{5}{n} - 3i$$

implying

$$\lim_{n\to\infty} \frac{3n^2 - 5in}{n^2 i} = -3i.$$

b) We have

$$\frac{(k - ik)^2 (2k + 5i)}{k^2 + (3 + ik)^3} = \frac{-4ik^3 + 10k^2}{27 - 8k^2 + i(27k - k^3)}$$

$$= \frac{-4i + \frac{10}{k}}{\frac{27}{k^3} - \frac{8}{k} + \frac{27i}{k} - i}$$

and it follows that

$$\lim_{k\to\infty} \frac{(k - ik)^2 (2k + 5i)}{k^2 + (3 + ik)^3} = 4.$$

c) First we note that

$$\sum_{k=4}^{N} z^k = \sum_{k=0}^{N} z^k - (1 + z + z^2 + z^3)$$

$$= \frac{1 - z^N}{1 - z} - (1 + z + z^2 + z^3).$$

Since $|z| < 1$ it follows that $\lim_{N\to\infty} z^N = 0$ and therefore

$$\lim_{N\to\infty} \sum_{k=4}^{N} z^k = \frac{1}{1 - z} - (1 + z + z^2 + z^3) = \frac{z^4}{1 - z}.$$

8. a) Since

$$\left| \frac{z^{k-1}}{(k+1)!} \frac{k!}{z^k} \right| = |z| \frac{1}{k+1} \to 0 \quad \text{as } k \to \infty,$$

by the ratio test the series $\sum_{k=0}^{\infty} \frac{z^k}{k!}$ converges for all $z \in \mathbb{C}$.

b) Since

$$\left| (-1)^k \frac{z^{2k}}{(2k)!} \right| \le \frac{|z|^{2k}}{(2k)!}$$

it follows that

$$\sum_{k=0}^{\infty} \left| (-1)^k \frac{z^{2k}}{(2k)!} \right| \le \sum_{l=0}^{\infty} \frac{|z|^l}{l!} = e^{|z|}$$

and by the comparison test the convergence of $\sum_{k=0}^{\infty} (-1)^k \frac{z^{2k}}{(2k)!}$ for all $z \in \mathbb{C}$ follows.

c) We argue as in part b) and observe first

$$\left|(-1)^{k-1}\frac{z^{2k-1}}{(2k-1)!}\right| \le \frac{|z|^{2k-1}}{(2k-1)!}$$

to conclude that

$$\sum_{k=1}^{\infty}\left|(-1)^k\frac{z^{2k-1}}{(2k-1)!}\right| \le e^{|z|}$$

and the convergence of $\sum_{k=1}^{\infty}(-1)^k\frac{z^{2k-1}}{(2k-1)!}$ for all $z \in \mathbb{C}$ follows again by the comparison test.

d) For $|z| < 1$ we find

$$\left|\frac{z^{k+1}}{k+1}\cdot\frac{k}{z}\right| = \frac{k}{k+1}|z| \le |z| < 1$$

and the ratio test yields the convergence of $\sum_{k=1}^{\infty}\frac{z^k}{k}$ for $|z| < 1$.

e) From part d) we know that for $|z| < 1$ the series $\sum_{k=1}^{\infty}\frac{|z|^k}{k}$ converges (which follows also from the convergence of the geometric series $\sum_{k=0}^{\infty}|z|^k$). Since

$$\sum_{k=1}^{\infty}\left|(-1)^k\frac{z^{2k-1}}{2k-1}\right| \le \sum_{l=1}^{\infty}\frac{|z|^l}{l} < \infty$$

the comparison test yields the convergence of $\sum_{k=1}^{\infty}(-1)^k\frac{z^{2k-1}}{2k-1}$ for all z, $|z| < 1$.

9. We can follow essentially the proof of Theorem I.29.21. We rewrite

$$\sum_{k=0}^{n}c_k = \sum_{k=0}^{n}a_{n-k}(B_k - B) + B\sum_{k=0}^{n}a_k \qquad (*)$$

where $B_k = \sum_{l=0}^{k}b_l$, $B = \lim_{k\to\infty}B_k$ and we also write $A = \sum_{k=0}^{\infty}a_k$. Thus, if we can prove

$$\lim_{n\to\infty}\sum_{k=0}^{n}a_{n-k}(B_k - B) = 0$$

it will follow that $\sum_{k=0}^{\infty}c_k = AB$.

Given $\epsilon > 0$ we can find $N = N(\epsilon)$ such that $k \ge N(\epsilon)$ implies $|B_k - B| < \epsilon$ and for $n > N(\epsilon)$ it follows that

$$\left|\sum_{k=0}^{n}a_{n-k}(B_k - B)\right| \le \sum_{k=0}^{N}|a_{n-k}||B_k - B| + \sum_{k=N+1}^{n}|a_{n-k}||B_k - B|$$

$$\le \max_{k\in\mathbb{N}}|B_k - B|\sum_{k=0}^{N}|a_{n-k}| + \epsilon\sum_{k=N+1}^{n}|a_{n-k}|$$

$$\le \max_{k\in\mathbb{N}}|B_k - B|\sum_{k=0}^{N}|a_{n-k}| + \epsilon\sum_{k=0}^{\infty}|a_k|.$$

657

For $n \to \infty$ it follows that $|a_n| \to 0$ since $\sum_{k=0}^{\infty} |a_k|$ converges. Therefore we find for every fixed N

$$\lim_{n \to \infty} \sum_{k=0}^{N} |a_{n-k}| = 0$$

implying that

$$0 \leq \limsup_{n \to \infty} \left| \sum_{k=0}^{n} a_{n-k}(B_k - B) \right| \leq \epsilon \sum_{k=0}^{\infty} |a_k|,$$

i.e. we find $\sum_{k=0}^{\infty} c_k = A \cdot B$.

Now we show the absolute convergence of $\sum_{k=0}^{\infty} c_k$:

$$\sum_{n=0}^{M} |c_k| = \sum_{n=0}^{M} \left| \sum_{k=0}^{n} a_{n-k} b_k \right| \leq \sum_{n=0}^{M} \sum_{k=0}^{n} |a_{n-k}||b_k|$$

$$= \left(\sum_{n=0}^{M} |a_n| \right) \left(\sum_{k=0}^{M} |b_k| \right) \leq \left(\sum_{n=0}^{\infty} |a_n| \right) \left(\sum_{k=0}^{\infty} |b_k| \right).$$

10. We form the Cauchy product of $\sum_{k=0}^{\infty} \frac{z_1^k}{k!}$ and $\sum_{k=0}^{\infty} \frac{z_2^k}{k!}$ to find

$$e^{z_1} e^{z_2} = \sum_{n=0}^{\infty} \left(\sum_{k=0}^{n} \frac{z_1^{n-k}}{(n-k)!} \frac{z_2^k}{k!} \right) = \sum_{n=0}^{\infty} \frac{1}{n!} \left(\sum_{k=0}^{n} \binom{n}{k} z_1^{n-k} z_2^k \right)$$

$$= \sum_{n=0}^{\infty} \frac{1}{n!} (z_1 + z_2)^n = e^{z_1 + z_2},$$

where we have used the binomial theorem, see Problem 4.

11.   a) Since $a_k = \frac{P_k}{P_{k-1}}$ we need to prove that $\lim_{k \to \infty} \frac{P_k}{P_{k-1}} = 1$. But this is trivial since $\lim_{k \to \infty} P_k = \lim_{k \to \infty} P_{k-1}$ implying that $\lim_{k \to \infty} \frac{P_k}{P_{k-1}} = \frac{\lim_{k \to \infty} P_k}{\lim_{k \to \infty} P_{k-1}} = 1$.

    b) We sketch once more the proof of Proposition I.30.7. Suppose that $\prod_{k=1}^{\infty} a_k$ converges to $a \neq 0$. By the Cauchy criterion applied to the sequence $\left(\prod_{k=1}^{N} a_k \right)_{N \in \mathbb{N}}$ we can find for $\epsilon > 0$ and $\eta > 0$ a number $N = N(\epsilon, \eta) \in \mathbb{N}$ such that $n > m > N(\epsilon, \eta)$ implies

$$\left| \prod_{k=1}^{n} a_k - \prod_{k=1}^{m} a_k \right| < \eta \epsilon$$

or

$$\left| \prod_{k=m+1}^{n} a_k - 1 \right| < \frac{\eta}{|\prod_{k=1}^{m} a_k|} \epsilon.$$

Since $\lim_{m \to \infty} \prod_{k=1}^{m} a_k = a \neq 0$ it follows for $m > N_0$ that $|\prod_{k=1}^{m} a_k| \geq \frac{|a|}{2}$ and therefore, if $n > m > \max(N_0, N(\epsilon, \eta))$, with $\eta = \frac{2}{|a|}$ we have

$$\left| \prod_{k=m+1}^{n} a_k - 1 \right| < \epsilon.$$

By inspecting the proof of Proposition I.30.7 that, as in the real case, the converse of the statement also holds, i.e. if the conclusion of the problem holds then the product converges.

12. Again we can argue as in the real case.

a) We note the estimate

$$|(1 + a_1) \cdot \ldots \cdot (1 + a_n) - 1| \leq (1 + |a_1|)(1 + |a_2|) \cdot \ldots \cdot (1 + |a_n|) - 1,$$

which follows as in the proof of Proposition I.30.9: For $n = 1$ we have $|(1+a_1)-1| = |a_1| = (1 + |a_1|) - 1$, and now we find

$$\begin{aligned} &|(1 + a_1)(1 + a_2) \cdot \ldots \cdot (1 + a_n)(1 + a_{n+1}) - 1| \\ &= |(1 + a_1)(1 + a_2) \cdot \ldots \cdot (1 + a_n + a_{n+1} + a_n a_{n+1}) - 1| \\ &\leq |(1 + |a_1|)(1 + |a_2|) \cdot \ldots \cdot (1 + |a_n + a_{n+1} + a_n a_{n+1}|) - 1|, \end{aligned}$$

but $1 + |a_n + a_{n+1} + a_n a_{n+1}| \leq (1 + |a_n|)(1 + |a_{n+1}|)$. Thus we have

$$\left| \prod_{k=m+1}^{n} (1 + a_k) - 1 \right| \leq \left| \prod_{k=m+1}^{n} (1 + |a_k|) - 1 \right|$$

and by the Cauchy criterion the convergence of $\prod_{k=1}^{\infty}(1 + a_k)$, follows from the convergence of $\prod_{k=1}^{\infty}(1 + |a_k|)$.

b) We argue as in the proof of Proposition I.30.10. Since

$$|a_1| + \cdots + |a_n| \leq (1 + |a_1|)(1 + |a_2|) \cdot \ldots \cdot (1 + |a_n|)$$

the absolute convergence of $\prod_{k=1}^{\infty}(1 + a_k)$ implies the absolute convergence of the series $\sum_{k=1}^{\infty} a_k$. On the other hand, for $x > 0$ we know that $1 + x \leq e^x$ and therefore

$$(1 + |a_1|)(1 + |a_2|) \cdot \ldots \cdot (1 + |a_n|) \leq e^{|a_1| + \cdots + |a_n|}.$$

If the series $\sum_{k=1}^{\infty} |a_k|$ converges then it follows that $\prod_{k=1}^{\infty}(1 + |a_k|)$ converges too since $\left(\prod_{k=1}^{N}(1 + |a_k|) \right)_{N \in \mathbb{N}}$ is an increasing sequence bounded from above.

Chapter 14

1.   a) Since $\langle z, z \rangle = \sum_{j=1}^{n} z_j \bar{z}_j = \sum_{j=1}^{n} |z_j|^2$ it follows that $\langle z, z \rangle \geq 0$ and if $\langle z, z \rangle = 0$ we must have $z_j = 0$ for all $j = 1, \ldots, n$, i.e. $z = 0 \in \mathbb{C}^n$. Moreover we find

$$\begin{aligned} \langle z, w \rangle = \sum_{j=1}^{n} z_j \bar{w}_j &= \overline{\sum_{j=1}^{n} \bar{z}_j w_j} \\ &= \overline{\sum_{j=1}^{n} w_j \bar{z}_j} = \overline{\langle w, z \rangle}. \end{aligned}$$

Finally we note for $\zeta_1, \zeta_2 \in \mathbb{C}$ that

$$\langle \zeta_1 z + \zeta_2 w, v \rangle = \sum_{j=1}^{n} (\zeta_1 z_j + \zeta_2 w_j) \overline{v}_j$$

$$= \zeta_1 \sum_{j=1}^{n} z_j \overline{v}_j + \zeta_2 \sum_{j=1}^{n} w_j \overline{v}_j$$

$$= \zeta_1 \langle z, v \rangle + \zeta_2 \langle w, v \rangle,$$

and we have proved that $\langle z, w \rangle$ is an unitary scalar product. Now we prove the Cauchy-Schwarz inequality. For $\langle z, w \rangle = 0$ nothing remains to be proved. From the definition of $\langle z, w \rangle$ we conclude for $\zeta \in \mathbb{C}$ that

$$0 \le \langle z - \zeta w, z - \zeta w \rangle$$

$$= \langle z, z \rangle - \zeta \langle w, z \rangle - \overline{\zeta} \langle z, w \rangle + \zeta \overline{\zeta} \langle w, w \rangle$$

$$= \|z\|^2 - 2\mathrm{Re}\left(\zeta \langle w, z \rangle\right) + |\zeta|^2 \|w\|^2.$$

With $\zeta = \frac{\|z\|^2}{\langle w, z \rangle}$, $\langle w, z \rangle \ne 0$, it follows that

$$0 \le \|z\|^2 - 2\|z\|^2 + \frac{\|z\|^4 \|w\|^2}{|\langle w, z \rangle|^2}$$

or

$$|\langle z, w \rangle|^2 \|z\|^2 \le \|z\|^4 \|w\|^2,$$

i.e. $|\langle z, w \rangle| \le \|z\| \|w\|$.

b) First let us make explicit the meaning of $\|u\|_\infty$. For $u : K \to \mathbb{C}$ we have the decomposition $u = a + ib$ where $a, b : K \to \mathbb{R}$ are continuous functions. Therefore $|u(x)| = \left(a^2(x) + b^2(x)\right)^{\frac{1}{2}}$ and

$$\|u\|_\infty = \sup_{x \in K} \left(a^2(x) + b^2(x)\right)^{\frac{1}{2}}.$$

Clearly $\|u\|_\infty \ge 0$ and $\|u\|_\infty = 0$ implies $a(x) = 0$ for all $x \in K$ and $b(x) = 0$ for all $x \in K$, thus $u = 0$. For $\zeta \in \mathbb{C}$ we find first $|\zeta u(x)| = |\zeta| |u(x)|$ and therefore

$$\|\zeta u\|_\infty = \sup_{x \in K} |\zeta u(x)| = |\zeta| \sup_{x \in K} |u(x)| = |\zeta| \|u\|_\infty.$$

Finally we derive the triangle inequality for $u, v \in C(K; \mathbb{C})$:

$$\|u + v\|_\infty = \sup_{x \in K} |u(x) + v(x)| \le \sup_{x \in K} |u(x)| + \sup_{x \in K} |v(x)| = \|u\|_\infty + \|v\|_\infty,$$

where we used that the triangle inequality holds for the modulus of complex numbers.

Next we turn to $\|u\|_{L^2}$ for $u \in C(K;\mathbb{C})$. Note that for $u = a + ib$, $a, b \in C(K;\mathbb{R})$ we have

$$\|u\|_{L^2} = \left(\int_K \left(|a(x)|^2 + |b(x)|^2 \right) \lambda^{(n)}(\mathrm{d}x) \right)^{\frac{1}{2}}.$$

Again it is easy to see that $\|u\|_{L^2} \geq 0$ and $\|\zeta u\|_{L^2} = |\zeta| \|u\|_{L^2}$. Suppose $\|u\|_{L^2} = 0$. Then both integrals $\int_K a^2(x)\,\mathrm{d}x$ and $\int_K b^2(x)\,\mathrm{d}x$ must vanish. Now a and b are continuous functions, then $a^2 \geq 0$ and $b^2 \geq 0$ are continuous functions and for a continuous function $f : K \to \mathbb{R}$, $f \geq 0$, we know that $\int_K f(x)\lambda^{(n)}(\mathrm{d}x) = 0$ implies that $f(x) = 0$ for all $x \in K$, see Proposition I.20.10. The triangle inequality can either be derived as in the next problem or it can be reduced to the real case: for $u, v \in C(K, \mathbb{C})$ we find

$$\left(\int_K |u + v|^2\,\mathrm{d}x \right)^{\frac{1}{2}} \leq \left(\int_K \big||u| + |v|\big|^2\,\mathrm{d}x \right)^{\frac{1}{2}}$$

and using the Cauchy-Schwarz inequality for real-valued functions we get

$$\int_K (|u| + |v|)^2\,\mathrm{d}x = \int_K |u|(|u| + |v|)\,\mathrm{d}x + \int_K |v|(|u| + |v|)\,\mathrm{d}x$$

$$\leq \left(\int_K |u|^2\,\mathrm{d}x \right)^{\frac{1}{2}} \left(\int_K (|u| + |v|)^2\,\mathrm{d}x \right)^{\frac{1}{2}} + \left(\int_K |v|^2\,\mathrm{d}x \right)^{\frac{1}{2}} \left(\int_K (|u| + |v|)^2\,\mathrm{d}x \right)^{\frac{1}{2}}$$

implying

$$\left(\int_K (|u| + |v|)^2\,\mathrm{d}x \right)^{\frac{1}{2}} \leq \left(\int_K |u|^2\,\mathrm{d}x \right)^{\frac{1}{2}} + \left(\int_K |v|^2\,\mathrm{d}x \right)^{\frac{1}{2}},$$

which yields

$$\|u + v\|_{L^2} \leq \|u\|_{L^2} + \|v\|_{L^2}.$$

2. It is possible to derive the Hölder inequality along the lines we proved in the previous problem the Cauchy-Schwarz inequality as we can reduce the new case to the case of real-valued functions which was treated in Chapter 5. However, since this chapter shall help to become some routine with complex-valued functions, we sketch a more direct proof here. We prove first Hölder's inequality $\mathbb{C}^n$, then we indicate the approximation process how to come from sums to integrals, and finally we use the standard argument to deduce from Hölder's inequality Minkowski's inequality. Recall that by Lemma I.23.11 for $p, q \in (1, \infty)$, $\frac{1}{p} + \frac{1}{q} = 1$, and $A, B \geq 0$ we have $A^{\frac{1}{p}} B^{\frac{1}{q}} \leq \frac{A}{p} + \frac{B}{q}$. Now assume for $z, w \in \mathbb{C}^N$ that at least one z_{j_0} and one w_{k_0}, $1 \leq j_0, k_0 \leq N$ is not zero. For $1 \leq j \leq N$ we replace in $A^{\frac{1}{p}} B^{\frac{1}{q}} \leq \frac{A}{p} + \frac{B}{q}$ now A by $\dfrac{|z_j|}{\left(\sum_{k=1}^{N} |z_k|^p \right)^{\frac{1}{p}}}$ and B by $\dfrac{|w_j|}{\left(\sum_{k=1}^{N} |w_k|^q \right)^{\frac{1}{q}}}$ to find

$$\frac{\sum_{j=1}^{N} |z_j| |w_j|}{\left(\sum_{k=1}^{N} |z_k|^p \right)^{\frac{1}{p}} \left(\sum_{k=1}^{N} |w_k|^q \right)^{\frac{1}{q}}} \leq \frac{1}{p} \frac{\sum_{j=1}^{N} |z_j|^p}{\sum_{k=1}^{N} |z_k|^p} + \frac{1}{q} \frac{\sum_{j=1}^{N} |w_j|^q}{\sum_{k=1}^{N} |w_k|^q} = 1$$

implying for $z, w \in \mathbb{C}^N$ Hölder's inequality

$$\left| \sum_{j=1}^{N} z_j w_k \right| \leq \left(\sum_{j=1}^{N} |z_j|^p \right)^{\frac{1}{p}} \left(\sum_{j=1}^{N} |w_j|^q \right)^{\frac{1}{q}}$$

as well as

$$\left| \sum_{j=1}^{N} z_j \overline{w}_j \right| \leq \left(\sum_{j=1}^{N} |z_j|^p \right)^{\frac{1}{p}} \left(\sum_{j=1}^{N} |w_j|^q \right)^{\frac{1}{q}}.$$

Minkowski's inequality in $\mathbb{C}^N$ follows now with the standard argument using Hölder's inequality

$$\sum_{j=1}^{N} |z_j + w_j|^p \leq \sum_{j=1}^{N} |z_j + w_j|^{p-1} |z_j| + \sum_{j=1}^{N} |z_j + w_j|^{p-1} |w_j|$$

$$\leq \left(\sum_{j=1}^{N} |z_j + w_j|^{q(p-1)} \right)^{\frac{1}{q}} \left(\sum_{j=1}^{N} |z_j|^p \right)^{\frac{1}{p}}$$

$$+ \left(\sum_{j=1}^{N} |z_j + w_j|^{q(p-1)} \right)^{\frac{1}{q}} \left(\sum_{j=1}^{N} |w_j|^p \right)^{\frac{1}{p}}.$$

For $\sum_{j=1}^{N} |z_j + w_j|^p \neq 0$ we now get Minkowski's inequality on $\mathbb{C}^N$ by dividing through $\left(\sum_{j=1}^{N} |z_j + w_j|^{q(p-1)} \right)^{\frac{1}{q}}$. The case where $\sum_{j=1}^{N} |z_j + w_j|^p = 0$ is trivial. Now we turn to the corresponding inequalities for integrals. For complex-valued step functions we get immediately

$$\left| \int_K f(x) g(x) \, dx \right| = \left| \sum_{j=1}^{N} (\alpha_j \beta_j) \lambda^{(n)}(K_j) \right|$$

$$\leq \left(\sum_{j=1}^{N} |\alpha_j| \left(\lambda^{(n)}(K_j) \right)^{\frac{1}{p}} \right) \left(\sum_{j=1}^{N} |\beta_j| \left(\lambda^{(n)}(K_j) \right)^{\frac{1}{q}} \right)$$

$$\leq \left(\sum_{j=1}^{N} |\alpha_j|^p \lambda^{(n)}(K_j) \right)^{\frac{1}{p}} \left(\sum_{j=1}^{N} |\beta_j|^q \lambda^{(n)}(K_j) \right)^{\frac{1}{q}}$$

$$= \left(\int_K |f(x)|^p \, dx \right)^{\frac{1}{p}} \left(\int_K |g(x)|^q \, dx \right)^{\frac{1}{q}}$$

where we assume that $f = \sum_{j=1}^{N} \alpha_j \chi_{K_j}$ and $g = \sum_{j=1}^{N} \beta_j \chi_{K_j}$ are already given with respect to the joint partition $(K_j)_{j=1,\dots,N}$ of K. The rest of the proof is now done by a standard approximation argument. In the case where K is Jordan measurable, i.e. $\lambda^{(n)}(\partial K) = 0$, we can approximate the integrals by Riemann sums. In the other case, i.e. the Borel set K is not Jordan measurable, we have to work with decompositions of f and g into real and imaginary parts and then decompose these into positive and negative parts.

For step functions Minkowski's inequality follows now from Minkowski's inequality in $\mathbb{C}^N$ and then again in the general case by approximation.

3. On $\mathbb{C}$ we take of course the Borel σ-field and since $\mathbb{C} \cong \mathbb{R}^2$ topologically the σ-field is $\mathcal{B}^{(2)}$. Thus when we are concerned with measurability we can switch from $f = u + iv$ to $f = \begin{pmatrix} u \\ v \end{pmatrix}$. Now the statement is that f is measurable if and only if $\mathrm{pr}_1(f) = u$ and $\mathrm{pr}_2(f) = v$ are measurable which is of course the case as proved in Lemma 4.1.

4.   a) If we choose $x = x_0 + he_k$ we find

$$f(x_0 + he_k) - f(x_0) = \langle a, he_k \rangle + \varphi_{x_0}(x_0 + he_k)$$

or

$$\frac{f(x_0 + he_k) - f(x_0)}{h} = a_k + \frac{\varphi_{x_0}(x_0 + he_k)}{h}$$

and therefore as $h \to 0$ we get $\frac{\partial f}{\partial x_k}(x_0) = a_k$ where we use that the limit on the right hand side exists by assumption. Note that $\frac{\partial f}{\partial x_k}(x_0) = \frac{\partial u}{\partial x_k}(x_0) + i\frac{\partial v}{\partial x_k}(x_0)$.

b) Suppose that F is differentiable at x_0. Then there exists a matrix $A = \begin{pmatrix} a_{11}(x_0) \ldots a_{n1}(x_0) \\ a_{21}(x_0) \ldots a_{2n}(x_0) \end{pmatrix}$ and a function $\psi_{x_0} : U(x_0) \to \mathbb{R}^2$, $\psi_{x_0} = \begin{pmatrix} \psi_{x_0,1} \\ \psi_{x_0,2} \end{pmatrix}$ defined in a neighbourhood of x_0 such that $\lim_{x \to x_0} \frac{\|\psi_{x_0}(x)\|}{\|x - x_0\|} = 0$ and $F(x) - F(x_0) = A(x - x_0) + \psi_{x_0}(x)$. With $a_k := a_{1k}(x_0) + ia_{2k}(x_0)$ as well as $a = (a_1, \ldots, a_n)$ and $\varphi_{x_0}(x) := \psi_{x_0,1}(x) + i\psi_{x_0,2}(x)$ we can rewrite $F(x) - F(x_0) = A(x - x_0) + \psi_{x_0}(x)$ as $f(x) - f(x_0) = \langle a, (x - x_0) \rangle + \varphi_{x_0}(x)$ and $\lim_{x \to x_0} \frac{\varphi_{x_0}(x)}{\|x - x_0\|} = 0$, i.e. the differentiability of F implies the differentiability of f. But now the converse is obvious: if f is differentiable define $a_{1k} := \mathrm{Re}\, a_k$, $a_{2k} := \mathrm{Im}\, a_k$ as well as $\psi_{x_0,1} := \mathrm{Re}\, \varphi_{x_0}$ and $\psi_{x_0,2} := \mathrm{Im}\, \varphi_{x_0}$.

Chapter 15

1.   a) We have $r = |z| = (16 + 16 \cdot 3)^{\frac{1}{2}} = \sqrt{64} = 8$ and $\varphi = \arctan\frac{4\sqrt{3}}{4} = \arctan\sqrt{3} = \frac{\pi}{3}$ which yields $z = 8e^{i\frac{\pi}{3}}$.

b) It follows that $r = |z| = (1 + 1)^{\frac{1}{2}} = \sqrt{2}$ and $\varphi = \frac{\pi}{2}\arctan 1 = \frac{3\pi}{4}$, i.e. $z = \sqrt{2}e^{i\frac{3\pi}{4}}$.

c) Now we find $r = |z| = (24 + 8)^{\frac{1}{2}} = \sqrt{32} = 4\sqrt{2}$ and further $\varphi = \pi + \arctan\frac{\sqrt{8}}{\sqrt{24}} = \pi + \arctan\frac{1}{\sqrt{3}} = \frac{7\pi}{6}$ which yields $z = 4\sqrt{2}e^{i\frac{7\pi}{6}}$.

2. a) Our starting point is the equality $\left(e^{i\varphi}\right)^5 = e^{5\varphi i}$ or

$$\cos 5\varphi + i\sin 5\varphi = (\cos\varphi + i\sin\varphi)^5 \qquad (*)$$

and by the binomial theorem we find

$$(\cos\varphi + i\sin\varphi)^5 = \binom{5}{0}\cos^5\varphi\,(i\sin\varphi)^0 + \binom{5}{1}\cos^4\varphi\,(i\sin\varphi)^1$$
$$+ \binom{5}{2}\cos^3\varphi\,(i\sin\varphi)^2 + \binom{5}{3}\cos^2\varphi\,(i\sin\varphi)^3 + \binom{5}{4}\cos\varphi\,(i\sin\varphi)^4$$
$$+ \binom{5}{5}(\cos\varphi)^0(i\sin\varphi)^5$$
$$= \cos^5\varphi - 10\cos^3\varphi\sin^2\varphi + 10\cos\varphi\sin^4\varphi$$
$$+ i(5\cos^4\varphi\sin\varphi - 10\cos^2\varphi\sin^3\varphi + \sin^5\varphi).$$

With the help of $(*)$ we now find by comparing real and imaginary parts

$$\cos 5\varphi = \cos^5\varphi - 10\cos^3\varphi\sin^2\varphi + 10\cos\varphi\sin^4\varphi$$
$$= \cos^5\varphi - 10\cos^3\varphi(1 - \cos^2\varphi) + 10\cos\varphi(1 - \cos^2\varphi)^2$$
$$= 16\cos^5\varphi - 20\cos^3\varphi + 5\cos\varphi.$$

b) We note that $8\cos^4(2\varphi) - 3 = \cos(8\varphi) + 4\cos(4\varphi)$ is equivalent to $\cos^4(2\varphi) = \frac{1}{8}\cos(8\varphi) + \frac{1}{2}\cos(2\varphi) + \frac{3}{8}$. Now we make use of

$$\cos^4(2\varphi) = \left(\frac{e^{2\varphi i} + e^{-2\varphi i}}{2}\right)^4$$
$$= \frac{1}{16}\left(e^{8\varphi i} + 4e^{4\varphi i} + 6 + 4e^{-4\varphi i} + e^{-8\varphi i}\right)$$
$$= \frac{1}{8}\left(\frac{e^{8\varphi i} + e^{-8\varphi i}}{2}\right) + \frac{1}{2}\left(\frac{e^{4\varphi i} + e^{-4\varphi i}}{2}\right) + \frac{3}{8}$$
$$= \frac{1}{8}\cos 8\varphi + \frac{1}{2}\cos 4\varphi + \frac{3}{8}.$$

c) Since $1 + \frac{1}{3}\sqrt{3}i = \frac{2}{\sqrt{3}}e^{i\frac{\pi}{6}}$ and $1 - \frac{1}{3}\sqrt{3}i = \frac{2}{\sqrt{3}}e^{i\frac{4\pi}{3}}$ we find

$$\left(\frac{1 + \frac{1}{3}\sqrt{3}i}{1 - \frac{1}{3}\sqrt{3}i}\right)^{12} = \left(\frac{\frac{2}{\sqrt{3}}e^{i\frac{\pi}{6}}}{\frac{2}{\sqrt{3}}e^{i\frac{4\pi}{3}}}\right)^{12} = \left(e^{-i\frac{7\pi}{6}}\right)^{12} = e^{-14\pi i} = 1.$$

3. a) Let $z_1, \ldots, z_n$ be the roots of $p(z) = \sum_{k=0}^{n} a_k z^k$. It follows that

$$a_n(z - z_1)(z - z_2)\cdot\ldots\cdot(z - z_n) = 0$$

or

$$a_n\left(z^n - (z_1 + \cdots + z_n)z^{n-1} + \cdots + (-1)^n z_1\cdots z_n\right) = 0$$

implying that $-a_n(z_1 + \cdots + z_n) = a_{n-1}$, i.e. $\sum_{k=1}^{n} z_k = -\frac{a_{n-1}}{a_n}$, and $(-1)^n$ $a_n(z_1 \cdots z_n) = a_0$ or $\prod_{k=1}^{n} z_k = (-1)^n \frac{a_0}{a_n}$.

b) The nth roots of unity are the roots of $z^n - 1$. In this polynomial the coefficient a_{n-1} is equal to 0, recall that we assume $n \geq 2$. Hence by part a) the sum of all n^{th} roots of unity for $n \geq 2$ is equal to 0.

4.   a) The locations of the 8th roots of unity are shown in the figure below:

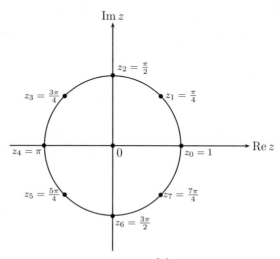

b) The elements of $E(m)$ are the numbers $e^{\frac{2\pi k}{m} i}$, $k = 0, \ldots, m-1$. Suppose that $n < m$ divides m, i.e. $m = pn$. We write $e^{\frac{2\pi k}{m} i} = e^{\frac{2\pi k}{np} i}$, $k = 0, \ldots, pn - 1$. Consider the quotient $\frac{k}{p}$ for $k = 0, \ldots, pn-1$. For the values $k = 0, p, 2p, \ldots, p(n-1) < pn-1$ it follows that

$$e^{\frac{2\pi k}{np} i} = e^{\frac{2\pi l}{n} i}, \quad l = 0, \ldots, n-1,$$

and the numbers $e^{\frac{2\pi l}{n} i}$, $l = 0, \ldots, n-1$ are exactly the elements of $E(n)$, hence $E(n) \subset E(m)$. Since for every group $E(j)$ the multiplication is multiplication in $\mathbb{C}$ it follows that $E(n)$ is a subgroup of $E(m)$.

5. Let $z \in A_{r,R}$ and $w \in S^1$. With T_w as in Example 15.5 we find $T_w z = wz$ and $|T_w z| = |z|$ implying that if $|z| = r_0$, $r < r_0 < R$, i.e. $z \in A_{r,R}$, then $|T_w z| = r_0 (= |z|)$ and therefore $T_w z \in A_{r,R}$, or $T_w(A_{r,R}) \subset A_{r,R}$. Since we can find for every $\zeta \in A_{r,R}$ a number $z \in A_{r,R}$ such that $T_w z = \zeta$, namely $z = w^{-1}\zeta$, it even follows that $T_w(A_{r,R}) = A_{r,R}$.

6. The subset $\mathcal{E}$ consists of all points in the plane with the property that their distance to the point $(3,0)$ and their distance to the point $(-3,0)$ always add up to a constant, namely 10. From this we deduce that $\mathcal{E}$ must be an ellipse with foci $(3,0)$ and $(-3,0)$. Replacing z by $-z$ and z by $\bar{z}$ we find further that this ellipse must be symmetric in the real axis as well as in the imaginary axis, hence its midpoint must be $(0,0)$. For $z \in \mathbb{R}$, $z = x$, we now find $x = 5$ or $x = -5$ depending whether

we are on the right of $(3,0)$ or on the left of $(-3,0)$. For $z = iy$, $y \in \mathbb{R}$ we find $|iy - 3| + |iy + 3| = 2(y^2 + y)^{\frac{1}{2}} = 10$ or $y = \pm 4$. Thus we can also give the equation of the ellipse with respect to the real coordinates x and y as $\frac{x^2}{25} + \frac{y^2}{16} = 1$.

7. With $c := \frac{z_2 - z_1}{|z_2 - z_1|}$ the equation according to (15.14) is given by $\mathrm{Im}\,(\bar{c}(z - z_1)) = 0$.
For $z_1 = -2 - i$ and $z_2 = 3 + 5i$ this yields $z_2 - z_1 = 3 + 5i + 2 + i = 5 + 6i$ and $c = \frac{5+6i}{\sqrt{61}}$. Therefore we find

$$0 = \mathrm{Im}\left(\frac{5 - 6i}{\sqrt{61}}(z + 2 + i)\right) = \frac{1}{\sqrt{61}}\,\mathrm{Im}\,((5 - 6i)z + 16 - 7i),$$

i.e. the equation for the line can be written as

$$\mathrm{Im}\,((15 - 6i)z + 16 - 7i) = 0.$$

If we write further $z = x + iy$ we find

$$0 = \mathrm{Im}\,((15 - 6i)(x + iy) + 16 - 7i) = 15y - 6x - 7$$

and we can give the equation of the line in Cartesian coordinates as $y = \frac{2}{5}x + \frac{7}{15}$.

8. Orthogonality in the plane refers of course to orthogonality of two vectors in $\mathbb{R}^2$. If $z = x + iy$ and $w = u + iv$ represent two vectors in $\mathbb{R}^2$, z corresponds to $\begin{pmatrix} x \\ y \end{pmatrix}$ and w corresponds to $\begin{pmatrix} u \\ v \end{pmatrix}$, then orthogonality means $xu + yv = 0$, or $\mathrm{Re}\,z\,\mathrm{Re}\,w + \mathrm{Im}\,z\,\mathrm{Im}\,w = 0$. The direction orthogonal to $\begin{pmatrix} x_2 - x_1 \\ y_2 - y_1 \end{pmatrix}$, $z_j = x_j + iy_j$, is of course $\begin{pmatrix} -(y_2 - y_1) \\ x_2 - x_1 \end{pmatrix}$ (or any non-zero real multiple of it). This vector corresponds to $i(z_2 - z_1)$, indeed

$$i(z_2 - z_1) = i((x_2 - x_1) + i(y_2 - y_1)) = i(x_2 - x_1) - (y_2 - y_1)$$

so $i(z_2 - z_1)$ is associated with $\begin{pmatrix} -(y_2 - y_1) \\ x_2 - x_1 \end{pmatrix}$. Thus a complex parametric form of the line orthogonal to $z = z_1 + t(z_2 - z_1)$ and passing through z_1 is $w = z_1 + t(i(z_2 - z_1))$. This is however not the form (15.10) but may be rewritten as $w = z_1 + t(z_1 + iz_2 - iz_1 - z_1)$. If we now replace c by ic in (15.14) we find a further form of the equation for the orthogonal line as $\mathrm{Im}\,(\overline{ic}(z - z_1)) = 0$, $c = \frac{z_2 - z_1}{|z_2 - z_1|}$.

9. The line L we consider in parametric form as $z(t) = z_0 t$, $|z_0| = 1$. In general for a point $z \in \mathbb{C}$ we find the position on the Riemann sphere by

$$\xi = \frac{z + \bar{z}}{z\bar{z} + 1}, \qquad \eta = \frac{i(\bar{z} - z)}{z\bar{z} + 1}, \qquad \zeta = \frac{z\bar{z} - 1}{z\bar{z} + 1}$$

which gives for $z = z(t) = z_0 t$

$$\xi = \frac{(z_0 + \bar{z}_0)t}{t^2 + 1} = \frac{2\operatorname{Re} z_0 t}{t^2 + 1},$$

$$\eta = \frac{i(\bar{z}_0 - z_0)t}{t^2 + 1} = \frac{2\operatorname{Im} z_0 t}{t^2 + 1},$$

$$\zeta = \frac{z_0 \bar{z}_0 t^2 - 1}{t^2 + 1} = \frac{t^2 - 1}{t^2 + 1}.$$

Thus on $S^2 \subset \mathbb{R}^3$ we have with T denoting the inverse of $\tilde{S}$

$$T(z(t)) = \frac{1}{1 + t^2} \begin{pmatrix} 2\operatorname{Re} z_0 t \\ 2\operatorname{Im} z_0 t \\ t^2 - 1 \end{pmatrix}.$$

Since with $\vec{n} := \begin{pmatrix} -\operatorname{Im} z_0 \\ \operatorname{Re} z_0 \\ 0 \end{pmatrix}$ we have

$$\left\langle \begin{pmatrix} 2\operatorname{Re} z_0 t \\ 2\operatorname{Im} z_0 t \\ t^2 - 1 \end{pmatrix}, \begin{pmatrix} -\operatorname{Im} z_0 \\ \operatorname{Re} z_0 \\ 0 \end{pmatrix} \right\rangle = 0$$

it follows that $T(L)$ lies in a plane orthogonal to $\vec{n}$. Furthermore we have $T(z(0)) = \begin{pmatrix} 0 \\ 0 \\ -1 \end{pmatrix}$, $T(z(1)) = \begin{pmatrix} \operatorname{Re} z_0 \\ \operatorname{Im} z_0 \\ 0 \end{pmatrix}$ and $\lim_{t \to \infty} T(z(t)) = \lim_{t \to -\infty} T(z(t)) = \begin{pmatrix} 0 \\ 0 \\ 1 \end{pmatrix}$. Iden-

tifying $T(z(1))$ with z_0 we have determined the plane and it follows that $T(L)$ is the circle on S^2 passing through $z_0 \left(\cong \begin{pmatrix} \operatorname{Re} z_0 \\ \operatorname{Im} z_0 \\ 0 \end{pmatrix} \right)$ the north pole $\begin{pmatrix} 0 \\ 0 \\ 1 \end{pmatrix}$ and the

south pole $\begin{pmatrix} 0 \\ 0 \\ -1 \end{pmatrix}$.

10. Given $R > 1$ and choose $N = N(R) = \left[\frac{\ln R}{\ln |z_0|} \right] + 1$. For $n > N$ we find

$$|z_0|^n = e^{n \ln |z_0|} \geq e^{\frac{\ln R}{\ln |z_0|} \ln |z_0|} = R$$

implying that $\lim_{n \to \infty} z_n = \lim_{n \to \infty} z_0^n = \infty$.

Now in the case of real numbers the situation is quite different. For $x_0 > 1$ we again find convergence for $(x_0^n)_{n \in \mathbb{N}}$, namely to $+\infty$, i.e. $\lim_{n \to \infty} x_0^n = +\infty$. However for $x_0 < -1$ we have $\lim_{n \to \infty} x_0^n = \lim_{n \to \infty} (-1)^n |x_0|^n$ does not exist.

This looks like a contradiction. In both cases "∞" denotes something different. In the complex case "∞" is the point at infinity of the one-point compactification, see

Appendix I. However in the case of the real line we have with "$+\infty$" and "$-\infty$" due to the order structure two points at infinity and the extended real-line $\overline{\mathbb{R}}$ is not a one-point compactification of $\mathbb{R}$ or a subset of the one-point compactification of $\mathbb{C}$.

Chapter 16

1.   a) Since $|n^4 + z^4| \geq n^4 - |z|^4 \geq n^4 - \frac{1}{4}n^4 = \frac{3}{4}n^4$ holds for $n \geq 4$ and $1 \leq |z| \leq 2$ we deduce that

$$\frac{1}{|n^4 + z^4|} \leq \frac{4}{3n^4} \qquad \text{for } n \geq 4 \text{ and } 1 \leq |z| \leq 2$$

and the Weierstrass M-test implies that the series $\sum_{n=1}^{\infty} \frac{1}{n^4 + z^4}$ converges in $1 \leq |z| \leq 2$ uniformly and absolutely.

    b) With $z = x + iy$ we find

$$\frac{\sin nz}{n^3} = \frac{1}{2in^3}\left(e^{i(x+iy)n} - e^{-i(x+iy)n}\right)$$
$$= \frac{e^{-ny}e^{ixn}}{2in^3} - \frac{e^{ny}e^{-ixn}}{2in^3}$$

implying

$$\left|\frac{\sin nz}{n^3}\right| \geq \frac{e^{n|y|}}{2n^3} - \frac{e^{-n|y|}}{2n^3} \geq \frac{e^{n|y|}}{2n^3} - \frac{1}{2n^3},$$

but $\sum_{n=1}^{\infty} \frac{e^{ny}}{2n^3}$ diverges for $|y| \neq 0$, note for this that for $a > 0$ we have that $\lim_{n\to\infty} \frac{e^{na}}{2n^3} = \infty$.

However, for $x \in \mathbb{R}$ we find

$$\left|\frac{\sin nx}{n^3}\right| \leq \frac{1}{n^3}$$

and the Weierstrass test implies the uniform and absolute convergence of $\sum_{n=1}^{\infty} \frac{\sin nx}{n^3}$ on $\mathbb{R}$. Note that $\sin$ is bounded on $\mathbb{R}$ but not on $\mathbb{C}$.

2.   a) We set $a_n := \left(\sqrt[n]{(n+2)(n+3)} - 1\right) > 0$ and find

$$(n+2)(n+3) = (1 + a_n)^n = \sum_{k=0}^{n}\binom{n}{k}a_n^k$$
$$\geq 1 + \binom{n}{3}a_n^3$$

which gives

$$a_n^3 \leq \frac{n^2 + 5n + 5}{\binom{n}{3}} = \frac{3(n^2 + 5n + 5)}{(n-2)(n-1)n},$$

i.e. $\lim_{n\to\infty} a_n = 0$ implying $\lim_{n\to\infty} \sqrt[n]{(n+2)(n+3)} = 1$ and by the Hadamard-Cauchy theorem it follows that the radius is 1.

b) We first estimate

$$\left| \frac{(-1)^{n-1}(n+1)}{3^n(n^2+1)^{\frac{1}{2}}} \right| \leq \frac{n+1}{3^n(n^2+1)^{\frac{1}{2}}}$$

and now study the series $\sum_{k=0}^{\infty} \frac{n+1}{3^n(n^2+1)^{\frac{1}{2}}} z^n$ by looking at

$$\sqrt[n]{\frac{1}{3^n} \frac{n+1}{(n^2+1)^{\frac{1}{2}}}} = \frac{1}{3} \sqrt[n]{\frac{n+1}{(n^2+1)^{\frac{1}{2}}}}.$$

As in part a) we see that $\lim_{n\to\infty} \sqrt[n]{\frac{n+1}{(n^2+1)^{\frac{1}{2}}}} = 1$ which first yields

$$\lim_{n\to\infty} \sqrt[n]{\frac{1}{3^n} \frac{n+1}{(n^2+1)^{\frac{1}{2}}}} = \frac{1}{3}$$

and then by the Hadamard-Cauchy theorem that the radius of convergence is 3.

c) Following the hint we write $n! = n^n e^{-n} r_n$ with $\lim_{n\to\infty} r_n^{\frac{1}{n}} = 1$. This implies

$$\left(\frac{(n!)^2}{(2n)!} \right)^{\frac{1}{n}} = \left(\frac{n^{2n} e^{-2n} r_n^2}{(2n)^{2n} e^{-2n} r_{2n}} \right)^{\frac{1}{n}}$$

$$= \left(\frac{r_n^2}{2^{2n} r_{2n}} \right)^{\frac{1}{n}} = \frac{1}{4} \left(\frac{r_n^2}{r_{2n}} \right)^{\frac{1}{n}}.$$

Since $\lim_{n\to\infty} \left(r_n^2 \right)^{\frac{1}{n}} = \lim_{n\to\infty} \left(r_{2n} \right)^{\frac{1}{n}} = 1$ we find that the radius of convergence is 4.

Now let us prove $n! = n^n e^{-n} r_n$, $\lim_{n\to\infty} r_n^{\frac{1}{n}} = 1$, using Stirling's formula, Theorem I.31.7. First we note that by Problem 10 of Chapter I.15 we have $\lim_{n\to\infty} (2\pi)^{\frac{n}{2}} = 1$ and by Example I.15.12 we know that $\lim_{n\to\infty} \sqrt[n]{n} = 1$. Now using (I.39.29) we find

$$\frac{(n!)^{\frac{1}{n}}}{(n^n e^{-n})^{\frac{1}{n}}} = \left(\frac{\sqrt{2\pi}\sqrt{n}(n!)}{\sqrt{2\pi}\sqrt{n}(n^n e^{-n})} \right)^{\frac{1}{n}}$$

which yields that $\lim_{n\to\infty} \frac{(n!)}{(n^n e^{-n})^{\frac{1}{n}}} = 1$, but

$$n! = n^n e^{-n} r_n, \qquad r_n = \frac{n!}{n^n e^{-n}}.$$

3. Since

$$\left| \frac{z^{k+1}}{\Gamma(\alpha k + 2)} \frac{\Gamma(\alpha k + 1)}{z^k} \right| = |z| \frac{(\alpha k)!}{(\alpha k + 1)!} = \frac{|z|}{\alpha k + 1},$$

the ratio test implies the uniform and absolute convergence in $\mathbb{C}$ for every $\alpha \in \mathbb{N}$ while for $\alpha = 0$ we have uniform and absolute convergence for $|z| < 1$. For $\alpha = 0$ we find $\Gamma(\alpha k + 1) = \Gamma(1) = 1$, i.e.

$$E_0(z) = \sum_{k=0}^{\infty} z^k = \frac{1}{1-z},$$

while for $\alpha = 2$ we have $\Gamma(2k + 1) = (2k)!$ and therefore

$$E_2(w) = \sum_{k=0}^{\infty} \frac{w^k}{(2k)!}.$$

With $w = z^2$ we obtain

$$E_2\left(z^2\right) = \sum_{k=0}^{\infty} \frac{z^{2k}}{(2k)!} = \cosh z$$

and

$$E_2\left(-z^2\right) = \sum_{k=0}^{\infty} \frac{\left(-z^2\right)^k}{(2k)!} = \sum_{k=0}^{\infty} (-1)^k \frac{z^{2k}}{(2k)!} = \cos z.$$

Also see the remark in the solution to Problem 4.

4. Again we use the ratio test and find first

$$\left| \frac{z^{k+1}}{\Gamma(\alpha k + \alpha + \beta)} \frac{\Gamma(\alpha k + \beta)}{z^k} \right| = |z| \frac{(\alpha k + \beta - 1)!}{(\alpha k + \alpha + \beta - 1)!} \le |z| \frac{1}{\alpha k + \beta}$$

since $\alpha, \beta \in \mathbb{N}$. Now it follows the convergence of $E_{\alpha\beta}(z)$ for all $z \in \mathbb{C}$. If we choose $\alpha = \beta = 1$ we have

$$E_{1,1}(z) = \sum_{k=0}^{\infty} \frac{z^k}{\Gamma(k+1)} = \sum_{k=0}^{\infty} \frac{z^k}{k!} = e^z.$$

For $\alpha = 1$ and $\beta = 2$ we have for $z \ne 0$

$$E_{1,2}(z) = \sum_{k=0}^{\infty} \frac{z^k}{\Gamma(k+2)} = \sum_{k=0}^{\infty} \frac{z^k}{(k+1)!}$$

$$= \frac{1}{z} \sum_{k=0}^{\infty} \frac{z^{k+1}}{(k+1)!} = \frac{1}{z} \left(\sum_{k=0}^{\infty} \frac{z^k}{k!} - 1 \right)$$

$$= \frac{e^z - 1}{z}.$$

Since $\lim_{z \to 0} \frac{e^z - 1}{z} = 1$ we have $E_{1,2}(z) = \frac{e^z - 1}{z}$ for all $z \in \mathbb{C}$. Finally for $\alpha = \beta = 2$ it follows that for $z \ne 0$

$$E_{2,2}\left(z^2\right) = \sum_{k=0}^{\infty} \frac{z^{2k}}{\Gamma(2k+2)} = \frac{1}{z} \sum_{k=0}^{\infty} \frac{z^{2k+1}}{(2k+1)!}$$

$$= \frac{1}{z} \sinh(z),$$

which extends continuously to $\mathbb{C}$.

Remark. *Denoting natural numbers by α or β is uncommon in our treatise. In fact the definitions of $E_\alpha(z)$ and $E_{\alpha,\beta}(z)$ extends to a much larger range of α and β. The functions $E_\alpha(z)$ can be defined for all $\alpha > 0$ and they are convergent in the plane $\mathbb{C}$ as are the functions $E_{\alpha,\beta}(z)$ for $\mathrm{Re}\,\alpha > 0$ and $\beta \in \mathbb{C}$, however many books also prefer to restrict $E_{\alpha,\beta}(z)$ to the range $\alpha > 0$ and $\beta > 0$. In order to prove the convergence we need the asymptotic expansion*

$$z^{b-a}\frac{\Gamma(z+a)}{\Gamma(z+b)} \sim 1 + \frac{(a-b)(a+b-1)}{2z} + \frac{1}{12}\binom{a-b}{2}\left(3(a+b-1)^2 - a+b-1\right)\frac{1}{z^2} + \cdots,$$

see [1], (6, 1.47) on page 257. Note that Mittag-Leffler functions play a crucial role in fractional calculus, see [31].

5.   a) Since $e^z = \sum_{k=0}^{\infty} \frac{z^k}{k!}$ we find

$$\frac{z}{e^z - 1} = \frac{z}{\sum_{k=1}^{\infty}\frac{z^k}{k!}} = \frac{1}{\sum_{k=0}^{\infty}\frac{z^k}{(k+1)!}},$$

which implies

$$1 = \left(\sum_{k=0}^{\infty}\frac{z^k}{(k+1)!}\right)\left(\sum_{l=0}^{\infty}\frac{B_l}{l!}z^l\right).$$

Taking the Cauchy product of the two power series we get

$$1 = \sum_{n=0}^{\infty}\left(\sum_{k=0}^{n}\frac{B_k}{(n-k+1)!k!}\right)z^n,$$

which yields

$$\sum_{k=0}^{N}\frac{B_k}{k!(N-k+1)!} = \begin{cases} 1, & N=0 \\ 0, & N>0 \end{cases},$$

i.e. for $N > 0$, summing up to $N - 1$, we get

$$0 = \sum_{k=0}^{N-1}\frac{B_k}{k!(N-k)!} = \frac{B_0}{0!N!} + \frac{B_1}{1!(N-1)!} + \cdots + \frac{B_{N-1}}{(N-1)!1!}$$

and multiplying by $N!$ we get

$$0 = \sum_{k=0}^{N-1}\frac{N!}{k!(N-k)!}B_k = \sum_{k=0}^{N-1}\binom{N}{k}B_k = \binom{N}{0}B_0 + \cdots + \binom{N}{N-1}B_{N-1}.$$

For B_0 we find of course the value $B_0 = 1$ and now it follows

$$\frac{B_0}{0!2!} + \frac{B_1}{1!1!} = 0 \quad \text{or} \quad B_1 = -\frac{B_0}{2} = -\frac{1}{2};$$

$$\frac{B_0}{0!3!} + \frac{B_1}{1!2!} + \frac{B_2}{2!1!} = 0 \quad \text{or} \quad B_2 = 2\left(-\frac{1}{6} + \frac{1}{4}\right) = \frac{1}{6};$$

$$\frac{B_0}{0!4!} + \frac{B_1}{1!3!} + \frac{B_2}{2!2!} + \frac{B_3}{3!1!} = 0 \quad \text{or} \quad B_3 = 6\left(-\frac{1}{24} + \frac{1}{12} - \frac{1}{24}\right) = 0;$$

and

$$\frac{B_0}{0!5!} + \frac{B_1}{1!4!} + \frac{B_2}{2!3!} + \frac{B_3}{3!2!} + \frac{B_4}{4!1!} = 0 \quad \text{or} \quad B_4 = 24 \left(-\frac{1}{120} + \frac{1}{48} - \frac{1}{72} \right) = -\frac{1}{30}.$$

b) Since $(*)$ holds for all $|z| < 2\pi$ it holds for all $x \in \mathbb{R}$, $|x| < 2\pi$. We consider the odd part of the function $g(x) = \frac{x}{e^x - 1} + \frac{x}{2}$, i.e.

$$2(g(x) - g(-x)) = \frac{x}{e^x - 1} + \frac{x}{2} - \left(\frac{-x}{e^x - 1} - \frac{x}{2} \right)$$

$$= x + x \left(\frac{-1 + e^{-x} + e^x - 1}{1 - e^x + e^{-x} - 1} \right) = 0.$$

Thus g is an even function implying that for $|x| < 2\pi$

$$\sum_{k=0}^{\infty} \frac{B_k}{k!} x^k + \frac{x}{2} = 1 - \frac{x}{2} + \sum_{k=2}^{\infty} \frac{B_k}{k!} x^k + \frac{x}{2}$$

$$= 1 + \sum_{k=2}^{\infty} \frac{B_k}{k!} x^k$$

must be an even function which yields $B_k = 0$ for $k = 2l + 1$, $l \in \mathbb{N}$.

6. In the solution to Problem 5 we have seen that for $|x| < 2\pi$, $x \in \mathbb{R}$, we have

$$\frac{x}{e^x - 1} + \frac{x}{2} = \sum_{k=0}^{\infty} \frac{B_{2k}}{(2k)!} x^{2k}.$$

From Theorem 22.12 we can conclude that this equality must hold also for all $z \in \mathbb{C}$, $|z| < 2\pi$. Thus we will justify the use of this identity for $|z| < 2\pi$ when proving Theorem 22.12. We look now at the function

$$f(z) = \frac{z}{e^z - 1} + \frac{z}{2}$$

and we find

$$f(z) = \frac{z}{2} \left(\frac{z}{e^z - 1} + 1 \right) = \frac{z}{2} \frac{e^z + 1}{e^z - 1} = \frac{z}{2} \frac{e^{\frac{z}{2}} + e^{-\frac{z}{2}}}{e^{\frac{z}{2}} - e^{-\frac{z}{2}}}.$$

With $z = 2\pi i \xi$ it follows that

$$f(\pi\xi) = f(2\pi i z) = -\frac{z}{2} i \frac{\cos \frac{z}{2}}{\sin \frac{z}{2}} = \pi\xi \cot(\pi\xi)$$

and replacing again ξ by z we arrive at

$$(\pi z) \cot(\pi z) = \sum_{k=0}^{\infty} \frac{B_{2k}}{(2k)!} (2\pi i z)^k = \sum_{k=0}^{\infty} (-1)^k \frac{(2\pi)^{2k} z^{2k}}{(2k)!}.$$

7. The first part is easy. By the definition of $\binom{a}{k}$, see (16.24) we have

$$\binom{a}{k} = \frac{a(a-1) \cdot \ldots \cdot (a-k+1)}{k!}$$

and in the case where $a \in \mathbb{N}_0$ it follows that the numerator becomes zero for $k > a$. The second part is the claim that for $|x| < 1$, $x \in \mathbb{R}$, and $a \in \mathbb{R}$ the Taylor expansion

$$(1+x)^a = \sum_{n=0}^{\infty} \binom{a}{n} x^n$$

holds. If $a \in \mathbb{N}$, this is just the binomial theorem

$$(1 = x)^a = \sum_{n=0}^{a} \binom{a}{n} x^n$$

in light of the first conclusion, i.e. the fact that $\binom{a}{n} = 0$ for $n > a$ if $a \in \mathbb{N}_0$. For the kth derivative of $x \mapsto (1+x)^a$, $a \in \mathbb{R}$, we find

$$f^{(k)}(x) = a(a-1) \cdot \ldots \cdot (a-k+1)(1+x)^{a-k}$$
$$= k! \binom{a}{k}(1+x)^{a-k}.$$

The Taylor coefficients are given by $\frac{f^{(n)}(0)}{n!}$ implying that the formal Taylor series is indeed given by

$$(1+x)^a = \sum_{n=0}^{\infty} \binom{a}{n} x^n.$$

We determine the domain of convergence by applying the ratio test and find

$$q_n := \left| \frac{\binom{a}{n+1} x^{n+1}}{\binom{a}{n} x^n} \right| = |x| \frac{\binom{a}{n+1}}{\binom{a}{n}} = |x| \left| \frac{a-n}{n+1} \right|$$

implying that $\lim q_n = |x| < 1$ for $|x| < 1$ which yields the convergence for $|x| < 1$. Now the equality

$$B_{a+b}(x) = (1+x)^{a+b} = (1+x)^a (1+x)^b = B_a(x) B_b(x)$$

which holds for $|x| < 1$ implies the final claim.

Note that the absolute convergence of $\sum_{n=0}^{\infty} \binom{a}{n} x^n$, $|x| < 1$, $x \in \mathbb{R}$, also implies the absolute convergence of the series for $|z| < 1$, $z \in \mathbb{C}$. We will see later in Theorem 22.12, that the identity $B_{a+b}(x) = B_a(x) B_b(x)$ for all $x \in (-1, 1)$ will extend to the identity $B_{a+b}(z) = B_a(z) B_b(z)$ for all $z \in \mathbb{C}$, $|z| < 1$.

8. Since $\binom{a}{k} = \frac{a-(k-1)}{k} \binom{a}{k-1}$ we find

$$\binom{a-1}{k} = \frac{(a-1)-(k-1)}{k} \binom{a-1}{k-1} = \frac{a-k}{k} \binom{a-1}{k-1}$$

673

and therefore

$$
\binom{a-1}{k} + \binom{a-1}{k-1} = \frac{a-k}{k}\binom{a-1}{k-1} + \binom{a-1}{k-1}
$$
$$
= \left(\frac{a-k}{k} + 1\right)\binom{a-1}{k-1} = \frac{a}{k}\binom{a-1}{k-1}
$$
$$
= \frac{a(a-1)\cdot\ldots\cdot(a-1-(k-1)+1)}{k(k-1)!}
$$
$$
= \frac{a(a-1)\cdot\ldots\cdot((a-k)+1)}{k!} = \binom{a}{k}.
$$

Furthermore, for $|z| < 1$ we have

$$
(1+z)B_{a-1}(z) = (1+z)\sum_{k=0}^{\infty}\binom{a-1}{k}z^k
$$
$$
= \sum_{k=0}^{\infty}\left(\binom{a-1}{k}z^k + \binom{a-1}{k}z^{k+1}\right)
$$
$$
= \sum_{k=0}^{\infty}\binom{a-1}{k}z^k + \sum_{k=1}^{\infty}\binom{a-1}{k-1}z^k
$$
$$
= \binom{a-1}{0}z^0 + \sum_{k=1}^{\infty}\left(\binom{a-1}{k} + \binom{a-1}{k-1}\right)z^k
$$
$$
= \binom{a-1}{0}z^0 + \sum_{k=1}^{\infty}\binom{a}{k}z^k = \sum_{k=0}^{\infty}\binom{a}{k}z^k = B_a(z),
$$

where we used the result of the previous consideration and the fact that $\binom{a-1}{0} = \binom{a}{0} = 1$.

9.   a) By definition we have

$$
(\alpha)_n = \alpha(\alpha+1)\cdot\ldots\cdot(\alpha+n-1)
$$

and the functional equation of the Γ-function yields

$$
\Gamma(\alpha+n) = (\alpha+n-1)\Gamma(\alpha+n-1)
$$
$$
= (\alpha+n-1)(\alpha+n-2)\Gamma(\alpha+n-2)
$$
$$
\vdots
$$
$$
= (\alpha+n-1)(\alpha+n-2)\cdot\ldots\cdot(\alpha+1)\alpha\Gamma(\alpha),
$$

i.e.

$$
\frac{\Gamma(\alpha+n)}{\Gamma(\alpha)} = (\alpha)_n.
$$

This leads now to

$$\frac{(\alpha)_n}{(\beta)_n} = \frac{\Gamma(\alpha+n)}{\Gamma(\alpha)} \frac{\Gamma(\beta)}{\Gamma(\beta+n)}$$

$$= \frac{\Gamma(\alpha+n)\Gamma(\beta-\alpha)}{\Gamma(n+\beta)} \cdot \frac{\Gamma(\beta)}{\Gamma(\alpha)\Gamma(\beta-\alpha)}$$

$$= \frac{B(\alpha+n, \beta-\alpha)}{B(\alpha, \beta-\alpha)}.$$

b) Since $(1)_k = k!$ we have

$$_2F_1(1,1;1;z) = \sum_{k=0}^{\infty} \frac{(1)_k(1)_k}{(1)_k k!} z^k$$

$$= \sum_{k=0}^{\infty} \frac{(1)_k}{k!} z^k = \sum_{k=0}^{\infty} z^k.$$

Moreover, since $(2)_k = 2 \cdot 3 \cdot \ldots \cdot (2+k-1) = (k+1)!$ it follows that

$$z \,_2F_1(1,1;2;-z) = \sum_{k=0}^{\infty} (-1)^k \frac{(1)_k(1)_k}{(2)_k k!} z^{k+1}$$

$$= \sum_{k=0}^{\infty} (-1)^k \frac{k!}{(k+1)!} z^{k+1} = \sum_{k=0}^{\infty} (-1)^k \frac{z^{k+1}}{k+1}$$

$$= \sum_{k=1}^{\infty} (-1)^{k+1} \frac{z^k}{k} = \tilde{\ln}(1+z).$$

10.   a) Recalling the definition of $_2F_1(\alpha, \beta; \gamma; z)$ we find

$$_2F_1(\alpha, \beta; \gamma; z) = \sum_{k=0}^{\infty} \frac{(\alpha)_k(\beta)_k}{(\gamma)_k k!} z^k$$

$$= \sum_{k=0}^{\infty} \frac{(\beta)_k(\alpha_k)}{(\gamma)_k k!} z^k = \,_2F_1(\beta, \alpha; \gamma; z).$$

b) We have

$$(\gamma - \alpha - \beta) \,_2F_1(\alpha, \beta; \gamma; z) + \alpha(1-z)_2F_1(\alpha+1, \beta; \gamma; z) - (\gamma - \beta)_2F_1(\alpha, \beta-1; \gamma; z)$$

$$= \sum_{k=1}^{\infty} \left((\gamma - \alpha - \beta)\frac{(\alpha)_k(\beta)_k}{(\gamma)_k k!} + \alpha\frac{(\alpha+1)_k(\beta)_k}{(\gamma)_k!} - (\gamma-\beta)\frac{(\alpha)_k(\beta-1)_k}{(\gamma)_k k!} \right.$$

$$\left. - \alpha\frac{(\alpha+1)_{k-1}(\beta)_{k-1}}{(\gamma)_{k-1}(k-1)!} \right) z^k$$

$$= \sum_{k=1}^{\infty} A_k z^k,$$

675

where for the coefficients A_k we find

$$A_k = \frac{(\alpha)_k(\beta)_k}{(\gamma)_k k!}\Big((\gamma - \alpha - \beta)(\beta + k - 1) + (\alpha + k)(\beta + k - 1) - (\gamma - \beta)(\beta - 1)$$

$$- (\gamma + k - 1)k\Big) = 0.$$

11.   a) For $k = 1, 2, 3$ we have

$$w - w_k = \frac{az + b}{cz + d} - \frac{az_k + b}{cz_k + d}$$
$$= \frac{(ad - bc)(z - z_k)}{(cz + d)(cz_k + d)}$$

and we find therefore

$$w - w_1 = \frac{(ad - bc)(z - z_1)}{(cz + d)(cz_1 + d)}$$
$$w - w_3 = \frac{(ad - bc)(z - z_3)}{(cz + d)(cz_3 + d)}$$
$$w_2 - w_1 = \frac{(ad - bc)(z_2 - z_1)}{(cz_2 + d)(cz_1 + d)}$$
$$w_2 - w_3 = \frac{(ad - bc)(z_2 - z_3)}{(cz_2 + d)(cz_3 + d)}.$$

Since we must assume that $ad - bc \neq 0$ we get further

$$\frac{(w - w_1)(w_2 - w_3)}{(w - w_3)(w_2 - w_1)} = \frac{(z - z_1)(z_2 - z_3)}{(z - z_3)(z_2 - z_1)}$$

which yields

$$w = \frac{az + b}{cz + d}$$

where

$$a = (w_2 - w_3)w_1(z_2 - z_1) + (w_1 - w_2)(z_3 - z_2)$$
$$b = (w_2 - w_3)w_1(z_2 - z_1) + (w_2 - w_1)z_1(z_2 - z_3)$$
$$c = (w_2 - w_3) - (z_2 - z_3)$$
$$d = (w_3 - w_2)z_3(z_2 - z_1) + z_1(z_2 - z_3).$$

b) Using part a) we find

$$w(z) = \frac{i(1 + i)z + 1 + i}{(2 - 2i)z + i(1 + i)}$$

676

12.   a) With $z = t \in \mathbb{R}$ we find

$$w(t) = \frac{t-i}{t+i} = \frac{t^2-1}{t^2+1} - \frac{2t}{t^2+1} i,$$

and

$$|w(t)|^2 = \left(\frac{t^2-1}{t^2+1}\right)^2 + \frac{4t^2}{(t^2+1)^2}$$
$$= \frac{t^4 - 2t^2 + 1 + 4t^2}{(t^2+1)^2} = \frac{t^4 + 2t^2 + 1}{(t^2+1)^2}$$
$$= 1.$$

Thus $w(\mathbb{R}) \subset S^1$. Clearly, $1 \notin w(\mathbb{R})$ since otherwise $1 = \frac{t^2-1}{t^2+1}$ or $t^2+1 = t^2-1$. Suppose that $w \in S^1\backslash\{1\}$. We can write w as $w = -\cos 2\varphi + i \sin 2\varphi$, $\varphi \in \left(-\frac{\pi}{2}, \frac{\pi}{2}\right)$, and expressing $\cos 2\varphi$ and $\sin 2\varphi$ by $\tan\varphi$, $\varphi \in \left(-\frac{\pi}{2}, \frac{\pi}{2}\right)$, we find

$$w = \frac{\tan^2\varphi - 1}{1 + \tan^2\varphi} - \frac{2\tan\varphi}{1 + \tan^2\varphi} i.$$

Thus the parametrization $t = \tan\varphi$, $\varphi \in \left(-\frac{\pi}{2}, \frac{\pi}{2}\right)$, allows us to find for every $w \in S^1\backslash\{1\}$ a $t = \tan\varphi \in \mathbb{R}$ such that $w = \frac{t^2-1}{t^2+1} - \frac{2t}{t^2+1} i$, i.e. $w(\mathbb{R}) = S^1\backslash\{1\}$.

b) For $z = x \in \mathbb{R}$ we have $w(x) = \frac{x-1}{x+1} \in \mathbb{R}\backslash\{1\}$. Indeed if for some x_0 we have $\frac{x_0-1}{x_0+1} = 1$, then this implies $x_0 - 1 = x_0 + 1$. Let $y \in \mathbb{R}\backslash\{1\}$ and consider the equation $\frac{x-1}{x+1} = y$ with solution $x = \frac{y+1}{1-y}$, $y \neq 1$. It follows that $w\left(\frac{y+1}{1-y}\right) = y$ i.e. $w(\mathbb{R}\backslash\{-1\}) = \mathbb{R}\backslash\{1\}$.

13. We have to solve the equation

$$\frac{a\zeta + b}{c\zeta + d} = \zeta$$

under the conditions $ac - bd \neq 0$ and $c \neq 0$. This leads to the quadratic equation

$$c\zeta^2 + d\zeta - a\zeta - b = 0$$

or

$$\zeta^2 + \frac{d-a}{c}\zeta - \frac{b}{c} = 0$$

which has two distinct solutions for $(d-a)^2 + 4bc \neq 0$. These solutions are given by

$$\zeta_{1,2} = \frac{d-a}{2c} \pm \sqrt{\left(\frac{d-a}{2c}\right)^2 + \frac{b}{c}} = \frac{d-a}{2c} \pm \sqrt{\frac{(d-a)^2 + 4bc}{4c^2}}.$$

Chapter 17

1. We can use the solution to Problem 2 of Chapter I.21. We use mathematical induction and the product rule, i.e. Leibniz's rule. For $m = 1$ it follows that

$$\frac{d}{dz}(f \cdot g)(z) = f'(z)g(z) + f(z)g'(z)$$

$$= \binom{1}{0}f^{(1)}(z)g^{(0)}(z) + \binom{1}{1}f^{(0)}(z)g^{(1)}(z).$$

Now suppose that for some $k \geq 1$

$$\frac{d^k}{dz^k}(f \cdot g)(z) = \sum_{l=0}^{k}\binom{k}{l}f^{(k-l)}(z)g^{(l)}(z)$$

holds and consider

$$\frac{d^{k+1}}{d^{k+1}}(f \cdot g)(z) = \frac{d}{dz}\left(\sum_{l=0}^{k}\binom{k}{l}f^{(k-l)}(z)g^{(l)}(z)\right)$$

$$= \sum_{l=0}^{k}\binom{k}{l}\frac{d}{dz}\left(f^{(k-l)}(z)g^{(l)}(z)\right)$$

$$= \sum_{l=0}^{k}\binom{k}{l}\left(f^{(k+1-l)}(z)g^{(l)}(z) + f^{(k-l)}(z)g^{(l+1)}(z)\right)$$

$$= f^{(k+1)}(z)g(z) + \sum_{l=1}^{k}\binom{k}{l}f^{(k+1-l)}(z)g^{(l)}(z)$$

$$+ \sum_{l=1}^{k}\binom{k}{l-1}f^{(k-(l-1))}(z)g^{(l)}(z) + f(z)g^{(k+1)}(z)$$

$$= \sum_{l=0}^{k+1}\binom{k+1}{l}f^{(k+1-l)}(z)g^{(l)}(z),$$

where we used the calculation of the solution to Problem 2 of Chapter I.21.

2. Suppose first that for $f = u + iv$ we have $\text{Im} f = v = c$. The Cauchy-Riemann differential equations $u_x = v_y$ and $u_y = -v_x$ imply $u_x = 0$ and $u_y = 0$. Since G is connected it follows that u must be constant on G, hence $f = u + iv$ must be constant on G.
Now assume that $|f| = (u^2 + v^2)^{\frac{1}{2}} = c$. If $c = 0$ it follows that $u = v = 0$. If $c \neq 0$

$$\frac{u_x + v_x}{(u^2 + v^2)^{\frac{3}{2}}} = 0 \quad \text{and} \quad \frac{u_y + v_y}{(u^2 + v^2)^{\frac{3}{2}}} = 0.$$

This implies that

$$0 = u_x^2 + 2u_x v_x + v_x^2 \tag{$*$}$$

and

$$0 = u_y^2 + 2u_y v_y + v_y^2. \tag{**}$$

Using the Cauchy-Riemann differential equations we find $2u_x v_x = -2u_x u_y$ and $2u_y v_y = 2u_x u_y$. Thus adding (*) and (**) yields

$$u_x^2 + v_x^2 + u_y^2 + v_y^2 = 0$$

which implies $u_x = u_y = v_x = v_y = 0$ and the result follows as before since G is connected.

3. Since $g(z) = u(z) + iv(z) = f(|z|^2)$ we find $v(z) = 0$ for all $z \in \mathbb{C}$ and by Problem 2 it follows that g is complex differentiable if and only if g, hence f, is constant.

4. Since passing from w to $\bar{w}$ is a reflection in the x-axis (or the real axis), G^* is obtained by reflecting G in the x-axis, see the figure below

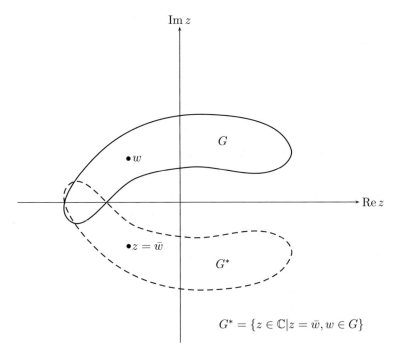

$$G^* = \{z \in \mathbb{C} | z = \bar{w}, w \in G\}$$

Now we look at $h(z) := \overline{f(\bar{z})}$, $z \in G^*$. For $z_0 \in G^*$ and $z \in G^*$ there exist $w_0, w \in G$

such that $w = \overline{z}$ and $w_0 = \overline{z_0}$ and we find

$$\frac{h(z) - h(z_0)}{z - z_0} = \frac{\overline{f(\overline{z})} - \overline{f(\overline{z_0})}}{z - z_0}$$

$$= \frac{\overline{f(w)} - \overline{f(w_0)}}{\overline{w} - \overline{w_0}}$$

$$= \overline{\frac{f(w) - f(w_0)}{w - w_0}}$$

$$= \left(\overline{\frac{f(w) - f(w_0)}{w - w_0}}\right).$$

Using the continuity of $z \mapsto \overline{z}$ it follows for $z \to z_0$ that,

$$h'(z_0) = \overline{h'(\overline{w}_0)} = \overline{f'(w_0)},$$

in particular h is differentiable.

5. For $z = x + iy$ we have $z^2 = x^2 - y^2 + 2ixy$ and $z^3 = x^3 - 3xy^2 + i(-y^3 + 3x^2y)$ which gives

$$e^{z^2} + z^3 = e^{x^2 - y^2} e^{i2xy} + x^3 - 3xy^2 + i(-y^3 + 3x^2y)$$

$$= e^{x^2 - y^2} \cos 2xy + ie^{x^2 - y^2} \sin 2xy + x^3 - 3xy^2 + i(-y^3 + 3x^2y)$$

$$= \left(e^{x^2 - y^2} \cos 2xy + x^3 - 3xy^2\right) + i\left(e^{x^2 - y^2} \sin 2xy - y^3 + 3x^2y\right)$$

$$=: u(x, y) + iv(x, y).$$

Now we find

$$u_x(x, y) = 2xe^{x^2 - y^2} \cos 2xy - 2ye^{x^2 - y^2} \sin 2xy + 6x^2 - 3y^2;$$

$$u_y(x, y) = -2ye^{x^2 - y^2} \cos 2xy - 2xe^{x^2 - y^2} \sin 2xy - 6xy;$$

$$v_x(x, y) = 2xe^{x^2 - y^2} \sin 2xy + 2ye^{x^2 - y^2} \cos 2xy + 6xy;$$

$$v_y(x, y) = -2ye^{x^2 - y^2} \sin 2xy + 2xe^{x^2 - y^2} \cos 2xy - 3y^2 + 6x^2$$

implying $u_x = v_y$ and $u_y = -v_x$.

6. We want to show for polar coordinates (r, φ) that

$$\frac{\partial u}{\partial r} = \frac{1}{r}\frac{\partial v}{\partial \varphi} \quad \text{and} \quad \frac{\partial v}{\partial r} = -\frac{\partial u}{\partial \varphi}$$

where $f(z) = u(z) + iv(z)$, $z = x + iy$. With $x = r\cos\varphi$, $y = r\sin\varphi$, $r = (x^2 + y^2)^{\frac{1}{2}}$ and $\varphi = \arctan\frac{y}{x}$ we find first

$$\frac{\partial r}{\partial x} = \frac{x}{r} = \cos\varphi, \qquad \frac{\partial r}{\partial y} = \frac{y}{r} = \sin\varphi$$

$$\frac{\partial \varphi}{\partial x} = \frac{-y}{x^2 + y^2} = -\frac{\sin\varphi}{r}, \qquad \frac{\partial \varphi}{\partial y} = \frac{x}{x^2 + y^2} = \frac{\cos\varphi}{r}$$

and now

$$\frac{\partial u}{\partial x} = \frac{\partial u(r\cos\varphi, r\sin\varphi)}{\partial x} = \frac{\partial u}{\partial r}\frac{\partial r}{\partial x} + \frac{\partial u}{\partial \varphi}\frac{\partial \varphi}{\partial x}$$

$$= \cos\varphi\frac{\partial u}{\partial r} - \frac{\sin\varphi}{r}\frac{\partial u}{\partial \varphi};$$

$$\frac{\partial u}{\partial y} = \frac{\partial u(r\cos\varphi, r\sin\varphi)}{\partial y} = \frac{\partial u}{\partial r}\frac{\partial r}{\partial y} + \frac{\partial u}{\partial \varphi}\frac{\partial \varphi}{\partial y}$$

$$= \sin\varphi\frac{\partial u}{\partial r} + \frac{\cos\varphi}{r}\frac{\partial u}{\partial \varphi};$$

$$\frac{\partial v}{\partial x} = \frac{\partial v(r\cos\varphi, r\sin\varphi)}{\partial x} = \frac{\partial v}{\partial r}\frac{\partial r}{\partial x} + \frac{\partial u}{\partial \varphi}\frac{\partial \varphi}{\partial x}$$

$$= \cos\varphi\frac{\partial v}{\partial r} - \frac{\sin\varphi}{r}\frac{\partial v}{\partial \varphi};$$

$$\frac{\partial v}{\partial y} = \frac{\partial v(r\cos\varphi, r\sin\varphi)}{\partial y} = \frac{\partial v}{\partial r}\frac{\partial r}{\partial y} + \frac{\partial v}{\partial \varphi}\frac{\partial \varphi}{\partial y}$$

$$= \sin\varphi\frac{\partial v}{\partial r} + \frac{\cos\varphi}{r}\frac{\partial v}{\partial \varphi}.$$

Using the Cauchy-Riemann differential equations we deduce that

$$\left(\frac{\partial u}{\partial r} - \frac{1}{r}\frac{\partial v}{\partial \varphi}\right)\cos\varphi - \left(\frac{\partial v}{\partial r} + \frac{1}{r}\frac{\partial u}{\partial \varphi}\right)\sin\varphi = 0 \qquad (*)$$

and

$$\left(\frac{\partial u}{\partial r} - \frac{1}{r}\frac{\partial v}{\partial \varphi}\right)\sin\varphi + \left(\frac{\partial v}{\partial r} + \frac{1}{r}\frac{\partial u}{\partial \varphi}\right)\cos\varphi = 0. \qquad (**)$$

If we multiply $(*)$ with $\cos\varphi$ and $(**)$ with $\sin\varphi$ and then add the resulting equalities we find

$$0 = \left(\frac{\partial u}{\partial r} - \frac{1}{r}\frac{\partial v}{\partial \varphi}\right)(\cos^2\varphi + \sin^2\varphi) = \frac{\partial u}{\partial r} - \frac{1}{r}\frac{\partial r}{\partial \varphi}.$$

Multiplying $(*)$ with $-\sin\varphi$ and $(**)$ with $\cos\varphi$ and adding these new equalities we get

$$0 = \left(\frac{\partial v}{\partial r} + \frac{1}{r}\frac{\partial u}{\partial \varphi}\right)(\sin^2\varphi + \cos^2\varphi) = \frac{\partial v}{\partial r} + \frac{1}{r}\frac{\partial u}{\partial \varphi},$$

and the result follows.

7. a) For $y = \operatorname{Im} z > 0$ we have the estimate

$$\left|\frac{e^{inz}}{n^4}\right| = \left|\frac{e^{inx}e^{-ny}}{n^4}\right| \leq \frac{1}{n^4}$$

implying that the series $\sum_{n=1}^{\infty}\frac{e^{inz}}{n^4}$ converges absolutely and uniformly, hence point-wise, for $\operatorname{Im} z > 0$. Consider now the series

$$\sum_{n=1}^{\infty}\frac{d}{dz}\frac{e^{inz}}{n^4} = \sum_{n=1}^{\infty}\frac{in}{n^4}e^{inz} = i\sum_{n=1}^{\infty}\frac{e^{inx}e^{-ny}}{n^3}.$$

681

For $\operatorname{Im} z > 0$ we have the estimate

$$\left| \frac{e^{inx}e^{-ny}}{n^3} \right| \leq \frac{1}{n^3}$$

implying that the differentiated series converges absolutely and uniformly for $\operatorname{Im} z > 0$. Hence we can justify the interchanging of summation and differentiation and it follows that for $\operatorname{Im} z > 0$ the function $z \mapsto \sum_{n=1}^{\infty} \frac{e^{inz}}{n^4}$ is holomorphic.

b) For $|z| > 2$, i.e. $\frac{2}{|z|} < 1$ we have the convergent geometric series $\sum_{n=1}^{\infty} \left(\frac{2}{|z|} \right)^n$ and therefore we find

$$\sum_{n=1}^{\infty} \frac{2^n}{|z|^n} = 1 - \frac{1}{1 - \frac{2}{z}} = \frac{2}{2 - z}$$

which is holomorphic for $z \neq 2$, in particular for all z with $|z| > 2$.

8. We know that for $z \neq -1$ the rational function $z \mapsto w(z) = \frac{z-1}{z+1}$ is holomorphic. We want to determine w^{-1} for which we solve the equation $\zeta = \frac{z-1}{z+1}$, which leads to $z = \frac{\zeta+1}{\zeta-1}$ provided $\zeta \neq 1$. Thus the image of $\mathbb{C}\backslash\{-1\}$ under w is $\mathbb{C}\backslash\{1\}$ and on $\mathbb{C}\backslash\{1\}$ the function $w^{-1}(\zeta) = \frac{\zeta+1}{\zeta-1}$ is holomorphic and inverse to w. Thus $w : \mathbb{C}\backslash\{-1\} \to \mathbb{C}\backslash\{1\}$ is a biholomorphic mapping.

9. If f is biholomorphic then f is holomorphic, bijective and since f^{-1} is also holomorphic it is continuous. Furthermore we have

$$z = f^{-1}(f(z)), \quad z \in G_1$$

which implies by the chain rule that

$$1 = \left(f^{-1} \right)' (f(z)) \cdot f'(z).$$

Hence $f'(z) \neq 0$ for $z \in G_1$ and with $w = f(z)$ we have

$$\left(f^{-1} \right)' (w) = \frac{1}{f'(z)}.$$

Now let f be holomorphic, bijective and assume that f^{-1} is continuous as well as that $f'(z) \neq 0$ for all $z \in G_1$. We want to show that $g := f^{-1} : G_2 \to G_1$ is holomorphic. With $w := f(z)$, $z \in G_1$, and $w_0 = f(z_0)$, $z_0 \in G_1$, we find

$$\frac{g(w) - g(w_0)}{w - w_0} = \frac{1}{\frac{w-w_0}{g(w)-g(w_0)}} = \frac{1}{\frac{f(z)-f(z_0)}{z-z_0}}$$

and since f' is continuous and $f'(z_0) \neq 0$ we find a neighbourhood of z_0 where $\frac{f(z)-f(z_0)}{z-z_0} \neq 0$. Thus we may pass to the limit as z tends to z_0 and we find

$$g'(w_0) = \frac{1}{f'(z_0)},$$

in particular $\left(f^{-1} \right)' (w_0)$ exists.

682

10. Clearly we have $|z| = r^2$ and

$$|\zeta - z|^2 = (\overline{\zeta - z})(\zeta - z) = \left(e^{-i\vartheta} - re^{-i\varphi}\right)\left(e^{i\vartheta} - re^{i\varphi}\right)$$
$$= 1 + r^2 - r\left(e^{-i(\vartheta - \varphi)} - e^{-(\varphi - \vartheta)}\right)$$
$$= 1 + r^2 - 2r\cos(\vartheta - \varphi)$$

implying that

$$P(\zeta, z) = \frac{1}{2\pi}\frac{1 - r^2}{1 + r^2 - 2r\cos(\vartheta - \varphi)}.$$

We know that the function $z \mapsto \frac{\zeta + z}{\zeta - z}$, $\zeta \in \partial B_1(0)$, $z \in B_1(0)$, is holomorphic, hence its real part is harmonic. Since

$$\frac{\zeta + z}{\zeta - z} = \frac{(\zeta + z)(\overline{\zeta} - \overline{z})}{|\zeta - z|^2} = \frac{\zeta\overline{\zeta} + z\overline{\zeta} - \zeta\overline{z} - z\overline{z}}{|\zeta - z|^2}$$
$$= \frac{1 - r^2}{1 - r^2 - 2r\cos(\vartheta - \varphi)} + i\frac{2r\sin(\vartheta - \varphi)}{1 - r^2 - 2r\cos(\vartheta - \varphi)},$$

where we used that $z\overline{\zeta} - \zeta\overline{z} = (2r\sin(\vartheta - \varphi))i$, it follows that $P(\zeta, (x, y))$ as a function of (x, y) is a harmonic function.

11. We have to find

$$\left(\frac{\partial^2}{\partial x^2} + \frac{\partial^2}{\partial y^2}\right)h(u(x, y), v(x, y)).$$

It follows that

$$\frac{\partial h}{\partial x} = \frac{\partial h}{\partial u}\frac{\partial u}{\partial x} + \frac{\partial h}{\partial v}\frac{\partial v}{\partial x},$$
$$\frac{\partial^2 h}{\partial x^2} = \frac{\partial^2 h}{\partial u^2}\frac{\partial u}{\partial x}\frac{\partial u}{\partial x} + \frac{\partial^2 h}{\partial v\partial u}\frac{\partial v}{\partial x}\frac{\partial u}{\partial x} + \frac{\partial h}{\partial u}\frac{\partial^2 u}{\partial x^2}$$
$$+ \frac{\partial^2 h}{\partial v^2}\frac{\partial v}{\partial x}\frac{\partial v}{\partial x} + \frac{\partial^2 h}{\partial v\partial u}\frac{\partial v}{\partial x}\frac{\partial u}{\partial x} + \frac{\partial h}{\partial v}\frac{\partial^2 v}{\partial x^2}$$
$$= h_{uu}u_x u_x + 2h_{uv}u_x v_x + h_{uv}v_x v_x + h_u u_{xx} + h_v v_{xx},$$

and analogously

$$\frac{\partial h}{\partial y} = \frac{\partial h}{\partial u}\frac{\partial u}{\partial y} + \frac{\partial h}{\partial v}\frac{\partial v}{\partial y},$$
$$\frac{\partial^2 h}{\partial y^2} = h_{uu}u_y u_y + 2h_{uv}u_y v_y + h_{vv}v_y v_y + h_u u_{yy} + h_v v_{yy},$$

which yields

$$\frac{\partial^2 h}{\partial x^2} + \frac{\partial^2 h}{\partial y^2} = h_{uu}(u_x^2 + u_y^2) + 2h_{uv}(v_x u_x + v_y u_y) + h_{vv}(v_x^2 + v_y^2)$$
$$+ h_u(u_{xx} + u_{yy}) + h_v(v_{xx} + v_{yy}).$$

Since f is holomorphic, u and v are harmonic, i.e. $u_{xx} + u_{yy} = 0$ and $v_{xx} + v_{yy} = 0$, and we find

$$\frac{\partial^2 h}{\partial x^2} + \frac{\partial^2 h}{\partial y^2} = h_{uu}(u_x^2 + u_y^2) + 2h_{uv}(v_x u_x + v_y u_y) + h_{vv}(v_x^2 + v_y^2).$$

The Cauchy-Riemann differential equations give $u_x = v_y$ and $u_y = -v_x$ and therefore we have

$$v_x u_x + v_y u_y = -u_x u_y + u_x u_y = 0,$$

and it follows that

$$\frac{\partial^2 h}{\partial x^2} + \frac{\partial^2 h}{\partial y^2} = h_{uu}(u_x^2 + u_y^2) + h_{vv}(v_x^2 + v_y^2).$$

However, again by the Cauchy-Riemann differential equations, we note that $u_x^2 + u_y^2 = v_x^2 + v_y^2$, i.e.

$$\left(\frac{\partial^2}{\partial x^2} + \frac{\partial^2}{\partial y^2} \right) h\left(u(xy), v(x, y) \right) = \left(h_{uu} + h_{vv} \right) \left(u_x^2 + u_y^2 \right)$$

and since by assumption h is harmonic we deduce that

$$\left(\frac{\partial^2}{\partial x^2} + \frac{\partial^2}{\partial y^2} \right) h\left(u(x, y), v(x, y) \right) = 0.$$

This result is interesting: we can obtain harmonic functions in G_1 once we know harmonic functions in G_2 and a holomorphic mapping from G_1 onto G_2. In the case where f is biholomorphic, hence $u_x^2 + u_y^2 \neq 0$, we obtain a one-to-one correspondence of harmonic functions on G_1 and harmonic functions on G_2.

Chapter 18

1. For $z = re^{i\varphi}$, $-\pi < \varphi < \pi$, we have

$$z^i = e^{i \log z} = e^{i(\ln r + i\varphi)} = e^{-\varphi}(\cos \ln r + i \sin \ln r).$$

Since $\frac{1}{2}\sqrt{2}e^{2\pi} + \frac{1}{2}\sqrt{2}e^{2\pi}i = e^{2\pi}e^{i\frac{\pi}{4}}$ it follows with $r = e^{2\pi}$ and $\varphi = \frac{\pi}{4}$ that

$$\left(\frac{1}{2}\sqrt{2}e^{2\pi} + \frac{1}{2}\sqrt{2}e^{2\pi}i \right)^i = e^{-\frac{\pi}{4}}(\cos 2\pi + i \sin 2\pi) = e^{-\frac{\pi}{4}}.$$

2. With $z_1 = e^{\frac{2\pi i}{3}}$ and $z_2 = e^{\frac{3\pi i}{4}}$ we find

$$\log z_1 = \frac{2\pi i}{3} \quad \text{and} \quad \log z_2 = \frac{3\pi i}{4}$$

leading to

$$\log z_1 + \log z_2 = \frac{17\pi i}{12}.$$

684

On the other hand we have

$$z_1 \cdot z_2 = e^{\frac{2\pi i}{3}} e^{\frac{3\pi i}{4}} = e^{-\frac{7\pi i}{12}}$$

and

$$\log(z_1 \cdot z_2) = -\frac{7\pi i}{12}.$$

Note however that

$$\log(z_1 \cdot z_2) + 2\pi i = \frac{17\pi i}{12} = \log z_1 + \log z_2.$$

3. From Problem 7 in Chapter 16 we know that for $x \in (-1, 1)$

$$\sum_{k=0}^{\infty} \binom{\beta}{k} x^k = (1+x)^{\beta}$$

holds and since for $x \in (-1, 1)$ the number $1 + x$ belongs to $\mathbb{C} \setminus (-\infty, 0]$ we find

$$\sum_{k=0}^{\infty} \binom{\beta}{k} x^k = (1+x)^{\beta} = e^{\beta \log(1+x)} = e^{s \log(1+x)} e^{it \log(1+x)}$$

$$= (1+x)^s \left(\cos(t \ln(1+x)) + i \sin(t \ln(1+x)) \right).$$

By Proposition 16.18 the series $\sum_{k=0}^{\infty} \binom{\beta}{k} z^k$ converges for all $|z| < 1$ and we know from Problem 7 of Chapter 16 that for $x \in (-1, 1)$ we have $\sum_{k=0}^{\infty} \binom{\beta}{k} x^k = (1+x)^{\beta}$. The uniqueness theorem for holomorphic functions which we will prove in Chapter 22, Theorem 22.16, now implies

$$\sum_{k=0}^{\infty} \binom{\beta}{k} z^k = (1+z)^{\beta} \quad \text{for} \quad z \in B_1(0) \quad \text{and} \quad \beta \in \mathbb{C}.$$

4. We know that $L_{a,b}(z) := b + \tilde{L}\left(\frac{z}{a}\right) = b + \sum_{k=1}^{\infty} \frac{(-1)^{k-1}}{k} \left(\frac{z}{a} - 1\right)^k$ converges in $B_{|a|}(a)$ and is continuous, in fact holomorphic. We want to use Theorem 18.4 in order to show that $L_{a,b}$ is a branch of the logarithmic function. For this we note that

$$L'_{a,b}(z) = \frac{d}{dz} \left(b + \sum_{k=1}^{\infty} \frac{(-1)^{k-1}}{k} \left(\frac{z}{a} - 1\right)^k \right) = \frac{d}{dz} \tilde{\ln}\left(\frac{z}{a} - 1\right)$$

$$= \frac{1}{a} \cdot \frac{1}{\frac{z}{a}} = \frac{1}{z}.$$

In addition it follows that

$$e^{L_{a,b}(a)} = e^b = a$$

since by assumption b is a logarithm of a and

$$L_{a,b}(a) = b + \sum_{k=1}^{\infty} (-1)^k \left(\frac{a}{a} - 1\right)^k = b.$$

5. The inverse of $w(z) = \frac{1+z}{1-z}$ is $\zeta(w) = \frac{w-1}{w+1}$. For $|z| < 1$ we find with $z = x + iy$

$$\frac{1+z}{1-z} = \frac{(1+z)(1-\bar{z})}{(1-x)^2+y^2} = \frac{1-(x^2+y^2)}{(1-x^2)+y^2} + i\frac{2y}{(1-x)^2+y^2}$$

implying $\operatorname{Re}\left(\frac{1+z}{1-z}\right) > 0$ for $|z| < 1$.

On the other hand for $w = u + iv$ the condition $\operatorname{Re} w = u > 0$ implies $|w-1| < |w+1|$ or $\left|\frac{w-1}{w+1}\right| < 1$. Thus we have proved that the bijective mapping $w : \mathbb{C}\backslash\{1\} \to \mathbb{C}\backslash\{-1\}$ maps $B_1(0)$ into $\{\zeta \in \mathbb{C}\,|\,\operatorname{Re}\zeta > 0\}$ and its inverse maps $\{\zeta \in \mathbb{C}\,|\,\operatorname{Re}\zeta > 0\}$ into $B_1(0)$, i.e. $w(B_1(0)) = \{\zeta \in \mathbb{C}\,|\,\operatorname{Re}\zeta > 0\}$.

Chapter 19

1.   a) Since in a discrete topological space every set is open and closed, a discrete topological space is connected if and only if its underlying set X has one element.

    b) Such a mapping cannot exist. A continuous mapping must map the connected space X onto a connected subset of Y. Since by assumption f is surjective, i.e. $f(X) = Y$, and Y is not connected such f cannot be continuous.

2. Take $(a_1, b_1), (a_2, b_2) \in A \times B$. The set $\{a_1\} \times B$ is homeomorphic to B and the set $A \times \{b_2\}$ is homeomorphic to A, hence $A \times \{b_2\}$ and $\{a_1\} \times B$ are connected in $\mathbb{R}^{n+m}$. Furthermore we have $\{a_1\} \times B \cap A \times \{b_2\} = \{a_1, b_2\}$, implying that $\{a_1\} \times B \cup A \times \{b_2\}$ is connected in $\mathbb{R}^{n+m}$ and therefore (a_1, b_1) and (a_2, b_2) belong to the same connectivity component of $A \times B$. However (a_1, b_1) and (a_2, b_2) are arbitrary points, hence $A \times B$ is connected.

3. We decompose $\operatorname{tr}(\gamma)$ into four curves $\gamma_1, \gamma_2, \gamma_3, \gamma_4$ such that

- $\gamma_1 : [0, \frac{1}{4}] \to \mathbb{C}$, $\operatorname{tr}(\gamma_1)$ is the half circle with centre $2 + i$ and radius 1 and $\gamma_1(0) = 2$ and $\gamma_1\left(\frac{1}{4}\right) = 2 + 2i$;

- $\gamma_2 : [\frac{1}{4}, \frac{1}{2}] \to \mathbb{C}$, $\operatorname{tr}(\gamma_2)$ is the line segment connecting $2 + 2i$ and $1 + 2i$, i.e. $\gamma_2\left(\frac{1}{4}\right) = 2 + 2i$ and $\gamma_2\left(\frac{1}{2}\right) = 1 + 2i$;

- $\gamma_3 : [\frac{1}{2}, \frac{3}{4}] \to \mathbb{C}$, $\operatorname{tr}(\gamma_3)$ is the line segment connecting $1 + 2i$ and 0, i.e. $\gamma_3\left(\frac{1}{2}\right) = 1 + 2i$ and $\gamma_3\left(\frac{3}{4}\right) = 0$;

- $\gamma_4 : [\frac{3}{4}, 1] \to \mathbb{C}$, $\operatorname{tr}(\gamma_4)$ is the line segment connecting 0 with 2, i.e. $\gamma_4\left(\frac{3}{4}\right) = 0$ and $\gamma_4(1) = 2$.

The corresponding curves are

$$\gamma_1(t) = (2+i) - ie^{4\pi it}, \quad t \in \left[0, \frac{1}{4}\right];$$

$$\gamma_2(t) = -4t + 3 + 2i, \quad t \in \left[\frac{1}{4}, \frac{1}{2}\right];$$

$$\gamma_3(t) = (-4 - 8i)t + 3 + 6i, \quad t \in \left[\frac{1}{2}, \frac{3}{4}\right];$$

$$\gamma_4(t) = 8t - 6i, \quad t \in \left[\frac{3}{4}, 1\right].$$

4. We identify $\mathbb{R}^2$ with $\mathbb{C}$ and we introduce polar coordinates (r, φ). Every $z \in \mathbb{C}\backslash\{0\} \cong \mathbb{R}^2\backslash\{0\}$ corresponds to $z = re^{i\varphi}$, $r > 0$, $\varphi \in [0, 2\pi)$. For $\rho > 0$ the mapping $R_\rho : \mathbb{C}\backslash\{0\} \to \mathbb{C}\backslash\{0\}$, $z \mapsto R_\rho(z) = \rho re^{i\varphi}$, $z = re^{i\varphi}$, is bijective, continuous with continuous inverse $R_{\frac{1}{\rho}}$. Finally, $f = R_{\frac{1}{2}}$ maps $\partial B_2(0)$ onto $\partial B_1(0)$ since for $z = 2e^{i\varphi}$ we have $R_{\frac{1}{2}}\left(2e^{i\varphi}\right) = e^{i\varphi}$ and conversely $2e^{i\varphi} = R_2\left(e^{i\varphi}\right)$.

5. Let $C \subset X$ be an arcwise connected set and $f : X \to Y$ a continuous mapping. For $y_1, y_2 \in f(C)$ we choose $x_1, x_2 \in C$ such that $f(x_1) = y_1$ and $f(x_2) = y_2$. Now let $\gamma : [0, 1] \to C$ be a continuous curve connecting x_1 and x_2, i.e. $\gamma(0) = x_1$ and $\gamma_1 = x_2$. We consider the arc $f \circ \gamma : [0, 1] \to f(C)$ which is continuous and $f \circ \gamma(0) = f(x_1) = y_1$ as well as $(f \circ \gamma)(1) = f(x_2) = y_2$. Hence $f(C)$ is arcwise connected.

6. Let $\gamma_j : [a_j, b_j] \to X$, $j = 1, 2, 3$, with $\gamma(b_1) = \gamma_2(a_2)$ and $\gamma_2(b_2) = \gamma_3(a_3)$. It follows that $\gamma_1 \oplus \gamma_2 : [a_1, b_1 + b_2 - a_2] \to X$,

$$(\gamma_1 \oplus \gamma_2)(t) = \begin{cases} \gamma_1(t), & t \in [a_1, b_1] \\ \gamma_2(t + a_2 - b_1), & t \in [b_1, b_1 + b_2 - a_2] \end{cases}$$

and $(\gamma_1 \oplus \gamma_2) \oplus \gamma_3 : [a_1, b_1 + b_2 - b_3 - a_2 - a_3] \to X$ where

$$((\gamma_1 \oplus \gamma_2) \oplus \gamma_3) = \begin{cases} (\gamma_1 \oplus \gamma_2)(t), & t \in [a_1, b_1 + b_2 - a_2] \\ \gamma_3(t + a_3 + a_2 - b_1 - b_2), & t \in [b_1 + b_2 - a_2, b_1 + b_2 + b_3 - a_2 - a_3] \end{cases}$$

$$= \begin{cases} \gamma_1(t), & t \in [a_1, b_1] \\ \gamma_2(t + a_2 - b_1), & t \in [b_1, b_1 + b_2 - a_2] \\ \gamma_3(t + a_2 + a_3 - b_1 - b_2), & t \in [b_1 + b_2 - a_2, b_1 + b_2 + b_3 - a_2 - a_3]. \end{cases}$$

On the other hand we have $\gamma_2 \oplus \gamma_3 : [a_2, b_2 + b_3 - a_3] \to X$ with

$$(\gamma_2 \oplus \gamma_3)(t) = \begin{cases} \gamma_2(t), & t \in [a_2, b_2] \\ \gamma_3(t + a_3 - b_2), & t \in [b_2, b_2 + b_3 - a_3] \end{cases}$$

and $\gamma_1 \oplus (\gamma_2 \oplus \gamma_3) : [a_1, b_1 + b_2 - a_2 - a_3] \to X$ where

$$(\gamma_1 \oplus (\gamma_2 \oplus \gamma_3))(t) = \begin{cases} \gamma_1(t), & t \in [a_1, b_1] \\ (\gamma_2 \oplus \gamma_3)(t + a_2 - b_1), & t \in [b_1, b_1 + b_2 + b_3 - a_2 - a_3] \end{cases}$$

$$= \begin{cases} \gamma_1(t), & t \in [a_1, b_1] \\ \gamma_2(t + a_2 - b_1), & t \in [b_1, b_1 + b_2 - a_2] \\ \gamma_3(t + a_2 + a_3 - b_1 - b_2), & t \in [b_1 + b_2 - a_2, b_1 + b_2 + b_3 - a_3], \end{cases}$$

i.e. we indeed have $(\gamma_1 \oplus \gamma_2) \oplus \gamma_3 = \gamma_1 \oplus (\gamma_2 \oplus \gamma_3)$.

7. Let $\gamma_1 \sim \tilde{\gamma}_1$ and $\gamma_2 \sim \tilde{\gamma}_2$, we have to prove that $\gamma_1 \oplus \gamma_2 \sim \tilde{\gamma}_1 \oplus \tilde{\gamma}_2$. With $\varphi_1 : [\tilde{a}_1, \tilde{b}_1] \to [a_1, b_1]$ and $\varphi_2 : [\tilde{a}_2, \tilde{b}_2] \to [a_2, b_2]$ we have $\tilde{\gamma}_1 = \gamma_1 \circ \varphi_1$ and $\tilde{\gamma}_2 = \gamma_2 \circ \varphi_2$. Furthermore it follows that $\gamma_1 \oplus \gamma_2 : [a_1, b_1 + b_2 - a_2] \to X$ where

$$(\gamma_1 \oplus \gamma_2)(t) = \begin{cases} \gamma_1(t), & t \in [a_1, b_1] \\ \gamma_2(t + a_2 - b_1), & t \in [b_1, b_1 + b_2 - a_2], \end{cases}$$

and $\tilde{\gamma}_1 \oplus \tilde{\gamma}_2 : [\tilde{a}_1, \tilde{b}_1 + \tilde{b}_2 - \tilde{a}_2] \to X$ where

$$(\tilde{\gamma}_1 \oplus \tilde{\gamma}_2)(t) = \begin{cases} \tilde{\gamma}_1(t), & t \in [\tilde{a}_1, \tilde{b}_1] \\ \tilde{\gamma}_2(t + \tilde{a}_2 - \tilde{b}_1), & t \in [\tilde{b}_1, \tilde{b}_1 + \tilde{b}_2 - \tilde{a}_2]. \end{cases}$$

With $\varphi : [\tilde{a}_1, \tilde{b}_1 + \tilde{b}_2 - \tilde{a}_2] \to [a_1, b_1 + b_2 - a_2]$ defined by

$$\varphi(\tilde{r}) = \begin{cases} \tilde{\varphi}_1(\tilde{r}), & \tilde{r} \in [\tilde{a}_1, \tilde{b}_1] \\ \tilde{\varphi}(\tilde{r} + \tilde{a}_2 - \tilde{b}_1), & \tilde{r} \in [\tilde{b}_1, \tilde{b}_1 + \tilde{b}_2 - \tilde{a}_2] \end{cases}$$

it follows that $\widetilde{\gamma_1 \oplus \gamma_2} = (\gamma_1 \oplus \gamma_2) \circ \varphi = (\gamma_1 \circ \varphi) \oplus (\gamma_2 \circ \varphi) = \tilde{\gamma}_1 \oplus \tilde{\gamma}_2$, and therefore the definition of $[\gamma_1] \oplus [\gamma_2]$ is independent of the parametrization. With the help of Problem 6 we now find

$$([\gamma_1] + [\gamma_2]) \oplus [\gamma_3] = [\gamma_1 \oplus \gamma_2] \oplus [\gamma_3] = [\gamma_1 \oplus \gamma_2 \oplus \gamma_3]$$
$$= [\gamma_1] \oplus [\gamma_2 \oplus \gamma_3] = [\gamma_1] \oplus ([\gamma_2] \oplus [\gamma_3]).$$

8. With $f = g = \mathrm{id}_X$ it follows that every topological space X is homotopically equivalent to itself. If X and Y are two topological spaces and X is homotopically equivalent to Y then there exists mappings $f : X \to Y$ and $g : Y \to X$ such that $g \circ f$ is homotopic to the identity id_X on X and $f \circ g$ is homotopic to the identity id_Y on Y. This implies that Y is homotopically equivalent to X just by switching the roles of X and Y as well as f and g. Now let X be homotopically equivalent to Y and Y homotopically equivalent to Z. Thus there exist continuous mappings $f_X : X \to Y$ and $g_Y : Y \to X$ such that $g_Y \circ f_X$ is homotopic to id_X and $f_X \circ g_Y$ is homotopic to id_Y. Further there exists continuous mappings $f_Y : Y \to Z$ and $g_Z : Z \to Y$ such that $g_Z \circ f_Y$ is homotopic to id_Y and $f_Y \circ g_Z$ is homotopic to id_Z. We consider the mappings $f_Y \circ f_X : X \to Z$ and $g_Y \circ g_Z : Z \to X$ which are continuous and from Theorem 19.19 we deduce that $(g_Y \circ g_Z) \circ (f_Y \circ f_X) = g_Y \circ (g_Z \circ f_Y) \circ f_X$ is homotopic to id_Z. Hence X is homotopically equivalent to Z.

9. Let $x_0 \in A_{\frac{1}{2},2}(0)$, $|x_0| = 1$. The path $\gamma : [0,1] \to A_{\frac{1}{2},2}(0)$ $\gamma(t) = e^{2\pi i t}$, has its trace in $A_{\frac{1}{2},2}(0)$, $1 \in \text{tr}(\gamma)$, but γ is not homotopic to the constant path $\gamma_1 : [0,1] \to A_{\frac{1}{2},2}(0)$, $\gamma_1(t) = 1$ for all $t \in [0,1]$.

Chapter 20

1. For $n \neq -1$ we find

$$\int_\gamma z^n \, dz = \int_0^{2\pi k} e^{int} i e^{it} \, dt = i \int_0^{2\pi k} e^{i(n+1)t} \, dt$$

$$= \frac{1}{n+1} \left(e^{2\pi k(n+1)} - 1 \right) = 0.$$

However for $n = -1$ we have

$$\int_\gamma z^{-1} \, dz = \int_0^{2\pi k} \frac{1}{e^{it}} i e^{it} \, dt = i \int_0^{2\pi k} 1 \, dt = 2\pi k i.$$

2. Recall that $\gamma : [0,\pi] \to \mathbb{C}$, $\gamma(t) = e^{i(\pi-1)}$, and $\sigma : [-1,1] \to \mathbb{C}$, $\sigma(t) = t$. Note that $\gamma(0) = \sigma(-1) = -1$ and $\gamma(\pi) = \gamma(1) = 1$. Further we have $\dot{\gamma}(t) = -ie^{i(\pi-t)}$ and $\dot{\sigma}(t) = 1$. On γ we find $\text{Re}\, z = \text{Re}\, e^{it} = \cos t$ and $\text{Im}\, z = \text{Im}\, e^{it} = \sin t$ whereas on σ we have $\text{Re}\, z = t$ and $\text{Im}\, z = 0$.
This yields

$$\int_\gamma \text{Re}\, z \, dz = \int_0^\pi (\cos t) \left(-ie^{i(\pi-t)} \right) \, dt$$

$$= -i \int_0^\pi \cos t \cos(\pi - t) \, dt + \int_0^\pi \cos t \sin(\pi - t) \, dt$$

$$= i \int_0^\pi \cos^2 t \, dt - \int_0^\pi \cos t \sin t \, dt$$

$$= i\frac{\pi}{2}$$

and

$$\int_\sigma \text{Re}\, z \, dz = \int_{-1}^1 t \, dt = \frac{t^2}{2} \Big|_{-1}^1 = 0.$$

Moreover we have

$$\int_\gamma \text{Im}\, z \, dz = \int_0^\pi \sin t \left(-ie^{i(\pi-t)} \right) \, dt$$

$$= -i \int_0^\pi \sin t \cos(\pi - t) \, dt - \int_0^\pi \sin^2 t \, dt$$

$$= -\frac{\pi}{2}$$

and

$$\int_\sigma \text{Im}\, z \, dz = \int_\sigma 0 \, dz = 0.$$

3. The parametrization is given by $\gamma : [0, 1] \to \mathbb{C}$ with

$$\gamma(t) = \begin{cases} 12t - 2, & t \in [0, \frac{1}{3}] \\ -(6 - 6i)t + 4 - 2i, & t \in [\frac{1}{3}, \frac{2}{3}] \\ -(6 + 6i)t + 4 + 6i, & t \in [\frac{2}{3}, 1] \end{cases}$$

with corresponding derivative

$$\dot{\gamma}(t) = \begin{cases} 12, & t \in [0, \frac{1}{3}] \\ -(6 - 6i), & t \in [\frac{1}{3}, \frac{2}{3}] \\ -(6 + 6i), & t \in [\frac{2}{3}, 1] . \end{cases}$$

Now we find with $\gamma_1 = \gamma\big|_{[0, \frac{1}{3}]}$, $\gamma_2 = \gamma\big|_{[\frac{1}{3}, \frac{2}{3}]}$ and $\gamma_3 = \gamma\big|_{[\frac{2}{3}, 1]}$ that

$$\int_\gamma z \, dz = \int_{\gamma_1} z \, dz + \int_{\gamma_2} z \, dz + \int_{\gamma_3} z \, dz$$

and further

$$\int_{\gamma_1} z \, dz = \int_0^{\frac{1}{3}} (12t - 2) \cdot 12 \, dt$$

$$= 144 \frac{t^2}{2} - 24t \Big|_0^{\frac{1}{3}} = 0$$

$$\int_{\gamma_2} z \, dz = \int_{\frac{1}{3}}^{\frac{2}{3}} (-(6 - 6i)t + (4 - 2i)) (-(6 - 6i)) \, dt$$

$$= (-6 + 6i)^2 \frac{t^2}{2} + (4 - 2i)(-6 + 6i)t \Big|_{\frac{1}{3}}^{\frac{2}{3}}$$

$$= -4$$

and

$$\int_{\gamma_3} z \, dz = \int_{\frac{2}{3}}^1 ((-6 - 6i)t + 4 + 6i) (-6 - 6i) \, dt$$

$$= (-6 - 6i)^2 \frac{t^2}{2} - (4 + 6i)(6 + 6i)t \Big|_{\frac{2}{3}}^1$$

$$= 4.$$

Thus we find indeed

$$\int_\gamma z \, dz = \int_{\gamma_1} z \, dz + \int_{\gamma_2} z \, dz + \int_{\gamma_3} z \, dz = 0 - 4 + 4 = 0.$$

4.   a) Since for every simply closed, piecewise continuously differentiable curve γ we have $\int_\gamma 1 \, dz = 0$ such a result is not possible.

b) From estimate (20.18) we deduce that

$$\left| \frac{1}{l_{\gamma_k}} \int_{\gamma_k} f(z)\,dz \right| \le \|f\|_{\infty,\mathrm{tr}\,\gamma_k} \le \|f\|_{\infty,K}.$$

Hence the sequence $\left(\left| \frac{1}{l_{\gamma_k}} \int_{\gamma_k} f(z)\,dz \right| \right)_{k\in\mathbb{N}}$ is bounded and therefore its lim sup must be finite.

5. With $z = x + iy$ we find when identifying $\mathbb{C}$ with $\mathbb{R}^2$ that $\gamma(t) = \begin{pmatrix} 4\cos t \\ 4\sin t \end{pmatrix} +$
$\begin{pmatrix} \cos t \\ -\sin t \end{pmatrix} = \begin{pmatrix} 5\cos t \\ 3\sin t \end{pmatrix}$ and for every $t \in [0, 2\pi]$ we find

$$\left(\frac{x}{5}\right)^2 + \left(\frac{y}{3}\right)^2 = \cos^2 t + \sin^2 t = 1.$$

Hence $\mathrm{tr}(\gamma)$ is the ellipse with centre 0 and principal axes being the real- and imaginary axes, respectively, passing through $(5,0), (-5,0), (0,3)$ and $(0,-3)$. Now we find with $\dot\gamma(t) = 4ie^{it} - ie^{-it}$ that

$$\int_\gamma z^3\,dz = \int_0^{2\pi} \left(4e^{it} + e^{-it}\right)^3 \left(4ie^{it} - ie^{-it}\right) dt$$
$$= \int_0^{2\pi} \left(256ie^{4it} + 128ie^{2it} - 8ie^{2it} - ie^{-4it}\right) dt$$
$$= 0,$$

where we used that for $n \in \mathbb{N}$ we have $\int_0^{2\pi} e^{nit}\,dt = 0$.

6. Let $f(z) = \sum_{k=0}^\infty a_k(z - z_0)^k$ be the power series representation of f in $B_R(z_0)$ and let $0 < r < R$. Since for $r < R$ it follows that $\sum_{k=0}^\infty |a_k| r^k$ converges, we deduce that $\sum_{k=0}^\infty \frac{|a_k|}{k+1} r^k$ and hence $\sum_{k=0}^\infty \frac{|a_k|}{k+1} r^{k+1}$ converges. This implies that for $z \in B_r(z_0)$, $0 < r < R$, the power series $F(z) := \sum_{k=0}^\infty \frac{a_k}{k+1}(z - z_0)^{k+1}$ converges absolute and uniformly and therefore we find that

$$F'(z) = \frac{d}{dz}\left(\sum_{k=0}^\infty \frac{a_k}{k+1}(z - z_0)^{k+1}\right) = \sum_{k=0}^\infty a_k(z - z_0)^k = f(z),$$

i.e. F is primitive of f in every disc $B_r(z_0)$, $0 < r < R$.

7.   a) We use the standard parametrization $\gamma : [0, 1] \to \mathbb{C}$, $\gamma(t) = (2+4\pi i)t - 1 - 2\pi i$

with $\dot{\gamma}(t) = 2 + 4\pi i$. Now we find

$$\int_{[-1-2\pi i, 1+2\pi i]} e^{2z}\, dz = \int_0^1 e^{2(2+4\pi i)t - 2 - 4\pi i}(2 + 4\pi i)\, dt$$

$$= (2 + 4\pi i)e^{-2}\int_0^1 e^{(4+8\pi i)t}\, dt$$

$$= \frac{2 + 4\pi i}{4 + 8\pi i}e^{-2}\left(e^{(4+8\pi i)t}\Big|_0^1\right)$$

$$= \frac{1}{2e^2}\left(e^{4+8\pi i} - 1\right) = \frac{1}{2}\left(e^2 - \frac{1}{e^2}\right).$$

b) It follows with the parametrization $\gamma : \left[0, \frac{3\pi}{2}\right] \to \mathbb{C}$, $\gamma(t) = 2e^{it}$ that $\dot{\gamma}(t) = 2ie^{it}$ and

$$\int_\gamma \sinh |z|\, dz = \int_0^{\frac{3\pi}{2}} \sinh\left|2e^{it}\right| 2ie^{it}\, dt$$

$$= 2i \sinh 2 \int_0^{\frac{3\pi}{2}} e^{it}\, dt$$

$$= 2\sinh 2 \left(e^{\frac{3\pi}{2}i} - 1\right)$$

$$= 2\sinh 2(-i - 1)$$

$$= -(2 + 2i)\sinh 2.$$

c) Once we know that $z \mapsto p(z)e^z$ admits a primitive in a neighbourhood of the closure of the interior of γ, it follows that $\int_\gamma p(z)e^z\, dz = 0$. We prove that $z \mapsto p(z)e^z$ has a convergent power series representation in $\mathbb{C}$ which in turn implies that the primitive exists, see Problem 6. Now let $p(z) = \sum_{k=0}^N a_k z^k$ and consider the Taylor expansion of the expontential function with centre 0, i.e.

$$e^z = \sum_{l=0}^\infty \frac{z^l}{l!}.$$

We find that

$$p(z)e^z = \sum_{k=0}^N a_k \sum_{l=0}^\infty \frac{z^{k+l}}{l!}$$

and the result follows if we can prove that for every $k \in \mathbb{N}_0$ the series $\sum_{l=0}^\infty \frac{z^{k+l}}{l!}$ converges in $\mathbb{C}$. Since

$$\sum_{l=0}^\infty \frac{z^{k+l}}{l!} = \sum_{m=k}^\infty \frac{z^m}{(m-k)!} \quad \text{and} \quad \frac{|z^{m-1}|}{(m+1-k)!} \cdot \frac{(m-k)!}{|z^m|} = \frac{|z|}{m+1-k},$$

we deduce from the ratio test that the series $\sum_{l=0}^\infty \frac{z^{k+l}}{l!}$ converges for all $z \in \mathbb{C}$.

8.   a) From Example 20.21 we know that $z \mapsto |z|$ has no primitive and Problem 2 yields that neither $z \mapsto \operatorname{Re} z$ nor $z \mapsto \operatorname{Im} z$ has a primitive.

b) With $\gamma : [0, \pi] \to \mathbb{C}$, $\gamma(t) = e^{i(\pi - t)}$ we find

$$\int_\gamma f(|z|) \, dz = \int_0^\pi f(|\gamma(t)|) \dot\gamma(t) \, dt$$

$$= f(1) \int_0^\pi \dot\gamma(t) \, dt$$

$$= f(1) \left(\gamma(\pi) - \gamma(0) \right) = 2 f(1).$$

On the other hand, with $\sigma : [-1, 1] \to \mathbb{C}$, $\sigma(t) = t$, we have

$$\int_\sigma f(|z|) \, dz = \int_{-1}^1 f(|t|) \, dt = \int_{-1}^0 f(-t) \, dt + \int_0^1 f(t) \, dt.$$

Now for $f(r) = r^k$ the first calculation yields

$$\int_\gamma |z|^k \, dz = 2.$$

For k odd we find $f(-t) = -f(t)$ and

$$\int_\sigma f(|z|) \, dz = -\int_{-1}^0 f(t) \, dt + \int_0^1 f(t) \, dt = 0,$$

whereas for k even it follows

$$\int_\sigma f(|z|) \, dz = 2 \int_0^1 f(t) \, dt = 2 \int_0^1 t^k \, dt = \frac{2}{k}.$$

In all cases we find $\int_\gamma f(|z|) \, dz \neq \int_\sigma f(|z|) \, dz$ and consequently $z \mapsto f(|z|) = |z|^k$ has no primitive.

Chapter 21

1. Since $f \cdot g$ is a primitive of $\frac{d}{dz}(f \cdot g)$ we find

$$\int_\gamma \frac{d}{dz}(f(z)g(z)) \, dz = f(\gamma(b))g(\gamma(b)) - f(\gamma(a))g(\gamma(a))$$

and on the other hand we have

$$\int_\gamma \frac{d}{dz}(f(z)g(z)) \, dz = \int_\gamma f(z)g'(z) \, dz + \int_\gamma f'(z)g(z) \, dz$$

implying the result. Since for a simply closed curve we have in addition $\gamma(b) = \gamma(a)$ the second statement follows too.

We have seen, compare with Theorem 21.12, that fg' and $f'g$ are complex differentiable too, hence we always have $\int_\gamma fg' \, dz = \int_\gamma gf' \, dz = 0$ and the second statement does not contain much useful new information.

693

2.   a) The curve γ is a simply closed differentiable curve. In fact it is the ellipse with centre $0 \in \mathbb{C}$, principal axes in the coordinate axes, i.e. the real and the imaginary axes passing through the points $\pm(a + b)$ and $\pm(a - b)i$, as we see from the representation

$$\gamma(t) = (a + b) \cos t + (a - b)i \sin t = x + iy$$

and

$$\frac{x^2}{(a + b)^2} + \frac{y^2}{(a - b)^2} = 1.$$

Since every polynomial is an entire function it follows that $\int_\gamma p(z) \, dz = 0$.

b) For $t \in [0, \pi]$ the trace of γ is just the graph of $\sin \big|_{[0, 2\pi]}$ while for $t \in [\pi, 2\pi]$ we deal with the line segment connecting π with 0 considered as points on the real axis.

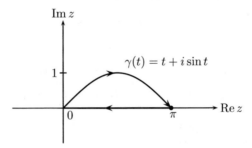

$$\gamma(t) = t + i \sin t$$

Since γ is a simply closed piecewise continuously differentiable curve we find, again by using the fact that polynomials as well as sin and cos are entire functions, that

$$\int_\gamma Q(\cos z, \sin z) \, dz = 0.$$

This implies in particular

$$\int_\pi^{2\pi} Q(\cos(2\pi - t), \sin(2\pi - t)) \, dt = \int_\pi^{2\pi} Q(\cos t, -\sin t) \, dt$$

$$= -\int_0^\pi Q(\cos(t + i \sin t), \sin(t + i \sin t))(1 + i \cos t) \, dt.$$

c) Since $\int_{|\zeta| = \rho} (g(\zeta) + h(\zeta)) \, d\zeta = \int_{|\zeta| = \rho} g(\zeta) \, d\zeta$ due to the holomorphy of h the estimate follows immediately from Proposition 20.13.B.

3.   a) Here is a sketch of the geometric situation

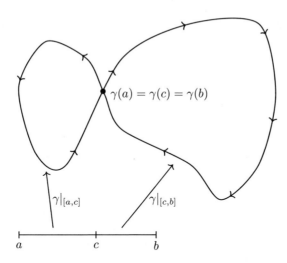

From our assumptions it follows that

$$\int_\gamma h(z)\,dt = \int_{\gamma_1} h(z)\,dz + \int_{\gamma_2} h(z)\,dz.$$

But each integral on the right hand side must be zero since it is an integral of a holomorphic function over a simply closed piecewise continuously differentiable curve.

b) First we look at Γ:

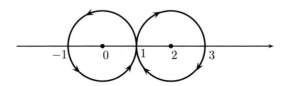

$$\eta : [0, 4\pi] \to \mathbb{C}, \quad \eta_1(t) := \eta\big|_{[0,2\pi]}(t) = e^{it} \quad \text{and} \quad \eta_2(t) := \eta\big|_{[2\pi,4\pi]}(t) = 2 - e^{it}.$$

It follows that $\eta(0) = 1 = \eta(2\pi) = \eta(4\pi)$ and further we find that $z \mapsto \frac{1}{z - \frac{3}{2}}$ is holomorphic in $B_{\frac{5}{4}}(0)$. Thus it follows that

$$\frac{1}{2\pi i} \int_\eta \frac{1}{z - \frac{3}{2}}\,dz = \frac{1}{2\pi i} \int_{\eta_1} \frac{1}{z - \frac{3}{2}}\,dz + \frac{1}{2\pi i} \int_{\eta_2} \frac{1}{z - \frac{3}{2}}\,dz$$

$$= \frac{1}{2\pi i} \int_{\eta_2} \frac{1}{z - \frac{3}{2}}\,dz.$$

695

We cannot apply directly Theorem 21.8 since η_2 is not given with the parametrization required. For this we have to change the orientation of the parametrization of η_2 and hence we arrive at

$$\frac{1}{2\pi i}\int_\eta \frac{1}{z-\frac{3}{2}}\,dz = -1.$$

4. We note that

$$\zeta^2 - \frac{5\pi}{4}\zeta - \frac{\pi^2}{4} = \left(\zeta - \frac{\pi}{4}\right)(\zeta - \pi)$$

which yields

$$\frac{\left(\zeta - \frac{\pi}{4}\right)\sin\zeta + (\zeta - \pi)\cos\zeta}{\zeta^2 - \frac{5\pi}{4} - \frac{\pi}{4}} = \frac{\sin\zeta}{\zeta - \pi} + \frac{\cos\zeta}{\zeta - \frac{\pi}{4}}$$

and next we note that $\pi \in B_1(4)$, $\frac{\pi}{4} \notin B_1(4)$ as well as $\pi \notin B_{\frac{5}{2}}\left(-\frac{1}{2}\right)$ but $\frac{\pi}{4} \in B_{\frac{5}{2}}\left(-\frac{1}{2}\right)$. By the Cauchy integral formula we find therefore

$$\frac{1}{2\pi i}\int_{|z-4|=1}\frac{\left(\zeta - \frac{\pi}{4}\right)\sin\zeta + (\zeta - \pi)\cos\zeta}{\zeta^2 - \frac{5\pi}{4}\zeta - \frac{\pi^2}{4}}\,d\zeta$$

$$= \frac{1}{2\pi i}\int_{|z-4|=1}\left(\frac{\sin\zeta}{\zeta - \pi} + \frac{\cos\zeta}{\zeta - \frac{\pi}{4}}\right)d\zeta$$

$$= \frac{1}{2\pi i}\int_{|z-4|=1}\frac{\sin\zeta}{\zeta - \pi}\,d\zeta = \sin 0 = 0,$$

where we have used that in $B_{\frac{5}{4}}(4)$ the function $\zeta \mapsto \frac{\cos\zeta}{\zeta-\frac{\pi}{4}}$ is holomorphic and for the remaining integral we applied the Cauchy integral formula. By a similar argument we find

$$\frac{1}{2\pi i}\int_{|z+\frac{1}{2}|=\frac{5}{2}}\frac{\left(\zeta-\frac{\pi}{4}\right)\sin\zeta+(\zeta-\pi)\cos\zeta}{\zeta^2-\frac{5\pi}{4}\zeta-\frac{\pi^2}{4}}\,d\zeta = \frac{1}{2\pi i}\int_{|z+\frac{1}{2}|=\frac{5}{2}}\frac{\cos\zeta}{\zeta-\frac{\pi}{4}}\,d\zeta$$

$$= \cos\frac{\pi}{4} = \frac{1}{2}\sqrt{2}.$$

5. Note that for all $k \in \mathbb{N}$ we have $0 < \ln(k^2 + k) < k^2$. Since $z \mapsto e^{-z}$ is an entire function we find

$$\frac{1}{2\pi i}\int_{|\zeta|=k^2}\frac{e^{-\zeta}}{\zeta - \ln(k^2+k)}\,d\zeta = e^{-\ln(k^2+k)} = \frac{1}{e^{\ln(k^2+k)}} = \frac{1}{k(k+1)}.$$

Using the result of Example I.16.3 we now derive

$$\sum_{k=1}^\infty \frac{1}{2\pi i}\int_{|\zeta|=k^2}\frac{e^{-\zeta}}{\zeta - \ln(k^2+k)}\,d\zeta = \sum_{k=1}^\infty \frac{1}{k(k+1)} = 1.$$

6. Since f, in fact f^k, satisfies the assumption of Theorem 21.8 we have for $k \in \mathbb{N}$ and $z \in B_r(z_0)$

$$f^k(z) = \frac{1}{2\pi i} \int_{|\zeta - z_0| = r} \frac{f^k(\zeta)}{\zeta - z} \, d\zeta$$

implying by (20.18)

$$|f(z)|^k = |f^k(z)| \leq \left(r \left\| \frac{1}{\cdot - z} \right\|_{\infty, \partial B_r(z_0)} \right) \|f^k\|_{\infty, \partial B_r(z_0)}$$

$$\leq \left(r \left\| \frac{1}{\cdot - z} \right\|_{\infty, \partial B_r(z_0)} \right) \|f\|_{\infty, \partial B_r(z_0)}$$

or

$$|f(z)| \leq \left(r \left\| \frac{1}{\cdot - z} \right\|_{\infty, \partial B_r(z_0)} \right)^{\frac{1}{k}} \|f\|_{\infty, \partial B_r(z_0)}.$$

The hint now implies the estimate $|f(z)| \leq \|f\|_{\infty, \partial B_r(z_0)}$ for all $z \in B_r(z_0)$. This nice argument is due to E. Landau, we refer to [67].

7.　　a) Since we have for f the mean-value equality

$$f(z_0) = \frac{1}{2\pi} \int_0^{2\pi} f(z_0 + re^{it}) \, dt$$

we may pass to the real and the imaginary part. Note that $u = \operatorname{Re} f$ and $v = \operatorname{Im} f$ are harmonic functions and we will see in Part 10 that harmonic functions are indeed characterised by the mean-value property.

　　b) For every $k \in \mathbb{N}$ and every ball $B_r(z_0) \subset \overline{B_r(z_0)} \subset G$ we have

$$f_k(z_0) = \frac{1}{2\pi} \int_0^{2\pi} f_k\left(z_0 + re^{it}\right) dt. \tag{$*$}$$

Since we have uniform convergence to f on all compact subsets of G we may pass in $(*)$ to the limit and obtain

$$f(z_0) = \frac{1}{2\pi} \int_0^{2\pi} f\left(z_0 + re^{it}\right) dt.$$

8. Given (21.19), i.e.

$$f(z) = \frac{1}{2\pi i} \int_{|\zeta - z_0| = r} \frac{f(\zeta)}{\zeta - z} \, d\zeta,$$

we have a starting point for our induction. Now suppose that for all $z \in B_r(z_0)$

$$f^{(n)}(z) = \frac{n!}{2\pi i} \int_{|\zeta - z_0| = r} \frac{f(\zeta)}{(\zeta - z)^{n+1}} \, d\zeta.$$

697

We are allowed to differentiate under the integral sign and therefore we find

$$f^{(n+1)}(z) = \frac{n!}{2\pi i} \frac{d}{dz} \int_{|\zeta-z_0|=r} \frac{f(\zeta)}{(\zeta-z)^{n+1}} \, d\zeta$$

$$= \frac{n!}{2\pi i} \int_{|\zeta-z_0|=r} \frac{d}{dz} \frac{f(\zeta)}{(\zeta-z)^{n+1}} \, d\zeta$$

$$= \frac{n!}{2\pi i} \int_{|\zeta-z_0|=r} \frac{n+1}{(\zeta-z)^{n+1}} f(\zeta) \, d\zeta$$

$$= \frac{(n+1)!}{2\pi i} \int_{|\zeta-z_0|=r} \frac{f(\zeta)}{(\zeta-z)^{n+1}} \, d\zeta.$$

9.   a) With $f(z) = \cos z^2$ the Cauchy integral formula for $n = 2$ reads

$$f^{(2)}(z) = \frac{2!}{2\pi i} \int_{|\zeta-2|=1} \frac{f(\zeta)}{(\zeta-\sqrt{\pi})^2} \, d\zeta = \frac{1}{\pi i} \int_{|\zeta-2|=1} \frac{\cos\zeta^2}{(\zeta-\sqrt{\pi})^2} \, d\zeta,$$

where we used that $1 < \sqrt{\pi} < 2$, i.e. $\sqrt{\pi} \in B_1(2)$. Since $\frac{d^2}{dz^2}(\cos\zeta^2) = -2\sin\zeta^2 - 4\zeta^2\cos\zeta^2$ it follows that

$$\int_{|\zeta-2|=1} \frac{\cos\zeta^2}{(\zeta-\sqrt{\pi})^2} \, d\zeta = \pi i(4\pi^2) = 4\pi^3 i.$$

b) Recall that for $\alpha, \beta \in \mathbb{R}, \gamma \in \mathbb{R}\setminus(-\mathbb{N}_0)$

$$_2F_1(z) = \sum_{k=0}^{\infty} \frac{(\alpha)_k(\beta)_k}{(\gamma)_k k!} z^k$$

converges for $|z| < 1$. Consequently we have

$$\frac{d^n}{dz^n} \, _2F_1(z) = \sum_{k=n}^{\infty} \frac{(\alpha)_n(\beta)_n}{(\gamma)_n k!} k(k-1)\cdots(k-n+1) z^{k-n}$$

and

$$\frac{d^n}{dz^n} \, _2F_1(0) = \frac{(\alpha)_n(\beta)_n}{(\gamma)_n}.$$

Now we find

$$\frac{1}{2\pi i} \int_{|z|=\frac{1}{2}} \frac{_2F_1(z)}{z^{n+1}} \, dz = \frac{1}{n!} \frac{d^n}{dz^n} \, _2F_1(0) = \frac{(\alpha)_n(\beta)_n}{(\gamma)_n n!}$$

which is of course the nth Taylor coefficient as expected.

10. The generalised Cauchy integral formula gives

$$f'(z) = \frac{1}{2\pi i} \int_{|\zeta - z_0| = 1} \frac{1}{(\zeta - z)^2} f(\zeta) \, d\zeta$$

$$f''(z) = \frac{2}{2\pi i} \int_{|\zeta - z_0| = 1} \frac{1}{(\zeta - z)^3} f(\zeta) \, d\zeta$$

$$f'''(z) = \frac{6}{2\pi i} \int_{|\zeta - z_0| = 1} \frac{1}{(\zeta - z)^4} f(\zeta) \, d\zeta$$

which yields

$$f'''(z) + 3f''(z) + 6f(z) = \frac{6}{2\pi i} \int_{|\zeta - z_0| = 1} \left(\frac{1}{(\zeta - z)^2} + \frac{1}{(\zeta - z)^3} + \frac{1}{(\zeta - z)^4} \right) f(\zeta) \, dz$$

$$= \frac{6}{2\pi i} \int_{|\zeta - z_0| = 1} \left(\frac{1 + \zeta - z + (\zeta - z)^2}{(\zeta - z)^4} \right) f(\zeta) \, d\zeta = 0.$$

Chapter 22

1. We have seen that a discrete set M is closed, hence a bounded discrete set is compact. Further, for every $x \in M$ we can find $r_x > 0$ such that $B_r(x) \cap (M \setminus \{x\}) = \emptyset$. Thus each $x \in M$ is exactly in one of these balls. They form an open covering of M and the compactness of M implies that finitely many of these balls will cover M, hence M itself must be finite.

2. We know from Problem 1 that M is a finite set, say $M = \{z_1, \ldots, z_n\} \subset G$. We choose $B_{r_1}(z_1) \subset G$ such that $B_{r_1}(z_1) \cap (M \setminus \{z_1\}) = \emptyset$. By Theorem 22.7 we can extend f as a holomorphic function f_1 to $B_{r_1}(z_1)$. The uniqueness theorem, Theorem 22.12, implies that $f_1|_{B_{r_1}(z_1) \setminus \{z_1\}} = f$ and hence we obtain a holomorphic function $\tilde{f}_1 : G \setminus \{z_2, \ldots, z_N\} \to \mathbb{C}$ with the property that $\tilde{f}_1|_G = f$ and $\tilde{f}_1$ is bounded in a neighbourhood of each point z_k, $2 \leq k \leq N$. This process can be iterated. If $\tilde{f}_k : G \setminus \{z_{k+1}, \ldots, z_N\} \to \mathbb{C}$, $k \leq N$, is already constructed as a holomorphic function such that $\tilde{f}_k|_G = f$ and $\tilde{f}_k$ is bounded in a neighbourhood of each z_l, $k + 1 \leq l \leq N$, then we choose $r_{k+1} > 0$ such that $B_{r_{k+1}}(z_{k+1}) \subset G$ and $B_{r_{k+1}}(z_{k+1}) \cap \{z_{k+2}, \ldots, z_n\} = \emptyset$. We can extend $\tilde{f}_k$ to a holomorphic function f_{k+1} defined $B_{r_{k+1}}(z_{k+1})$ and then we apply once more the uniqueness theorem to find that this gives rise to a holomorphic function $\tilde{f}_{k+1} : G \setminus \{z_{k+2}, \ldots, z_N\}$ extending to f. Eventually the result follows.

3. Since $\sum_{k=0}^{\infty} a_{2k+1} z^{2k+1}$ converges uniformly and absolutely in $B_r(0)$ it follows that $\sum_{k=0}^{\infty} a_{2k+1}(2k + 1) z^{2k}$ converges uniformly and absolutely in $B_r(0)$, i.e. $\sum_{k=0}^{\infty} |a_{2k+1}|(2k+1)|z|^{2k}$ converges uniformly in $B_r(0)$ implying that $\sum_{k=0}^{\infty} |a_{2k+1}| |z|^{2k}$ also converges uniformly in $B_r(0)$, hence $\sum_{k=0}^{\infty} a_{2k+1} z^{2k}$ converges uniformly and absolutely in $B_r(0)$. Now, on $B_r(0) \setminus \{0\}$ we have

$$g(z) = \frac{f(z)}{z} = \sum_{k=0}^{\infty} a_{2k+1} z^{2k}$$

699

which is bounded in a neighbourhood of $z = 0$. Therefore g has at $z = 0$ a removable singularity and can be extended to $B_r(0)$ by $g(0) = a_1$.

Since $\sin z = \sum_{k=0}^{\infty}(-1)^k \frac{z^{2k+1}}{(2k+1)!}$ for all $z \in \mathbb{C}$ we first conclude that $z \mapsto \frac{\sin z}{z}$ must have an entire extension and at $z = 0$ this extension has the value 1.

4. We know that $h(z_0) = 0$ as well as $h^{(l)}(z_0) = 0$ for $l = 1, \ldots, k-1$, but $h^{(k)}(z_0) \neq 0$. We claim that h^2 has at z_0 a zero of order $2k$. The Leibniz rule gives

$$\frac{\mathrm{d}^l}{\mathrm{d}z^l}\left(h^2(z)\right) = \frac{\mathrm{d}^l}{\mathrm{d}z^l}\left(h(z)h(z)\right)$$

$$= \sum_{j=0}^{l}\binom{l}{j}h^{(l-j)}(z)h^{(j)}(z).$$

If $l < 2k-1$ at least one term in the product $h^{(l-j)}(z_0)h^{(j)}(z_0)$ must vanish, however, if $l = 2k$ all products $h^{(2k-j)}(z_0)h^{(j)}(z_0)$ vanish except the term $h^{(k)}(z_0)h^{(k)}(z_0)$. Therefore we have $\left(\frac{\mathrm{d}^l}{\mathrm{d}z^l}h^2\right)(z_0) = 0$ for $l \leq 2k-1$, but $\left(\frac{\mathrm{d}^{2k}}{\mathrm{d}z^{2k}}h^2\right)(z_0) \neq 0$ implying that h^2 has at z_0 a zero of order $2k$.

5. We know that $z \mapsto e^{az}$, $a \in \mathbb{C}$, has no zeroes. This implies that $p(z)e^{az} = 0$ if and only if $p(z) = 0$ and as a polynomial of order k we know that $p(z)$ can have zeroes only of order less than or equal to k. Indeed, if $p(z) = a_k z^k + q(z)$, $a_k \neq 0$, and $q(z) = \sum_{l=0}^{k-1} a_l z^l$, we find $\frac{\mathrm{d}^k}{\mathrm{d}z^k}p(z) = k!a_k \neq 0$.

6. We proceed as in our considerations leading to $e^{z+w} = e^z e^w$. The functions $z \mapsto \cos z$ and $z \mapsto \sin z$ are entire functions. For $y \in \mathbb{R}$ fixed we consider the two functions $x \mapsto \cos(x - y)$ and $x \mapsto \cos z \cos y + \sin x \sin y$. Both functions have a holomorphic extension to $\mathbb{C}$ and by the uniqueness theorem we must have for all $z \in \mathbb{C}$ that $\cos(z - y) = \cos z \cos y + \sin z \sin y$. Now we fix $z \in \mathbb{C}$ and consider the functions $y \mapsto \cos(z-y)$ and $y \mapsto \cos z \cos y + \sin z \sin y$. Again, both functions have a holomorphic extension to $\mathbb{C}$ and we find once more by the uniqueness theorem $\cos(z - w) = \cos z \cos w + \sin z \sin w$ for all $z, w \in \mathbb{C}$.

7. Since $\mathrm{tr}(\gamma)$ is compact we can cover $\mathrm{tr}(\gamma)$ with finitely many balls $B_{r_{z_{0j}}}(z_{0j})$, $j = 1, \ldots, M$, with the property that $f|_{B_{r_{z_{0j}}}(z_{0j})}$ has a convergent power series represenation. Hence the function f has for every point $w \in \bigcup_{j=1}^{M} B_{r_{z_{0j}}}(z_{0j})$ a convergent power series representation $\sum_{k=0}^{\infty} b_k(z - w)^k$ with some positive radius of convergence. Therefore f is in $\bigcup_{j=0}^{M} B_{r_{z_{0j}}}(z_{0j})$ a holomorphic function. Note that since f is a priori given in G we cannot run into problems as we could in the consideration following Corollary 22.13.

8.   a) The series is a geometric series, namely

$$\sum_{k=0}^{\infty} z^{2k} = \sum_{k=0}^{\infty} (z^2)^k = \frac{1}{1 - z^2}.$$

700

We note further that

$$\frac{d}{dz}\left(\frac{1}{1-z^2}\right) = \frac{2z}{(1-z^2)^2}$$

and

$$\frac{d^2}{dz^2}\left(\frac{1}{1-z^2}\right) = \frac{2+6z^2}{(1-z^2)^3}.$$

It follows that

$$\frac{2+6z^2}{(1-z^2)^3} = \frac{d^2}{dz^2}\sum_{k=0}^{\infty} z^{2k}$$

$$= \frac{d}{dz}\sum_{k=0}^{\infty} 2kz^{2k-1}$$

$$= \sum_{k=0}^{\infty}(2k)(2k-1)z^{2k-2}$$

$$= \sum_{k=1}^{\infty}(2k+2)(2k+1)z^{2k}.$$

b) First note that $(1)_k = k!$ and further

$$\left(\frac{1}{2}\right)_k = \frac{1}{2}\cdot\frac{3}{2}\cdot\frac{5}{2}\cdot\ldots\cdot\frac{2k-1}{2}$$

and

$$\left(\frac{3}{2}\right)_k = \frac{3}{2}\cdot\frac{5}{2}\cdot\frac{7}{2}\cdot\ldots\cdot\frac{2k+1}{2}.$$

Therefore we find

$$x\,{}_2F_1\left(\frac{1}{2},1;\frac{3}{2};-x^2\right) = x\sum_{k=0}^{\infty}\frac{\left(\frac{1}{2}\right)_k(1)_k}{k!\left(\frac{3}{2}\right)_k}\left(-x^2\right)^k$$

$$= \sum_{k=0}^{\infty}\frac{1}{2k+1}(-1)^k x^{2k+1} = \arctan x.$$

Since

$$\frac{d}{dx}\arctan x = \frac{1}{1+x^2}$$

it follows that

$$\frac{1}{1+x^2} = \frac{d}{dx}\sum_{k=0}^{\infty}\frac{(-1)^k}{2k+1}x^{2k+1} = \sum_{k=0}^{\infty}(-1)^k x^{2k},$$

a result which we of course expect when looking at $\frac{1}{1+x^2}$ as limit of a geometric series.

701

9. Since $u(x) = \sum_{k=0}^{\infty} a_k(x - x_0)^k$ converges uniformly in $(-\eta + x_0, x_0 + \eta)$, we can extend u to a holomorphic function $h(z) := \sum_{k=0}^{\infty} a_k(z - x_0)^k$ with a uniformly and absolutely convergent power series in $B_\eta(x_0) \subset \mathbb{C}$. We claim that h satisfies $h''(z) + z^2 h'(z) + z^4 h(z) = 0$ in $B_\eta(x_0)$. The function $z \mapsto w(z) := h''(z) + z^2 h'(z) + z^4 h(z)$ is in $B_\eta(x_0)$ holomorphic since h is. On the open interval $(-\eta + x_0, x_0 + \eta)$ this function coincides with $x \mapsto u''(x) + x^2 u'(x) + x^4 u(x)$, but $u''(x) + x^2 u'(x) + x^4 u(x) = 0$ for all $x \in (a, b)$, in particular for all $x \in (-\eta + x_0, x_0 + \eta)$. Now the uniqueness theorem for holomorphic functions implies $w(z) = 0$ for all $z \in B_\eta(x_0)$, i.e. $h''(z) + z^2 h'(z) + z^4 h(z) = 0$ for all $z \in B_\eta(x_0)$.

10. We start to prove the identity for $x \in \mathbb{R}$. Recall the definition of the beta-function, (I.31.31),

$$B(a, b) = \int_0^1 t^{a-1}(1 - t)^{b-1}\, dt, \qquad a > 0,\ b > 0,$$

and the relation to the Γ-function, Theorem I.31.11,

$$B(a, b) = \frac{\Gamma(a)\Gamma(b)}{\Gamma(a + b)}.$$

It follows that

$$\frac{\Gamma\left(k + \tfrac{1}{2}\right)\Gamma\left(l + \tfrac{1}{2}\right)}{\Gamma(k + l + 1)} = \int_0^1 s^{k-\frac{1}{2}}(1 - s)^{l-\frac{1}{2}}\, ds$$

$$= 2 \int_0^1 t^{2k-1}(1 - t^2)^{l-\frac{1}{2}} t\, dt$$

$$= 2 \int_0^1 t^{2k}(1 - t^2)^{l-\frac{1}{2}}\, dt$$

$$= \int_{-1}^1 t^{2k}(1 - t^2)^{l-\frac{1}{2}}\, dt.$$

Thus we have

$$\frac{1}{\Gamma(k + l + 1)} = \frac{1}{\Gamma\left(k + \tfrac{1}{2}\right)\Gamma\left(l + \tfrac{1}{2}\right)} \int_{-1}^1 t^{2k}(1 - t^2)^{l-\frac{1}{2}}\, dt.$$

For the Bessel function J_l, $l \in \mathbb{N}$, this yields

$$J_l(x) = \sum_{k=0}^{\infty} \frac{(-1)^k \left(\frac{x}{2}\right)^{l+2k}}{\Gamma(k + 1)\Gamma(k + l + 1)}$$

$$= \sum_{k=0}^{\infty} \frac{(-1)^k \left(\frac{x}{2}\right)^{l+2k}}{\Gamma(k + 1)} \frac{1}{\Gamma\left(k + \tfrac{1}{2}\right)\Gamma\left(l + \tfrac{1}{2}\right)} \int_{-1}^1 t^{2k}(1 - t^2)^{l-\frac{1}{2}}\, dt$$

$$= \frac{\left(\frac{x}{2}\right)^l}{\Gamma\left(l + \tfrac{1}{2}\right)} \int_{-1}^1 (1 - t^2)^{l-\frac{1}{2}} \sum_{k=0}^{\infty} \frac{(-1)^k}{2^{2k}} \frac{(xt)^{2k}}{\Gamma(k + 1)\Gamma\left(k + \tfrac{1}{2}\right)}\, dt.$$

702

Now we use Theorem I.31.12, i.e. the doubling formula for the Γ-function

$$2^{2k}\Gamma(k+1)\Gamma\left(k+\frac{1}{2}\right) = \Gamma\left(\frac{1}{2}\right)\Gamma(2k+1) = \Gamma\left(\frac{1}{2}\right)(2k)!$$

and we find

$$J_l(x) = \frac{\left(\frac{x}{2}\right)^l}{\Gamma\left(1+\frac{1}{2}\right)} \int_{-1}^{1}(1-t^2)^{l-\frac{1}{2}} \sum_{k=0}^{\infty} \frac{(-1)^k}{2^{2k}} \frac{(xt)^{2k}}{\Gamma(k+1)\Gamma\left(k+\frac{1}{2}\right)} \, dt$$

$$= \frac{\left(\frac{x}{2}\right)^l}{\Gamma\left(l+\frac{1}{2}\right)} \int_{-1}^{1}(1-t^2)^{l-\frac{1}{2}} \sum_{k=0}^{\infty} \frac{(-1)^k (xt)^{2k}}{\Gamma\left(\frac{1}{2}\right)\Gamma(2k+1)} \, dt$$

$$= \frac{\left(\frac{x}{2}\right)^l}{\Gamma\left(l+\frac{1}{2}\right)} \int_{-1}^{1}(1-t^2)^{l-\frac{1}{2}} \sum_{k=0}^{\infty} \frac{(-1)^k}{(2k)!} (xt)^{2k} \, dt,$$

or for all $x \in \mathbb{R}$

$$J_l(x) = \frac{\left(\frac{x}{2}\right)^l}{\Gamma\left(\frac{1}{2}\right)\Gamma\left(l+\frac{1}{2}\right)} \int_{-1}^{1}(1-t^2)^{l-\frac{1}{2}}\cos(xt)\, dt. \qquad (*)$$

We know that J_l is an entire function. The complex differentiability of the function $z \mapsto \int_{-1}^{1}(1-t^2)^{l-\frac{1}{2}}\cos(zt)\,dt$ for $z \in B_R(0)$, $R > 0$, follows by using standard arguments for differentiating parameter depending integrals. Hence $(*)$ holds for all $z \in B_R(0)$, $R > 0$, therefore $(*)$ holds in $\mathbb{C}$.

Note that the result also holds for $l = 0$, but for this case we must be a bit more careful when arguing that $z \mapsto \int_{-1}^{1}(1-t^2)^{-\frac{1}{2}}\cos(zt)\,dt$ is complex differentiable due to the (weak) singularity of the integral at $t = \pm 1$.

11. We prove $\frac{1}{(2k)!} = \frac{1}{4^k\left(\frac{1}{2}\right)_k k!}$ by induction. For $k=1$ we have $\frac{1}{2} = \frac{1}{2!} \stackrel{!}{=} \frac{1}{4^1\left(\frac{1}{2}\right)_1 1!} = \frac{1}{4\cdot\frac{1}{2}\cdot 1} = \frac{1}{2}$. Now if the statement holds for k we find

$$\frac{1}{2(k+1)!} = \frac{1}{(2k)!(2k+1)(2k+2)} = \frac{1}{4^k\left(\frac{1}{2}\right)_k k!(2k+1)(2k+2)}$$

and we long to prove

$$\frac{1}{4^k\left(\frac{1}{2}\right)_k k!(2k+1)(2k+2)} \stackrel{!}{=} \frac{1}{4^{k+1}\left(\frac{1}{2}\right)_{k+1}(k+1)!}$$

or

$$\frac{1}{(2k+1)(2k+2)} \stackrel{!}{=} \frac{1}{4(k+1)\left(\frac{1}{2}+k\right)}$$

where we used that $\left(\frac{1}{2}\right)_{k+1} = \left(\frac{1}{2}\right)_k \left(\frac{1}{2}+k\right)$. Since

$$(2k+1)(2k+2) = 4k^2 + 6k + 2 = 4(k+1)\left(\frac{1}{2}+k\right),$$

703

the equality follows.

Analogously the second identity follows: For $k = 1$ we have

$$\frac{1}{6} = \frac{1}{3!} \overset{!}{=} \frac{1}{4\left(\frac{3}{2}\right)_1} = \frac{1}{6}.$$

Furthermore we find assuming the statement for k that

$$\frac{1}{(2(k+1)+1)!} = \frac{1}{(2k+3)!} = \frac{1}{(2k+1)!(2k+2)(2k+3)}$$

$$= \frac{1}{4^k \left(\frac{3}{2}\right)_k k!(2k+2)(2k+3)}$$

$$\overset{!}{=} \frac{1}{4^{k+1} \left(\frac{3}{2}\right)_{k+1} (k+1)!},$$

which reduces to the claim

$$\frac{1}{(2k+2)(2k+3)} \overset{!}{=} \frac{1}{4(k+1)\left(\frac{3}{2}+k\right)}$$

which follows from $(2k+2)(2k+3) = 4k^2 + 10k + 6 = 4(k+1)\left(\frac{3}{2}+k\right)$. With these identities we find now

$$_0F_1\left(\frac{1}{2}; -\frac{1}{4}z^2\right) = \sum_{k=0}^{\infty} \frac{1}{k!\left(\frac{1}{2}\right)_k}\left(-\frac{1}{4}z^2\right)^k$$

$$= \sum_{k=0}^{\infty} \frac{(-1)^k}{4^k k! \left(\frac{1}{2}\right)_k} z^{2k}$$

$$= \sum_{k=0}^{\infty} \frac{(-1)^k}{(2k)!} z^{2k} = \cos z,$$

and analogously

$$_0F_1\left(\frac{1}{2}; \frac{1}{4}z^2\right) = \sum_{k=0}^{\infty} \frac{z^{2k}}{(2k)!} = \cosh z.$$

Chapter 23

1.   a) The exponential function $z \mapsto e^z$ is an entire function without any zeroes and $(e^z)' = e^z$. However it is a periodic function with periodic $2\pi i$ and therefore it is not globally bijective. Indeed all our problems of defining and understanding the logarithmic function are due to the non-injectivity of $z \mapsto e^z$.

b) Since

$$f'(z) = g'(z)\sin z + g(z)\cos z$$

we have $f'(0) = g(0)$. If $g(0) \neq 0$ then by Theorem 23.7 there exists an open neighbourhood U of 0 such that $f : U \to f(U)$ is biholomorphic, i.e. f is locally biholomorphic at 0.

2. For $w \in B_\delta(f(z_0))$, i.e. $|w - f(z_0)| < \delta$, it follows for all $z \in \partial B_R(z_0)$ that

$$|f(z) - w| \geq |f(z) - f(z_0)| - |w - f(z_0)| > \delta$$

and therefore

$$\min_{z \in \partial B_R(z_0)} |f(z) - w| > |f(z_0) - w|.$$

By applying Lemma 23.1 to $z \mapsto f(z) - w$, we conclude that there exists some $z_1 \in B_R(z_0)$ such that $f(z_1) = w$, i.e. $w \in f(B_R(z_0))$.

3. We consider on $G = \{z \in \mathbb{C} \mid \operatorname{Re} z > 0, \operatorname{Im} z > 0\}$ the function $f(z) = e^{-iz^4}$. Since $\partial G = ([0, \infty) + 0i) \cup (0 + i[0, \infty))$ i.e. ∂G is the union of the closed positive real and positive imaginary axis, we find for f on ∂G that $z = x + iy : z \in [0, \infty) + i0$ implies $z^4 = x^4$ and $f(z) = e^{ix^4}$, hence $|f(z)| = 1$. Then for $z \in 0 + i[0, \infty)$ we have $z^4 = (iy)^4 = y^4$ and it follows that $f(z) = e^{iy^4}$ and we find again $|f(z)| = 1$. Thus $\sup_{z \in \partial G} |f(z)| = 1$. However for $z = r\left(\frac{1}{2}\sqrt{2} + i\frac{1}{2}\sqrt{3}\right)$ we have $z^4 = r^4\left(-\frac{5}{16} - i\frac{\sqrt{6}}{8}\right)$ and therefore $f(z) = e^{\frac{\sqrt{6}r^4}{8}} e^{-\frac{5r^4}{16}i}$ implying that $|f(z)| = e^{\frac{\sqrt{6}r^4}{8}}$ and for r tending to $+\infty$ it yields that $|f(z)|$ is unbounded, i.e. the estimate $|f(z)| \leq \sup_{z \in \partial G} |f(z)| = 1$ cannot hold. This example shows that the boundedness of G is essential for (23.3) to hold.

4. Since $\lim_{k \to \infty} \|f_k - f\|_{\infty, \partial G} = 0$ it follows that the sequence $\left(f_k|_{\partial G}\right)_{k \in \mathbb{N}}$ is a Cauchy sequence and the boundary maximum principle implies that $\left(f_k|_{\overline{G}}\right)_{k \in \mathbb{N}}$ is a Cauchy sequence too, i.e. $\lim_{l, k \to \infty} \|f_k - f_l\|_{\infty, \overline{G}} = 0$. Since $(f_k(z))_{k \in \mathbb{N}}$ converges for all $z \in \overline{G}$ to $f(z)$ we find that $(f_k)_{k \in \mathbb{N}}$ converges uniformly on $\overline{G}$ to f, i.e. $\lim_{k \to \infty} \|f_k - f\|_{\infty, \overline{G}} = 0$, also compare with Theorem II.14.2. As a uniform limit of holomorphic functions f_k the function f is on G holomorphic too.

5. Since $g, h \in \mathcal{F}$ implies for $\lambda, \mu \in \mathbb{C}$ that $\lambda g + \mu h \in \mathcal{F}$, i.e. $\mathcal{F}$ is vector space over $\mathbb{C}$. Clearly $\|f\|_\mathcal{F} \geq 0$ and if $\|f\|_\mathcal{F} = 0$ then by the boundary maximum principle we have $|f(z)| \leq \max_{z \in \partial B_R(0)} |f(z)| = \|f\|_\mathcal{F} = 0$ for all $z \in B_R(0)$ which yields $f \equiv 0$. For $\lambda \in \mathbb{C}$ we find further that

$$\|\lambda f\|_\mathcal{F} = \max_{z \in \partial B_R(0)} |\lambda f(z)| = |\lambda| \max_{z \in \partial B_R(0)} |f(z)| = |\lambda| \|f\|_\mathcal{F}$$

and the triangle inequality follows from

$$\|f + g\|_\mathcal{F} = \max_{z \in \partial B_R(0)} |f(z) + g(z)|$$
$$\leq \max_{z \in \partial B_R(0)} (|f(z)| + |g(z)|)$$
$$\leq \|f\|_\mathcal{F} + \|g\|_\mathcal{F}.$$

Thus $(\mathcal{F}, \|.\|_\mathcal{F})$ is a normed space. Now let $(f_k)_{k \in \mathbb{N}}$ be a sequence in $\mathcal{F}$ which is a Cauchy sequence with respect to $\|.\|_\mathcal{F}$. By boundary maximum principle it follows that $\left(f_k|_{\overline{B_R(0)}}\right)$ is a Cauchy sequence with respect to the norm $\|.\|_{\infty, \overline{B_R(0)}}$.

From Theorem II.14.2 we deduce that $(f_k)_{k\in\mathbb{N}}$ has a limit in $C\left(\overline{B_R(0)}\right)$ and now Problem 4 implies that f must be holomorphic in $B_R(0)$. Hence $(\mathcal{F}, \|.\|_{\mathcal{F}})$ is complete.

6. Suppose that for $0 < r_2 < r_1 < R$ we have $M(r_1) < M(r_2)$ i.e. $\max_{z\in\partial B_{r_1}(0)} |f(z)| < \max_{z\in B_{r_2}(0)} |f(z)|$. By the maximum boundary principle this implies

$$\max_{z\in \overline{B_{r_1}(0)}} |f(z)| \le \max_{z\in\partial B_{r_1}(0)} |f(z)| < \max_{z\in\partial B_{r_2}(0)} |f(z)|$$

but $\partial B_{r_2}(0) \subset \overline{B_{r_1}(0)}$ which leads to a contradiction.

7. If necessary by using the transformation $z \mapsto f_k(z) - f(z_0)$ we may assume that $f(z_0) = 0$. Since $(f_k)_{k\in\mathbb{N}}$ converges locally and uniformly to f the function f is holomorphic and not constant equal to 0. Therefore we can find some disc $B_R(z_0)$ such that $f|_{\overline{B_R(z_0)}\setminus\{z_0\}}$ has no zeroes. (Otherwise $f|_{\overline{B_R(z_0)}}$ would have countably many zeroes in the compact set $\overline{B_R(z_0)}$ and by the uniqueness theorem it would be identically zero in $B_R(z_0)$.) Further, the locally uniform convergence of f_k to f implies the uniform convergence of $f_k|_{\partial B_R(z_0)\cup\{z_0\}}$ to f, and since $f(z_0) = 0$ there exists $N(z_0)$ such that for $n \ge N(z_0)$

$$|f_n(z_0)| < \min\left\{f_n(z) \,\middle|\, \partial B_R(z_0)\right\}.$$

By Lemma 23.1 each of the functions f_n, $n \ge N(z_0)$, has a zero z_n, i.e. $f_n(z_n) = 0$. We claim $\lim_{n\to\infty} z_n = z_0$. Suppose that $(z_n)_{n\in\mathbb{N}}$ does not converge to z_0. Since this sequence is bounded it would have a subsequence $(z_{n_l})_{l\in\mathbb{N}}$ converging to some $w \in \overline{B_R(z_0)}\setminus\{z_0\}$. This however would yield $0 = \lim_{l\to\infty} f_{n_l}(z_{n_l}) = f(w)$ which is a contradiction.

8. Let $(w_k)_{k\in\mathbb{N}}$ be a sequence in the range of p converging in $\mathbb{C}$, i.e. $\lim_{k\to\infty} w_k = w$. We want to prove that $w \in R(p)$. Consider the sequence $(z_k)_{k\in\mathbb{N}}$, $z_k \in \mathbb{C}$, $p(z_k) = w_k$. From (23.9) it follows that

$$|z_k| \le 2|a_n||p(z_k)| = 2|a_n||w_k| \le 2|a_n|R$$

where R is a bound for the sequence $(w_k)_{k\in\mathbb{N}}$. Hence the sequence $(z_k)_{k\in\mathbb{N}}$ is bounded and must have a convergent subsequence $(z_{k_l})_{l\in\mathbb{N}}$ with limit z_0. For this subsequence we find

$$p(z_0) = \lim_{l\to\infty} p(z_{k_l}) = \lim_{l\to\infty} w_{k_l} = w,$$

i.e. $w \in R(p)$.

Since $\mathbb{C}$ is a region $p(\mathbb{C})$ is for every non-constant polynomial an open, non-empty and connected set. Furthermore it is closed, i.e. $p(\mathbb{C}) = \mathbb{C}$. Thus $0 \in R(p)$, or for some $\zeta \in \mathbb{C}$ we have $p(\zeta) = 0$, i.e. p has a zero.

9. Suppose that $f : \mathbb{C} \to B_1(0)$ is holomorphic. Then f is an entire function and $\sup_{z\in\mathbb{C}} |f(z)| \le 1$ which by the Liouville theorem is not possible. Thus such a mapping cannot exist.

10. The two complex numbers a and b span a parallelogram $P(a,b)$ in the plane $\mathbb{C}$, see the figure below:

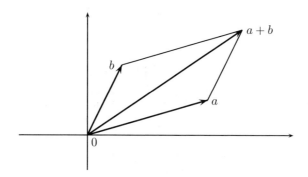

On $\overline{P(a,b)}$ the continuous function f is bounded. For $z \in \mathbb{C}$ we can find $n, m \in \mathbb{Z}$ such that $z - na - mb \in \overline{P(a,b)}$. Thus for all $z \in \mathbb{C}$ we have $|f(z)| \leq \max_{\zeta \in \overline{P(a,b)}} |f(\zeta)| = M < \infty$. Again the Liouville theorem implies that if f is holomorphic on $\mathbb{C}$ then it must be constant.

11.   a) This statement follows as Remark I.30.3: If $\prod_{k=1}^{\infty} c_k$ converges then we find with $P_N := \prod_{k=1}^{N} c_k$ that

$$\lim_{N \to \infty} c_N = \lim_{N \to \infty} \frac{P_N}{P_{N-1}} = \frac{\lim_{N \to \infty} P_N}{\lim_{N \to \infty} P_{N-1}} = 1.$$

b) We follow the considerations made in Proposition I.30.7: The infinite product $\prod_{k=1}^{\infty} c_k$, $c_k \in \mathbb{C}$, converges if and only if for every $\epsilon > 0$ there exists $N = N(\epsilon) \in \mathbb{N}$ such that $n > m > N(\epsilon)$ implies $\left| \prod_{k=m+1}^{n} c_k - 1 \right| < \epsilon$.

Proof. Suppose first that $\prod_{k=1}^{\infty} c_k = c \neq 0$. The Cauchy criterion applied to $\left(\prod_{k=1}^{N} c_k \right)_{N \in \mathbb{N}}$ states: for every $\epsilon > 0$ and $\eta > 0$ there exists $N = N(\epsilon, \eta) \in \mathbb{N}$ such that $n > m > N(\eta, \epsilon)$ implies

$$\left| \prod_{k=1}^{n} c_k - \prod_{k=1}^{m} c_k \right| < \eta \epsilon,$$

or

$$\left| \prod_{k=m+1}^{n} c_k - 1 \right| < \frac{\eta}{\left| \prod_{k=1}^{m} c_k \right|} \epsilon.$$

Since $\lim_{m \to \infty} \prod_{k=1}^{m} c_k = c \neq 0$ it follows that for $m \geq N_0$ we have $\left| \prod_{k=1}^{m} c_k \right| \geq \frac{|c|}{2} \neq 0$, hence we have for $\eta = \frac{2}{|c|}$ that $n > m > \max \left(N_0, N\left(\epsilon, \frac{2}{|c|} \right) \right)$ implies

$$\left| \prod_{k=m+1}^{n} c_k - 1 \right| < \frac{\frac{2}{|c|}}{\left| \prod_{k=1}^{m} c_k \right|} \epsilon \leq \epsilon.$$

Now we prove the converse. For $\epsilon = \frac{1}{2}$ there exists $N_1 \in \mathbb{N}$ such that $n > m > N_1$ implies

$$\left| \prod_{k=m}^{n} c_k - 1 \right| < \frac{1}{2}$$

which yields

$$\frac{1}{2} < \left| \prod_{k=m}^{n} c_k \right| < \frac{3}{2}.$$

In particular $c_l \neq 0$ for all $l > N_1$. For $N > N_1$ fixed and $0 < \epsilon < \frac{1}{2}$ there exists by assumption $N(\epsilon) > N$ such that $n > m > N(\epsilon)$ implies

$$\left| \frac{\prod_{k=N}^{n} c_k}{\prod_{k=N}^{m} c_k} - 1 \right| = \left| \prod_{k=m}^{n} c_k - 1 \right|$$

$$= |c_{m+1} \cdot c_{m+2} \cdot \ldots \cdot c_n - 1| < \frac{2}{3}\epsilon,$$

or

$$\left| \prod_{k=N}^{n} c_k - \prod_{k=N}^{m} c_k \right| < \left| \prod_{k=N}^{m} c_k \right| \frac{2}{3}\epsilon < \epsilon$$

implying that $\left(\prod_{k=1}^{M} c_k \right)_{M \in \mathbb{N}}$ is a Cauchy sequence in $\mathbb{C}$ and therefore convergent.

We use Theorem 23.34 to find

$$\pi \cot \pi z = \frac{(\sin \pi z)'}{\sin \pi z}$$

$$= \frac{\left(\pi z \prod_{k=1}^{\infty} \left(1 - \frac{z^2}{k^2} \right) \right)'}{\pi z \prod_{k=1}^{\infty} \left(1 - \frac{z^2}{k^2} \right)}$$

$$= \frac{\pi \prod_{k=1}^{\infty} \left(1 - \frac{z^2}{k^2} \right)}{\pi z \prod_{k=1}^{\infty} \left(1 - \frac{z^2}{k^2} \right)} + \frac{\left(\prod_{k=1}^{\infty} \left(1 - \frac{z^2}{k^2} \right) \right)'}{\prod_{k=1}^{\infty} \left(1 - \frac{z^2}{k^2} \right)}$$

$$= \frac{1}{z} + \sum_{k=1}^{\infty} \frac{\left(1 - \frac{z^2}{k^2} \right)'}{\left(1 - \frac{z^2}{k^2} \right)}$$

$$= \frac{1}{z} + \sum_{k=1}^{\infty} \frac{-2z}{k^2 - z^2} = \frac{1}{z} + \sum_{k=1}^{\infty} \frac{2z}{z^2 - k^2}.$$

$\square$

Chapter 24

1. For $w \in B_{\frac{1}{r}}(0) \setminus \{0\}$ the functions $h(w) = z_0 + \frac{1}{w}$ is holomorphic and with $z = h(w) = z_0 + \frac{1}{w}$ we find $w = h^{-1}(z) = \frac{1}{z - z_0}$ which is holomorphic on $\mathbb{C} \setminus \{z_0\}$. It

remains to prove that $h\left(B_{\frac{1}{r}}(0)\backslash\{0\}\right) = \mathbb{C}\backslash\overline{B_r(z_0)}$. We note that $w \in B_{\frac{1}{r}}(0)\backslash\{0\}$ if and only if $0 < |w| < \frac{1}{r}$ implying that $|w| = \frac{1}{z-z_0} < \frac{1}{r}$ or $r < |z - z_0|$, i.e. we find $h\left(B_{\frac{1}{r}}(0)\backslash\{0\}\right) \subset \mathbb{C}\backslash\overline{B_r(z_0)}$. On the other hand for $z \in \mathbb{C}\backslash\overline{B_r(z_0)}$, i.e. $|z - z_0| > r$, and $w = \frac{1}{z - z_0}$, we find $w \in B_{\frac{1}{r}}(0)\backslash\{0\}$ and $z = z_0 + \frac{1}{w}$, i.e. $\mathbb{C}\backslash\overline{B_r(z_0)} \subset h\left(B_{\frac{1}{r}}(0)\backslash\{0\}\right)$.

2.　a) We have

$$\frac{z - \frac{z^3}{6} - \sin z}{z^5} = \frac{z - \frac{z^3}{6} - \left(z - \frac{z^3}{3!} + \frac{z^5}{5!} - \frac{z^7}{7!} + \cdots\right)}{z^5}$$

$$= \frac{-\frac{z^5}{5!} + \frac{z^7}{7!} - \frac{z^9}{9!} \pm \cdots}{z^5}$$

$$= -\frac{1}{5!} + \frac{z^2}{7!} - \frac{z^4}{9!} \pm \cdots$$

$$= \sum_{k=0}^{\infty}(-1)^{k+1}\frac{z^{2k}}{(2k+5)!}$$

and therefore $z \mapsto \frac{z - \frac{z^3}{6} - \sin z}{z^5}$ has a removable singularity at $z_0 = 0$.

b) With $z + 3 = w$, i.e. $z = w - 3$, it follows that

$$(z-4)\sin\frac{1}{z+3} = (w-z)\sin\frac{1}{w}$$

$$= (w-z)\left(\frac{1}{w} - \frac{1}{3!w^3} + \frac{1}{5!w^5} \pm \cdots\right)$$

$$= 1 - \frac{7}{w} - \frac{1}{3!w^2} + \frac{7}{3!w^3} + \frac{1}{5!w^4} - \frac{7}{5!w^5} \pm \cdots$$

$$= 1 - \frac{7}{z+3} - \frac{1}{3!(z+3)^2} + \frac{7}{3!(z+3)^3} + \frac{1}{5!(z+3)^4} - \frac{7}{5!(z+3)^5} \pm \cdots$$

and therefore $z \mapsto (z-4)\sin\frac{1}{z+3}$ has an essential singularity at $z_0 = -3$.

c) With $z - \frac{\pi}{4} = w$, i.e. $z = w + \frac{\pi}{4}$, we find

$$\frac{\cos 2z}{\left(z - \frac{\pi}{4}\right)^3} = \frac{\cos\left(\frac{\pi}{2} + 2w\right)}{w^3} = \frac{-\sin 2w}{w^3}$$

and further

$$\frac{-\sin 2w}{w^3} = -8\frac{\sin 2w}{(2w)^3} = -\frac{8}{(2w^3)}\left(2w - \frac{(2w)^3}{3!} + \frac{(2w)^5}{5!} - \frac{(2w)^7}{7!} \pm \cdots\right)$$

$$= -\frac{8}{(2w)^2} + \frac{8}{3!} - \frac{8(2w)^2}{5!} + \frac{8(2w)^4}{7!} \pm \cdots$$

$$= -\frac{8}{(2w)^2} + 8\sum_{k=0}^{\infty}(-1)^k\frac{(2w)^{2k}}{(2k+3)!}$$

709

or

$$\frac{\cos 2z}{\left(z - \frac{\pi}{4}\right)} = -\frac{8}{\left(2z - \frac{\pi}{2}\right)^2} + 8\sum_{k=0}^{\infty}(-1)^k\frac{\left(2z - \frac{\pi}{2}\right)^{2k}}{(2k+3)!}$$

$$= -\frac{2}{\left(z - \frac{\pi}{4}\right)^2} + 8\sum_{k=0}^{\infty}(-1)^k\frac{2^{2k}\left(z - \frac{\pi}{4}\right)^{2k}}{(2k+3)!}$$

implying that $z \mapsto \frac{\cos 2z}{\left(z - \frac{\pi}{4}\right)^3}$ has a pole of order 2 at $z_0 = \frac{\pi}{4}$.

3. First we note that we can decompose $f(z) = \frac{1}{(z+2)(z+4)}$ by using partial fractions:

$$\frac{1}{(z+2)(z+4)} = \frac{1}{2(z+2)} - \frac{1}{2(z+4)}.$$

a) For $A_{2,4}(0) = B_4(0)\backslash\overline{B_2(0)}$ we find: since $|z| > 2$

$$\frac{1}{2(z+2)} = \frac{1}{2z\left(1 + \frac{2}{z}\right)} = \frac{1}{2z}\sum_{k=0}^{\infty}(-1)^k\frac{2^k}{z^k}$$

and since $|z| < 4$

$$\frac{1}{2(z+4)} = \frac{1}{8\left(1 + \frac{z}{4}\right)} = \frac{1}{8}\sum_{k=0}^{\infty}(-1)^k\frac{z^k}{4^k}$$

it follows for the Laurent series of $f(z) = \frac{1}{(z+2)(z+4)}$ in $A_{2,4}(0)$ that

$$f(z) = \sum_{k=0}^{\infty}(-1)^k\frac{2^{k-1}}{z^{k+1}} - \frac{1}{8}\sum_{k=0}^{\infty}(-1)^k\frac{z^k}{4^k}$$

$$= \sum_{k=0}^{\infty}(-1)^k\frac{2^{k-1}}{z^{k+1}} + \sum_{k=0}^{\infty}(-1)^{k+1}\frac{z^k}{2 \cdot 4^{k+1}}.$$

b) For $z \in A_{4,\infty}(0)$ we can use a partial result from the previous part to find

$$f(z) = \frac{1}{2(z+2)} - \frac{1}{2(z+4)} = \sum_{k=0}^{\infty}(-1)^k\frac{2^{k-1}}{z^{k+1}} - \frac{1}{2(z+4)}.$$

But for $|z| > 4$ we have

$$\frac{1}{2(z+4)} = \frac{1}{2z\left(1 + \frac{4}{z}\right)} = \frac{1}{2z}\sum_{k=0}^{\infty}(-1)^k\frac{4^k}{z^k}$$

$$= \sum_{k=0}^{\infty}(-1)^k\frac{2 \cdot 4^{k-1}}{z^{k+1}},$$

and therefore

$$f(z) = \sum_{k=0}^{\infty} \frac{(-1)^k 2^{k-1}}{z^{k+1}} - \sum_{k=0}^{\infty} \frac{(-1)^k 2 \cdot 4^{k-1}}{z^{k+1}}$$

$$= \sum_{k=0}^{\infty} (-1)^k \frac{2^{k-1} \left(1 - 2^k\right)}{z^{k+1}}$$

for $z \in A_{4,\infty}(0)$.

c) Note that $z \in A_{1,2}(-2)$ means $1 < |z+2| < 2$ and with $z + 2 = w$ we find

$$\frac{1}{(z+2)(z+4)} = \frac{1}{w(w+2)} = \frac{1}{2w\left(1 + \frac{w}{2}\right)}$$

$$= \frac{1}{2w} \sum_{k=0}^{\infty} (-1)^k \frac{w^k}{2^k} = \frac{1}{2(z+2)} + \sum_{k=1}^{\infty} (-1)^k \frac{(z+2)^{k-1}}{2^{k+1}}$$

$$= \frac{1}{2(z+2)} + \sum_{k=0}^{\infty} (-1)^{k+1} \frac{(z+2)^k}{2^{k+2}}.$$

d) Again using the decomposition $f(z) = \frac{1}{2(z+2)} - \frac{1}{2(z+4)}$ we get for $z \in B_2(0)$, i.e. $|z| < 2$, that

$$\frac{1}{2(z+2)} = \frac{1}{4\left(1 + \frac{z}{2}\right)} = \frac{1}{4} \sum_{k=0}^{\infty} (-1)^k z^k,$$

and for the second term we may use the result of part a), i.e.

$$\frac{1}{2(z+4)} = \frac{1}{8} \sum_{k=0}^{\infty} (-1)^k \frac{z^k}{4^k},$$

which yields in $B_2(0)$

$$f(z) = \frac{1}{4} \sum_{k=0}^{\infty} (-1)^k z^k - \frac{1}{8} \sum_{k=0}^{\infty} (-1)^k \frac{z^k}{4^k}$$

$$= \frac{1}{4} \sum_{k=0}^{\infty} (-1)^k \left(1 - \frac{1}{2 \cdot 4^k}\right) z^k$$

$$= \sum_{k=0}^{\infty} (-1)^k \left(\frac{2 \cdot 4^k - 1}{2 \cdot 4^{k+1}}\right) z^k,$$

hence in $B_2(0)$ we obtain the Taylor series.

4.   a) We have to look at the zeroes of $z \mapsto 4\sin z - 2$, i.e. we have to solve the equation $\sin z = \frac{1}{2}$. For z real we obtain the points $\frac{\pi}{4} + 2k\pi$ and $\frac{5\pi}{4} + 2k\pi$, $k \in \mathbb{Z}$, and since $\sin' = \cos$ these are simple zeroes. Consequently f has at the points $\frac{\pi}{4} + 2k\pi$ and $\frac{5\pi}{4} + 2k\pi$, $k \in \mathbb{Z}$, a pole of order 2. Using the representation

$\sin z = \frac{1}{2i}\left(e^{iz} - e^{-iz}\right)$ we first deduce that $\sin z = \frac{1}{2}$ cannot have a purely imaginary solution iy, and in the general case, i.e. $z = x + iy$, for $\sin z = \frac{1}{2}$ we must have $\sin x = \frac{1}{e^{-y}+e^{y}} > 1$.

b) We first look at the equation $e^{w} - 1$, $w \in \mathbb{C}$. With $w = u+iv$ it follows that $e^{u}e^{iv} = 1$ must hold, or $e^{u} = e^{-iv}$. Since $\left|e^{-iv}\right| = 1$ we deduce that $e^{u} = 1$ or $u = 0$. Thus $e^{w} = 1$, $w = u+iv$ implies $w = iv$, and we have to solve $\cos v + i \sin v = 1$ which implies $v = 2k\pi$. With $w = \frac{1}{2z}$ we now deduce $z_k = \frac{1}{4\pi i k} = -\frac{1}{4\pi k}i$, $k \in \mathbb{Z}\setminus\{0\}$. So for these values of z_k the function $\frac{z}{e^{\frac{1}{2z}}-1}$ has a pole of first order. Whereas for $z = 0$ we obtain that Laurent expansion

$$\frac{z}{e^{\frac{1}{2z}} - 1} = \sum_{k=1}^{\infty} \frac{1}{k!}\frac{1}{2^k}z^{-k+1}$$

implying that the function has an essential singularity at $z_0 = 0$.

5. Consider the function $g(z) = \frac{\sin \frac{\pi}{2}z}{((\cos \pi(z-1))-1)^2}$. For $z_0 = 1$ we have $\sin \frac{\pi}{2}z_0 = 1$ and $\cos(\pi(z_0 - 1)) - 1 = 0$. Moreover $z_0 = 1$ is a simple zero of $\cos \pi(z - 1) - 1$, hence it is a zero of order two for $(\cos \pi(z - 1) - 1)^2$ implying that g has a pole of order two at $z_0 = 1$.

6. If $z \to z_0$ then $|z - z_0| \to 0$ and therefore $e^{\frac{1}{|z-z_0|}} \to \infty$. This implies that f is not bounded in a neighbourhood of z_0, hence it cannot have a removable singularity. If f had a pole at z_0 we must have for some $N \in \mathbb{N}$ with $\kappa_0 > 0$ that

$$\kappa_0 e^{\frac{1}{|z-z_0|}} \leq |f(z)| \leq \kappa_1 \frac{1}{|z - z_0|^N}$$

holds in a neighbourhood of z_0. This however yields that

$$|z - z_0|^N e^{\frac{1}{|z-z_0|}} \leq \frac{\kappa_1}{\kappa_0},$$

i.e. $\limsup_{r \to 0} r^N e^{\frac{1}{r}} < \infty$, but we know that $\lim_{r \to 0} r^N e^{\frac{1}{r}} = \infty$. Hence f must have an essential singularity at z_0.

7. We consider the equation $\exp \frac{1}{z} = w_0$, $w_0 \neq 0$, for $0 < |z| < \frac{1}{2}$. With $\zeta = \frac{1}{z}$ this is equivalent to $\exp \zeta = w_0$, $w_0 \neq 0$, $2 < |\zeta| < \infty$. With $\zeta = u + iv$ and $w_0 = re^{i\varphi}$ it follows now that $e^{u}e^{iv} = re^{i\varphi}$. With $u = \ln r$ we arrive at $e^{i(v-\varphi)} = 1$, i.e. $\cos(v - \varphi) + i\sin(v - \varphi) = 1$ or $v - \varphi = 2k\pi$, $k \in \mathbb{Z}$. Hence we have in $0 < |z| < \frac{1}{2}$ the countable many zeroes $z_k = \frac{1}{\zeta_k} = \frac{1}{\ln r + i(\varphi + 2\pi k)}$, $k \in \mathbb{Z}$, of $e^{\frac{1}{z}} = w_0$.

8. If f and g have a pole of order N at z_0 we can find an open neighbourhood of z_0 such that with $a_{-N} \neq 0$ and $b_{-N} \neq 0$ we have

$$f(z) = a_{-N}(z - z_0)^{-N} + \sum_{l=1}^{N-1} a_{-l}(z - z_0)^{-l} + \sum_{k=0}^{\infty} a_k(z - z_0)^k$$

and

$$g(z) = b_{-N}(z - z_0)^{-N} + \sum_{l=1}^{N-1} b_{-l}(z - z_0)^{-l} + \sum_{k=0}^{\infty} b_k(z - z_0)^k$$

or

$$f(z) = a_{-N}(z - z_0)^{-N}\left(1 + \sum_{l=1}^{N-1} \frac{a_{-l}}{a_{-N}}(z - z_0)^{N-l} + \sum_{k=0}^{\infty} \frac{a_k}{a_{-N}}(z - z_0)^{k+N}\right)$$

and

$$g(z) = b_{-N}(z - z_0)^{-N}\left(1 + \sum_{l=1}^{N-1} \frac{b_{-l}}{b_{-N}}(z - z_0)^{N-l} + \sum_{k=0}^{\infty} \frac{b_k}{b_{-N}}(z - z_0)^{k+N}\right),$$

which yields for $z \neq z_0$ that

$$\begin{aligned}
\frac{f(z)}{g(z)} &= \frac{a_{-N}}{b_{-N}}\left(\frac{1 + \sum_{k=-N+1}^{\infty} \frac{a_k}{a_{-N}}(z - z_0)^{k+N}}{1 + \sum_{k=-N+1}^{\infty} \frac{b_k}{b_{-N}}(z - z_0)^{k+N}}\right) \\
&= \frac{a_{-N}}{b_{-N}} \cdot \frac{F(z)}{G(z)}.
\end{aligned}$$

For $z \to z_0$ it follows that $\lim_{z \to z_0} F(z) = 1$ and $\lim_{z \to z_0} G(z) = 1$ which first of all implies that $\frac{f(z)}{g(z)}$ is bounded in a neighbourhood of z_0, hence the singularity at z_0 is removable, and further we find

$$\lim_{z \to z_0} \frac{f(z)}{g(z)} = \frac{a_{-N}}{b_{-N}}.$$

9. First we choose $R > 0$ such that $|p(z)| \geq 1$ for $|z| \geq R$. Now we note that

$$\begin{aligned}
\frac{p'(z)}{p(z)} &= \frac{nz^{n-1} + a_1(n-1)z^{n-1} + \cdots + a_{n-1}}{z^n + a_1 z^{n-1} + \cdots + a_0} \\
&= \frac{n}{z} + h(z),
\end{aligned}$$

where $h(z)$ is a sum of terms of the type $\frac{A_\nu}{z^\nu}$, $\nu \geq 2$, as we can deduce by applying polynomial division. Integrating over $\partial B_R(0)$ given in standard parametrization yields, recall that for $\nu \geq 2$ we have

$$\int_{\partial B_R(0)} \frac{1}{z^\nu}\, dz = 0,$$

that

$$\frac{1}{2\pi i}\int_{\partial B_R(0)} \frac{p'(z_0)}{p(z)}\, dz = \frac{1}{2\pi i}\int_{\partial B_R(0)} \frac{n}{z}\, dz = n.$$

Since $p(z)$ has no pole in $B_R^{\complement}(0)$ the argument principle implies $\sum_{k=1}^{M} \alpha_k = n$ where $\alpha_1, \ldots, \alpha_M$ are the multiplicities of the zeroes of p.

713

10. First we decompose $h(z) = z^5 - 2z^3 + 10$ according to $h(z) = f(z) + g(z)$ where $f(z) = 10$ and $g(z) = z^5 - 2z^3$. We note that

$$|g(z)| = |z^5 - 2z^3| \le |z|^5 + 2|z|^3 \le 3 < 10$$

for $z \in \overline{B_1(0)}$. This implies by Rouché's theorem that $f(z) + g(z) = z^5 - 2z^2 + 10$ and $f(z) = 10$ have the same number of zeroes in $B_1(0)$, but $f(z)$ has no zero. Now we decompose $h(z) = f(z) + g(z)$ with $f(z) = z^5$ and $g(z) = 10 - 2z^3$. On $\partial B_2(0)$ we find

$$|g(z)| = |10 - 2z^3| \le 10 + 2|z|^3 \le 26 < 2^5 = 32.$$

Thus $h(z) = z^5 - 2z^3 + 10$ has the same numbers of zeroes inside $B_2(0)$ as $f(z) = z^5$, i.e. all zeroes of $h(z)$ are inside $B_2(0)$ implying now that all zeroes of $h(z) = z^5 - 2z^3 + 10$ belong to $A_{1,2}(0)$.

Chapter 25

1. We can consider ∂K as the trace of the N line segments $[A_j, A_{j+1}]$, $j = 1, \ldots, N$, $A_{N+1} = A_1$, i.e. we have

$$\partial K = \text{tr}\left([A_1, A_2] \oplus [A_2, A_3] \oplus \cdots \oplus [A_{N-1}, A_N] \oplus [A_N, A_1]\right),$$

which implies that ∂K is a cycle provided that the line segments are parametrized in the standard way.

2. Using standard parametrization we introduce the two curves $\gamma : [A, B] \oplus [B, C] \oplus [C, A]$ and $\eta := [D, E] \oplus [E, F] \oplus [F, D]$. These are two simply closed curves which we can add to a chain $\gamma \tilde{+} \eta$. Since $\text{tr}(\gamma) \cap \text{tr}(\eta) = \emptyset$ and both curves are simply closed $\gamma \tilde{+} \eta$ is indeed a cycle.

3. We denote the simply closed curve with trace $\kappa_{\sqrt{2}}(z_0)$ by γ_1 and that with trace being the rectangle $ABCD$ by γ_2. We already know that both curves are cycles. Moreover, both are null-homologous in $G = B_2(z_0)$ and since

$$\text{ind}_{\gamma_1 - \gamma_2}(z) = \text{ind}_{\gamma_1}(z) - \text{ind}_{\gamma_2}(z)$$

we have for $z \in G^{\complement}$ that $\text{ind}_{\gamma_1 - \gamma_2}(z) = 0$ implying that γ_1 and γ_2 are homologous.

4. For $z \in \overline{\Delta}^{\complement}$ we already know that $\text{ind}_\gamma(z) = 0$ since in this case $\zeta \mapsto \frac{1}{z - \zeta}$ is holomorphic in a neighbourhood of $\overline{\Delta}$. Next let $z \in \Delta$ and consider the figure below

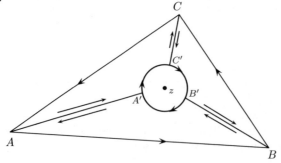

$$714$$

The Cauchy integral theorem yields

$$\int_\gamma \frac{1}{\zeta - z}\, d\zeta = \int_{\partial B_r(z)} \frac{1}{\zeta - z}\, dz = 1.$$

5. Let z_0 be a pole of order k. Then we have by the translation invariance of the differentiation

$$\operatorname{res}(g, z_0) = \lim_{z \to z_0} \frac{1}{(k-1)!} \frac{d^{k-1}}{dz^{k-1}} \left((z - z_0)^k g(z) \right)$$

$$= \lim_{z \to z_0} \frac{1}{(k-1)!} \frac{d^{k-1}}{dz^{k-1}} \left((z + a - z_0)^k g(z + a) \right)$$

$$= \operatorname{res}(g \circ \tau_a, z_0 - a).$$

6. a) We have a simple pole at $z = 2i$, $z = -2i$ and a pole of order two at $z = -1$.
Therefore we find with $f(z) = \frac{z^3 - z^2 + 2z}{(z+1)^2(z^2+4)}$

$$\operatorname{res}(f, 2i) = \lim_{z \to 2i} \left((z - 2i) \left(\frac{z^3 - z^2 + 2z}{(z+1)^2(z-2i)(z+2i)} \right) \right)$$

$$= \lim_{z \to 2i} \frac{z^3 - z^2 + 2z}{(z+1)^2(z+2i)} = \frac{1-i}{-4-3i},$$

$$\operatorname{res}(f, -2i) = \lim_{z \to -2i} \left((z + 2i) \frac{z^3 - z^2 + 2z}{(z+1)^2(z-2i)(z+2i)} \right)$$

$$= \lim_{z \to -2i} \frac{z^3 - z^2 + 2z}{(z+1)^2(z-2i)} = \frac{1+i}{-4+3i}$$

$$\operatorname{res}(f, -1) = \lim_{z \to -1} \frac{1}{1!} \frac{d}{dz} \left((z+1)^2 \frac{z^3 - z^2 + 2z}{(z+1)^2(z^2+4)} \right)$$

$$= \lim_{z \to -1} \frac{d}{dz} \left(\frac{z^3 - z^2 + 2z}{z^2+4} \right)$$

$$= \lim_{z \to -1} \frac{(3z^2 - 2z + 2)(z^2 + 4) - (z^3 - z^2 + 2z)(2z)}{(z^2+4)^2}$$

$$= \frac{34}{25}.$$

b) At $z = k\pi$, $k \in \mathbb{Z}$, we have poles of order two and with $g(z) = \frac{e^{2z}}{\sin^2 z}$ we have

$$\operatorname{res}(g, k\pi) = \lim_{k \to k\pi} \frac{1}{1!} \frac{d}{dz} \left((z - k\pi)^2 \frac{e^{2z}}{\sin 2z} \right)$$

$$= \lim_{z \to k\pi} 2e^{2z} \left(\frac{(z - k\pi)^2 \sin z + 2(z - k\pi) \sin z - 2(z - k\pi)^2 \cos z}{\sin^3 z} \right)$$

and the substitution $w = z - k\pi$ yields

$$\text{res}(g, k\pi) = 2 \lim_{w \to 0} e^{2w+2k\pi} \left(\frac{w^2 \sin w + 2w \sin w - 2w^2 \cos w}{\sin^3 w} \right)$$

$$= 2e^{2k\pi} \lim_{w \to 0} \left(\left(\frac{w^2 \sin w + 2w \sin w - 2w^2 \cos w}{w^3} \right) \left(\frac{w^3}{\sin^3 w} \right) \right)$$

$$= 2e^{2k\pi} \lim_{w \to 0} \frac{w^2 \sin w + 2w \sin w - 2w^2 \cos w}{w^3}$$

where we used that $\lim_{w \to 0} \frac{\sin w}{w} = 1$. Applying the rules of l'Hospital three times we find

$$\lim_{w \to 0} \frac{w^2 \sin w + 2w \sin w - 2w^2 \cos w}{w^3} = \lim_{w \to 0} \frac{(2w^2 + 2w + 2) \sin w + (w^2 - 2w) \cos w}{3w^2}$$

$$= \lim_{w \to 0} \frac{(-w^2 + 6w + 2) \sin w + (2w^2 + 4w) \cos w}{6w}$$

$$= \lim_{w \to 0} \frac{(-2w^2 + 2w + 6) \sin w + (-w^2 + 10w + 6) \cos w}{6}$$

$$= 1,$$

hence we have $\text{res}(g, k\pi) = 2e^{2k\pi}$.

7. Consider the path $\gamma_R = \kappa_r^+(0) \oplus [-R, R]$ as below

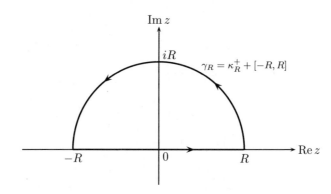

where $R > 0$ and all poles $c_1, \ldots, c_K$ of $f = \frac{p}{q}$ lie in the interior of γ_R. By the residue theorem we have

$$\int_{\gamma_R} f(z) \, dz = \int_{\gamma_R} \frac{p(z)}{q(z)} \, dz = 2\pi i \sum_{j=1}^{K} c_j.$$

We split the integral according to

$$\int_{\gamma_R} f(z) \, dz = \int_{\kappa_R^+(0)} f(z) \, dz + \int_{-R}^{R} f(z) \, dz.$$

716

For the first integral on the right hand side we find

$$\int_{\kappa_R^+(0)} f(z)\,dz = \int_0^\pi f\left(Re^{i\varphi}\right) \cdot iRe^{i\varphi}\,d\varphi$$

and since $M \geq N + 2$ we can find a constant $d > 0$ such that

$$\left|f\left(Re^{i\varphi}\right) iRe^{i\varphi}\right| \leq \frac{d}{R}$$

implying

$$\left|\int_{\kappa_R^+(0)} f(z)\,dz\right| \leq \frac{d\pi}{R}.$$

Now we find

$$2\pi i \sum_{j=1}^K c_j = \lim_{R\to\infty}\left(\int_{\kappa_R^+(0)} f(z)\,dz + \int_{-R}^R f(z)\,dz\right)$$

$$= \lim_{R\to\infty}\int_{\kappa_R^+(0)} f(z)\,dz + \int_{-\infty}^\infty f(z)\,dz$$

and since

$$\lim_{R\to\infty}\int_{\kappa_R^+(0)} f(z)\,dz = 0$$

the result follows.

8. The integral is of the type discussed in Problem 7. The poles of $z \mapsto \frac{1}{z^6+1}$ are the zeroes of $z^6 + 1 = 0$ which yields that the poles in the upper half plane $\operatorname{Im} z > 0$ are at the points $e^{\frac{\pi i}{6}}$, $e^{\frac{\pi i}{2}}$ and $e^{\frac{5\pi i}{6}}$, and these are simple poles. Now let z_0 be any zero of $z^6 + 1$. It follows that

$$\frac{z - z_0}{z^6 + 1} = \frac{1}{\sum_{k=0}^5 z^{5-k} z_0^k}$$

and therefore we find for the residues of $z \mapsto \frac{1}{z^6+1}$ by Proposition 25.22

$$\operatorname{res}\left(\frac{1}{z^6+1}, z_0\right) = \lim_{z\to z_0}\left((z - z_0)\frac{1}{z^6+1}\right) = \lim_{z\to z_0}\frac{1}{\sum_{k=0}^5 z^{5-k} z_0^k} = \frac{1}{6z_0^5}.$$

This now implies

$$\int_0^\infty \frac{1}{x^6+1}\,dx = \frac{1}{2}\int_{-\infty}^\infty \frac{1}{x^6+1}\,dx = \frac{1}{2} \cdot 2\pi i\left(\frac{e^{-5\pi i}6}{6} + e^{-\frac{5\pi i}{2}} + e^{-\frac{25\pi i}{6}}\right)$$

$$= \frac{\pi}{3}.$$

717

9. The meromorphic function $f(z) = \frac{1}{(z^2+1)^k}$ has in the upper half-plane $\operatorname{Im} z > 0$ the pole $z_0 = i$ only and this pole is of order k. Therefore we have

$$
\begin{aligned}
\operatorname{res}(f, i) &= \lim_{z \to i} \frac{d^{k-1}}{dz^{k-1}} \left(\frac{1}{(k-1)!} (z-i)^k \frac{1}{(z^2+1)^k} \right) \\
&= \lim_{z \to i} \frac{d^{k-1}}{dz^{k-1}} \left(\frac{1}{(k-1)!} \frac{(z-i)^k}{(z-i)^k (z+i)^k} \right) \\
&= \lim_{z \to i} \frac{d^{k-1}}{dz^{k-1}} \left(\frac{1}{(k-1)!} \frac{1}{(z+i)^k} \right) \\
&= \frac{-i}{2^{2k-1}} \frac{(2k-2)!}{((k-1)!)^2},
\end{aligned}
$$

which implies by the residue theorem (in form of Problem 7)

$$
\int_{-\infty}^{\infty} \frac{1}{(s^2+1)^k} \, ds = \frac{\pi}{2^{2k-2}} \frac{(2k-2)!}{((k-1)!)^2}.
$$

We note the first four integrals explicitly:

$$
\int_{-\infty}^{\infty} \frac{1}{s^2+1} \, ds = \pi, \qquad \int_{-\infty}^{\infty} \frac{1}{(s^2+1)^2} \, ds = \frac{\pi}{2},
$$
$$
\int_{-\infty}^{\infty} \frac{1}{(s^2+1)^3} \, ds = \frac{3\pi}{8}, \qquad \int_{-\infty}^{\infty} \frac{1}{(s^2+1)^4} \, ds = \frac{5\pi}{16}.
$$

10. Again we can apply Problem 7. The polynomial $q(z) = 1 + z^{2\nu}$ has in $\operatorname{Im} z > 0$ the simple zeroes $z_l = e^{\frac{(2l+1)\pi i}{2\nu}}$, $l = 0, \ldots, \nu - 1$. Using Proposition 25.22, also compare with Problem 8, we find

$$
\begin{aligned}
\operatorname{res}\left(\frac{z^{2\mu}}{1+z^{2\nu}}; z_l \right) &= \lim_{z \to z_l} \left((z - z_l) \frac{z^{2\mu}}{1+z^{2\nu}} \right) \\
&= \frac{1}{2\nu} z_l^{2\mu-2\nu+1} = -\frac{1}{2\nu} z_l^{2\mu+1}.
\end{aligned}
$$

718

where in the last step we used that $z_l^{-2\nu} = -1$. Thus we find

$$\int_{-\infty}^{\infty} \frac{t^{2\mu}}{1+t^{2\nu}}\, dt = -\frac{2\pi i}{2\nu} \sum_{l=0}^{\nu-1} z_l^{2\mu+1}$$

$$= -\frac{i\pi}{\nu} \sum_{l=0}^{\nu-1} e^{\frac{\pi i}{2\nu}(2l+1)(2\mu+1)}$$

$$= -\frac{i\pi}{\nu} e^{\frac{(2\nu_1)\pi i}{2\nu}} \sum_{l=0}^{\nu-1} \left(e^{\frac{\pi i(2\mu+1)}{\nu}} \right)^l$$

$$= -\frac{i\pi}{\nu} e^{\frac{(2\mu+1)\pi i}{2\nu}} \sum_{l=0}^{\nu-1} \left(e^{\frac{\pi i(2\mu+1)}{\nu}} \right)^l$$

$$= -\frac{i\pi}{\nu} e^{\frac{(2\mu+1)\pi i}{2\nu}} \frac{1 - e^{(2\mu+1)\pi i}}{1 - e^{\frac{(2\mu+1)\pi i}{\nu}}}$$

$$= -\frac{i\pi}{\nu} \frac{i}{\sin\left(\frac{(2\mu+1)\pi}{2\nu} \right)} = \frac{\pi}{\nu \sin\left(\frac{(2\mu+1)\pi}{\nu} \right)}.$$

11. The integral $\int_0^{2\pi} \frac{d\varphi}{1-2r\cos\varphi+r^2}$ is of the type (25.32) and with $R(\cos\varphi, \sin\varphi) = \frac{1}{1-2r\cos\varphi+r^2}$ we find $h(z) = \frac{1}{(z-r)(1-rz)}$ which has in the unit circle one simple pole, either r if $|r| < 1$ or $\frac{1}{r}$ if $|r| > 1$. Using Proposition 25.22 to calculate the corresponding residue we arrive at

$$\int_0^{2\pi} \frac{d\varphi}{1 - 2r\cos\varphi + r^2} = \begin{cases} \frac{2\pi}{1-r^2}, & |r| < 1 \\ \frac{2\pi}{r^2-1}, & |r| > 1. \end{cases}$$

12. We consider the integral $\int_\gamma \frac{z^{\alpha-1}}{1+z}\, dz$ along the curve given in the hint where $0 < \epsilon < 1$ and $0 < r < 1 < R$. In the interior of γ the function $g(z) = \frac{z^{\alpha-1}}{1+z}$ has a simple pole at $z_0 = -1$ and the residue at z_0 is

$$\lim_{z\to -1} \left((z+1)\frac{z^{\alpha-1}}{1+z} \right) = e^{(\alpha-1)\pi i}.$$

This implies

$$\int_\gamma \frac{z^{\alpha-1}}{1+z}\, dz = 2\pi i e^{(\alpha-1)\pi i}.$$

We now split the integral and we obtain

$$\int_\gamma g(z)\, dz = \int_{[A,B]} g(z)\, dz + \int_{\widehat{BF}} g(z)\, dz + \int_{[F,G]} g(z)\, dz + \int_{\widehat{GA}} g(z)\, dz,$$

where as usual $[A, B]$ and $[F, G]$ are the line segments from A to B and from F to G, respectively, and $\widehat{BF}$ denotes the arc of the circle with centre 0 and radius

719

R that goes through B, C, D, E and F, whereas $\overset{\frown}{GA}$ is the arc of the circle with centre 0 and radius r that goes through G, H and A.

First we note that with $B = Re^{i\varphi_\epsilon}$

$$\left| \int_{\overset{\frown}{BF}} g(z)\,dz \right| = \left| \int_{\varphi_\epsilon}^{2\pi - \varphi_\epsilon} \frac{R^{\alpha-1}e^{i\varphi(\alpha-1)}iRe^{i\varphi}}{1 + Re^{i\varphi}}\,d\varphi \right|$$

$$= \int_{\varphi_\epsilon}^{2\pi - \varphi_\epsilon} \frac{R^\alpha}{|1 + Re^{i\varphi}|}\,d\varphi \leq \int_0^{2\pi} \frac{R^\alpha}{R-1}\,d\varphi$$

implying that

$$\lim_{\epsilon \to 0}\lim_{R \to 0} \int_{\overset{\frown}{BF}} g(z)\,dz = \lim_{R \to \infty}\lim_{\epsilon \to 0} \int_{\overset{\frown}{BF}} g(z)\,dz = 0.$$

Furthermore we have $A = re^{i\psi_\epsilon}$ that

$$\left| \int_{\overset{\frown}{GA}} g(z)\,dz \right| = \left| \int_{2\pi - \psi_\epsilon}^{\psi_\epsilon} \frac{r^{\alpha-1}e^{i\psi(\alpha-1)}ire^{i\varphi}}{1 + re^{i\varphi}}\,d\varphi \right|$$

$$\leq \int_{\psi_\epsilon}^{2\pi - \psi_\epsilon} \frac{r^\alpha}{1-r}\,d\varphi \leq \int_0^{2\pi} \frac{r^\alpha}{1-r}\,d\varphi$$

which yields

$$\lim_{\epsilon \to 0}\lim_{r \to 0} \int_{\overset{\frown}{GA}} g(z)\,dz = \lim_{r \to 0}\lim_{\epsilon \to 0} \int_{\overset{\frown}{GA}} g(z)\,dz = 0.$$

Now we use the second part of the hint, i.e.

$$\lim_{\epsilon \to 0}(x + i\epsilon)^\beta = x^\beta \quad \text{and} \quad \lim_{\epsilon \to 0}(x - i\epsilon)^\beta = x^\beta e^{2\pi i\beta},$$

$x > 0$, which simply follows from $z^\alpha = e^{\alpha \log z}$ and $\lim_{\epsilon \to 0}\log(x + i\epsilon) = \ln x$ and $\lim_{\epsilon \to 0}\log(x - i\epsilon) = \ln x + 2\pi i$. For $x \in [r, R]$ we find that

$$\lim_{\epsilon \to 0} \int_{[A,B]} \frac{z^{\alpha-1}}{1+z}\,dz = \int_r^R \frac{x^{\alpha-1}}{1+x}\,dx,$$

and

$$\lim_{\epsilon \to 0} \int_{[F,G]} \frac{z^{\alpha-1}}{1+z}\,dz = -e^{2\pi i\alpha} \int_r^R \frac{x^{\alpha-1}}{1+x}\,dx,$$

or

$$\lim_{R \to \infty}\lim_{r \to 0}\lim_{\epsilon \to 0} \left(\int_{[A,B]} \frac{z^{\alpha-1}}{1+z}\,dz + \int_{[F,G]} \frac{z^{\alpha-1}}{1+z}\,dz \right) = (1 - e^{2\pi i\alpha}) \left. \frac{x^{\alpha-1}}{1+x} \right|_0^\infty dx.$$

Finally we arrive at

$$-2\pi i e^{\alpha\pi i} = e^{(\alpha-1)\pi i} = \int_\gamma \frac{z^{\alpha-1}}{1+z}\,dz = (1 - e^{2\pi i\alpha}) \int_0^\infty \frac{x^{\alpha-1}}{1+x}\,dx$$

or

$$\int_0^\infty \frac{x^{\alpha-1}}{1+x}\,dx = -\frac{2\pi i e^{\alpha\pi i}}{1 - e^{2\pi i\alpha}} = -\frac{2\pi i}{e^{-\alpha\pi i} - e^{\alpha\pi i}} = \frac{\pi}{\sin \alpha\pi}.$$

13. It turns out that some calculations become easier when first noting that

$$\int_{-\infty}^{\infty} \frac{e^{-2\pi i x \xi}}{\cosh \pi x} \, dx = \frac{1}{\cosh \pi \xi}$$

will imply our result. This follows from the substitutions $\sqrt{2\pi}x = y$ and $\sqrt{2\pi}\xi = \eta$. In the remaining part we argue along the lines of [83]. We integrate $f(z) = \frac{e^{-2\pi i z \xi}}{\cosh \pi z}$ over the curve indicated in the limit. The denominator of this function vanishes if $\cosh \pi z = \frac{e^{\pi z} + e^{-\pi z}}{2} = 0$, i.e. $e^{\pi z} = -e^{-\pi z}$ or $e^{2\pi z} = -1$ implying that in the interior of γ_R we have two simple poles $\frac{i}{2}$ and $\frac{3i}{2}$. We now use Proposition 25.22 to find the residue at $z_0 \in \{\frac{i}{2}, \frac{3i}{2}\}$:

$$(z - z_0)f(z) = e^{-2\pi i z \xi} \frac{2(z - z_0)}{e^{\pi z} + e^{-\pi z}}$$

$$= 2e^{-2\pi i z \xi} e^{\pi z} \frac{1}{\frac{e^{2\pi z} - e^{-2\pi z_0}}{z - z_0}},$$

where we have used that $e^{\pi i} = e^{2\pi\left(\frac{i}{2}\right)} = e^{-2\pi\left(\frac{3i}{2}\right)} = 1$. We note further that

$$\lim_{z \to z_0} \frac{1}{\frac{e^{2\pi z} - e^{2\pi z_0}}{z - z_0}} = \frac{1}{\left(\frac{d}{dz} e^{2\pi z}\right)(z_0)} = \frac{1}{2\pi e^{2\pi z_0}}.$$

Therefore it follows that

$$\lim_{z \to z_0} (z - z_0)f(z) = \lim_{z \to z_0} \left(\left(2e^{-2\pi i z \xi} e^{\pi z}\right) \frac{1}{\frac{e^{2\pi z} - e^{-2\pi z_0}}{z - z_0}} \right)$$

$$= \frac{1}{\pi} e^{-2\pi i z_0 \xi} e^{-\pi z_0},$$

which yields

$$\operatorname{res}\left(f, \frac{i}{2}\right) = -\frac{i}{\pi} e^{\pi \xi}, \qquad \operatorname{res}\left(f, \frac{3i}{2}\right) = \frac{i}{\pi} e^{3\pi \xi}.$$

So far we have proved

$$\int_{\gamma_R} e^{-2\pi i z \xi} \frac{1}{\cosh \pi z} = 2\pi i \left(-\frac{i}{\pi} e^{\pi \xi} + \frac{i}{\pi} e^{3\pi \xi} \right)$$

$$= 2\left(e^{\pi \xi} - e^{3\pi \xi} \right).$$

Now we split the integral $\int_{\gamma_R} f(z) \, dz$ in the following way:

$$\int_{\gamma_R} f(z) \, dz = \int_R^{R+2i} f(z) \, dz + \int_{R+2i}^{-R+2i} f(z) \, dz + \int_{-R+2i}^{-R} f(z) \, dz + \int_{-R}^{R} f(z) \, dz.$$

721

We observe that

$$\left|\cosh \pi z\right| = \left|\frac{e^{\pi z} + e^{-\pi z}}{2}\right| \geq \frac{1}{2}\left|\left|e^{\pi z}\right| - \left|e^{-\pi z}\right|\right|$$

and for $z = R + iy$, $0 \leq y \leq 2$ we get

$$\left|\cosh \pi z\right| \geq \frac{1}{2}\left(e^{\pi R} - e^{-\pi R}\right)$$

whereas

$$\left|e^{-2\pi i z \xi}\right| = \left|e^{-2\pi i R \xi} e^{2\pi y \xi}\right| \leq e^{4\pi \xi}$$

implying

$$\lim_{R \to \infty} \left|\frac{e^{-2\pi i (R+iy)\xi}}{\cosh \pi (R + iy)\xi}\right| = 0, \qquad \text{for } 0 \leq y \leq 2,$$

i.e.

$$\lim_{R \to \infty} \int_{R}^{R+2i} f(z)\, dz = 0.$$

Replacing $z = R + iy$ by $z = -R + iy$, $0 \leq y \leq 2$, yields similar estimates and consequently

$$\lim_{R \to \infty} \int_{-R+2i}^{-R} f(z)\, dz = 0.$$

On the other hand we find $e^{-2\pi i (x+2i)} = e^{-2\pi i x \xi} e^{4\pi \xi}$ and $\cosh(\pi x) = \cosh(\pi(x+2i))$ since $\cosh(z + 2\pi i) = \cosh z$. It follows that

$$\int_{-R+2i}^{R+2i} e^{-2\pi i z \xi} \frac{1}{\cosh \pi z} = \int_{-R+2i}^{R+2i} e^{-2\pi i (x+2i)\xi} \frac{1}{\cosh \pi (x + 2i)}\, dx$$

$$= e^{4\pi \xi} \int_{-R}^{R} e^{-2\pi i x \xi} \frac{1}{\cosh \pi x}\, dx,$$

or

$$\int_{R+2i}^{-R+2i} e^{-2\pi i z \xi} \frac{1}{\cosh \pi z}\, dz = -e^{4\pi \xi} \int_{-R}^{R} e^{-2\pi i x \xi} \frac{1}{\cosh \pi x}\, dx.$$

This implies

$$\int_{R+2i}^{-R+2i} e^{-2\pi i z \xi} \frac{1}{\cosh \pi z}\, dz + \int_{-R}^{R} e^{-2\pi i x \xi} \frac{1}{\cosh \pi x}\, dx = \left(1 - e^{4\pi \xi}\right) \int_{-R}^{R} f(z)\, dz$$

and for $R \to \infty$ we finally arrive at

$$2\left(e^{\pi \xi} - e^{3\pi \xi}\right) = \lim_{R \to \infty} \int_{\gamma_R} e^{-2\pi i z \xi} \frac{1}{\cosh \pi z}\, dz = \left(1 - e^{4\pi \xi}\right) \int_{-\infty}^{\infty} e^{-2\pi i x \xi} \frac{1}{\cosh \pi x}\, dx$$

or

$$\int_{-\infty}^{\infty} e^{-2\pi x \xi} \frac{1}{\cosh \pi x}\, dx = \frac{2\left(e^{\pi \xi} - e^{3\pi \xi}\right)}{1 - e^{4\pi \xi}}.$$

Since $1 - e^{4\pi\xi} = -e^{2\pi\xi}\left(e^{\pi\xi} - e^{-\pi\xi}\right)$ and $e^{\pi\xi} - e^{3\pi\xi} = -e^{2\pi\xi}\left(e^{\pi\xi} - e^{-\pi\xi}\right)$ we eventually obtain

$$\int_{-\infty}^{\infty} e^{-2\pi i x\xi} \frac{1}{\cosh \pi x} = \frac{2}{e^{\pi\xi} - e^{-\pi\xi}} = \frac{1}{\cosh \pi z}.$$

Chapter 26

1.   a) From (26.11) we deduce for $k \in \mathbb{N}$ that

$$\Gamma\left(-k + \frac{1}{2}\right) = \frac{1}{\Gamma\left(k + \frac{1}{2}\right)} \frac{\pi}{\sin \pi(k+1)} = (-1)^k \frac{\pi}{\Gamma\left(k + \frac{1}{2}\right)}.$$

Furthermore we know that $\Gamma\left(\frac{1}{2}\right) = \sqrt{\pi}$ and a short induction yields

$$\Gamma\left(k + \frac{1}{2}\right) = \frac{1 \cdot 3 \cdot 5 \cdot \ldots \cdot (2k-1)}{2^k} \Gamma\left(\frac{1}{2}\right).$$

Indeed, for $k = 0$ the result is known, and assuming the statement for k we find

$$\Gamma\left(k + 1 + \frac{1}{2}\right) = \Gamma\left(\left(k + \frac{1}{2}\right) + 1\right) = \left(k + \frac{1}{2}\right)\Gamma\left(k + \frac{1}{2}\right)$$
$$= \frac{2k+1}{2}\Gamma\left(k + \frac{1}{2}\right) = \frac{1 \cdot 2 \cdot 3 \cdot \ldots \cdot (2k-1)(2k+1)}{2^{k+1}}\Gamma\left(\frac{1}{2}\right).$$

Note that

$$\frac{1 \cdot 3 \cdot 5 \cdot \ldots \cdot (2k-1)}{2^k} = \frac{(2k)!}{4^k k!}$$

and therefore we find

$$\Gamma\left(-k + \frac{1}{2}\right) = \frac{(-1)^k (2k)!}{4^k k!}\sqrt{\pi}.$$

   b) From (26.11) we deduce

$$\frac{\pi}{\sin \pi z} = \Gamma(z)\Gamma(1-z) = -z\Gamma(z)\Gamma(-z) \tag{$*$}$$

or

$$\Gamma(z)\Gamma(-z) = -\frac{\pi}{z \sin \pi z}.$$

Since by Lemma 26.1 we have $\Gamma(\overline{z}) = \overline{\Gamma(z)}$ we find for $z = iy$, $y \in \mathbb{R}$, that

$$\Gamma(iy)\Gamma(-iy) = \Gamma(iy)\overline{\Gamma(iy)} = |\Gamma(iy)|^2$$

implying that

$$|\Gamma(iy)|^2 = -\frac{\pi}{iy \sin \pi(iy)} = \frac{\pi}{y \sinh y}.$$

723

where we used that $-i\sin(iy) = \sinh y$.

Analogously we now find when taking $z + \frac{1}{2}$ instead of z in $(*)$ that

$$\Gamma\left(z + \frac{1}{2}\right)\Gamma\left(\frac{1}{2} - z\right) = \frac{\pi}{\sin \pi \left(z + \frac{1}{2}\right)} = \frac{\pi}{\cos \pi z}$$

and with $z = iy$, $y \in \mathbb{R}$, it follows that

$$\Gamma\left(\frac{1}{2} + iy\right)\Gamma\left(\frac{1}{2} - iy\right) = \left|\Gamma\left(\frac{1}{2} + iy\right)\right|^2 = \frac{\pi}{\cos \pi iy} = \frac{\pi}{\cosh \pi y},$$

since $\cos iw = \cosh w$, $w \in \mathbb{R}$.

2. We know from (I.31.35) that this identity holds for all $x > 0$ and that Γ is in $\operatorname{Re} z > 0$ a holomorphic function. The uniqueness theorem for holomorphic functions implies now that the identity must hold for all $z \in \mathbb{C}$, $\operatorname{Re} z > 0$.

3. By Theorem I.31.11 we have the equality

$$B(x, y) = \int_0^1 t^{x-1}(1-t)^{y-1}\, dt = \frac{\Gamma(x)\Gamma(y)}{\Gamma(x+y)}$$

for all $x, y > 0$. We can extend $y \mapsto \frac{\Gamma(x)\Gamma(y)}{\Gamma(x+y)}$ for fixed $x > 0$ to a holomorphic function $w \mapsto \frac{\Gamma(x)\Gamma(w)}{\Gamma(x+w)}$ on $\operatorname{Re} w > 0$ as we can extend $\int_0^1 t^{x-1}(1-t)^{y-1}\, dt$ for $x > 0$ fixed to a holomorphic function $w \mapsto \int_0^1 t^{x-1}(1-t)^{w-1}\, dt$ on $\operatorname{Re} w > 0$. By the uniqueness theorem for holomorphic functions it follows that for all $x > 0$ and $w \in \mathbb{C}$, $\operatorname{Re} w > 0$, we have

$$\int_0^1 t^{x-1}(1-y)^{w-1}\, dt = \frac{\Gamma(x)\Gamma(w)}{\Gamma(x+w)}. \tag{$*$}$$

Now we fix w, $\operatorname{Re} w > 0$, and extend both sides in $(*)$ to holomorphic functions in $\operatorname{Re} z > 0$, i.e. $z \mapsto \int_0^1 t^{z-1}(1-t)^{w-1}\, dt$ and $z \mapsto \frac{\Gamma(z)\Gamma(w)}{\Gamma(z+w)}$, and again the uniqueness result for holomorphic function yields that $(*)$ must hold for $\operatorname{Re} z > 0$ and $\operatorname{Re} w > 0$.

4. Once we have proved

$$\Gamma(x) = 2\int_0^\infty t^{2x-1}e^{-t^2}\, dt$$

for all $x > 0$ we can argue as in Problem 3 to extend this equality to $\operatorname{Re} z > 0$. The substitution $s = t^2$ gives

$$\Gamma(x) = \int_0^\infty s^{x-1}e^{-s}\, ds = \int_0^\infty \left(t^2\right)^{x-1}e^{-t^2}2t\, dt$$

$$= 2\int_0^\infty t^{2x-1}e^{-t^2}\, dt.$$

5. We know that $z \mapsto \frac{1}{\Gamma(z)}$ is an entire function and Γ is a meromorphic function in $\mathbb{C} \backslash (-\mathbb{N}_0)$ with simple poles at $-n, n \in \mathbb{N}_0$. Thus $\psi(z) := \frac{\Gamma'(z)}{\Gamma(z)}$ is a meromorphic function with poles at $-k, k \in \mathbb{N}_0$. Since about $-k$ we have the Laurent expansion (compare with (26.10))

$$\Gamma(z) = \frac{(-1)^k}{k!} \frac{1}{z+k} + T(z+b)$$

where T is the regular part, we find

$$\Gamma'(z) = \frac{(-1)^{k+1}}{k!} \frac{1}{(z+k)^2} + T'(z+k),$$

and it follows that $z \mapsto \frac{\Gamma'(z)}{\Gamma(z)}$ has a simple pole at $-k$. Using the functional equation of the Γ-function for $x > 0$ we find

$$\frac{d}{dx} \ln \Gamma(x+1) = \frac{d}{dx} \ln (x\Gamma(x))$$

or

$$\psi(x+1) = \frac{d}{dx} \ln \Gamma(x+1) = \frac{1}{x\Gamma(x)} \frac{d}{dx} (x\Gamma(x))$$

$$= \frac{1}{x\Gamma(x)} (\Gamma(x) + x\Gamma'(x)) = \frac{1}{x} + \frac{\Gamma'(x)}{\Gamma(x)} = \frac{1}{x} + \psi(x).$$

This identity extends to the half plane $\operatorname{Re} z > 0$. Similarly we have $x \in (0,1)$

$$\frac{d}{dx} \ln (\Gamma(x)\Gamma(1-x)) = \frac{d}{dx} \left(\ln \frac{\pi}{\sin \pi x} \right)$$

or

$$\frac{1}{\Gamma(x)\Gamma(1-x)} \frac{d}{dx} (\Gamma(x)\Gamma(1-x)) = \frac{1}{\frac{\pi}{\sin \pi x}} \frac{d}{dx} \frac{\pi}{\sin x}.$$

For the left hand side we find

$$\frac{1}{\Gamma(x)\Gamma(1-x)} \frac{d}{dx} (\Gamma(x)\Gamma(1-x)) = \frac{\Gamma'(x)\Gamma(1-x) - \Gamma(x)\Gamma'(1-x)}{\Gamma(x)\Gamma(1-x)}$$

$$= \frac{\Gamma'(x)}{\Gamma(x)} - \frac{\Gamma'(1-x)}{\Gamma(1-x)} = \psi(x) - \psi(1-x),$$

and the right hand side gives

$$\frac{1}{\frac{\pi}{\sin \pi x}} \frac{d}{dx} \frac{\pi}{\sin \pi x} = \pi \cot \pi x.$$

Thus we have for $x \in (0,1)$

$$\psi(x) - \psi(1-x) = \pi \cot \pi x$$

which by the uniqueness theorem extends to the holomorphy domain of both sides.

6. We use the product representation of $\Gamma(z)$, i.e. (26.16), and for $x > 0$ we obtain

$$\psi(x) = \frac{d}{d} \ln \Gamma(x) = \frac{d}{dx} \ln \left(\frac{e^{-\gamma x}}{x} \prod_{k=1}^{\infty} \frac{e^{\frac{x}{k}}}{1 + \frac{x}{k}} \right).$$

We note that

$$\ln \left(\frac{e^{\gamma x}}{x} \prod_{k=1}^{N} \frac{e^{\frac{x}{k}}}{1 + \frac{x}{k}} \right) = -\gamma x - \ln x + \sum_{k=1}^{N} \frac{x}{k} - \sum_{k=1}^{N} \ln \left(1 + \frac{x}{k} \right)$$

and differentiation yields

$$\frac{d}{dx} \ln \left(\frac{e^{-\gamma x}}{x} \prod_{k=1}^{N} \frac{e^{\frac{x}{k}}}{1 + \frac{x}{k}} \right) = -\gamma - \frac{1}{x} + \sum_{k=1}^{N} \frac{1}{k} - \sum_{k=1}^{N} \frac{1}{x+k}$$

$$= -\gamma - \frac{1}{x} - \sum_{k=1}^{N} \left(\frac{1}{x+k} - \frac{1}{k} \right).$$

Since $\sum_{k=1}^{N} \left(\frac{1}{x+k} - \frac{1}{k} \right) = - \sum_{k=1}^{N} \frac{x}{k^2 + kx}$ we have convergence for $x > 0$ and hence we arrive at

$$\psi(x) = -\gamma - \frac{1}{x} - \sum_{k=1}^{\infty} \left(\frac{1}{x+k} - \frac{1}{k} \right)$$

which again extends to the half plane $\operatorname{Re} z > 0$.

7. a) This is trivial. From the definition we deduce that if some $\nu_j > 1$ then $\nu_j = 2 + \alpha_j$, $\alpha_j \geq 0$, and $p_j^{\nu_j} = p_j^2 p_j^{\alpha_j}$.

  b) Let m and n be relative prime. In the case where m or n has a prime square factor then mn also has a prime square factor. Hence $\mu(m \cdot n) = 0$ as well as $\mu(m)\mu(n) = 0$. If both m and n have no square factor, then we can write $m = p_1 \cdot \ldots \cdot p_r$ and $n = q_1 \cdot \ldots \cdot q_s$ with distinct primes $p_1, \ldots, p_r, q_1, \ldots, q_s$. Consequently we have $\mu(m) = (-1)^r$, $\mu(n) = (-1)^s$ and since $m \cdot n = p_1 \cdots p_r q_1 \cdots q_s$ we find $\mu(mn) = (-1)^{r+s} = \mu(m)\mu(n)$, i.e. μ is multiplicative. The example $\mu(2)\mu(2) = 1$ but $\mu(4) = 0$ shows that μ is not completely multiplicative.

8. (Following [5]) With $f = f_1 \cdot f_2$ we have to prove that $f(n) = 0$ for all $n \in \mathbb{N}$. Let n_0 be the smallest natural number with $f(n) \neq 0$. It follows that

$$D_f(s) = \sum_{n=n_0}^{\infty} \frac{f(n)}{n^s} = \frac{f(n_0)}{n_0^s} + \sum_{n=n_0+1}^{\infty} \frac{f(n)}{n^s}$$

or

$$f(n_0) = n_0^s D_f(s) - n_0^s \sum_{n=n_0+1}^{\infty} \frac{f(n)}{n^s}.$$

726

For $s = s_k$ we have $D_f(s_k) = 0$, hence

$$f(n_0) = -n_0^{s_k} \sum_{n=n_0+1}^{\infty} \frac{f(n)}{n^{s_k}}.$$

Using the estimate from the hint we find for k such that $\sigma_k > c > \sigma_a$

$$|f(n_0)| \leq n_0^{\sigma_k} (n_0 + 1)^{-\sigma_k - c} \sum_{n=n_0+1}^{\infty} |f(n)| n^{-c}$$

$$\leq \left(\frac{n_0}{n_0 + 1} \right)^{\sigma_k} \kappa,$$

where κ is independent of k. Since $\lim_{k \to \infty} \left(\frac{n_0}{n_0+1} \right)^{\sigma_k} = 0$ we arrive at $f(n_0) = 0$
which is a contradiction, i.e. $f(n) = 0$ for all $n \in \mathbb{N}$.
Finally we prove the estimate given in the hint. We have

$$\left| \sum_{n=k}^{\infty} \frac{g(n)}{n^s} \right| \leq \sum_{n=k}^{\infty} \frac{|g(n)|}{n^c} n^{-(\sigma-c)}$$

$$\leq k^{-(\sigma-c)} \sum_{n=k}^{\infty} \frac{|g(n)|}{n^c}.$$

9. Note that $n^s = n^{\sigma} n^{it} = e^{\sigma \ln n} e^{it \ln n}$ for $s = \sigma + it$. We start with

$$\int_{-R}^{R} \left| \sum_{n=1}^{N} \frac{a_n}{n^s} \right|^2 dt = \int_{-R}^{R} \sum_{n=1}^{N} \sum_{k=1}^{N} \frac{a_n \bar{a}_k}{n^s \overline{k^s}} dt$$

$$= \int_{-R}^{R} \sum_{n=1}^{N} \sum_{k=1}^{N} a_k \bar{a}_k e^{-\sigma \ln n} e^{-\sigma \ln k} e^{it \ln n} e^{-it \ln k} dt$$

$$= \sum_{n=1}^{N} \sum_{k=1}^{N} a_n \bar{a}_k e^{-\sigma \ln n} e^{-\sigma \ln k} \int_{-R}^{R} e^{it \ln n} e^{-t \ln k} dt.$$

For $n = k$ it follows that

$$a_n \bar{a}_k e^{-\sigma \ln n} e^{-\sigma \ln k} \int_{-R}^{R} e^{it \ln n} e^{-i \ln k} dt = 2T |a_n|^2 e^{-2\sigma \ln} = \frac{2R |a_n|^2}{n^{2\sigma}}.$$

However for $n \neq k$ we find

$$\int_{-R}^{R} e^{it(\ln n - \ln k)} dt = \frac{1}{i \ln \frac{n}{k}} \left(e^{iR \ln \frac{n}{k}} - e^{-iR \ln \frac{n}{k}} \right)$$

$$= \frac{2 \sin R \ln \frac{n}{k}}{\ln \frac{n}{k}}.$$

Thus we find

$$\frac{1}{2R} \int_{-R}^{R} \left| \sum_{n=1}^{N} \frac{a_n}{n^s} \right|^2 dt = \frac{1}{2R} \sum_{n=1}^{N} \frac{2R|a_n|^2}{n^{2\sigma}}$$

$$+ \frac{1}{2R} \sum_{\substack{n \neq k}}^{N} \frac{a_n \bar{a}_k}{n^\sigma k^\sigma} \frac{2\sin\left(R\ln\frac{n}{k}\right)}{\ln\frac{n}{k}}$$

$$= \sum_{n=1}^{N} \frac{|a_n|^2}{n^{2\sigma}} + \sum_{\substack{n \neq k}}^{N} \frac{a_n \bar{a}_k}{n^\sigma k^\sigma} \frac{\sin\left(R\ln\left(\frac{n}{k}\right)\right)}{R\ln\left(\frac{n}{k}\right)}.$$

Since $\sigma > 1$ and $|a_n| \leq M$ we obtain in the limit $N \to \infty$ and $R \to \infty$ that

$$\lim_{R \to \infty} \frac{1}{2R} \int_{-R}^{R} \left| \sum_{n=1}^{\infty} \frac{a_n}{n^{\sigma+it}} \right|^2 dt = \sum_{n=1}^{\infty} \frac{|a_n|^2}{n^{2\sigma}}.$$

10. For $0 < x < \frac{\pi}{2}$ we have $\sin x < x < \tan x$ or $\cot x < x < \frac{1}{\sin x}$, which gives $\cot^2 x < \frac{1}{x^2} < \frac{1}{\sin^2 x} = 1 + \cot^2 x$. For $x = \frac{k\pi}{2m+1}$, $k = 1, \ldots, m$, it follows that

$$\sum_{k=1}^{m} \cot^2\left(\frac{k\pi}{2m+1}\right) \leq \frac{(2m+1)^2}{\pi^2} \sum_{k=1}^{m} \frac{1}{k^2} < m + \sum_{k=1}^{m} \cot^2\left(\frac{k\pi}{2m+1}\right). \qquad (*)$$

We claim that

$$\sum_{k=1}^{m} \cot^2 \frac{\pi k}{2m+1} = \frac{m(2m-1)}{3}.$$

For $y \in \left(0, \frac{\pi}{2}\right)$ we have

$$(\cos ny + i\sin ny) = (\cos y + i\sin y)^n = \sin^n y \, (i + \cot y)^n$$

$$= \sin^n y \sum_{k=0}^{n} \binom{n}{k} i^k \cot^{n-k} y.$$

With $n = 2m+1$ we find for the imaginary parts

$$\sin(2m+t)y = \sin^{2m+1} y P_m\left(\cot^2 y\right) \qquad (**)$$

with the polynomial

$$P_m(y) = \binom{2m+1}{1} y^m - \binom{2m+1}{3} y^{m-1} + \ldots \pm 1.$$

Choosing $y = \frac{k\pi}{2m+1}$ we deduce further from $(**)$ that

$$P_m\left(\cot^2 \frac{k\pi}{2m+1}\right) = 0.$$

728

Thus the zeroes of P_m are giving by $z_k = \cot\frac{k\pi}{2m+1}$, $k = 1,\ldots,m$, which yields

$$\sum_{n=1}^{m} \cot^2 \frac{k\pi}{2m+1} = \frac{\binom{2m+1}{3}}{\binom{2m+1}{1}} = \frac{m(2m-1)}{3}.$$

Combined with $(*)$ this implies

$$\frac{m(2m-1)}{3} < \frac{(2m+1)^2}{\pi^2} \sum_{k=1}^{m} \frac{1}{k^2} < m + \frac{m(2m-1)}{3}$$

or

$$\frac{\pi^2(2m-1)m}{3(2m+1)^2} < \sum_{k=1}^{\infty} \frac{1}{k^2} < \frac{\pi^2 m}{(2m+1)^2} + \frac{\pi^2(2m-1)m}{3(2m+1)^2}$$

which gives in the limit, note that $\sum_{k=1}^{\infty} \frac{1}{k^2} = \zeta(2)$,

$$\frac{\pi^2}{6} \le \sum_{k=1}^{\infty} \frac{1}{k^2} = \zeta(2) \le \frac{\pi^2}{6},$$

i.e. $\zeta(2) = \frac{\pi^2}{6}$.

Chapter 27

1. Since $\dot{\gamma}(t) = (a\cos t, -b\sin t)$ we find

$$l_{\gamma|[0,t]} = \int_0^t \|\dot{\gamma}(s)\| \, ds = \int_0^t \left(a^2 \cos^2 s + b^2 \sin^2 s\right)^{\frac{1}{2}} \, ds$$

$$= a \int_0^t \left(1 - e^2 \sin^2 s\right)^{\frac{1}{2}} \, ds$$

where $\epsilon = \frac{\sqrt{a^2-b^2}}{a}$ is the excentricity of the ellipse. In particular we find

$$l_{\gamma|\left[0,\frac{\pi}{2}\right]} = a \int_0^{\frac{\pi}{2}} \left(1 - e^2 \sin^2 t\right)^{\frac{1}{2}} \, ds = aE(\epsilon)$$

where $E(k) = \int_0^{\frac{\pi}{2}} \left(1 - k^2 \sin\varphi\right)^{\frac{1}{2}} \, d\varphi$, compare with (27.7). This now implies that

$$l_\gamma = l(\epsilon) = 4aE(\epsilon).$$

2. Using the equation $\left(x^2 + y^2\right)^2 = x^2 - y^2$ we arrive at the following figure showing the lemniscate

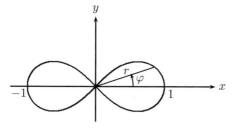

729

The length of the lemniscate is four times the length of the curve $r^2 = 2\cos 2\varphi$, $0 \le \varphi \le \frac{\pi}{4}$. Since $\frac{dr}{d\varphi} = -\frac{2\sin 2\varphi}{r}$ and consequently $\sqrt{r^2 + \left(\frac{dr}{d\varphi}\right)^2} = \frac{2}{r} = \frac{\sqrt{2}}{\sqrt{\cos 2\varphi}}$, we find for the arclengths $l(\vartheta)$, $0 < \vartheta < \frac{\pi}{4}$,

$$l(\vartheta) = \sqrt{2} \int_0^\vartheta \frac{d\varphi}{\sqrt{1 - 2\sin^2\varphi}}.$$

For $0 \le \vartheta < \frac{\pi}{4}$ we choose the substitution $2\sin^2\varphi = \sin^2\psi$ to find

$$\sqrt{2} \int_0^\vartheta \frac{d\varphi}{\sqrt{1 - 2\sin^2\varphi}} = \int_0^{2\vartheta} \frac{d\psi}{\sqrt{1 - \frac{1}{2}\sin^2\psi}}.$$

Passing to the limit $\vartheta \to \frac{\pi}{4}$ we arrive at

$$l = 4\int_0^{\frac{\pi}{2}} \frac{d\varphi}{\sqrt{1 - \frac{1}{2}\sin^2\psi}} = 4K\left(\frac{1}{\sqrt{2}}\right).$$

3. With the substitution $t = \sin\vartheta$ and $x = \sin\varphi$ we find $\frac{d\vartheta}{dt} = \frac{1}{\sqrt{1-t^2}}$ and consequently

$$\int_0^\varphi \frac{d\vartheta}{\sqrt{1 - k^2\sin^2\vartheta}} = \int_0^x \frac{dt}{\sqrt{1 - t^2}\sqrt{1 - k^2 t^2}}.$$

Analogously we can prove (27.7) and (27.8).

4. First we note that with the substitution $x = \frac{\pi}{2} - y$ we find

$$\int_0^{\frac{\pi}{2}} \frac{dx}{\sqrt{\sin x}} = \int_0^{\frac{\pi}{2}} \frac{dy}{\sqrt{\cos y}}.$$

Now we set $\cos y = \cos^2\varphi$ to find $\frac{dy}{d\vartheta} = \frac{2\cos\varphi}{\sqrt{1 + \cos^2\varphi}}$ and therefore

$$\int_0^{\frac{\pi}{2}} \frac{dx}{\sqrt{\sin x}} = \int_0^{\frac{\pi}{2}} \frac{dy}{\sqrt{\cos y}} = 2\int_0^{\frac{\pi}{2}} \frac{d\varphi}{\sqrt{1 + \cos^2\varphi}}$$

$$= 2\int_0^{\frac{\pi}{2}} \frac{d\varphi}{\sqrt{2 - \sin^2\varphi}} = \sqrt{2}\int_0^{\frac{\pi}{2}} \frac{d\varphi}{\sqrt{1 - \frac{1}{2}\sin^2\varphi}}$$

$$= \sqrt{2}K\left(\sqrt{\frac{1}{2}}\right).$$

On the other hand we know, compare with Problem 11 of Chapter I.28 and Theorem I.31.11, that

$$\int_0^{\frac{\pi}{2}} \frac{1}{\sqrt{\sin x}}\, dx = \frac{1}{2}B\left(\frac{1}{4}, \frac{1}{2}\right) = \frac{1}{2}\frac{\Gamma\left(\frac{1}{4}\right)\Gamma\left(\frac{1}{2}\right)}{\Gamma\left(\frac{3}{4}\right)}$$

$$= \frac{\sqrt{\pi}\,\Gamma\left(\frac{1}{4}\right)}{2\,\Gamma\left(\frac{3}{4}\right)}.$$

730

implying that

$$K\left(\frac{1}{2}\right) = \frac{\sqrt{\pi}\,\Gamma\left(\frac{1}{4}\right)}{2\sqrt{2}\,\Gamma\left(\frac{3}{4}\right)}.$$

5. We start with the substitution $\vartheta = 2\vartheta$ and using $\cos 2\varphi = 1 - 2\sin^2\varphi$ we have

$$J = \int_0^\pi \frac{d\vartheta}{\sqrt{a^2 + 2ab\cos\vartheta + b^2}} = \frac{2}{a+b} \int_0^{\frac{\pi}{2}} \frac{d\varphi}{\sqrt{1 - \frac{4ab}{(a+b)^2}\sin^2\varphi}}$$

or $J = \frac{2}{a+b} K\left(\frac{2\sqrt{ab}}{a+b}\right)$.

Now we use the substitution $a\sin(\vartheta - \psi) = b\sin\psi$ which yields

$$a\cos(\vartheta - \psi)\frac{d\vartheta}{d\psi} = a\cos(\vartheta - \psi) + b\cos\psi.$$

A study of the following figure gives a geometric interpretation of this substitution:

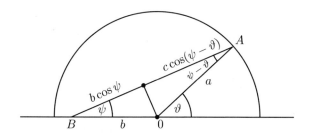

$$\overline{AB} = b + \cos\phi + a\cos(\psi - \vartheta) \quad \text{(from the figure)}$$
$$\overline{AB} = \sqrt{a^2 + 2ab\cos\vartheta + b^2} \quad \text{(from the cosine theorem)}.$$

Thus we find with $r = \overline{AB}$

$$a\cos(\vartheta - \psi)\frac{d\vartheta}{d\psi} = a\cos(\vartheta - \psi) + b\cos\psi = r$$

which implies

$$J = \int_0^\pi \frac{d\vartheta}{\sqrt{a^2 + 2ab\cos\psi + b^2}} = \int_0^\pi \frac{d\vartheta}{r}$$
$$= \frac{1}{a}\int_0^\pi \frac{d\psi}{\cos(\vartheta - \psi)} = \frac{1}{a}\int_0^\pi \frac{d\psi}{\sqrt{1 - \sin^2(\vartheta - \psi)}}$$
$$= \frac{2}{a}\int_0^{\frac{\pi}{2}} \frac{d\psi}{\sqrt{1 - \frac{b^2}{a^2}\sin^2\psi}},$$

731

where we used in the last step again $a \sin(\vartheta - \psi) = b \sin \psi$. Thus we arrive at

$$J = \frac{2}{a} K \left(\frac{b}{a} \right).$$

Combining these results we find

$$K \left(\frac{b}{a} \right) = \frac{a}{a+b} K \left(\frac{2\sqrt{ab}}{a+b} \right),$$

and with $k = \frac{2\sqrt{ab}}{a+b}$ and $k' = \sqrt{1 - b^2}$ it follows that $K(k) = \frac{2}{1+k} K \left(\frac{1-k'}{1+k} \right)$. (Our solution makes good use of the presentation in [72].)

6. Since $\operatorname{dn} z = \sqrt{1 - k^2 \operatorname{sn}^2 z}$ and $\operatorname{cn} z = \sqrt{1 - \operatorname{sn}^2 z}$ we find first

$$\operatorname{dn}^2 z - k^2 \operatorname{cn}^2 z = 1 - k^2 \operatorname{sn}^2 z - k^2 - k \operatorname{sn}^2 z = 1 - k^2 = k'^2.$$

Furthermore, using (27.17) and (27.18) we get

$$\operatorname{cn} 2z = \frac{\operatorname{cn}^2 z - \operatorname{sn}^2 z \operatorname{dn}^2 z}{1 - k^2 \operatorname{sn}^4 z} \quad \text{and} \quad \operatorname{dn} 2z = \frac{\operatorname{dn}^2 z - k^2 \operatorname{sn}^2 z \operatorname{cn}^2 z}{1 - k^2 \operatorname{sn}^4 z}$$

which yields

$$1 - \operatorname{cn} 2z = \frac{1 - k^2 \operatorname{sn}^4 z - \operatorname{cn}^2 z + \operatorname{sn}^2 z \operatorname{dn}^2 z}{1 - k^2 \operatorname{sn}^4 z}$$

and

$$1 + \operatorname{dn} 2z = \frac{1 - k^2 \operatorname{sn}^4 z + \operatorname{dn}^2 z - k^2 \operatorname{sn}^2 z \operatorname{cn}^2 z}{1 - k^2 \operatorname{sn}^4 z}.$$

Consequently we have

$$\begin{aligned}
\frac{1 - \operatorname{cn} 2z}{1 + \operatorname{dn} 2z} &= \frac{1 - k^2 \operatorname{sn}^4 z - \operatorname{cn}^2 z + \operatorname{sn}^2 z \operatorname{dn}^2 z}{1 - k^2 \operatorname{sn}^4 z + \operatorname{dn}^2 z - k^2 \operatorname{sn}^2 z \operatorname{cn}^2 z} \\
&= \frac{\operatorname{sn}^2 z - k^2 \operatorname{sn}^4 z + \operatorname{sn}^2 z \left(1 - k^2 \operatorname{sn}^2 z \right)}{1 - k^2 \operatorname{sn}^4 z + 1 - k^2 \operatorname{sn}^2 z - k^2 \left(\operatorname{sn}^2 z \left(1 - \operatorname{sn}^2 z \right) \right)} \\
&= \frac{2 \operatorname{sn}^2 z - 2k^2 \operatorname{sn}^4 z}{2 - k^2 \operatorname{sn}^4 z - k^2 \operatorname{sn}^2 z - k^2 \operatorname{sn}^2 z + k^2 \operatorname{sn}^4 z} \\
&= \frac{\operatorname{sn}^2 z \left(2 - 2k^2 \operatorname{sn}^2 z \right)}{2 - 2k^2 \operatorname{sn}^2 z} = \operatorname{sn}^2 z.
\end{aligned}$$

7. For z_1, z_2 fixed we want to prove

$$\operatorname{sn}(z_1 + z_2) = \frac{\operatorname{sn} z_1 \operatorname{cn} z_2 \operatorname{dn} z_2 + \operatorname{cn} z_1 \operatorname{sn} z_2 \operatorname{dn} z_1}{1 - k^2 \operatorname{sn}^2 z_1 \operatorname{sn}^2 z_2}$$

which by the uniqueness theorem for holomorphic functions will hold if we can prove it for z_1, z_2 in an open non-empty interval in $\mathbb{R}$. Since z_1 and z_2 are fixed we find that $\alpha := z_1 + z_2$ is a constant and we can consider z_2 as a function of z_1 with $\frac{dz_2}{dz_1} = -1$. We consider the function $g(z_1) := \operatorname{sn} z_1$ and $h(z_1) = \operatorname{sn} z_2$, i.e. $h(z_1) = \operatorname{sn}(\alpha - z_1)$. We find

$$g'(z_1) = \frac{dg}{dz_1}(z_1) = \operatorname{cn} z_1 \operatorname{dn} z_1$$

and

$$h'(z_1) = \frac{dh(\alpha - z_1)}{dz_1} = \frac{dh(z_2)}{dz_2}\frac{dz_2}{dz_1} = -\operatorname{cn} z_2 \operatorname{dn} z_2$$

implying

$$g'^2 = \left(1 - g^2\right)\left(1 - k^2 g^2\right) \qquad \text{and} \qquad h'^2 = \left(1 - h^2\right)\left(1 - k^2 h^2\right)$$

where we used (27.10) and (27.11). For the second derivatives we find

$$g'' = 2k^2 g^3 - \left(1 + k^2\right) g \tag{a}$$

and

$$h'' = 2k^2 h^3 - \left(1 + k^2\right) h. \tag{b}$$

Multiplying (a) by h and (b) by g and taking the difference we obtain

$$g'' h - g h'' = 2k^2 g h \left(g^2 - h^2\right) \tag{c}$$

as well as

$$g'^2 h^2 - g^2 h'^2 = \left(1 - k^2 g^2 h^2\right)\left(h^2 - g^2\right), \tag{d}$$

i.e.

$$g' h - g h' = \frac{\left(1 - k^2 g^2 h^2\right)\left(h^2 - g^2\right)}{g' h + g h'}. \tag{e}$$

Dividing (c) by (e) we arrive at

$$\frac{g'' h - g h''}{g' h - g h'} = -\frac{2k^2 g h \left(g' h + g h'\right)}{1 - k^2 g^2 h^2}. \tag{f}$$

Now we note that $g'' h - g h'' = \frac{d}{dz_1}\left(g' h - g h'\right)$ and that $-2k^2 g h \left(g' h + g h'\right) = \frac{d}{dz_1}\left(1 - k^2 g^2 h^2\right)$, which implies

$$\frac{\frac{d}{dz_1}\left(g' h - h g'\right)}{g' h - g h'} = \frac{\frac{d}{dz_1}\left(1 - k^2 g^2 h^2\right)}{1 - k^2 g^2 h^2} \tag{g}$$

or

$$\frac{d}{dz_1} \ln\left(g' h - h g'\right) = \frac{d}{dz_1} \ln\left(1 - k^2 g^2 h^2\right),$$

which yields

$$\frac{g' h - g h'}{1 - k^2 g^2 h^2} = c, \qquad c \in \mathbb{R}.$$

733

Using the definition of g and h we find with $\alpha = z_1 + z_2$ that

$$\frac{\operatorname{sn} z_1 \operatorname{cn} z_2 \operatorname{dn} z_2 + \operatorname{cn} z_1 \operatorname{sn} z_2 \operatorname{dn} z_1}{1 - k^2 \operatorname{sn}^2 z_1 \operatorname{sn}^2 z_2} = c$$

solves the above differential equation as does $z_2 = \alpha - z_1$. From the general theory of differential equations (compare with Part 9) we now deduce

$$\frac{\operatorname{sn} z_1 \operatorname{cn} z_2 \operatorname{dn} z_2 + \operatorname{cn} z_1 \operatorname{sn} z_2 \operatorname{dn} z_2}{1 - k^2 \operatorname{sn}^2 z_1 \operatorname{sn}^2 z_2} = f(z_1 + z_2)$$

with some function f. But for $z_2 = 0$ it follows that $f(z_1) = \operatorname{sn} z_1$ implying finally

$$\operatorname{sn}(z_1 + z_2) = \frac{\operatorname{sn} z_1 \operatorname{cn} z_2 \operatorname{dn} z_2 + \operatorname{cn} z_1 \operatorname{sn} z_2 \operatorname{dn} z_2}{1 - k^2 \operatorname{sn}^2 z_1 \operatorname{sn}^2 z_2}.$$

As already mentioned, we follow [80] closely in this solution.

8. From $\operatorname{dn}^2 z = 1 - k^2 \operatorname{sn}^2 z$ we deduce $\operatorname{sn}^2 z = \frac{\operatorname{dn}^2 z - 1}{k^2}$ and $\operatorname{cn}^2 z = \frac{\operatorname{dn}^2 z - 1 + k^2}{-k^2} = \frac{k'^2 - \operatorname{dn}^2 z}{k^2}$. Now we use the differential equations (27.13) - (27.15) to find

$$(\operatorname{sn}' z)^2 = \operatorname{cn}^2 z \operatorname{dn}^2 z = \left(1 - \operatorname{sn}^2 z\right)\left(1 - k^2 \operatorname{sn}^2 z\right);$$

$$(\operatorname{cn}' z)^2 = \operatorname{sn}^2 z \operatorname{dn}^2 z = \left(1 - \operatorname{cn}^2 z\right)\left(1 - k^2 \operatorname{sn}^2 z\right)$$
$$= \left(1 - \operatorname{cn}^2 z\right)\left(1 - k^2 + k^2 \operatorname{cn}^2 z\right)$$
$$= \left(1 - \operatorname{cn}^2 z\right)\left(k'^2 + k^2 \operatorname{cn}^2 z\right);$$

$$\left(\operatorname{dn}' z\right)^2 = k^4 \operatorname{sn}^2 z \operatorname{cn}^2 z$$
$$= k^4 \operatorname{sn}^2 z \left(1 - \operatorname{sn}^2 z\right)$$
$$= k^4 \left(\frac{\operatorname{dn}^2 z - 1}{k^2}\right)\left(1 - \frac{\operatorname{dn}^2 z - 1}{k^2}\right)$$
$$= -\left(1 - \operatorname{dn}^2 z\right)\left(k'^2 - \operatorname{dn}^2 z\right).$$

9. We start with the observation that

$$\sum_{(k_1,k_2)\in\mathbb{Z}^2\setminus\{0\}} \frac{1}{\left(k_1^2 + k_2^2\right)^\alpha} = \sum_{k_1\in\mathbb{Z}\setminus\{0\},k_2=0} \frac{1}{\left(k_1^2 + k_2^2\right)^\alpha} + \sum_{k_1=0,\,k_2\in\mathbb{Z}\setminus\{0\}} \frac{1}{\left(k_1^2 + k_2^2\right)^\alpha}$$

$$+ \sum_{k_1,k_2\in\mathbb{Z}\setminus\{0\}} \frac{1}{\left(k_1^2 + k_2^2\right)^\alpha}$$

$$= 4 \sum_{k_1\in\mathbb{N}} \frac{1}{k_1^{2\alpha}} + 4 \sum_{k_1,k_2\in\mathbb{N}} \frac{1}{\left(k_1^2 + k_2^2\right)^\alpha}.$$

The first sum on the right hand side compares with the integral $\int_1^\infty \frac{1}{x^{2\alpha}}\,dx$ and already gives the condition $2\alpha > 1$. For $N \in \mathbb{N}$ we find

$$\sum_{j,l=1}^N \frac{1}{\left((j+1)^2 + (l+1)^2\right)^\alpha} < \int_{[1,N]^2} \frac{1}{(x^2 + y^2)^\alpha}\,dx\,dy < \sum_{j,l=1}^{N-1} \frac{1}{\left(j^2 + l^2\right)^\alpha}. \tag{$*$}$$

Here we used for $[1, N]$ the partition $\{1, 2, \ldots, N\}$. Thus the left hand side is a Riemann sum for the integral with selected points $((j + 1), (l + 1))$ and the right hand side is a Riemann sum of the integral with selected points (j, l). Note that the two dimensional volume of a square of this partition is $\mathrm{vol}_2\left([j, j + 1] \times [l, l + 1]\right) = 1$. Now it follows that the convergence of the original series implies the convergence of the integral $\int_{[1,N]^2} \frac{1}{(x^2+y^2)} \, dx \, dy$ as well as of the integral $\int_1^N \frac{1}{x^{2\alpha}} \, dx$. It remains to note that

$$\int_K \frac{1}{(x^2 + y^2)^\alpha} \, dx \, dy = 4 \int_{[1,N]^2} \frac{1}{(x^2 + y^2)^\alpha} \, dx \, dy + 4 \int_1^N \left(\int_{-1}^1 \frac{1}{(x^2 + y^2)^\alpha} \, dy \right) dx$$

and that

$$\int_1^N \int_{-1}^1 \frac{1}{(x^2 + y^2)^\alpha} \, dy \, dx \leq 2 \int_1^N \frac{1}{x^{2\alpha}} \, dx$$

and we can indeed conclude that the convergence of the integral $\int_K \frac{1}{(x^2+y^2)^\alpha} \, dx \, dy$ is equivalent to the convergence of the series $\sum_{(k_1,k_2) \in \mathbb{Z}^2 \setminus \{0\}} \frac{1}{(k_1^2 + k_2^2)^\alpha}$.

Although a different proof (or an induction) is needed, we shall expect that the convergence of

$$\sum_{(k_1,\ldots,k_n) \in \mathbb{Z}^n \setminus \{0\}} \frac{1}{(k_1^2 + \cdots + k_n^2)^\alpha}$$

is equivalent to the convergence of the integral

$$\int_{W_n} \frac{1}{(x_1^2 + \cdots + x_n^2)^\alpha} \, dx_1 \ldots dx_n$$

where $W_n := \{x = (x_1, \ldots, x_n) \in \mathbb{R}^n \mid \|x\|_\infty \geq 1\}$. However it is easier to work with the integral

$$\int_{\mathbb{R}^n \setminus B_1(0)} \frac{1}{(x_1^2 + \cdots + x_n^2)^\alpha} \, dx_1 \cdots dx_n$$

and to use spherical polar coordinates in $\mathbb{R}^n$ to derive as equivalent condition for the convergence of the series the convergence of the integral

$$\int_1^\infty r^{-2\alpha} r^{n-1} \, dr$$

or $\alpha > \frac{n}{2}$.

10. Using (27.55) we find for $\lambda > 0$ that

$$\wp(\lambda z; \lambda w_1, \lambda w_2) = \frac{1}{(\lambda z)^2} + \sum_{w \in L \setminus \{0\}} \left(\frac{1}{(\lambda z - \lambda w)^2} - \frac{1}{(\lambda w)^2} \right)$$

$$= \frac{1}{\lambda^2} \left(\frac{1}{z^2} + \sum_{w \in L \setminus \{0\}} \left(\frac{1}{(z - w)^2} - \frac{1}{w^2} \right) \right)$$

$$= \frac{1}{\lambda^2} \wp(z; w_1, w_2) = \lambda^{-2} \wp(z; w_1, w_2).$$

11. Note that in our situation with $w_1 = 1$ and $w_2 = \tau$ we have

$$G_n(\tau) = G_n(1, \tau) = \sum_{(k_1, k_2) \in \mathbb{Z}^2 \setminus \{0\}} \frac{1}{(k_1 + k_2 \tau)^n}. \qquad (*)$$

Since $k_1 + k_2(\tau + 1) = k_1 + k_2 + k_2\tau$ a rearrangement of the summation in $(*)$ with $k_1 + k_2$ replacing k_1 we see that $G_n(\tau)$ has period 1, i.e. $G_n(\tau + 1) = G_n(\tau)$. In addition we have

$$\left(k_1 + k_2 \left(\frac{-1}{\tau}\right)\right)^n = \tau^{-1}(k_1\tau - k_2)^n$$

and replacing in $(*)$ the summation over (k_1, k_2) by the summation over $(-k_2, k_1)$ we obtain the relation $G_n(\tau) = \tau^{-n} G_n\left(-\frac{1}{\tau}\right)$.

Chapter 28

1. The mapping W_1 is continuous on $\mathbb{C} \setminus \{-i\}$, in particular it is continuous when restricted to $\partial \mathbb{H}$ and $\partial \mathbb{H}$ can be identified by $\mathbb{R}$, recall $\mathbb{H} = \{z \in \mathbb{C} \mid \operatorname{Im} z > 0\}$. For $z = x \in \mathbb{R}$ we have $|x - i| = |x + i|$ which implies $|W_1(x)| = \frac{|i-x|}{|i+w|} = 1$ implying that $W_1(\partial \mathbb{H}) \subset \partial \mathbb{D}$. Note that with some extra effort, see [83], it is possible to show that W_1 maps $\partial \mathbb{H}$ bijectively onto $\partial \mathbb{D} \setminus \{-1\}$.

2. We decompose $w(z) = \frac{1+z}{1-z}$ into its real and imaginary parts, i.e. with $z = x + iy$ we have

$$w(z) = w(x, y) = \frac{(1+z)(\overline{1-z})}{|1-z|^2}$$

$$= \frac{1 - (x^2 + y^2)}{(1-x)^2 + y^2} + \frac{2y}{(1-x)^2 + y^2}i.$$

For $z \in \mathbb{D} \cap \mathbb{H} = \{x + iy \in \mathbb{C} \mid x^2 + y^2 < 1, y > 0\}$ we find $\operatorname{Re} w(z) = \frac{1-(x^2+y^2)}{(1-x)^2+y^2} > 0$ and $\operatorname{Im} w(z) = \frac{2y}{(1-x)^2+y^2} > 0$, i.e. $w(\mathbb{D} \cap \mathbb{H}) \subset \mathbb{K} := \{x + iy \in \mathbb{C} \mid x > 0, y > 0\}$. The inverse Möbius transformation of $w : \mathbb{C} \setminus \{1\} \to \mathbb{C} \setminus \{-1\}$ is $v(\zeta) = \frac{\zeta-1}{\zeta+1}$ and since for $\zeta = \xi + i\eta$, $\xi > 0$ and $\eta > 0$, we have $|\zeta + 1| > |\zeta - 1|$, see the figure below, we conclude that $|v(\zeta)| < 1$, i.e. $v(\mathbb{K}) \subset \mathbb{D}$.

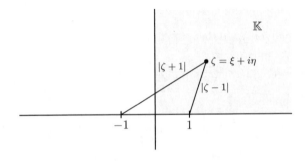

736

For $v(\zeta)$ we have the decomposition

$$v(\xi, \eta) = v(\zeta) = \frac{(\zeta - 1)(\overline{\zeta + 1})}{|\zeta + 1|^2} = \frac{\xi^2 + \eta^2 - 1}{(\xi + 1)^2 + \eta^2} + \frac{2\eta}{(\xi + 1)^2 + \eta^2} i$$

implying that $\operatorname{Im} v(\zeta) > 0$ for $(\xi, \eta) \in \mathbb{K}$. Hence we have proved that $v(\mathbb{K}) \subset \mathbb{D} \cap \mathbb{H}$ and together with the first part we conclude that $w(\mathbb{D} \cap \mathbb{H}) = \mathbb{K}$. It remains to remark that $w : \mathbb{C} \backslash \{1\} \to \mathbb{C} \backslash \{-1\}$ and $v : \mathbb{C} \backslash \{-1\} \to \mathbb{C} \backslash \{1\}$ are holomorphic functions.

3. We can adopt the arguments from the proof of Lemma 28.9. Let $g : G_1 \to G_2$ be a biholomorphic mapping with biholomorphic inverse $g^{-1} : G_2 \to G_1$ and define for $W \in \operatorname{Aut}(G_1)$

$$\Phi : \operatorname{Aut}(G_1) \to \operatorname{Aut}(G_2)$$
$$\Phi(W) := g \circ W \circ g^{-1}.$$

We claim that Φ is a group isomorphism. First we note for $W, V \in \operatorname{Aut}(G_1)$ that

$$\Phi\,(W \circ V) = g \circ (W \circ V) \circ g^{-1}$$
$$= g \circ W \circ g^{-1} \circ g \circ V \circ g^{-1}$$
$$= \Phi(W) \circ \Phi(V),$$

and further

$$\Phi\,(\operatorname{id}_{G_1}) = g \circ \operatorname{id}_{G_1} \circ g^{-1} = g \circ g^{-1} = \operatorname{id}_{G_2}.$$

Next we note that

$$\Phi(W)^{-1} = \left(g \circ W \circ g^{-1}\right)^{-1}$$
$$= g^{-1} \circ W^{-1} \circ g$$

and both $\Phi(W)$ as well as $\Phi(W)^{-1}$ are biholomorphic mappings. Hence Φ is a group isomorphism with inverse $\Phi^{-1} : \operatorname{Aut}(G_2) \to \operatorname{Aut}(G_1)$ given by $\Phi^{-1}(V) = g^{-1} \circ V \circ g$.

4. We decompose $f_A(z)$ into its real and imaginary part and find

$$f_A(z) = \frac{ac|z|^2 + (bc + ad)\operatorname{Re} z + bd}{|cz + d|^2} + i\frac{(ad - bc)\operatorname{Im} z}{|cz + d|^2}$$

and since $ad - bc = 1$ we get $\operatorname{Im} f_A(z) > 0$ for $z \in \mathbb{H}$, i.e. $f_A(\mathbb{H}) \subset \mathbb{H}$. Since this holds for all $A \in SL(2; \mathbb{R})$ it follows already that $f_A(\mathbb{H}) = \mathbb{H}$. Next we note that by Lemma 16.24 we have for $A, A' \in SL(2; \mathbb{R})$ the relations $f_A \circ f_{A'} = f_{A \circ A'}$, and $f_{A^{-1}} = (f_A)^{-1}$. Hence f_A is for every $A \in SL(2; \mathbb{R})$ an automorphism of $\mathbb{H}$.

5. According to Theorem 28.13 it is sufficient to show for every compact set $K \subset G$ that

$$\sup_{f \in \mathcal{H}} \left\| f^{(b)} \right\|_{\infty, K} \leq M_{K,k} < \infty. \tag{$*$}$$

737

Of course $(*)$ will follows if we can prove the estimate for all closed balls $\overline{B_{2r}(\zeta)} \subset G$. Since $\mathcal{H}$ is normal we can find $M_{\overline{B_{2r}(\zeta)}}$ such that

$$\sup_{f \in H} \|f\|_{\infty, \overline{B_{2r}(\zeta)}} \leq M_{\overline{B_{2r}(\zeta)}}.$$

The standard estimates, i.e. Theorem 21.15, imply

$$\left\| f^{(k)} \right\|_{\infty, \overline{B_r(\zeta)}} \leq \frac{2M_{\overline{B_{2r}(\zeta)}}k!}{r^k}$$

for all $f \in \mathcal{H}$, hence

$$\sup_{f \in \mathcal{H}} \left\| f^{(k)} \right\|_{\infty, \overline{B_r(\xi)}} \leq \frac{2M_{\overline{B_{2r}(\zeta)}}k!}{r^k}$$

and Theorem 28.13 implies the normality of $\mathcal{H}_k$.

6. Since for all $k \in \mathbb{N}$ the functions $f_k := W_k - a$ are injective and $W_k(z_1) = a$, we can find $\eta > 0$ such that with z_2 as in the proof of Theorem 28.17 it follows that

$$\inf_{k \in \mathbb{N}} \inf_{z \in \partial B_r(z_2)} |f_k(z)| \geq \eta > 0.$$

Since $g_k := (J - a) - (W_k - a)$ converges uniformly on $\overline{B_r(z_2)}$ to zero we can find $k_0 \in \mathbb{N}$ such that

$$|g_{k_0}(z)| < |f_{k_0}(z)| \quad \text{for all } z \in \partial B_r(z_2).$$

This implies by the theorem of Rouché, Theorem 24.26, that $f_{k_0} = W_{k_0} - a$ and $f_{k_0} + g_{k_0} = J - a$ have in $B_r(z_2)$ the same number of zeroes, i.e. $J - a$ has no zero in $B_r(z_2)$.

7. With $W_1(z) = W_1(x, y) = u(x, y) + iv(x, y)$ we find

$$g(x, y) = h(u(x, y), v(x, y)) = u^2(x, y) - v^2(x, y).$$

Since $W_1(z) = \frac{i-z}{i+z}$ we find further

$$W_1(x, y) = u(x, y) + iv(x, y) = \frac{1 - (x^2 + y^2)}{x^2 + (y + 1)^2} + i\frac{2x}{x^2 + (y + 1)^2}$$

implying that

$$g(x, y) = \left(\frac{1 - (x^2 - y^2)}{x^2 + (y + 1)^2} \right)^2 - \frac{4x^2}{(x^2 + (y + 1)^2)^2} = \frac{(x^2 + y^2)^2 - 6x^2 - 2y^2 + 1}{(x^2 + (y + 1)^2)^2}$$

is harmonic in $\{(x, y) \in \mathbb{R}^2 \mid y > 0\}$.

Chapter 29

1. The convergence of $\left(\sum_{|\alpha| \leq N} |c_\alpha| \right)_{N \in \mathbb{N}}$ implies the convergence of $\sum_{|\alpha| \leq N} c_\alpha = C$.
   We have to prove that $\lim_{M \to \infty} \sum_{k=0}^{M} c_{\varphi(k)} = C$. For $\epsilon > 0$ we know the existence
   of $N_0 \in \mathbb{N}$ such that $\sum_{|\alpha| \geq N_0} |c_\alpha| < \frac{\epsilon}{2}$ which yields

$$\left| C - \sum_{|\alpha| \leq N_0 - 1} c_\alpha \right| = \left| \sum_{|\alpha| \geq N_0} c_\alpha \right| \leq \sum_{|\alpha| \geq N_0} |c_{|\alpha|}| < \frac{\epsilon}{2}.$$

Now we take N such that $\{\alpha \,|\, |\alpha| \leq N_0 - 1\} \subset \{\varphi(\beta) \,|\, |\beta| \leq N\}$. For $m \geq N$ we
find

$$\left| \sum_{k=0}^{m} c_{\varphi(k)} - C \right| \leq \left| \sum_{k=0}^{m} c_{\varphi(k)} - \sum_{|\alpha| \leq N_0 - 1} c_\alpha \right| + \left| \sum_{|\alpha| \leq N_0 - 1} c_\alpha - C \right|$$

$$\leq \sum_{|\alpha| \geq N_0} |c_\alpha| + \left| \sum_{|\alpha| \leq N_0 - 1} c_\alpha - C \right|$$

$$\frac{\epsilon}{2} + \frac{\epsilon}{2} = \epsilon$$

and the result is proved.

2.   a) We have $\sum_{\alpha \in \mathbb{N}_0^2} z^\alpha = \sum_{\alpha_1, \alpha_2 \in \mathbb{N}_0} z_1^{\alpha_1} z_2^{\alpha_2} = \left(\sum_{\alpha_1 \in \mathbb{N}_0} z_1^{\alpha_1} \right) \left(\sum_{\alpha_2 \in \mathbb{N}_0} z_2^{\alpha_2} \right)$ and
   for $|z_1| < 1$, $|z_2| < 1$ we find $\sum_{\alpha_1 \in \mathbb{N}_0} z_1^{\alpha_1} = \frac{1}{1 - z_1}$ as well as $\sum_{\alpha_2 \in \mathbb{N}_0} z_2^{\alpha_2} = \frac{1}{1 - z_2}$
   implying that

$$\sum_{\alpha \in \mathbb{N}_0^2} z^\alpha = \frac{1}{(1 - z_1)(1 - z_2)}$$

for $(z_1, z_2) \in B_1(0) \times B_1(0)$.

   b) We note that $\sum_{(1, \alpha_2) \in \mathbb{N}_0^2} z^\alpha = \sum_{\alpha_2 \in \mathbb{N}_0} z_1 z_2^{\alpha_2} = z_1 \sum_{\alpha_2 \in \mathbb{N}_0} z_2^{\alpha_2}$ and therefore
   we obtain for $(z_1, z_2) \in \mathbb{C} \times B_1(0)$ that

$$\sum_{(1, \alpha) \in \mathbb{N}_0^2} z^\alpha = \frac{z_1}{1 - z_2}.$$

   c) Since $\sum_{\nu \in \mathbb{N}_0} z_1^\nu z_2^\nu = \sum_{\nu \in \mathbb{N}_0} (z_1 z_2)^\nu$ we deduce that for $|z_1 z_2| < 1$ the series
   converges and has limit $\frac{1}{1 - z_1 z_2}$, i.e.

$$\sum_{\nu \in \mathbb{N}_0} z_1^\nu z_2^\nu = \frac{1}{1 - z_1 z_2}.$$

739

References

[1] Abramowitz, M., Stegun, J. A., (eds.), *Handbook of Mathematical Functions.* 7th printing. Dover Publications, New York 1970.

[2] Agricola, J., Friedrich, T., *Elementary Geometry.* American Mathematical Society, Providence R.I. 2008.

[3] Ahlfors, L.V., *Complex Analysis.* 2nd ed. McGraw-Hill Book Company, New York 1966.

[4] Ahlfors, L., *Conformal Invariants. Topics in Geometric Function Theory.* McGraw-Hill Book Company, New York 1973.

[5] Apostol, T.M., *Introduction to Analytic Number Theory.* Springer Verlag, Berlin 1976.

[6] Appell, J., *Analysis in Beispielen und Gegenbeispielen.* Springer Verlag, Berlin · Heidelberg 2009.

[7] Appell, J., Väth, M., *Elemente der Funktionalanalysis.* Vieweg Verlag, Wiesbaden 2005.

[8] Armstrong, M.,A., *Basic Topology.* McGraw-Hill Book Company (U.K.), London 1979.

[9] Bartoszynski, T., Judah, H., *Set Theory. On the Structure of the Real Line.* A. K. Peters, Wellesly MA 1995.

[10] Bauer, H., *Wahrscheinlichkeitstheorie und Grundzüge der Maßtheorie.* 2. Aufl. Walter de Gruyter Verlag, Berlin 1974.
English: *Probability Theory and Elements of Measure Theory.* 2nd ed. Academic Press, London · New York, 1981.

[11] Bauer, H., *Maß-und Integrationstheorie.* Walter de Gruyter Verlag, Berlin 1990.
English: *Measure and Integration Theory.* Walter de Gruyter Verlag, Berlin 2001.

[12] Beals, R., and Wong, R., *Special Functions. A Graduate Text.* Cambridge University Press, 2010.

[13] Behrends, E., *Maß-und Integrationstheorie.* Springer Verlag, Berlin 1987.

[14] Benedetto, J.J., Czaja, W., *Integration and Modern Analysis.* Birkhäuser Verlag, Boston · Basel · Berlin 2009.

[15] Billingsley, P., *Probability and Measure.* 3rd ed. John Wiley & Sons, New York 1995.

[16] Carathéodory, C., *Funktionentheorie.* 1. Band. 2. Aufl. Birkhäuser Verlag, Basel · Stuttgart 1960.
English: *Theory of Functions of a Complex Variable* Volume 1. 2nd ed. Chelsea Publishing Company, New York 1964.

[17] Carathéodory, C., *Funktionentheorie.* 2. Band. 2. Aufl. Birkhäuser Verlag, Basel · Stuttgart 1961.
English: *Theory of Functions of a Complex Variable.* Volume 2. 2nd ed. Chelsea Publishing Company, New York 1960.

[18] Cerny, J., *Foundations of Analysis in the Complex Domain*. Academia, Praha 1992.

[19] Deimling, K., *Nichtlineare Gleichungen und Abbildungsgrade*. Springer Verlag, Berlin 1974.

[20] Dinghas, A., *Vorlesungen über Funktionentheorie*. Springer Verlag, Berlin 1961.

[21] Dudley, R.M., *Real Analysis and Probability*. Wadsworth & Brooks, Pacific Grove CA 1989.

[22] Duren, P., *Invitation to Classical Analysis*. American Mathematical Society, Providence R.I. 2012.

[23] Falconer, K.J., *Fraktale Geometrie*. Spektrum Akademischer Verlag, Heidelberg 1993. English: *Fractal Geometry*. Mathematical Foundations and Applications. John Wiley & Sons, New York 1990.

[24] Falconer, K.J., *The Geometry of Fractal Sets*. Cambridge University Press, paperback reprint, Cambridge 1995.

[25] Federer. H., *Geometric Measure Theory*. Reprint: Classics in Mathematics. Springer-Verlag, Berlin 1996.

[26] Fischer, W., Lieb, I., *Funktionentheorie*. 5. Aufl. Vieweg Verlag, Braunschweig · Wiesbaden 1988.

[27] Folland, G.B., *Real Analysis. Modern Techniques and Their Applications*. 2nd ed. John Wiley & Sons, New York 1999.

[28] Freitag, E., Busam, R., *Funktionentheorie 1*. 4th ed. Springer Verlag, Berlin 2006.

[29] Fritsche, K., Grauert, H., *From Holomorphic Functions to Complex Manifolds*. Springer Verlag, Berlin 2002.

[30] Gaier, D., *Vorlesungen über Approximation im Komplexen*. Birkhäuser Verlag, Basel · Boston · Stuttgart 1980.

[31] Gorenflo, R., Kilbas, A.A., Mainardi, F., Rogosini, S.V., *Mittag-Leffler Functions, Related Topics and Applications*. Springer Verlag, Berlin 2014.

[32] Hardy, G.H., Wright, E.M., *An Introduction to the Theory of Numbers*. 5th ed. Clarendon Press, Oxford 1979, several corrected reprints.

[33] Heins, M., *Selected Topics in the Classical Theory of Functions of a Complex Variable*. Holt, Rinehart and Winston, New York 1962.

[34] Heins, M., *Complex Function Theory*. Academic Press, New York 1968.

[35] Hewitt, E., Stromberg, K., *Real and Abstract Analysis*. Springer Verlag, 1978.

[36] Hille, E., *Analytic Function Theory. Volume I*. Ginn and Company, Boston MA 1959.

[37] Hille, E., *Analytic Function Thoery. Volume II*. Ginn and Company, Boston MA 1962.

[38] Hirzebruch, F., Scharlau, W., *Einführung in die Funktionalanalysis*. B. I.-Wissenschaftsverlag, Mannheim · Wien · Zürich 1971.

[39] Ingham, A.E., *The Distribution of Prime Numbers*. Reprinted in "Cambridge Mathematical Library", Cambridge University Press, Cambridge 1995.

[40] Jacob, N., Knopova, V., Landwehr, S., Schilling, R.L., *A Geometric Interpretation of the Transition Density of a Symmetric Lévy Process*. Sci. China Math. 55(2012), 1099-1126.

[41] Jacobi, C.G.J., *Fundamenta nova theoria functionum elliptarium*. Gebrüder Borntträger, Königsberg 1829.

[42] Jacobs, K., *Measure and Integral*. Academic Press, New York 1978.

[43] Kaczor, W.J., and Nowak, M.T., *Problems in Mathematical Analysis I*. American Mathematical Society, Providence R.I., 2000.

[44] Kaczor, W.J., and Nowak, M.T., *Problems in Mathematical Analysis II*. American Mathematical Society, Providence R.I., 2001.

[45] Kaczor, W.J., and Nowak, M.T., *Problems in Mathematical Analysis III*. American Mathematical Society, Providence RI 2003.

[46] Kharazishvili, A.B., *Strange Functions in Real Analysis*. 2nd ed. Chapman & Hall, Boca Raton FL 2006.

[47] Knapp, A. W., *Basic Real Analysis*. Birkhäuser Verlag, Boston · Basel · Berlin 2005.

[48] Koecher, M., Krieg, A., *Elliptische Funktionen und Modulformen*. Springer Verlag, Berlin 1998.

[49] Kolmogoroff, A.N., *Grundbegriffe der Wahrscheinlichkeitsrechnung*. Springer Verlag, Berlin 1933.

[50] Kolmogorov, A.N., Fomin, S.V., *Reelle Funktionen und Funktionalanalysis*. VEB Deutscher Verlag der Wissenschaften, Berlin 1975.

[51] Krantz, S.G., Parks, H.R., *A Primer of Real Analytic Functions*. Birkhäuser Verlag, Basel · Boston · Berlin 1992.

[52] Kuttler, K.L., *Modern Analysis*. CRC Press, Boca Raton FL 1998.

[53] Lang, S., *Elliptic Functions*. 2nd ed. Springer Verlag, Berlin 1987.

[54] Lebesgue, H., *Intégrale, Longueur, Aire*. These Paris 1902. Also: Ann. Mat. Pura Appl. 7(1902), 231-359.

[55] Magnus, W., Oberhettinger, F., Soni, R.P., *Formulas and Theorems for Special Functions of Mathematical Physics*. 3rd ed. Springer Verlag, Berlin 1966.

[56] Malliavin, P., *Integration and Probability*. Springer Verlag, Berlin 1995.

[57] Milnor, J.W., *Topology from the Differential Viewpoint*. The University Press of Virginia, Charlottesville, 5th printing, 1978.

743

[58] Narasimhan R., *Analysis on Real and Complex Manifolds*. North-Holland Publishing Company, Amsterdam, 3rd printing, 1985.

[59] Natanson, J.P., *Theorie der Funktionen einer reelen Veränderlichen*. 4. Aufl. Verlag Harri Deutsch, Zürich · Frankfurt · Thun 1975.

[60] Nevanlinna, R., *Uniformisierung*. Springer Verlag, Berlin 1953.

[61] Nevanlinna, R., Paatero, V., *Einführung in die Funktionentheorie*. Birkhäuser Verlag, Basel · Stuttgart 1965.
English: *Introduction to Complex Analysis*. Addison-Wesley Publishing Company, Reading MA 1969.

[62] Nielsen, N., *Traité élémentaire des nombres de Bernoulli*. Gauthier-Villars, Paris 1923.

[63] Patterson, S.J., *An Introduction to the Theory of the Riemann Zeta-Function*. Cambridge University Press, Paperback Edition, Cambridge 1995.

[64] Peschl, E., *Funktionentheorie*. Band 1. B.I.-Verlag, Mannheim · Wien · Zürich 1967.

[65] Querenburg, Boto von, *Mengentheoretische Topologie*. Springer Verlag, Berlin 1973.

[66] Range, R.M., *Holomorphic Functions and Integral Representations in Several Complex Variables*. Springer Verlag, Berlin 1986.

[67] Remmert, R., *Funktionentheorie 1*. 3. Aufl. Springer Verlag, Berlin 1992.
English: *Theory of Complex Functions*. Springer Verlag, Berlin 1991.

[68] Remmert, R., *Funktionentheorie 2*. 2. Aufl. Springer Verlag, Berlin 1995.
English: *Classical Topics in Complex Function Theory*. Springer Verlag, Berlin 1998.

[69] Rinow, W., *Topologie*. VEB Deutscher Verlag der Wissenschaften, Berlin 1975.

[70] Royden, H.L., *Real Analysis*. Macmillan Company, New York 1963.

[71] Rudin, W., *Real and Complex Analysis*. 2nd ed. McGraw-Hill Book Company, New York 1974.

[72] Rühs, F., *Funktionentheorie*. 2 Aufl. VEB Deutscher Verlag der Wissenchaften, Berlin 1971.

[73] Saks, S., Zygmund, A., *Analytic Functions*. Nakladem Polskiego Towarzystwa Matematycznego, Warszawa 1952.

[74] Schechter, M., *An Introduction to Nonlinear Analysis*. Cambridge University Press, Cambridge 2004.

[75] Schilling, R.L., *Measures, Integrals and Martingales*. Cambridge University Press, Cambridge 2005. (2nd ed. to be published in 2017.)

[76] Schilling, R.L., *Mass und Integral*. Walter de Gruyter Verlag, Berlin 2015.

[77] Schubert, H., *Topologie*. 4. Aufl. Teubner Verlag, Stuttgart 1976.

[78] Schwartz, J., *Nonlinear Functionalanalysis*. Gordon and Breach Science Publishers, New York 1969.

[79] Shen, A., Vereshchagin, N.K., *Basic Set Theory*. American Mathematical Society, Providence R.I. 2002.

[80] Spiegel, M.R., *Theory and Problems of Complex Variables*. (SI(metric) edition.) McGraw-Hill Book Company, New York 1974.

[81] Srivastara, S.M., *A Course on Borel Sets*. Springer Verlag, Berlin 1998.

[82] Stein, E.M., Shakarchi, R., *Princeton Lectures in Analysis I. Fourier Analysis: An Introduction*. Princeton University Press, Princeton NJ 2003.

[83] Stein, E.M., Shakarchi, R., *Princeton Lectures in Analysis II. Complex Analysis*. Princeton University Press, Princeton NJ 2003.

[84] Stein, E.M., Shakarchi, R., *Princeton Lectures in Analysis III. Real Analysis*. Princeton University Press, Princeton NJ 2005.

[85] Stroock, D.W., *A Concise Introduction to the Theory of Integration*. 3rd ed. Birkhäuser Verlag, Boston · Basel · Berlin 1999.

[86] Titchmarsh, E.C., *The Theory of the Riemann Zeta-Function*. Oxford at the Clarendon Press, Oxford 1951.

[87] Torchinsky, A., *Real Variables*. Perseus Books, Reading MA 1995.

[88] Tricomi, F., Krafft, M., *Elliptische Funktionen*. Akademische Verlagsgesellschaft, Geest & Partig, Leipzig 1948.

[89] Triebel, H., *Höhere Analysis*. VEB Deutscher Verlag der Wissenschaften, Berlin 1972. English: *Higher Analysis*. J. A. Barth Verlag, Leipzig 1992.

[90] Voigt, A., Wloka, J., *Hilberträume und elliptische Differentialoperatoren*. B.-I Verlag, Mannheim · Wien · Zürich 1975.

[91] Weil, A., *Elliptic Functions According to Eisenstein and Kronecker*. Springer Verlag, Berlin 1976. (Reprinted as Springer Classics in Mathematics 1999.)

[92] Weyl, H., *Die Idee der Riemannschen Fläche*. Teubner Verlag 1913, 2nd corrected ed. 1923.

[93] Whittaker, E.T., Watson, G.N., *A Course of Modern Analysis*. 4th ed., reprinted in "Cambridge Mathematical Library", Cambridge University Press, Cambridge 1996.

[94] Wheeden, R.L., Zygmund, A., *Measure and Integral. An Introduction to Real Analysis*. Marcel Dekker Inc., New York · Basel 1977.

Mathematicians Contributing to Analysis (Continued)

Ahlfors, Lars Valerian (1907-1996).

Alexandrov, Pawel Sergejevich (1896-1982).

Bernoulli, Johann I (1667-1748).

Carathéodory, Constantin (1873-1950).

Casorati, Felice (1835-1890).

Chebyshev, Pafnuti Lwowich (1821-1894).

Dynkin, Eugene Borisovich (1924-2014).

Egorov, Dmitri Fedorovich (1869-1931).

Eisenstein, Ferdinand Gotthold Max (1823-1852).

Fatou, Pierre Joseph Louis (1878-1829).

Federer, Herbert (1920-2010).

Fischer, Ernst (1875-1954).

Fresnel, Augustin Jean (1788-1827).

Friedrichs, Kurt-Otto (1901-1982).

Goursat, Edouard Jean-Baptiste (1858-1936).

Grauert, Hans (1930-2011).

Hardy, Godefrey Harold (1877-1947).

Kolmogorov, Andrej Nikolajevich (1903-1987).

Laurent, Pierre Alphonse (1813-1854).

Levi, Beppo (1875-1961).

Liouville, Joseph (1809-1882).

Littlewood, John Edensor (1885-1977).

Lusin, Nikolaj Nikolajevich (1883-1950).

Möbius, August Ferdinand (1790-1868).

Marcinkiewicz, Jozef (1910-1940).

Markov, Andrej Andrejevich (1856-1922).

Mittag-Leffler, Magnus Gösta (1846-1927).

Montel, Paul Antoine Aristide (1876-1975).

Morera, Giacinto (1856-1909).

Nevanlinna, Rolf Hermann (1895-1980).

Nielsen, Niels (1865-1931).

Nikodym, Otton Martin (1887-1974).

Pochhammer, Leo August (1841-1920).

Radon, Johann Karl August (1887-1956).

Reinhardt, Karl August (1895-1941).

Riesz, Frigyes (1880-1956).

Rouché, Eugene (1832-1910).

Runge, Carl David Tolmé (1856-1927).

Sard, Arthur (1909-1980).

Tonelli, Leonida (1885-1946).

Tricomi, Francesco Giacomo Filippo (1897-1978).

Urysohn, Pawel Samuilovich (1898-1924).

Viète (Vieta), François (1540-1603).

Vitali, Giuseppe (1875-1932).

Weil, André (1906-1998).

Weyl, Hermann Klaus Hugo (1885-1955).

Wiener Norbert (1894-1964).

Zygmund Antoni (1900-1992).

Subject Index

749

世界著名数学家 R. L. Wilder 曾指出：

然而看来有理由期望，一旦一个能够提供研究沃土的数学新分支存在于足够多的数学家头脑之中，这个分支就将沿着与推进该分支的数学家的个人气质无关（除去纯细节方面）的方向发展，先是出现许多重复性的结果，该学科的活动达到了最热烈的程度，继而兴趣开始衰落. 衰落可能是因为最重要的、最有兴趣的问题已经解决，也许是因为这个分支已经被包容更大领域的数学新分支所吸收. 最后只有分散的工作者滞留在原分支中，而主要的年轻人则忙于新理论的工作.

他还说过：

数学中出现了各种各样的时兴方式，这些方式所遵循的进程与文化中其他部分各种方式的进程并无不同. 一个处于 20 世纪前期的数学家可能会对今日的数学感到吃惊. 就像在人类事物的其他领域一样，长者对年轻人所喜爱的新方式总是兴叹.

本书是一部论及数学经典理论的教程，中文书名或可译为《分析学教程. 第 3 卷，测度与积分理论，复变量的复值函数》.

本书的作者有两位：一位是尼尔斯·雅各布（Niels Jacob），英国数学家，英国斯旺西大学教授；另一位是克里斯蒂安·P. 埃文斯（Kristian P. Evans），英国数学家，英国斯旺西大学教授.

复分析,特别是 Cauchy 的复分析的真正来源是定积分,这些定积分是实变函数的实积分上、下限的积分,但是用通常方法不能求出来. 从 Euler 起, D'Alembert, Lagrange, Laplace, Legendre 以及后来的 Poisson 等都遇到过这种问题,他们都曾经运用由实到复的过渡来解决问题. Leibniz 及 John Bernoulli 都曾应用"虚变换"来解积分问题,但系统地应用这种方法的是 Euler. Euler 在 1776—1777 及 1781 年曾向彼得堡科学院宣读过 9 篇论文,他的方法是把积分

$$\int Z(z)\,\mathrm{d}z = \Delta z$$

(Δz 表示 z 的函数)中的实变元换成复变元.

正如本书的两位作者在前言中所介绍:

　　本书是我们的《分析学教程》著作的第 3 卷,涵盖了每个数学家都必须要研究的两个主题,无论他们是否专门研究分析学. 第 6 部分讨论了 Lebesgue 的积分理论和实变量的实值函数理论中的第一个结果. 第 7 部分介绍了一个复变量的复值函数理论,习惯上简称"函数理论".

　　实值函数、Fourier 分析、函数分析、动力系统理论、偏微分方程或变分法都是基于 Lebesgue 积分理论的课题. 但除此之外,概率论、几何学的部分内容和更多的应用学科,如信息论或优化,都需要对 Lebesgue 积分有一个正确的理解,而且,许多数学家将会认同实线结构如果不是最复杂的结构,那么也是最复杂的结构之一,我们必须去研究它. 许多问题使我们对 $\mathbb{R}$ 有了一个更好的理解,事实上,当我们将 Euclid 度量所引起的 $\mathbb{R}$ 的拓扑结构与定义和确定 $\mathbb{R}$ 的子集大小的问题联系起来时,会导致数学基础(数学逻辑和集合论)的深入发展. 通常,集合论的基础模型决定了拓扑学和测度理论之间的关系. 在我们的论文中,总是假设 ZF 是集合论的基础模型,但我们没有花太多时间去研究上述问题,它们是不同的和更高级的课程的主题.

　　下面我们来更详细地描述第 6 部分的内容,我们想在介绍中提到[11]的影响. H. Bauer 的专著[11]的先驱,即[10],是几十年来德国测度理论的标准教科书. 当[10]被分成两本书时,第一作者参与了大量校对和讨论材料的工作.

　　函数理论是分析学中所有其他高级理论的关键,同时它也是许多

760

其他应用数学学科的工具,如力学、电气工程或物理学等.但更多的是全纯函数,即复变量的复可微复值函数进入了纯数学的许多领域,如数论、代数几何、表示论、微分几何或组合学等.公平地说,如果没有适当的函数理论知识,纯数学或应用数学的本科教育就不是完整的.下面是对第7部分的更详细的讨论.

在讨论细节之前我们先补充两点.第一,最初我们还计划在第3卷第8部分介绍 Fourier 分析.在编写第6部分和第7部分时,我们发现一个更好的策略是在第6部分和第7部分已经涵盖的两个理论中添加更多(高级的或专业的)内容,而不是在后面需要它们的部分中再提到一点儿关于它们的内容,例如,实值函数的微分性质、Sard 定理、L^p – 空间中的稠密子集和 Friedrichs 光滑化算子,或超几何函数、椭圆积分与椭圆函数.这当然会导致之前安排整个材料的计划有所改变.第二,读者会注意到引用文献的不同方式.我们现在有时会用完全不同的方法和表示来处理问题.显然,我们会受到之前处理过这些材料的作者的影响,但我们必须在适当的情况下给予这些材料公平的评价.当我们在一些证明中将简单计算过程留给读者完成时,我们同时也考虑了更高级的内容这一事实,这是我们在前两卷中严格避免的.书中的一些问题越复杂,一些解法越简洁,这再次反映了我们的书面向的是更多(数学上)成熟的读者.

第1章介绍了 σ 域及其生成元和测度.第2章致力于研究 Carathéodory 扩展定理.第3章讨论了 Lebesgue-Borel 测度和 Hausdorff 测度,也作为 Hausdorff 维数,特别详细地讲解了 Cantor 集合.可测映射是第4章的主题.定义测度的 Lebesgue 积分的标准方法在第5章中进行了阐述.第6章介绍了带稠密性的测度,我们用证明 Lebesgue 积分的变换理论的方法证明了 Radon-Nikodym 定理.第7章讨论了测度零集的作用,介绍了与主要收敛结果有关的几乎所有的知识,尤其是控制收敛定理.我们也考虑了 p 幂可积函数空间.这些结果会在第8章中用来证明关于极限交换的典型定理,比如连续性或参数相关积分的可微性.我们证明了 Jensen 不等式,讨论了 Legendre 积分和 Riemann 积分的关系.通过在第2卷中处理更高维的 Riemann 积分我们填补了某些空白.这个讨论包括反常积分,特别地介绍了 L^p – 空间.乘积积分与 Tonelli 和 Fubini 的大部分定理是第9章的主要内容.作为一个重要的应用,我们研究了关于分布函数的积分,同时给出了一些具有理论意义的例子.我们也提供了 Minkowski 积分不等式的完整证明.从我们的观点来看,第 1~9 章的内容构成了任意测度模的核心,同时这些内容也适用于以分析为目的的积分理论和概率论.第 10~12 章研究的主题很有趣,但它们也是 Fourier 分析、泛函分析或(偏)微分方程等领域的主要工具.第 10 章致力于研究函数和测度的卷积,我们证明了 $L^p(\mathbb{R}^n)(1 \leqslant p < \infty)$ 空间中具有紧支连续函数的稠密性.此外,我们研究了 Friedrichs 光滑化算子,在以后的许多计算中它将成为一流的工具.最后,我们先来看看卷积算子.Legendre 的微分理论是我们第 11 章的主题.在证明了 Vitali 覆盖定理之后,我们引入了绝对连续函数和有界变分函数,并研究了它们的关系.我们得到的主要结果是微积分基本定理和 Lebesgue 微分定理的一个新版本,其证明是利用 Hardy-Littlewood 极大函数给出的.最后,在第 12 章中,我们讨论三个特殊结果:关于可微映射的临界点测度的 Sard 定理的一个版本、Lusin 定理、L^p – 空间中表征相对紧集的 Kolmogorov-Riesz 定理.

我们对复变量的复值函数的处理是以对复数的简短回忆开始的,该复数包括序列和级数的收敛.在第 14 章中,我们简述了一些题外内容,并总结了定义在任意集合上的复值函数的明显的性质.第 15 章专门讨论了平面几何和复数.书中还包括 Riemann 球面和立体投影的内容.接下来是关于复变量的复值函数的一章,我们研究了连续性、收敛性和一致收敛,并证明了 Abel 定理以及幂级数的 Cauchy-Hadamard 定理.幂级数示例包括二项级数和 Gauss 超几何级数.我们以 Möbius 变换来结束本章.第 17 章介绍了复微分,给出了所有的标准结果,包括与 Cauchy-Riemann 微分方程相关的结果.关键的概念是全纯函数的概念,我们也第一次接触双全纯函数与调和函数.接下来讨论了一些重要的全纯函数,例如指数函数、三角函数及其逆函数、双曲函数及其逆函数.第 18 章中最重要的内容是对对数函数的研究,比如连通性、简单连通性或同调等拓扑概念在函数论中起着至关重要的作用,第 19 章在第 20 章介绍线积分之前研究了一些相关问题.对于线积分的定义,我们可以依据第 2 卷第 15 章中的内容,就像在推导它们的基本性质一样.通常我们只需要改变符号,即使用复数.一旦完成这个步骤,在(简单)闭合曲线上的线积分的背景下,复变量的复值函数理论可以有许多新结果.

我们引入了原数概念,并将线积分的性质与原数的存在联系起来.第 21 章是该理论的核心章节,详细讨论了 Cauchy 积分定理、Cauchy 积分公式及其许多应用,包括全纯函数导数的标准估计,以及它们与 Taylor 展开式的关系.在第 22 章中我们继续研究幂级数、全纯和全纯域,以及微分方程的应用.一个重要的结果是关于可移奇点的 Riemann 定理和全纯函数的唯一性定理.在第 23 章我们通过讨论边界最大值原理、Schwarz 引理、Liouville 定理和 Runge 逼近定理进一步研究了该理论.第 24 章介绍了亚纯函数和 Laurent 扩展,包括它们的奇点分类和残差定理的第一个版本,以及论证原理和 Rouché 定理.在第 25 章中,我们给出了残差定理的通用版本.这种理论性的研究具有许多非常具体的应用,例如某些(反常)积分的评估.我们从第 1 卷开始讨论 Γ – 函数,现在在第 26 章中把 $\Gamma(z)$ 看作一个亚纯函数.我们还介绍了 Dirichlet 级数,特别是 Riemann ζ – 函数,并证明了它们的一些性质.这使我们能够最终表述 Riemann 假设,相信读者会理解,在这里不提供对它的证明.数学家可能在他们接受教育的后期才会遇到椭圆积分,物理学家很早就遇到了.我们在第 27 章中讨论了椭圆积分和椭圆函数,从积分开始,然后转向 Jacobi 椭圆函数.我们证明了它们是双周期函数,同时开始通过证明三个 Liouville 定理来研究双周期函数.本章虽然是第一个介绍此内容,但它涵盖的内容(与前一章一样)比通常包含在函数理论第一门课程中的内容要多.第 28 章提供了 Riemann 映射定理的证明,该结果必须包含在每门课程中,但是我们不会讨论保角映射的一些应用,例如 Schwarz-Christoffel 映射.第 7 部分的最后一章专门讨论了多复变量的幂级数.其目的是指出在多个(实和复)变量中,幂级数的收敛域是一个比在一维(实或复)中更加微妙的问题.

两个附录介绍了一些拓扑问题,一部分是关于测度理论的.附录Ⅲ中,我们更进一步讨论了 Möbius 变换和扩展平面的几何内容.最后一个附录是关于我们之前在许多地方都遇到过的 Bernoulli 数.

像之前的几卷一样,我们提供了约 275 个问题的解答.此外,特别是在第 7 部分中,我们需要相当多的数字(大约 90 个).所有这些数字都是由第二位作者使用 LaTeX 完成的.最后参考卷 1 或卷 2 的注释:定理Ⅱ.4.20 表示卷 2 中的定理 4.20,(Ⅰ.3.12) 表示卷 1 中的公式 (3.12).标有 $*$ 的问题更具有挑战性.

本书的目录为：

第 6 部分：测度和积分论

第 7 部分：复变量的复值函数

25. 留数定理

26. Γ – 函数,ζ – 函数与 Dirichlet 级数

27. 椭圆积分与椭圆函数

28. Riemann 映射定理

29. 多变量中的幂级数

本书共分成两大部分,首先是测度论和积分论,相当于我国大学教材中的实变函数论. 它是数学分析的继续. 比如 §8 中收敛理论的应用,在经典的分析中讨论积分时,我们也用到收敛的理论.

举个例子,在分析中有一个定理,如下:

定理　若 $\iint\limits_{D} f(x,y)\,\mathrm{d}x\mathrm{d}y$ 收敛,则 $\iint\limits_{D} f(x,y)\,\mathrm{d}x\mathrm{d}y$ 也绝对收敛.

证明　用反证法. 参见 [菲]《微积分学教程》三卷一分册(第 223 页).

取 $\{D_n\}: D_1 \subset D_2 \subset \cdots \subset D_n \subset \cdots$ 且 $\bigcup\limits_{n=1}^{\infty} D_n = D$,而

$$\lim_{n\to+\infty} \iint\limits_{D_n} \mid f(x,y) \mid \mathrm{d}x\mathrm{d}y = +\infty$$

不失一般性,可假定对每一个 n,不等式

$$\iint\limits_{D_{n+1}} \mid f(x,y) \mid \mathrm{d}x\mathrm{d}y > 3\iint\limits_{D_n} \mid f(x,y) \mid \mathrm{d}x\mathrm{d}y + 2n$$

成立. 否则可取子列,记 $C_n = D_{n+1} - D_n$,则有

$$\iint\limits_{C_n} \mid f(x,y) \mid \mathrm{d}x\mathrm{d}y > 2\iint\limits_{D_n} \mid f(x,y) \mid \mathrm{d}x\mathrm{d}y + 2n$$

但 $\mid f(x,y) \mid = f_+(x,y) - f_-(x,y)$,故

$$\iint\limits_{C_n} \mid f_+(x,y) \mid \mathrm{d}x\mathrm{d}y = \iint\limits_{C_n} f_+(x,y)\,\mathrm{d}x\mathrm{d}y + \iint\limits_{C_n} f_-(x,y)\,\mathrm{d}x\mathrm{d}y$$

设 $\iint\limits_{C_n} f_+(x,y)\,\mathrm{d}x\mathrm{d}y > \iint\limits_{C_n} f_-(x,y)\,\mathrm{d}x\mathrm{d}y$,则

$$\iint\limits_{C_n} f_+(x,y)\,\mathrm{d}x\mathrm{d}y > \iint\limits_{D_n} f(x,y)\,\mathrm{d}x\mathrm{d}y + n$$

将左端的二重积分以 Darboux 下和代替,成立

$$\sum_{i\in\mathbf{Z}} m_n^{(i)} C_n^{(i)} > \iint\limits_{D_n} \mid f(x,y) \mid \mathrm{d}x\mathrm{d}y + n$$

在和式中只留下与 $m_n^{(i)} > 0$ 相对应的项,以 $\widetilde{C}_n$ 表示对应元素 $C_n^{(i)}$ 的集合,有

$$\iint\limits_{C_n} f(x,y)\,\mathrm{d}x\mathrm{d}y = \iint\limits_{C_n} f_+(x,y)\,\mathrm{d}x\mathrm{d}y > \iint\limits_{D_n} |f(x,y)|\,\mathrm{d}x\mathrm{d}y + n$$

记 $\widetilde{D}_n = D_n \cup \widetilde{C}_n$,由于 $\iint\limits_{D_n} f(x,y)\,\mathrm{d}x\mathrm{d}y > -\iint\limits_{D_n} |f(x,y)|\,\mathrm{d}x\mathrm{d}y$,将上述

两个不等式相加,得 $\iint\limits_{\widetilde{D}_n} f(x,y)\,\mathrm{d}x\mathrm{d}y > n$. 可考虑改变 $\widetilde{C}_n$ 及 $\widetilde{D}_n$,使得有一

连通区域 D_n',其面积与 $\widetilde{D}_n$ 相差之小,使得不等式 $\iint\limits_{D_n'} f(x,y)\,\mathrm{d}x\mathrm{d}y > n$ 成

立. 这样反常重积分 $\iint f(x,y)\,\mathrm{d}x\mathrm{d}y$ 就不收敛了.

由定理可知,二重积分绝对收敛与收敛是等价的.
瑞典著名数学家 L. 戈丁曾这样介绍:

1　$\mathbb{R}^n$ 中的积分和测度[①]

在 $\mathbb{R}^n$ 中的一个区间上对函数 1 进行积分,结果就得出这个区间的 n 维体积,而进行变量更换就引出一个行列式来. 如果把积分看作是从 C_0 函数映到数上的函数,那么积分便是线性而且单调递增的.

(1) $\mathbb{R}^n$ 中的 Riemann 积分,变量的更换.

假设 f 是 n 维区间 $I: a_1 \leqslant x_1 \leqslant b_1, \cdots, a_n \leqslant x_n \leqslant b_n$ 上的连续函数. 我们可以通过 n 次积分而得出 f 在 I 上的积分,例如,当 $n = 3$ 时

$$\int_I f(\boldsymbol{x})\,\mathrm{d}\boldsymbol{x} = \int_{a_1}^{b_1} \Big[\int_{a_2}^{b_2} \Big(\int_{a_3}^{b_3} f(x_1,x_2,x_3)\,\mathrm{d}x_3 \Big) \mathrm{d}x_2 \Big] \mathrm{d}x_1 \tag{1}$$

一般情形也是一样. 这里左边只是右边的一种记号. 这个公式是合理的,公式右边的值与进行积分的顺序无关. 当 $f = 1$ 时,积分值就等于 $(b_1 - a_1)\cdots(b_n - a_n)$,假如采用标准正交化的坐标系,那么这就是 I 的 n 维体积($n = 2$ 时就是面积),可得

---

① 摘自《数学概观》,L. 戈丁著,胡作玄译,科学出版社,2001.

$$f \leqslant g \Rightarrow \int_I f(\boldsymbol{x}) \, \mathrm{d}\boldsymbol{x} \leqslant \int_I g(\boldsymbol{x}) \, \mathrm{d}\boldsymbol{x} \tag{2}$$

$$\int_I (A f(\boldsymbol{x}) + B g(\boldsymbol{x})) \, \mathrm{d}\boldsymbol{x} = A \int_I f(\boldsymbol{x}) \, \mathrm{d}\boldsymbol{x} + B \int_I g(\boldsymbol{x}) \, \mathrm{d}\boldsymbol{x} \tag{3}$$

换句话说,积分是被积函数 f 的单调递增的线性函数. 如果 f 是 C_0 函数,即 f 连续而且当和式 $|x_1| + \cdots + |x_n|$ 充分大时 $f(\boldsymbol{x}) = 0$,那么积分(1)的值就与 I 无关,但设 I 很大,使得在 I 外面 $f = 0$. 于是我们就可以把所述积分写成 $\int_{\mathbb{R}^n} f(\boldsymbol{x}) \, \mathrm{d}\boldsymbol{x}$,或者 $\int f(\boldsymbol{x}) \, \mathrm{d}\boldsymbol{x}$. 此时性质(2)和(3)仍然成立.

自然我们也可在并不是区间的其他区域上进行积分. 我们要在 $\mathbb{R}^n$ 的开子集 V 上对连续函数进行积分. 我们分两步来定义相应的积分. 第一步,设 $C_0(V)$ 为在 V 的紧子集外面等于零的 C_0 函数所构成的集合,并且当 $f \geqslant 0$ 是连续函数时,定义

$$\int_V f(\boldsymbol{x}) \, \mathrm{d}\boldsymbol{x} = \sup \int g(\boldsymbol{x}) \, \mathrm{d}\boldsymbol{x}$$

其中

$$g \in C_0(V), 0 \leqslant g \leqslant f \tag{4}$$

如果 f 是 $C_0(V)$ 中的函数,则当 $g = f$ 时即可达到上确界,从而并不能得到新结论. 在一般情形下,右边可以是正无穷大.

第二步,我们对可以变号的函数 $f(\boldsymbol{x})$ 进行积分,不过它必须是绝对可积的,也就是 $\int_V |f(\boldsymbol{x})| \, \mathrm{d}\boldsymbol{x} < \infty$. 此时令

$$\int_V f(\boldsymbol{x}) \, \mathrm{d}\boldsymbol{x} = \int_V f_+(\boldsymbol{x}) \, \mathrm{d}\boldsymbol{x} - \int_V f_-(\boldsymbol{x}) \, \mathrm{d}\boldsymbol{x} \tag{5}$$

式中 $f_+(\boldsymbol{x}) = \max\{f(\boldsymbol{x}), 0\}$ 是 $f(\boldsymbol{x})$ 的正值部分,而 $f_-(\boldsymbol{x}) = \max\{-f(\boldsymbol{x}), 0\}$ 是 $f(\boldsymbol{x})$ 的负值部分再变号. 因为 $|f| = f_+ + f_-$,所以右边两项皆为有限;又因为 $f = f_+ - f_-$,所以(5)对于 $C_0(V)$ 函数成立. 不难证明,积分(5)具有性质(2)和(3).

按照我们上面选取的记号,在(1)中进行变量代换以后的公式,看起来就像单个变量情形中的相应公式,只不过现在加上一个表示行列式的记号"det",有

$$\int_{h(V)} f(\boldsymbol{x}) \, \mathrm{d}\boldsymbol{x} = \int_V f \circ h(\boldsymbol{y}) \, |\det h'(\boldsymbol{y})| \, \mathrm{d}\boldsymbol{y} \tag{6}$$

式中 h 是由 V 到 $h(V)$ 的一个 C^1 双映射,其 Jacobi 矩阵为 $h'(\boldsymbol{y}) =$

$(\partial_k h_j(\boldsymbol{y}))$. 下面我们只能简要地叙述一下这个相当麻烦的证明. 假如 h 只改变一个变量, 也就是说, 除一个 k 之外(比如 $k = n$), 都有 $h_k(\boldsymbol{y}) = y_k$, 而 $V = \mathbb{R}^n$. 我们把右边写成

$$\cdots\left(\int f\circ h(\boldsymbol{y}) \mid \partial_n h_n(\boldsymbol{y}) \mid \mathrm{d}y_n\right)\cdots$$

然后在内层积分中把变量换成 $x_n = h_n(\boldsymbol{y})$, 即得所需的结果. 如果 h 正好把变量作一个置换, 那么公式(6)也成立. 重复进行这两类变量更换, 那么至少当 $f\circ h$ 在 V 上任意选定的某一小块的外面等于零的情况下, 可以推出一般的公式. 因此, 如果把使得已知函数恒等于零的最大开区域的余集称为函数的支集, 那么我们现在已经知道, 对于具有小支集的函数, (6)成立. 再进一步, 我们就要用单位分解了. 所谓单位分解, 就是指一串 $C_0(V)$ 函数 $g_1, g_2, \cdots$, 其和为1, 并且 V 的每个紧子集都至多和有限个这种函数的支集相交. 这个概念的要点在于, 它可以和 V 的开覆盖联系起来. V 的开覆盖的定义, 就是一些开集 W 所成的族 (W), 这些开集合并起来覆盖 V. 我们说一个单位分解从属于某一开覆盖, 如果对于每个 g_k, 都存在族中的一个开集 W_k, 使得 g_k 属于 $C_0(W_k)$, 即 g_k 在 W_k 的一个紧子集外面等于0, 如图1所示.

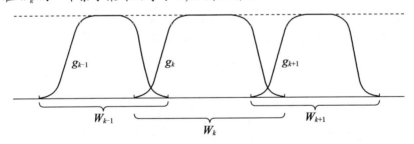

图1　$\mathbb{R}$ 上的单位分解, 和 $g_1 + g_2 + \cdots$ 处处等于1,
每个 g_k 在相应的区间的一个紧子集 W_k 外等于零

我们有:

定理(单位分解)　$\mathbb{R}^n$ 中一个开集的任何开覆盖, 都有一个从属的单位分解.

证明并不困难, 但是证明过程太长, 所以这里不讲了. 现在我们来看一下怎么应用这个定理. 假设我们已经证明, V 的每个点都具有一个开邻域 W, 使得当 $f\circ h$ 是 $C_0(W)$ 函数时, (6)成立, 那么我们就有一个 V

的开覆盖. 假设 $g_1, g_2, \cdots$ 是它的从属单位分解, 并令 $f \in C_0(V)$, 我们考虑公式(6). 把 f 用 $f_k(\boldsymbol{x}) = g_k(h^{-1}(\boldsymbol{x}))f(\boldsymbol{x})$ 来代替, 则 $f_k \circ h(\boldsymbol{y}) = g_k(\boldsymbol{y})f \circ h(\boldsymbol{y})$ 有一个很小的支集, 使得(6)对于 f_k 成立. 把相应的公式加在一起, 并且注意到 $f(\boldsymbol{x}) = f_1(\boldsymbol{x}) + f_2(\boldsymbol{x}) + \cdots$, 其中至多只有有限项不等于零, 这样就对 f 证明了(6).

公式(6)要用到微分, 可知

$$\mathrm{d}x_1 \wedge \cdots \wedge \mathrm{d}x_n = \det h'(\boldsymbol{y})\mathrm{d}y_1 \wedge \cdots \wedge \mathrm{d}y_n$$

取绝对值, 并令

$$\mathrm{d}\boldsymbol{x} = \mathrm{d}x_1 \cdots \mathrm{d}x_n = | \mathrm{d}x_1 \wedge \cdots \wedge \mathrm{d}x_n | \tag{7}$$

则得 $\mathrm{d}\boldsymbol{x} = | \det h'(\boldsymbol{y}) | \mathrm{d}\boldsymbol{y}$. 这样就把(6)归结为一个恒等式.

现在令 y 表示二维和三维极坐标且应用公式(6). 首先, 令 $y = r, \theta$ 为平面上的极坐标: $x_1 = r\cos\theta, x_2 = r\sin\theta$, 并用 V 代表区间 $\theta_1 < \theta < \theta_2$. 于是区域 $U = h(V)$ 就是圆环 $r_1^2 < x_1^2 + x_2^2 < r_2^2$ 中的一个扇形, 从而得

$$\int_U f(x_1, x_2)\mathrm{d}x_1\mathrm{d}x_2 = \int_V f(r\cos\theta, r\sin\theta)r\mathrm{d}r\mathrm{d}\theta$$

当 $f = 1$ 时, 上式右边等于 $(\theta_2 - \theta_1)(r_2^2 - r_1^2)/2$, 这就是扇形的面积. 由空间的极坐标可得一个相应的公式

$$\int_U f(x_1, x_2, x_3)\mathrm{d}x_1\mathrm{d}x_2\mathrm{d}x_3$$

$$= \int_V f(r\cos\varphi\cos\theta, r\cos\theta\sin\varphi, r\sin\theta)r^2\cos\theta\mathrm{d}r\mathrm{d}\theta\mathrm{d}\varphi$$

假如 V 是区间 $r_1 < r < r_2, \theta_1 < \theta < \theta_2, \varphi_1 < \varphi < \varphi_2$, 而且 $f = 1$, 则上式右边就成为

$$\frac{1}{3}(r_2^3 - r_1^3)(\sin\theta_2 - \sin\theta_1)(\varphi_2 - \varphi_1)$$

这就是球壳 $r_1^2 < x_1^2 + x_2^2 + x_3^2 < r_2^2$ 上的相应一块的体积. 最后, 我们来计算 $\mathbb{R}^n$ 中的单位球体 $E: x_1^2 + \cdots + x_n^2 < 1$ 的 n 维体积, 也就是积分

$$\int_E \mathrm{d}x_1 \cdots \mathrm{d}x_n = 2^n\int_V \mathrm{d}x_1 \cdots \mathrm{d}x_n$$

其中 V 是 E 中满足 $x_1 > 0, \cdots, x_n > 0$ 的部分. 把变量更换为 $t_k = x_k^2$, 就得出 $2\mathrm{d}x_k = t_k^{-1/2}$. 从而右边等于

$$\int_U (t_1 \cdots t_n)^{-1/2}\mathrm{d}t_1 \cdots \mathrm{d}t_n$$

其中 U 是区域 $t_1 > 0, \cdots, t_n > 0, t_0 = 1 - (t_1 + \cdots + t_n) > 0$. 这是积分

$$\int_U t_0^{\alpha_0-1} t_1^{\alpha_1-1} \cdots t_n^{\alpha_n-1} \mathrm{d}t_1 \cdots \mathrm{d}t_n = \frac{\Gamma(\alpha_0 + \cdots + \alpha_n)}{\Gamma(\alpha_0) \cdots \Gamma(\alpha_n)} \tag{8}$$

的特殊情形,其中 $\alpha_0 > 0, \cdots, \alpha_n > 0$,这个公式是 Euler 公式的推广. 所以要求的体积就等于

$$\Gamma\left(\frac{1}{2}\right)^n \Big/ \Gamma\left(\frac{n}{2} + 1\right)$$

要证明(8),可用归纳法,作变量替换使得 $t_1 = s(1 - t_2 - \cdots - t_n)$,然后对 s 进行积分.

(2) Stieltjes 积分及测度.

我们已经指出,把 Riemann 积分看作是由 C_0 函数到数的映射 $f \rightarrow L(f)$ 时,它是单调递增且是线性的. 换句话说,对于所有数 A,B 及 C_0 函数 f,g,都有

$$f \geqslant 0 \Rightarrow L(f) \geqslant 0, L(Af + Bg) = AL(f) + BL(g) \tag{9}$$

如果我们进一步追问,具有这种一般性质的任意映射 L 是否有任何有趣的性质,那么我们对于什么是积分就会有更好的理解. 倘若我们回到公式(4)而利用它使 L 定义一个测度,那么我们就得到肯定的答案. 说得更确切些,假设给定了具有性质(9)的 L,并令

$$m(E) = \sup L(f),\ \text{其中}\ 0 \leqslant f \leqslant \chi_E\ \text{且}\ f\ \text{是}\ C_0\ \text{函数} \tag{10}$$

这里 E 是 $\mathbb{R}^n$ 中的任何开集,χ_E 是 E 的示性函数,它在 E 上取值 1 而在 E 外面取值 0. 这样就定义了一个由开集到大于或等于 0 的数(包含 $+\infty$)的函数 $E \rightarrow m(E)$,称为测度,它显然具有下列三条性质(箭头 $\uparrow$ 表示"递增地趋近于极限")

$$m(E) \geqslant 0, E\ \text{有界} \Rightarrow m(E) < \infty \tag{11a}$$

$$E' \uparrow E \Rightarrow m(E') \uparrow m(E) \tag{11b}$$

$$E \cap E' = \varnothing \Rightarrow m(E \cup E') = m(E) + m(E') \tag{11c}$$

对于任意的集合 F,定义

$$m(F) = \inf m(E),\ \text{其中}\ E\ \text{是包含}\ F\ \text{的开集}. \tag{12}$$

现在反过来,假设已经给定一个测度 m,它具有性质(11). 我们要定义 C_0 函数关于 m 的积分. 此时先把 $\mathbb{R}^n$ 分成不相交的有界区间 $I_1, I_2, \cdots$ 的族 (I),使得任何紧集都至多与其中有限个区间相交,然后考虑相应的 Riemann 和

$$\sum f(\xi_k) m(I_k),\ \text{对于所有的}\ k, \xi_k\ \text{都属于}\ I_k \qquad (13)$$

因为 f 具有紧支集,所以这里只有有限个项不等于零. 正如 Riemann 积分的情形一样,我们证明当这一族区间中的最大距离趋近于零时,这个和有一个极限,这个极限用

$$\int f(\boldsymbol{x})\,\mathrm{d}m(\boldsymbol{x})$$

表示,称为 f 关于 m 的 Stieltjes 积分. 显然它具有性质(9),此时我们可以问:假如测度由(10)定义,那么这个积分是否等于我们开始所讲的那个映射 L?回答是肯定的,下面的结果主要归功于 F. 黎斯.

定理 具有性质(9)的任何线性映射,都是关于由 L 生成的测度(10)的 Stieltjes 积分.

证明并不困难,至少 $n=1$ 时不难. 考虑以所有整数为下标的严格递增数列 x_k,把 $\mathbb{R}$ 分成 x_k 与 x_{k+1} 之间的区间 I_k,点 x_k 属于 I_{k-1} 或 I_k. 然后构造出单位分解 (g_k),使得 g_k 在 I_k 外部以及不属于 I_k 的端点处等于零,而在其他端点处 $g_k=1$. 由(10)与(12)可以推知,这种单位分解存在,并且我们可以这样选取这种分解,使得对于所有的 $k, L(g_k)$ 都与(10)和(12)所定义的 $m(I_k)$ 任意接近. 因此,Riemann 和(13)与 $\sum f(\xi_k) L(g_k)$ 的差可以取得任意小. 另外,因为 $L(f) = \sum L(fg_k)$,所以

$$|L(f) - \sum f(\xi_k) L(g_k)| \leqslant \sum |L(fg_k) - f(\xi_k) L(g_k)| \leqslant \sum{}' \delta_k L(g_k)$$

其中 δ_k 是 f 在 g_k 的支集上的最大值与最小值的差,最后的和取遍使得 g_k 的支集与 f 的支集相交的 k. 再作更细致的分割,使得 g_k 的支集与 I_k 充分接近,最大的 δ_k 趋近于零而 $\sum{}' L(g_k)$ 保持有界. 这样就在 $n=1$ 的情形下证明了定理. 当 $n>1$ 时,证明也是类似的.

在我们讲完这部分相当抽象的数学之后,现在该是构造一些测度和 Stieltjes 积分的例子的时候了. 假设 $g(\boldsymbol{x}) \geqslant 0$ 是连续函数,$\{\boldsymbol{x}^{(k)}\}$ 是 $\mathbb{R}^n$ 中的点列,而 $\{a_k\}$ 是一串大于或等于 0 的数,使得只要 $\boldsymbol{x}^{(k)}$ 属于 $\mathbb{R}^n$ 中的某个紧集,相应的和 $\sum a_k$ 就总是有限的,那么

$$L(f) = \int f(\boldsymbol{x}) g(\boldsymbol{x})\,\mathrm{d}\boldsymbol{x} + \sum a_k f(\boldsymbol{x}^{(k)})$$

便是 Stieltjes 积分. 显然,此时对应的测度是

$$m(E) = \int_E g(\boldsymbol{x})\,\mathrm{d}\boldsymbol{x} + \sum{}' a_k$$

其中的和取遍所有使得 $\boldsymbol{x}^{(k)}$ 属于 E 的 k. 这个例子虽然不是最一般的,却非常有代表性. 当 $g = 0$ 时,我们就说上述测度是离散的;当 $a_k = 0$ 时,就说它具有密度 g. 当 $n = 1$ 时,每个测度 m 都可以由一个这种非增函数 $h(x)$ 来表示:当 $x \geqslant 0$ 时,它定义为 $h(x) = m(I_x)$,而当 $x < 0$ 时,则定义为 $h(x) = -m(J_x)$,这里 I_x 是区间 $0 < t \leqslant x$, J_x 是区间 $x < t \leqslant$ 0. 于是 Stieltjes 积分 $\int f(x)\,\mathrm{d}m(x)$ 便具有 Riemann 和

$$\sum f(\xi_j)\big[h(x_{j+1}) - h(x_j)\big]$$

它也可以写成 $\int f(x)\,\mathrm{d}h(x)$,这是 Stieltjes 所使用的形式. 黎斯定理是 1909 年以后黎斯发现的,他所考虑的是一个区间 $a \leqslant x \leqslant b$ 上所有连续实函数 f 所成的线性空间 C,以及由 C 到实数的线性映射 $f \to L(f)$,黎斯定理满足:对于所有的 f 及某个与 f 无关的 c, $|L(f)| \leqslant c \max |f(x)|$ 成立. 黎斯证明,每一个这样的 L 都是两个 Stieltjes 积分的差

$$L(f) = \int_a^b f(x)\,\mathrm{d}h_1(x) - \int_a^b f(x)\,\mathrm{d}h_2(x) \tag{14}$$

其中 h_1, h_2 为不减的函数,如果当 $f \geqslant 0$ 时有 $L(f) \geqslant 0$,那么便可以取 $h_2 = 0$. 我们注意一下后面这种情形,此时条件 $|L(f)| \leqslant c|f|$ 成立,其中 $|f| = \max |f(x)|$, $c = L(1)$,这是因为,由 $|f| \geqslant f(x) \geqslant -|f|$ 可以推出 $|f|L(1) \geqslant L(f) \geqslant -|f|L(1)$. 用现代的术语来陈述,黎斯定理就是说,Banach 空间 C 上的任何连续线性泛函都有(14)的形式,反过来也是如此. 这样陈述的定理,是泛函分析中最早的也是最基本的定理之一.

(3)Lebesgue 积分及其他各种积分.

到现在为止,为了简单起见,我们大多都只是求了连续函数的积分,但我们也可以走得更远一些. 对于 Stieltjes 积分,可以施行单调地取极限的步骤,这样就得出 Lebesgue-Stieltjes 积分. 或者,倘若从 Riemann 积分出发,那么通过上述步骤就得出 Lebesgue 积分. 说得更确切些,我们从 C_0 函数集上的性质(9)的映射 $L(f)$ 出发,当 f 是 C_0 函数的递增序列(递减序列)$\{f_k\}$ 的逐点极限时,就令 $L(f) = \lim L(f_k)$,这样便推广了积分 L,此时 f 称为下(上)半连续函数. 任意一个函数 f 称为可积函

772

数,假如下面的条件成立:f 介于两个半连续函数之间(一样一个,其中比较大的一个为下半连续,另外一个为上半连续),而且它们的积分可以相差得任意小.此时积分 $L(f)$ 便自然而然有定义.事实证明,可积函数空间是线性空间,而将 L 看作这个空间上的映射时,它仍然具有性质 (9).这种积分的表述是 Lebesgue(1900) 所做的,它的优点在于:为使控制收敛定理成立,只须逐点收敛的条件成立就够了,而不需要局部一致收敛的条件.

积分还有其他的方面,例如,我们可以从所谓微积分的基本定理出发,这个定理可以表述为

$$f(b) - f(a) = \int_a^b f'(x)\,\mathrm{d}x \tag{15}$$

如果 $a = x_0 < x_1 < \cdots < x_n = b$ 是一个划分,而且 f' 存在,则由中值定理即知

$$f(b) - f(a) = \sum_{j=0}^{n} (f(x_{j+1}) - f(x_j))$$
$$= \sum_{j=0}^{n} f'(\xi_j)(x_{j+1} - x_j)$$

是式(15)右边的 Riemann 和,因此,当 f' 是 Riemann 可积时,等式(15)成立.但导数 f' 可以是相当复杂的函数,甚至当它有界的时候也是这样,此时它不一定是 Riemann 可积的,因而(15)的右边没有意义.另外,在 Lebesgue 的理论中,有界导数总是可积的,而且(15)成立.但是,即便如此也并不是万事大吉.例如,倘若 f 单调而且连续,除一个非常小的集合——所谓零集——以外,f' 都存在,而且 f' 按照 Lebesgue 意义是可积的,但是(15)的左边可以比右边大(当 f 是递增函数时).存在这样的积分,它能使(15)成立,或者干脆以(15)为定义.但 Lebesgue 积分是数学分析中的标准积分,它将来肯定仍然会保持这种地位.

2   流形上的积分

在导致微积分产生的问题当中,有曲线弧长和曲面面积的计算.这两个问题都归结为积分的问题.事实证明,弧 $y = f(x)$ 在 $x = a$ 及 $x = b$ 之间的长度公式为

$$\int_a^b \mathrm{d}s \tag{16}$$

其中 $\mathrm{d}s = (\mathrm{d}x^2 + \mathrm{d}y^2)^{1/2} = (1 + (f'(x))^2)^{1/2}\mathrm{d}x$ 是所谓弧元素. 它的思想出发点是: 非常小的一段弧十分接近于直线段, 从而可以对它应用 Pythagoras 定理. Leibniz 的巧妙的记号, 把这个公式同我们的直觉完善地协调起来. 而后来人们所做的事, 只不过是把弧长定义为逼近所给弧的折线长度的上确界.

下面我们要把公式(16)推广到高维情形以及流形上, 但我们事先做一点准备: 在流形上求密度的积分, 另外还要对 Riemann 几何做一点说明. 然后, 我们要谈到定向流形上的微分形式的积分, 并且作为应用, 给出 Green 公式和 Stokes 公式的证明. 尽管这些公式表面看来颇为惊人, 但实际上它们只不过是微积分基本定理的推广, 所谓基本定理, 无非就是对于 C^1 函数 $f(x)$, 我们有

$$\int_a^b f'(x)\,\mathrm{d}x = f(b) - f(a)$$

(1) 密度的积分.

在讲 Stieltjes 积分时, 我们已经把 $\mathbb{R}^n$ 中的连续密度 $g(x) \geqslant 0$ 的积分定义为 C_0 函数集上的映射 $f \to L(fg) = \int f(x)g(x)\,\mathrm{d}x$. 现在也要考虑 p 流形 M 上的连续密度的积分. 所谓 f 是 M 上的 C_0 函数, 就是指 f 是连续函数并且具有紧支集, 也就是说, 在 M 的一个紧子集外部等于零. 我们考虑 C_0 函数集上的这种线性映射 $f \to L(f)$, 使得对于 M 的每个图 h, V, 如果 f 在附图区域 $h(V)$ 的一个紧子集外部等于 0, 那么

$$L(f) = \int f_V(t) H_V(t)\,\mathrm{d}t,\ \mathrm{d}t = \mathrm{d}t_1 \cdots \mathrm{d}t_p$$

这里, 当 $x = h(t)$ 时 $f_V(t) = f(t)$, 而且 $H_V(t) \geqslant 0$ 是连续函数. 当 f 的支集属于两个附图区域的交集 $h(V) \cap h'(V')$ 时, 我们就有

$$\int f_V(t) H_V(t)\,\mathrm{d}t = \int f_{V'}(t') H_{V'}(t')\,\mathrm{d}t'$$

其中当 $x = h(t) = h'(t')$ 时, 即有 $f_V(t) = f_{V'}(t') = f(x)$. 为了使这个等式能够适合积分中的变量更换公式(6)和(7), 对于所有的图 h, V 和 h', V', H_V 和 $H_{V'}$ 必须通过下面的公式相联系

$$h(t) = h'(t') \Rightarrow H_V(t)\,|\,\mathrm{d}t_1 \wedge \cdots \wedge \mathrm{d}t_p\,|$$
$$= H_{V'}(t')\,|\,\mathrm{d}t'_1 \wedge \cdots \wedge \mathrm{d}t'_p\,| \tag{17}$$

因此,流形上的密度便由大于或等于 0 的连续函数集合 (H_V) 所给出,使得每个图都有一个连续函数,使 (17) 成立. 特别地,如果 $\omega(\boldsymbol{x})$ 是 M 上的 p 形式,那么便可以令 $H_V(\boldsymbol{t}) = |\, g_V(\boldsymbol{t}) \,|$,此处 $\omega(\boldsymbol{x})$ 用图 (h, V) 的变量表示时,即有 $\omega(\boldsymbol{x}) = g_V(\boldsymbol{t}) \mathrm{d}t_1 \wedge \cdots \wedge \mathrm{d}t_p$.

由于密度可以相加,因此不难构造密度. 当 h, V 是已给的图时,由任意的 $C_0(h(V))$ 函数 H_V 都可以通过一定条件给出一个密度,为此,只要当 $h(V) \cap h'(V')$ 是空集或者当 $h'(t') = h(t)$ 且 $H_V(t) = 0$ 时,令 $H_V(t) = 0$ 即可. 同样的方法有时也用来构造微分形式. 为了对已知密度 (H_V) 构造积分 $L(f)$,我们要用单位分解,也就是可数个大于或等于 0 的 C_0 函数所成的集 (g_k),这些 g_k 的和为 1,使得每个 g_k 的支集都包含在一个附图区域之中,并且最多只有有限个支集与 M 的给定紧集相交. 我们以前的 $\mathbb{R}^n$ 中的单位分解定理,对于流形也成立. 但我们不得不略去证明. 特别地,每个图册都有一个从属的单位分解,于是一个已知密度的积分就定义为

$$L(f) = \sum L(g_k f)$$

其中 (g_k) 是一个单位分解,g_k 的支集包含在某个附图区域 $h_k(V_k)$ 中,而且

$$L(g_k f) = \int (g_k f)_{V_k}(\boldsymbol{t}) H_{V_k}(\boldsymbol{t}) \mathrm{d}\boldsymbol{t}$$

因为按照假设 f 具有紧支集,所以和式中至多有有限个项不等于 0. 如果 (g'_j) 是另外一个单位分解,而 $L'(f) = \sum L'(g'_j f)$ 为相应的积分,则由 (17) 即可证明,对于所有的 j 与 k,有 $L(g_k g'_j f) = L'(g_k g'_j f)$. 这样,如果对 j 求和,则因 $\sum g'_j = 1$,所以 $L(g_k f) = L'(g_k f)$;再对 b 求和,则因 $\sum g_k = 1$,所以 $L(f) = L'(f)$. 因此,积分就不依赖于我们所用的单位分解. 下面讲几个古典的密度的例子.

(2) 弧元和面积元.

令 $\boldsymbol{x} = (x_1, \cdots, x_n)$ 为 $\mathbb{R}^n$ 中的标准正交坐标系,因而 $\boldsymbol{x}, \boldsymbol{y}$ 两点的距离平方便是 $|\, \boldsymbol{x} - \boldsymbol{y} \,|^2 = (x_1 - y_1)^2 + \cdots + (x_n - y_n)^2$. $\mathbb{R}^n$ 中一条 C^1 曲线 γ 的弧元,是由

$$\begin{aligned} \mathrm{d}s &= (\mathrm{d}x_1^2 + \cdots + \mathrm{d}x_n^2)^{1/2} = |\, \boldsymbol{x}'(t) \,| \, \mathrm{d}t \\ &= ((x_1'(t))^2 + \cdots + (x_n'(t))^2)^{1/2} \mathrm{d}t \end{aligned} \tag{18}$$

所定义的密度. 这里第一个等式是纯粹记号上的, 关于图 $t \rightarrow \boldsymbol{x}(t)$ 的密度 $|\boldsymbol{x}'(t)|$ 出现于最后一项中. 注意, 假如 $s \rightarrow \boldsymbol{y}(s) = \boldsymbol{x}(t)$ 是另外一个图, 则 $|\boldsymbol{x}'(t)| = |\boldsymbol{y}'(s)| \left| \dfrac{\mathrm{d}s}{\mathrm{d}t} \right|$, 所以 $|\boldsymbol{x}'(t)|$ 的确是一个密度. 为了看出 $\int_I \mathrm{d}s$ 就是 γ 的弧 I 的长度, 我们注意到, 假如 I 是联结 $\boldsymbol{y}$ 与 $\boldsymbol{z}$ 的直线段, 其方程为 $\boldsymbol{x}(t) = \boldsymbol{y} + t(\boldsymbol{z} - \boldsymbol{y})(0 \leqslant t \leqslant 1)$, 则 $\mathrm{d}s = |\boldsymbol{z} - \boldsymbol{y}| \, \mathrm{d}t$, 因而积分就等于 $|\boldsymbol{z} - \boldsymbol{y}|$.

作为例子, 我们考虑 xOy 平面上的椭圆 $\dfrac{x^2}{a^2} + \dfrac{y^2}{b^2} = 1$. 用图 $x = a(1 - t^2), y = bt(-1 < t < 1)$ 来表示, 它的弧元就是公式

$$F(u) = \int_0^u \left[(a^2 - b^2)t^2 + b^2 \right]^{1/2} (1 - t^2)^{-1/2} \mathrm{d}t$$

中积分号下的表达式, 而这个积分本身就是所给椭圆在 $0 < t < u$ 上的一段的弧长. 利用极坐标 $x = a\cos\theta, y = b\sin\theta$, 这个公式可以改写为

$$F(u) = \int_0^{\arcsin u} (a^2\sin^2\theta + b^2\cos^2\theta)^{1/2} \mathrm{d}\theta$$

当椭圆退化为圆, 即 $a = b$ 时, 我们当然可以明显地算出这个积分而得到 $F(u) = a\arcsin u$. 但是对于真正的椭圆来说, 弧长 $F(u)$ 不能够用传统的初等函数来表示. 与面积 πab 不一样, 椭圆的全长 $F(u)$ 不是 u, b 和 π 的简单表达式. 通过 Euler, Jacobi 和 Abel 的工作, 像 F 这种函数的研究已经促使数学中产生了一个完整的分支, 这就是椭圆函数论.

在标准正交坐标系 x, y, z 之下, $\mathbb{R}^3$ 中的 C^1 曲面 Γ 的面积元素为

$$\mathrm{d}S = ((\mathrm{d}y \wedge \mathrm{d}z)^2 + (\mathrm{d}z \wedge \mathrm{d}x)^2 + (\mathrm{d}x \wedge \mathrm{d}y)^2)^{1/2} \qquad (19)$$

把图 $x = x(u,v), y = y(u,v), z = z(u,v)$ 代入, 则得

$$\mathrm{d}S = |\boldsymbol{J}| \, \mathrm{d}u\mathrm{d}v, \ |\boldsymbol{J}| = |J_x^2 + J_y^2 + J_z^2|^{1/2}$$

这里 J_x, J_y, J_z 都是 Jacobi 行列式, 它们由 $\mathrm{d}y \wedge \mathrm{d}z = J_x\mathrm{d}u \wedge \mathrm{d}v, \cdots$ 来定义. 向量 $\boldsymbol{J} = (J_x, J_y, J_z)$ 是曲面的法线. Jacobi 行列式的性质表明, $|\boldsymbol{J}|$ 是密度. 倘若 Γ 是一个平行四边形, 其参数表示为 $(x, y, z) = (x_0 + ux_1 + vx_2, y_0 + uy_1 + vy_2, z_0 + uz_1 + vz_2)(0 \leqslant u, v \leqslant 1)$, 因而它的四个顶点就对应于 $(u,v) = (0,0), (1,0), (0,1), (1,1)$, 那么 $|J_x| = |y_2z_3 - y_3z_2|$ 就是 Γ 在 yOz 平面上的投影的面积, 其他的 Jacobi 行列式 $|J_y|, |J_z|$ 也是一样. 因此由初等几何即知, $|\boldsymbol{J}| = \int_\Gamma \mathrm{d}S$ 就是 Γ 的面

积. 这就是在一般情形下, 我们也把 $\int_{\Gamma} dS$ 作为 Γ 的面积的动机.

当 Γ 是单位球面 $x^2 + y^2 + z^2 = 1$ 时, 让我们来计算 dS. 选取具有坐标 z 与 θ 的图, 使得

$$x = r\cos\theta, y = r\sin\theta, r = (1 - z^2)^{1/2}$$

其中 $|z| < 1, 0 < \theta < 2\pi$, 我们就得出

$$dy \wedge dz = r\cos\theta d\theta \wedge dz$$
$$dz \wedge dx = r\sin\theta d\theta \wedge dz$$
$$dx \wedge dy = z d\theta \wedge dz$$

所以 $dS = dz d\theta$. 这个结果等价于 Archimedes 的结果: 两个相距为 h 的平行平面所截出的那部分球面, 其面积等于 $2\pi h$. 现在我们要求读者证明, 如果选取 x, y 为坐标, 那么抛物面 $2z = x^2 + y^2$ 的面积元素等于 $(1 + x^2 + y^2)^{1/2} dx dy$, 并且把抛物面满足 $x^2 + y^2 < R^2$ 的那块面积明显地计算出来.

最后让我们提一下, $\mathbb{R}^n$ 中 $n - 1$ 维的 C^1 流形的 $(n - 1)$ 维面积元, 由下面的公式给出

$$dS = (\sigma_1^2(\boldsymbol{x}) + \cdots + \sigma_n^2(\boldsymbol{x}))^{1/2} \qquad (20)$$

这个公式和 (19) 相似, 其中 $\sigma_1, \cdots, \sigma_n$ 是由

$$\sigma_k(\boldsymbol{x}) = (-1)^k dx_1 \wedge \cdots \wedge dx_{k-1} \wedge dx_{k+1} \wedge \cdots \wedge dx_n$$

所定义的 $(n - 1)$ 微分形式.

(3) Riemann 几何学.

假设 $M \subset \mathbb{R}^n$ 是 p 维的 C^1 流形, 而 h, V 为一个图. 将 $dx_i = \sum_{j=1}^{p} \partial_j h_i(\boldsymbol{t}) dt_j$ 代入 $\mathbb{R}^n$ 的弧元 $(dx_1^2 + \cdots + dx_n^2)^{1/2}$ 中去, 写出平方项, 然后求和, 再把最后的结果写成乘积形式 $dt_j dt_k = dt_k dt_j$ 的线性组合, 我们就得出流形 M 上相应的弧元, 即

$$ds = (\sum g_{jk}(\boldsymbol{t}) dt_j dt_k)^{1/2} \qquad (21)$$

其中 $(g_{jk}(\boldsymbol{t}))$ 是正定对称矩阵. 对于由 V 中的曲线 $u \to t(u)$ 提升到 M 上的曲线 $u \to h(t(u))$, 它的弧元可以由 ds 得出, 只要令 $dt_j = t'_j(u) du$ 即可. Riemann 在他 1854 年的著名论文《论几何学的基本假设》(On the Basic Hypotheses of Geometry) 中的思想, 就是通过事先给定的度量或者弧元 (21) 来测量流形上的距离. 在公式 (21) 中, 右端的括号内的式子

与所用的图有关,但是在两个叠交的图中的对应点上,要求

$$\sum g_{jk}(\boldsymbol{t})\,\mathrm{d}t_j\mathrm{d}t_k = \sum g'_{jk}(\boldsymbol{t}')\,\mathrm{d}t'_j\mathrm{d}t'_k$$

成立. 应用单位分解,通过简单的论据即可证明,任何流形都具有 Riemann 度量. 不难看出,对任意选取的度量,$(\det g_{jk}(\boldsymbol{t}))^{1/2}\mathrm{d}t_1\cdots\mathrm{d}t_p$ 都是一个密度. 我们也能够定义度量流形 M 的曲率的量,而不考虑它在某一 $\mathbb{R}^n$ 中的任何嵌入. 甚至于当度量是无定度量,亦即 (g_{jk}) 仅仅是一个对称矩阵时,这些公式仍然成立. 这就是广义相对论的数学上的出发点.

(4) 定向和微分形式的积分.

按照定义,当 a 和 b 互换位置时,积分

$$\int_a^b f(\boldsymbol{x})\,\mathrm{d}\boldsymbol{x}$$

改变符号. 到现在为止,在区间上的积分公式 (1) 中,我们一直假设 $a_1 < b_1,\cdots,a_n < b_n$,所以并没有应用这个规定. 我们的密度也假设大于或等于 0,因而当 $f \geqslant 0$ 时,积分 $L(f) \geqslant 0$. 在公式 (7) 中,通过在 $\mathrm{d}x_1 \wedge \cdots \wedge \mathrm{d}x_n$ 的两边加上绝对值的符号来表示. 现在我们要去掉这些符号而考虑区间 $I \subset \mathbb{R}^n$ 上的积分

$$L(f) = \int_I f(\boldsymbol{x})\,\mathrm{d}x_1 \wedge \cdots \wedge \mathrm{d}x_n$$

此时必须有一个约定,来固定积分的符号. 一种方法是用下面两种规则的某一种使区间定向,这就是在 I 上

$$\mathrm{d}x_1 \wedge \cdots \wedge \mathrm{d}x_n > 0 \text{ 或 } < 0 \qquad (22)$$

可以解释为,对于正的 f,分别有 $L(f) > 0$ 和 $L(f) < 0$. 在 (22) 中,改变因子的次序或者乘以不等于 0 的函数,就会使不等式的符号有相应的改变. 例如:$\mathrm{d}x_1 \wedge \mathrm{d}x_2 > 0, \mathrm{d}x_2 \wedge \mathrm{d}x_1 < 0$ 和 $-(1+x_1^2)\,\mathrm{d}x_1 \wedge \mathrm{d}x_2 < 0$,这三种定向是完全一样的.

p 维流形 M 可以利用在 M 上到处都不为零的 p 微分形式 τ 来定向,所谓到处都不为零就是指:如果 h, V 是一个图,而 $\tau(\boldsymbol{x}) = g(\boldsymbol{t})\mathrm{d}t_1\cdots\mathrm{d}t_p$,那么对所有的 $\boldsymbol{t}$,有 $g(\boldsymbol{t}) \neq 0$. 于是 $\tau(\boldsymbol{x}) > 0$ 这个规则就使得每个图都有定向,同时使整个流形定向. 如果流形 M 已经定向,我们就可以在它上面求 p 微分形式 ω 的积分. 如果 ω 的支集包含在附图区域 $h(V)$ 内,而且 $\omega = f(\boldsymbol{t})\mathrm{d}t_1 \wedge \cdots \wedge \mathrm{d}t_p$,我们就令

$$\int_M \omega(\boldsymbol{x}) = \int_V f(\boldsymbol{t})\mathrm{d}t_1 \wedge \cdots \wedge \mathrm{d}t_p$$

此处 V 通过 $\tau(\boldsymbol{x}) > 0$ 来定向,从而采用上面的记号时,有 $g(\boldsymbol{t})\,\mathrm{d}t_1 \wedge \cdots > 0$. 对于具有紧支集的 p 微分形式,我们可以用单位分解,其论证方法和密度的情形是一样的. 于是积分即为定义在 M 上具有紧支集的 p 微分形式所成的空间上的线性函数 $\omega \to \displaystyle\int_M \omega(\boldsymbol{x})$.

在一维、二维和三维的情形下,定向是一种非常直觉的观念,它可以通过许多方式来表示. 现在举几个例子.

C^1 曲线 γ 可以通过规则 $\mathrm{d}g > 0$ 来定向,这里 g 是曲线上的一个实 C^1 函数,使得到处都有 $\mathrm{d}g \neq 0$. 这可以在曲线上沿着 g 的增长方向画箭头来表示,如图 2 所示. 令 $\omega(\boldsymbol{x}) = f_1(\boldsymbol{x})\,\mathrm{d}x_1 + \cdots + f_n(\boldsymbol{x})\,\mathrm{d}x_n$ 为 γ 上的具有紧支集的连续 1 微分形式,并设 γ 已经定向. 此时线积分 $\displaystyle\int_\gamma \omega$ 有定义. 令 $\mathrm{d}s$ 为弧元素,选取坐标 $t \to \boldsymbol{x}(t)$ 使 $\mathrm{d}t > 0$,于是等式 $\boldsymbol{e}\,\mathrm{d}s = \mathrm{d}\boldsymbol{x} = \boldsymbol{x}'(t)\,\mathrm{d}t$ 便定义了曲线上的长度等于 1 的连续切向量. 我们也可以把积分写成密度的积分

$$\int_\partial (f_1 \mathrm{d}x_1 + \cdots + f_n \mathrm{d}x_n) = \int_\gamma (f_1 e_1 + \cdots + f_n e_n)\,\mathrm{d}s \qquad (23)$$

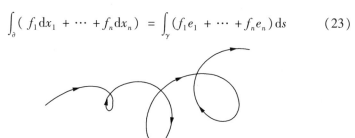

图 2 曲线的定向

在 $\mathbb{R}^2$ 中,我们可以利用长度为 1 的连续法向量 $\boldsymbol{v} = (v_1, v_2)$,把线积分表示为密度的积分,其公式为

$$\int_\gamma (P\mathrm{d}y - Q\mathrm{d}x) = \int_\partial (Pv_1 + Qv_2)\,\mathrm{d}s \qquad (24)$$

因为 $e_1 = v_2, e_2 = -v_1$ 定义了 γ 的一个切向量,故由 (23) 可以推知,除可能差一个符号以外,(24) 成立. 如果用 $v_1\mathrm{d}y - v_2\mathrm{d}x > 0$ 使 γ 定向,那么符号也是对的,这是因为,如果 $P = hv_1, Q = hv_2$,而且 $h \geqslant 0$,那么 (24) 的两边都大于或等于 0.

在 $\mathbb{R}^3$ 中,曲面 Γ 上的 2 微分形式的积分,也可以写成密度的积分,其

公式为

$$\int_{\Gamma} (Pdy \wedge dz + Qdz \wedge dx + Rdx \wedge dy) = \int_{\Gamma} (Pv_1 + Qv_2 + Rv_3) dS$$

(25)

此式与(24)很相像. 这里 dS 是面积元, $v = (v_1, v_2, v_3)$ 是 Γ 上的连续单位法向量, Γ 如此定向, 使得 $v_1 dy \wedge dz + \cdots \geqslant 0$. 利用单位分解, 只需在附图区域 $h(V)$ 之外 P, Q, R 等于零的情形下证明这个公式就行了. 但此时由(19)可知, 公式两边都等于

$$\int_{V} (PJ_x + QJ_y + RJ_z) dudv$$

其中可能差一个符号. 这里 $x = x(u,v)$ 是一个图, $dy \wedge dz = J_x dudv, \cdots$, 因而 $J = (J_x, J_y, J_z)$ 是 Γ 的法向量. 如果我们选取 (P,Q,R) 为 v 的非负倍数, 那么便可以推知(25)的符号也是对的.

最后, 我们对于 $\mathbb{R}^n$ 中的超曲面 M 写下相当于(25)的公式. 令 $(f, \sigma) = f_1\sigma_1 + \cdots + f_n\sigma_n$, 而 $(f, v) = f_1 v_1 + \cdots + f_n v_n$, 则公式为

$$\int_M (f, \sigma) = \int_M (f, v) dS$$

(26)

这里 $f = (f_1, \cdots, f_n)$ 的分量是 C_0 函数, $v = (v_1, \cdots, v_n)$ 是连续单位法向量, dS 由(20)给出, 而 $\sigma = (\sigma_1, \cdots, \sigma_n)$ 的分量是由(20)后面的公式所定义的 $(n-1)$ 微分形式. M 的定向是 $v_1\sigma_1 + \cdots + v_n\sigma_n > 0$. 证明完全和 $n = 3$ 的情形一样.

(5) Creen 公式和 Stokes 公式.

Green 公式(1827)(也称为 Gauss 公式)用传统的记号可以写成

$$\int_V \mathrm{div}\, f dV = \int_S (f, v) dS$$

(27)

这里 V 是 $\mathbb{R}^n$ 中的有界开子集, 其边界 S 是 $n-1$ 维的 C^1 流形, dV 和 dS 分别是相应的体积元和面积元. $v = (v_1, \cdots, v_n)$ 是 S 的外单位法向量, 向量 $f = (f_1, \cdots, f_n)$ 的分量是 C^1 函数, (f, v) 是标量积 $f_1 v_1 + \cdots + f_n v_n$, div f 是函数 $\partial_1 f_1 + \cdots + \partial_n f_n$. 这个公式适用于所有的维数, 如图 3 所示.

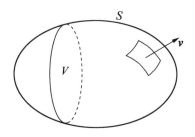

图 3　Green 公式的图示

假如把 f 想象成某种物质流，而 (f, v) 是在单位时间内，通过 S 上具有单位 $(n-1)$ 维面积的一块表面，沿着其法方向 v 的流量，则 (27) 右边就是单位时间内由 V 经过 S 向外流出的物质流，而 (27) 左边就是单位时间内由某种源头加给 V 的物质. 于是单位体积内的源头的密度是 $\operatorname{div} f$. 当 $\operatorname{div} f = 0$ 时，流就称为无源的.

为了证明 (27)，我们把公式的两边都写成微分形式的积分，根据 (26)，右边是微分形式

$$\omega(x) = f_1(x)\sigma_1(x) + \cdots + f_n(x)\sigma_n(x)$$

在 S 上的积分，通过简单的计算可以证明 $\mathrm{d}\omega(x) = \operatorname{div} f(x) \cdot \tau(x)$，其中 $\tau(x) = \mathrm{d}x_1 \wedge \cdots \wedge \mathrm{d}x_n$. 因此我们可以把 Green 公式写成

$$\int_V \mathrm{d}\omega(x) = \int_S \omega(x) \tag{28}$$

其中 V 和 S 分别由 $\tau(x) > 0$ 和 $(v, \sigma(x)) > 0$ 来定向. 其次，我们用两种图 $h(t)$，W 在包含紧区域 $V \cup S$ 的开集上构造一个图册，其中一种图满足 $h(W) \subset V$，另一种图满足在 $h(W) \cap V$ 上 $t_1 \leqslant 0$，而在 S 上 $t_1 = 0$. 这里 W 表示 $\mathbb{R}^n$ 中的开集. 利用单位分解，显然我们只需对 ω 在这种附图区域之外等于零的情形证明 (28) 就够了. 如果 $\omega = g_1\sigma_1(t) + \cdots + g_n\sigma_n(t)$，则 $\mathrm{d}\omega = (\operatorname{div} g)\tau(t)$. 这是因为，根据 Leibniz 引理，$d$ 与代换可以交换. 因此整个定理的证明就归结为，对于在 W 上及 W 的子集 $t^1 \leqslant 0$ 上具有紧支集的 C^1 函数，分别证明

$$\int (\partial_1 g_1(t) + \cdots + \partial_n g_n(t)) \mathrm{d}t_1 \cdots \mathrm{d}t_n = 0 \tag{29}$$

$$\int_{t_1 < 0} (\partial_1 g_1(t) + \cdots + \partial_n g_n(t)) \mathrm{d}t_1 \cdots \mathrm{d}t_n = \int_{t_1 = 0} g_1(t) \mathrm{d}t_2 \cdots \mathrm{d}t_n \tag{30}$$

首先对 t_k 积分 $\partial_k g_k(\boldsymbol{t})$，然后由关于单变量 C^1 函数的事实 $\int_a^b f'(\boldsymbol{x})\,\mathrm{d}x = f(b) - f(a)$ 就可以推出这两个公式. 除一个细节, 也就是 (30) 的符号之外, 我们已经证毕. 但这个符号也是对的, 因为当 $g_1 \geqslant 0$ 而且随着 t_1 一同增大时, (30) 的两边都大于或等于 0.

平面上的 Green 公式通常写成

$$\int_\Omega (Q_x - P_y)\,\mathrm{d}x\mathrm{d}y = \int_\gamma P\mathrm{d}x + Q\mathrm{d}y \tag{31}$$

其中 Ω 是具有光滑边界 γ 的有界开区域, γ 由 $e_1\mathrm{d}x + e_2\mathrm{d}y > 0$ 来定向, 这里 (e_1, e_2) 是 γ 的切向量, 它如此选取, 使得 $(e_2, -e_1)$ 是外法向量. 因为 $\mathrm{d}(P\mathrm{d}x + Q\mathrm{d}y) = (Q_x - P_y)\mathrm{d}x \wedge \mathrm{d}y$, 所以公式 (31) 可由 (28) 推出.

Stokes 公式 (1840) 是 (31) 的推广, 这个公式讨论的是流 $\boldsymbol{f}(\boldsymbol{x}) = (f_1(\boldsymbol{x}), f_2(\boldsymbol{x}), f_3(\boldsymbol{x}))$ 在 $\mathbb{R}^3$ 中的曲面及其边界上的积分. 用传统的记号, 它可以写成

$$\int_\Omega (\mathrm{rot}\,\boldsymbol{f}, \boldsymbol{v})\,\mathrm{d}\Omega = \int_\gamma (\boldsymbol{f}, \boldsymbol{e})\,\mathrm{d}\gamma$$

其中 $\mathrm{d}\Omega$ 是面积元, $\mathrm{d}\gamma$ 是弧元, $\boldsymbol{v} = (v_1, v_2, v_3)$ 是 Ω 的连续单位法向量, $\boldsymbol{e} = (e_1, e_2, e_3)$ 是 γ 的连续单位切向量, 而 $\boldsymbol{f}$ 的旋量 $\mathrm{rot}\,\boldsymbol{f}$ 是一个向量, 其分量为

$$(\partial_2 f_3 - \partial_3 f_2, \partial_3 f_1 - \partial_1 f_3, \partial_1 f_2 - \partial_2 f_1)$$

此外, 我们如此选取 $\boldsymbol{v}$ 和 $\boldsymbol{e}$, 使得向量积 $\boldsymbol{e} \times \boldsymbol{v}$ 指向 Ω 外部, $\boldsymbol{e} \times \boldsymbol{v}$ 的分量是 $(e_2 v_3 - e_3 v_2, e_3 v_1 - e_1 v_3, e_1 v_2 - e_2 v_1)$. 如图 4 所示, 从直观上来看, 右边是环绕 γ 的质量流, 而左边是更抽象的涡旋密度在 Ω 上的积分. 因为

$$\omega = f_1\mathrm{d}x_1 + f_2\mathrm{d}x_2 + f_3\mathrm{d}x_3 \Rightarrow \mathrm{d}\omega = (\partial_2 f_3 - \partial_3 f_2)\mathrm{d}x_2 \wedge \mathrm{d}x_3 + \cdots$$

所以可将 Stokes 公式写成

$$\int_\Omega \mathrm{d}\omega = \int_\gamma \omega \tag{32}$$

其中 Ω 与 γ 分别由 $v_1\mathrm{d}x_2 \wedge \mathrm{d}x_3 + \cdots > 0$ 与 $e_1\mathrm{d}x_1 + \cdots > 0$ 来定向. 这个公式可以归结为它的特殊情形 (31); 但是利用我们证明 Green 公式的方法, 也就是利用 Ω 上的图 $x = h(t_1, t_2)$, 最后归结成 (29) 与 (30)($n = 2$), 也不难给出证明. 事实上, 由这个证明可知, 当 Ω 是具有 C^1 边界 γ 的 C^1 类 p 维流形, 而 Ω 与 γ 都适当地定向时, 式 (32) 对于 $(p-1)$ 微分形式也成立. 这种一般性的公式, 也称为 Stokes 公式.

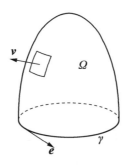

图 4　Stokes 公式的图示

本书的第二个主题是复变函数论,它是研究复变数的函数的性质及应用的一门学科,是分析学的一个重要分支,简介如下:

形如 $x + iy(x, y$ 为实数,i 是虚数单位,满足 $i^2 = -1)$ 的数被称为复数.复数早在 16 世纪就已经出现,它起源于求代数方程的根.在相当长的一段时间内,复数不被人们所接受.直到 19 世纪,才阐明复数是从已知量确定出的数学实体.以复数为自变量的函数叫作复变函数.

对复变函数的研究是从 18 世纪开始的.18 世纪 30 ~ 40 年代,Euler 曾利用幂级数详细讨论过初等复变函数的性质,并得出了著名的 Euler 公式 $e^{ix} = \cos x + i\sin x$.1752 年,D'Alembert 在论述流体力学的论文中,考虑复函数 $f(z) = u + iv$ 的导数存在的条件,导出了关系式

$$\frac{\partial u}{\partial x} = \frac{\partial v}{\partial y}, \frac{\partial u}{\partial y} = -\frac{\partial v}{\partial x} \tag{1}$$

Euler 在 1777 年提交给圣彼得堡科学院的一篇论文中,利用实函数计算复函数的积分,也得到了关系式(1).因此,式(1)有时被称为 D'Alembert-Euler 方程,但后来更多地被称为 Cauchy-Riemann 方程.在这一时期,Laplace 也研究过复函数的积分.但是以上三人的工作都存在着本质上的局限性,因为他们把 $f(z)$ 的实部和虚部分开考虑,没有把它看成一个基本实体.

复变函数论的全面发展是在 19 世纪.首先,Cauchy 的工作为单复变函数论的发展奠定了基础.他从 1814 年开始致力于复变函数的研究,完成了一系列重要论著.他把一个复变函数 $f(z)$ 看作复变数 z 的一元函数来研究.他首先证明复数的代数运算与极限运算的合理性,引进了复

函数连续性的概念,接着给出了复函数可导的充分必要条件(即 Cauchy-Riemann 方程). 他定义了复函数的积分,得到复函数在无奇点的区域内积分值与积分路径无关的重要定理,从而导出著名的 Cauchy 积分公式

$$f(z) = \frac{1}{2\pi i}\int_{\Gamma}\frac{f(s)}{\zeta - z}\mathrm{d}s$$

Cauchy 还给出了复函数在极点处的留数的定义,建立了计算留数的定理. 他还研究了多值函数,为 Riemann 面的创立提供了理论依据.

紧接着,Abel 和 Jacobi 创立了椭圆函数理论(1826),给复变函数论带来了新的生机. 1851 年,Riemann 的博士论文《单复变函数的一般理论基础》第一次给出单值解析函数的定义,指出实函数与复函数导数的基本差别. 他把单值解析函数推广到多值解析函数,阐述了现被称为 Riemann 面的概念,开辟了多值函数研究的方向. Riemann 最重要的工作是建立了保角映射的基本定理,解析函数与保角映射的联结不仅奠定了复变函数几何理论的基础,还开辟了求解许多数学物理问题的新途径. Riemann 的工作为 19 世纪复变函数论的全面发展开辟了道路.

Weierstrass 与 Cauchy,Riemann 不同,他摆脱了复函数的几何直观,从研究幂级数出发来发展复变函数理论. 按照他的定义,一个函数 $w = f(z)$ 如果在区域 D 内的每一个点 z_0 的邻域内都能展开为幂级数

$$w = f(z) = \sum_{k=0}^{\infty} c_k(z - z_0)^k$$

则称 $f(z)$ 在 D 内是解析的. 他还提出了复函数的解析开拓理论,引入完全解析函数的概念. 他在椭圆函数论方面也有很重要的工作.

19 世纪后期,复变函数论得到迅速发展. 在相当一段时间内,Cauchy,Riemann,Weierstrass 这三位主要奠基人的工作被他们各自的追随者继续研究. 后来,Cauchy 和 Riemann 的思想被融合在一起,而 Weierstrass 的方法逐渐从 Cauchy,Riemann 的观点推导出来. 人们发现,Weierstrass 的研究途径不是本质的,因此不再强调从幂级数出发考虑问题,这是 20 世纪初的事.

20 世纪以来,复变函数论又有很大的发展,形成了一些专门的研究领域. 在这方面做出较多工作的有瑞典数学家米塔-列夫勒,法国数学家 Poincaré,Picard,Borel, 芬兰数学家 Nevanlinna, 德国数学家

Bieberbach, 苏联数学家韦夸、拉夫连季耶夫, 以及中国数学家杨乐、张广厚等.

复变函数论研究的主要对象为解析函数, 我们讨论具有复系数的单复变量幂级数. 这就意味着, 我们把函数改为

$$f(z) = \sum_{k=0}^{\infty} c_k (z - a)^k$$

其中 a 是某个复数, 而 z 是一个复变量. 此时 r 仍旧和之前一样定义, 也可以推知, 当 $|z - a| < r$ 时, 上述级数收敛, 而当 $|z - a| > 4$ 时, 这个级数发散. 此处我们的收敛区间就成为复数平面上的收敛圆盘, 这大概就是实数和复数情形的唯一区别. 前面的定理及其证明仍然逐字逐句地成立, 导数 f' 是

$$f'(z) = \lim_{h \to 0} \frac{f(z + h) - f(z)}{h} \tag{2}$$

其中 $h \to 0$ 是通过复数值的. 以前的论据还证明, 当 x 改为 z 时, 在以 b 为中心的圆盘 $|z - b| < r - |a - b|$ 中仍旧成立, 这个圆盘包含在圆盘 $|z - a| < r$ 内, 并与之相切.

所谓复平面的开子集 A 上的复值函数 f 在 A 中是解析的, 就是指在 A 的每一点附近, f 都等于它以此点为中心的 Taylor 级数. 例如, 由一个幂级数定义的函数在其收敛圆盘中是解析函数; 多项式是处处解析的函数; 在实轴的区间 $|x - a| < r$ 中定义的函数, 可以扩张成在圆盘 $|z - a| < r$ 中解析的函数; 函数 z^{-1} 在原点解析, 因为 $z^{-1} = (z - a + a)^{-1} = \sum_{k=0}^{\infty} (-1)^k a^{-k-1} (z - a)^k$, 当 $|z - a| < |a|$ 时, 这个级数收敛.

这里解析函数是仿照 Weierstrass(大约 1870 年) 定义的, 它具有许多突出的性质, 在数学分析的所有分支中都是极为重要的. 这里举几个例子: 解析函数的线性组合, 乘积、分母不为零的商以及复合函数仍旧是解析函数. 复平面的开子集 A 上的复值函数为解析函数, 当且仅当其导数在这个开集中到处存在. 如果 A 是连通的, 那么这样的函数便由它在 A 的任意一点的 Taylor 级数的系数唯一决定. 如果 C 是 A 的紧子集, 它由一个开集 Ω 及其光滑的边界曲线 γ 组成, 那么 f 在 Ω 中的值可以由 f 在 γ 上的值通过 Cauchy 积分公式

$$f(z) = (2\pi i)^{-1} \int_{\gamma} (w - z)^{-1} f(w) \, dw$$

计算出来,其中的积分是通常的线积分,γ 如此定向,使得 Ω 处于 γ 的左侧.

本书第二部分复变函数中的重点有三个,如下:

一是所谓留数(residue),又称残数,复变函数论中的一个重要概念.解析函数 $f(z)$ 在孤立奇点 $z = z_0$ 处的 Laurent 展开式

$$f(z) = \sum_{n=-\infty}^{+\infty} c_n (z - z_0)^n$$

中,$(z - z_0)^{-1}$ 项的系数 c_{-1} 被称为 $f(x)$ 在 $z = z_0$ 处的留数.

留数的概念由法国数学家 Cauchy 提出.他在 1814 年于巴黎科学院宣读的论文《关于定积分理论的报告》(1827 年发表)中,已经接触到这个概念.对留数较完整的论述是在他 1825 年的论文《关于积分线为虚数的定积分的报告》(1874 年发表)中给出的.第二年他提出了积分留数的术语,并指出 $f(z)$ 在 z_0 处的留数就是 $f(z)$ 在 z_0 处的 Laurent 展开式中 $(z - z_0)^{-1}$ 项的系数.到 1841 年,他建立了留数的积分表达式

$$F(z_0) = E[f(z)]z_0 = \frac{1}{2\pi i} \int_\Gamma f(z)\,\mathrm{d}z$$

其中积分路径 Γ 表示以 z_0 为中心的小圆. 1846 年,Cauchy 又指出,如果曲线 Γ 包围着一些极点,那么积分 $\int_\Gamma f(z)\,\mathrm{d}z$ 的值等于 $f(z)$ 在这些极点上的留数之和的 $2\pi i$ 倍,即

$$\int_\Gamma f(z)\,\mathrm{d}z = 2\pi i E[f(z)]$$

其中 $E[f(z)]$ 是 Cauchy 用以表示留数之和的记号,这个结果被称为留数定理.留数定理在复变函数论中有着广泛的应用,例如利用它可计算一些较复杂的定积分.

二是一个重要的定理:Cauchy 积分定理,法国数学家 Cauchy 在研究复变函数的积分时所得到的基本定理.

Cauchy 在 1825 年完成的论文《关于积分限为虚数的定积分的报告》(1874 年发表)中叙述了这个定理:若 $f(x + iy)$ 在区域 $x_0 \leqslant x \leqslant X$,$y_0 \leqslant y \leqslant Y$ 中有界并连续,则积分

$$\int_{x_0+iy_0}^{x+iy} f(z)\,\mathrm{d}z$$

的值与 $x = \varphi(t)$ 和 $y = \psi(t)$ 的形式无关. Cauchy 在这篇论文中给出的证明并不十分严谨,他在 1846 年的论文中给出了这个定理的一个新证明. Gauss 在 1811 年致 Bessel 的信中也表述了上述积分定理,称它为"一条非常美妙的定理",说他"将在适当时候给出它的一个不难的证明",但一直没有发表.

由 Cauchy 积分定理可导出 Cauchy 积分公式. Cauchy 在他 1831 年的论文(1841 年发表)中给出的结果相当于现在所称的 Cauchy 积分公式,并通过幂级数证明了这个公式. 直到 1884 年,法国数学家 Goursat 才给出了 Cauchy 积分定理第一个完整的证明. Cauchy 积分定理所表达的解析函数的特性完全刻画了这类函数,因而解析函数的所有基本性质都可以由这个定理推出. Cauchy 积分定理在多复变量解析函数的情形由 Poincaré 推广,被称为 Cauchy-Poincaré 积分定理.

第三个稍微有些专门化,即单叶函数,复变函数中一类重要的解析函数. 若对于复区域 D 上单值的解析函数 $f(z)$,当 z_1 不同于 z_2 时有 $f(z_1) \neq f(z_2)$,则称 $f(z)$ 为 D 上的单叶函数. 单叶函数及与其相关的单叶映射等课题是复变函数论最重要的研究内容之一.

德国数学家 Koebe 和 Bieberbach 等人最早对单叶函数做出重要贡献. 例如,Bieberbach 从 1916 年开始对单位圆内全纯单叶函数进行了定量研究,他讨论了单叶的半纯函数,建立了面积原理,由此导出 Koebe 掩盖定理,最后证明了重要的偏差定理. 苏联数学家 Goluzin 研究了单叶函数的几何性质,证明了有趣的回转定理. Bieberbach 与芬兰数学家 Nevanlinna 共同建立了单位圆内单叶函数的系统理论. 1916 年 Bieberbach 提出了一个著名猜想,被称为 Bieberbach 猜想. 它曾经是单叶函数研究的中心问题,吸引过许多著名数学家. 围绕这个猜想所做的工作推动了复变函数论的发展.

本书所述内容是经典数学中比较完美的理论,既有应用价值,又同时具有审美价值;既可供目前正在求学的青年人攻读,亦可供具有一定数学修养的老年人把玩.

张岱说:"人无癖不可与交,以其无深情也;人无疵不可与交,以其无真气也". 其实,何止是交友需有癖、有疵,人活着更是离不开这两样德行. 人有癖好,精神才会有寄托,专注于自己的精神世界,可以做到风雨不动安如山;有疵,性情才会真诚,可以在无聊中觅得欢喜.

"南山松随感"中有一句话说得挺对:现实总是逼仄得很,外界总是很喧闹,风雨随时会侵袭每一个人,有癖好的痴情人,他总是以一个高级玩家的姿态面对世界,无论风雨,无论阴晴,他都会陶醉在自己的世界中,忘乎所以.

数学博大精深,品味数学是一个雅好,入此坑者多半有癖、有疵,所以祝各位各美其美、自得其乐.

刘培杰

2023 年 2 月 20 日

于哈工大

刘培杰数学工作室
已出版(即将出版)图书目录——原版影印

书　名	出版时间	定　价	编号
数学物理大百科全书.第1卷(英文)	2016—01	418.00	508
数学物理大百科全书.第2卷(英文)	2016—01	408.00	509
数学物理大百科全书.第3卷(英文)	2016—01	396.00	510
数学物理大百科全书.第4卷(英文)	2016—01	408.00	511
数学物理大百科全书.第5卷(英文)	2016—01	368.00	512
zeta函数,q-zeta函数,相伴级数与积分(英文)	2015—08	88.00	513
微分形式:理论与练习(英文)	2015—08	58.00	514
离散与微分包含的逼近和优化(英文)	2015—08	58.00	515
艾伦·图灵:他的工作与影响(英文)	2016—01	98.00	560
测度理论概率导论,第2版(英文)	2016—01	88.00	561
带有潜在故障恢复系统的半马尔柯夫模型控制(英文)	2016—01	98.00	562
数学分析原理(英文)	2016—01	88.00	563
随机偏微分方程的有效动力学(英文)	2016—01	88.00	564
图的谱半径(英文)	2016—01	58.00	565
量子机器学习中数据挖掘的量子计算方法(英文)	2016—01	98.00	566
量子物理的非常规方法(英文)	2016—01	118.00	567
运输过程的统一非局部理论:广义波尔兹曼物理动力学,第2版(英文)	2016—01	198.00	568
量子力学与经典力学之间的联系在原子、分子及电动力学系统建模中的应用(英文)	2016—01	58.00	569
算术域(英文)	2018—01	158.00	821
高等数学竞赛:1962—1991年的米洛克斯·史怀哲竞赛(英文)	2018—01	128.00	822
用数学奥林匹克精神解决数论问题(英文)	2018—01	108.00	823
代数几何(德文)	2018—04	68.00	824
丢番图逼近论(英文)	2018—01	78.00	825
代数几何学基础教程(英文)	2018—01	98.00	826
解析数论入门课程(英文)	2018—01	78.00	827
数论中的丢番图问题(英文)	2018—01	78.00	829
数论(梦幻之旅):第五届中日数论研讨会演讲集(英文)	2018—01	68.00	830
数论新应用(英文)	2018—01	68.00	831
数论(英文)	2018—01	78.00	832

刘培杰数学工作室
已出版(即将出版)图书目录——原版影印

书　名	出版时间	定　价	编号
湍流十讲(英文)	2018—04	108.00	886
无穷维李代数:第3版(英文)	2018—04	98.00	887
等值、不变量和对称性(英文)	2018—04	78.00	888
解析数论(英文)	2018—09	78.00	889
《数学原理》的演化:伯特兰·罗素撰写第二版时的手稿与笔记(英文)	2018—04	108.00	890
哈密尔顿数学论文集(第4卷):几何学、分析学、天文学、概率和有限差分等(英文)	2019—05	108.00	891
偏微分方程全局吸引子的特性(英文)	2018—09	108.00	979
整函数与下调和函数(英文)	2018—09	118.00	980
幂等分析(英文)	2018—09	118.00	981
李群、离散子群与不变量理论(英文)	2018—09	108.00	982
动力系统与统计力学(英文)	2018—09	118.00	983
表示论与动力系统(英文)	2018—09	118.00	984
分析学练习.第1部分(英文)	2021—01	88.00	1247
分析学练习.第2部分,非线性分析(英文)	2021—01	88.00	1248
初级统计学:循序渐进的方法:第10版(英文)	2019—05	68.00	1067
工程师与科学家微分方程用书:第4版(英文)	2019—07	58.00	1068
大学代数与三角学(英文)	2019—06	78.00	1069
培养数学能力的途径(英文)	2019—07	38.00	1070
工程师与科学家统计学:第4版(英文)	2019—06	58.00	1071
贸易与经济中的应用统计学:第6版(英文)	2019—06	58.00	1072
傅立叶级数和边值问题:第8版(英文)	2019—05	48.00	1073
通往天文学的途径:第5版(英文)	2019—05	58.00	1074
拉马努金笔记.第1卷(英文)	2019—06	165.00	1078
拉马努金笔记.第2卷(英文)	2019—06	165.00	1079
拉马努金笔记.第3卷(英文)	2019—06	165.00	1080
拉马努金笔记.第4卷(英文)	2019—06	165.00	1081
拉马努金笔记.第5卷(英文)	2019—06	165.00	1082
拉马努金遗失笔记.第1卷(英文)	2019—06	109.00	1083
拉马努金遗失笔记.第2卷(英文)	2019—06	109.00	1084
拉马努金遗失笔记.第3卷(英文)	2019—06	109.00	1085
拉马努金遗失笔记.第4卷(英文)	2019—06	109.00	1086
数论:1976年纽约洛克菲勒大学数论会议记录(英文)	2020—06	68.00	1145
数论:卡本代尔1979:1979年在南伊利诺伊卡本代尔大学举行的数论会议记录(英文)	2020—06	78.00	1146
数论:诺德韦克豪特1983:1983年在诺德韦克豪特举行的Journees Arithmetiques数论大会会议记录(英文)	2020—06	68.00	1147
数论:1985—1988年在纽约城市大学研究生院和大学中心举办的研讨会(英文)	2020—06	68.00	1148

刘培杰数学工作室
已出版(即将出版)图书目录——原版影印

书 名	出版时间	定 价	编号
数论:1987年在乌尔姆举行的 Journees Arithmetiques 数论大会会议记录(英文)	2020—06	68.00	1149
数论:马德拉斯 1987:1987 年在马德拉斯安娜大学举行的国际拉马努金百年纪念大会会议记录(英文)	2020—06	68.00	1150
解析数论:1988 年在东京举行的日法研讨会会议记录(英文)	2020—06	68.00	1151
解析数论:2002 年在意大利切特拉罗举行的 C. I. M. E. 暑期班演讲集(英文)	2020—06	68.00	1152
量子世界中的蝴蝶:最迷人的量子分形故事(英文)	2020—06	118.00	1157
走进量子力学(英文)	2020—06	118.00	1158
计算物理学概论(英文)	2020—06	48.00	1159
物质,空间和时间的理论:量子理论(英文)	2020—10	48.00	1160
物质,空间和时间的理论:经典理论(英文)	2020—10	48.00	1161
量子场理论:解释世界的神秘背景(英文)	2020—07	38.00	1162
计算物理学概论(英文)	2020—06	48.00	1163
行星状星云(英文)	2020—10	38.00	1164
基本宇宙学:从亚里士多德的宇宙到大爆炸(英文)	2020—08	58.00	1165
数学磁流体力学(英文)	2020—07	58.00	1166
计算科学:第 1 卷,计算的科学(日文)	2020—07	88.00	1167
计算科学:第 2 卷,计算与宇宙(日文)	2020—07	88.00	1168
计算科学:第 3 卷,计算与物质(日文)	2020—07	88.00	1169
计算科学:第 4 卷,计算与生命(日文)	2020—07	88.00	1170
计算科学:第 5 卷,计算与地球环境(日文)	2020—07	88.00	1171
计算科学:第 6 卷,计算与社会(日文)	2020—07	88.00	1172
计算科学:别卷,超级计算机(日文)	2020—07	88.00	1173
多复变函数论(日文)	2022—06	78.00	1518
复变函数入门(日文)	2022—06	78.00	1523
代数与数论:综合方法(英文)	2020—10	78.00	1185
复分析:现代函数理论第一课(英文)	2020—07	58.00	1186
斐波那契数列和卡特兰数:导论(英文)	2020—10	68.00	1187
组合推理:计数艺术介绍(英文)	2020—07	88.00	1188
二次互反律的傅里叶分析证明(英文)	2020—07	48.00	1189
旋瓦兹分布的希尔伯特变换与应用(英文)	2020—07	58.00	1190
泛函分析:巴拿赫空间理论入门(英文)	2020—07	48.00	1191
卡塔兰数入门(英文)	2019—05	68.00	1060
测度与积分(英文)	2019—04	68.00	1059
组合学手册.第一卷(英文)	2020—06	128.00	1153
* 一代数、局部紧群和巴拿赫 * 一代数丛的表示.第一卷,群和代数的基本表示理论(英文)	2020—05	148.00	1154
电磁理论(英文)	2020—08	48.00	1193
连续介质力学中的非线性问题(英文)	2020—09	78.00	1195
多变量数学入门(英文)	2021—05	68.00	1317
偏微分方程入门(英文)	2021—05	88.00	1318
若尔当典范性:理论与实践(英文)	2021—07	68.00	1366
伽罗瓦理论.第 4 版(英文)	2021—08	88.00	1408
R 统计学概论	2023—03	88.00	1614
基于不确定静态和动态问题解的仿射算术(英文)	2023—03	38.00	1618

刘培杰数学工作室
已出版(即将出版)图书目录——原版影印

书　名	出版时间	定　价	编号
典型群,错排与素数(英文)	2020—11	58.00	1204
李代数的表示:通过 gln 进行介绍(英文)	2020—10	38.00	1205
实分析演讲集(英文)	2020—10	38.00	1206
现代分析及其应用的课程(英文)	2020—10	58.00	1207
运动中的抛射物数学(英文)	2020—10	38.00	1208
2—纽结与它们的群(英文)	2020—10	38.00	1209
概率,策略和选择:博弈与选举中的数学(英文)	2020—11	58.00	1210
分析学引论(英文)	2020—11	58.00	1211
量子群:通往流代数的路径(英文)	2020—11	38.00	1212
集合论入门(英文)	2020—10	48.00	1213
酉反射群(英文)	2020—11	58.00	1214
探索数学:吸引人的证明方式(英文)	2020—11	58.00	1215
微分拓扑短期课程(英文)	2020—10	48.00	1216
抽象凸分析(英文)	2020—11	68.00	1222
费马大定理笔记(英文)	2021—03	48.00	1223
高斯与雅可比和(英文)	2021—03	78.00	1224
π 与算术几何平均:关于解析数论和计算复杂性的研究(英文)	2021—01	58.00	1225
复分析入门(英文)	2021—03	48.00	1226
爱德华·卢卡斯与素性测定(英文)	2021—03	78.00	1227
通往凸分析及其应用的简单路径(英文)	2021—01	68.00	1229
微分几何的各个方面.第一卷(英文)	2021—01	58.00	1230
微分几何的各个方面.第二卷(英文)	2020—12	58.00	1231
微分几何的各个方面.第三卷(英文)	2020—12	58.00	1232
沃克流形几何学(英文)	2020—11	58.00	1233
彷射和韦尔几何应用(英文)	2020—12	58.00	1234
双曲几何学的旋转向量空间方法(英文)	2021—02	58.00	1235
积分:分析学的关键(英文)	2020—12	48.00	1236
为有天分的新生准备的分析学基础教材(英文)	2020—11	48.00	1237
数学不等式.第一卷.对称多项式不等式(英文)	2021—03	108.00	1273
数学不等式.第二卷.对称有理不等式与对称无理不等式(英文)	2021—03	108.00	1274
数学不等式.第三卷.循环不等式与非循环不等式(英文)	2021—03	108.00	1275
数学不等式.第四卷.Jensen 不等式的扩展与加细(英文)	2021—03	108.00	1276
数学不等式.第五卷.创建不等式与解不等式的其他方法(英文)	2021—04	108.00	1277

刘培杰数学工作室
已出版(即将出版)图书目录——原版影印

书　名	出版时间	定　价	编号
冯·诺依曼代数中的谱位移函数:半有限冯·诺依曼代数中的谱位移函数与谱流(英文)	2021－06	98.00	1308
链接结构:关于嵌入完全图的直线中链接单形的组合结构(英文)	2021－05	58.00	1309
代数几何方法.第1卷(英文)	2021－06	68.00	1310
代数几何方法.第2卷(英文)	2021－06	68.00	1311
代数几何方法.第3卷(英文)	2021－06	58.00	1312

书　名	出版时间	定　价	编号
代数、生物信息和机器人技术的算法问题.第四卷,独立恒等式系统(俄文)	2020－08	118.00	1199
代数、生物信息和机器人技术的算法问题.第五卷,相对覆盖性和独立可拆分恒等式系统(俄文)	2020－08	118.00	1200
代数、生物信息和机器人技术的算法问题.第六卷,恒等式和准恒等式的相等 问题、可推导性和可实现性(俄文)	2020－08	128.00	1201
分数阶微积分的应用:非局部动态过程,分数阶导热系数(俄文)	2021－01	68.00	1241
泛函分析问题与练习:第2版(俄文)	2021－01	98.00	1242
集合论、数学逻辑和算法论问题:第5版(俄文)	2021－01	98.00	1243
微分几何和拓扑短期课程(俄文)	2021－01	98.00	1244
素数规律(俄文)	2021－01	88.00	1245
无穷边值问题解的递减:无界域中的拟线性椭圆和抛物方程(俄文)	2021－01	48.00	1246
微分几何讲义(俄文)	2020－12	98.00	1253
二次型和矩阵(俄文)	2021－01	98.00	1255
积分和级数.第2卷,特殊函数(俄文)	2021－01	168.00	1258
积分和级数.第3卷,特殊函数补充:第2版(俄文)	2021－01	178.00	1264
几何图上的微分方程(俄文)	2021－01	138.00	1259
数论教程:第2版(俄文)	2021－01	98.00	1260
非阿基米德分析及其应用(俄文)	2021－03	98.00	1261
古典群和量子群的压缩(俄文)	2021－03	98.00	1263
数学分析习题集.第3卷,多元函数:第3版(俄文)	2021－03	98.00	1266
数学习题:乌拉尔国立大学数学力学系大学生奥林匹克(俄文)	2021－03	98.00	1267
柯西定理和微分方程的特解(俄文)	2021－03	98.00	1268
组合极值问题及其应用:第3版(俄文)	2021－03	98.00	1269
数学词典(俄文)	2021－01	98.00	1271
确定性混沌分析模型(俄文)	2021－06	168.00	1307
精选初等数学习题和定理.立体几何.第3版(俄文)	2021－03	68.00	1316
微分几何习题:第3版(俄文)	2021－05	98.00	1336
精选初等数学习题和定理.平面几何.第4版(俄文)	2021－05	68.00	1335
曲面理论在欧氏空间 E_n 中的直接表示(俄文)	2022－01	68.00	1444
维纳－霍普夫离散算子和托普利兹算子:某些可数赋范空间中的诺特性和可逆性(俄文)	2022－03	108.00	1496
Maple 中的数论:数论中的计算机计算(俄文)	2022－03	88.00	1497
贝尔曼和克努特问题及其概括:加法运算的复杂性(俄文)	2022－03	138.00	1498

刘培杰数学工作室
已出版（即将出版）图书目录——原版影印

书　名	出版时间	定　价	编号
复分析:共形映射(俄文)	2022—07	48.00	1542
微积分代数样条和多项式及其在数值方法中的应用(俄文)	2022—08	128.00	1543
蒙特卡罗方法中的随机过程和场模型:算法和应用(俄文)	2022—08	88.00	1544
线性椭圆型方程组:论二阶椭圆型方程的迪利克雷问题(俄文)	2022—08	98.00	1561
动态系统解的增长特性:估值、稳定性、应用(俄文)	2022—08	118.00	1565
群的自由积分解:建立和应用(俄文)	2022—08	78.00	1570
混合方程和偏差自变数方程问题:解的存在和唯一性(俄文)	2023—01	78.00	1582
拟变量空间分析:存在和逼近定理(俄文)	2023—01	108.00	1583
二维和三维流形上函数的拓扑性质:函数的拓扑分类(俄文)	2023—03	68.00	1584
齐次马尔科夫过程建模的矩阵方法:此类方法能够用于不同目的的复杂系统研究、设计和完善(俄文)	2023—03	68.00	1594
周期函数的近似方法和特性:特殊课程(俄文)	2023—04	158.00	1622
扩散方程解的矩函数:变分法(俄文)	2023—03	58.00	1623

书　名	出版时间	定　价	编号
狭义相对论与广义相对论:时空与引力导论(英文)	2021—07	88.00	1319
束流物理学和粒子加速器的实践介绍:第2版(英文)	2021—07	88.00	1320
凝聚态物理中的拓扑和微分几何简介(英文)	2021—05	88.00	1321
混沌映射:动力学、分形学和快速涨落(英文)	2021—05	128.00	1322
广义相对论:黑洞、引力波和宇宙学介绍(英文)	2021—06	68.00	1323
现代分析电磁均质化(英文)	2021—06	68.00	1324
为科学家提供的基本流体动力学(英文)	2021—06	88.00	1325
视觉天文学:理解夜空的指南(英文)	2021—06	68.00	1326
物理学中的计算方法(英文)	2021—06	68.00	1327
单星的结构与演化:导论(英文)	2021—06	108.00	1328
超越居里:1903年至1963年物理界四位女性及其著名发现(英文)	2021—06	68.00	1329
范德瓦尔斯流体热力学的进展(英文)	2021—06	68.00	1330
先进的托卡马克稳定性理论(英文)	2021—06	88.00	1331
经典场论导论:基本相互作用的过程(英文)	2021—07	88.00	1332
光致电离量子动力学方法原理(英文)	2021—07	108.00	1333
经典域论和应力:能量张量(英文)	2021—05	88.00	1334
非线性太赫兹光谱的概念与应用(英文)	2021—06	68.00	1337
电磁学中的无穷空间并矢格林函数(英文)	2021—06	88.00	1338
物理科学基础数学.第1卷,齐次边值问题、傅里叶方法和特殊函数(英文)	2021—07	108.00	1339
离散量子力学(英文)	2021—07	68.00	1340
核磁共振的物理学和数学(英文)	2021—07	108.00	1341
分子水平的静电学(英文)	2021—08	68.00	1342
非线性波:理论、计算机模拟、实验(英文)	2021—06	108.00	1343
石墨烯光学:经典问题的电解决解决方案(英文)	2021—06	68.00	1344
超材料多元宇宙(英文)	2021—07	68.00	1345
银河系外的天体物理学(英文)	2021—07	68.00	1346
原子物理学(英文)	2021—07	68.00	1347
将光打结:将拓扑学应用于光学(英文)	2021—07	68.00	1348
电磁学:问题与解法(英文)	2021—07	88.00	1364
海浪的原理:介绍量子力学的技巧与应用(英文)	2021—07	108.00	1365
多孔介质中的流体:输运与相变(英文)	2021—07	68.00	1372
洛伦兹群的物理学(英文)	2021—08	68.00	1373
物理导论的数学方法和解决方法手册(英文)	2021—08	68.00	1374

刘培杰数学工作室
已出版(即将出版)图书目录——原版影印

书　　名	出版时间	定　价	编号
非线性波数学物理学入门(英文)	2021—08	88.00	1376
波:基本原理和动力学(英文)	2021—07	68.00	1377
光电子量子计量学.第1卷,基础(英文)	2021—07	88.00	1383
光电子量子计量学.第2卷,应用与进展(英文)	2021—07	68.00	1384
复杂流的格子玻尔兹曼建模的工程应用(英文)	2021—08	68.00	1393
电偶极矩挑战(英文)	2021—08	108.00	1394
电动力学:问题与解法(英文)	2021—09	68.00	1395
自由电子激光的经典理论(英文)	2021—08	68.00	1397
曼哈顿计划——核武器物理学简介(英文)	2021—09	68.00	1401
粒子物理学(英文)	2021—09	68.00	1402
引力场中的量子信息(英文)	2021—09	128.00	1403
器件物理学的基本经典力学(英文)	2021—09	68.00	1404
等离子体物理及其空间应用导论.第1卷,基本原理和初步过程(英文)	2021—09	68.00	1405
磁约束聚变等离子体物理:理想MHD理论(英文)	2023—03	68.00	1613
相对论量子场论.第1卷,典范形式体系(英文)	2023—03	38.00	1615
涌现的物理学(英文)	2023—05	58.00	1619
量子化旋涡:一本拓扑激发手册(英文)	2023—04	68.00	1620
非线性动力学:实践的介绍性调查(英文)	2023—05	68.00	1621
拓扑与超弦理论焦点问题(英文)	2021—07	58.00	1349
应用数学:理论、方法与实践(英文)	2021—07	78.00	1350
非线性特征值问题:牛顿型方法与非线性瑞利函数(英文)	2021—07	58.00	1351
广义膨胀和齐性:利用齐性构造齐次系统的李雅普诺夫函数和控制律(英文)	2021—06	48.00	1352
解析数论焦点问题(英文)	2021—07	58.00	1353
随机微分方程:动态系统方法(英文)	2021—07	58.00	1354
经典力学与微分几何(英文)	2021—07	58.00	1355
负定相交形式流形上的瞬子模空间几何(英文)	2021—07	68.00	1356
广义卡塔兰轨道分析:广义卡塔兰轨道计算数字的方法(英文)	2021—07	48.00	1367
洛伦兹方法的变分:二维与三维洛伦兹方法(英文)	2021—08	38.00	1378
几何、分析和数论精编(英文)	2021—08	68.00	1380
从一个新角度看数论:通过遗传方法引入现实的概念(英文)	2021—07	58.00	1387
动力系统:短期课程(英文)	2021—08	68.00	1382
几何路径:理论与实践(英文)	2021—08	48.00	1385
论天体力学中某些问题的不可积性(英文)	2021—07	88.00	1396
广义斐波那契数列及其性质(英文)	2021—08	38.00	1386
对称函数和麦克唐纳多项式:余代数结构与Kawanaka恒等式(英文)	2021—09	38.00	1400
杰弗里·英格拉姆·泰勒科学论文集:第1卷.固体力学(英文)	2021—05	78.00	1360
杰弗里·英格拉姆·泰勒科学论文集:第2卷.气象学、海洋学和湍流(英文)	2021—05	68.00	1361
杰弗里·英格拉姆·泰勒科学论文集:第3卷.空气动力学以及落弹数和爆炸的力学(英文)	2021—05	68.00	1362
杰弗里·英格拉姆·泰勒科学论文集:第4卷.有关流体力学(英文)	2021—05	58.00	1363

刘培杰数学工作室
已出版(即将出版)图书目录——原版影印

书　名	出版时间	定　价	编号
非局域泛函演化方程:积分与分数阶(英文)	2021—08	48.00	1390
理论工作者的高等微分几何:纤维丛、射流流形和拉格朗日理论(英文)	2021—08	68.00	1391
半线性退化椭圆微分方程:局部定理与整体定理(英文)	2021—07	48.00	1392
非交换几何、规范理论和重整化:一般简介与非交换量子场论的重整化(英文)	2021—09	78.00	1406
数论论文集:拉普拉斯变换和带有数论系数的幂级数(俄文)	2021—09	48.00	1407
挠理论专题:相对极大值,单射与扩充模(英文)	2021—09	88.00	1410
强正则图与欧几里得若尔当代数:非通常关系中的启示(英文)	2021—10	48.00	1411
拉格朗日几何和哈密顿几何:力学的应用(英文)	2021—10	48.00	1412

书　名	出版时间	定　价	编号
时滞微分方程与差分方程的振动理论:二阶与三阶(英文)	2021—10	98.00	1417
卷积结构与几何函数理论:用以研究特定几何函数理论方向的分数阶微积分算子与卷积结构(英文)	2021—10	48.00	1418
经典数学物理的历史发展(英文)	2021—10	78.00	1419
扩展线性丢番图问题(英文)	2021—10	38.00	1420
一类混沌动力系统的分歧分析与控制:分歧分析与控制(英文)	2021—11	38.00	1421
伽利略空间和伪伽利略空间中一些特殊曲线的几何性质(英文)	2022—01	68.00	1422
一阶偏微分方程:哈密尔顿—雅可比理论(英文)	2021—11	48.00	1424
各向异性黎曼多面体的反问题:分段光滑的各向异性黎曼多面体反边界谱问题:唯一性(英文)	2021—11	38.00	1425

书　名	出版时间	定　价	编号
项目反应理论手册.第一卷,模型(英文)	2021—11	138.00	1431
项目反应理论手册.第二卷,统计工具(英文)	2021—11	118.00	1432
项目反应理论手册.第三卷,应用(英文)	2021—11	138.00	1433
二次无理数:经典数论入门(英文)	2022—05	138.00	1434
数,形与对称性:数论,几何和群论导论(英文)	2022—05	128.00	1435
有限域手册(英文)	2021—11	178.00	1436
计算数论(英文)	2021—11	148.00	1437
拟群与其表示简介(英文)	2021—11	88.00	1438
数论与密码学导论:第二版(英文)	2022—01	148.00	1423

刘培杰数学工作室
已出版(即将出版)图书目录——原版影印

书　名	出 版 时 间	定　价	编号
几何分析中的柯西变换与黎兹变换:解析调和容量和李普希兹调和容量、变化和振荡以及一致可求长性(英文)	2021—12	38.00	1465
近似不动点定理及其应用(英文)	2022—05	28.00	1466
局部域的相关内容解析:对局部域的扩展及其伽罗瓦群的研究(英文)	2022—01	38.00	1467
反问题的二进制恢复方法(英文)	2022—03	28.00	1468
对几何函数中某些类的各个方面的研究:复变量理论(英文)	2022—01	38.00	1469
覆盖、对应和非交换几何(英文)	2022—01	28.00	1470
最优控制理论中的随机线性调节器问题:随机最优线性调节器问题(英文)	2022—01	38.00	1473
正交分解法:涡流流体动力学应用的正交分解法(英文)	2022—01	38.00	1475

书　名	出 版 时 间	定　价	编号
芬斯勒几何的某些问题(英文)	2022—03	38.00	1476
受限三体问题(英文)	2022—05	38.00	1477
利用马利亚万微积分进行 Greeks 的计算:连续过程、跳跃过程中的马利亚万微积分和金融领域中的 Greeks(英文)	2022—05	48.00	1478
经典分析和泛函分析的应用:分析学的应用(英文)	2022—03	38.00	1479
特殊芬斯勒空间的探究(英文)	2022—03	48.00	1480
某些图形的施泰纳距离的细谷多项式:细谷多项式与图的维纳指数(英文)	2022—05	38.00	1481
图论问题的遗传算法:在新鲜与模糊的环境中(英文)	2022—05	48.00	1482
多项式映射的渐近簇(英文)	2022—05	38.00	1483

书　名	出 版 时 间	定　价	编号
一维系统中的混沌:符号动力学,映射序列,一致收敛和沙可夫斯基定理(英文)	2022—05	38.00	1509
多维边界层流动与传热分析:粘性流体流动的数学建模与分析(英文)	2022—05	38.00	1510
演绎理论物理学的原理:一种基于量子力学波函数的逐次置信估计的一般理论的提议(英文)	2022—05	38.00	1511
R^2 和 R^3 中的仿射弹性曲线:概念和方法(英文)	2022—08	38.00	1512
算术数列中除数函数的分布:基本内容、调查、方法、第二矩、新结果(英文)	2022—05	28.00	1513
抛物型狄拉克算子和薛定谔方程:不定常薛定谔方程的抛物型狄拉克算子及其应用(英文)	2022—07	28.00	1514
黎曼-希尔伯特问题与量子场论:可积重正化、戴森-施温格方程(英文)	2022—08	38.00	1515
代数结构和几何结构的形变理论(英文)	2022—08	48.00	1516
概率结构和模糊结构上的不动点:概率结构和直觉模糊度量空间的不动点定理(英文)	2022—08	38.00	1517

刘培杰数学工作室
已出版(即将出版)图书目录——原版影印

书 名	出版时间	定 价	编号
反若尔当对:简单反若尔当对的自同构(英文)	2022－07	28.00	1533
对某些黎曼－芬斯勒空间变换的研究:芬斯勒几何中的某些变换(英文)	2022－07	38.00	1534
内谐零流形映射的尼尔森数的阿诺索夫关系(英文)	2023－01	38.00	1535
与广义积分变换有关的分数次演算:对分数次演算的研究(英文)	2023－01	48.00	1536
强子的芬斯勒几何和吕拉几何(宇宙学方面):强子结构的芬斯勒几何和吕拉几何(拓扑缺陷)(英文)	2022－08	38.00	1537
一种基于混沌的非线性最优化问题:作业调度问题(英文)	2023－03	38.00	1538
广义概率论发展前景:关于趣味数学与置信函数实际应用的一些原创观点(英文)	2023－03	48.00	1539
纽结与物理学:第二版(英文)	2022－09	118.00	1547
正交多项式和 q－级数的前沿(英文)	2022－09	98.00	1548
算子理论问题集(英文)	2022－09	108.00	1549
抽象代数:群、环与域的应用导论:第二版(英文)	2023－01	98.00	1550
菲尔兹奖得主演讲集:第三版(英文)	2023－01	138.00	1551
多元实函数教程(英文)	2022－09	118.00	1552
球面空间形式群的几何学:第二版(英文)	2022－09	98.00	1566
对称群的表示论(英文)	2023－01	98.00	1585
纽结理论:第二版(英文)	2023－01	88.00	1586
拟群理论的基础与应用(英文)	2023－01	88.00	1587
组合学:第二版(英文)	2023－01	98.00	1588
加性组合学:研究问题手册(英文)	2023－01	68.00	1589
扭曲、平铺与镶嵌:几何折纸中的数学方法(英文)	2023－01	98.00	1590
离散与计算几何手册:第三版(英文)	2023－01	248.00	1591
离散与组合数学手册:第二版(英文)	2023－01	248.00	1592
分析学教程.第1卷,一元实变量函数的微积分分析学介绍(英文)	2023－01	118.00	1595
分析学教程.第2卷,多元函数的微分和积分,向量微积分(英文)	2023－01	118.00	1596
分析学教程.第3卷,测度与积分理论,复变量的复值函数(英文)	2023－01	118.00	1597
分析学教程.第4卷,傅里叶分析,常微分方程,变分法(英文)	2023－01	118.00	1598

联系地址:哈尔滨市南岗区复华四道街 10 号　哈尔滨工业大学出版社刘培杰数学工作室
网　　址:http://lpj.hit.edu.cn/
邮　　编:150006
联系电话:0451－86281378　　13904613167
E-mail:lpj1378@163.com